T0318660

The unique characteristic of this book is that it considers the theory of partial differential equations in mathematical physics as the language of continuous processes, that is to say, as an interdisciplinary science that treats the hierarchy of mathematical phenomena as reflections of their physical counterparts. Special attention is drawn to tracing the development of these mathematical phenomena in different natural sciences, with examples drawn from continuum mechanics, electrodynamics, transport phenomena, thermodynamics, and chemical kinetics. At the same time, the authors trace the interrelation between the different types of problems – elliptic, parabolic, and hyperbolic – as the mathematical counterparts of stationary and evolutionary processes. This interrelation is traced through study of the asymptotics of the solutions of the respective initial boundary-value problems both with respect to time and the governing parameters of the problem.

This combination of mathematical comprehensiveness and natural scientific motivation represents a step forward in the presentation of the classical theory of PDEs, one that will be appreciated by both graduate students and researchers alike.

Partial differential equations in classical mathematical physics

Partial differential equations in classical mathematical physics

Isaak Rubinstein
Jacob Blaustein Institute for Desert Research
Department of Mathematics and Computer Science
Ben Gurion University of the Negev

Lev Rubinstein
Professor Emeritus
School of Applied Science and Technology
The Hebrew University of Jerusalem

CAMBRIDGE
UNIVERSITY PRESS

CAMBRIDGE UNIVERSITY PRESS
Cambridge, New York, Melbourne, Madrid, Cape Town, Singapore, São Paulo

Cambridge University Press
The Edinburgh Building, Cambridge CB2 2RU, UK

Published in the United States of America by Cambridge University Press, New York

www.cambridge.org
Information on this title: www.cambridge.org/9780521410588

First published 1993
First paperback edition 1998

A catalogue record for this publication is available from the British Library

ISBN-13 978-0-521-41058-8 hardback
ISBN-10 0-521-41058-4 hardback

ISBN-13 978-0-521-55846-4 paperback
ISBN-10 0-521-55846-8 paperback

Transferred to digital printing 2007

Contents

Preface

This book represents an attempt to implement a general approach that in essence views the theory of partial differential equations (PDEs) of mathematical physics as the language of continuous processes, that is, an interdisciplinary science that considers the hierarchy of mathematical phenomena as a reflection of their physical counterparts. A comprehensive, mathematically rigorous account of the classical theory of PDEs in mathematical physics is thus inseparably bound with the features of the corresponding natural continuum objects. We shall therefore endeavor to trace the simultaneous origins of some basic mathematical objects in different natural contexts (continuum mechanics, electrodynamics, transport phenomena, thermodynamics, and chemical kinetics). In parallel, we shall trace the interrelation between different types of problems (elliptic, parabolic, and hyperbolic) as mathematical counterparts of their natural prototypes: steady-state and evolutionary processes (dissipative and conservative). This will be done by an asymptotic analysis of the behavior of these processes in time and their dependence on the relevant governing parameters.

In view of the almost complete absence of a physics background in undergraduate and graduate curricula of mathematics and applied mathematics, it seems important, in a course of mathematical physics, to provide an introduction to the basic concepts of different natural sciences and their relation with PDEs in terms of certain typical boundary-value problems that recur in different scientific contexts. Chapters 1 and 2 are therefore addressed primarily to students of mathematics. On the other hand, a rigorous and systematic exposition of classical methods of mathematical physics is undoubtedly necessary for future workers in the applied and engineering sciences, particularly in view of the growing sophistication of industry and the increasing use of mathematical methods in natural sciences. In addition, since modern mathematical education virtually ignores certain efficient classical methods of mathematical physics (such as potential theory and the use of Fredholm and Volterra integral equations in solving boundary-value problems), some attention to these topics is highly desirable.

To summarize, this book is addressed to graduate students of applied mathematics and of the natural and engineering sciences. Selected parts may form the basis of an undergraduate course in applied mathematics for mathematics, physics, or engineering students. Most of the material in this book was featured in courses given by one of us (L.R.) for graduate students in mathematics and applied mathematics and by the other (I.R.) for undergraduate students of electrical engineering and physics.

Besides the bulk of the text, which is addressed primarily to graduate students, some chapters are addressed to specialists and are rather monographic in nature (i.e., Chapters 16–18).

1

Introduction

1. Mathematical physics

Mathematical physics is an interdisciplinary science which, on the basis of the fundamental (mostly phenomenological) laws of physics, uses mathematical methods to study processes evolving in material media. Its purpose is to formulate equations describing a process to within a reasonable degree of idealization (i.e., disregarding details that are not essential for its qualitative and quantitative essences), to develop methods for solution of the resulting problem, and to analyze the qualitative and quantitative properties of the solutions. In this latter respect mathematical physics borders on numerical analysis and mathematical simulation, but in its most important aspects it borders on the theoretical and even experimental natural sciences.

In this book we shall restrict our attention to phenomena of the "macro" world – more precisely, to processes evolving in continuous media. At this point some elaboration of the very notion of a continuous medium is desirable, since at first sight it might seem incompatible with an atomistic view of the universe. The notion of continuous medium is related to the following notion of a physical element of volume. Consider some process evolving in a region $D \subset \mathbb{R}_3$ and let $K \subset D$ be a subset of positive 3-dimensional measure with diameter d:

$$d = \max_{p \in K} \max_{q \in K} r_{pq} \tag{1.1}$$

where r_{pq} is the distance between points p and q. Fixing $p \in K$, consider it as a representative of K. Assume that d is much smaller than the characteristic size d_0 of D (e.g., the upper limit of the diameters of all spheres contained in D), but that the number of individual material particles in K is very large and their maximum size is very small compared with d. Let us consider some physical characteristic F of particles in K (e.g., the velocity $\mathbf{v}$ at time t of the individual particles in K). Let $\hat{F}(p,t)$ denote the value of F averaged over all particles in K. The medium in D may be called *continuous with*

1

respect to F if $\hat{F}(p,t)$ is a continuous function of p and t everywhere in D, except for finitely many sets of points of zero 3-dimensional measure (i.e., except for finitely many surfaces, lines, or separate points). If the medium is continuous with respect to all physical parameters of relevance for the process in question, one can speak of the medium as simply continuous.

Processes in nature may be divided, roughly speaking, into three groups: (1) stationary processes, in which the state of the system is independent of time; (2) dissipative time-dependent evolution processes; and (3) conservative evolution processes. Although the processes in group (3) may be viewed as the processes of group (2) in the limit of vanishing dissipation, we shall see in due course that this is a "singular" limit, in the sense that the processes of this group exhibit very different features.

There is a similarity between the fundamental (phenomenological) laws governing processes of the same group. For example, Fourier's law of heat conduction, Fick's law of diffusion, and Darcy's law of liquid percolation through porous media are identical, up to renaming of the variables. Indeed, *Fourier's law* reads: The amount of heat flowing in an isotropic homogeneous thermally conductive body through a surface element $d\tilde{\sigma}$ in the direction of the normal **n** to $d\tilde{\sigma}$ in time $d\tilde{t}$ is

$$d\tilde{q} = -\tilde{\lambda}\frac{\partial}{\partial n}\tilde{T}d\tilde{\sigma}d\tilde{t} \tag{1.2}$$

where $\tilde{T}$ is the temperature and the minus sign indicates that the heat is flowing in the direction of decreasing temperature, so that the coefficient $\tilde{\lambda}$ of thermal conductivity may be assumed positive.

Now *Fick's law* reads as follows: The mass of a solute transferred by diffusion in an isotropic solution through a surface element $d\sigma$ in the direction of the normal **n** to $d\tilde{\sigma}$ in time $d\tilde{t}$ is

$$d\tilde{q} = -\tilde{D}\frac{\partial}{\partial n}\tilde{C}d\tilde{\sigma}d\tilde{t} \tag{1.3}$$

where $\tilde{C}$ is the solute concentration.

Finally, *Darcy's law* reads: The mass of liquid percolating through the pore space of a homogeneous porous medium through a surface element $d\tilde{\sigma}$ in direction of the normal **n** to $d\tilde{\sigma}$ in time $d\tilde{t}$ is

$$d\tilde{q} = -\tilde{K}\frac{\partial}{\partial n}\tilde{p}d\tilde{\sigma}d\tilde{t} \tag{1.4}$$

where $\tilde{p}$ is the pore pressure and $\tilde{K}$ the percolation coefficient. The tilde "~" indicates that variables are dimensional.

All these phenomenological laws are of the same form; written in terms of dimensionless variables, they are indistinguishable. It is this possibility of simultaneously describing processes of a different physical nature, but belonging to the same group, that makes mathematical physics a universal language of the continuum, a connecting link between different disciplines of physics, chemistry, biology, and so on. The interrelation between the various properties of partial differential equations (PDEs) and

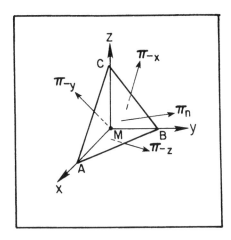

Figure 1.1. Sketch of stresses on the tetrahedron $MABC$.

their natural prototypes is very profound and helpful for research. Quite frequently, previously unknown mathematical phenomena are discovered by looking for the explanation of a natural phenomenon and vice versa, a natural phenomenon may be predicted by analyzing the properties of the corresponding mathematical object.

Before going on to a systematic study of the classical methods of mathematical physics, we shall present some preliminary information about the natural systems that will be referred to later. This includes some basic facts from thermodynamics, continuum mechanics, electrodynamics, and chemical kinetics. For a systematic presentation of these topics the reader is referred to the relevant specialized texts.

2. Basic concepts of continuum mechanics [47]

A. Mass (volume) forces and surface forces (stresses)

Mass (or volume) forces, such as forces of gravity and inertia (e.g., centrifugal force), are additive functions of the mass (volume) of a material body and are thus characterized by their densities per unit mass (or volume).

Surface forces (e.g., forces due to molecular interaction) are additive functions of the body's surface area and are thus characterized by their surface density (stress). Consider some domain D in a continuous medium. Let Σ be a two-sided smooth surface dividing D into two parts D_i and D_e, and let the normal $\mathbf{n}$ to Σ be directed into D_e. The surface force with which D_e attracts D_i, called the *stress*, is characterized by its density $\boldsymbol{\Pi}_n(P)$ (force per unit area of Σ at point P).

Let (x_1, x_2, x_3) be an orthogonal cartesian coordinate system and $\mathbf{i}_k$, $k = 1, 2, 3$, the basis vectors of the system (i.e., unit vectors directed along the x_k axis). Let us express $\boldsymbol{\Pi}_n$ in terms of $\boldsymbol{\Pi}_k$, $k = 1$, 2, 3. ($\boldsymbol{\Pi}_k$ is a stress on the surface with a normal $\mathbf{i}_k$). Imagine an infinitesimal tetrahedron $MABC$ cut out of the medium (see Figure 1.1). Let $d\sigma_1$, $d\sigma_2$, $d\sigma_3$, and

$d\sigma_n$ denote the surface areas of the faces MBC, MAC, MAB, and ABC, respectively. If h_n is the altitude of the tetrahedron dropped from its vertex M, $\alpha_1, \alpha_2, \alpha_3$ are the direction cosines of $\mathbf{n}$ and $d\omega$ is the volume of $MABC$, then

$$d\sigma_i = \alpha_i \cdot d\sigma_n, \quad d\omega = h_n \cdot d\sigma_n/3, \quad i = 1, 2, 3 \qquad (1.5)$$

The surface forces acting on the tetrahedron at its faces MBC, MAC, MAB, and ABC are, respectively,

$$-\mathbf{\Pi}_1 d\sigma_1, \quad -\mathbf{\Pi}_2 d\sigma_2, \quad -\mathbf{\Pi}_3 d\sigma_3, \quad \mathbf{\Pi}_n d\sigma_n \qquad (1.6)$$

The mass force F acting on the tetrahedron is

$$\mathbf{F} = \mathbf{f}\rho d\omega \qquad (1.7)$$

where $\mathbf{f}$ is the mass force density and ρ the density of the medium. Finally, the force of inertia is

$$\mathbf{F} = -\rho\frac{d}{dt}\mathbf{v}d\omega \qquad (1.8)$$

Thus, by Newton's second law applied to the tetrahedron $MABC$, formulas (1.5)–(1.8) yield

$$\mathbf{\Pi}_n - \sum_{i=1}^{3} \alpha_i \mathbf{\Pi}_i - h_n\rho\left(\mathbf{f} + \frac{d}{dt}\mathbf{v}\right) = 0 \qquad (1.9)$$

Hence, going to the limit of $h_n \downarrow 0$ we obtain condition of local mechanical equilibrium

$$\mathbf{\Pi}_n = \sum_{i=1}^{3} \alpha_i \mathbf{\Pi}_i \qquad (1.10)$$

Now let

$$\Pi_{nj}, \ \Pi_{ij}, \quad i, j = 1, 2, 3 \qquad (1.11)$$

be the jth components of the vectors $\mathbf{\Pi}_n$ and $\mathbf{\Pi}_i$, respectively. Then by (1.10),

$$\Pi_{nj} = \sum_{i=1}^{3} \alpha_i \cdot \Pi_{ij}, \quad j = 1, 2, 3 \qquad (1.12)$$

According to (1.12), the stress on an arbitrarily oriented surface with normal $\mathbf{n}(\alpha_1, \alpha_2, \alpha_3)$ is determined by a tensor

$$\tilde{\mathbf{\Pi}} = (\Pi_{ij}) \qquad (1.13)$$

known as the *stress tensor*. Reasoning as in (1.5)–(1.10) for the momenta of the surface forces on the tetrahedron $MABC$, one sees that the tensor $\tilde{\mathbf{\Pi}}$ is symmetric, that is

$$\Pi_{ij} = \Pi_{ji}, \quad i, j = 1, 2, 3 \qquad (1.14)$$

The diagonal components Π_{ii} of $\tilde{\Pi}$ are known as *normal* stresses, the nondiagonal ones *tangential* or *shear* stresses.

Define $\tilde{p}$ to be minus one third of the trace of the stress tensor

$$\tilde{p} = -\frac{1}{3} \sum_{i=1}^{3} \Pi_{ii} \tag{1.15}$$

This quantity is known as the *pressure*. Let $\tilde{I}$ be the unit tensor, that is, the tensor whose components are

$$I_{jk} = \delta_{ij} = \begin{cases} 1 & \text{if } i = j, \ j = 1, 2, 3 \\ 0 & \text{if } i \neq j, \ i, j = 1, 2, 3 \end{cases} \tag{1.16}$$

Finally, let $\tilde{\tau}$ be the deviator of the tensor $\tilde{\Pi}$:

$$\tilde{\tau} = \tilde{\Pi} + \tilde{p}\tilde{I} \tag{1.17}$$

A continuum for which $\tilde{\tau}$ vanishes identically (admits no shear stresses at any point) and for which the only normal stress is pressure (with a minus sign) is occasionally called *perfect* (*ideal*).

B. *Deformation of continuous medium. Strain tensor and strain velocity tensor*

The deformation of a continuum is described as follows. Let (x_1, x_2, x_3) be a rectangular cartesian coordinate system, $p = (x_1, x_2, x_3)$ some material point in the medium, and $q = (\xi_1, \xi_2, \xi_3)$ the position of the same material point in the deformed state. Let us assume that the Jacobian

$$\frac{D(x_1, x_2, x_3)}{D(\xi_1, \xi_2, \xi_3)} \neq 0 \tag{1.18}$$

so that the x_i may be considered as implicit functions of ξ_j, $i, j = 1, 2, 3$. Consider a line element

$$\mathbf{ds} = \sum_{i=1}^{3} dx_i \mathbf{e}_i \tag{1.19}$$

in the undeformed body, which transforms into a line element $\mathbf{d\tilde{s}}$ as result of the deformation:

$$\mathbf{d\tilde{s}} = \sum_{i=1}^{3} d\xi_i \mathbf{e}_i \tag{1.20}$$

The vector

$$\mathbf{r}_{pq} = \sum_{i=1}^{3} V_i \mathbf{e}_i \tag{1.21}$$

where

$$V_i = \xi_i - x_i, \quad i = 1, 2, 3 \tag{1.22}$$

is known as the *displacement vector*. Let ds and $d\tilde{s}$ be the absolute values of $\mathbf{ds}$ and $\mathbf{d\tilde{s}}$, respectively. The relative extension of $\mathbf{ds}$ is

$$e = \frac{d\tilde{s}}{ds} - 1. \tag{1.23}$$

By (1.19)–(1.22),

$$d\xi_1 = \left(1 + \frac{\partial}{\partial x_1}V_1\right)dx_1 + \frac{\partial}{\partial x_2}V_1 dx_2 + \frac{\partial}{\partial x_3}V_1 dx_3 \tag{1.24}$$

$$d\xi_2 = \frac{\partial}{\partial x_1}V_2 dx_1 + \left(1 + \frac{\partial}{\partial x_2}V_2\right)dx_2 + \frac{\partial}{\partial x_3}V_2 dx_3 \tag{1.25}$$

$$d\xi_3 = \frac{\partial}{\partial x_1}V_3 dx_1 + \frac{\partial}{\partial x_2}V_3 dx_2 + \left(1 + \frac{\partial}{\partial x_3}V_3\right)dx_3 \tag{1.26}$$

Let l_i and $\bar{l}_i$, $i = 1, 2, 3$, be the direction cosines of the line elements $\mathbf{ds}$ and $\mathbf{d\tilde{s}}$, respectively, so that

$$ds = \sum_{i=1}^{3} l_i dx_i, \qquad d\tilde{s} = \sum_{i=1}^{3} \tilde{l}_i d\xi_i \tag{1.27}$$

Since

$$\sum_{i=1}^{3} l_i^2 = 1, \qquad \sum_{i=1}^{3} \tilde{l}_i^2 = 1 \tag{1.28}$$

equalities (1.24)–(1.26) imply that

$$\left(\frac{d\tilde{s}}{ds}\right)^2 - 1 = e(e+2) = 2F(l_1, l_2, l_3) \tag{1.29}$$

where

$$F(l_1, l_2, l_3) = \sum_{i=1}^{3}\sum_{j=1}^{3} \varepsilon_{ij} l_i l_j \tag{1.30}$$

Here

$$\varepsilon_{ij} = e_{ij} + e_{ij}^* \tag{1.31}$$

$$e_{ij} = \frac{1}{2}\left(\frac{\partial V_i}{\partial x_j} + \frac{\partial V_j}{\partial x_i}\right) \tag{1.32}$$

$$e_{ij}^* = \frac{1}{2}\frac{\partial V}{\partial x_i}\frac{\partial V}{\partial x_j} \tag{1.33}$$

Thus the quadratic form F consists of two parts: one part that is linear in the derivatives of the displacement vector,

$$f(l_1, l_2, l_3) = \sum_{i=1}^{3}\sum_{j=1}^{3} e_{ij} l_i l_j \tag{1.34}$$

and a bilinear part,

$$f^*(l_1, l_2, l_3) = \sum_{i=1}^{3} \sum_{j=1}^{3} e_{ij}^* l_i l_j \tag{1.35}$$

The quadratic forms F and f generate two tensors

$$\tilde{\varepsilon} = (\varepsilon_{ij}) \tag{1.36}$$
$$\tilde{e} = (e_{ij}) \tag{1.37}$$

$\tilde{\varepsilon}$ is known as the tensor of finite deformations, and $\tilde{e}$ as the tensor of small deformations or the strain tensor.[1] If the process under consideration is time dependent, the tensors

$$\tilde{\varepsilon}' = (\dot{\varepsilon}_{ij}), \quad \tilde{e}' = (\dot{e}_{ij}) \tag{1.38}$$

are known, respectively, as the rate of finite deformation and rate of strain tensors. (The dot designates partial differentiation with respect to time.)

C. Volumetric dilatation

Volumetric dilatation is defined as

$$\Theta = \frac{\omega_1 - \omega}{\omega} \tag{1.39}$$

where ω and ω_1 are the specific volumes of the medium before and after deformation, evaluated at the same material point. Let ρ_1 and ρ be the densities of the medium in the deformed and undeformed states, respectively. Then

$$\rho = \frac{1}{\omega} \Rightarrow \Theta = \frac{\rho - \rho_1}{\rho_1} \tag{1.40}$$

Let M be the mass of some fixed set of material particles, occupying an arbitrarily chosen simply connected region D at some time, so that

$$M = \int_D \rho(q) dx \tag{1.41}$$

where dx denotes the volume element

$$dx = dx_1 dx_2 dx_3 \tag{1.42}$$

After deformation, the same set of material points occupies a deformed region D_1, so that

$$M = \int_{D_1} \rho_1 d\xi \tag{1.43}$$

where

$$\xi_i = x_i + V_i(x_1, x_2, x_3) \tag{1.44}$$

---

[1] In what follows we consider only small deformations. Accordingly, we use the term "strain tensor" only for small deformations.

and V_i are the components of the displacement vector.

We have

$$M = \int_{D_1} \rho_1 d\xi = \int_D \rho_1 \left| \frac{D(\xi_1, \xi_2, \xi_3)}{D(x_1, x_2, x_3)} \right| dx \qquad (1.45)$$

where

$$\frac{D(\xi_1, \xi_2, \xi_3)}{D(x_1, x_2, x_3)} = \begin{vmatrix} 1 + \dfrac{\partial}{\partial x_1} V_1 & \dfrac{\partial}{\partial x_2} V_2 & \dfrac{\partial}{\partial x_3} V_1 \\[2mm] \dfrac{\partial}{\partial x_1} V_2 & 1 + \dfrac{\partial}{\partial x_2} V_2 & \dfrac{\partial}{\partial x_3} V_2 \\[2mm] \dfrac{\partial}{\partial x_1} V_3 & \dfrac{\partial}{\partial x_2} V_3 & 1 + \dfrac{\partial}{\partial x_3} V_3 \end{vmatrix} = 1 + \text{div } V + A \quad (1.46)$$

with A containing all quadratic terms in the components of the displacement vector $\mathbf{V}$. Since only small deformations are considered, the Jacobian (1.46) is positive. Therefore, there is no need for absolute-value signs in (1.45).

Equalities 1.41 and 1.45 yield

$$\Theta = \int_D [\rho - \rho_1[1 + \text{div } V + A]] \, dx \qquad (1.47)$$

Since D is an arbitrary simply connected region and we are assuming that the integrand is continuous, it follows from (1.47) that

$$\rho - \rho_1[1 + \text{div } V + A] = 0 \qquad (1.48)$$

or, up to highest order terms in V,

$$\Theta \overset{\text{def}}{=} \frac{\rho - \rho_1}{\rho_1} = \text{div } V \qquad (1.49)$$

D. Substantial (material) derivatives and continuity equation

Assume that the deformation takes place during a time dt, so that

$$\mathbf{V} = \mathbf{v} dt \qquad (1.50)$$

where $\mathbf{v}$ is the velocity of small deformations. Then

$$\rho_1 = \rho + \sum_{i=1}^{3} \frac{\partial \rho}{\partial x_i} v_i dt = \rho + \mathbf{v} \cdot \text{grad } \rho dt \qquad (1.51)$$

which implies

$$\Theta = -\frac{\frac{d}{dt}\rho dt}{\rho + \frac{d}{dt}\rho dt} \approx -\frac{1}{\rho}\frac{d}{dt}\rho dt \qquad (1.52)$$

Comparing (1.52) with (1.49), we obtain

$$\frac{d}{dt}\rho + \rho \, \text{div } \mathbf{v} = 0 \qquad (1.53)$$

On the other hand, ρ_1 is the density of the medium at the point $(\xi_1, \xi_2, \xi_3, t + dt)$,

$$\rho_1 = \rho(\xi_1, \xi_2, \xi_3, t + dt) \tag{1.54}$$

so that

$$\frac{d}{dt}\rho = \frac{\partial}{\partial t}\rho + \sum_{i=1}^{3}\frac{\partial\rho}{\partial x_i}v_i \equiv \frac{\partial}{\partial t}\rho + \mathbf{v} \cdot \operatorname{grad}\rho \tag{1.55}$$

The operator

$$\frac{d}{dt} = \frac{\partial}{\partial t} + \mathbf{v} \cdot \operatorname{grad} \tag{1.56}$$

is called the *substantial* (or *material*) *differentiation* operator; equation (1.53), that is, the local form of the law of mass conservation, is known as the *equation of continuity*[2]

E. Rheological state of continuous medium

The rheological state of a continuous medium is determined by the relationship between the stress tensor and the strain and rate of strain tensors. In the general case,

$$\tilde{\mathbf{\Pi}} = F(\tilde{\varepsilon}, \tilde{\varepsilon}') \tag{1.57}$$

The rheological equation of state of a linearly elastic isotropic body is given by a phenomenological law – Hook's law – which, as formulated by Lamé, reads

$$\Pi_{ij} = \lambda \operatorname{div} \mathbf{V}\delta_{ij} + 2\mu e_{ij} \tag{1.58}$$

where δ_{ij} is the Kronecker delta[3] and λ and μ are known as Lamé coefficients.

Similarly, the rheological equation of state of a real (Newtonian viscous) liquid is given by the phenomenological Navier law, in the form

$$\Pi_{ij} = -\left(p - \left(\frac{2}{3} \cdot \mu - \mu'\right)\operatorname{div}\mathbf{v}\right)\delta_{ij} + 2\mu\dot{e}_{ij} \tag{1.59}$$

If $\tilde{\tau}$ is the deviator of the stress tensor and $-\tilde{p}$ one third of the trace, then it follows from (1.15), (1.37), and (1.32) that

$$p = \tilde{p} - \mu'\operatorname{div}\mathbf{v} \tag{1.60}$$

The quantity p defined by (1.60) is the pressure in the liquid in a state of local thermodynamic equilibrium. If the liquid is incompressible, or if its

---

[2] The equation of continuity will be derived again, in a different way, in Section 3 of Chapter 2.

[3] Recall that the Kronecker delta is defined by $\delta_{ii} = 1$, $\delta_{ij} = 0$, $i, j = 1, 2, 3$, $i \neq j$.

compressibility is small enough, the pressure p is minus one third the trace of the stress tensor. The coefficients μ and μ' are known, respectively, as the coefficients of dynamic (or Newtonian, or shear) viscosity and volume viscosity. In most hydrodynamic situations, the effect of volume viscosity is negligible. It must be taken into account, however, in the case of a compressible medium with a long relaxation process, for example, a gas with a slow chemical reaction.

The rheological equation of state for an inviscid (perfect) liquid is

$$\tilde{\Pi} = -p\tilde{\mathbf{I}} \tag{1.61}$$

that is, there are no shear stresses in an inviscid liquid.

F. Kinematics of fluid motion. Inviscid fluid flow

The following definitions and theorems are basic for the kinematics of any liquid.

Definition 1.2.1. Let $\mathbf{v}$ be the velocity vector of fluid motion. The motion of the fluid is said to be *rotational*, if there exists a vector $\mathbf{\Omega}$, called the *vector velocity potential*, such that

$$\mathbf{v} = \operatorname{rot} \mathbf{\Omega} \tag{1.62}$$

Obviously, the velocity of rotational motion is solenoidal, in the sense that

$$\operatorname{div} \mathbf{v} = \operatorname{div} \operatorname{rot} \mathbf{\Omega} \equiv 0 \tag{1.63}$$

Definition 1.2.2. The motion of a fluid motion is said to be *potential* if there exists a scalar φ, called the *velocity potential*, such that

$$\mathbf{v} = \operatorname{grad} \varphi \tag{1.64}$$

Recall that any continuously differentiable vector function $\mathbf{v}$ may be represented as the sum of the rotor of a vector potential $\mathbf{\Omega}$ and the gradient of a scalar potential φ; that is, in general,

$$\mathbf{v} = \operatorname{rot} \mathbf{\Omega} + \operatorname{grad} \varphi \tag{1.65}$$

Definition 1.2.3. Let $\mathbf{A}(p, t)$ be a vector field. A curve Γ is called a *vector line* if it is tangent to $\mathbf{A}$ at each of its points.

In particular, if $\mathbf{A}(p, t)$ is a field of vorticity ω, defined as

$$\omega = \operatorname{rot} \mathbf{v} \tag{1.66}$$

then the vector line of ω is called a *vortex line*.

Definition 1.2.4. Let $L \subset \mathbb{R}_3$ be a closed curve and let S be the surface composed of the vector lines Γ of $\mathbf{A}$, passing through L. Then S is called

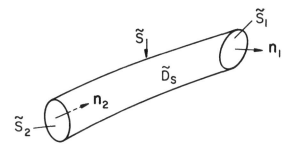

Figure 1.2. Section $\tilde{D}_s$ of a vector tube D by
plane S_1 and S_2 with outward and inward normals
$\mathbf{n}_1$ and $\mathbf{n}_2$, respectively.

the *surface of a vector tube*, and the part D_s of the space $\mathbb{R}_3$ bounded by S
is called a *vector tube*.[4] A vector tube generated by vortex lines is known
as a *vortex tube*.

Let $\tilde{D}_s$ be the part of a vortex tube D_s bounded by S and by two planes
S_1 and S_2 without common points in a closure of $\tilde{D}_s$ and not tangent to S.
Let $\mathbf{n}_1$ and $\mathbf{n}_2$ be the normals to S_1 and S_2, directed outward and inward
to $\tilde{D}_s$ respectively (see Figure 1.2). Let

$$\tilde{S} = S \cap \tilde{D}_s, \quad \tilde{S}_i = S_i \cap \tilde{D}_s, \qquad i = 1,2 \tag{1.67}$$

Consider the integral

$$I = \int_{\tilde{S}} \boldsymbol{\omega} \cdot \mathbf{d\sigma} + \int_{\tilde{S}_1} \boldsymbol{\omega} \cdot \mathbf{d\sigma} - \int_{\tilde{S}_2} \boldsymbol{\omega} \cdot \mathbf{d\sigma} \tag{1.68}$$

By the Gauss divergence theorem,

$$I = \int_{\tilde{D}_S} \operatorname{div} \boldsymbol{\omega} dx = 0 \tag{1.69}$$

since $\boldsymbol{\omega}$ is a solenoidal vector. Besides,

$$\int_{\tilde{S}} \boldsymbol{\omega} \cdot \mathbf{d\sigma} = 0 \tag{1.70}$$

It follows from (1.68)–(1.70) that

$$\int_{\tilde{S}_1} \boldsymbol{\omega} \cdot \mathbf{d\sigma} = \int_{\tilde{S}_2} \boldsymbol{\omega} \cdot \mathbf{d\sigma} \tag{1.71}$$

Definition 1.2.5. The integral

$$\mathbf{J} = \int_{\tilde{S}_1} \boldsymbol{\omega}(q,t) \mathbf{d\sigma}_q \tag{1.72}$$

is called the *strength* of the vortex tube.

---

[4]It is assumed that any part of D_s bounded by S and two planes, not tangent to S,
does not contain a point at infinity.

Equality 1.71 means that the following theorem is proved:

Theorem 1.2.1. *The strength of any vortex tube is constant (i.e., independent of the choice of sections $\tilde{S}_1$).*

Corollary. *A vortex line ends either on the solid surfaces bounding the fluid or at infinity.*

Definition 1.2.6. A force field $\mathbf{F}$ is said to be *conservative* if it is independent of time and has a potential U; that is, if $\mathbf{F} = \operatorname{grad} U$.

Definition 1.2.7. The vector line is said to be *frozen into the fluid* if it always consists of the same material particles. The same definition applies to vector tubes.

We may now formulate Friedman's theorem on rotational motion of a fluid.

Theorem 1.2.2 (Friedman [22]). *Let $\mathbf{A}$ be a continuous vector field associated with a fluid in motion. Assume that $\mathbf{A}$ does not vanish inside the fluid. Then the vector lines and vector tubes are frozen into the fluid if and only if*

$$\frac{d}{dt}\mathbf{A} - (\mathbf{A} \cdot \operatorname{grad})\mathbf{q} + \mathbf{A} \cdot \operatorname{div} \mathbf{q} = 0 \tag{1.73}$$

where $\mathbf{q}$ is the velocity of fluid particles.

Corollary (Helmholtz theorem). *If a barotropic (i.e., with density dependent on pressure only) ideal fluid is flowing in a conservative force field, the vortex lines and vortex tubes are frozen into fluid.*

This means that if at some initial moment the motion of an inviscid fluid is irrotational, it will remain so always.

3. Elements of electrostatics [20]

Electrostatics is that part of electrodynamics which deals with electrical equilibrium (i.e., steady states without electric current).

Coulomb's law, electric potential, conductors, dielectrics

Let the space $\mathbb{R}_3$ be occupied by a homogeneous continuous medium. Assume that a point charge e_0 is placed at a point p and another charge e is placed at a point q. Then the charge e_0 acts upon e with a force

$$\mathbf{F} = \frac{e_0 e}{4\pi \varepsilon r_{pq}^2} \mathbf{r}_{pq}^0 \tag{1.74}$$

where $\mathbf{r}_{pq}^0$ is the unit vector in the direction from p to q. In the Gaussian Electrostatic System of units (GES) ε is a positive dimensionless constant, characterizing the medium in D; it is known as the *dielectric permeability*. ε is equal to 1 in vacuum and greater than 1 for any dielectrically polarizable medium. Occasionally we shall refer to a unit charge as a *test charge*, assuming whenever possible that $\varepsilon = 1$ independently of the material in D. According to (1.74), a charge e_0 placed at point $p \in \mathbb{R}_3$ generates in $\mathbb{R}_3$ a radially symmetric field of strength (intensity) $\mathbf{E}$,

$$\mathbf{E}(r) = \frac{e_0}{4\pi\varepsilon r_{pq}^2}\mathbf{r}^0 \tag{1.75}$$

By (1.74) and (1.75), the charge e at q experiences a force

$$\mathbf{F} = e\mathbf{E} \tag{1.76}$$

The function

$$V(r) \stackrel{\text{def}}{=} - \int_\infty^r \mathbf{F} \cdot d\mathbf{r} = -e \int_\infty^r E(\rho)d\rho = \frac{e_0 e}{4\pi r_{pq}} \tag{1.77}$$

determines the work necessary to move the charge e from infinity to the surface of a sphere of radius r. The quantity

$$\varphi = V(r)/e \tag{1.78}$$

is known as the *electric potential* of the electric field due to a charge e_0 placed at the center of the sphere. From this and (1.76) we have

$$\mathbf{E} = - \operatorname{grad} \varphi \tag{1.79}$$

The differential equations for the vector lines of $\mathbf{E}$ are

$$\frac{dx_1}{E_1} = \frac{dx_2}{E_2} = \frac{dx_3}{E_3} \tag{1.80}$$

Here (x_1, x_2, x_3) are Cartesian coordinates and E_1, E_2, E_3 are the components of $\mathbf{E}$. The force lines of the electrostatic field begin and end at electric charges or at infinity. The electric field generated by a system of charges is the superposition of the fields generated by the various charges and the same is true of its potential.

Among materials, one distinguishes between conductors and dielectrics. At equilibrium, there is no electric field in the bulk of a charged conductor, so that all charges are concentrated at the surface. (A nonvanishing field inside a conductor would generate electric current, violating the equilibrium.)

In dielectrics, besides free charges there exist so-called linked charges, due to the polarization of the medium. These charges may be viewed as resulting from the arrangement of dipoles. By a *dipole* we mean a pair of two equal charges of opposite signs, separated by a distance vector $\mathbf{d}$. The dipole moment $\mathbf{N}$ is defined as

$$\mathbf{N} = e\mathbf{d} \tag{1.81}$$

Thus, the molecules of a dielectric may be viewed as dipoles of the following two types: those for which the absolute value of the dipole moment is fixed and for which orientation is affected by the local electric field (orientational dielectric polarization); and those for which the dipole moment is proportional to the intensity of the local electric field (elastic polarization).

In a dielectrically polarizable isotropic medium we can define two vectors,

$$\mathbf{D} = \varepsilon\mathbf{E}, \qquad \mathbf{P} = (\varepsilon - 1)\mathbf{E} \equiv \mathbf{D} - \mathbf{E} \tag{1.82}$$

known respectively as the *electric displacement*[5] and *electric polarization*. It can be deduced from (1.75) and (1.82) that

$$\operatorname{div}\mathbf{D} = \rho \tag{1.83}$$

where ρ is the volume density of the free charges.

If Σ is a surface separating two dielectric media, then the normal component of D experiences a jump at Σ equal to the surface density of the free charges distributed over Σ; the tangential components of $\mathbf{E}$ are, however, continuous at Σ (see Chapter 2, Section 10).

4. Elements of electrodynamics [20]

A flow of free charges is called a *conduction current*. According to Ohm's phenomenological law, the conduction current density $\mathbf{j}$ is proportional to $\mathbf{E}$. Thus

$$\mathbf{j} = \sigma\mathbf{E} \tag{1.84}$$

In the simplest case – a linear homogeneous conductor – the conductivity σ is a constant.

Conduction current and the variation in time of electric displacement generate a magnetic field. The relation between the magnetic field intensity $\mathbf{H}$ and the former is established by the generalized Ampère law; in GES system,

$$\operatorname{rot}\mathbf{H} = \frac{4\pi}{c}\mathbf{j} + \frac{1}{c}\frac{\partial}{\partial t}\mathbf{D} \tag{1.85}$$

where $c = 10^{10}$ cm/sec is the speed of light. The quantity $(1/4\pi)(\partial\mathbf{D}/\partial t)$ is known as the *displacement current density*. The total current density is defined as

$$\mathbf{C} = \mathbf{j} + \frac{1}{4\pi}\frac{\partial\mathbf{D}}{\partial t} \tag{1.86}$$

so that the generalized Ampère law reads

$$\operatorname{rot}\mathbf{H} = \frac{4\pi}{c}\mathbf{C} \tag{1.87}$$

---

[5]Generally ε depends on the field, but for moderate $\mathbf{E}$ in most media it may be considered with a reasonable accuracy as constant.

The magnetic field in a magnetically polarizable medium is characterized, besides **H**, by the magnetic induction (displacement) vector **B**, related to **H** by

$$\mathbf{B} = \mu \mathbf{H} \tag{1.88}$$

Here μ is the magnetic permeability of the medium, which is a positive constant equal to unity in a vacuum, smaller than unity in what are known as diamagnetic media, and greater than unity in para- or ferromagnetic media. From now on we shall assume whenever possible that $\mu = 1$. Magnetic induction is a solenoidal vector; this reflects the absence of magnetic charges in nature. Thus

$$\operatorname{div} \mathbf{B} = 0 \tag{1.89}$$

Variations of magnetic induction in time generate an electric field; according to Faraday's law,

$$\operatorname{rot} \mathbf{E} = -\frac{1}{c}\frac{\partial \mathbf{B}}{\partial t} \tag{1.90}$$

Equations (1.83), (1.86), and (1.90), together with (1.82), (1.88), (1.84), and (1.85), constitute Maxwell's system of equations for electrodynamics (see Chapter 2, Section 7).

5. Elements of chemical kinetics [3]

Variables of reactive system. Reaction rate

The macroscopic variables of a reactive system are chosen according to the specific context. Besides the usual thermodynamic variables of temperature, volume, and pressure, and such hydrodynamic variables as velocity and stresses, these usually include the chemical composition of the reactive mixture. The composition of a mixture is characterized by the concentrations of its components.

Reactions evolving in a single-phase system are said to be *homogeneous*, as opposed to *heterogeneous* reactions, which evolve in a multiphase system.

It is customary to distinguish between simple and complex kinetics. Simple kinetics involves a single irreversible reaction

$$A + B + C \cdots \to P \tag{1.91}$$

The rate of production in the course of reaction (1.91) is determined by the mass action law,

$$\frac{d}{dt}C_p = k \cdot C_a^{\alpha} \cdot C_b^{\beta} \cdot \cdots \cdot C_c^{\gamma} \tag{1.92}$$

where $C_a, C_b, \ldots, C_c$ are the molar concentrations of $A, B, \ldots, C$, respectively, and $\alpha, \beta, \ldots, \gamma, k$ are positive constants. Of these, k is known as the *rate constant*. The reaction is said to be of order $\alpha, \beta, \ldots, \gamma$ with respect to species $A, B, \ldots, C$, respectively. The *total order of the reaction* (or simply *reaction order*) is defined as $\alpha + \beta + \cdots + \gamma$. Equations of type (1.92) are written for all components.

Consider the stochiometric equation of a simple reaction (i.e., one including the number of reacting molecules of each reactant in a single reaction act),

$$\alpha C_a + \beta C_b + \cdots \Rightarrow pC_p + qC_q + \cdots \tag{1.93}$$

The rate R of the simple reaction is then defined as

$$R = -\frac{1}{\alpha} \cdot \frac{d}{dt}C_a = -\frac{1}{\beta} \cdot \frac{d}{dt}C_b = \cdots = \frac{1}{p} \cdot \frac{d}{dt}C_p = \frac{1}{q} \cdot \frac{d}{dt}C_q = \cdots \tag{1.94}$$

In complex kinetics, each component of the reactive mixture may participate in several reactions. Several types of complex kinetics may be distinguished as follows.

(1) *Reversible reactions.* In a reversible reaction the products of the direct reaction participate in a reverse reaction whose products are the components of the direct reaction. In kinetics of this type, the reactive system may have a nontrivial dynamic equilibrium.

(2) *Parallel reactions.* For kinetics of this group, each species may react in several different patterns. The total reaction rate is then the sum of rates of all the parallel competing reactions.

(3) *Successive reactions* [73]. In successive reaction kinetics, the reaction products of the first reaction become the components of a subsequent reaction, and so on.

Here the following general remark is due. *The limiting stage of a sequence of consecutive processes (e.g., reactions) is that with the slowest rate.* In particular, the limiting stage of a heterogeneous reaction evolving in a multiphase systems may either be related to developments at the phase interfaces or depend on the mass transfer from one phase to another. Examples of processes of the second kind are heterogeneous catalysis and physical sorption (adsorption). The first group includes chemosorption (absorption). In either of these systems the rate of reaction is determined by two factors, the rate of supply of reagents to the reaction site and the true reaction rate.

Consider a heterogeneous reaction evolving in a bed of porous grains, for example, of activated coal. A distinction must be made between the external surface of the grains and the walls of the pores inside the grains. In activated coal, the specific surface of the pores of grains may be of the order of 10^6–10^7 cm^2/cm^3, whereas the specific external surface of the grains may be of the order of 10^2–10^3 cm^2/cm^3. The reaction rate is determined by the slowest of the subprocesses involved, which are the rate of supply of reagent to the external boundaries of the grains from the bulk of the solution (gas or liquid) filling the porous space between the grains, the rate of transfer of reagent to the surfaces of the pores of the grains, and the true reaction rate per unit reactive surface area. The following terminology is used. [21]

- A heterogeneous reaction is evolving in the *outer diffusion* region if the rate of transfer of reagent from the bulk of the solution to the external surfaces of the grains is the slowest subprocess.
- A heterogeneous reaction is evolving in the *inner diffusion* region if the rate of transfer of reagent to the surfaces of the inner pores of the grains is the slowest subprocess.
- A heterogeneous reaction is evolving in the *kinetic* region if the limiting subprocess is the true reaction rather than mass transfer.

Naturally, cases are possible in which one cannot select the slowest subprocess, since different subprocesses evolve at comparable rates. The mathematical description of the process depends essentially on which of the aforementioned kinetic regimes the reaction is evolving in.

To illustrate, we shall now develop kinetic equations for the simplest case of adsorption by a sorbent bed from a percolating solution, assuming that the limiting stages are sorptive diffusion toward the surface of the sorbent grains and a reversible adsorption reaction. (By *sorbent* we mean the phase at which adsorption occurs; by *sorptive*, the species which is adsorbed.)

For simplicity, let us assume that the concentration of sorptive A in the percolating solution is constant, equal to c_0 (mol/cm³) everywhere in the porous space of the bed, except for a narrow layer of solution of thickness δ (cm) adjacent to a sorbent grain. Regarding this boundary layer, called diffusion or unstirred layer or Nernst film, let us assume that the fluid in it is motionless and that the only sorptive transport mechanism in it is molecular diffusion with diffusivity D (cm²/sec). This crude idealization, which usually yields qualitatively correct results, has proved useful in numerous kinetic contexts. Let the reversible adsorption reaction be

$$A + S \underset{k_2}{\overset{k_1}{\rightleftharpoons}} R \tag{1.95}$$

Here A, S, R, k_1, and k_2 stand for the sorptive, sorbent, adsorbed species, and the direct (adsorption) and reverse (desorption) reaction rate constants, respectively. Let the respective concentrations be c_w (mol/cm³) for the sorptive concentration at the reactive surface Σ (characterized by a unit normal vector **n**), s (mol/cm²) for surface concentration of the unoccupied reactive sites, and a (mol/cm²) for the surface concentration of the adsorbed species. Note that (1.95) is based on the assumption that each reactive site reacts with only one sorptive molecule.

The kinetic equation corresponding to (1.95) is

$$\frac{\partial}{\partial t} a = k_1 c_w s - k_2 a \tag{1.96}$$

On the other hand, recall that the rate of adsorption da/dt should be equal to the rate of diffusive supply of sorptive towards the reactive surface Σ. Thus, by mass balance,

$$\frac{\partial}{\partial t} a = -D \frac{\partial}{\partial n} c \bigg|_{\Sigma} \tag{1.97}$$

Recall that $-D(\partial c/\partial n)\,|_\Sigma$ is the sorptive diffusion flux at Σ, where c is the sorptive concentration in the diffusion layer. Let us approximate this flux by the expression

$$-D\frac{\partial}{\partial n}c\,\bigg|_\Sigma = -\frac{D}{\delta}(c_w - c_0) \tag{1.98}$$

Returning to equation (1.96), we note that

$$s = a_\infty - a \tag{1.99}$$

Here a_∞ (mol/cm^2) is the maximum available surface concentration of the adsorption sites. Substitution of (1.99) into (1.96) and of (1.98) into (1.97), respectively, yields

$$\frac{\partial}{\partial t}a = k_1 c_w(a_\infty - a) - k_2 a \tag{1.100}$$

$$\frac{\partial}{\partial t}a = -\frac{D}{\delta}(c_w - c_0) \tag{1.101}$$

Equations (1.100) and (1.101) represent the final dimensional form of the kinetic adsorption–diffusion equations. Note that the dimension of the rate constants k_1 and k_2 are, respectively,

$$[k_1] = \frac{\text{cm}^3}{\text{mol}\cdot\text{sec}}, \qquad [k_2] = \frac{1}{\text{sec}} \tag{1.102}$$

Let us introduce in (1.98)–(1.101) dimensionless variables (indicated by a "tilde"), defined as

$$\tilde{a} = \frac{a}{a_\infty}, \quad \tilde{c}_e = \frac{c_w}{c_0}, \quad \tilde{t} = \frac{t}{t_0} \tag{1.103}$$

Here t_0 is some time scale, arbitrary so far. Substitution of (1.103) into (1.100) and (1.101) yields

$$\frac{\partial \tilde{a}}{\partial \tilde{t}} = \alpha[\tilde{c}_e(1 - \tilde{a}) - K\tilde{a}] \tag{1.104}$$

$$\frac{\partial \tilde{a}}{\partial \tilde{t}} = \beta[1 - \tilde{c}_e] \tag{1.105}$$

Here α, β, and k are dimensionless parameters, defined as

$$\alpha = t_0 k_1 c_0, \quad \beta = \frac{t_0 D c_0}{\delta a_\infty}, \quad K = \frac{k_2}{k_1 c_0} \tag{1.106}$$

Let us assume that if t_0 is suitably chosen then $\partial \tilde{a}/\partial \tilde{t}$ is of the order of unity. Note that both $(k_1 c_0)^{-1}$ and $(Dc_0/\delta a_\infty)^{-1}$ have the dimension of time. Denote

$$\tau_1 = (k_1 c_0)^{-1}, \qquad \tau_2 = (Dc_0/\delta a_\infty)^{-1} \tag{1.107}$$

These quantities τ_1, τ_2 define the reactive and diffusion time scales, respectively, intrinsic to the problem. Chose t_0 to be the slowest of them. We consider two limiting cases.

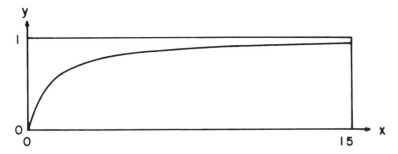

Figure 1.3. Langmuir's isotherm; $y = x/(1+x)$, $x = \tilde{c}_e/K$, $y = \tilde{a}/\tilde{a}_\infty$.

Case 1. $\tau_1 \ll \tau_2 \, (= t_0) \to$ outer diffusion kinetics (outer diffusion region). In this case, according to (1.104), (1.105), and (1.107),

$$\beta = 1, \qquad \alpha \gg 1 \tag{1.108}$$

whence, noting that

$$\frac{\partial \tilde{a}}{\partial \tilde{t}} = O(1) \tag{1.109}$$

we obtain

$$\tilde{c}_e(1 - \tilde{a}) - K\tilde{a} = 0 \tag{1.110}$$

$$\frac{\partial \tilde{a}}{\partial \tilde{t}} = 1 - \tilde{c}_e \tag{1.111}$$

The relation (1.110), implying local equilibrium between the sorptive unadsorbed at the reactive surface and that in the adsorbed state, is known as the Langmuir adsorption isotherm; K is called the adsorption equilibrium (dissociation) constant. Equalities (1.110) and (1.111) are the kinetic equations for the outer diffusion region. Solving (1.110) for $\tilde{c}_e$ and substituting the result into (1.111), we can rewrite the latter as

$$\frac{\partial \tilde{u}}{\partial \tilde{t}} = 1 - K\frac{\tilde{a}}{1 - \tilde{a}} \tag{1.112}$$

Alternatively, resolving (1.110) with respect to $\tilde{a}$ we find

$$a(p,t) = \frac{a_\infty \tilde{c}_e}{K + \tilde{c}_e}, \quad K = \frac{k_2}{k_1 c_0} \tag{1.113}$$

In Figure 1.3 $\tilde{a}/a_\infty$ is plotted as a function of $\tilde{c}_e/K$. This equilibrium plot has a plateau, corresponding to the limit $\tilde{c}_e \uparrow \infty$, and a linear part

$$\tilde{a}(p,t) = \Gamma\tilde{c}_e(p,t) \tag{1.114}$$

corresponding to $\tilde{c}_e \ll K$. The relation (1.114) is known as the Henry isotherm. The simplest case of adsorption evolving in the outer diffusion region is considered in subsequent chapters (see Chapter 2, Section 8, and Chapter 6, Section 3).

Case 2. $\tau_2 \ll \tau_1 \ (= t_0) \to$ reaction kinetics (kinetic region). In this limit $\alpha = 1$ and $\beta \gg 1$. This, together with (1.108) and (1.109), implies that

$$\tilde{c}_e = 1, \qquad \frac{\partial \tilde{a}}{\partial \tilde{t}} = \tilde{c}_e (1 - \tilde{a}) - K\tilde{a} \qquad (1.115)$$

Relations (1.115) represent the adsorption kinetics equations for the kinetic region. Whenever the bulk concentration is not constant $(c \neq c_0)$, as it was in the simplest case considered here, but rather determined by a transport process (e.g., convective diffusion, dispersion, etc.), an appropriate modification of the left side of (1.115) is necessary.

Finally, another remark is due concerning the thermal effect taking place in a reaction. A reaction is said to be *endothermal* if it evolves with absorption of heat, and *exothermal* if heat is released in the course of reaction. An endothermal reaction may evolve only if a sufficient amount of heat is supplied to the reaction vessel; it may stop if the heat supply is inadequate. The very opposite is the case with an exothermal reaction. Usually the rate of reaction increases with the temperature. If the rate of removal of the heat released in an exothermal reaction is less than the rate of heat release itself, then the temperature may increase unboundedly, resulting in what is known as *thermal explosion* (see Chapter 17, Section 3).

6. Elements of equilibrium thermodynamics [60]

Thermodynamics studies the thermal properties of macroscopic systems at equilibrium. We shall outline a few basic thermodynamic concepts.

Let D be a domain in a continuum in $\mathbb{R}_3$, bounded by a surface Σ. The physical system corresponding to this domain may be either isolated, closed, or open, depending on the conditions at the surface Σ. The system is said to be *isolated* if there is no exchange of material and heat between D and the external space. If there is no exchange of material but an influx or efflux of heat may occur, the system is said to be *closed*. If both material and heat may be exchanged with the external space, the system is said to be *open*.

A system is said to be *at equilibrium* if it is independent of time and there are in it no macroscopic fluxes. Any isolated system, not initially at equilibrium, eventually reaches its equilibrium state. The time necessary for the system to reach its equilibrium is known as its *relaxation time*. If the passage from one equilibrium state of the system to another, due to an infinitesimally slow change of external conditions, occurs at infinitesimal speed, then the system may be viewed as passing through an infinite sequence of equilibrium states in a reversible manner, which means that reversion of the external conditions will invert the sequence of equilibrium states experienced by the system during its evolution.

A. First law of thermodynamics and related concepts

1. Functions of state: internal energy, entropy, enthalpy

Assume that the material system in a region D contains n components A_i at molar concentration c_i, $i = 1, 2, \ldots, n$.

The system at equilibrium may be characterized by various parameters, such as its volume V, pressure p, and temperature T, the concentrations c_i, and so forth. One such equilibrium characteristic is the *internal energy E*, defined by the first law of thermodynamics (the postulate of total energy balance) as

$$dE = \delta Q - \delta W - \sum_{i=1}^{n} \mu_i \cdot dc_i \qquad (1.116)$$

where δQ is the amount of heat supplied to the system, δW is the work performed by the system, c_i the molar concentrations of the ith component, and μ_i its chemical potential.

If a change of state occurs at constant volume, then

$$\delta Q = C_v dT \qquad (1.117)$$

If the change occurs at constant pressure, then

$$\delta Q = C_p dT \qquad (1.118)$$

Here C_v and C_p are the total heat capacities of the system at constant volume and constant pressure, respectively.[6]

According to the fundamentals of equilibrium thermodynamics, p, V, c_i, and μ_i form conjugated pairs, so that any equilibrium characteristic of a state may be expressed as a function of single members of these pairs together with temperature as independent variables.

The quantity δQ is not a full differential of thermodynamic parameters (V, c_i, T) or (P, μ_i, T), or (V, μ_i, T), etc. However, by the first part of the second law of thermodynamics, use of the integrating factor $1/T$ yields the total differential of a function of state S, known as the *entropy*:

$$dS = \frac{\delta Q}{T} \qquad (1.119)$$

Along with entropy and internal energy, a number of additional functions of state, called "potentials," are introduced, depending on the independent variables chosen. These are *free energy* $F(T, V, c_i)$, *enthalpy* $H(S, p, c_i)$, and so on. Thermodynamic potentials are related to each other through the

---

[6] C_v and C_p refer to heat capacities of the system of an arbitrary mass. If the mass of the system is unity, C_v and C_p become the specific heat capacities c_v and c_p, respectively, at constant volume and pressure. An analogous distinction between capital and lowercase letters is used in all relevant cases.

so-called Legendre transformation. Thus, for example, the enthalpy of a monocomponent system is related to the internal energy $E(S, V)$ by

$$dH = dE + d(pV) \tag{1.120}$$

On the other hand, by (1.116) and (1.119),

$$dE = TdS + pdV \tag{1.121}$$

so that for a monocomponent system

$$dH = TdS - Vdp \tag{1.122}$$

Similarly, specific internal energy and specific enthalpy are related by the equality

$$dh = de + d(p/\rho) \tag{1.123}$$

where ρ is the density of the medium.

2. Caloric coefficients of monocomponent system. Coefficient of thermal expansion

As mentioned previously, only two of the parameters (T, p, V) determining the state of a thermodynamic system are independent. Let these be T and V. The caloric coefficients c_v, c_p defined in the previous subsection may be redefined through the equalities

$$c_V \stackrel{\text{def}}{=} \left(\frac{\partial}{\partial T} e \right)_V , \qquad c_p \stackrel{\text{def}}{=} \left(\frac{\partial}{\partial T} h \right)_p \tag{1.124}$$

which yield the specific heat capacities at constant volume and at constant pressure, respectively. (The subscripts v, p imply, in accordance with usual thermodynamic convention, that the respective variables are kept constant during differentiation.) The coefficient

$$h_T \stackrel{\text{def}}{=} \left(\frac{\partial}{\partial p} h \right)_T - \frac{1}{\rho} = -\frac{\alpha}{\rho} T \tag{1.125}$$

is known as the *latent heat of isothermal pressure variation*, and α is the *coefficient of thermal expansion*.

3. Chemical potential in multicomponent system

Similarly, the chemical potential μ_i of a component A_i of a multicomponent system may be rewritten as

$$\mu_i = \rho \left(\frac{\partial}{\partial c_i} e \right)_{S,V} = \rho \left(\frac{\partial}{\partial c_i} h \right)_{S,p} , \qquad i = 1, 2, \ldots, n \tag{1.126}$$

Let Ω_i be the partial molar volume of A_i, and let

$$x_i = \frac{c_i}{c_1 + c_2 + \cdots + c_n} \tag{1.127}$$

be the molar fraction of A_i. In "perfect" systems (e.g., an ideal gas),

$$\mu_i(p,t) = \mu_i^0(p,T) + \Omega_i p - RT \ln(x_i), \quad i = 1,2,\dots,n \qquad (1.128)$$

for arbitrary concentration c_i. In nonperfect (real) systems, equality 1.128 holds only in the limit of infinite dilution – that is, only when the molar fractions of all but one of the components, known as *solutes*, are very low; the exceptional component is the *solvent*.

4. One-phase and multiphase systems

The domain D in an equilibrium continuum may be divided into a finite number of subdomains, so that passage from one subdomain to another is associated with discontinuity of at least one thermodynamic parameter such as T, V, p, or c_i. If there are no surfaces of discontinuity in the relevant domain, the system is said to be a *one-phase* system. In a multiphase system at equilibrium, the temperature and chemical potentials of all components capable of penetrating the phase interfaces must have the same values in all phases. In addition, mechanical equilibrium implies that the normal stress components must be continuous at the interfaces. In ideal gaseous or liquid systems this implies the same value of pressure in all phases.

B. Second law of thermodynamics

A process is said to be *adiabatic* if it is not accompanied by an influx (efflux) of heat to (from) the system. The second part of the second law of thermodynamics asserts that every irreversible process evolving in an isolated (i.e., closed adiabatic) system is accompanied by an increase of entropy. Conversely, any reversible process in an isolated system must evolve at a constant entropy.

7. Integral laws of conservation of extensive parameters

All thermodynamic parameters characterizing a continuum may be divided into two groups: intensive and extensive parameters. Intensive parameters are point functions of the spatial coordinates and time, whereas extensive parameters characterize properties dependent on the mass or volume of a set of material particles, or on the area of the surface or length of the line along which they are distributed. Thus, extensive parameters are set-additive functions. If $\mathfrak{F}(t)$ is an extensive parameter, $\mathfrak{f}(q,t)$ its mass density (the intensive parameter related to $\mathfrak{F}(t)$), and $\rho(q,t)$ the density of the material system, then

$$\mathfrak{F}(t) = \int_{D_t} \rho(q,t)\mathfrak{f}(q,t)d\omega_q \qquad (1.129)$$

where D_t is the region related to the material system with boundary Σ_t. Thus D_t may vary in time. In particular, if $M(t)$ is the mass in D_t then

$$M(t) = \int_{D_t} \rho(q,t)d\omega_q \qquad (1.130)$$

Underlying most of the equations of mathematical physics considered in this book are laws of conservation, which, in their integral form, determine the rate of change of an extensive parameter and are identical in structure, irrespective of the physical origin of the problem or the tensor character of the extensive parameter in question. Integral conservation laws are of the following general structure.

Let us evaluate the rate of change of $\mathfrak{F}(t)$, assuming (to fix ideas) that $\mathfrak{F}$ is a scalar parameter and that the medium is moving at velocity $\mathbf{v}(p,t)$ in a laboratory frame of reference. We have

$$\frac{d}{dt}\mathfrak{F}(t) = \lim_{\tau\downarrow 0}\frac{1}{\tau}\left(\int_{D_{t+\tau}}\rho(q,t+\tau)\mathfrak{f}(q,t+\tau)d\omega_q - \int_{D_t}\rho(q,t)\mathfrak{f}(q,t)d\omega_q\right)$$
$$= I_1 + I_2 \tag{1.131}$$

where

$$I_1 = \int_{D_t}\frac{\partial}{\partial t}[\rho(q,t)\mathfrak{f}(q,t)]d\omega_q \tag{1.132}$$

and

$$I_2 = \lim_{\tau\downarrow 0}\frac{1}{\tau}\left(\int_{D_{t+\tau}} - \int_{D_t}\right)\rho(q,t)\mathfrak{f}(q,t)d\omega_q \tag{1.133}$$

The integral I_1 is called the *total rate of local change in D_t*.

To evaluate and identify the second integral I_2, we proceed as follows. Let $q \in \Sigma_t$ and let n_q be the outward normal to Σ_t at q. Further, let q^* be the point of $\Sigma_{t+\tau}$ nearest to q on n_q and δn_q the length of the segment $[q, q^*]$, taken with a plus sign if $q^* \notin D_t$ and with a minus sign if $q^* \in D_t$. It is obvious that, up to infinitesimal terms of higher order in τ (i.e., up to infinitesimal terms of higher order in δn_q),

$$\delta n_q = \tau \mathbf{v}(q,t)\cdot\mathbf{n}_q^0 \tag{1.134}$$

where $\mathbf{n}_q^0$ is the unit vector of the normal $\mathbf{n}_q$.

Let

$$D = D_{t+\tau}\cap D_t, \quad D_+ = D_{t+\tau}\setminus\bar{D}_t, \quad D_- = D_t\setminus\bar{D}_{t+\tau},$$
$$\Sigma_t^+ = \Sigma_t\cap\bar{D}_+, \quad \Sigma_t^- = \Sigma_t\cap D_- \tag{1.135}$$

(see Figure 1.4), so that

$$\Sigma_t = \Sigma_t^+\cup\Sigma_t^-, \qquad \int_{D_{t+\tau}} - \int_{D_t} = \int_{D_+} - \int_{D_-} \tag{1.136}$$

Evidently, to the leading order in τ[7]

$$\int_{D_-}\rho\mathfrak{F}d\omega = -\tau\int_{\Sigma_t^+}\rho\mathfrak{f}\cdot\mathbf{v}\cdot\mathbf{n}_q^0\cdot d\sigma, \qquad \int_{D_+}\rho\mathfrak{F}d\omega = \tau\int_{\Sigma_t^+}\rho\mathfrak{f}\cdot\mathbf{v}\cdot\mathbf{n}_q^0\cdot d\sigma \tag{1.137}$$

---

[7]This means up to the infinitesimal terms of higher order in τ.

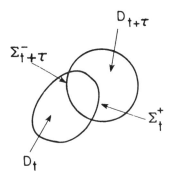

$D_{t+\tau}$

$\Sigma^{-}_{t+\tau}$

Σ^{+}_{t}

D_{t}

Figure 1.4.   Regions D_t and $D_{t+\tau}$.

Since $\Sigma_t = \Sigma_+ + \Sigma_-$, we finally obtain

$$I_2 = \int_{\Sigma_t} \rho(q,t)\mathfrak{f}(q,t)\mathbf{v}(q,t)\mathbf{d\sigma}_q \qquad (1.138)$$

where

$$\mathbf{d\sigma}_q = d\sigma_q \cdot \mathbf{n}^0 \qquad (1.139)$$

Note that $\rho\mathfrak{f}$ is the density of $\mathfrak{F}$ per unit volume. Thus, I_2 is the *total convective influx* of $\mathfrak{F}$ into D_t through its boundary. (By "convective" we mean a quantity related to the motion of the medium.)

Combining (1.131), (1.132), and (1.138), we obtain the identity

$$\frac{d}{dt}\mathfrak{F}(t) = \int_{D_t} \frac{\partial}{\partial t}[\rho(q,t)\mathfrak{f}(q,t)]d\omega_q + \int_{\Sigma_t} \rho(q,t)\mathfrak{f}(q,t)\mathbf{v}(q,t)\mathbf{d\sigma}_q \qquad (1.140)$$

We can now omit the subscript t in the symbols D_t and Σ_t without any risk of confusion. Note that the left-hand side of (1.140) refers to a property of a fixed set of material particles at time t, whereas the right-side refers to a property of material particles located at time t inside a fixed geometrical region D. Naturally, since the medium is in motion, the material contents of any region fixed in space vary in time. In this connection we point out two different approaches that have developed in the mechanics of continuous media, one due to Euler and the other to Lagrange. According to Euler, one traces the state of medium in a fixed geometrical region (i.e., deals with a medium of variable material contents); Lagrange, however, considers a fixed set of moving material particles. Equation (1.140) illustrates the relation between these two equivalent, though methodologically different, approaches.

We summarize our result as follows. The rate of change of any extensive property of a continuous medium dependent on mass or volume is composed of two components: the total rate of local change, and the influx through the surface Σ bounding the region D due to convective motion of the system.

By general arguments of balance, the total rate of change of $\mathfrak{F}$ in D must be equal to the total production of $\mathfrak{F}$ by sources or sinks distributed within D and by the nonconvective influx of $\mathfrak{F}$ into D through its surface Σ:

$$\frac{d}{dt}\mathfrak{F}(t) \equiv \int_D \frac{\partial}{\partial t}[\rho(q,t)\mathfrak{f}(q,t)]d\omega_q + \int_\Sigma \rho(q,t)\mathfrak{f}(q,t)\mathbf{v}(q,t)\mathbf{d\sigma}_q$$

$$= \int_D \rho(q,t)f(q,t)d\omega_q - \int_\Sigma \mathbf{J}(q,t\mid\mathfrak{f})\mathbf{d\sigma}_q \qquad (1.141)$$

where $f(q,t)$ is the mass density of inner sources of the property $\mathfrak{F}$ and $\mathbf{J}$ the surface density of the nonconvective flux of $\mathfrak{F}$. Though we have proved the identity (1.141) for a scalar extensive parameter, it is evidently true for an extensive parameter of any tensor nature.

The flux $\mathbf{J}$ occurring in (1.141) may sometimes be determined theoretically, by means of physical statistics or physical kinetics, but in most cases it is defined purely phenomenologically. Two kinds of flux may be distinguished: flux generated by regular (ordered) processes, and flux resulting from the average effect of random microscopic processes such as thermal movement of molecules resulting in diffusion.

The flux generated by a regular phenomenon may be exemplified by the process of sedimentation under gravity of colloidal particles in viscous liquid. If the interaction of different particles can be disregarded, their rate of sedimentation is determined by Stokes's theory, which has been confirmed experimentally (for details see Chapter 2, Section 9).

The macroscopic flux generated by a random process depends on the gradients of the densities of the quantities involved, averaged over a physical element of volume. To understand this statement, consider the following simple situation. Suppose we are observing a random walk of particles in the x direction, with the number of particles walking to the left and to the right proportional to their local concentration, and the probability of a particle moving to the left equal to the probability of its moving to the right. It is obvious that this random walk cannot generate any average flux if there is no gradient of particle concentration. On the contrary, if there is a gradient then the random walk will obviously generate a flux of particles in the direction of decreasing concentration. This is an intuitive explanation of the character of such phenomenological laws as those of Fourier, Fick, and so on, which were formulated on the basis of experimental observations and subsequently generalized in the thermodynamics of irreversible processes.[8]

---

[8] All our results remain true in the 2-dimensional case, where D is a plane region and the density of the extensive parameter is independent of the coordinate normal to the plane. In the general case of an extensive parameter defined on a surface in $\mathbb{R}_3$ or on a line of double curvature, the computations are slightly more complicated although the results are quite analogous (see Problem 1.7.1 at the end of this chapter).

8. Elements of thermodynamics of irreversible processes [32]

The thermodynamics of irreversible processes studies processes evolving (generally speaking) in open systems that are in a state near equilibrium, although the very concept of "nearness" of equilibrium remains undefined, and the only criterion for the validity of the description is the adequacy of the results. Thus, for example, propagating shock waves in gas dynamics, or even blast waves, prove to be sufficiently near equilibrium.

The basic *concept* of irreversible thermodynamics is local thermodynamic equilibrium. We say that there is a *local thermodynamic equilibrium* at a point q at time t if there exist a neighborhood V_q of q and a time interval $[t, t + \delta t)$ such that the state of the system in V_q may be treated as an equilibrium state at any moment $\tau \in [t, t + \delta t)$. Thus all the concepts of thermodynamics, understood locally, remain in force in irreversible thermodynamics, except for the formulation of the second law. In the thermodynamics of irreversible processes, the entropy production dS splits into two parts:

$$dS = d_i S + d_e S \qquad (1.142)$$

where $d_e S$ is generated by the exchange (material and thermal) of the open system under consideration with the external medium, and $d_i S$ is the entropy generated within the system by irreversible processes. The second law of thermodynamics now reads

$$d_i S \geq 0 \qquad (1.143)$$

Let D be some region. Then the rate of generation of entropy inside D is

$$\frac{d_i}{dt} S = \int_D \sigma d\omega \qquad (1.144)$$

where σ is the generation of entropy per unit volume and unit time. In the thermodynamics of irreversible process, one finds, using the conservation laws, that σ is a bilinear function of all the independent macroscopic fluxes J_i and the so-called thermodynamic forces X_i related to the spatial gradients of the appropriate variables

$$\sigma = \sum_{i=1}^{n} J_i X_i \qquad (1.145)$$

The fluxes J_i in turn are assumed to be linear functions of the thermodynamic forces:

$$J_i = \sum L_{ik} X_k \qquad (1.146)$$

The coefficients L_{ik} are called *phenomenological coefficients*. Inserting (1.146) into (1.145), one sees that σ is a bilinear form in the thermodynamic forces:

$$\sigma = \sum_{i=1}^{n} \sum_{k=1}^{n} L_{ik} X_i X_k \qquad (1.147)$$

As already mentioned, thermodynamic forces are defined as linear combinations of gradients of the thermodynamic parameters involved,[9] such as concentration, temperature, pressure, and so on. According to *Curie's rule*, the fluxes and forces occurring in (1.146) cannot be of different tensor types.

In particular, in a monocomponent nonisothemal incompressible medium the thermodynamic force is the gradient of temperature T, so that

$$\sigma = L_T \operatorname{grad}^2 T \qquad (1.148)$$

Similarly, if the medium is a bicomponent incompressible isothermal solution, then the thermodynamic force is the gradient of the solute concentration c, so that

$$\sigma = L_c \operatorname{grad}^2 c \qquad (1.149)$$

Assuming that the phenomenological coefficients L_T and L_c are constant, one can introduce a suitable scaling to make both L_T and L_c equal to 1. *Onsager's rule* states that the phenomenological coefficients are symmetric functions of their indices:

$$L_{ik} = L_{ki} \qquad (1.150)$$

The thermodynamics of irreversible processes provides a universal way to derive the equations of mathematical physics, which otherwise seem to be no more than a generalization of empirical observations; it may even predict the existence of new effects. For example, a nonzero temperature gradient in a binary solution generates a flux of matter which may be in the same or the opposite direction to the flux generated by the concentration gradient (this is known as the thermodiffusion effect); conversely, a concentration gradient may generate a nonzero temperature gradient. As a matter of fact, thermal diffusion in liquid solutions was first discovered experimentally, whereas the existence of thermal diffusion in gases was predicted theoretically [31].

Problems

P1.5.1. Derive the equations of kinetics of adsorption evolving in the inner diffusion region.

P1.7.1. Derive an expression for the rate of change of an extensive parameter of the matter distributed over a surface $S \subset \mathbb{R}_3$, taking into account possible exchange with the space surrounding S and the convection along S.

---

[9]See the end of Section 7.

2

Typical equations of mathematical physics. Boundary conditions

1. Laws of conservation and continuity. Three prototypic second-order equations of mathematical physics

Let $\mathfrak{F}(t)$ be an extensive parameter, characterizing some property of a continuous medium filling a region D at time t. Recall the expression for its rate of change as determined by the integral conservation law (see Chapter 1, Section 7):

$$\frac{d}{dt}\mathfrak{F}(t) = \int_D \frac{\partial}{\partial t}[\rho(q,t)\mathfrak{f}(q,t)]d\omega_q + \int_\Sigma \rho(q,t)\mathfrak{f}(q,t)\mathbf{v}(q,t)\mathbf{d}\boldsymbol{\sigma}_q$$

$$= \int_D \rho(q,t)f(q,t)d\omega_q - \int_\Sigma \mathbf{J}(q,t)\mathbf{d}\boldsymbol{\sigma}_q \qquad (2.1)$$

where ρ is the density of the medium, $\mathfrak{f}$ mass density of $\mathfrak{F}$, $\mathbf{v}$ the velocity of motion of the medium in the laboratory coordinate system, f the mass density of distributed inner sources of $\mathfrak{F}$, and $\mathbf{J}$ the nonconvective flux of $\mathfrak{F}$. As we have seen, $\mathbf{J}$ is generally the superposition of two components: $\mathbf{J}_1$, due to ordered motion of the relevant component of the medium, such as particles settling under gravity; and a diffusive component $\mathbf{J}_2$, due to the average transport effect of random motions in the system (see Chapter 1, Section 7 and this chapter, Section 9). The diffusive component $\mathbf{J}_2$ is related to the rate of spatial or temporal variations of the intensive parameters in the medium, as reflected in phenomenological laws.

Applying Gauss's divergence theorem,

$$\int_\Sigma \mathbf{A}\mathbf{d}\boldsymbol{\sigma}_q = \int_D \operatorname{div}\mathbf{A}d\omega_q \qquad (2.2)$$

we bring equality (2.1) to the form

$$\int_D \left(\frac{\partial}{\partial t}(\rho\mathfrak{f}) + \operatorname{div}(\rho\mathfrak{f}\mathbf{v}) - \rho f + \operatorname{div}\mathbf{J}\right)d\omega_q = 0 \qquad (2.3)$$

Now assume that D is an arbitrary simply connected subregion of the region D_0 occupied by the medium, and that the integrand in (2.3) is a function continuous everywhere in D_0. We recall the following lemma.

Lemma 2.1. *Let $F(q)$ be a function, continuous everywhere in D_0 such that*

$$\int_D F(q)d\omega_q = 0 \tag{2.4}$$

for any simply connected subregion D of D_0. Then

$$F(p) = 0 \quad \forall p \in D_0 \tag{2.5}$$

Using this lemma, we obtain the local form of the conservation law for the density of any scalar intensive parameter $\mathfrak{F}$:

$$\frac{\partial}{\partial t}(\rho\mathfrak{f}) + \operatorname{div}(\rho\mathfrak{f}\mathbf{v}) - \rho f + \operatorname{div}\mathbf{J} = 0 \tag{2.6}$$

Consider, in particular, the law of mass conservation in a medium without chemical reactions or any other distributed sources or sinks of mass. Then equation (2.6) becomes the *equation of continuity,*

$$\frac{\partial}{\partial t}\rho + \operatorname{div}(\rho\mathbf{v}) = 0 \tag{2.7}$$

Using the identity

$$\operatorname{div}(\rho\mathbf{v}) = \rho\operatorname{div}\mathbf{v} + \mathbf{v}\cdot\operatorname{grad}\rho \tag{2.8}$$

we see that the equation of continuity may be rewritten as

$$\frac{d}{dt}\rho + \rho\operatorname{div}\mathbf{v} = 0 \tag{2.9}$$

where

$$\frac{d}{dt} \overset{\text{def}}{=} \frac{\partial}{\partial t} + \mathbf{v}\cdot\operatorname{grad} \tag{2.10}$$

is the *substantial differentiation* operator. (Compare with the derivation of (1.56).)

Note that by (2.10) the time derivative of any extensive parameter $\mathfrak{F}$ in a medium without inner sources of $\mathfrak{F}$ may be rewritten as

$$\frac{d}{dt}\mathfrak{F} = \int_D \rho\frac{d}{dt}\mathfrak{f}\,d\omega_q \tag{2.11}$$

so that substantial differentiation of an extensive parameter characterizing a property of a fixed system of material particles does not affect the density of the medium.

We shall now apply this local conservation law to derive three prototypical equations of mathematical physics that will be referred to repeatedly in the rest of this chapter.

A. Equation of heat conduction in immobile solid without inner sources of heat

The density of the extensive parameter $\mathfrak{F}$ under consideration is the specific internal energy

$$\mathfrak{f} = c\rho u \tag{2.12}$$

where $c = \text{const}$ and $\rho = \text{const}$ are the specific heat capacity and the mass density of the solid, respectively, and u is the temperature. The flux $\mathbf{J}$ is related to the temperature gradient by Fourier's law

$$\mathbf{J} = -k \operatorname{grad} u \tag{2.13}$$

where $k = \text{const}$ is the coefficient of thermal conductivity. Substitution of (2.12)–(2.13) into (2.6) yields

$$a^2 \Delta u = \frac{\partial}{\partial t} u \tag{2.14}$$

where Δ is the Laplace operator and $a^2 = k/c\rho$ the coefficient of thermal diffusivity. We have thus obtained Fourier's equation of heat conduction, mentioned in Chapter 1, Section 1. In steady-state conditions, Fourier's equation becomes the Laplace equation

$$\Delta u = 0 \tag{2.15}$$

B. Equation of small longitudinal vibrations of a string (1-dimensional wave equation)

We now consider small longitudinal vibrations of a homogeneous string. The relevant extensive parameter is the momentum of the string

$$\mathfrak{F}(t \mid x_1, x_2) = \int_{x_1}^{x_2} \rho v(x, t) dx \tag{2.16}$$

where $v(x, t)$ is the velocity of a point of the string and ρ the linear density of the string. The momentum flux J is the stress in the string, which by Hooke's law is given by the equality

$$J = E \frac{\partial}{\partial x} u \tag{2.17}$$

where E is the modulus of elasticity. On the other hand,

$$v = \frac{\partial}{\partial t} u \tag{2.18}$$

where u is the displacement of points in the string. Hence, in view of (2.17), equation (2.6) yields the 1-dimensional wave equation

$$a^2 \frac{\partial^2}{\partial x^2} u = \frac{\partial^2}{\partial t^2} u, \quad a^2 = \frac{E}{\rho} \tag{2.19}$$

The corresponding 3-dimensional wave equation is

$$a^2 \Delta u = \frac{\partial^2}{\partial t^2} u \tag{2.20}$$

In what follows (see Sections 2–10) we shall derive the equations of mathematical physics in different natural scientific contexts. It will be shown that the fundamental similarity of the phenomenological laws determining the flux **J** in different physical processes implies a similarity of the corresponding partial differential equations, in keeping with our claim in Chapter 1 that the partial differential equations of mathematical physics constitute the universal language of the natural sciences. Moreover, we shall demonstrate that the study of properties of the wave, Laplace, and Fourier operators is the primary task of mathematical physics from the natural–scientific point of view. Later (in Chapter 4) we shall show that this task is also of paramount importance from a purely mathematical standpoint.

The rest of this chapter (Sections 10–12) is devoted to a description of the characteristic boundary and initial conditions occurring in mathematical physics, and to a rigorous formulation of the problems to be studied. There is also a short description of some problems that, though of interest for modern mathematical physics, are beyond the scope of this book but still deserve mention in order to provide a perspective for further studies.

2. Equations of continuity. Convective and diffusion flux in nonelectrolyte solutions in presence of chemical reactions. Fick's equation of diffusion in binary solutions. Diffusion of electrolytes. Nernst–Planck equation

A. Diffusion of nonelectrolyte accompanied by chemical reactions

Suppose now that the region $D \subset \mathbb{R}_3$ is occupied by a homogeneous solution of n nonelectrolyte species $a_1, a_2, \ldots, a_n$, moving at volume velocities $\mathbf{v}_i$, $i = 1, 2, \ldots, n$, relative to the laboratory coordinate system. Let c_i, ρ_i, M_i, and Ω_i denote the molar concentration, partial density, molar weight, and partial molar volume, respectively, of a_i. Assume for the sake of simplicity that the solution is a perfect mixture of incompressible liquid components, so that we have additivity of volumes;[1] that is,

$$\sum_{i=1}^{n} \Omega_i c_i = 1, \quad \Omega_i = \text{const}, \quad i = 1, 2, \ldots, n \qquad (2.21)$$

Define the total density of the solution by

$$\rho = \sum_{i=1}^{n} \rho_i \qquad (2.22)$$

---

[1] Additivity of volumes determines the simplest thermodynamic equation of state for a mixture; the partial molar volume of any component of a solution will equal the product of its molar volume and its molar concentration only in a system with such an equation of state. In most real solutions, particularly in salt solutions, this is not the case. The general definition of the partial molar volumes (and all other partial extensive parameters) may be found in any treatise on chemical thermodynamics (see e.g. Prigogine and Defay [60]).

The motion of the solution as a whole may be characterized by different
kinds of velocity, of which we shall refer to the following two: the average
volume velocity $\mathbf{w}$ and the center of mass velocity $\mathbf{v}$, defined respectively
by

$$\mathbf{w} = \sum_{i=1}^{n} \Omega_i c_i \mathbf{v}_i, \qquad \mathbf{v} = \frac{1}{\rho} \sum_{i=1}^{n} \rho_i \mathbf{v}_i \qquad (2.23)$$

In accordance with these definitions, we have two different definitions of
diffusion flux of a_i: the diffusion flux $\mathbf{J}_i$ in the frame of reference of the
average-volume velocity, and the diffusion flux $\tilde{\mathbf{J}}_i$ in frame of reference of
the center-of-mass velocity:

$$\mathbf{J}_i = c_i(\mathbf{v}_i - \mathbf{w}), \qquad \tilde{\mathbf{J}}_i = \rho_i(\mathbf{v}_i - \mathbf{v}) \qquad (2.24)$$

These fluxes $\mathbf{J}_i$ and $\tilde{\mathbf{J}}_i$ possess dimensions $\mathrm{mol}\,l^{-2}t^{-1}$ and $\mathrm{m}\,l^{-2}t^{-1}$,
respectively. Since

$$\rho_i = M_i c_i, \quad i = 1, 2, \ldots, n \qquad (2.25)$$

it follows from (2.23)–(2.25) that

$$\mathbf{v} = \mathbf{w} + \frac{1}{\rho} \sum_{i=1}^{n} M_i \mathbf{J}_i, \qquad \tilde{\mathbf{J}}_i = \frac{1}{\rho}[M_i \mathbf{J}_i - M_i c_i (\mathbf{v} - \mathbf{w})] \qquad (2.26)$$

Let f_i, $i = 1, 2, \ldots, n$, be the specific overall rate of production of a_i in
moles per unit volume due to all chemical reactions present. The equation
of continuity for a_i is

$$\frac{\partial}{\partial t} c_i + \mathrm{div}(c_i \mathbf{v}_i) - f_i = 0, \quad i = 1, 2, \ldots, n \qquad (2.27)$$

or, after separating the convection at a rate equal to the average-volume
velocity $\mathbf{w}$ from the corresponding diffusion flux $\mathbf{J}_i$,

$$\frac{\partial}{\partial t} c_i + \mathrm{div}(c_i \mathbf{w}) - f_i = -\,\mathrm{div}\,\mathbf{J}_i, \quad i = 1, 2, \ldots, n \qquad (2.28)$$

Multiplying (2.28) by Ω_i, summing over i, and using the assumption that
Ω_i is constant and the definitions (2.23) and (2.24), we find

$$\mathrm{div}\,\mathbf{w} = 0 \qquad (2.29)$$

where we have used the equalities

$$\sum_{i=1}^{n} \Omega_i \mathbf{J}_i = 0, \quad \sum_{i=1}^{n} \Omega_i f_i = 0, \qquad (2.30)$$

expressing mass conservation in chemical reactions. Equations (2.26) and
(2.29) imply that the mixture behaves like an incompressible liquid in
the frame of reference of average volume velocity, whereas it does like a
compressible liquid in the frame of reference of the center-of-mass velocity.
This distinction may be important, since in multicomponent solutions the
distinction between average volume velocity and center of mass velocity, and

hence also the distinction between the corresponding diffusion fluxes, may be essential even in diluted solutions.

The dependence of diffusion flux on concentration gradients, pressure, and temperature must be prescribed phenomenologically, in accordance with the rules of the thermodynamics of irreversible processes (see Chapter 1, Section 8) or as purely experimental laws; alternatively, it may be derived from nonequilibrium statistical mechanics. In the simplest case of a binary system with no temperature and pressure gradients there exists only one diffusion coefficient D, and the diffusion flux is determined by the classical Fick law

$$\mathbf{J} = -D \operatorname{grad} c \tag{2.31}$$

where c is the concentration of solute in the solution, which is usually assumed to be dilute.

Inserting (2.31) into (2.28) (specialized to a binary system), one obtains the Fick equation for convective diffusion accompanied by a chemical reaction:

$$\frac{\partial}{\partial t} c + \operatorname{div}(c\mathbf{w}) = D\Delta c + f \tag{2.32}$$

In a ternary system with solute concentrations c_1 and c_2, under isothermal conditions equation (2.32) must be replaced by the system of equations

$$\frac{\partial}{\partial t} c_i + \operatorname{div}(c_i\mathbf{w}) = D_{ii}\Delta c_i + D_{ij}\Delta c_j, \quad i,j = 1,2, \; j \neq i \tag{2.33}$$

where, by Onsager's theorem on the symmetry of phenomenological coefficients, $D_{ij} = D_{ji}$ (see Chapter 1, Section 8).

Note the resemblance between (2.32)–(2.33) and (2.14) of the previous section. This resemblance is due to the similarity of the underlying phenomenological equations and the identical general structure of the conservation laws.

B. Free and forced convective diffusion

There are two types of convection: free and forced. Free convection is caused by forces acting inside the fluid (e.g., free gravitational convection due to the dependence of the density of the solution on the concentrations of its components). In such cases the equation of diffusion is not independent but must rather be considered simultaneously with the equations of hydrodynamics. Forced convection is imposed externally, the flow velocity being independent of the process evolving in the fluid. For example, consider the flow of an aqueous solution of some material along a pipe of finite length L, caused by injection of the solution at a given rate (discharge) into the pipe. Assume that the viscosity is independent of concentration of the solute. Then the flow velocity distribution in the pipe will be independent of the concentration distribution; hence the diffusion equation will be independent of the momentum equation. Thus, if no chemical reactions take place inside the pipe, the equation of convective diffusion in the pipe will be

$$\frac{\partial}{\partial t}c = D\frac{\partial^2}{\partial x^2}c - v\frac{\partial}{\partial x}c \qquad (2.34)$$

where the flow velocity field v is given or calculated independently.

If the maximal flow velocity $\mathbf{v}_{max}$ of the solution is much smaller than D/L then the convective term $v(\partial/\partial x)c$ is negligible compared with the diffusion term $D(\partial^2/\partial x^2)c$, so that equation (2.34) can be replaced by the 1-dimensional Fick equation. If, on the contrary, the velocity is uniformly large enough, the diffusion term may be disregarded so that equation (2.34) becomes

$$\frac{\partial}{\partial t}c + v\frac{\partial}{\partial x}c = 0 \qquad (2.35)$$

The external force may also act only on the solute, so that its velocity $\mathbf{v}$ will be determined by the balance of this force and the force of friction between solute and solvent. If this friction force is proportional to the velocity of the solute $\mathbf{v}$ then the latter will become proportional to the external force, with the mobility B of the solute as proportionality factor

$$\mathbf{v} = B\mathbf{k} \qquad (2.36)$$

Here $\mathbf{k}$ is the density of external force per unit volume and $B > 0$ is a scalar constant. Accordingly, equation (2.34) becomes

$$\frac{\partial}{\partial t}c = D\Delta c - B\,\mathrm{div}(c\mathbf{k}) \qquad (2.37)$$

C. Diffusion of electrolytes. Nernst–Planck equation

In diffusion of electrolytes one must take into account that electrically neutral molecules dissociate in solutions (a water solution for instance) into positively and negatively charged ions, cations and anions, respectively, generally with different mobilities. Since cations and anions are electrically charged, the fact that they are moving at different velocities generally implies a violation of strict local electrical neutrality in the system. This in turn implies the appearance of a gradient of electric potential. As a result it becomes necessary simultaneously to determine the concentrations of the ions c_+ and c_- and the electric potential φ. The establishment of an electric potential at a given ionic concentration is a very fast process, compared with the redistribution of these concentrations by convective diffusion. Hence the potential φ is determined by Poisson's equation only (see Section 7 of this chapter and Chapter 12, Section 6), whereas the concentrations must satisfy the equations of diffusion in a force field. Assume, for the sake of simplicity, that we are dealing with a strong (completely dissociated) univalent electrolyte (e.g., NaCl) and that there are no chemical reactions. Then the molar concentrations c_+ and c_- of cations and anions must satisfy the equation (2.37) of forced convective diffusion. In this case the force $\mathbf{k}$

is the gradient of electrostatic potential, with a minus sign, times the ionic charge (see Chapter 1, Section 6). The velocity $\mathbf{w}$ for each ionic species will thus be $\pm B_\pm e \cdot \operatorname{grad} \varphi$,[2] so that

$$\frac{\partial}{\partial t} c_+ = D_+ \Delta c_+ + B_+ e \operatorname{div}(c_+ \operatorname{grad} \varphi) \tag{2.38}$$

$$\frac{\partial}{\partial t} c_- = D_- \Delta c_+ - B_- e \operatorname{div}(c_- \operatorname{grad} \varphi) \tag{2.39}$$

and

$$\Delta \varphi = \frac{e}{\varepsilon}(c_- - c_+)N \tag{2.40}$$

where $B_\pm$ and $D_\pm$ are positive constants, e the charge of the electron, ε the dielectric permeability of the solution and N the Avogadro number [20]. Einstein's equality establishes the following relation between $B_\pm$ and $D_\pm$:

$$B_\pm = D_\pm/(kT) \tag{2.41}$$

where k is the Boltzmann constant and T is the absolute temperature.[3] System (2.38)–(2.40) is known as the Nernst–Planck–Poisson equations.

3. Equation of motion of continuous medium

The equation of motion of a continuous medium in its integral form is the law of conservation of momentum. Recall that momentum evolves due to forces of two kinds: the mass forces, characterized by their mass density $\mathbf{f}$, and the forces acting on the surface of the body, known as stresses, which can be characterized by their surface density (see Chapter 1, Section 2):

$$\mathbf{\Pi}_n = \sum_{i=1}^{3} \Pi_{ni} \mathbf{e}_i \tag{2.42}$$

In the terminology of the previous section and of Chapter 1, the total stress acting on the body is the nonconvective part of the total momentum flux through the surface.

Recall also that, by (1.12),

$$\Pi_{ni} = \sum_{j=1}^{3} \Pi_{ji} \alpha_j \tag{2.43}$$

where α_j, $j = 1, 2, 3$, are the direction cosines of the normal $\mathbf{n}$ and where Π_{ij} are the components of the stress tensor. The latter is thus the nonconvective component of the momentum flux.

---

[2] The resistance of the solution to the motion of the ions, which balances the electric field force, is proportional to the "migration" velocity $\mathbf{v}$.

[3] Recall that Avogadro's number is the number of molecules in one mole: $N = 6.022 \cdot 10^{23}$ mol^{-1}; Boltzmann's constant is $k = R/N$, where R is the universal gas constant: $k = 1.3807 \cdot 10^{-23}$ J K^{-1} (J denotes Joule, K denotes degrees Kelvin).

The law of conservation of the momentum components reads

$$\frac{d}{dt}\int_D \rho v_i d\omega_q = \int_D \rho f_i d\omega_q + \sum_{j=1}^{3}\int_\Sigma \Pi_{ji}\alpha_j d\sigma_q \qquad (2.44)$$

Since

$$\alpha_j d\sigma_q = \mathbf{e}_j \mathbf{d}\boldsymbol{\sigma}_q \qquad (2.45)$$

equation (2.44) may be rewritten as

$$\frac{d}{dt}\int_D \rho v_i d\omega_q = \int_D \rho f_i d\omega_q + \sum_{j=1}^{3}\int_\Sigma (\Pi_{ji}\mathbf{e}_j)\mathbf{d}\boldsymbol{\sigma}_q \qquad (2.44)$$

so that applying the Gauss divergence theorem and equality (2.3) we obtain

$$\int_D \left(\rho(q,t)\left(\frac{d}{dt}v_i(q,t) - f_i(q,t)\right) - \sum_{j=1}^{3}\operatorname{div}(\Pi_{ji}\mathbf{e}_j)\right)d\omega_q = 0 \qquad (2.45)$$

Since D is an arbitrarily chosen region within the material medium, Lemma 2.1 may be applied, on the assumption that the integrand in (2.45) is continuous. This gives the local form of the equations of motion for any continuous medium,

$$\rho\left(\frac{d}{dt}v_i - f_i\right) - \sum_{j=1}^{3}\frac{\partial}{\partial x_j}\Pi_{ji} = 0, \quad i = 1, 2, 3 \qquad (2.48)$$

or, in vector notation,

$$\rho\frac{d}{dt}\mathbf{v} = \rho\mathbf{f} + \sum_{j=1}^{3}\frac{\partial}{\partial x_j}\mathbf{\Pi}_j \qquad (2.49)$$

To these equations we must add the continuity equation and the phenomenological equations of the rheological state of the medium (see Chapter 1, Section 2), which include for example the following.

A. Equations of motion of inviscid incompressible liquid (Euler equations)

The rheological relation for an inviscid liquid is

$$\Pi_{jj} = -p, \quad \Pi_{ji} = 0, \quad i, j = 1, 2, 3, \ i \neq j \qquad (2.50)$$

where p is hydrostatic pressure. Since the liquid is incompressible,

$$\operatorname{div}\mathbf{v} = 0 \qquad (2.51)$$

Hence, by (2.48),

$$\frac{d}{dt}v_i + \frac{1}{\rho}\frac{\partial}{\partial x_i}p = f_i, \quad i = 1, 2, 3 \qquad (2.52)$$

or, in vector notation,

$$\frac{d}{dt}\mathbf{v} + \frac{1}{\rho}\operatorname{grad}p = \mathbf{f} \tag{2.53}$$

Recall that d/dt denotes substantial differentiation:

$$\frac{d}{dt} = \frac{\partial}{\partial t} + (\mathbf{v}\cdot\operatorname{grad}) \tag{2.54}$$

Once again, we note the similarity between (2.53) and (2.35), due to the absence of "diffusive" components in the expressions for the relevant fluxes, as dictated by the underlying phenomenological laws.

B. Equation of motion of inviscid compressible liquid

Equation (2.53) still holds in this case, but the equation of continuity becomes

$$\frac{\partial}{\partial t}\rho + \operatorname{div}(\rho\mathbf{v}) = 0 \tag{2.55}$$

System (2.53), (2.55) is not closed; it consists of one vector equation or (what is the same) three scalar equations of motion plus the equation of continuity. On the other hand, it must serve for determination of five variables: the three components of the velocity vector, the density, and the pressure. In order to complete the system, one must add the thermodynamic equation of state, whose general form is

$$\rho = R(p, T) \tag{2.56}$$

where T is the temperature. Hence, in order to complete the system of equations, it is generally necessary to add the heat conduction equation, which will be derived in the next section. However, there are certain liquids, known as *barotropic* liquids, in which density is independent of temperature,

$$\rho = R(p) \tag{2.57}$$

In this case the system of equations (2.53), (2.55), and (2.56) is closed, so that the velocity and pressure fields can be determined independently of the temperature distribution.

C. Navier–Stokes equations of motion of Newtonian viscous liquid

The equation of rheological state is

$$\Pi_{ij} = (-p + {}^2\!/_3\mu\operatorname{div}\mathbf{v})\delta_{ij} + 2\mu\dot{e}_{ij} \tag{2.58}$$

where the components of the strain velocity tensor $\tilde{e}'$ are determined by (1.32) and (1.38), and the effect of volume viscosity is disregarded (see (1.59)). One notes the similarity of the phenomenological relations (2.58) to the Fourier and Fick laws (2.13) and (2.31). Hence the equation of motion of a viscous liquid is similar to the heat conduction equation. Indeed, substitution of (2.58) into (2.49) yields

$$\frac{d}{dt}\mathbf{v} + \frac{1}{\rho}\operatorname{grad} p = \mathbf{f} + \frac{2}{3}\nu \operatorname{grad}\operatorname{div}\mathbf{v} + \nu\Delta\mathbf{v} \qquad (2.59)$$

where

$$\nu = \mu/\rho \qquad (2.60)$$

is the *kinematic viscosity*. Note that the dimension of kinematic viscosity ν coincides with that of the coefficient of thermal diffusivity and diffusion (in this connection, see Chapter 1, Section 1).

When the liquid is incompressible and the motion is so slow that all inertia terms may be disregarded, the Navier–Stokes equations become the Stokes equations:

$$\frac{1}{\rho}\frac{\partial}{\partial x_i}p = f_i + \nu\Delta v_i, \quad i = 1,2,3 \qquad (2.61)$$

D. Equations of linear theory of elasticity (Lamé equations)

In linear elasticity theory all displacements are assumed to be small, which allows us to disregard the distinction between local and material derivatives. This means that in the linear theory the coordinates (x_1, x_2, x_3) refer to a point of the body in its initial (usually unstressed) state. This approximation implies (see (1.53))

$$\frac{d}{dt}\ln(\rho) \approx \frac{\partial}{\partial t}\ln(\rho) = -\operatorname{div}\mathbf{v} \qquad (2.62)$$

The acceleration vector $d\mathbf{v}/dt$ in this approximation is

$$\mathbf{a} = \frac{\partial^2}{\partial t^2}\mathbf{V} \qquad (2.63)$$

where $\mathbf{V}$ is the displacement vector (i.e., in the framework of linear elasticity theory, the displacement vector of a fixed point of the body in its initial state).

Hence, by (1.58), the non–steady-state equations of motion of the linear theory of elasticity in terms of displacements (Cauchy's equations) are

$$\rho\frac{\partial^2}{\partial t^2}V_k = \rho f_k + (\lambda + \mu)\frac{\partial}{\partial x_k}\operatorname{div}\mathbf{V} + \mu\Delta V_k, \quad k = 1,2,3 \qquad (2.64)$$

where it is assumed that the Lamé coefficients λ and μ are constants – an assumption that is valid for homogeneous elastic bodies, provided that the temperature stresses may be neglected (e.g., under isothermal conditions).

The similarity of (2.64) to the wave equation (2.20) should be noted. This similarity is obviously due to the identity of the underlying conservation arguments and phenomenological laws.

4. Equation of heat conduction in continuous media. Heat conduction in moving homogeneous compressible fluid

Let ε and e be the densities of total and internal energy per unit mass of a fluid, so that

$$\varepsilon = e + \tfrac{1}{2}v^2 \tag{2.65}$$

where $v^2/2$ is the kinetic energy density per unit mass. Let $\mathcal{E}$ be the total energy of the moving medium filling a region D at time t. By the law of conservation of energy (first law of thermodynamics),

$$\frac{d}{dt}\mathcal{E} = \int_D \rho \frac{d}{dt}\left[e + \frac{v^2}{2}\right] d\omega_q = Q + W \tag{2.66}$$

where Q is the sum of the total rate of heat production in D and the amount of heat flowing into D through its surface Σ; W is the total power (work done on the system per unit time) of the bulk and surface forces. We have

$$Q = Q_1 + Q_2, \quad W = W_1 + W_2 \tag{2.67}$$

where

$$Q_1 = \int_\Sigma k \operatorname{grad} T \cdot \mathbf{d\sigma}_q \tag{2.68}$$

is the "diffusive" heat influx according to Fourier's law, and

$$Q_2 = \int_D \rho F d\omega_q \tag{2.69}$$

is the rate of heat production by distributed internal sources of heat (e.g., the rate of release of Joule heat by electric current) per unit mass. Finally,

$$W_1 = \int_D \rho \mathbf{f} \cdot \mathbf{v} d\omega_q \tag{2.70}$$

is the power of the mass force and

$$W_2 = \int_\Sigma \mathbf{\Pi}_n \cdot \mathbf{v} d\sigma_q \tag{2.71}$$

the power of the surface force.

We have

$$\mathbf{v} \cdot \frac{d}{dt}\mathbf{v} = \mathbf{v} \cdot \left(\frac{\partial}{\partial t}\mathbf{v} + (\mathbf{v} \cdot \operatorname{grad})\mathbf{v}\right) = \frac{1}{2}\left(\frac{\partial}{\partial t}v^2 + \mathbf{v} \cdot \operatorname{grad} v^2\right) = \frac{1}{2}\frac{d}{dt}v^2 \tag{2.72}$$

and (see (2.49))

$$\rho \mathbf{v} \cdot \frac{d}{dt}\mathbf{v} = \rho(q,t)\mathbf{v} \cdot \mathbf{f}(q,t) + \sum_{i=1}^{3} \mathbf{v} \cdot \frac{\partial}{\partial x_i}\mathbf{\Pi}_i$$

$$\equiv \rho(q,t)\mathbf{f}(q,t) + \sum_{i=1}^{3} \frac{\partial}{\partial x_i}(\mathbf{v} \cdot \mathbf{\Pi}_i) - \sum_{i=1}^{3} \mathbf{\Pi}_i \cdot \frac{\partial}{\partial x_i}\mathbf{v} \tag{2.73}$$

This yields

$$\frac{1}{2}\int_D \rho \frac{d}{dt}\mathbf{v}^2 d\omega_q = \int_D \rho \mathbf{v}\cdot \mathbf{f}\omega_q + \int_D \sum_{i=1}^{3}\frac{\partial}{\partial x_i}(\mathbf{v}\cdot\mathbf{\Pi}_i)d\omega_q - \int_D \sum_{i=1}^{3}\mathbf{\Pi}_i\frac{\partial}{\partial x_i}\mathbf{v}\,d\omega_q$$

$$(2.74)$$

By the Gauss divergence theorem, using (1.10),

$$\int_D \sum_{i=1}^{3}\frac{\partial}{\partial x_i}(\mathbf{v}\cdot\mathbf{\Pi}_i)d\omega_q = \int_\Sigma \mathbf{v}\cdot\mathbf{\Pi}_n d\sigma_q \qquad (2.75)$$

Subtracting (2.74) from (2.66) and using (2.75) and (2.67)–(2.71), we obtain

$$\int_D \rho \frac{d}{dt}e\,d\omega_q = \int_\Sigma k\operatorname{grad} T\cdot \mathbf{d\sigma} + \int_D \rho F d\omega_q + \int_D \sum_{i=1}^{3}\mathbf{\Pi}_i\frac{\partial}{\partial x_i}\mathbf{v}\,d\omega_q \quad (2.76)$$

Again using the Gauss divergence theorem in order to convert the first integral on the right into a volume integral, and applying Lemma 2.1, we obtain a local differential equation determining the rate of change of specific internal energy in a moving compressible continuous medium:

$$\rho \frac{d}{dt}e = k\Delta T + \rho F + \sum_{i=1}^{3}\mathbf{\Pi}_i\frac{\partial}{\partial x_i}\mathbf{v} \qquad (2.77)$$

Defining the hydrodynamics pressure as the trace of the stress density tensor, we find that

$$\sum_{i=1}^{3}\mathbf{\Pi}_i\cdot\frac{\partial}{\partial x_i}\mathbf{v} = -p\operatorname{div}\mathbf{v} + \sum_{i=1}^{3}\tau_i\cdot\frac{\partial}{\partial x_i}\mathbf{v} \qquad (2.78)$$

where τ_i is the stress density on a surface with normal $\mathbf{i}$, generated by the deviator $\tilde{\tau}$ of the stress tensor.[4] Thus

$$\rho \frac{d}{dt}e = k\Delta T + \rho F - p\operatorname{div}\mathbf{v} + \sum_{i=1}^{3}\tau_i\frac{\partial}{\partial x_i}\mathbf{v} \qquad (2.79)$$

We now recall the relationship between the specific internal energy e, the specific enthalpy h, and the caloric coefficients (see Chapter 1, Section 6):

$$h = e + \frac{p}{\rho} \qquad (2.80)$$

$$\left(\frac{dh}{dT}\right)_p = c_p, \quad \left(\frac{dh}{dp}\right)_T = -\frac{\alpha}{\rho}T \Rightarrow \frac{d}{dt}h = c_p\frac{d}{dt}T - \frac{\alpha}{\rho}T\frac{d}{dt}p \qquad (2.81)$$

---

[4]Recall (see Chapter 1, Section 2) that identification of the hydrostatic pressure thus defined with the equilibrium pressure occurring in the thermodynamic equation of state amounts to neglecting the bulk viscosity.

Equality (2.81), the equation of continuity (1.53), and definition of the material derivative imply

$$\rho\frac{d}{dt}e = \rho c_p\frac{d}{dt}T - (1+\alpha T)\frac{d}{dt}p - p\operatorname{div}\mathbf{v} \tag{2.82}$$

Comparing (2.78) and (2.82) and using the identity

$$\operatorname{div}(p\mathbf{v}) = \mathbf{v}\cdot\operatorname{grad}p + p\operatorname{div}\mathbf{v} \tag{2.83}$$

we obtain[5]

$$\rho c_p\frac{d}{dt}T = k\Delta T + \rho F + (1+\alpha T)\frac{d}{dt}p + \sum_{i=1}^{3}\tau_i\frac{\partial}{\partial x_i}\mathbf{v} \tag{2.84}$$

The last term on the right of this equation represents the volumetric density of heat release due to viscous dissipation of energy; the second term is the density of the internal sources of heat. Finally, the third term on the right of (2.84) represents the rate of release of heat of compression. If the viscous dissipation of energy and the rate of change in the pressure may be neglected, equation (2.84) becomes the equation of convective diffusion.

The equation of energy dissipation in an elastic body is quite similar to that in a viscous liquid.

5. Potential motion of inviscid incompressible liquid. Equations of vibrations of elastic body and of slightly compressible inviscid liquid

In certain key situations, the basic general equations of hydrodynamics and elasticity theory yield simpler prototypical equations of classical mathematical physics, such as Laplace and wave equations.

A. Potential motion of inviscid incompressible fluid in conservative field of mass forces

Recall that the flow of any liquid is said to be *potential* if its velocity vector $\mathbf{v}$ is the gradient of a scalar φ, called the velocity potential. Since the vortex lines are frozen in an barotropic inviscid liquid (in particular, in an incompressible liquid), the absence of vorticity at the initial time is a necessary condition for the flow to be potential at all later times (see Chapter 1, Section 2). We shall now show that if the motion of an inviscid incompressible liquid in a conservative field of mass forces (i.e., in a potential field of force independent of time) is irrotational, then the assumption that the flow is potential is compatible with the Euler equation of motion. Indeed, let U be the force potential in the liquid. The flow is governed by Euler's equation

$$\rho\frac{d}{dt}\mathbf{v} + \operatorname{grad}(p+U) = 0 \tag{2.85}$$

---

[5]In thermodynamics, temperature is measured in absolute (Kelvin) degrees.

and by the equation of continuity which, in the absence of inner sources or sinks of liquid, reads

$$\operatorname{div} \mathbf{v} = 0 \tag{2.86}$$

Assume that there exists a velocity potential, so that

$$\mathbf{v} = \operatorname{grad} \varphi \tag{2.87}$$

Substitution of (2.87) into the continuity equation (2.86) yields

$$\operatorname{div} \operatorname{grad} \varphi \equiv \Delta \varphi = 0 \tag{2.88}$$

so that the velocity potential satisfies the Laplace equation, that is, is a harmonic function.

We now substitute (2.87) into the Euler equation (2.85). It is easy to check that

$$\frac{d}{dt} \operatorname{grad} \varphi = \operatorname{grad} \left(\frac{\partial}{\partial t} \varphi + \frac{1}{2} \operatorname{grad}^2 \varphi \right) \tag{2.89}$$

Equations (2.85) and (2.87) imply

$$\operatorname{grad} \left(\frac{\partial}{\partial t} \varphi + \frac{1}{2} \operatorname{grad}^2 \varphi + \frac{p+U}{\rho} \right) = 0 \tag{2.90}$$

Hence

$$\frac{\partial}{\partial t} \varphi + \frac{1}{2} \operatorname{grad}^2 \varphi + \frac{p+U}{\rho} = f(t) \tag{2.91}$$

where $f(t)$ is a time-dependent constant of integration.

Assume that the velocity potential has been determined and that the pressure p is prescribed at some point as a function of time. Then (2.91) shows that the pressure p in all other points is uniquely determined in terms of the derivatives of the velocity potential. This proves that assuming the existence of a velocity potential is compatible with the system of governing equations. The results may be summarized as follows.

- If there are no inner sources or sinks of mass (e.g., due to injection of material or chemical reactions), then the velocity potential is a harmonic function:

$$\Delta \varphi = 0 \tag{2.92}$$

- If the liquid contains distributed sources or sinks of mass of volume density f, then the velocity potential is a solution of the Poisson equation

$$\Delta \varphi + f = 0 \tag{2.93}$$

- If, moreover, the pressure p is prescribed as a function of time at some point in space, then a knowledge of the velocity potential completely determines the state variables for irrotational flow in a conservative force field.

Thus the Laplace and Poisson equations are the principal equations of the theory of potential motion of incompressible inviscid fluid.

B. Vibration of elastic body and slightly compressible inviscid barotropic liquids

Consider now the Lamé equations (2.64) of elasticity theory. In vector notation, they are

$$\rho\frac{\partial^2}{\partial t^2}\mathbf{V} = \rho\mathbf{f} + (\lambda+\mu)\,\text{grad}\,\text{div}\,\mathbf{V} + \mu\sum_{i=1}^{3}\Delta V\cdot\mathbf{e}_i \qquad (2.94)$$

Calculating the divergence of (2.94) and remembering that

$$\text{div}\,\text{grad} = \Delta \qquad (2.95)$$

we obtain

$$\text{div}\left(\rho\frac{\partial^2}{\partial t^2}\mathbf{V}\right) = \text{div}(\rho\mathbf{f}) + (\lambda+\mu)\Delta(\text{div}\,\mathbf{V}) + \mu\,\text{div}(\Delta\mathbf{V}) \qquad (2.96)$$

where we have used the identity

$$\Delta\mathbf{V} = \sum_{i=1}^{3}\Delta V_i\mathbf{e}_i \qquad (2.97)$$

Let Θ be the volumetric dilatation of the body

$$\Theta = \text{div}\,\mathbf{V} \qquad (2.98)$$

Using the commutability of differentiation and disregarding the quadratic terms

$$\frac{\partial^2}{\partial t^2}\mathbf{V}\cdot\text{grad}\,\rho, \quad \mathbf{f}\cdot\text{grad}\,\rho \qquad (2.99)$$

which are assumed to be small, we obtain the *wave equation*

$$a^2\Delta\Theta + \mathfrak{F} = \frac{\partial^2}{\partial t^2}\Theta \qquad (2.100)$$

where

$$\mathfrak{F} = \text{div}\,\mathbf{f}, \quad a^2 = \frac{\lambda+2\mu}{\rho} \qquad (2.101)$$

The constant a has the dimension of velocity.

Now let us consider a slightly compressible barotropic inviscid liquid in a potential force field. The coefficient of compressibility is defined by

$$\beta = \frac{1}{\rho}\frac{d\rho}{dp} \qquad (2.102)$$

In a slightly compressible liquid the pressure is assumed to vary by a small amount about its average value $p_0 = 0$. Similarly, let ρ_0 be the "average" density, so that $|\delta\rho| \overset{\text{def}}{=} |\rho - \rho_0| \ll \rho_0$. Then (2.102) will be replaced by

$$\frac{d}{dp}\rho = \beta\rho_0 = \text{const} \Rightarrow \rho = \rho_0(1 + \beta\cdot p) \qquad (2.103)$$

and the equation of continuity

$$\frac{\partial}{\partial t}\rho + \mathrm{div}(\rho \mathbf{v}) = 0 \tag{2.104}$$

(owing to the "smallness" of the motion, βp is negligible compared with 1) may be replaced by its approximation

$$\beta \frac{\partial}{\partial t}p + \mathrm{div}\,\mathbf{v} = 0 \tag{2.105}$$

Similarly, the Euler equation

$$\frac{d}{dt}\mathbf{v} + \frac{1}{\rho}\,\mathrm{grad}\,p = -\,\mathrm{grad}\,U \tag{2.106}$$

may be replaced by the linearized equation

$$\frac{\partial}{\partial t}\mathbf{v} + \frac{1}{\rho_0}\,\mathrm{grad}\,p + \mathrm{grad}\,U = 0 \tag{2.107}$$

It follows from (2.105) and (2.107) that

$$\beta \frac{\partial^2}{\partial t^2}p + \mathrm{div}\left(\frac{\partial}{\partial t}\mathbf{v}\right) = 0 \tag{2.108}$$

and

$$\mathrm{div}\left(\frac{\partial}{\partial t}\mathbf{v}\right) + \frac{1}{\rho_0}\Delta p + \Delta U = 0 \tag{2.109}$$

Eliminating $\mathrm{div}(\partial \mathbf{v}/\partial t)$, we obtain

$$a^2 \Delta p + \mathfrak{F} = \frac{\partial^2}{\partial t^2}p \tag{2.110}$$

where

$$\mathfrak{F} = \frac{1}{\beta}\Delta U \tag{2.111}$$

and

$$a = (\beta \rho_0)^{-1/2} \tag{2.112}$$

Thus the pressure in the "small" motion of a slightly compressible liquid satisfies the wave equation with velocity of propagation a.

It is easily shown that in this case the density ρ also satisfies the wave equation

$$a^2 \Delta \rho + \Psi = \frac{\partial^2}{\partial t^2}\rho \tag{2.113}$$

with the same wave velocity a and the free term $\Psi = \beta \rho_0 \mathfrak{F}$.

In the case under consideration, the velocity vector may be expressed as the gradient of the velocity potential. It is easily seen that the velocity potential satisfies the wave equation with the same wave velocity.

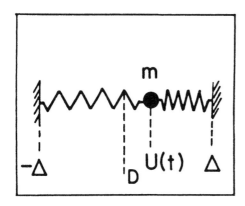

Figure 2.1.   Spring loaded by mass M.

6. Chain of springs oscillating in medium with friction. Wave equation

Assume that a weightless spring of length 2Δ, with its ends rigidly attached at points $x = \pm\Delta$, is oscillating in a fluid medium, together with a point mass m fixed at the middle of the undeformed spring. Denote the displacement of m from equilibrium by $U(t)$ (see Figure 2.1). Let $k = E/2\Delta$ (E is the elastic constant of the material) be the coefficient of stiffness of the spring and assume that the friction per unit length of the spring is proportional to the velocity of the mass m; the surrounding liquid is assumed to remain motionless. Since the resultant of the forces of contraction and dilatation of the parts $[U(t), \Delta]$ and $[-\Delta, U(t)]$ (respectively) acting on the mass m is $-2kU$, the momentum equation for m becomes

$$m\frac{d^2}{dt^2}U(t) = -2kU(t) - 2\beta\Delta\frac{d}{dt}U(t) \qquad (2.114)$$

where β is the friction coefficient per unit length of the spring and $U(t)$ the displacement of m. Note that the transition from large values of $m^{1/2}k^{1/2}/(\beta\Delta)$ to small values corresponds to a transition from inertia-dominated, oscillatory behavior of the solution of (2.114) to a friction-dominated solution depending monotonically on time.

Now consider a chain of identical springs connected in series, with masses m. Let $U_i(t)$ denote the displacement of mass i from its equilibrium position, corresponding to the nondeformed state of all the springs (see Figure 2.2). Accordingly, equation (2.114) will be replaced by

$$m\frac{d^2}{dt^2}U_i(t) = -k[(U_i - U_{i-1}) - (U_{i+1} - U_i)] - 2\beta\Delta\frac{d}{dt}U_i(t), \quad i = 0, \pm1, \pm2, \ldots$$
$$(2.115)$$

Evidently, this system of equations can be viewed as a finite difference approximation to the partial differential equation for longitudinal vibrations of a homogeneous spring in a medium with friction. Indeed, let us replace the discrete index i by a continuous space variable x. Let $u(x, t)$ be the

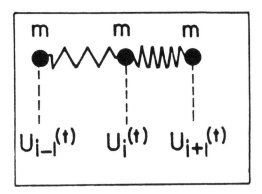

Figure 2.2.   Chain of loaded springs.

displacement of the point x of the spring at time t. Letting $\Delta \downarrow 0$ in (2.115) and defining the linear density of the spring as $\rho = m/2\Delta$, we have

$$\rho \frac{\partial^2}{\partial t^2} u(x,t) + \beta \frac{\partial}{\partial t} u(x,t) = E \frac{\partial^2}{\partial x^2} u(x,t) \qquad (2.116)$$

or setting

$$a^2 \stackrel{\text{def}}{=} \frac{E}{\beta}, \qquad \tau \stackrel{\text{def}}{=} \frac{\rho}{\beta} \qquad (2.117)$$

we obtain

$$a^2 \frac{\partial^2}{\partial x^2} u(x,t) = \tau \frac{\partial^2}{\partial t^2} u(x,t) + \frac{\partial}{\partial t} u(x,t) \qquad (2.118)$$

Note that a^2 and τ have the same dimension as the diffusion coefficient and the time, respectively.

Remark. The transition to predominance of dissipation or friction in the *partial* differential equation (2.118) (i.e., neglect of the inertia term) leading to the Fourier equation (2.14), may be compared to neglect of the inertia term in the *ordinary* differential equation (2.114).

Equation (2.118), transformed to dimensionless variables, is identical with the equation for 1-dimensional electromagnetic oscillations (see Section 7) or the hyperbolic equation of diffusion (see Section 9 and Chapter 13, Section 2).

7. Maxwell's equations of electrodynamics

A. Maxwell's equations in integral form

The phenomenological integral laws of classical electrodynamics are as follows.

1. Generalized Ampère law

The circulation of the magnetic field intensity vector $\mathbf{H}$ around a smooth closed contour L is equal to the flux of the total electric current C through any smooth two-sided surface Σ bounded by L, so that in the Gaussian absolute system of units

$$\oint_L \mathbf{H} \cdot \mathbf{ds} = \frac{4\pi}{c} \int_\Sigma \mathbf{J} \cdot \mathbf{d\sigma} + \frac{1}{c}\frac{\partial}{\partial t} \int_\Sigma \mathbf{D} \cdot \mathbf{d\sigma} \qquad (2.119)$$

where the sense of description of L and the direction of the normal $\mathbf{n}$ to Σ satisfy the definition of a right-handed system (see Appendix 1). The first integral in the right-hand side is the total conductive current; the second is the total displacement current through Σ.

2. Faraday's law of electromagnetic induction

The circulation of the electric field intensity around L is proportional to the rate of change of the flux of magnetic induction $\mathbf{B}$ through Σ, so that

$$\oint_L \mathbf{E} \cdot \mathbf{ds} = -\frac{1}{c}\frac{\partial}{\partial t} \int_\Sigma \mathbf{B} \cdot \mathbf{d\sigma} \qquad (2.120)$$

3. Generalized Coulomb's law

The flux of of the electric displacement vector $\mathbf{D}$ through an arbitrary closed surface Σ bounding a region $G \subset \mathbb{R}_3$ is equal to the electric charge in the region, so that

$$\oint_\Sigma \mathbf{D} \cdot \mathbf{d\sigma} = 4\pi \int_G \rho d\omega \qquad (2.121)$$

where ρ is the volume charge density.

4. Flux of magnetic induction through any closed surface is equal to zero

$$\oint_\Sigma \mathbf{B} \cdot \mathbf{d\sigma} = 0 \qquad (2.122)$$

Of these four fundamental laws, only the last two are conservation (continuity) laws of the type discussed in Section 1, which form the basis for the description the transport processes. The first two laws are of a different (circulation) type, which we have not encountered so far among the fundamental components from which the basic differential equations of mathematical physics are derived.

B. *Maxwell's differential equations of electrodynamics*

Let as assume that the Gauss divergence theorem and the Stokes theorem are applicable to the appropriate integrals in (2.119)–(2.122), and that the integrands in all the integrals obtained after the application of these two theorems are continuous functions. Then, applying Lemma 2.1, one obtains Maxwell's differential equations of electrodynamics:

$$\operatorname{rot} \mathbf{H} = \frac{4\pi}{c}\mathbf{J} + \frac{1}{c}\frac{d}{dt}\mathbf{D}, \quad \operatorname{div} \mathbf{D} = 4\pi\rho \tag{2.123}$$

$$\operatorname{rot} \mathbf{E} = -\frac{1}{c}\frac{d}{dt}\bar{\mathbf{B}}, \quad \operatorname{div} \mathbf{B} = 0 \tag{2.124}$$

These equations must be complemented by the following phenomenological equations:

$$\mathbf{D} = \varepsilon\mathbf{E}, \quad \mathbf{B} = \mu\mathbf{H}, \quad \mathbf{J} = \sigma\mathbf{E} \tag{2.125}$$

The dielectric permeability ε and magnetic permeability μ are material constants, equal to 1 in a vacuum,[6] and σ is the electric conductance of the medium.[7]

As we shall be concerned with electrodynamics in an stationary medium only, the material differentiation d/dt will be replaced everywhere by partial differentiation $\partial/\partial t$.

C. Differential equations of electrostatics, magnetostatics, and direct electric current

These are Maxwell's equations for steady processes; one obtains the equations of electrostatics by equating all the time derivatives in Maxwell's equations to zero. On the other hand, the Laplace and Poisson equations (i.e., the fundamental equations of electrostatics) may be obtained directly as corollaries of Coulomb's law, without any reference to the Maxwell's equations (for details see Chapter 12, Sections 4 and 6). They are

$$\Delta\varphi = 0 \tag{2.126}$$

and

$$\Delta\varphi + \rho/\varepsilon = 0 \tag{2.127}$$

where the Laplace equation (2.126) holds in any region in vacuum without free charges, and the Poisson equation (2.127) holds in any region occupied by a dielectric of permeability ε and free charges of volume density ρ.

The *equation of direct electric (conduction) current* is a consequence of the local form of the law of conservation of electric charge (which is itself a linear combination of the mass conservation laws for the different charged components in the system) and of the relation between the flux $\mathbf{J}$ of the

---

[6] $\varepsilon = 1$ and $\mu = 1$ in a vacuum in the absolute Gaussian system of units. $\varepsilon, \mu > 0$ in all media; media with $\varepsilon < 1$ are unknown. μ may be less or greater than 1. Both ε and μ may vary within a given medium, with change in temperature and pressure.

[7] (a) In some cases the equality $\mathbf{J} = \sigma\mathbf{E}$ must be replaced by $\mathbf{J} = \sigma\mathbf{E} + \mathbf{J}_e$, where $\mathbf{J}_e$ is the current density due to nonelectric sources (e.g., diffusion). (b) For the definitions of electric and magnetic forces, see Chapter 1, Section 3.

electric charge to the driving force – that is, the electric field intensity in the conductor – as expressed by Ohm's law,

$$\mathbf{J} = -\sigma \operatorname{grad} \varphi \tag{2.128}$$

Thus the equation of conduction current (charge conservation) becomes

$$\frac{\partial}{\partial t}\rho = \sigma \Delta \varphi \tag{2.129}$$

where ρ is the charge density, $\sigma = \text{const} > 0$ is again the electric conductance of the conductor $(r = 1/\sigma$ is the specific resistance), and φ is the electric potential.

D. Equation of propagation of electromagnetic oscillations in homogeneous isotropic conductive medium at constant temperature and pressure. Dissipative wave equation

Let us assume that the medium under consideration is homogeneous, isotropic, and in an isothermal, isobaric state at rest. Then $\varepsilon = \text{const}$, $\mu = \text{const}$, and $d/dt = \partial/\partial t$, so that the first of equations (2.123)–(2.124) may be rewritten in the form

$$\operatorname{rot} \mathbf{H} = \frac{\varepsilon}{c}\frac{\partial}{\partial t}\mathbf{E} + \frac{4\pi\sigma}{c}\mathbf{E} \tag{2.130}$$

$$\operatorname{rot} \mathbf{E} = -\frac{\mu}{c}\frac{\partial}{\partial t}\mathbf{H} \tag{2.131}$$

We can eliminate $\mathbf{H}$ from this system. Indeed, by (2.130)–(2.131),

$$\mu\frac{\partial}{\partial t}\operatorname{rot}\mathbf{H} = \frac{\varepsilon\mu}{c}\frac{\partial^2}{\partial t^2}\mathbf{E} + \frac{4\pi\sigma\mu}{c}\frac{\partial}{\partial t}\mathbf{E} \tag{2.132}$$

$$\mu\frac{\partial}{\partial t}\operatorname{rot}\mathbf{H} = -c\operatorname{rot}\operatorname{rot}\mathbf{E} \tag{2.133}$$

so that

$$\frac{\varepsilon\mu}{c}\frac{\partial^2}{\partial t^2}\mathbf{E} + \frac{4\pi\sigma\mu}{c}\frac{\partial}{\partial t}\mathbf{E} = -c\operatorname{rot}\operatorname{rot}\mathbf{E} \tag{2.134}$$

Applying the identity

$$\operatorname{rot}\operatorname{rot}\mathbf{E} = \operatorname{grad}\operatorname{div}\mathbf{E} - \Delta\mathbf{E} \tag{2.135}$$

to (2.134), we obtain

$$\frac{\varepsilon\mu}{c}\frac{\partial^2}{\partial t^2}\mathbf{E} + \frac{4\pi\sigma\mu}{c}\frac{\partial}{\partial t}\mathbf{E} = c\Delta\mathbf{E} - c\operatorname{grad}\operatorname{div}\mathbf{E} \tag{2.136}$$

Use of equations (2.123)–(2.124) yields the following equations for the components of $\mathbf{E}$

$$\Delta E_k - 4\pi\frac{\partial}{\partial x_k}\rho = \frac{\varepsilon\mu}{c^2}\frac{\partial^2}{\partial t^2}E_k + \frac{4\pi\sigma\mu}{c^2}\frac{\partial}{\partial t}E_k, \quad k = 1,2,3 \tag{2.137}$$

Let us assume that the electric field E has only one component and that the free charge density ρ is prescribed as a function of x and t. Then, using appropriate scaling, we reduce (2.137) to a 1-dimensional dissipative wave equation:

$$a^2 \frac{\partial^2}{\partial x^2} u + f(x,t) = \tau \frac{\partial^2}{\partial t^2} u + \alpha \frac{\partial}{\partial t} u \qquad (2.138)$$

This equation, also known as the Telegraph equation (see Problem 2.7.1), has already been obtained for a chain of springs coupled in series in a viscous medium. It also appears in the theory of diffusion of Brownian particles if the contribution due to the inertia of the particles is not neglected (see the Langevin equation and the so-called hyperbolic diffusion or heat-conduction equation; see Section 9 of this chapter as well as Chapter 13, Section 2 and Chapter 21, Section 1). If $\alpha = \text{const} > 0$ is sufficiently small (i.e., if viscous friction or electric conductance is negligible) then the term $\alpha \partial u / \partial t$ may be omitted and equation (2.138) becomes an inhomogeneous wave equation. If $\tau = \text{const} > 0$ is sufficiently small (i.e., if the inertia or electromagnetic polarizability of the medium is negligible) then the term $\tau \partial^2 u / \partial t^2$ may be omitted and (2.138) becomes the standard equation of heat conduction or diffusion. (In this connection we recall the well-known electromagnetic–mechanical analogy [76;77].) Finally, if this term is negligible and simultaneously $\alpha = \text{const} < 0$, equation (2.138) becomes the equation of heat conduction with reversed time, which is an important object of study in the theory of historical climatology (see Section 1 of this chapter).

We can also eliminate **E** from system (2.130)–(2.131) and obtain an equation for the magnetic displacement vector **H** that is similar to equation (2.137) except for the absence of the source term. (This results from the equality $\text{div}\,\mathbf{H} = 0$ for $\mu = \text{const}$.) Thus we find that **H** satisfies the equation

$$\Delta \mathbf{H}_k = \frac{\varepsilon\mu}{c^2} \frac{\partial^2}{\partial t^2} H_k + \frac{4\pi\sigma\mu}{c^2} \frac{\partial}{\partial t} H_k, \quad k = 1, 2, 3 \qquad (2.139)$$

In particular, in the 1-dimensional case,

$$\frac{\partial^2}{\partial x^2} H = \frac{\varepsilon\mu}{c^2} \frac{\partial^2}{\partial t^2} H + \frac{4\pi\sigma\mu}{c^2} \frac{\partial}{\partial t} H \qquad (2.140)$$

In the case of axial symmetry, equation (2.139) yields

$$\frac{1}{r} \frac{\partial}{\partial r}\left(r \frac{\partial}{\partial r} H\right) = \frac{\varepsilon\mu}{c^2} \frac{\partial^2}{\partial t^2} H + \frac{4\pi\sigma\mu}{c^2} \frac{\partial}{\partial t} H \qquad (2.141)$$

Note that the first term in the right-hand side of (2.139) is due to the displacement current, the second to the conductance current. If the displacement current is small compared with the conductance current (i.e., if the oscillation frequency is sufficiently small) then the term $((\varepsilon\mu)/c^2)\partial^2 H/\partial t^2$ may be dropped, so that equation (2.139) becomes

$$\Delta \mathbf{H}_k = \frac{4\pi\sigma\mu}{c^2} \frac{\partial}{\partial t} H_k, \quad k = 1, 2, 3 \qquad (2.142)$$

and once again the 1-dimensional dissipative wave equations (2.140) and (2.141) become the magnetic field "diffusion equation"

$$\frac{\partial^2}{\partial x^2} H = \frac{4\pi\sigma\mu}{c^2} \frac{\partial}{\partial t} H \qquad (2.143)$$

Correspondingly, in the axially symmetric case we have

$$\frac{1}{r} \frac{\partial}{\partial r} \left(r \frac{\partial}{\partial r} H \right) = \frac{4\pi\sigma\mu}{c^2} \frac{\partial}{\partial t} H \qquad (2.144)$$

We reiterate that these equations are formally identical with Fick's 1-dimensional (planar) and cylindrically symmetric diffusion equations. This transition from the dissipative wave equation to the diffusion equation corresponds to the so-called quasistatic current approximation in electrodynamics.

8. Theory of percolation of multicomponent liquids [68]

Consider a liquid moving through the pore space of a porous medium. A porous medium is a multiphase medium with randomly distributed particles of different phases.[8] We assume that one of the phases, known as the solid matrix, is composed of grains of solid particles, bonded together by some kind of mineral cement or located within some vessel obstructing their motion; alternatively, they may be in close contact under high pressure, which obstructs their motion. The space between the solid particles, which is assumed to be permeable to liquid, is called the pore space. The liquid (or gas) phases may be either monocomponent (such as pure water) or a mixture, such as a solution of several components in a liquid solvent (e.g., oil containing a large number of hydrocarbons, heavy and light, which may change their phase state in the process of percolation and go from one liquid phase into another, or be selectively adsorbed by the solid grains). Because of the great variety of different processes that may accompany percolation, the mathematical approaches to the subject are differentiated into a large number of particular theories. In what follows we shall restrict our consideration to the simplest situations.

A. Basic assumptions. Two modes of averaging

1.

As mentioned in Chapter 1, a porous medium is usually described as a fictitious continuous medium. This passage from a real, heterogeneous medium to a fictitious, homogeneous one is based on the use of a number of assumptions, outlined in this section.

---

[8]Recall that a material medium is a multiphase medium if, at equilibrium, at least one of the physical parameters characterizing its state is a discontinuous piecewise constant function of the space coordinates. The surfaces of discontinuity of such parameters are called "phase interfaces" (see Chapter 1, Section 6).

To fix ideas, let us consider a porous medium composed of $n+1$ phases A_i, where $i = 0$ refers to the solid matrix, $i = 1, 2, \ldots, k$ to immiscible monocomponent liquid phases, not involved in mass exchange with the other phases, and $i = k+1, k+2, \ldots, k+r = n$ to perfect solutions of n_i components in a liquid solvent a_i. The components with $k+1 \le i \le n$ may be involved in a mass exchange with other phases due to evaporation, absorption or adsorption at the phase interfaces, or surface chemical reactions. When dealing with concrete problems, for the sake of simplicity we shall here assume $k = 1$, $r = 1$, thus restricting our attention to a 3-phase system with two immiscible liquid nonelectrolyte phases, A_1 and A_2, percolating through the pore space of the heterogeneous medium. We shall also assume that phase A_2 is a perfect solution of a single component b in a solvent a_2, which may be adsorbed at the surfaces of the grains composing the solid matrix A_0 but cannot be transferred from phase A_2 into the first liquid phase A_1. The particles making up each phase are assumed to be large enough to permit application of the equations of continuum mechanics, heat conduction, and so on, derived in previous sections of this chapter. At the same time, the particles are very small when compared with the characteristic size of the region occupied by the porous medium. Finally, we assume that the number of particles of at least two phases within any physical volume element is very large. The number, location, orientation, and shape of the phase interfaces are unknown, and therefore the process under consideration cannot be described as evolving in a real heterogeneous medium. Instead, it is necessary to consider a fictitious homogeneous medium. The passage to such a medium is based on the process of averaging, as described briefly in Section 1 of Chapter 1.

2.

Let $D \subset \mathbb{R}_3$ be the simply connected region occupied by the porous medium under consideration, and let K_M^R be a sphere of radius R with center at $M \in D$; we shall treat this sphere as a physical element of volume. Further, let $D_{it} \subset D$ be the set of points occupied by phase A_i at time t. We shall write $\Delta\omega(M)$ instead of K_M^R and treat it as a geometrically infinitesimal element of arbitrary shape. To avoid possible misunderstandings, we shall denote the set of points itself by $\Delta\omega(M)$ and its volume by $\operatorname{mes}\Delta\omega(M)$. Let M be an arbitrary point in D. Define $\Delta_i\omega(M,t)$ by

$$\Delta_i\omega(M,t) = D_{it} \cap \Delta\omega(M), \quad i = 0, 1, \ldots, n \qquad (2.145)$$

so that $\Delta_i\omega(M,t)$ is the set of all points of A_i belonging at time t to $\Delta\omega(M)$ and $\operatorname{mes}\Delta_i\omega(M,t)$ is the volume of this set.[9] Now define the volume

---

[9]We may obviously assume that $\Delta_i\omega(M,t)$ is a union of finitely many regions bounded by piecewise smooth surfaces. It is therefore legitimate to say that $\operatorname{mes}\Delta_i\omega(M,t)$ is the volume of $\Delta_i\omega(M,t)$.

concentration $m_i(M, t)$ of A_i at point M at moment t by

$$m_i(M, t) = \frac{\text{mes } \Delta_i \omega(M, t)}{\text{mes } \Delta \omega(M, t)}, \quad i = 0, 1, \ldots, n \qquad (2.146)$$

Similarly, let Σ be a smooth two-sided surface such that $M \in \Sigma$ and let **n** be a vector normal to Σ at M. Let

$$\Delta\sigma(M) = \sum \cap \Delta\omega(M), \quad \text{mes } \Delta\sigma(M) = \text{mes}_2 \, \Delta\sigma(M) \qquad (2.147)$$

so that $\text{mes}_2 \, \Delta\sigma(M)$ is the surface measure of $\Delta\sigma(M)$, and let

$$\Delta_i\sigma(M, t) = \Delta\sigma(M) \cap D_{it}, \quad \text{mes } \Delta_i\sigma(M, t) = \text{mes}_2 \, \Delta_i\sigma(M, t) \qquad (2.148)$$

Finally, define

$$\mu_i(M, t) = \frac{\text{mes } \Delta_i\sigma(M, t)}{\text{mes } \Delta\sigma(M)} \qquad (2.149)$$

We shall call $\mu_i(M, t)$ the *surface concentration* of phase A_i.

Let us assume that the real heterogeneous medium under consideration is isotropic, so that all its geometrical and physical characteristics are independent of the direction of the normal **n**. Then $\mu_i(M, t)$ is independent of **n** and, moreover,[10]

$$\mu_i(M, t) \equiv m_i(M, t), \quad i = 0, 1, \ldots, n \qquad (2.150)$$

Define the *phase function* $\delta(M, t)$ by

$$\delta_i(M, t) = \begin{cases} 1, & \text{if } M \in D_{it} \\ 0, & \text{if } M \notin D_{it} \end{cases} \quad i = 0, 1, \ldots, n \qquad (2.151)$$

and consider the volume density $\mathfrak{f}$ of an extensive parameter $\mathfrak{F}$ characterizing some property of the real porous medium at M at time t. The function $\mathfrak{f}(M, t)$ or its derivatives may have discontinuities inside D_{it}. Taking this into account, we define

$$\mathfrak{f}^i(M, t) = \mathfrak{f}(M, t)\delta_i(M, t), \quad i = 0, 1, \ldots, n \qquad (2.152)$$

so that instead of one field we now have n fields of the same name, since any of n phases composing the medium may be that actually found at M.

Let $\Delta\omega(M, t) \subset D_t$ be an arbitrarily chosen physical element of volume around M. The following two modes of averaging may be introduced:

$$\mathfrak{F}_i^{**}(M, t) = \frac{1}{\text{mes } \Delta\omega(M)} \int_{\Delta\omega(M)} \mathfrak{f}^i(q, t)d\omega_q$$

or   (2.153)

$$\mathfrak{F}_i^{*}(M, t) = \frac{1}{\text{mes } \Delta_i\omega(M, t)} \int_{\Delta_i\omega(M, t)} \mathfrak{f}^i(q, t)d\omega_q$$

----

[10]The fact that the surface phase concentration is equal to the volume phase concentration in a homogeneous isotropic porous medium may either be proved by the methods of the theory of probability or considered as an experimental fact.

Here $\mathfrak{F}_i^{**}(M,t)$ is the average value of $\mathfrak{f}^i(M,t)$ per unit volume of the whole medium, whereas $\mathfrak{F}_i^*(M,t)$ is the average value of $\mathfrak{f}^i(M,t)$ per unit volume of the ith phase.

Note that, by virtue of the identity

$$\mathfrak{f}^i(q,t) \equiv 0 \quad \forall q \in \Delta\omega(M) \setminus \Delta_i(M), \ \forall i, \ i = 1,2,\dots,n \qquad (2.154)$$

it follows from (2.153) that

$$\int_{\Delta\omega(M)} \mathfrak{f}^i(q,t)d\omega_q = \int_{\Delta_i\omega(M,t)} \mathfrak{f}^i(q,t)d\omega_q \qquad (2.155)$$

so that, by (2.146),

$$\mathfrak{f}_i^{**}(M,t) = m_i(M,t)\mathfrak{f}_i^*(M,t) \qquad (2.156)$$

In particular, let $\mathbf{v}^i(M,t)$ be the velocity of the real ith phase in the laboratory coordinate system. Then $\mathbf{v}_i^{**}(M,t)$ is what hydrologists call the *velocity of percolation* of the liquid phase A_i and $\mathbf{v}_i^*$ the *true velocity* (or simply velocity) of the phase. Thus, the velocity of percolation is related to the true velocity by the equality

$$\mathbf{v}_i^{**} = m_i\mathbf{v}_i^*, \quad i = 1,2,\dots,n \qquad (2.157)$$

Let us consider two different fields $\mathfrak{f}$ and ψ. In general, the average of their product is not equal to the product of the respective averages. Let us assume, however, that the fields possess the following properties: Let $\mathfrak{M}$ be a set of field functions $f^i(M,t)$ such that, if $f^i \in \mathfrak{M}$ then there exists a positive constant N such that, for all $\Delta\omega(M) \subset D_t$ and $t \geq 0$,

$$\frac{1}{\operatorname{mes}\Delta_i\omega(M,t)} \int_{\Delta\omega(M)} [f^i(q,t)]^2 d\omega_q < N \qquad (2.158)$$

Furthermore, assume that for a suitable choice of a fixed small $\varepsilon > 0$,

$$\frac{1}{\operatorname{mes}\Delta_i\omega(M,t)} \int_{\Delta\omega(M)} [f^*(q,t) - f^i(q,t)]^2 d\omega_q < \frac{\varepsilon^2}{N} \qquad (2.159)$$

Then, for every $\mathfrak{f}^i, \psi^i \in \mathfrak{M}$, the average of the product $\mathfrak{f} \cdot \psi$ is equal to the product of the averages of $\mathfrak{f}$ and ψ, to within an error not exceeding ε; that is,

$$\left| \mathfrak{f}^*(M,t)\psi^*(M,t) - \frac{1}{\operatorname{mes}\Delta_i\omega(M,t)} \int_{\Delta\omega(M)} \mathfrak{f}^i(q,t)\psi^i(q,t)d\omega_q \right| < \varepsilon \qquad (2.160)$$

Indeed, by the Cauchy–Schwarz inequality, if $\mathfrak{f}, \psi \in \mathfrak{M}$ then

$$\left| \mathfrak{f}_i^*(M,t)\psi_i^*(M,t) - \frac{1}{\operatorname{mes}\Delta_i\omega(M,t)} \int_{\Delta\omega(M,t)} \mathfrak{f}^i(q,t)\psi^i(q,t)d\omega_q \right|$$

$$= \left| \frac{1}{\operatorname{mes}\Delta_i\omega(M,t)} \int_{\Delta_i\omega(M,t)} \mathfrak{f}^i(q,t)[\psi_i^*(M,t) - \psi^i(q,t)]d\omega_q \right|$$

$$\leq \left| \frac{1}{\operatorname{mes} \Delta_i \omega(M,t)} \int_{\Delta_i \omega(M,t)} f^i(q,t)^2 d\omega_q \right|^{1/2}$$

$$\left| \frac{1}{\operatorname{mes} \Delta_i \omega(M,t)} \int_{\Delta_i \omega(M,t)} [\psi_i^*(M,t) - \psi^i(q,t)]^2 d\omega_q \right|^{1/2}$$

$$\leq \left[N \cdot \frac{\varepsilon^2}{N} \right]^{1/2} = \varepsilon \qquad (2.161)$$

In what follows, we shall always replace the average of a product by the product of averages without any additional stipulations. Incidentally, this assumption is used tacitly in the theory of percolation.

We now consider an integral of the form

$$\mathfrak{F}_i(t) = \int_D f^i(q,t) \cdot d\omega_q \qquad (2.162)$$

where D is an arbitrarily chosen simply connected region bounded by a smooth surface Σ. Let us represent D as the sum of n nonintersecting physical elements of volume $\Delta\omega(M_k)$:

$$D = \bigcup_1^n \Delta\omega(M_k) \qquad (2.163)$$

so that

$$\mathfrak{F}_i(t) = \sum_{k=1}^{n} \int_{\Delta\omega(M_k)} f^i(q,t) \cdot d\omega_q \qquad (2.164)$$

By the definition (2.153),

$$\mathfrak{F}_i(t) = \sum_{k=1}^{n} f_i^{**}(M_k,t) \cdot \operatorname{mes} \Delta\omega(M_k) \qquad (2.165)$$

The sum in the right-hand side of (2.165) is an integral sum, so that in the limit

$$\mathfrak{F}_i(t) = \int_D f_i^{**}(q,t) \, d\omega_q = \int_D f_i^*(q,t) m_i(q,t) \, d\omega_q \qquad (2.166)$$

Surface integrals may be treated in similar fashion. In particular, if

$$\Psi_i(t) = \int_\Sigma f^i(q,t) \psi^i(q,t) d\sigma_q \qquad (2.167)$$

then, replacing the average of the product by the product of the averages, we have

$$\Psi_i(t) = \sum_{k=1}^{n} f_i^*(M_k,t) \psi_i^*(M_k,t) m_i(M_k,t) \operatorname{mes} \Delta\sigma(M_k) \qquad (2.168)$$

Once more, after the passage from the integral sum to the integral, we obtain

$$\Psi_i(t) = \int_\Sigma f_i^*(q,t)\psi_i^*(q,t)m_i(q,t)d\sigma_q \qquad (2.169)$$

or, using (2.156),

$$\Psi_i(t) = \int_\Sigma f_i^{**}(q,t)\psi_i^*(q,t)d\sigma_q \qquad (2.170)$$

Let us now consider the averaging of derivatives with respect to the spatial coordinates. Denote

$$J = \frac{\partial}{\partial x}\int_D f^i(q,t)d\omega_q, \quad I = \int_D f_x^i(q,t)d\omega_q \qquad (2.171)$$

where f_x stands for $\partial f/\partial x$. Using (2.166), we find that

$$I = \int_D f_i^*(q,t)_x m_i(q,t)d\omega_q \qquad (2.172)$$

$$J = \frac{\partial}{\partial x}\int_D f_i^*(q)m_i(q,t)d\omega_q = \int_D \frac{\partial}{\partial x}[f_i^*(q)m_i(q,t)]d\omega_q \qquad (2.173)$$

Here $f_i^*(q,t)_x$ stands for h^* where $h \equiv f_x$. Thus, derivatives with respect to the spatial coordinates of an averaged field are not equal to the average of the derivatives of the field, unless the volumetric concentration of the appropriate phase is constant.

We can now proceed to deriving the laws of conservation in a moving porous medium.

B. Law of mass conservation. Equation of continuity

Let $\rho_i^*(M,t)$ be the partial density of the ith phase A_i of the fictitious homogeneous medium, so that

$$\rho_i^*(M,t) = \frac{1}{\text{mes }\Delta_i(M,t)}\int_{\Delta\omega(M)} \rho^i(q,t)d\omega q \qquad (2.174)$$

Further, let $M_i(t)$ be the mass of phase A_i of the fixed simply connected region $D \subset \mathbb{R}_3$, so that

$$M_i(t) = \int_D \rho^i(q,t)d\omega_q \qquad (2.175)$$

Then, according to (2.146), (2.153), and (2.164),

$$M_i(t) = \int_D \rho_i(q,t)m_i(q,t)d\omega_q \qquad (2.176)$$

where ρ_i is written instead of ρ_i^*, so that ρ_i is the density of the ith phase A_i of the real heterogeneous medium under consideration.

In the same way, the rate of variation of $M_i(t)$ is a sum of three terms:

$$\frac{d}{dt}M_i(t) = I_1 + I_2 + I_3 \qquad (2.177)$$

where

$$I_1 = \int_D \frac{\partial}{\partial t}[\rho_i(q,t)m_i(q,t)]d\omega_q \tag{2.178}$$

is the total rate of change of density ρ_i due to phase compressibility and thermal dilatation. Furthermore,

$$I_2 = -\int_\Sigma \rho_i(q,t)\mathbf{v}_i^{**} \cdot \mathbf{d\sigma}_q \tag{2.179}$$

is the mass increase due to the influx of A_i into D through Σ, at rate of percolation $\mathbf{v}_i^{**}$. Finally

$$I_3 = \int_D \rho_i(q,t)f_i^{**}(q,t)d\omega_q \tag{2.180}$$

is the contribution of the internal sources with mass density f_i.

On the assumption that Gauss's divergence theorem is applicable and the resulting integrand continuous, we obtain, omitting stars, the local continuity equation of percolation theory

$$\frac{\partial}{\partial t}[\rho_i(q,t)m_i(q,t)] + \mathrm{div}(\rho_i\mathbf{v}_i) = \rho_i(q,t)f_i(q,t) \tag{2.181}$$

C. Law of momentum conservation. Darcy's law

Let us consider the case in which all the phases A_i, $i = 1, 2, \ldots, n$, are viscous immiscible liquids. In a real, heterogeneous medium with chaotically distributed particles in different phases, these particles are subjected not only to viscous but also to frictional forces between the different phases. To describe all the processes in a real heterogeneous medium in terms of those taking place in a fictitious, homogeneous medium, we must replace the real forces of friction by fictitious volume forces, characterized by their mass density, whose contribution to the momentum equations must be equivalent to the average contribution of the real forces of friction.

Thus, let $\mathbf{f}^i$ be the density of the real mass forces acting on the ith phase, $\mathbf{R}^i$ the mass density of the fictitious mass forces of resistance, p^i the pressure, $\mathbf{v}^i$ the velocity vector, and ρ^i the density of the phase A_i. Then the rate of change of the momentum $\mathbf{Q}$ of this phase in a fixed region D is

$$\frac{d}{dt}\mathbf{Q} \equiv \int_D \frac{\partial}{\partial t}(\rho^i\mathbf{v}^i)d\omega + \int_\Sigma \rho^i\mathbf{v}^i(\mathbf{v}^i \cdot \mathbf{d\sigma}) = \int_D \rho^i(\mathbf{f}^i + \mathbf{R}^i)d\omega - \int_\Sigma p^i\mathbf{n}^0(\mathbf{n}^0 \cdot \mathbf{d\sigma}) \tag{2.182}$$

"Homogenization" in (2.182) yields

$$\int_D \frac{\partial}{\partial t}(\rho_i\mathbf{v}_i^{**})d\omega + \int_\Sigma \rho_i\mathbf{v}_i^{**}(\mathbf{v}_i^* \cdot \mathbf{d\sigma}) + \int_\Sigma p_i^*\mathbf{n}^0(\mathbf{n}^0 \cdot \mathbf{d\sigma}) = \int_D \rho_i(\mathbf{f}_i^* + \mathbf{R}_i^*)d\omega \tag{2.183}$$

or, assuming that the Gauss divergence theorem is applicable and the resulting integrands are continuous, we obtain the local form of the equation of motion:

$$\frac{\partial}{\partial t}(\rho_i \mathbf{v}_i^{**}) + (\mathbf{v}_i^* \cdot \mathrm{grad})(\rho_i \mathbf{v}_i^{**}) + p_i^* \, \mathrm{grad} \, m_i = m_i \rho_i (\mathbf{f}_i^* + \mathbf{R}_i^*) - m_i \, \mathrm{grad} \, p_i^*$$
$$(2.184)$$

Experimental studies of water percolation through sand-filled tubes have led to the following phenomenological rule, known as Darcy's law: The volumetric density of the fictitious force of resistance is proportional to the dynamic viscosity of the liquid and to its rate of percolation relative to the speed of motion of the solid matrix, so that

$$\rho_i \mathbf{R}_i^* = m_i \frac{\mu_i}{\kappa_i}(\mathbf{v}_i^* - \mathbf{v}_0^*)$$
$$(2.185)$$

Here κ_i is a positive constant with the dimension of area, dependent on the pore space geometry and the volume concentrations of all the liquid phases, and μ_i is the dynamic viscosity of phase i. For a slightly more accurate formulation of Darcy's law, the speed of matrix motion $\mathbf{v}_0^{**}$ must be replaced by the average volume velocity of all the phases except the ith.

In applications of theory of percolation, the inertia terms and the term $p^* \, \mathrm{grad} \, m_i$ on the left of (2.184) are usually neglected. This yields an equation of motion in the form

$$(\mathbf{v}_i^{**} - \mathbf{v}_0^{**}) = -\frac{\kappa_i}{\mu_i} \cdot \mathrm{grad}(p_i^*) + \rho_i f_i^{**}$$
$$(2.186)$$

If there is only one liquid phase, the coefficient κ_i is denoted simply by k and called the *coefficient of permeability*. Experimental studies of the percolation of different homogeneous liquids have proved that k is dependent solely on the internal geometry of the matrix. Therefore κ_i is usually expressed as

$$\kappa_i = k f_i(m_1, m_2, \ldots, m_n)$$
$$(2.187)$$

The functions f_i that depend on the volume concentrations of all the liquid phases are known as the *coefficients of phase permeability*; they must be determined experimentally. Since the force of interaction of the any of the liquid phases with the others is added to the forces of friction with the solid skeleton particles, the coefficients of phase permeability are positive and smaller than 1.

D. Equation of piezoconductivity

Consider the slow, isothermal percolation of a homogeneous, slightly compressible liquid, saturating the pore space of a slightly compressible porous stratum under the action of ground pressure p_0 and the pore pressure p. The initial pore pressure is assumed to be zero. (In our notation the pore pressure coincides with p^*.) Suppose that the equation of state of the liquid saturating the pore space is

$$\rho = \rho_0(1 + \beta_0 p)$$
$$(2.188)$$

where β_0 is the coefficient of compressibility and ρ_0 the density at the initial (zero) pressure. Let

$$m = 1 - m_0 \tag{2.189}$$

be the porosity of the medium. The decrease in pore pressure in the process of percolation affects the porosity according to the following semiempirical equation of state:

$$m = m^0(1 - \beta^0 p) \tag{2.190}$$

The equation of continuity, in the absence of sources, reads

$$\frac{\partial}{\partial t}(\rho m) + \operatorname{div}(\rho \mathbf{v}) = 0 \tag{2.191}$$

where $\mathbf{v}$ is the rate of percolation. Equation (2.191) may be rewritten as

$$m\frac{\partial}{\partial t}\rho + \rho\frac{\partial}{\partial t}m + \rho \operatorname{div}\mathbf{v} + \mathbf{v}\cdot\operatorname{grad}\rho = 0 \tag{2.192}$$

so that the equations of state (2.188) and (2.190) yield

$$[m^0(1 - \beta^0 p)\rho_0\beta_0 - \rho_0(1 + \beta_0)m^0\beta^0]\frac{\partial}{\partial t}p + \rho_0(1 + \beta_0 p)\operatorname{div}\mathbf{v} + \mathbf{v}\cdot\operatorname{grad}p = 0 \tag{2.193}$$

Neglecting all the quadratic terms as small, we obtain

$$m^0(\beta^0 - \beta_0)\frac{\partial}{\partial t}p + \operatorname{div}\mathbf{v} = 0 \tag{2.194}$$

Here $\mathbf{v}$ is determined from the equation of motion

$$\mathbf{v} = -\frac{k}{\mu}\operatorname{grad}p \tag{2.195}$$

Ignoring the possible variation of the coefficient of permeability due to the decrease in pore pressure, we find that

$$\kappa\Delta p = \frac{\partial}{\partial t}p \tag{2.196}$$

where

$$\kappa = \frac{k}{\mu m^0(\beta^0 - \beta_0)} \tag{2.197}$$

The coefficient k is known as the *coefficient of piezoconductivity*; equation (2.196) is the equation of piezoconductivity.[11]

E. Isothermal percolation of two immiscible incompressible liquids through immobile porous medium. Rapoport–Leas and Buckley–Leverett equations [7;63]

Let us consider a porous medium with a rigid solid skeleton of porosity m, saturated with two immiscible liquids L_1 and L_2; for convenience, we

---

[11] Equation (2.196) is derived in the theory of oil recovery, and is often called the "equation of the elastic regime of oil recovery."

shall refer to these liquids as "water" and "oil," respectively. Let m_1 be the volume concentration of water and m_2 the volume concentration of oil. The quotient

$$\sigma = \frac{m_1}{m} \tag{2.198}$$

is called the water saturation and $1 - \sigma$ the oil saturation. We treat both water and oil as incompressible liquids; their equations of continuity are

$$m\frac{\partial}{\partial t}\sigma + \operatorname{div} \mathbf{v}_1 = 0, \qquad -m\frac{\partial}{\partial t}\sigma + \operatorname{div} \mathbf{v}_2 = 0 \tag{2.199}$$

where $\mathbf{v}_1$ and $\mathbf{v}_2$ are the rates of percolation of water and oil, respectively. Summing equations (2.199), we obtain

$$\operatorname{div} \mathbf{v} = 0 \tag{2.200}$$

where

$$\mathbf{v} = \mathbf{v}_1 + \mathbf{v}_2 \tag{2.201}$$

Let us assume that $\mathbf{v}$ is the gradient of a harmonic function U:

$$\mathbf{v} = \operatorname{grad} U, \quad \Delta U = 0 \tag{2.202}$$

We shall assume that the effect of the mass forces may be disregarded.[12] The equations of motion of water and oil are

$$\mathbf{v}_1 = -\frac{k}{\mu_1}f_1(\sigma)\operatorname{grad} p_1, \qquad \mathbf{v}_2 = -\frac{k}{\mu_2}f_2(\sigma)\operatorname{grad} p_2 \tag{2.203}$$

where the phase permeabilities f_1 and f_2 are functions of the form shown in Figure 2.3. They may be approximated by the equations

$$f_1(\sigma) = \begin{cases} a_1(\sigma - \sigma_1)^{b_1} & \text{if } 0 < \sigma_1 \le \sigma \\ 0 & \text{if } 0 \le \sigma \le \sigma_1 \end{cases} \tag{2.204}$$

$$f_2(\sigma) = \begin{cases} 0 & \text{if } \sigma_2 < \sigma \\ a_2(\sigma_2 - \sigma)^{b_2} & \text{if } 0 \le \sigma < \sigma_2 \end{cases} \tag{2.205}$$

where

$$a_1 > 0, \quad a_2 > 0, \quad b_1 > 1, \quad b_2 > 1 \tag{2.206}$$

We are thus dealing with an underdetermined system of four equations in five unknown functions. In order to "close" the system one either adds an

---

[12]The theory of percolation of two immiscible liquids was created in the theory of oil recovery, on the assumption that the percolation takes place in a productive layer of thickness much smaller than its length. If, in addition, the angle of dip of the layer is small, then the influence of the only active mass force (i.e., the force of gravity) may be disregarded.

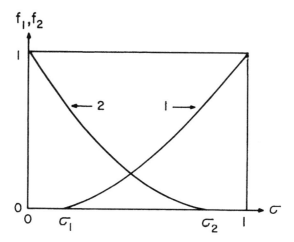

Figure 2.3. Dependence of phase permeabilities f_i on σ, $i = 1, 2$; $f_1(\sigma) = 1.276(\sigma - \sigma_1)^{1.5}$, $f_2(\sigma) = 1.340(\sigma_2 - \sigma)^{1.8}$, $\sigma_1 = 0.15$, $\sigma_2 = 0.85$.

experimentally determined equation, expressing the dependence of capillary pressure $p_c(\sigma)$ on water saturation σ,

$$p_1 - p_2 \stackrel{\text{def}}{=} p_c(\sigma) \tag{2.207}$$

or makes certain simplifying assumptions. Capillary pressure depends not only on water saturation but also on temperature. Nevertheless, as we are concerned here with isothermal percolation, we may use experimental data obtained under isothermal conditions only. The typical shape of an experimentally defined plot of p_c vs. σ is shown in Figure 2.4. Curves of this shape may be approximated to satisfactory accuracy by equations of the form

$$p_c(\sigma) = B\frac{(\sigma_2 - \sigma)^{\gamma_2}}{(\sigma - \sigma_1)^{\gamma_1}} \tag{2.208}$$

where σ_1 and σ_2 are as in (2.204) and (2.205) and B, γ_1, and γ_2 are positive constants. In particular, let us consider 1-dimensional percolation in the x-direction. The system of governing equations becomes

$$v_1 = -\frac{k}{\mu_1}f_1(\sigma)\frac{\partial}{\partial x}p_1, \quad v_2 = -\frac{k}{\mu_2}f_2(\sigma)\frac{\partial}{\partial x}p_2 \tag{2.209}$$

$$p_1 - p_2 = p_c(\sigma) \tag{2.210}$$

$$\frac{\partial}{\partial t}\sigma + \frac{\partial}{\partial x}v_1 = 0, \; v = v_1 + v_2, \; \frac{\partial}{\partial x}v \equiv 0 \Rightarrow v \equiv q(t) \tag{2.211}$$

Let us assume that $q(t)$ is given. Then equations (2.209)–(2.211) may be reduced to a single equation for the water saturation σ. Indeed, a

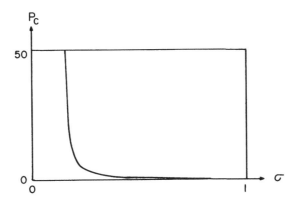

Figure 2.4. Capillary pressure p_c plot vs. water saturation σ; $p_c(\sigma) = B(\sigma_2 - \sigma)^{\gamma_2}/(\sigma - \sigma_1)^{\gamma_1}$, $B = 0.5$, $\sigma_1 = 0.15$, $\sigma_2 = 0.85$.

straightforward computation yields

$$m\frac{\partial}{\partial t}\sigma + q(t)\frac{\partial}{\partial x}\Phi(\sigma) = \frac{\partial}{\partial x}\left[\lambda(\sigma)\frac{\partial}{\partial x}\sigma\right] \tag{2.212}$$

where

$$\Phi(\sigma) = \frac{1}{1 + f(\sigma)\mu}, \quad \mu = \frac{\mu_1}{\mu_2}, \quad f(\sigma) = \frac{f_2(\sigma)}{f_1(\sigma)}, \quad \lambda(\sigma) = -\frac{k}{\mu_2}f_2(\sigma)\Phi(\sigma)\frac{dp_c}{d\sigma} \tag{2.213}$$

Note that $\Phi(\sigma)$, known as the Buckley–Leverett function, increases monotonically from 0 to 1, with a point of inflection at some $\sigma = \sigma_0 \in (\sigma_1, \sigma_2)$ (see Figure 2.5). Equation (2.212) was derived by Rapoport and Leas in 1953 [63] in order to improve the earlier (1942) theory of Buckley and Leverett [7], who disregarded the effect of capillary pressure so that instead of the equation (2.212) they had

$$m\frac{\partial}{\partial t}\sigma + q(t)\frac{\partial}{\partial x}\Phi(\sigma) = 0 \tag{2.214}$$

known as the Buckley–Leverett equation. In spite of its better physical foundation, the practical impact of the Leas–Rapoport theory has been much smaller than the Buckley–Leverett theory, since the latter has produced reasonable and easy predictions in oil-layer processing and this cannot be said of the Leas–Rapoport theory. Moreover, the Buckley–Leverett equation is of mathematical interest as a typical representative of a broad class of quasilinear first-order partial differential equations for which the concept of discontinuous (generalized) solutions becomes indispensable; in addition, the point of inflection of the function $\mathfrak{F}(\sigma)$ is a very essential feature (see Chapter 3, Section 4).

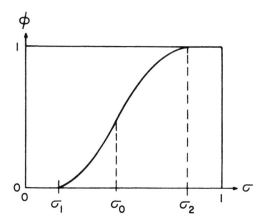

Figure 2.5. The Buckley–Leverett function $\Phi(\sigma)$; $\Phi(\sigma) = (1 + \mu f(\sigma))^{-1}$, $\mu = \mu_2/\mu_1$, $f(\sigma) = f_2(\sigma)/f_1(\sigma)$. Data as in Figure 2.3; $\mu = 1$.

F. Dynamics of adsorption in convective diffusion percolation of solution through porous adsorbent layer

Our purpose is now to derive an equation for the adsorption of a nonelectrolyte from a solution in the process of convective diffusion percolation through the pore space of a nondeformable porous adsorbent, on the assumption that the adsorption is taking place in the outer diffusion region (kinetics) (see Chapter 1, Section 5).

Let $S(M,t)$ be the specific (external) surface of the adsorbent:

$$S(M,t) = \frac{\tilde{\sigma}_0(M,t)}{\Delta\omega(N)} \qquad (2.215)$$

where $\tilde{\sigma}_0$ is the total external surface area of the grains composing the solid skeleton of the porous adsorbent in a physical volume element $\Delta\omega(M)$. Assuming that the medium is homogeneous and rigid, we conclude that the specific surface S is a positive constant, characteristic of the entire porous medium. Let the porosity be a constant ($m = \text{const} > 0$). Let $\mathbf{v}(p,t)$ be the rate of percolation of the solution and $c(p,t)$ the molar concentration of the solute per unit volume of solution. Let $a(p,t)$ be the molar concentration of the adsorbed solute per unit adsorbent surface. As indicated in Chapter 1, the rate of adsorption in the outer diffusion region is proportional to the difference between the bulk concentration of the solute and its equilibrium concentration c^e, that is,

$$\frac{\partial}{\partial t}a(p,t) = b[c(p,t) - c^e(p,t)] \qquad (2.216)$$

where the kinetic coefficient b is assumed to be a positive constant.

The relation between the adsorbed concentration $a(p,t)$ and the equilibrium concentration $c^e(p,t)$ is given by the adsorption isotherm (see

Chapter 1, Section 5):

$$a = f(c^e) \tag{2.217}$$

Let D be an arbitrary simply connected region and Q the total amount of sorptive in its free and adsorbed state:

$$Q = \int_D [mc(q,t) + Sa(q,t)]d\omega_q \tag{2.218}$$

so that, assuming there are no inner sources of solute, we have for the total rate of change of Q

$$\frac{d}{dt}Q = \int_D \frac{\partial}{\partial t}[mc(q,t) + Sa(q,t)]d\omega_q + \int_\Sigma [c(q,t)\mathbf{v}(q,t) + m\mathbf{J}(q,t)] \cdot d\boldsymbol{\sigma}_q = 0 \tag{2.219}$$

Here $\mathbf{v}$ is the rate of percolation and $\mathbf{J}$ is the flux of solute per unit pore area. Localizing the conservation law (2.219), we obtain

$$\frac{\partial}{\partial t}[mc(q,t) + Sa(q,t)] + \mathrm{div}[c(q,t)\mathbf{v}(q,t) + m\mathbf{J}(q,t)] = 0 \tag{2.220}$$

Hence, using Fick's law for the flux, on the assumption that m and S are constants and the solution is an incompressible liquid, we have

$$m\frac{\partial}{\partial t}c + S\frac{\partial}{\partial t}a + \mathrm{div}[c\mathbf{v}(q,t)] = mD\Delta c \tag{2.221}$$

In particular, if the process is 1-dimensional and evolving in the x direction at a constant velocity of forced convection, then

$$\frac{\partial}{\partial t}c + s\frac{\partial}{\partial t} \cdot a + \alpha v\frac{\partial}{\partial x}c = D\frac{\partial^2}{\partial x^2}c, \quad s = \frac{S}{m}, \quad \alpha = \frac{1}{m} \tag{2.222}$$

There are two interesting limiting cases: (1) the limit of a negligibly small rate of percolation, corresponding to purely diffusive mass transfer

$$\frac{\partial}{\partial t}c + s\frac{\partial}{\partial t}a = D\frac{\partial^2}{\partial x^2}c \tag{2.223}$$

and (2) the limit of a negligibly small contribution of diffusion compared with that of forced convection, so that the equation (2.222) becomes a first-order equation

$$\frac{\partial}{\partial t}c + s\frac{\partial}{\partial t}a + \alpha v\frac{\partial}{\partial x}c = 0 \tag{2.224}$$

Thus, in the second case, the 1-dimensional "outer diffusion" sorption of the nonelectrolyte solute by the porous adsorbent from the incompressible solution is described by the following system of equations:

- Mass conservation equation

$$\frac{\partial}{\partial t}c + s\frac{\partial}{\partial t}a + \alpha v\frac{\partial}{\partial x}c = 0 \tag{2.225}$$

- Kinetic equation

$$\frac{\partial}{\partial t}a(p,t) = b[c(p,t) - c^e(p,t)] \tag{2.226}$$

- Adsorption isotherm

$$a = f(c^e) \tag{2.227}$$

and, in particular, for the Langmuir isotherm:

$$a = \frac{a^\infty c^e}{k + c^e} \tag{2.228}$$

and for the Henry isotherm:

$$a = \Gamma c^e \tag{2.229}$$

9. Brownian motion. Langevin's equation [10] and hyperbolic diffusion equation

The purpose of this section is to provide a motivation for the hyperbolic diffusion equation mentioned in Section 7 and discussed in Chapter 13, Section 2 and Chapter 21, Section 1. This will be done in a purely heuristic manner, with reference to some well-known concepts and results from the theory of stochastic processes and some facts about the slow (Stokes) motion of a spherical solid particle in a viscous liquid under the influence of gravity. We shall make no attempt to reproduce the derivation of these results or even to elaborate upon their intuitive origin. Our sole purpose is to demonstrate that an equation of the type of (2.138) may appear directly in the treatment of transport processes. Specifically, we shall need the following assertions.

(a) Consider particles with concentration c moving along random trajectories under the influence of a random force $\mathbf{A}(t)$ which is independent of the trajectories. Then the average value $\langle c\mathbf{A}(t) \rangle$ of $c\mathbf{A}(t)$ over the ensemble of all admissible random trajectories is proportional to the gradient of c average:

$$\langle c\mathbf{A}(t) \rangle = -k \operatorname{grad}\langle c \rangle \tag{2.230}$$

where k is a constant of proportionality.

(b) Consider a spherical particle of radius a and density ρ, falling under the force of gravity in an unbounded region filled with a homogeneous liquid of dynamic viscosity η and density ρ_0. Then the resistance of the liquid to the motion of the particle (i.e., the frictional force decelerating the particle), which is proportional to the velocity $\mathbf{v}$ of the latter, will ultimately cancel out the accelerating effect of gravitation whereafter the particle will move with a constant velocity v_s proportional to the force of gravity. It can be shown that the coefficient of proportionality β^{-1} is

$$\beta^{-1} = \frac{1}{6\pi a\eta}, \quad v_s = \beta^{-1}m'g, \quad m' = m\left(1 - \frac{\rho_0}{\rho}\right) \tag{2.231}$$

where m is the mass of the particle.

Let us now consider Brownian motion of a colloidal particle immersed in a viscous liquid. Brownian motion is the motion of a particle due to its random collisions with the molecules of the surrounding liquid. The frequency of these collisions is of the order of 10^{12} times per second. The size of the colloidal particles may vary over a broad range, from particles so small that their motion may be treated as inertia-free, to a size so large that the inertia of the particles must be taken into account. Langevin's equation is concerned precisely with the inertial motion of colloidal particles. It is in fact Newtonian's law of momentum for a material particle subjected to two kinds of forces: a deterministic force, representing the surrounding liquid's resistance to the motion of the particle, which is proportional to the velocity of the particle; and a random force, due to the particle's collisions with the molecules of the liquid. Thus, let $\mathbf{v}(p,t)$ be the velocity of a Brownian particle, viewed as a sphere of radius a and mass m in a liquid of dynamic viscosity η, and let $\mathbf{A}(t)$ be the stochastic force per particle. Langevin's equation is

$$m\frac{d}{dt}\mathbf{v} = -\beta\mathbf{v} + \mathbf{A}(t) \tag{2.232}$$

It has been assumed here that the liquid at a great distance from the Brownian particle is quiescent. The coefficient β in this theory is taken as equal to that in the Stokes solution for uniform motion of a sphere in a viscous liquid; that is, it is determined by (2.231).

A rigorous stochastic analysis of Brownian motion leads to the Fokker–Planck equation of diffusion in 6-dimensional (coordinate–velocity) phase space [10; 20]. For our purpose it is sufficient to bypass the Fokker–Planck equation, restricting ourselves to the following observations.

Let c be the concentration of Brownian particles per unit volume of the medium. Since the "average" velocity of the fluid motion is assumed to be zero, we have

$$\langle c\mathbf{v}\rangle = \mathbf{J} \tag{2.233}$$

where $\mathbf{J}$ is the diffusion flux. By multiplying equation (2.232) by c we obtain (after averaging)

$$m\left\langle c\frac{\partial}{\partial t}\cdot\mathbf{v}\right\rangle = -\beta\cdot\langle c\mathbf{v}\rangle + \langle c\mathbf{A}(t)\rangle \tag{2.234}$$

The following hypotheses are being used (see (2.158)–(2.161))

$$\left\langle \mathbf{v}\frac{\partial c}{\partial t}\right\rangle = \langle\mathbf{v}\rangle\left\langle\frac{\partial c}{\partial t}\right\rangle, \quad \left\langle\frac{\partial c}{\partial t}\right\rangle = \frac{\partial}{\partial t}\langle c\rangle, \quad \langle\mathbf{v}\rangle \approx \frac{\mathbf{J}}{\langle c\rangle} \tag{2.235}$$

Since the particles are conserved, we have

$$\left\langle\frac{\partial c}{\partial t}\right\rangle = -\operatorname{div}\mathbf{J} \tag{2.236}$$

By (2.235),

$$\left\langle c\frac{\partial}{\partial t}\mathbf{v}\right\rangle = \frac{\partial}{\partial t}\mathbf{J} - \langle\mathbf{v}\rangle\left\langle\frac{\partial}{\partial t}c\right\rangle \tag{2.237}$$

It follows from (2.233), (2.237), and (2.230) that

$$m\frac{\partial}{\partial t}\mathbf{J} - m\langle\mathbf{v}\rangle\left\langle\frac{\partial}{\partial t}c\right\rangle = -\beta\mathbf{J} + k\operatorname{grad}\langle c\rangle \tag{2.238}$$

Let us omit the symbol $\langle\ \rangle$, so that nonbracketed expressions will refer to averages. Then, neglecting the product of averages on the left of (2.238) (which, by (2.235) and (2.236), means that we are neglecting the term quadratic in the flux), we obtain the desired equation for the rate of change of the diffusive flux:

$$m\frac{\partial}{\partial t}\mathbf{J} = -\beta\mathbf{J} - k\operatorname{grad}c \tag{2.239}$$

Equation (2.239) implies

$$m\frac{\partial}{\partial t}\operatorname{div}\mathbf{J} = -\beta\operatorname{div}\mathbf{J} - k\Delta c, \quad \frac{\partial^2}{\partial t^2}c + \frac{\partial}{\partial t}\operatorname{div}\mathbf{J} = 0$$

$$\Rightarrow \tau\frac{\partial^2}{\partial t^2}c + \frac{\partial}{\partial t}c = D\Delta c \tag{2.240}$$

where

$$\tau = \frac{m}{\beta}, \quad D = \frac{k}{\beta} \tag{2.241}$$

Note that, by (2.231), β is independent of m; hence, by (2.241), the *relaxation time* τ is proportional to m, whereas the *diffusion coefficient* D is independent of m. Hence, in the limit of $m \downarrow 0$, it follows from (2.240) that

$$\frac{\partial}{\partial t}u = D\Delta u \tag{2.242}$$

that is, the inertia-free limit $(m = 0)$ is the limit as $\tau \downarrow 0$ of the hyperbolic diffusion equation, and we obtain the usual Fick equation.

10. Boundary and initial conditions

Every differential equation has an infinite set of solutions, from which one must select a particular solution: one describing the physical process, the mathematical simulation of which led to the particular problem of mathematical physics being considered. The state and evolution of the material medium in a region D of space depends on the interaction between this medium and the medium in the complement CD of D, and thus depends also on processes in CD. If one wishes to restrict the detailed considerations to D only, one models the process in CD in some very crude fashion, so as to infer the effect of CD on D through the boundary Σ of D. Essentially, this means that the processes in CD prescribe certain features of the solution in D at the surface Σ. The totality of these "prescriptions" is known as *boundary conditions*. As a rule, the boundary conditions follow from two physical principles: local thermodynamic equilibrium, and continuity of the fluxes of the physical variables involved. Frequently, however, appropriate conditions may be derived from the specific features of the process under consideration. This is particularly true in the case of multiphase systems.

If one is concerned with an evolution process, then it is usually necessary to prescribe, along with the boundary conditions, suitable initial conditions. (We say "usually" because one can have non–steady-state evolution problems without initial conditions.) Thus, consider a function $u(p,t)$ that is to be determined inside a region D. Let us indicate some of the most typical and most commonly used of the infinitely many different types of possible boundary conditions.

1. Conditions of the first kind (also called Dirichlet conditions)

The value of u is prescribed on Σ

$$u(p,t) = f(p,t) \quad \forall p \in \sum, \ \forall t \geq 0 \tag{2.243}$$

where f is a given function.

2. Conditions of the second kind (Neumann conditions)

The normal derivative of u on Σ is a prescribed function φ:

$$\frac{\partial}{\partial n} u(p,t) = \varphi(p,t) \quad \forall p \in \sum, \ \forall t \geq 0 \tag{2.244}$$

3. Conditions of the third kind (also known as *irradiation* conditions in the theory of heat conduction or *elastic fixation* conditions in the theory of elasticity)

The normal derivative of u on Σ is a prescribed function of u and of the independent variables

$$\frac{\partial}{\partial n} u(p,t) = \psi[u(p,t),p,t] \tag{2.245}$$

4. Periodicity conditions

If the solution of a boundary value problem is expected to be spatially periodic with a known period, owing to some spatial periodicity in the formulation of the problem, then the boundary conditions will dictate that the solution repeats over the period.

For example, in the spatially 1-dimensional case, if the element of periodicity is an infinite layer of thickness ℓ,

$$n\ell < x < (n+1)\ell, \quad -\infty < y, z < \infty \tag{2.246}$$

then the periodicity conditions are

$$\frac{\partial}{\partial x} u(n\ell + 0, y, z, t) = \frac{\partial}{\partial x} u((n+1)\ell - 0, y, z, t) \tag{2.247}$$

$$u[n\ell + 0, y, z, t] = u[(n+1)\ell - 0], \quad n = 0, 1 \pm 0, 2 \pm 0 \tag{2.248}$$

Periodicity conditions may be imposed even when the problem is being solved in a bounded domain D, if a similar problem can be formulated in the whole space in such a manner that the solution satisfies all the conditions originally prescribed in D (see Chapter 5, Section 2).

5. Boundary conditions in multiphase medium problems

Recall that when a multiphase medium is occupying some region G, there are in G certain surfaces across which at least one of the dependent variables (e.g., temperature, pressure, velocity of motion, etc.) or at least one of the physical parameters (e.g., thermal conductivity, density, viscosity, conductance, etc.) that characterize the medium is discontinuous. To handle processes evolving in a multiphase medium, one must prescribe some kind of matching conditions at the phase interfaces. Here, too, such conditions typically account for local thermodynamic equilibrium and conservation of the appropriate variables at the surfaces of discontinuity, as implied by the balance in narrow layers adjacent to the surfaces. Such boundary conditions are sometimes called dynamic compatibility conditions.

Local thermodynamic equilibrium. This may imply nothing more than continuity of the unknown function (temperature, pressure, concentration, etc.) or it may yield a more complex condition, involving continuity of the chemical or electrochemical potential of any component of the medium for which the phase interface is permeable.

For example, consider the process of heat conduction in a solid D, composed of two parts D_1 and D_2 in a perfect contact along a surface Σ, with different thermal conductivities k_1 and k_2, respectively. Despite the discontinuity of the physical property in question (thermal conductivity), the temperature must be continuous across Σ by virtue of the local equilibrium, so that

$$u_1 = u_2 \quad \forall p \in \sum, \ \forall t > 0 \tag{2.249}$$

As a second example, consider isothermal osmotic water transfer through a surface Σ. Suppose that the physical meaning of Σ is an infinitely thin, semipermeable membrane, separating two water solutions of the same nonelectrolyte solute A at two different concentrations, filling regions D_1 and D_2. Denote these (molar) concentrations of A c_{a1} and c_{a2}, respectively, and the respective water concentrations c_{w1} and c_{w2}. Let p_1 and p_2 be the pressures in D_1 and D_2, T the absolute temperature, Ω the molar volume of water, and R the universal gas constant. Then the condition of local thermodynamic equilibrium at Σ reads

$$RT \ln \left(\frac{c_{w1}}{c_{w1} + c_{a1}} \right) + \Omega p_1 = RT \ln \left(\frac{c_{w2}}{c_{w2} + c_{a2}} \right) + \Omega p_2 \tag{2.250}$$

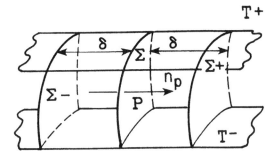

Figure 2.6. Sketch of the domain $\bar{D} = \bar{D}^+ \cup \bar{D}^-$.

Dynamic compatibility conditions.[13] Two cases must be distinguished:

(1) The surface of discontinuity is prescribed in space and time.
(2) The surface Σ of discontinuity is not prescribed but must be determined in the course of solution; in this case Σ is called a *free boundary.*

Let $\mathfrak{f}$ be the density of some extensive parameter $\mathfrak{F}$ in a region $\tilde{D}$ of space, whose material contents may vary. Let Σ be a smooth two-sided surface dividing $\tilde{D}$ into two parts $\tilde{D}^+$ and $\tilde{D}^-$, so that the normal $\mathbf{n}$ to Σ is directed into $\tilde{D}^+$. Consider a point P on Σ and take $\delta > 0$ small and $h > 0$. Now let D be the region bounded by two planes T^+ and T^-, both parallel to the normal $\mathbf{n}$ at P at a distance h from P, and by the two surfaces Σ^+ and Σ^- obtained by shifting Σ for a distance δ in the outward and inward direction, respectively (see Figure 2.6). The rate of change $\mathfrak{F}$ in D is:

$$\frac{d}{dt}\mathfrak{F}(t) = \int_D \frac{\partial}{\partial t}\mathfrak{f}(q,t)d\omega_q + \int_{\partial D}\mathfrak{f}(q,t)\tilde{\mathbf{v}}(q,t)d\boldsymbol{\sigma}_q \qquad (2.251)$$

Here (in contrast to (2.3)) $\tilde{\mathbf{v}}$ is the relative velocity of motion of the medium relative to Σ, that is, the difference between the rate $\mathbf{v}$ of the medium in the laboratory frame of reference and the velocity $\delta n/\delta t$ at which Σ moves in the direction of its normal $\mathbf{n}$ in the same frame:[14]

$$\tilde{\mathbf{v}} = \mathbf{v} - \frac{\delta \mathbf{n}}{\delta t} \qquad (2.252)$$

We now include the effect of volume and surface sources of $\mathfrak{f}$, with volumetric and surface densities F and f, respectively. Let $\mathbf{J}$ denote the

---

[13]This term originated in gas dynamics [64].

[14]Let $F(p,t) = 0$ be the equation of Σ at the point P at time t. Then

$$\frac{\delta n}{\delta t} = -\frac{\partial}{\partial t}F \bigg/ |\operatorname{grad} F|$$

flux of $\mathfrak{f}$ through Σ. Then the integral law of conservation of $\mathfrak{F}$ in D reads

$$\int_D \frac{\partial}{\partial t} \mathfrak{f}(q,t) d\omega_q + \int_{\partial D} \mathfrak{f}(q,t) \tilde{\mathbf{v}}(q,t) \cdot d\boldsymbol{\sigma}_q \qquad (2.253)$$

$$= -\int_{\partial D} \mathbf{J}(q,t) \cdot d\boldsymbol{\sigma}_q + \int_D F(q,t) d\omega_q + \int_{\partial D} f(q,t) d\sigma_q$$

Now let $\delta \downarrow 0$. The volume integrals in (2.253) tend to zero, while the surfaces Σ^+ and Σ^- approach the positive and negative sides (respectively) of the discontinuity surface Σ. Let

$$[F] = F^+ - F^- \qquad (2.254)$$

denote the jump of any function F at the points of Σ. We have

$$\int_\Sigma [\mathfrak{f}(q,t)\tilde{v}_n(q,t)] d\sigma_q = \int_\Sigma f(q,t) d\sigma_q - \int_\Sigma [J_n(q,t)] d\sigma_q \qquad (2.255)$$

Assuming that the integrands are continuous, we can apply Lemma 2.1 (which is possible since h is arbitrarily small) to obtain the local form of the dynamic compatibility condition, written for any intensive scalar variable $\mathfrak{f}$:

$$[\mathfrak{f}(q,t)\tilde{v}_n(q,t)] = f(q,t) - [J_n(q,t)] \qquad (2.256)$$

To get the dynamic compatibility condition for an intensive vector variable, simply apply this condition to each component separately.

Let us apply this derivation of matching conditions in a multiphase medium to the equation

$$\text{div } \boldsymbol{\Psi} = 0 \qquad (2.257)$$

for an arbitrary vector field. Repeating the above arguments, we find that

$$[\Psi_n] = 0 \qquad (2.258)$$

In particular, if $\boldsymbol{\Psi}$ is the average volume velocity of a solution flowing through the phase interface Σ,

$$\boldsymbol{\Psi} = \mathbf{w} \qquad (2.259)$$

then we find that the normal component of $\mathbf{w}$ varies continuously. The corresponding electrodynamical compatibility conditions at the boundaries between two media are:

$$[\mathbf{n}^0 \times \mathbf{H}] = \frac{4\pi}{c} i, \quad [n^0 \cdot D] = 4\pi\delta, \quad [n^0 \times E] = 0, \quad [n^0 \cdot B] = 0 \qquad (2.260)$$

Here " $\times$ " and "." denote signs of the vector and scalar products, [ ] symbolizes the jump upon passage through the surface of discontinuity in the direction of the normal $\mathbf{n}$ to the surface, $\mathbf{n}^0$ is the unit vector in the direction of $\mathbf{n}$, and $\mathbf{i}$ and δ are the surface densities of the conductive current and true electric charges, respectively. Thus the normal component of $\mathbf{B}$ and tangential component of $\mathbf{E}$ are continuous, and the normal component of $\mathbf{D}$ and tangential component of $\mathbf{H}$ experience jumps as indicated.

The distinction between the two cases (1) and (2) is essential. In case (1), the local thermodynamic equilibrium and dynamic compatibility conditions are the only boundary conditions necessary. In contrast, for a free boundary-value problem (case (2)), the region in which the function $u(p, t)$ has to be determined is itself not known in advance but must be determined in the course of solution. For this to be possible, an additional boundary condition, depending on the specific features of the problem, must be prescribed (see Section 11). This is in contrast to dynamic compatibility conditions, which essentially possess the same very general, universal structure common to all systems.

A. Concentrated capacity conditions

All the boundary conditions considered so far contain derivatives of at most first order, because the partial differential equations in question are of second order (as was the case in viscous hydrodynamics, theory of elasticity, electrodynamics, theory of diffusion or heat conduction, etc.). In certain situations, however, it is natural to impose boundary conditions of the same or even higher order than that of the differential equation to be solved. This class of boundary conditions is known as concentrated capacity conditions. Such conditions usually appear in the treatment of processes evolving in multilayered regions, where the time of relaxation in some layers is much shorter than in others. To illustrate, we shall consider two examples from the theory of heat conduction.[15]

1. Let us consider a slab S of length L and normal cross-sectional area s, made of a thermally conductive material. Assume that the slab is thermally insulated along its lateral boundary. Furthermore, it is attached at one end ($x = 0$) to another slab S_1 of a very high thermal conductivity, with the same cross-sectional area s but length ℓ, which is also thermally insulated along its lateral boundary and at its free end $x = -\ell$. The temperature distribution in both slabs obeys the heat conduction equation

$$k \frac{\partial^2}{\partial x^2} u(x, t) = c\rho \frac{\partial}{\partial t} u(x, t), \quad 0 < x < L, \ t > 0 \tag{2.261}$$

$$k_1 \frac{\partial^2}{\partial x^2} u_1(x, t) = c_1 \rho_1 \frac{\partial}{\partial t} u_1(x, t), \quad -\ell < x < 0, \ t > 0 \tag{2.262}$$

subject to compatibility conditions at $x = 0$,

$$u(+0, t) = u_1(-0, t), \quad k \frac{\partial}{\partial x} u(+0, t) = k_1 \frac{\partial}{\partial x} u_1(-0, t) \tag{2.263}$$

the condition of thermal insulation at $x = -\ell$,

$$\frac{\partial}{\partial x} u(-\ell, t) = 0 \tag{2.264}$$

and certain boundary condition at $x = L$, as well as some initial conditions at $t = 0$ that are of no relevance here.

---

[15] Concentrated capacity conditions were first introduced by A. Tichonov [84].

If $k(\partial/\partial x)u(+0,t)$ is of order unity and $k_1 \uparrow \infty$, then by (2.263)

$$\frac{\partial}{\partial x}u_1(-0,t) \downarrow 0 \tag{2.265}$$

Thus, if k_1 is very large compared with k then the temperature gradient in S_1 is very small. This and equation (2.264) show that, except for a very small neighborhood of the initial time, the temperature of the slab S_1 depends on t only.

Owing to the heat influx into S_1 from S, equal to $ks(\partial/\partial x)u(+0,t)$, the slab S_1 heats up, so that

$$c_1\rho_1 s\ell\frac{\partial}{\partial t}u_1(-0,t) = ks\frac{\partial}{\partial x}u(+0,t) \tag{2.266}$$

Hence, by the left side of (2.263),

$$\frac{k}{c_1\rho_1\ell}\frac{\partial}{\partial x}u(+0,t) = \frac{\partial}{\partial t}u(+0,t) \tag{2.267}$$

so that, by (2.261),

$$\frac{\partial^2}{\partial x^2}u(0,t) = \alpha\frac{\partial}{\partial x}u(0,t), \quad \alpha = \frac{c\rho}{c_1\rho_1\ell} \tag{2.268}$$

We can now consider the temperature in S without referring to S_1, since the latter has been completely eliminated from the argument. This slab S_1 is called the *concentrated thermal capacity*, and condition (2.268) is a boundary condition of the concentrated capacity type.

2. In our second example we shall be concerned with a three-layered medium occupying a region $D = D_1 \cup D_2 \cup D_3$, where

$$D_1 = \{0 < r < \infty, \ 0 < \varphi \leq 2\pi, \ -2h < z < 0\}$$
$$D_2 = \{0 < r < \infty, \ 0 < \varphi \leq 2\pi, \ 0 < z < \infty\} \tag{2.269}$$
$$D_3 = \{0 < r < \infty, \ 0 < \varphi \leq 2\pi, \ -\infty < z < -2h\}$$

and (r,φ,z) is a cylindrical coordinate system. Assume that D_1 is a homogeneous porous stratum of constant porosity and small thickness $2h$, while D_2 and D_3 represent the overlying and underlying impermeable strata. Let us consider the temperature distribution in D_2 and D_3. We may assume that the temperature is initially identically equal to zero and varies owing to injection into the porous layer D_1 of hot water at a constant temperature ϑ and a constant rate of injection.

Assuming also that the plane $z = -h$ is a plane of symmetry, we restrict ourselves to the axially symmetric case. Then the temperature in D_1 and D_2 is a solution of the following problem:

$$a^2\left(\frac{\partial^2}{\partial r^2} + \frac{1-2\nu}{r}\frac{\partial}{\partial r} + \frac{\partial^2}{\partial z^2}\right)u = \frac{\partial}{\partial t}u,$$

$$0 < r < \infty, \quad -h < z < 0 \ (\nu = \text{const} > 0) \tag{2.270}$$

$$a_1^2\left(\frac{\partial^2}{\partial r^2} + \frac{1}{r}\frac{\partial}{\partial r} + \frac{\partial^2}{\partial z^2}\right)u = \frac{\partial}{\partial t}u, \quad 0 < r < \infty, \ 0 < z < \infty \tag{2.271}$$

$$\frac{\partial}{\partial z} u(r, -h, t) = 0 \tag{2.272}$$

$$u(r, -0, t) = u(r, +0, t), \quad \frac{\partial}{\partial z} u(r, -0, t) = \lambda \frac{\partial}{\partial z} u(r, +0, t) \tag{2.273}$$

$$u(r, z, 0) = 0, \quad 0 < r < \infty, \quad -h < z < \infty, \quad u(0, z, t) = \vartheta \tag{2.274}$$

Here the constants a^2, a_1^2 are the respective heat diffusivities, λ–ratio of heat conductivities and the positive constant ϑ is related to the injection rate (below, for the sake of simplicity, a_1^2 is equated to a^2). Integrating equation (2.270) with respect to z and using the continuity conditions (2.273), we obtain

$$a^2 \left(\frac{\partial^2}{\partial r^2} + \frac{1 - 2\nu}{r} \frac{\partial}{\partial r} \right) \tilde{u} + k \frac{\partial}{\partial z} u \bigg|_{z=+0} = \frac{\partial}{\partial t} \cdot \tilde{u}, \quad 0 < r < \infty, \ t > 0 \tag{2.275}$$

where $k = a^2 \lambda / h$ and $\tilde{u}$ is the value of u, averaged with respect to z over $z \in (-h, 0)$;

$$\tilde{u}(r, t) = \frac{1}{h} \int_{-h}^{0} u(r, z, t) dz \tag{2.276}$$

Now, disregarding the distinction between $\tilde{u}(r, t)$ and $u(r, -0, t) \equiv u(r, +0, t)$, we obtain

$$a^2 \left(\frac{\partial^2}{\partial r^2} + \frac{1 - 2\nu}{r} \frac{\partial}{\partial r} \right) u + k \frac{\partial}{\partial z} u = \frac{\partial}{\partial t} u, \quad 0 < r < \infty, \ z = 0, \ t > 0 \tag{2.277}$$

The term $\partial u / \partial t$ may be eliminated from (2.277) by using (2.271), and the result is a boundary condition of the form

$$\frac{\partial^2}{\partial z^2} u + \frac{\alpha}{r} \frac{\partial}{\partial r} u + \beta \frac{\partial}{\partial z} u = 0, \quad 0 < r < \infty, \ z = 0, \ t > 0, \alpha, \beta = \text{const} \tag{2.278}$$

Again, the need to consider the process in a two-layered medium has been eliminated by introducing a boundary condition of the concentrated capacity type. This approach to problems in multilayered media, in which there are very thin layers together with thick layers, is quite efficient for numerical or asymptotic analysis. This is because direct application of any numerical method (finite difference, finite elements, variational methods, etc.) to a multilayered problem in its full formulation, such as (2.270)–(2.274) is a very complicated task, even with modern computers.

B. Initial conditions. Cauchy problem. Non–steady-state problems without initial conditions

A non–steady-state process may evolve in all of space or in a bounded or unbounded subregion. A non–steady state problem considered in all of space is known as a Cauchy problem. In such problems, to be able to select a unique solution from the infinite set of possible solutions one must add initial conditions, or Cauchy data as they are sometimes called. (In most

Cauchy problems there is also an analog of the boundary conditions, e.g., the prescribed rate at which the solution varies at infinity.) The number and nature of the initial conditions are determined by the highest order of the derivative with respect to time in the partial differential equation to be solved. Indeed, consider a partial differential equation

$$\frac{\partial^r u}{\partial t^r} = F\left(t, x, y, z, u, \frac{\partial u}{\partial x}, \dots, \frac{\partial u}{\partial z}, \frac{\partial u}{\partial t}, \dots, \frac{\partial^{i+k+m+n} u}{\partial x^i \partial y^k \partial z^m \partial t^n}, \dots, \frac{\partial^{r-1} u}{\partial t^{r-1}}\right)$$
(2.279)

Let $u = u(x, y, z, t)$ be a definite solution of this equation, and introduce the notation

$$u_k = \frac{\partial^k}{\partial t^k} u, \quad k = 0, 1, 2, \dots, r - 1$$
(2.280)

Suppose we consider F as a function $\mathfrak{F}(t, u_0, u_1, \dots, u_{r-1} \mid x, y, z)$ with x, y, z treated as parameters, so that equation (2.279) is replaced by a system of r first-order ordinary differential equations

$$\frac{d}{dt} u_0 = u_1, \quad \frac{d}{dt} u_1 = u_2, \quad \dots, \quad \frac{d}{dt} u_{r-1} = \mathfrak{F}$$
(2.281)

The solution of this system is uniquely determined by prescribing Cauchy data

$$u_k(0 \mid x, y, z) = u_{k0}(x, y, z), \quad k = 0, 1, 2, \dots, r - 1$$
(2.282)

This heuristic argument indicates that, in order to specify a definite solution of a partial differential equation of the form (2.279), one must prescribe, in addition to the boundary conditions, the initial values of all the derivatives with respect to time up to order $r - 1$. This conjecture may be justified by a rigorous theory, as we shall see in due course.

Nevertheless, there exists one class of non–steady-state problems that can be formulated and solved without initial conditions. Such problems arise in situations where one is interested in the state of the system well away from the initial time, and where it is known in advance that initial irregularities in the solution decay in time, so that after a sufficiently long time the time-dependent solutions of any two problems, with different initial conditions but identical boundary conditions, become practically indistinguishable. Therefore, in non–steady-state problems with no initial conditions, the solution is unique in an asymptotic sense only.

A typical example of a non–steady-state problem without initial conditions is the propagation of temperature waves. This problem, originating in climatology, is formulated as follows. Let $u(x, t)$ be the temperature beneath the earth's surface ($x = 0$), determined by the equation of heat conduction

$$a^2 \frac{\partial^2}{\partial x^2} u = \frac{\partial}{\partial t} u, \quad 0 < x < \infty$$
(2.283)

and the boundary condition

$$u(0, t) = \alpha e^{ik^2 t}$$
(2.284)

This boundary condition simulates seasonal temperature fluctuations at the surface.

The solution u may be sought in the form of steady oscillations with the amplitude $v(x)$:

$$u(x,t) = v(x)e^{ik^2t} \qquad (2.285)$$

The amplitude $v(x)$ is then a solution of the equation of stationary oscillations (the Helmholtz equation)

$$a^2 \frac{d^2}{dx^2} v(x) = ik^2 v(x) \qquad (2.286)$$

which is bounded at infinity and satisfies the boundary condition

$$v(0) = \alpha \qquad (2.287)$$

The solution to problem (2.283) and (2.284) bounded at infinity and of the form (2.285) is easily found to be

$$u(x,t) = \alpha e^{-\sigma x} \cos(kt - \sigma x), \quad \sigma = \frac{k}{a\sqrt{2}} \qquad (2.288)$$

so that temperature fluctuations rapidly decay in space as one moves away from the surface of the earth. On the other hand, it can be shown that any solution of an initial boundary-value problem for (2.283) and (2.284) satisfying an arbitrary initial condition and decaying at infinity tends to (2.288) as t increases.

11. Examples of typical free boundary-value problems

A. Thermal two-phase Stefan problem

Let $D \subset \mathbb{R}_3$ be a doubly connected region bounded by a surface

$$\Sigma = \Sigma_e \cup \Sigma_i \qquad (2.289)$$

(see Figure 2.7). Assume that it is filled with some monocomponent material in two phases: solid and liquid that occupy subregions D_t^s and D_t^f, separated by a smooth interface Σ_t. Let

$$\Sigma_t^s = \Sigma_e \cup \Sigma_t, \quad \Sigma_t^f = \Sigma_i \cup \Sigma_t \qquad (2.290)$$

be the boundaries of the solid and liquid phases, respectively. The melting temperature is assumed to be $u = 0$, and prescribed temperatures $\vartheta_s(p,t) < 0$ and $\vartheta_f(p,t) > 0$, respectively, are maintained on Σ_e and Σ_i. Under these conditions the transfer of heat is accompanied by a phase transition.

Let us assume that thermal expansion is negligible, so that the densities ρ_s and ρ_f are constant and equal. Moreover, we shall assume that there occurs no undercooling of the liquid phase or superheating of the solid phase during the phase transition.[16] Under these assumptions, in the absence of

---

[16] The theoretical treatment (mathematical and physical) of phase transitions accompanied by superheating and undercooling is a fascinating unsolved problem.

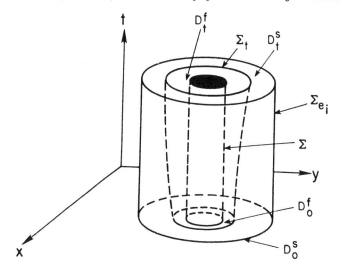

Figure 2.7. Free boundary D_t in a monocomponent material.

forced convection, the temperature distribution in both phases obeys the Fourier heat conduction equation:

$$a_s^2 \Delta u_s = \frac{\partial}{\partial t} u_s \quad \forall p \in D_t^s, \ \forall t > 0 \qquad (2.291)$$

$$a_f^2 \Delta u_f = \frac{\partial}{\partial t} u_f \quad \forall p \in D_t^f, \ \forall t > 0 \qquad (2.292)$$

$$u_s(p,t) = \vartheta_s(p,t) \quad \forall p \in \Sigma_e, \quad u_f(p,t) = \vartheta_f(p,t) \quad \forall p \in \Sigma_e, \ \forall t > 0 \quad (2.293)$$

$$u_s(p,0) = u_{s0}(p) \quad \forall p \in D_0^s, \quad u_f(p,0) = u_{f0}(p) \quad \forall p \in D_0^f \qquad (2.294)$$

$$u_s(p,t) = u_f(p,t) = 0 \quad \forall p \in \Sigma_t, \ \forall t > 0 \qquad (2.295)$$

$$\gamma \frac{\delta n}{\delta t} = k_s \frac{\partial}{\partial n_p} u_s - k_f \frac{\partial}{\partial n_p} u_f \quad \forall p \in \Sigma_t, \ \forall t > 0 \qquad (2.296)$$

The left-hand side of the Stefan condition (2.296) represents the rate of release of latent heat of fusion, which may be viewed as the density of the surface source of heat (see (2.256)). It is assumed that the initial location and shape of the phase interface Σ_t are known, the specific (per unit of volume) latent heat of fusion γ is a positive constant, and the initial temperature satisfies the conditions

$$u_{s0}(p) < 0 \quad \forall p \in D_0^s, \quad u_{f0}(p) > 0 \quad \forall p \in D_0^f \qquad (2.297)$$

B. Diffusion Stefan Problem. Dissolution of rock salt

Let us consider the isothermal dissolution of the rock salt bottom of a basin filled with a water solution of the same salt. (To simplify matters, we shall treat the dissolved salt as a nonelectrolyte, e.g., sugar.) The redistribution of the salt concentration inside the solution is assumed to be purely diffusional. In other words, no forced convection is involved and neither is any free convection, since the density of the solution increases with the increased concentration and the dissolution of salt takes place at the bottom.

The x axis is assumed to point vertically upward, and the solution fills the half-space

$$D = \{x: s(t) < x < \infty\} \quad \forall t \geq 0 \tag{2.298}$$

Here $x = s(t)$ is the free boundary, the determination of whose location is our main concern.

Let us denote the volume concentration of the salt by $c(x,t)$ and the concentration of the saturated solution by $c_\infty < 1$, with a unit "concentration" adopted for the rock salt. Assume for the sake of simplicity that the initial concentration is zero, $c(x,0) \equiv 0$. Under these conditions, the problem to be solved is the following one-phase diffusion Stefan problem:

$$D\frac{\partial^2}{\partial x^2}c(x,t) = \frac{\partial}{\partial t}c(x,t), \quad x > s(t), \ t > 0 \tag{2.299}$$

$$c(x,0) = 0 \quad \forall x > s(0) = 0 \tag{2.300}$$

$$c[s(t),t] = c_\infty \quad \forall t > 0 \tag{2.301}$$

$$(1 - c_\infty)\frac{d}{dt}s(t) = D\frac{\partial}{\partial x}c[s(t),t] \quad \forall t > 0 \tag{2.302}$$

The left-hand side in the last condition may be viewed as the density of a surface source of salt.

C. Wind drive of sea water in vicinity of vertical shore

Let us represent the deep sea in the vicinity of a vertical shore by a region

$$D_t = \{x,y,z: 0 < x < \infty, \ -\infty < y < \infty, \ -\infty < z < s(x,t)\} \quad \forall t > 0 \tag{2.303}$$

filled with water of constant density ρ and dynamic viscosity μ. Let

$$\Sigma_t = \{x,z: z = s(x,t)\}, \quad \forall t > 0 \tag{2.304}$$

denote the free surface. Suppose that a wind is blowing in the x direction at constant velocity v_0. This generates a water drive in the sea toward the shore (due to the constant tangential stress $T_{n\tau}$ along the free surface Σ_t), which in turn causes the water to circulate in a plane normal to the shoreline (y-axis). The problem is thus 2-dimensional.

Let $\mathbf{v}(x,z,t)$ be the velocity vector of the water with components (u,v). Rigorously speaking, the drive of water affects the wind velocity in the vicinity of the surface, but we shall use the approximation $v_0 = \text{const}$. The air pressure at the surface is assumed to be negligible, so that the only influence to be taken into account is that of tangential stresses. Under these conditions the motion of the water is described by the following system of governing equations.

$$\frac{\partial}{\partial t}u + u\frac{\partial}{\partial x}u + v\frac{\partial}{\partial z}u = \frac{\partial}{\partial x}p + \nu\Delta u \tag{2.305}$$

$$\frac{\partial}{\partial t}v + u\frac{\partial}{\partial x}v + v\frac{\partial}{\partial z}v = -\frac{\partial}{\partial z}p + \rho g + \nu\Delta v \tag{2.306}$$

$$\frac{\partial}{\partial x}u + \frac{\partial}{\partial z}v = 0, \quad (x,z) \in D_t, \ t > 0 \tag{2.307}$$

$$u = v = 0, \quad x = 0 \quad \forall z < 0, \ \forall t \geq 0 \tag{2.308}$$

$$p(x, s(x,t), t) = 0 \quad \forall x > 0, \ \forall t > 0 \tag{2.309}$$

$$\frac{\delta n}{\delta t} = v_n \quad \forall p = (x, z) \in \Sigma_t, \ \forall t > 0 \tag{2.310}$$

$$T_{n\tau} = A = \text{const} \quad \forall p = (x, z) \in \Sigma_t, \ \forall t > 0 \tag{2.311}$$

$$s(x, 0) = 0, \quad 0 < x < \infty \tag{2.312}$$

Here, as before, $\delta n/\delta t$ is the rate of displacement of the free boundary in the normal direction, $T_{n\tau}$ the tangential component of the viscous stress in the water at the free boundary, and $\nu = \mu/\rho$ the kinematic viscosity of water. The corresponding steady-state problem may be obtained from (2.303)–(2.309) by setting $\delta n/\delta t = 0$ and omitting the initial conditions (2.308) and (2.312).

The free-boundary-value problem (2.305)–(2.312) may be viewed as typical of a very broad class of problems concerning the motion of a fluid with a "free" surface. In particular, this includes the classical problem of water waves.

D. Stationary dam problem [14]

Consider an earth dam occupying a region

$$D = \{x, y, z\colon 0 < x < L, \ -\infty < y < \infty, \ 0 < z < H\} \tag{2.313}$$

that separates two infinitely large water basins

$$D_1 = \{x, y, z\colon -\infty < x < 0, \ -\infty < y < \infty, \ 0 < z < h_0\} \tag{2.314}$$
$$D_2 = \{x, y, z\colon L < x < \infty, \ -\infty < y < \infty, \ 0 < z < h_1\} \tag{2.315}$$

where

$$0 < h_1 < h_0 < H \tag{2.316}$$

Water percolates through the dam from D_1 into D_2 under the force of gravity. As a result of percolation and evaporation, water saturates the region

$$D_0 = \{x, z\colon 0 \leq x \leq L, \ 0 < z < Z(x)\} \tag{2.317}$$

where $Z(x)$ is a monotonically decreasing function such that

$$Z(0) = h_0, \quad Z(L) = h \tag{2.318}$$

with

$$h_1 \le h \le h_0 \tag{2.319}$$

The problem is to determine the free boundary

$$S = \{x, z \colon z = Z(x), \ 0 \le x \le L\} \tag{2.320}$$

together with the pressure distribution $p(x, z)$ and percolation rate $v(x, z)$.

In the simplest formulation of the dam problem, evaporation from the free boundary is disregarded, so that the pressure p there is equated to zero; that is,

$$p(x, z(x)) = 0 \tag{2.321}$$

One also adopts the simplifying assumption that a "toe drain" cannot occur; that is, the end of the free boundary cannot touch dam bottom at $x < L$.

The rate of seepage $\mathbf{v}$ must satisfy the following conditions:

(1) The Darcy law:

$$\mathbf{v} = -\operatorname{grad} \varphi \tag{2.322}$$

where the velocity potential is defined as

$$\varphi = \frac{k}{\mu}(p + \rho g z) \tag{2.323}$$

(2) Equation of continuity:

$$\operatorname{div} \mathbf{v} = 0 \tag{2.324}$$

(3) Free boundary conditions:

$$p = 0, \quad \frac{\partial}{\partial n}\varphi = 0 \quad \forall(x, z) \subset S \tag{2.325}$$

(4) Continuity of pressure at the boundaries $x = 0$, $x = L$:

$$p(-0, z) = p(+0, z) \quad \forall z \in [0, h_1], \quad p(L - 0, z) = p(L + 0, z) \quad \forall z \in [0, h_2] \tag{2.326}$$

Since the reservoirs D_1 and D_2 are infinitely large, the velocity of the water in them is negligibly small compared with the percolation rate $\mathbf{v}$, so that hydrostatic equilibrium may be assumed at the dam boundaries, which yields

$$p(-0, z) = g\rho(h_1 - z) \quad \forall z \in [0, h_1], \quad p(L + 0, z) = g\rho(h_2 - z) \quad \forall z \in [0, h_2] \tag{2.327}$$

(5) Impermeability of the dam bottom:

$$\frac{\partial}{\partial z}\varphi(x, 0) = 0 \quad \forall x \in [0, L] \tag{2.328}$$

(6) Condition at the seepage surface $x = L$, $h_0 < z < h$

$$\frac{\partial}{\partial x}\varphi(L, z) \le 0 \tag{2.329}$$

This inequality expresses the simple fact that no water flows into the dam from the basin D_2, whereas there is an efflux of water from the dam at the seepage surface.

(7) Unboundedness of dZ/dx:

$$\lim_{x \to L-0} \frac{d}{dx} Z(x) = -\infty \tag{2.328}$$

This condition follows from the fact that the seepage surface $x = L$, which is a stream line, may be treated as a continuation of the free boundary $z = Z(x, t)$, so that its slope becomes vertical at $x = L$. Note that h is unknown in advance and must be determined in the course of solution.

E. Penetration of a magnetic field into a superconductor [43]

Superconductivity is the property of certain conducting materials (superconductors) to have infinitely high conductivity (zero ohmic resistance) at temperatures below a certain very (and recently not so) low critical temperature T_c. (For example, T_c for Hg is approximately 4°K and for Ru 0.47°K; whereas the maximal known T_c is about 150°K, for newly discovered ceramic "high-temperature" superconductors.) There is a critical value $\mathbf{H}_c$ of the intensity of the magnetic field applied to a superconductor; this value is related to the critical temperature T_c. The defining feature of $\mathbf{H}_c$ is that a magnetic field of intensity lower than $\mathbf{H}_c$ ($\mathbf{H} < \mathbf{H}_c$) does not penetrate the superconductor. Actually, the field penetrates a very thin (typically 10^{-6}–10^{-5} cm) boundary layer at the surface of the conductor. On the other hand, when a magnetic field of intensity higher than $\mathbf{H}_c$ is applied, the superconductor loses its superconductive properties and thus becomes permeable to a magnetic field, like an ordinary conductor.

Thus, consider a quasistationary magnetic field $\mathbf{H}$ in a cylindrical superconductor A of radius R at a temperature below the critical temperature T_c. We shall assume that $\mathbf{H}$ has a radial component H only, prescribed at the boundary $r = R$ and at the initial time, so that

$$H(R, t) = H_0 > H_c \tag{2.331}$$

$$H(r, 0) = \begin{cases} H^0(r) & \forall r \in (S(0), R) \\ 0 & \forall r \leq S(0) \end{cases} \tag{2.332}$$

where $S(0)$ is the depth to which the magnetic field penetrates at the initial moment. Since H is assumed to be quasistatic, it satisfies the following equation (see (2.144)):

$$\frac{D}{r} \frac{\partial}{\partial r} \left(r \frac{\partial}{\partial r} H \right) - \frac{\partial}{\partial t} H = 0 \quad \text{for } S(t) < r < R, \ \forall t > 0 \tag{2.333}$$

On the other hand,

$$H \equiv 0 \quad \text{for } 0 < r < S(t) \tag{2.334}$$

Here

$$S(0) \in [0, R) \qquad (2.335)$$

is a given quantity. The compatibility conditions on the circle $r = S(t)$ are

$$H[S(t), t] = H_c \qquad (2.336)$$

$$H_c \frac{d}{dt} S(t) = -D \frac{\partial}{\partial r} H[S(t), t] > 0 \qquad (2.337)$$

where

$$D = \frac{c^2}{4\pi\sigma\mu} \qquad (2.338)$$

Thus the problem of a magnetic field penetrating a superconductor is an axially symmetric version of the one-phase diffusion Stefan problem.

12. Well-posedness in Hadamard's sense. Examples of ill-posed problems

In the previous sections of this chapter we have derived a number of differential equations of mathematical physics together with boundary and initial conditions. However none of these problems was formulated rigorously. Indeed, nowhere was it indicated in what sense the differential equations or boundary and initial conditions must be satisfied. Moreover, the total number of boundary and initial conditions was determined on the basis of physical intuition rather than the rigorous mathematical requirements for each particular equation. A rigorous formulation of the requirements that must be satisfied by a solution of a problem of mathematical physics was presented in 1923 by J. Hadamard in a classical monograph [34]. According to Hadamard, the problem is formulated correctly or (as we now say) is *well-posed* if the following three conditions hold:

(1) a solution exists;
(2) the solution is unique;
(3) the solution is stable with respect to small perturbations of all the input data.

The meaning of the last condition is as follows. Let

$$Q = (q_1, q_2, \ldots, q_m) \qquad (2.339)$$

be a vector-valued function representing all the input data of the problem (the boundary and initial functions and the scalar parameters entering into the formulation of the problem). Let u be the solution of the problem with these input data. Define the norm of Q by

$$\|Q\| = \max(\sup |q_1|, \ \sup |q_2|, \ldots, \sup |q_m|) \qquad (2.340)$$

Let $\tilde{Q}$ be the analogous vector-valued function representing the perturbed input data, with $\tilde{u}$ representing the corresponding solution of the problem.

Then the solution u is said to be stable with respect to perturbations of the input data if, for every $\varepsilon > 0$, there exists $\delta > 0$ such that

$$\|Q - \tilde{Q}\| < \delta \Rightarrow \|u - \tilde{u}\| < \varepsilon \qquad (2.341)$$

The solution is unstable if there exists $N > 0$ such that for all $\varepsilon > 0$ there exists $\tilde{Q}$ such that

$$\|u - \tilde{u}\| > N \wedge \|Q - \tilde{Q}\| < \varepsilon \qquad (2.342)$$

The motivation of Hadamard's first condition is obvious. Underlying the second condition is the deterministic philosophy of classical science of that time, according to which – knowing the initial state of the system and the external conditions affecting the process – one can predict the state of the system at any subsequent time. This, of course, necessarily implies uniqueness of the solution. In this respect scientific psychology has changed appreciably, and the loss of uniqueness (bifurcation) may be considered one of the most interesting topics in connection with the deterministic, classical problems of mathematical physics. Unfortunately, due to space limitations this exciting issue is beyond the scope of this text. Finally, the stability condition reflects the same deterministic philosophy, which held that in principle the measurement in any experiment may be carried out with absolute accuracy, and that accordingly it should be possible to predict the evolution of the process to the same degree of accuracy.

These conditions imply, first, that the formulation of the problem should include a precise definition of the class of functions to which the solution belongs. For example, consider the following Dirichlet problem: Find a function satisfying the Laplace equation

$$\Delta u = 0 \qquad (2.343)$$

inside the sphere

$$K = \left\{ x, y \colon x^2 + y^2 + z^2 < 1 \right\} \qquad (2.344)$$

and equal to unity at surface S of the sphere:

$$S \overset{\text{def}}{=} \partial K = \left\{ x, y \colon x^2 + y^2 + z^2 = 1 \right\} \qquad (2.345)$$

Evidently, any function

$$u(x, y, z) = \begin{cases} \alpha x + \beta y + \gamma z & \forall (x, y, z) \in K \\ 1 & \forall (x, y, z) \in S \end{cases} \qquad (2.346)$$

where α, β, γ are constants, satisfies both the Laplace equation and the boundary conditions. However, if we add the requirement that u is continuous in the closure of K, then the only admissible solution of the problem turns out to be

$$u(x, y, z) \equiv 1 \qquad (2.347)$$

The need to define the class of functions to which the solution should belong is thus quite clear. A much more difficult question, though absolutely essential for understanding the essence of a problem of mathematical physics, is to determine the number and type of boundary conditions compatible with stability. The following example by Hadamard illustrates this.

Consider the following Cauchy problem: Find a function $u(x, y)$, continuous in the closure of the region

$$D \stackrel{\text{def}}{=} \{x, y: -\infty < x < \infty, \ 0 < y < \infty\} \tag{2.348}$$

harmonic inside D,

$$\frac{\partial^2}{\partial x^2} u(x, y) + \frac{\partial^2}{\partial y^2} u(x, y) = 0 \quad \forall(x, y) \in D \tag{2.349}$$

and satisfying the Cauchy conditions

$$u(x, 0) = u_0(x), \quad \frac{\partial}{\partial y} u(x, 0) = u_1(x) \quad \forall x \in (-\infty, \infty) \tag{2.350}$$

where

$$u_0 \equiv 0, \quad u_1(x) = 0 \tag{2.351}$$

Now let $\tilde{u}(x, y)$ correspond to perturbed input data

$$\tilde{u}(x, 0) = 0, \quad \frac{\partial}{\partial y} \tilde{u}(x, 0) = \frac{\sin(nx)}{n} \tag{2.352}$$

where $n > 0$ is arbitrary. The solutions u and $\tilde{u}$ of these problems are, respectively,

$$u(x, y) \equiv 0, \quad \tilde{u}(x, y) = n^{-2} \sin(nx) \sinh(ny) \tag{2.353}$$

We have

$$Q = \{0, 0\}, \quad \tilde{Q} = \{0, \sin(nx)/n\} \Rightarrow \|Q - \tilde{Q}\| = 1/n \tag{2.354}$$

Hence

$$\lim_{n \uparrow \infty} \|Q - \tilde{Q}\| = 0 \tag{2.355}$$

At the same time, for any constant positive y,

$$\lim_{n \uparrow \infty} \|u - \tilde{u}\| = \infty \tag{2.356}$$

so that the solution of the Cauchy problem for the Laplace equation is unstable. However, it can (and will) be shown that solutions of the Cauchy problem for the heat-conduction (Fourier) equation and for the wave equation, or the solutions of the first, second, and third boundary-value problems formulated for the Laplace, Fourier, and wave equations, as well as the parallel problems in multiphase regions, are stable.

Hadamard's principle of well-posedness has played a very important role in mathematical physics. However, as already mentioned, well-posed problems are by no means the only important ones from the scientific, engineering, or purely mathematical standpoint. There is a broad class of problems, under intensive study in present-day mathematical physics, whose formulations and solutions are ill-posed in Hadamard's sense. This class includes, in particular, all problems that can be reduced to the solution of integral equations of the first kind, among them what are known as inverse problems.

The general idea behind inverse problems is as follows. Assume that the function $u(p,t)$ (scalar or vector-valued) is a solution of some boundary-value or Cauchy problem. Assume further that certain characteristics of $u(p,t)$ are known inside the region under consideration, for example, the value of $u(p,t)$ on a surface that is in the region at a given time (in the case of a time-dependent process). The goal of the inverse problem is to determine the input data to which the function $u(p,t)$ corresponds. Let us illustrate this with two examples.

A. Inverse gravimetric problem

The gravimetric field of the earth is determined by the distribution of gravitational mass beneath the earth's surface. The regional gravimetric field is defined as the mean field of the force of gravity characteristic for the region in question. A deviation of the local gravimetric field from the regional field may result from the presence of some underground body whose mass density differs from the average density of the mass distribution determining the regional field. The aim of gravimetric exploration is to determine the shape, location, and density of the body causing the anomaly. This is the inverse problem of the theory of Newton potential. The simplest inverse gravimetry problem may be formulated as follows.

Suppose that the gravity potential u and its derivative $\partial u/\partial z$ are known at the boundary $z = 0$ of the half-space

$$D = \{x, y, z: \ -\infty < x, y < \infty, \ 0 < z < \infty\} \tag{2.357}$$

bounded by the plane

$$T = \{x, y, z: \ -\infty < x, y < \infty, \ z = 0\} \tag{2.358}$$

(the axis z is directed vertically downward):

$$u = u_0, \quad \frac{\partial}{\partial z}u = u_1 \quad \forall p \in T \tag{2.359}$$

Suppose that the distribution of the potential u is due to a body Ω of density ρ located beneath the plane $z = 0$. The density ρ_0 of the mass surrounding Ω is assumed to be known. Let $U(p)$ denote the gravity potential of the masses distributed in D with density ρ_0, and let

$$V = \frac{\partial}{\partial z}U \tag{2.360}$$

where U is a solution of the problem

$$\Delta U + \rho_0 = 0 \quad \forall p \in D, \quad U = U_0 \quad \forall p \in T \tag{2.361}$$

with U_0 known. This problem has a unique solution that may be evaluated, for example, by methods of potential theory (see Chapter 12, Sections 4–8). Hence it is also possible to find the normal derivative of U at the boundary. This yields

$$V_0(p) = V(p) \quad \forall p \in T \tag{2.362}$$

Let us denote

$$v = u - U, \quad w = \frac{\partial}{\partial z}v, \quad \rho - \rho_0 = f \tag{2.363}$$

In this notation,

$$v_0(p) = v, \quad w_0(p) = w(p) \quad \forall p \in T \tag{2.364}$$

are known. The potential v is a bounded function, satisfying the Poisson equation

$$\Delta v(p) = F(p) \quad \forall p \in D \tag{2.365}$$

where

$$F = \begin{cases} 0 & \forall p \in D \setminus \Omega \\ f & \forall p \in \Omega \end{cases} \tag{2.366}$$

The boundary conditions satisfied by $v(p)$ are

$$v(p) = v_0(p) \quad \forall p \in T \tag{2.367}$$

$$\frac{\partial}{\partial z}v(p) = w_0(p) \quad \forall p \in T \tag{2.368}$$

If the location and shape of the body Ω were known together with its density ρ, problem (2.365) (2.367) would have a unique solution v. The inverse gravimetric problem is to determine the location, shape, and density of Ω in such a way that v will also satisfy condition (2.368). Thus stated, the problem does not have a unique solution. However, it may be simplified by first prescribing ρ and the shape and dimension of the body.[17] Thus the simplest inverse gravimetric problem may be formulated as follows. Let the anomaly v of the derivative w of the gravimetric potential be prescribed on a plane T. The body Ω that causes this gravitational anomaly is a sphere K of radius R and density $\rho = \text{const}$. The problem is to determine the location of the center of K.

The use of the potential theory makes it possible to reduce this problem to the solution of a Fredholm integral equation of the first kind. This is a problem whose solution is not stable, that is, does not satisfy the Hadamard criterion of well-posedness.

---

[17] The density ρ may be determined experimentally by other geophysical methods (e.g., from seismological data). The shape and size of Ω may also be roughly estimated from these data.

B. Inverse problem of historical climatology

Again, let D and T be the half-space $z > 0$ and its boundary $z = 0$. Assume that the thermal conductivity and heat capacity of the rocks occupying the region are known. The simplest problem of historical climatology is to determine the distribution of temperature beneath the earth's surface at some time $t = -\tau_0$ ($\tau_0 > 0$) if the temperature at the plane T at the moment $t = 0$ is known. Setting $\tau = -t$, we find that the problem is to find the solution of the heat conduction equation

$$a^2 \Delta u + \frac{\partial}{\partial \tau} u = 0 \tag{2.369}$$

with *reversed time*, satisfying the initial condition

$$u(p, 0) = f(p) \quad \forall p \in T \tag{2.370}$$

where $f(p)$ is a given function.

The solution of this problem may be reduced to the solution of a Volterra integral equation of the first kind. This again implies that boundary-value problems for the heat-conduction equation with reversed time are ill-posed in Hadamard's sense.

There are a great many solved and even more unsolved inverse problems of considerable applied and theoretical significance. They are the subject of mathematical control theory, a rapidly developing branch of the theories of ordinary and partial differential equations. However, as classical mathematical physics was not concerned with this type of problems, we restrict ourselves to these brief remarks and merely refer the reader to the appropriate literature.

13. Terminology. Concluding remark. Notation

Before going on to a systematic study of the classical methods of mathematical physics, we introduce a preliminary classification of partial differential equations.

Consider a partial differential equation of the form

$$F\left(p, u, \frac{\partial u}{\partial x_1}, \ldots, \frac{\partial u}{\partial x_n}, \ldots, \frac{\partial^m u}{\partial x_1^m}, \ldots, \frac{\partial^{k_1 + \cdots + k_n} u}{\partial x_1^{k_1} \ldots \partial x_n^{k_n}}, \ldots, \frac{\partial^m u}{\partial x_n^m}\right) = 0 \tag{2.371}$$

where

$$p = (x_1, x_2, \ldots, x_n) \wedge k = k_1 + k_2 + \cdots + k_n \leq m \tag{2.372}$$

This equation is called a nonlinear partial differential equation of the mth order in n independent variables. Thus, the order of the partial differential equation is determined by the highest order of the derivatives of u occurring in the function F. If F is linear with respect to the totality of all the derivatives of the highest order,

$$F = \sum_{k_1 + \cdots + k_n = m} a_{k_1, \ldots, k_n} \frac{\partial^m u}{\partial x_1^{k_1} \ldots \partial x_n^{k_n}} + \mathfrak{F}\left((x_1, \ldots, x_n, u, \ldots, \frac{\partial^{m-1} u}{\partial x^{m-1}}\right)$$

$$\tag{2.373}$$

where the coefficients $a_{k_1,\ldots,k_n}$ may depend on all independent variables, on u, and on all its partial derivatives up to $(m-1)$th order, then the equation $F=0$ is called a quasilinear partial differential equation of mth order. If all these coefficients are independent of u and its derivatives, the equation $F=0$ is called a partial differential equation of mth order with linear principal part. Analogous terminology applies to a system of partial differential equations.

The equations derived in this chapter originated in different natural sciences: mechanics of continuous media, electrostatics and electrodynamics, theory of heat conduction and diffusion, industrial chemistry, and chemical kinetics. Most of them deal with at least one of the following three operators: the Laplace operator Δ, the heat conduction operator $\Delta - \partial/\partial t$ and the wave operator $\Delta - \partial^2/\partial t^2$. The appearance of the Laplace operator in the formulation of a problem is a characteristic property of many steady-state processes, as we have shown. The heat-conduction operator is usually present in problems concerned with dissipative inertia-free processes, whereas the wave operator appears in all problems dealing with the propagation of oscillations. In accordance with the observation that the Laplace, heat-conduction, and wave operators play a crucial role in the formulation of problems covering most processes of continuum physics, these three operators are the main objects of study in the theory of partial differential equations of classical mathematical physics. As will be demonstrated in the following chapters, the interest in these three operators is not only justified by their importance in physics, but is motivated by the fundamental fact that their properties are typical for a broad class of much more general quasilinear and even nonlinear operators.

There is, moreover, a close relationship between steady-state, dissipative, and oscillatory processes. As may be observed from several of the previous examples, one type of problem may be transformed quite naturally into a problem of another type. This transition depends on the range of physical parameters involved in the formulation of the problem, or on the duration of the evolution of the process. These questions will be treated rigorously in what follows (see Chapter 17, Section 2 and Chapter 21, Sections 1 and 3).

In what follows, $F(x,\ldots,t \mid f,\varphi,\ldots,h)$ denotes any operator acting on functions $f,\varphi,\ldots,h$ defined in a space of arguments $x,\ldots,t$. Vectors are designated by bold letters $(\mathbf{a},\mathbf{v})$; $\mathbf{i},\mathbf{j},\mathbf{k}$ denote everywhere the basis vectors of a rectangular cartesian coordinate system. Any tensor is denoted by a bold letter with tilde (e.g., $\tilde{\mathbf{\Pi}}$, $\tilde{\boldsymbol{\varepsilon}}$, $\ldots$). Other notations are explained in a due place.

Problems

P2.2.1. Formulate a problem describing nonelectrolyte diffusion in a solution accompanied by a second-order chemical reaction with an instantaneous removal of the reaction product from the solution. (The reaction product is assumed to be insoluble.)

P2.6.1. Derive the equations of small 1-dimensional longitudinal and transversal vibrations of an elastic medium from the general 3-dimensional wave equation (2.94).[18] Determine the wave speeds of longitudinal and transversal vibrations and explain the source of their difference.

Hint: Evaluate all the components of the stress tensor in both types of vibrations.

P2.6.2. Is the assumption that the vibrations of an elastic body are 1-dimensional compatible with the assumption that the stress tensor has only one nonzero component?

P2.6.3. Derive the equation of small transversal vibrations of an initially stressed elastic string that is nonresistant to bending.

P2.6.4. Derive the equation of transversal vibrations of a membrane that is not resistant to bending and shearing.[19]

Hint: Consider the infinitesimal element of the membrane surface cut out of it by a parallelepiped with lateral boundaries parallel to the coordinate planes in (x, y, z) space. Let $z = u(x, y, t)$ be the equation of the membrane surface at time t, and use the dynamic equilibrium condition.

P2.6.5. Formulate the equation of transversal vibrations of a string immersed in a viscous liquid. Assume that the viscous friction resists the vibrations with a force proportional to the velocity of displacements of the points of the string. Allow for the effect of the force of gravity. Consider, in particular, a string consisting of two parts with different moduli of elasticity.

P2.7.1. Derive a system of equations for the electric current and potential in a cable, given ohmic resistance R, inductance L, capacity C, and loss factor α, all per unit length of the cable. The meaning of the loss factor α is given by the equality

$$q = \alpha[u(x, t) - u_0]$$

where q is the linear leakage current density (the charge leaving the cable in unit time per unit length of the cable), $u(x, t)$ the local potential in the cable at position x at time t, and u_0 the (given) potential outside the cable. Show that the resulting system of equations (the so-called telegraph system)

---

[18] The idea of the transition from the 3-dimensional theory of elasticity to the theory of a string is as follows. Consider an elastic cylinder with a stress-free lateral surface. Take a normal cross section of the cylinder and identify the total average stress on this section with the stress at the corresponding point of the string.

[19] Elastic membrane not resisting bending and shearing is an elastic film where the forces of tension, applied to the contour bounding the membrane, are vectors lying in the plane tangent to the membrane surface at points under consideration and directed along the normal to that contour.

is identical to the 1-dimensional hyperbolic diffusion equation (telegraph equation).

P2.8.1. Formulate a problem describing the loss of heat during the injection of hot water into a thin horizontal layer of porous medium. View this layer as a concentrated thermal capacity in the surrounding impermeable rocks. Assume that the hot water is injected into an infinitely long gallery at a constant rate per unit length of the gallery. (A "gallery" is a channel of infinitely small width, cutting the permeable layer through all its thickness. The water injected into the layer percolates at an equal rate, independent of the altitude, on both sides of the gallery.)

P2.8.2. Formulate the problem on nonelectrolyte diffusion in a two-component water solution accompanied by a second-order chemical reaction. Assume that the solid product of the reaction is insoluble and that it precipitates instantaneously, creating a layer of a sediment of variable porosity. The porosity of the layer is assumed to decrease linearly with increasing thickness. The solution percolates through the porous space of the sediment with a constant diffusivity (dispersion coefficient). The effect of the chemical reaction inside the pores of sediment may be disregarded; the initial concentrations of both components of the solution are prescribed; the initial thickness of the sediment layer is assumed to be zero; and the initial thickness of the solution layer is prescribed and the initial concentrations of both components are given constants. No influx of reagents into the solution occurs. The process may be considered as 1-dimensional.

P2.9.1. Consider problem (1) of Section 10, assuming that the layers $(-\ell, 0)$ and $(0, L)$ are separated by an infinitely thin film of a finite thermal conductivity and finite (total) heat capacity. What condition should be imposed in this case instead of condition (2.268)?

3

Cauchy problem for first-order partial differential equations

1. Local Cauchy problem for quasilinear equation with two independent variables

Let D be a simply connected region in $\mathbb{R}_3$ and Γ a smooth curve

$$\Gamma = \{x, y, z : x = x_0(s), \ y = y_0(s), \ z = z_0(s), \ \alpha^* < \alpha \leq s \leq \beta < \beta^*\} \in D \tag{3.1}$$

without self-intersections. Let $P(x, y, z)$, $Q(x, y, z)$, and $R(x, y, z)$ be continuously differentiable functions defined inside D such that

$$P^2 + Q^2 \neq 0 \quad \forall (x, y, z) \in D \tag{3.2}$$

and

$$\Delta_0 \stackrel{\text{def}}{=} \begin{vmatrix} P & Q \\ \dfrac{dx_0}{ds} & \dfrac{dy_0}{ds} \end{vmatrix} \neq 0 \quad \forall (x, y, z) \in \Gamma \tag{3.3}$$

The local Cauchy problem is: *In a neighborhood U of Γ, find a continuously differentiable function $z(x, y)$ such that*

$$Pp + Qq = R \quad \forall (x, y, z) \in U \tag{3.4}$$

$$z = z_0(s), \quad \frac{dz_0}{ds} = p_0(s)\frac{dx_0}{ds} + q_0(s)\frac{dy_0}{ds}, \quad (x, y, z) \in \Gamma \tag{3.5}$$

We have used the abbreviations

$$p(x, y) = \frac{\partial}{\partial x} z, \quad q(x, y) = \frac{\partial}{\partial y} z \quad \forall (x, y, z) \in U \tag{3.6}$$

and

$$p_0(s) = p[x_0(s), y_0(s)], \quad q_0(s) = q[x_0(s), y_0(s)] \tag{3.7}$$

to simplify the notation.

If $z(x, y)$ is a solution of this problem, the surface

$$\Sigma = \{x, y, z(x, y)\} \tag{3.8}$$

is called an *integral surface*. Let Σ be an integral surface and T the plane tangent to Σ at a point $[x, y, z(x, y)]$ of Σ so that

$$-(X - x)p(x, y) - (Y - y)q(x, y) + Z - z(x, y) = 0 \quad \forall [x, y, z(x, y)] \in \Gamma \quad (3.9)$$

where X, Y, Z are the cartesian coordinates of any point in T. This means that if $\mathbf{i}$, $\mathbf{j}$, $\mathbf{k}$ are basis vectors then the vector

$$\mathbf{n} = -p\mathbf{i} - q\mathbf{j} + \mathbf{k} \quad (3.10)$$

is a vector normal to Σ at (x, y, z). By (3.4) the vector

$$\boldsymbol{\tau} = P\mathbf{i} + Q\mathbf{j} - R\mathbf{k} \quad (3.11)$$

is orthogonal to $\mathbf{n}$, that is, is parallel to T; hence the straight line

$$L = \left\{ \frac{X - x}{P} = \frac{Y - y}{Q} = \frac{z - Z}{R} \right\} \in T \quad (3.12)$$

lies in T, so that its infinitesimal element

$$dx = X - x, \quad dy = Y - y, \quad dz = Z - z \quad (3.13)$$

lies in Σ. Hence the system of ordinary differential equations

$$\frac{dx}{P(x, y, z)} = \frac{dy}{Q(x, y, z)} = \frac{dz}{R(x, y, z)} \quad (3.14)$$

defines a curve that belongs to Σ.

These geometrical considerations indicate the close connection between the local Cauchy problem (3.4), (3.5) and the Cauchy problem

$$\frac{dx}{P(x, y, z)} = \frac{dy}{Q(x, y, z)} = \frac{dz}{R(x, y, z)} = dt \quad (3.15)$$

$$x = x_0(s), \quad y = y_0(s), \quad z = z_0(s) \quad \text{at } t = 0, \quad \alpha < s < \beta \quad (3.16)$$

for a system of ordinary differential equations. Equations (3.14) are called the *characteristic equations* and their integral curves $\{x(t, s), y(t, s), z(t, s)\}$ the *characteristic curves*. The projections of characteristic curves on the plane $z = 0$ are called the *characteristic curve projections*.

In what follows we assume that the functions $x_0(s)$, $y_0(s)$, and $z_0(s)$ are continuously differentiable everywhere along Γ. Then – since P, Q, and R are smooth – there exists a 1-parametric system of characteristic curves, continuously differentiable with respect to the parameter s.

Theorem 3.1.1. *If the carrier Γ of the Cauchy data is not tangent to characteristic curves at each of its points, then a local solution of the Cauchy problem (3.4) and (3.5) exists and is unique.*

PROOF. Let $[x(t, s), y(t, s), z(x, s)]$ be a solution of the system of characteristic equations. Then the Jacobian

$$\Delta \stackrel{\text{def}}{=} \begin{vmatrix} P & Q \\ \dfrac{\partial x}{\partial s} & \dfrac{\partial y}{\partial s} \end{vmatrix}_\Gamma = \Delta_0 \quad (3.17)$$

is not equal to zero. This means that in the neighborhood U of Γ,

$$\Delta \Big|_U \neq 0 \tag{3.18}$$

so that, by the implicit function theorem, t and s may be defined as functions of x and y:

$$t = t(x, y), \quad s = s(x, y), \quad (x, y) \in U \tag{3.19}$$

and hence, along the characteristic curves,

$$z \equiv z[t(x, y), s(x, y)] \stackrel{\text{def}}{=} Z(x, y) \tag{3.20}$$

Evidently, the surface

$$\Sigma = (x, y, Z(x, y)) \tag{3.21}$$

is an integral surface. Indeed, by (3.14),

$$\frac{\partial x}{\partial t} = P, \frac{\partial y}{\partial t} = Q, \frac{\partial z}{\partial t} = \frac{\partial Z}{\partial x} \cdot \frac{\partial x}{\partial t} + \frac{\partial Z}{\partial y} \cdot \frac{\partial y}{\partial t} = R \tag{3.22}$$

so that, by (3.6),

$$Pp + Qq \equiv R(x, y, z) \in U \tag{3.23}$$

Thus a solution of the local Cauchy problem (3.4) and (3.5) indeed exists. It is easy to prove that it is unique. Consider an integral curve of the system of equations

$$\frac{dx}{P[x, y, Z(x, y)]} = \frac{dy}{Q[x, y, Z(x, y)]} = dt \tag{3.24}$$

or, equivalently, of the system

$$\frac{\partial x}{\partial t} = P[x, y, Z(x, y)], \quad \frac{\partial y}{\partial t} = P[x, y, Z(x, y)] \tag{3.25}$$

By the uniqueness theorem of the theory of ordinary differential equations, this integral curve is uniquely determined by any of its points, considered as an initial point. In other words, if one point of the characteristic belongs to the integral surface Σ, then the entire characteristic belongs to Σ. Since the family of characteristic curves emanating from points of Γ is uniquely determined, this means that the solution of the local Cauchy problem is unique. **Q.E.D.**

Remark. The implicit function theorem is local in nature. Therefore the result just proved is also valid only locally.[1]

---

[1] The existence and uniqueness of a global solution of the Cauchy problem (3.4)–(3.5) are considered in Section 3 of this chapter.

2. Local Cauchy problem for nonlinear first-order partial differential equation

Consider now a nonlinear first-order partial differential equation with n independent variables:

$$F(x, z, p) = 0 \qquad (3.26)$$

where

$$x = (x_1, x_2, \ldots, x_n), \quad p = (p_1, p_2, \ldots, p_n), \quad p_i = \frac{\partial z}{\partial x_i}, \quad i = 1, 2, \ldots, n \qquad (3.27)$$

Assume that F is a twice continuously differentiable function of all its arguments, defined in some simply connected region $D^* \in \mathbb{R}_{2n+1}$ and let D be the projection of D^* on $R_n = \{x\}$. Let Γ_0 be an $(n-1)$-dimensional manifold in D, defined as

$$\Gamma_0 = \{x_i = x_{i0}(s_1, s_2, \ldots, s_{n-1})\} \qquad (3.28)$$

Assume that x_{i0}, $i = 1, 2, \ldots, n$, are continuously differentiable functions of all their arguments and that the rank of the matrix

$$\mathfrak{M} = \begin{pmatrix} (x_{10})_1 & (x_{10})_2 & \cdots & (x_{10})_{n-1} \\ (x_{20})_1 & (x_{20})_2 & \cdots & (x_{20})_{n-1} \\ \vdots & \vdots & & \vdots \\ (x_{n0})_1 & (x_{n0})_2 & \cdots & (x_{n0})_{n-1} \end{pmatrix} \qquad (3.29)$$

is $n - 1$.

In the case of a quasilinear equation $F = 0$, the Cauchy problem under consideration would be to determine, in a neighborhood U of the manifold Γ_0, a solution $z(t)$ taking the prescribed values

$$z = z_0(s_1, s_2, \ldots, s_{n-1}) \qquad (3.30)$$

on Γ_0. In the nonlinear case, the Cauchy data must be supplemented by prescribed values

$$p_j = p_{j0}(s_1, s_2, \ldots s_{n-1}), \quad j = 1, 2, \ldots, n \qquad (3.31)$$

satisfying the conditions

$$\frac{\partial z}{\partial x_k} = \sum_{m=1}^{n} p_{m0} \frac{\partial x_{m0}}{\partial x_k}, \quad k = 1, 2, \ldots, n - 1 \qquad (3.32)$$

Denote

$$\frac{\partial}{\partial x_j} F = X_j, \quad \frac{\partial}{\partial z} F = Z, \quad \frac{\partial}{\partial p_j} F = P_j, \quad j = 1, 2, \ldots, n \qquad (3.33)$$

Assume that the Cauchy data (3.30) and (3.31) satisfy conditions (3.32) and make the equation

$$F(x_0, z_0, p_0) = 0 \qquad (3.34)$$

an identity everywhere on Γ_0. Differentiation of F with respect to x_i yields

$$X_i + Zp_i + \sum_{k=1}^{n} P_k \frac{\partial}{\partial x_i} p_k = 0, \quad i = 1, 2, \ldots, n \tag{3.35}$$

Since for all i, k

$$\frac{\partial}{\partial x_i} p_k = \frac{\partial}{\partial x_k} p_i \tag{3.36}$$

by the continuity of the right- and left-hand sides, equation (3.35) may be rewritten as

$$X_i + Zp_i + \sum_{k=1}^{n} P_k \frac{\partial}{\partial x_k} p_i = 0, \quad i = 1, 2, \ldots, n \tag{3.37}$$

Let $Z(x)$ be a twice continuously differentiable solution of the Cauchy problem under consideration, so that

$$p_k = \frac{\partial z}{\partial x_k}, \quad k = 1, 2, \ldots, n \tag{3.38}$$

Then

$$F[x, z(x), p(x)] \equiv 0 \tag{3.39}$$

so that (3.37) is identically valid.

We can now consider z and p_{jk}, $j \neq k$, as known functions of $x_1, x_2, \ldots, x_n$, and treat (3.37) as a quasilinear equation for p_k.

Now, in the space (x, p_k) (as in Section 1), consider the vector

$$\mathbf{n} = (-p_{k1}, -p_{k2}, \ldots, -p_{kn}, 1); \quad p_{km} = \frac{\partial}{\partial x_m} p_k \tag{3.40}$$

along the normal $\mathbf{n}$ to the integral surface Σ and the vector

$$\mathbf{T} = (P_1, P_2, \ldots, P_n, X_k + Zp_k) \tag{3.41}$$

Then the left-hand side of (3.37) is the scalar product of $\mathbf{n}$ and $\mathbf{T}$:

$$X_i + Zp_i + \sum_{k=1}^{n} P_k \frac{\partial}{\partial x_k} p_i = \mathbf{n} \cdot \mathbf{T} \tag{3.42}$$

so that $\mathbf{T}$ is orthogonal to $\mathbf{n}$. Hence the straight line

$$\frac{X_1 - x_1}{P_1} = \frac{X_2 - x_2}{P_2} = \cdots = \frac{X_n - x_n}{P_n} = -\frac{\tilde{p}_k - p_k}{X_k + Zp_k} \tag{3.43}$$

lies in the tangent plane to Σ so that the infinitesimal element of the curve

$$\frac{dx_1}{P_1} = \frac{dx_2}{P_2} = \cdots = \frac{dx_n}{P_n} = -\frac{dp_k}{X_k + Zp_k} = dt \tag{3.44}$$

lies in Σ. Recall that here k is any of $1, 2, \ldots, n$. Since

$$dz = \sum_{k=1}^{n} p_k dx_k \tag{3.45}$$

it follows from (3.44) that

$$dz \left(\sum_{k=1}^{n} P_k p_k \right)^{-1} = dt \qquad (3.46)$$

Again, the system of ordinary differential equations

$$\frac{dx_1}{P_1} = \frac{dx_2}{P_2} = \cdots = \frac{dx_n}{P_n} = -\frac{dp_k}{X_k + Zp_k} = dz \left(\sum_{k=1}^{n} P_k p_k \right)^{-1} = dt \qquad (3.47)$$

is called the *system of characteristic equations*, and its integral curves, called the characteristic curves, constitute an $(n-1)$-parametric family

$$\{x, p\} = \{x(s_1, s_2, \ldots, s_{n-1}, t), p(s_1, s_2, \ldots, s_{n-1}, t)\} \qquad (3.48)$$

The projections of characteristic curves on the space $\mathbb{R}_n = \{x\}$ are called *characteristic projections*.

As in the case of a quasilinear equation with two independent variables (treated in the previous section), the following theorem is true.

Theorem 3.2.1. *If the Jacobian*

$$\Delta \equiv \frac{D(x_1, x_2, \ldots, x_{n-1}, x_n)}{D(s_1, s_2, \ldots, s_{n-1}, t)} \qquad (3.49)$$

is not equal to zero everywhere along the manifold Γ_0 (see (3.28)), then the system of characteristic curves emanating from the points of Γ_0 form a uniquely determined integral surface Σ of the equation $F = 0$ in the neighborhood U of Γ_0.

PROOF. By the continuity of all derivatives occurring in the expression of Δ,

$$\Delta \neq 0 \quad \forall (x) \in U \qquad (3.50)$$

Hence, by the implicit function theorem, $s_1, s_2, \ldots, s_{n-1}$ and t can be uniquely represented in U as continuous functions of $x_1, x_2, \ldots, x_n$. Since (3.47) implies

$$dz = \sum_{k=1}^{n} P_k p_k dt = \sum_{k=1}^{n} p_k dx_k \qquad (3.51)$$

we have

$$p_k = \frac{\partial z}{\partial x_k}, \quad k = 1, 2, \ldots, n \qquad (3.52)$$

Hence Σ is indeed an integral surface, since by (3.52) and (3.47)

$$X_i + Zp_i + \sum_{k=1}^{n} P_k \frac{\partial p_i}{\partial x_k} = 0 \qquad (3.53)$$

This integral surface contains the manifold Γ_0 by construction; hence $F \equiv 0$ along Σ, so that (x, z, p) is indeed a solution of the Cauchy problem (3.26)–(3.28), (3.30)–(3.32). Since the input data satisfy the conditions of the uniqueness theorem for solutions of systems of ordinary differential equations, it follows that a characteristic curve emanating from any point of Γ_0 is uniquely determined by any of its points. Thus the integral surface Σ is determined uniquely. **Q.E.D.**

3. Global Cauchy problem for quasilinear partial differential first-order equation with two independent variables. Need for broader class of generalized (discontinuous) solutions

Let us return to the Cauchy problem for a quasilinear equation

$$Pp + Qq = R \tag{3.54}$$

where the curve

$$\Gamma = \{x, y, z : x = x_0(s),\ y = y_0(s),\ z = z_0(s),\ \alpha^* < \alpha \leq S \leq \beta < \beta^*\} \tag{3.55}$$

is the carrier of the Cauchy data. By Theorem 3.1.1, we know that the system of characteristic curves

$$x = (t, s), y = y(t, s), z = z(t, s) \tag{3.56}$$

emanating from a point of Γ can be solved for t and s in a neighborhood of Γ. Thus $z(t, s)$ becomes a continuously differentiable function $z(x, y)$, which is a solution of the original Cauchy problem. Thus the possibility of solving this Cauchy problem everywhere inside some region D crossed by Γ depends entirely on whether the implicit function theorem is applicable everywhere in D. There is an essential distinction between two cases, according as P and Q are independent of z or not. Indeed, if P and Q are independent of z then the system of equations

$$\frac{dx}{P(x,y)} = \frac{dy}{Q(x,y)} = dt, \quad x\Big|_{t=0} = x_0(s),\quad y\Big|_{t=0} = y_0(s) \tag{3.57}$$

determines x and y as continuously differentiable functions of t and s, so that by the implicit function theorem s and t can be represented as continuous functions of x and y everywhere in the projection $\tilde{D}$ of D on the plane $z = 0$ (see Figure 3.1a). This means that the integral surface

$$\Sigma = \Big\{x, y, z : (x, y) \in \tilde{D},\ z = z(x, y)\Big\} \tag{3.58}$$

is uniquely defined globally. But if P and Q depend on z then the characteristic projections may intersect at some point $x = \tilde{x}$, $y = \tilde{y}$, whereas the characteristic curves $x = x(t, s)$, $y = y(t, s)$, and $z = z(t, s)$ are uniquely determined (see Figure 3.1b). As an example, consider the following Cauchy problem:

$$\frac{\partial z}{\partial y} + z\frac{\partial z}{\partial x} = 0,\ z(x, 0) = -\tanh(x) \tag{3.59}$$

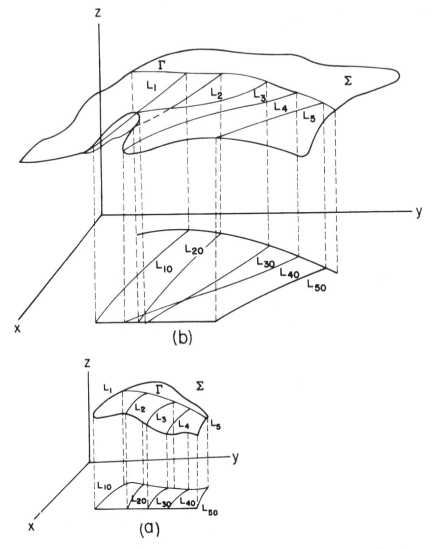

Figure 3.1. Sketch of the integral surface and its projection on the plane $z = 0$; Cauchy data carrier Γ, characteristic curves L_i, and characteristic curve projections L_{iO}. (a) quasilinear equation with linear principal part; $P = P(x, y)$, $Q = Q(x, y)$. (b) quasilinear equation; $P = P(x, y, z)$, $Q_1 = Q(x, y, z)$.

The characteristic system is

$$\frac{dx}{z} = \frac{dy}{1} = \frac{dz}{0} \qquad (3.60)$$

with initial data

$$x \Big|_{y=0} = \xi, \quad z \Big|_{y=0} = -\tanh(\xi) \qquad (3.61)$$

Integration yields

$$z = -\tanh(\xi), x = \xi - y\tanh(\xi) \qquad (3.62)$$

Hence, for all $y \geq 0$

$$\frac{dx}{d\xi} = 1 - \frac{y}{\cosh^2(\xi)} \geq 1 - y \qquad (3.63)$$

Thus, for all $y < 1$, ξ can be determined as a function of x, so that the Cauchy problem under consideration has a unique solution for every $y < 1$. However, for every pair of different values of ξ the characteristic curves intersect. Indeed, given ξ_1 and ξ_2 such that $\xi_2 > \xi_1$, define

$$y_{12} = \frac{\xi_2 - \xi_1}{\tanh(\xi_2) - \tanh(\xi_1)} \qquad (3.64)$$

Then, by (3.62),

$$x(y_{12}, \xi_1) = x(y_{12}, \xi_2) \qquad (3.65)$$

Thus we see that a global solution to the Cauchy problem need not necessarily exist in the class of smooth functions. Therefore the need arises to introduce a broader class of functions within which Cauchy problems always have global solutions. Naturally this makes it necessary to change the very definition of a solution to the Cauchy problem.

4. Necessary conditions of discontinuity. Problem of decay of arbitrary discontinuity. Gelfand's heuristic theory [24]

In what follows we restrict ourselves to consideration of the Cauchy problem for the model "conservation law"

$$\frac{\partial}{\partial t}u + \frac{\partial}{\partial x}\Phi(u) = 0 \quad \forall t > 0, \ \forall x \in (-\infty, \infty) \qquad (3.66)$$

$$u(x, 0) = f(x) \qquad (3.67)$$

where $f(x)$ is a piecewise continuous function in $(-\infty, \infty)$ having at most a finite number of points of discontinuity, and Φ is the smooth function of u. This is the form of the Buckley–Leverett equation, describing the percolation of two immiscible liquids through a porous medium (see Chapter 2, Section 8).

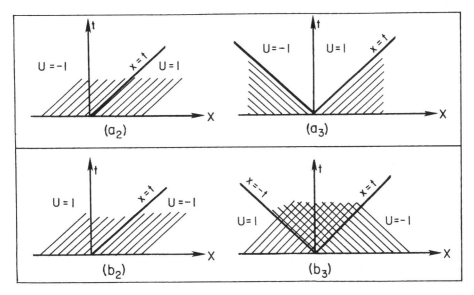

Figure 3.2. Initial conditions and characteristics in Cauchy problems (3.68a)–(3.79b). (a_2) Initial conditions (3.69a), characteristics (3.72a)–(3.75a). (a_3) Initial conditions (3.69b), characteristics (3.72b)–(3.75b). (b_2) Initial conditions (3.70a), characteristics (3.76a)–(3.79a). (b_3) Initial conditions (3.70b), characteristics (3.76b)–(3.79b).

We begin with consideration of the following two examples in parallel where $\Phi(u) = u$ and $\Phi(u) = u^2/2$ in the first and the second example, respectively (see Figure 3.2).

Linear equation		*Quasilinear equation*	
$\dfrac{\partial u}{\partial t} + \dfrac{\partial u}{\partial x} = 0$	(3.68a)	$\dfrac{\partial u}{\partial t} + u\dfrac{\partial u}{\partial x} = 0$	(3.68b)

Initial condition   *Initial condition*

a: $u\big|_{t=0} = \begin{cases} -1 & \forall x < 0 \\ 1 & \forall x > 0 \end{cases}$ (3.69a) a: $u\big|_{t=0} = \begin{cases} -1 & \forall x < 0 \\ 1 & \forall x > 0 \end{cases}$ (3.69b)

b: $u\big|_{t=0} = \begin{cases} 1 & \forall x < 0 \\ -1 & \forall x > 0 \end{cases}$ (3.70a) b: $u\big|_{t=0} = \begin{cases} 1 & \forall x < 0 \\ -1 & \forall x > 0 \end{cases}$ (3.70b)

Characteristic equations   *Characteristic equations*

$$\frac{dt}{1} = \frac{dx}{1} = \frac{du}{0} \quad (3.71a) \qquad \frac{dt}{1} = \frac{dx}{u} = \frac{du}{0} \quad (3.71b)$$

Initial data   *Initial data*

a: a:

$$x\big|_{t=0} = \xi \quad (3.72a) \qquad x\big|_{t=0} = \xi \quad (3.72b)$$

$$u\big|_{t=0} = \begin{cases} -1 & \xi \le 0 \\ 1 & \xi > 0 \end{cases} \quad (3.73a) \qquad u\big|_{t=0} = \begin{cases} -1 & \xi \le 0 \\ 1 & \xi > 0 \end{cases} \quad (3.73b)$$

b: b:

$$x\Big|_{t=0} = \xi \qquad\qquad (3.74a)$$ $$x\Big|_{t=0} = \xi \qquad\qquad (3.74b)$$

$$u\Big|_{t=0} = \begin{cases} 1 & \xi \le 0 \\ -1 & \xi > 0 \end{cases} \qquad (3.75a)$$ $$u\Big|_{t=0} = \begin{cases} 1 & \xi \le 0 \\ -1 & \xi > 0 \end{cases} \qquad (3.75b)$$

Characteristic curves   *Characteristic curves*

a: $$x = \xi + t \qquad (3.76a)$$ a: $$x = \begin{cases} \xi - t & \xi < 0 \\ \xi + t & \xi > 0 \end{cases} \qquad (3.76b)$$

$$u = \begin{cases} -1 & \xi < 0 \\ 1 & \xi > 0 \end{cases} \qquad (3.77a)$$ $$u\Big|_{t=0} = \begin{cases} -1 & \xi < 0 \\ 1 & \xi > 0 \end{cases} \qquad (3.77b)$$

b: $$x = \xi + t \qquad (3.78a)$$ b: $$x = \begin{cases} \xi + t & \xi < 0 \\ \xi - t & \xi > 0 \end{cases} \qquad (3.78b)$$

$$u = \begin{cases} 1 & \xi < 0 \\ -1 & \xi > 0 \end{cases} \qquad (3.79a)$$ $$u = \begin{cases} 1 & \xi < 0 \\ -1 & \xi > 0 \end{cases} \qquad (3.79b)$$

We see that in the linear case *a* discontinuity propagates along a characteristic going from the origin (0,0) in both cases *a* and *b*. The quasilinear case, however, exhibits a different picture. In case *a* all the characteristic curves emanating from points $\xi < 0$, $t = 0$ and $\xi > 0$, $t = 0$ are directed into the half-plane $x < 0$, $t > 0$ and $x > 0$, $t = 0$, respectively, so that the angle

$$\alpha = \{x, t : 0 < t < \infty, \ -t < x < t\} \qquad (3.80)$$

between the characteristic curves emanating from points $\xi \le -0$, $t = 0$ and $\xi \ge +0$, $t = 0$ remains unfilled by characteristic curves. In contrast, in case *b* all the characteristic curves emanating from points of the half-axis $\xi < 0$ are directed to the right, whereas characteristic curves emanating from points of the half-axis $\xi > 0$ are directed leftward. Therefore the angle α is covered by two sets of characteristic curves: those emanating from points $\xi < 0$, $t = 0$ and from points $\xi > 0$, $t = 0$.

In both cases, the Cauchy problem is not solved by reduction to a Cauchy problem for the system of characteristic equations with Cauchy data prescribed on the two half-axes $t = 0$, $|\xi| > 0$. However, in case *a* there exists a continuous solution of equation (3.68b): the extension by continuity of the solution of the Cauchy problem constructed outside the angle α. This solution can be constructed as follows. Consider all possible characteristic curves emanating from the origin (0,0). They are defined as solutions of the system of characteristic equations

$$\frac{dt}{1} = \frac{dx}{u} = \frac{du}{0} = ds \qquad (3.81)$$

carrying values

$$t = \frac{x}{k}, \quad u = k, \quad k = \text{const} \qquad (3.82)$$

When k varies in the interval $(-1, 1)$ the characteristic curves emanating from the origin (0,0) cover the angle α and coincide at the boundaries of

α with the characteristic curves emanating from points $(x = -0,\ t = 0)$ and $(x = +0,\ t = 0)$, respectively, which carry values $u = -1$ and $u = 1$, respectively. Evidently it follows from (3.82) that

$$u = \frac{x}{t}, \quad (x,t) \in \alpha \qquad (3.83)$$

so that u is indeed a solution of equation (3.68) which is the extension by continuity of the solution of the original Cauchy problem constructed outside the angle α.

In the case b, no continuous solution of the Cauchy problem can exist within the angle α since in this case α is covered by two different families of characteristic curves. Hence it is precisely in such cases that the concept of discontinuous solutions of a quasilinear partial differential equation must be introduced.

Naturally we can assume that the points of discontinuity exist at the initial time $t = 0$, since otherwise the classical solution of the Cauchy problem exists locally and we may therefore consider it as known up to the first time of the discontinuity appearance. In what follows we restrict ourselves to consideration of the simplest case, where only one point of discontinuity $(x = 0)$ exists at $t = 0$ and $u(x,0)$ is piecewise constant. Nevertheless, before proceeding to the construction of a generalized solution to this problem, we must define the sense of the concept of "generalized solution" for a general case and derive the necessary condition for a curve to be the curve of discontinuity.

We now proceed to study the model problem (3.66), (3.67). Assume first that $u(x,t)$ is continuously differentiable everywhere in a simply connected region $D \subset (-\infty < x < \infty, t > 0)$ bounded by a piecewise smooth contour Γ. Integration of (3.66) over D and the use of Green's formula yield

$$0 \equiv \int_D \left(\frac{\partial}{\partial t} u + \frac{\partial}{\partial x} \Phi(u) \right) dx dt = - \int_\Gamma u dx - \Phi(u) dt \qquad (3.84)$$

The identity

$$\int_\Gamma u dx - \Phi(u) dt \equiv 0 \qquad (3.85)$$

will serve as the basis for definition of a generalized solution of the Cauchy problem (3.66), (3.67).

Definition 3.4.1. A function $u(x,t)$ is called a *generalized solution* of the problem

$$\frac{\partial}{\partial t} u + \frac{\partial}{\partial x} \Phi(u) = 0 \quad \forall t > 0, \ \forall x \in (-\infty, \infty), \quad u(x,0) = f(x) \qquad (3.86)$$

if the following conditions are satisfied.

(α) $u(x,t)$ is continuous everywhere in a half-plane $t > 0$ except for a finite number of smooth curves L_k, $k = 1, 2, \ldots, n$, without self-intersections, that intersect one another at most at a finite number of points.

(β) If x_0 is a point of continuity of $f(x)$, then

$$\lim_{t \downarrow 0} u(x(t), t) = f(x_0) \tag{3.87}$$

if $x(t)$ tends to x_0 along a path not tangent to the axis $t = 0$.

(γ) If Γ is an arbitrary smooth contour in the half-plane $t > 0$ intersected by curves of discontinuity of u at most at a finite number points, then

$$\int_\Gamma \Phi(u) dt - u dx = 0 \tag{3.88}$$

(δ) If L is a curve of discontinuity of u then there exist one-sided limits u^+ and u^- of u that are continuously differentiable along L.

Theorem 3.4.1. *The curve*

$$L = \{x, t : x = x(t), \ t > 0\} \tag{3.89}$$

can be the curve of discontinuity of $u(x, t)$ only if along it[2]

$$\frac{dx}{dt} = \frac{[\Phi(u)]}{[u]} \tag{3.90}$$

PROOF. Let L be a segment of a curve of discontinuity,

$$L = \{x, t : x = x(s), \ t = t(s), \ s_1 < s < s_2\} \tag{3.91}$$

and Γ the piecewise smooth contour $MPQNRSM$, where

$$\begin{aligned} M &= [x(s_2), t(s_2)], & P &= [x(s_2) - \varepsilon, t(s_2)], & Q &= [x(s_1) - \varepsilon, t(s_1)] \\ N &= [x(s_1), t(s_1)], & R &= [x(s_1) + \varepsilon, t(s_1)], & S &= [x(s_2) + \varepsilon, t(s_2)] \end{aligned} \tag{3.92}$$

Assume that $s_2 - s_1 > 0$ and $\varepsilon > 0$ are so small that no other curve of discontinuity cuts the region D bounded by the contour $MPQNRSM$. Then, by condition γ,

$$\int_\Gamma \Phi(u) dt - u dx = 0 \tag{3.93}$$

Hence, letting $\varepsilon \to 0$, by virtue of condition (δ), we see that

$$\int_{s_1}^{s_2} \left([\Phi(u)] \frac{dt}{ds} - [u] \frac{dx}{ds} \right) ds = 0 \tag{3.94}$$

Here the integrand is a continuous function of s because of condition (δ). Besides, $s_2 - s_1$ is arbitrarily small. Hence

$$[\Phi(u)] \frac{dt}{ds} - [u] \frac{dx}{ds} = 0 \quad \forall (x, t) \in L \tag{3.95}$$

---

[2] (a) Recall that if L is a smooth curve that divides a region D into two parts D^+ and D^-, where $p^+ \in D^+$, $p^- \in D^-$, $p \in L$, and p^+, p^- tend to p along any path not tangent to L, then

$$[f(p)] = \lim [f(p^+) - f(p^-)]$$

(b) The equation analogous to equation (3.90) is known in gas dynamics as the Rankine–Hugoniot condition.

If

$$\frac{dt}{ds} = 0, \quad \frac{dx}{ds} \neq 0 \quad \forall (x,t) \in L \tag{3.96}$$

then

$$[u] = 0 \quad \forall (x,t) \in L \tag{3.97}$$

If

$$\frac{dx}{ds} = 0, \quad \frac{dt}{ds} \neq 0 \quad \forall (x,t) \in L \tag{3.98}$$

then (3.97) is true, since

$$\Phi[u] = 0 \Rightarrow [u] = 0 \tag{3.99}$$

This means that along any curve of discontinuity

$$0 < \left| \frac{dx}{dt} \right| < \infty \quad \forall (x,t) \in L \tag{3.100}$$

so that (3.95) implies

$$\frac{dx}{dt} = \frac{[\Phi(u)]}{[u]} \quad \forall (x,t) \in L \quad \textbf{Q.E.D.} \tag{3.101}$$

Remark. Equality (3.90) shows that along a curve of discontinuity $x = x(t)$ the slope dx/dt of the tangent line in the (x,t) plane coincides with the slope of the secant line to the curve $\Phi = \Phi(u)$ in the (u, Φ) plane, which passes trough those values of u between which the jump occurs.

Note that in the linear case this condition means that a curve of discontinuity is a characteristic curve. Indeed, if

$$\Phi(u) = \varphi(x,t)u \tag{3.102}$$

where $\varphi(x,t)$ is continuous in the half-plane $t > 0$, then

$$[\Phi(u)] = \varphi(x,t)[u] \tag{3.103}$$

so that (3.101) becomes

$$\frac{dx}{\varphi(x,t)} = \frac{dt}{1} \tag{3.104}$$

But the system of characteristic equations in this case is

$$\frac{dx}{\varphi(x,t)} = \frac{dt}{1} = \frac{du}{0} \tag{3.105}$$

verifying our statement.

By borrowing terminology from gas dynamics, we shall say that if there exists an angle with its vertex at a point of discontinuity of u that is not covered by characteristic curves emanating from points located to the left and right of the vertex of the angle, then a *rarefaction wave* appears within that angle. In this case there exists a solution of the equation continuous in *the neighborhood of* the point of discontinuity if $\Phi(u)$ has no points of inflection for u between the values carried by the edge characteristic curves. If a curve of discontinuity emanates from the vertex then we shall say that a *shock wave* occurs. In what follows, we show that whenever $\Phi(u)$ has an inflection point in the relevant range of u, appearance of a shock is imminent.

It is easy to see that the necessary condition (3.101) for L to be a curve of discontinuity is not sufficient for a unique determination of L. Indeed, consider two examples, the first of which is the Cauchy problem for the Buckley–Leverett equation:

$$m\frac{\partial}{\partial t}\sigma + q(t)\frac{\partial}{\partial x}\Phi(\sigma) = 0, \quad \sigma(x,0) = \begin{cases} \sigma^- = \text{const} & \text{if } x < 0 \\ \sigma^+ = \text{const} & \text{if } x > 0 \end{cases} \quad (3.106)$$

where $\Phi(\sigma)$ is a monotonically increasing, twice continuously differentiable function such that

$$\Phi(\sigma) \equiv 1 \ \forall \sigma \geq \sigma_2, \quad \Phi(\sigma) \equiv 0 \ \forall \sigma \leq \sigma_1, \quad 0 < \sigma_1 < \sigma_2 < \sigma^- \leq 1 \quad (3.107)$$

$$\frac{d^2}{d\sigma^2}\Phi(\sigma) = 0 \quad \text{at } \sigma = \sigma_0 = \text{const} \in (\sigma_1, \sigma_2) \quad (3.108)$$

Let us assume that

$$\sigma^- = 1, \quad \sigma^+ = \text{const} \in (\sigma_1, \sigma_2) \quad (3.109)$$

In physical terms, the Cauchy problem (3.106)–(3.109) models the recovery of oil from a porous stratum of constant porosity m, where the total liquid percolation rate is equal to $q(t)$ and $\sigma(x,t)$ is the water saturation of the porous space.[3] It is assumed that the region $x < 0$ represents a water reservoir adjoining the productive layer and σ^+ is the initial water saturation of the latter.

Introduce a new time variable by setting

$$\tau = m^{-1}\int_0^t q(s)ds \quad (3.110)$$

Then the problem (3.106)–(3.109) becomes

$$\frac{\partial}{\partial \tau}\sigma + \frac{\partial}{\partial x}\Phi(\sigma) = 0 \quad (3.111)$$

$$\sigma(x,0) = \begin{cases} \sigma^- = 1 & \text{if } x < 0 \\ \sigma^+ & \text{if } x > 0 \end{cases} \quad (3.112)$$

---

[3]If v_w and v_o are the percolation rates of water and oil, respectively, then

$$q = v_w\sigma + (1-\sigma)v_o$$

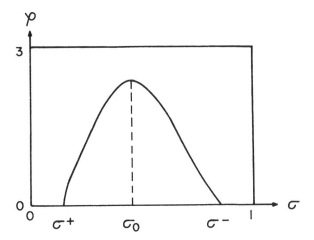

Figure 3.3. Dependence of $\varphi(\sigma) \stackrel{\text{def}}{=} d\Phi/d\sigma$ on σ. Data as in Figure 2.5.

that is, the model problem just considered. The corresponding Cauchy problem for the system of characteristic equations is[4]

$$\frac{dt}{1} = \frac{dx}{\Phi'(\sigma)} = \frac{d\sigma}{0} \tag{3.113}$$

$$x\bigg|_{t=0} = s, \sigma\bigg|_{t=0} = \begin{cases} \sigma^- & \forall s < 0 \\ \sigma^+ & \forall s > 0 \end{cases} \tag{3.114}$$

Hence the equations of the characteristic curves are

$$x = \begin{cases} s + \varphi(\sigma^-)t & \forall s < 0 \\ s + \varphi(\sigma^+)t & \forall s > 0 \end{cases} \tag{3.115}$$

where $\varphi(\sigma) \stackrel{\text{def}}{=} d\Phi(\sigma)/d\sigma$ (see Figure 3.3). The angle α, bounded by the characteristic curves $x = \Phi(\sigma^-)t$ and $x = \Phi(\sigma^+)t$, is not covered by characteristic curves emanating from half-lines $t=0$, $s < 0$ and $t=0$, $s > 0$ (see Figure 3.4). Hence, one could expect that there should exist a rarefaction wave determined, similarly to (3.81)–(3.83), by the equation

$$\frac{dx}{dt} \equiv \frac{x}{t} = \varphi(\sigma) \tag{3.116}$$

so that the slope of such a wave in (x,t) plane would be that of the tangent line to the curve $\Phi = \Phi(\sigma)$ in the (Φ,σ) plane. However, the solution, continuous everywhere within the angle α and varying between the edges $\sigma = \sigma^-$ and $\sigma = \sigma^+$, does not exist. Indeed, Φ has a point σ_0 of inflection

---

[4]Here we write t instead of τ and Φ' instead of $d\Phi/dx$.

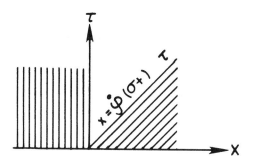

Figure 3.4.   Characteristics in Buckley–Leverett's equation. Data as in Figure 2.5.

between σ^- and σ^+, so that $\varphi(\sigma)$ has its maximum between σ^- and σ^+. Since equation (3.116) must hold along the characteristic lines that emanate from the point of discontinuity $x = 0$ and determine the rarefaction wave, two different values of σ correspond to the same value of x/t (see Figure 3.3), which is impossible for a single-valued function $\sigma(x/t)$. This means that rarefaction and shock waves must alternate. If a shock wave represents a jump between saturations σ^* and σ^{**} and the transition from an rarefaction wave to a shock wave (or vice versa) occurs at σ^*, then simultaneously

$$\lim_{\sigma \to \sigma^* + 0} \frac{dx}{dt} = \varphi(\sigma^*), \quad \lim_{\sigma \to \sigma^* - 0} \frac{dx}{dt} = \frac{\Phi(\sigma^{**}) - \Phi(\sigma^*)}{\sigma^{**} - \sigma^*},$$

$$\lim_{\sigma \to \sigma^* - 0} \frac{dx}{dt} = \lim_{\sigma \to \sigma^* + 0} \frac{dx}{dt} = \frac{dx}{dt}\bigg|_{\sigma = \sigma^*}$$

(3.117)

(See the remark following equation (3.101).). Obviously, in our examples there are two (and only two) ways to satisfy conditions (3.117).

(1) A rarefaction wave within the angle with edges $x/t = 0$ and $x/t = \varphi(\tilde{\sigma}^-)$, with $\tilde{\sigma}^- > \sigma_0$ followed by a jump from $\sigma = \tilde{\sigma}^-$ to σ^+.
(2) A jump from the saturation σ^- to some value $\sigma = \tilde{\sigma}^+ < \sigma_0$ followed by the rarefaction wave filling the angle between the edges $x/t = 0$ and $x/t = \varphi(\sigma^+)$.

Let us reiterate that, within each option, the points $\tilde{\sigma}^-, \tilde{\sigma}^+$ are those for which the secants with one end at σ^+ or σ^- (respectively), become simultaneously tangent to the curve $\Phi = \Phi(\sigma)$ (see Figure 3.5). The water saturation corresponding to these two options is shown in Figure 3.6. Thus, if the function $\Phi(\sigma)$ in our model equation has a point of inflection, the necessary condition of Theorem 3.4.1 is not sufficient for a unique determination of the location of a shock wave.

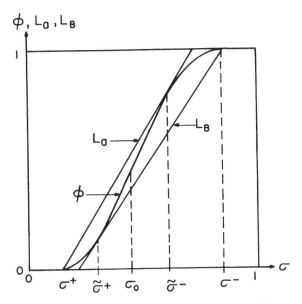

Figure 3.5. Two options satisfying conditions (3.101). Data as in Figure 3.5.

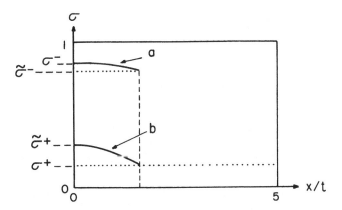

$$z = \wp(\tilde{\sigma}^-) = \wp(\tilde{\sigma}^+)$$

Figure 3.6. Solution of Buckley–Leverett's equation; $\sigma = \sigma(z/t)$. (a) Jump from σ^- to the wave of expansion; (b) jump from the wave of expansion to σ^+. Data as in Figure 3.5.

As a second example, consider the Cauchy problem (3.68b)–(3.70b), that is, the problem

$$\frac{\partial u}{\partial t} + u\frac{\partial u}{\partial x} = 0 \tag{3.118}$$

$$u\Big|_{t=0} = \begin{cases} 1 & \forall x < 0 \\ -1 & \forall x > 0 \end{cases} \tag{3.119}$$

Take an arbitrary $a \geq 1$ and define

$$u(x,t) = \begin{cases} 1 & \text{for } -\infty < x < \dfrac{1-a}{2}t \\[2mm] -a & \text{for } \dfrac{1-a}{2}t < x < 0 \\[2mm] a & \text{for } 0 < x < \dfrac{1+a}{2}t \\[2mm] -1 & \text{for } \dfrac{1+a}{2}t < x < \infty \end{cases} \tag{3.120}$$

Obviously, $u(x,t)$ so defined is a generalized solution to the Cauchy problem (3.118), (3.119). Lines of its discontinuity are

$$x = \frac{1-a}{2}t, \quad x = 0, \quad x = \frac{1+a}{2}t \tag{3.121}$$

Along these curves we have, respectively,

$$u^+ = \begin{cases} -a \\ a \\ -1 \end{cases}, \quad u^- = \begin{cases} 1 \\ -a \\ a \end{cases}, \quad \Phi^+ = \begin{cases} a^2/2 \\ a^2/2 \\ 1/2 \end{cases} \quad \Phi^- = \begin{cases} 1/2 \\ a^2/2 \\ a^2/2 \end{cases} \tag{3.122}$$

Hence, along all these three lines of discontinuity,

$$\frac{dx}{dt} = \frac{\Phi^+ - \Phi^-}{u^+ - u^-} \tag{3.123}$$

so that the necessary condition (3.90) holds. Note that in this case $\Phi(u)$ has no points of inflection, so that their occurrence (as in the previous example), is not the only source of insufficiency of condition (3.90) for uniquely determining the position of the shock.[5]

Thus there is need for a condition, additional to (3.90), for a unique determination of the position of shock waves. The required additional condition for unique determination of curves of discontinuity is provided by the method of *vanishing viscosity* [24] (see Chapter 21, Section 1), which we expose here by following the heuristic arguments of Gelfand. Thus we consider, together with the model problem

$$\frac{\partial}{\partial t}u + \frac{\partial}{\partial x}\Phi(u) = 0, \quad x \in (-\infty, \infty), \ t > 0, u(x,0) = f(x) \tag{3.124}$$

---

[5] The assumption that $\Phi(u)$ has no points of inflection was essential in the papers of O. Oleinik, one of the creators of the theory of generalized solutions of quasilinear first-order partial differential equations (see [50;51;52]).

the Cauchy problem

$$\frac{\partial}{\partial t}u + \frac{\partial}{\partial x}\Phi(u) = \varepsilon\frac{\partial^2}{\partial x^2}u, \quad x \in (-\infty, \infty), \ t > 0 \tag{3.125}$$

$$u(x,0) = f(x) \tag{3.126}$$

Existence and uniqueness theorems for this kind of Cauchy problem will be proved in Section 3 of Chapter 17, where we shall prove the existence and uniqueness of a global solution to the Cauchy problem

$$\frac{\partial}{\partial t}u + \varphi(u)\frac{\partial}{\partial x}u = \frac{\partial^2}{\partial x^2}u, \quad x \in (-\infty, \infty), \ t > 0 \tag{3.127}$$

$$u(x,0) = f(x) \tag{3.128}$$

Evidently, by use of the scaling

$$x_1 = \varepsilon x, \quad t_1 = \varepsilon t \tag{3.129}$$

we can reduce this problem to the form (3.125) and (3.126) if

$$\varphi(u) = \frac{d}{du}\Phi(u) \tag{3.130}$$

Thus there exists a unique global solution to the problem (3.125) and (3.126), satisfying the initial condition at every point of continuity of $f(x)$. In what follows we denote the solution of the problem (3.125) and (3.126) by $u(x,t \mid \varepsilon)$. It is natural to consider the solution of problem (3.124) as the limit of $u(x,t \mid \varepsilon)$ as $\varepsilon \downarrow 0$. This idea is natural because there is a close connection between the physical content of problems (3.125), (3.126) and (3.124). Namely, in the case

$$\Phi(u) = \frac{u^2}{2} \tag{3.131}$$

the left side of equation (3.124) can be treated as the equation of 1-dimensional motion of an inviscid liquid not affected by any external forces when the pressure is independent of x, whereas equation (3.125) can be treated as the equation of motion of a viscous liquid. Thus, in terms of the method of vanishing viscosity, one defines

$$u(x,t) = \lim_{\varepsilon \downarrow 0} u(x,t \mid \varepsilon) \tag{3.132}$$

This selects a unique solution of the model problem (3.124) from the entire set of solutions, satisfying the necessary condition (3.90).

The preceding physical interpretation is not the only possible one. For example, equation (3.68a) could be interpreted as the right side of the Rapoport–Leas equation (the right side of equation 2.211) if the capillary pressure $\lambda(u)$ were assumed identical constant

$$\lambda(u) \equiv \varepsilon = \text{const} \tag{3.133}$$

In the context of this interpretation, a solution of the Buckley–Leverett equation can be treated as a solution of the Rapoport–Leas equation with vanishing capillary pressure.

Since Gelfand's purpose was to establish a necessary and sufficient condition for determination of a shock wave, he did not prove the method of vanishing viscosity to be applicable in general, but restricted his attention to the particular case of the problem of decay of an arbitrary initial discontinuity, which is the same as problem (3.124) if

$$f(x) \equiv \begin{cases} u^- = \text{const} & \forall x < 0 \\ u^+ = \text{const} & \forall x > 0 \end{cases} \tag{3.134}$$

Thus, following Gelfand, consider problem (3.125), (3.134). The change of variables

$$y = \frac{x}{\varepsilon}, \quad \tau = \frac{t}{\varepsilon}, \quad v(y, \tau \mid \varepsilon) = u(x, t \mid \varepsilon) \tag{3.135}$$

transforms problem (3.127), (3.134) into the problem

$$\frac{\partial}{\partial \tau} v(y, \tau \mid \varepsilon) + \frac{\partial}{\partial y} \Phi[v(y, \tau \mid \varepsilon)] = \frac{\partial^2}{\partial y^2} v(y, \tau \mid \varepsilon) \tag{3.136}$$

$$v(y, \tau \mid \varepsilon) = \begin{cases} u^- & \forall y < 0 \\ u^+ & \forall y > 0 \end{cases} \tag{3.137}$$

Note that

$$\lim_{\varepsilon \downarrow 0} y = \begin{cases} -\infty & \forall x = \text{const} < 0, \ \forall t \geq 0 \\ \infty & \forall x = \text{const} > 0, \ \forall t \geq 0 \end{cases} \tag{3.138}$$

Since ε does not occur in (3.136) and (3.137), the solution of the problem is independent of ε and this means, by (3.138), that we must seek a solution to the equation

$$\frac{\partial}{\partial \tau} u + \frac{\partial}{\partial y} \Phi(u) - \frac{\partial^2}{\partial y^2} u \quad \forall y \in (-\infty, \infty), \ \forall \tau > 0 \tag{3.139}$$

satisfying the boundary conditions

$$\lim_{u \downarrow -\infty} u(y, \tau) = u^-, \quad \lim_{y \uparrow \infty} u(y, \tau) = u^+ \tag{3.140}$$

Let ω be the expected rate of the shock wave

$$\omega = \frac{dx}{dt}\bigg|_{t=0} = \frac{dy}{d\tau}\bigg|_{\tau=0} \tag{3.141}$$

We seek a solution of problem (3.140) and (3.141) in the form of a traveling wave:

$$u = u(z), \quad z = y - \omega\tau \tag{3.142}$$

Substitution of (3.142) into (3.139) yields

$$\frac{d^2}{dz^2} u = \omega \frac{d}{dz} \Phi(u) - \omega \frac{d}{dx} u \tag{3.143}$$

$$u \bigg|_{-\infty} = u^-, \quad u_\infty = u^+ \tag{3.144}$$

Integration, taking into account that

$$\lim_{z \downarrow -\infty} \Phi(u) = \Phi(u^-) \stackrel{\text{def}}{=} \Phi^-, \quad \lim_{z \uparrow \infty} \Phi(u) = \Phi(u^+) \stackrel{\text{def}}{=} \Phi^+, \quad \lim_{|z| \uparrow \infty} \frac{d}{dz} u = 0$$

$$(3.145)$$

yields

$$\omega = \frac{\Phi^+ - \Phi^-}{u^+ - u^-} \qquad (3.146)$$

which coincides with (3.101). However, we can now derive a sufficient condition too.

Lemma 3.4.1. *A discontinuous solution of the model problem may be the limit of a solution of the corresponding problem with vanishing viscosity if u^- and u^+ are neighboring roots of the function*

$$\Psi(u) \stackrel{\text{def}}{=} \Phi(u) - \Phi^- - \omega(u - u^-) \qquad (3.147)$$

PROOF. Integration of (3.143) yields

$$\int_{-\infty}^{z} \frac{d^2}{ds^2} u(s) \, ds = \frac{d}{dz} u(z) = \Psi[u(z)] \qquad (3.148)$$

Assume that

$$\exists u_0 \in (u^-, u^+) \quad \text{such that} \quad \Psi(u_0) = 0 \qquad (3.149)$$

and

$$u(z_0) = u_0 \qquad (3.150)$$

By the uniqueness theorem for solutions of ordinary differential equations, the solution of the Cauchy problem

$$\frac{d}{dz} u(z) = \Psi(z), \quad u(z_0) = u_0 \qquad (3.151)$$

is

$$u(z) \equiv u_0 \qquad (3.152)$$

But an integral curve of an ordinary differential equation is uniquely determined by any of its points. This contradicts the assumption that

$$\lim_{z \downarrow -\infty} u = u^- \neq u_0 \neq \lim_{z \uparrow \infty} u = u^+ \qquad (3.153)$$

(see Figure 3.7). Thus u^- and u^+ must indeed be the neighboring roots of $\psi(u)$. **Q.E.D.**

Remark. The statement "u^- and u^+ are neighboring roots of $\psi(u)$" is equivalent to the statement:

$$\text{sign } \Psi(u) = \text{sign}(u^+ - u^-) \quad \forall u \in (u^-, u^+) \qquad (3.154)$$

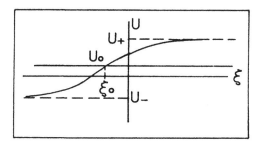

Figure 3.7. Sketch of an integral curve of
the equation $du/dz = \Phi[u(z)]$.

Indeed, integration of (3.151) yields

$$u^+ - u^- = \int_{-\infty}^{\infty} \Psi[u(s)]ds \tag{3.155}$$

Hence, if $\Psi(u)$ is not of constant sign then there is a root of Ψ between u^- and u^+. On the other hand, if $\Psi(u)$ is of a constant sign then this is precisely the sign of $u^+ - u^-$.

Condition (3.154) admits a convenient geometrical interpretation: Let

$$\Gamma = \{u, \Phi : \Phi = \Phi(u), \ u \in (u^-, u^+)\} \tag{3.156}$$

and

$$L = \left\{ u, \varphi : \frac{\varphi - \Phi^-}{u - u^-} = \frac{\Phi^+ - \Phi^-}{u^+ - u^-}, \ u \in (u^-, u^+) \right\} \tag{3.157}$$

Then in the interval (u^-, u^+) the straight line L is located under the curve Γ if $u^- < u^+$ and above it if $u^- > u^+$.

Collecting our results, we let

$$\frac{\partial}{\partial t}u(x,t) + \frac{\partial}{\partial x}\Phi[u(x,t)] = 0 \tag{3.158}$$

$$u(x,0) = \begin{cases} u_- = \text{const} & \forall x < 0 \\ u_+ = \text{const} & \forall x > 0 \end{cases} \tag{3.159}$$

$$y = \frac{x}{\varepsilon}, \quad \tau = \frac{t}{\varepsilon}, \quad \varepsilon = \text{const} \downarrow 0 \tag{3.160}$$

$$z = \omega(y - \tau), \quad \omega = \text{const} \tag{3.161}$$

$$-\omega\frac{d}{dz}U(z) + \frac{d}{dz}\Phi[U(z)] = \frac{d^2}{dz^2}U(z) \tag{3.162}$$

$$U\Big|_{z=-\infty} = u_- = \text{const}, \quad U\Big|_{z=\infty} = u_+ = \text{const} \tag{3.163}$$

Definition 3.4.2. $u(x,t)$ is called an *admissible solution* of the problem (3.158) and (3.159) if

$$u(x,t) = \lim_{\varepsilon\downarrow0} U(z) \tag{3.164}$$

We have proved the following theorem.

Theorem 3.4.2. *If $u(x,t)$ is an admissible solution of problem (3.158) and (3.159), then a smooth curve $\tilde{L} = \{x, t : t > 0, x = x(t)\}$ may be a curve of propagation of a shock wave if and only if*

(a) *The rate dx/dt of motion of $\tilde{L}$ is*

$$\frac{dx}{dt} = \frac{\Phi^+ - \Phi^-}{u^+ - u^-} \tag{3.165}$$

(b) *If $\Phi(u)$ has no points of inflection between u_+ and u_- then*

$$u^+ = u_+, \quad u^- = u_- \tag{3.166}$$

(c) *If $\Phi(u)$ has a finite number of points of inflection*

$$\frac{d^2}{du^2}\Phi(u_k) = 0, \quad k = 1, 2, \ldots, n \tag{3.167}$$

between u_+ and u_-, where $u_-, u_1, u_2, \ldots, u_n, u_+$ is a monotone sequence, then the transition from u_- to u_+ occurs by alternation of rarefaction waves and shock waves such that:

(1) *If $u_1 > u_-$ then the transition from $u = u_-$ to $u = u_1^-$ occurs by a shock wave, where u_1^- is the nearest point to u such that the straight line L_1 tangent to the curve $\Gamma = [u, \Phi(u)]$ is located under Γ.*
(2) *If $u_1 < u_-$ then the straight line defined in (1) is located above Γ.*
(3) *If $u_n > u_+$ then the passage from $u = u_n^+$ to u_+ occurs by a shock wave, where u_n^+ is the nearest point to u_+ such that the straight line L_n tangent to the curve $\Gamma = [u, \Phi(u)]$ is located above Γ.*
(4) *If $u_n < u_+$ then the straight line defined in (3) is located under Γ.*
(5) *Between u_1^- and u_n^+ there alternate rarefaction waves and shock waves connecting the points u_m^- and u_m^+, $u_{m+1}^- > u_m^+$, so that the straight line L_m connecting points u_m^- and u_m^+ is tangent to the curve Γ at both these points; moreover, between u_m^- and u_m^+, L_m is located under Γ if $u_m^- < u_m^+$ and above Γ if $u_m^- > u_m^+$*

Problems

P3.1.1. Solve the following Cauchy problems:
$$yp + xq = 0$$
(a)
$$\Gamma = \left\{x, y, z : y = \alpha x, \ z = f(x), \ a^2 \neq 1\right\}$$
$$xp + yq = x + y + z$$
(b)
$$\Gamma = \{x, y, z : x = \cos(\varphi), \ y = \sin(\varphi), \ z = 0, \ 0 \le \varphi \le 2\pi\}$$

P3.1.2. General solution for a n-th order partial differential equation in m variables is a family of particular solutions dependent on n arbitrary functions of $m-1$ variables [28]. In particular, general solution of equation (3.4) may be defined by the equality

$$\beta(x,y,z) = \mathbf{F}(\alpha(x,y,z))$$

where $\mathbf{F}$ is an arbitrary continuously differentiable function of one variable and $\alpha(x,y,z)$, $\beta(x,y,z)$ are two independent first integrals of the characteristic system (3.14). Show that $z(x,y)$, defined by (*), indeed satisfies equation (3.4).

P3.2.1.   Solve the Cauchy problem

$$p^2 + q^2 = x + y$$
$$\Gamma = \{x,y,z : -\infty < x \le \infty,\ y = 0,\ z = 0\}$$

P3.3.1.   Consider the Cauchy problem

$$\frac{\partial}{\partial t}u + u\frac{\partial}{\partial x}u = 0, \quad -\infty < x < \infty,\ t > 0$$
$$u(x,0) = \tanh(x)$$

Does there exist a global solution? If so, prove its existence; if not, estimate the interval of t values in which the problem has a local solution.

P3.3.2.   Solve Problem 3.3.1 if $u(x,0) = \ln(1 + x^2)$.

P3.4.1.   Determine the solutions of problem (3.106) in view of Theorem 3.4.2. Take as the input data those in Figures 2.3 and Figures 3.3–3.6.

P3.4.2.   Explain why the discontinuity $\sigma(x,t) \equiv 1 \forall x < 0, \forall t > 0$, $\sigma(x,t) \equiv \sigma^- \ \forall x > 0, \forall t > 0$ corresponding to the data of Problem 3.4.1 does not propagate.

P3.4.3.   Determine the solutions of problems (3.118) and (3.119) in view of Theorem 3.4.2.

P3.4.4.   Draw graphs $(x/t, u)$ of solutions to: (a) problems (3.68b) and (3.69b); (b) problems (3.68b) and (3.70b); (c) the model problems with $u_- = 0$, $u_+ = 1$, and $\Phi(u) = \sin(2\pi u)$ or $\Phi(u) = -\sin(2\pi u)$.

4

Classification of second-order partial differential equations with linear principal part. Elements of the theory of characteristics

1. Classification of second-order partial differential equations

A number of equations and systems of equations, describing different phenomena, were derived in Chapter 2. The classical theory of partial differential equations is concerned with the simplest and the most typical equations, such as the Laplace and Poisson, Fourier, and wave equations. As we have seen, these equations describe physical processes of very different origins. For example, the Laplace and Poisson equations appear in the theories of gravimetry, electrostatics, and flow of inviscid incompressible liquids. The Fourier equation appears in the theories of thermal conductivity, diffusion and percolation of elastic liquid through a porous medium, and in electrodynamics. The wave equation describes processes of transversal and longitudinal vibrations in solids and liquids and field propagation in electrodynamics.

This diversity of phenomena described by the same type of second-order partial differential equations is not accidental, from the purely mathematical point of view; it reflects the essential properties of these equations. We shall demonstrate this through consideration of the possible classification of these equations and the derivation of their canonical forms.

We begin with the classification of second-order partial differential equations with linear principal part:

$$\Phi^* \equiv \sum_{k,m}^{n} a_{km}(x_1, x_2, \ldots, x_n) \frac{\partial^2 u}{\partial x_k \partial x_m} + F(x, u^0, u^1, \ldots, u^n) = 0 \qquad (4.1)$$

where u^k, $k = 0, 1, \ldots, n$, denotes the set of all kth partial derivatives of u. The classification can be accomplished through several different approaches, which will be presented in this chapter. We begin with an approach based on consideration of invariants under all possible nondegenerate coordinate transformations:

$$\xi_j = \xi_j(x_1, \ldots, x_n), \quad j = 1, 2, \ldots, n \qquad (4.2)$$

117

By "nondegenerate" we mean that the Jacobian of the transformation does not vanish in the region where equation (4.1) is being considered:

$$\frac{D(\xi_1, \xi_2, \ldots, \xi_n)}{D(x_1, x_2, \ldots, x_n)} \neq 0 \tag{4.3}$$

By the implicit function theorem, inequality (4.3) means that the system (4.2) can be solved for $x_1, \ldots, x_n$, so that

$$x_k = x_k(\xi_1, \ldots, \xi_n), \quad k = 1, 2, \ldots, n \tag{4.4}$$

We have

$$\frac{\partial}{\partial x_i} = \sum_{k=1}^{n} \frac{\partial \xi_k}{\partial x_i} \cdot \frac{\partial}{\partial \xi_k} \tag{4.5}$$

$$\frac{\partial^2}{\partial x_i \partial x_j} = \sum_{k=1}^{n} \frac{\partial \xi_k}{\partial x_i} \sum_{m=1}^{n} \frac{\partial \xi_m}{\partial x_j} \cdot \frac{\partial^2}{\partial \xi_k \partial \xi_m} + \sum_{k=1}^{n} \frac{\partial^2 \xi_k}{\partial x_i \partial x_j} \cdot \frac{\partial}{\partial \xi_k} \tag{4.6}$$

Hence

$$\Phi^* = \Phi + F^* \tag{4.7}$$

where

$$\Phi = \sum_{i=1}^{n} \sum_{j=1}^{n} a_{ij} \sum_{k=1}^{n} \sum_{m=1}^{n} \frac{\partial^2 u}{\partial \xi_k \partial \xi_m} \cdot \frac{\partial \xi_k}{\partial x_i} \cdot \frac{\partial \xi_m}{\partial x_j} \tag{4.8}$$

and

$$F^* \left(\xi_1, \ldots, \xi_n, u, \frac{\partial u}{\partial \xi_1}, \ldots, \frac{\partial}{\partial \xi_n} \right) = F + \sum_{i-1}^{n} \sum_{j=1}^{n} a_{ij} \sum_{k=1}^{n} \frac{\partial u}{\partial \xi_k} \cdot \frac{\partial^2 \xi_k}{\partial x_i \partial x_j} \tag{4.9}$$

Changing the order of summation yields

$$\Phi = \sum_{k=1}^{n} \sum_{m=1}^{n} a_{km}^* \frac{\partial^2 u}{\partial \xi_k \partial \xi_m} \tag{4.10}$$

where

$$a_{km}^* = \sum_{i=1}^{n} \sum_{j=1}^{n} a_{ij} \frac{\partial \xi_k}{\partial x_i} \frac{\partial \xi_m}{\partial x_j} \tag{4.11}$$

Let us associate with the differential form

$$f = \sum_{i=1}^{n} \sum_{j=1}^{n} a_{ij} \frac{\partial^2}{\partial x_i \partial x_j} \tag{4.12}$$

a quadratic form

$$\varphi = \sum_{i=1}^{n} \sum_{j=1}^{n} a_{ij} x_i x_j \tag{4.13}$$

and with the transformation (4.4) a linear transformation

$$x_i = \sum_{k=1}^{n} \alpha_{ik} \xi_k \tag{4.14}$$

which transforms φ into

$$\varphi = \sum_{i=1}^{n} \sum_{j=1}^{n} \tilde{a}_{ij}^{*} \xi_i \xi_j \tag{4.15}$$

where

$$\tilde{a}_{ij}^{*} = \sum_{k=1}^{n} \sum_{m=1}^{n} a_{km} \alpha_{ki} \alpha_{mj} \tag{4.16}$$

Let us compare the forms (4.12) and (4.15). If at a fixed point $(x_1, \dots, x_n)$

$$\frac{\partial \xi_i}{\partial x_k} = \alpha_{ki}, \qquad \frac{\partial \xi_j}{\partial x_m} = \alpha_{mj} \tag{4.17}$$

then at that point

$$f = \varphi \quad \text{and} \quad a_{ij}^{*} = \tilde{a}_{ij}^{*} \quad \forall(i,j) \tag{4.18}$$

This means that the invariants of transformation (4.4) coincide with those of the transform (4.14), so that the classification of quadratic forms may be used for the classification of partial differential equations with linear principal part.

Quadratic forms are classified by reducing the latter to canonical form and using the so-called law of inertia. Namely, it is proved in the theory of quadratic forms that any form φ can be reduced by infinitely many nondegenerate transformations (4.15) to a canonical form, that is, such that the matrix of the transformed form contains only diagonal terms:

$$\begin{bmatrix} a_{11} & a_{12} & \cdots & a_{1n} \\ a_{21} & a_{22} & \cdots & a_{2n} \\ \vdots & & \ddots & \vdots \\ a_{n1} & a_{n2} & \cdots & a_{nn} \end{bmatrix} = A \begin{bmatrix} \lambda_1 & 0 & \cdots & 0 \\ 0 & \lambda_2 & \cdots & 0 \\ \vdots & & \ddots & \vdots \\ 0 & 0 & \cdots & \lambda_n \end{bmatrix} \tag{4.19}$$

Here A is a coefficient of proportionality, depending on the transformation;

$$\lambda_k = \begin{cases} 1 & \text{for } k = 1, 2, \dots, p \\ -1 & \text{for } k = p+1, \ p+2, \dots, p+q \\ 0 & \text{for } k = p+q+1, \ p+q+2, \dots, \ p+q+r = n \end{cases} \tag{4.20}$$

and the numbers $p \geq q$ and r are invariants of the transformation, that is, are independent of the choice of transformation reducing the quadratic form to its canonical form.[1] This means that if a transformation (4.15) reduces φ to the form

$$\varphi^* = \sum_{k=1}^{p} \xi_k^2 - \sum_{k=1}^{q} \xi_{k+p}^2 \tag{4.21}$$

then the partial differential equation (4.1) will be transformed at the fixed point under consideration into the equation

$$\sum_{k=1}^{p} \frac{\partial^2}{\partial \xi_k^2} u - \sum_{k=1}^{q} \frac{\partial^2}{\partial \xi_{k+p}^2} u + F^* = 0 \tag{4.22}$$

In what follows we shall assume that only the three simplest cases may occur:[2]

$$\begin{array}{llll}
C_1: & p = n, & q = 0, & r = 0 \\
\end{array} \tag{4.23}$$
$$\begin{array}{llll}
C_2: & p = n - 1, & q = 1, & r = 0 \\
\end{array} \tag{4.24}$$
$$\begin{array}{llll}
C_3: & p = n - 1, & q = 0, & r = 1 \\
\end{array} \tag{4.25}$$

In $\mathbb{R}_3$,

$$\varphi(x_1, x_2, x_3) = 0 \tag{4.26}$$

is the equation of a quadratic surface: in case C_1 it is an ellipsoid, in case C_2 a hyperboloid, and in case C_3 a paraboloid. Generalizing this terminology, the theory of quadratic forms calls the form φ *elliptic*, *hyperbolic*, and *parabolic* in case C_1, C_2, and C_3, respectively. This terminology is also used in the theory of second-order partial differential equations with linear principal part, so that if the form φ is elliptic, hyperbolic, or parabolic at the fixed point $(x_1, x_2, \ldots, x_3)$, then the differential form f is also called elliptic, hyperbolic, or parabolic, (respectively), at this point. Similarly, the equation $\Phi^* = 0$ is said to be elliptic, hyperbolic, or parabolic at that point. Clearly, the type of the equation may remain unchanged in part or all of the region under consideration. Thus the Poisson equation

$$\Delta u + f(p) = 0 \tag{4.27}$$

is elliptic everywhere, the wave equation

$$a^2 \Delta u + f(p, t) = \frac{\partial^2}{\partial t^2} u \tag{4.28}$$

---

[1]The assumption $p \geq q$ is not restrictive because, if it turns out after reducing φ to diagonal form with new independent variables $y_1, y_2, \ldots, y_n$ that $p < q$, then the substitution $z_k = \sqrt{-1} y_k$, $k = 1, 2, \ldots, n$ replaces p by q and q by p.

[2]The general cases $p < n-1$, $q > 0$, and $r \geq 0$ (when the type of the equation remains unchanged within some region) are not considered in the classical theory of partial differential equations of second order, but have been subject to intensive study in the modern theory. In the case $p = q = 0$, the second-order partial differential equation degenerates into a first-order equation. Partial differential equations may also degenerate into an ordinary differential equation, so that the presence of partial derivatives in an equation not in canonical form does not reflect its properties.

is hyperbolic, and the heat conduction (Fourier) equation

$$a^2 \Delta u + f(p,t) = \frac{\partial}{\partial t} u \qquad (4.29)$$

is parabolic everywhere.[3]

2. Reduction of second-order equation to canonical form

We have shown that a second-order partial differential equation with a linear principal part may be transformed into canonical form in any fixed point of the region where the equation is defined. We now consider the possibility of reducing an equation to canonical form not in a fixed point but everywhere within the region where the type of the equation remains unchanged. In other words, we seek a nondegenerate transformation of coordinates under which the nondiagonal elements of the matrix (a_{km}^*) will be equated to 0 and its diagonal elements to ± 1 or 0:

$$a_{km}^* = 0 \quad \text{if } k \neq m;$$

$$a_{kk}^* = 1 \quad \text{if } k \leq p, a_{kk}^* = -1 \quad \text{if } p+1 \leq k \leq p+q, \qquad (4.30)$$

$$a_{kk}^* = 0 \quad \text{if } p+q+1 \leq k \leq n$$

This means that n functions $\xi_k(x_1, \ldots, x_n)$, $k = 1, 2, \ldots, n$, must satisfy $n(n-1)$ differential equations. Since the matrix (a_{km}) is symmetric,[4] so is the matrix (a_{km}^*). Hence, for all the nondiagonal elements of the matrix (a_{km}^*) to vanish, we must satisfy $n(n-1)/2$ equations. If $n > 3$ then $n(n-1)/2 > n$ so that the problem is overdetermined. If $n = 3$ then $n(n-1)/2 = n$, so that it should be possible to equate all the nondiagonal elements to 0. However, it will be impossible to make $p+q$ diagonal elements of this matrix equal to ± 1, so we cannot expect the equation to be reducible to canonical form in an entire region of $\mathbb{R}_3$ (except, naturally, when the principal part of the linear equation has constant coefficients when a suitable scaling of the independent variables do this job). Therefore the problem of reducing the equation to canonical form in an entire region must be restricted to consideration of the 2-dimensional case.

Thus, let us consider an equation

$$\Phi^* \equiv \sum_{k,m}^{2} a_{km}(x_1, x_2) \frac{\partial^2 u}{\partial x_1 \partial x_2} + F\left(x_1, x_2, u, \frac{\partial u}{\partial x_1}, \frac{\partial u}{\partial x_2}\right) = 0 \qquad (4.31)$$

---

[3]Recall that by appropriate scaling the coefficient of diffusivity a^2 in the wave and Fourier equations may be made equal to 1, so that these equation may be considered as reduced to canonical form.

[4]By assumption, second derivatives of u are continuous, so that

$$\partial^2 u / \partial x_m \partial x_k \equiv \partial^2 u / \partial x_k \partial x_m.$$

Hence the terms $a_{km} \partial^2 u / \partial x_k \partial x_m$ and $a_{mk} \partial^2 u / \partial x_m \partial x_k$ can always be joint, and a symmetric coefficient $\tilde{a}_{km}$ defined as $\tilde{a}_{am} = 1/2(a_{km} + a_{mk})$ may be introduced.

whose type is preserved everywhere in some simply connected region D of $\mathbb{R}_2$. Recall that the type of a 2-dimensional quadratic form

$$\varphi = a_{11}x_1^2 + 2a_{12}x_1x_2 + a_{22}x^2 \qquad (4.32)$$

is determined by the sign of its discriminant:

$$\delta = a_{12}^2 - a_{11} \cdot a_{22} \qquad (4.33)$$

- $\delta > 0$ then $\varphi = 0$ is the equation of a hyperbola;
- $\delta < 0$ then $\varphi = 0$ is the equation of an ellipse;
- $\delta = 0$ then $\varphi = 0$ is the equation of a parabola.

Hence the characteristic equation

$$a_{11}x_1^2 + 2a_{12}x_1x_2 + a_{22}x_2^2 = 0 \qquad (4.34)$$

has

- two different real roots r_1 and r_2 in the hyperbolic case;
- two complex conjugate roots $r_1 = \alpha + i\beta$, and $r_2 = \alpha - i\beta$ in the elliptic case;
- a double root r in the parabolic case.

Let us begin with an equation that is *hyperbolic* in D. Recall that the equation of a hyperbola may be reduced to two canonical forms:

$$x_1x_2 = 1/4 \qquad (4.35)$$

or

$$x_1^2 - x_2^2 = 1 \qquad (4.36)$$

Hence a hyperbolic equation in $\mathbb{R}_2$ may be reduced to two canonical forms:

$$\frac{\partial^2}{\partial x_1 \partial x_2}u + \frac{1}{4}F = 0 \qquad (4.37)$$

or

$$\frac{\partial^2}{\partial x_1^2}u - \frac{\partial^2}{\partial x_2^2}u + F = 0 \qquad (4.38)$$

Note that the substitution

$$\xi_1 = \frac{x_1 + x_2}{2}, \quad \xi_2 = \frac{x_1 - x_2}{2} \qquad (4.39)$$

transforms the first canonical form (4.35) of a hyperbola, or the first canonical form (4.37) of the corresponding hyperbolic partial differential equation, into their second canonical forms. Therefore it is sufficient to reduce the general hyperbolic equation to its first canonical form.

Note, first of all, that in the hyperbolic case we may assume that one of coefficients a_{11} or a_{22} does not vanish everywhere in the region D. Indeed, if $a_{11} \equiv a_{22} \equiv 0$ then $a_{12} \neq 0$ everywhere in D, since if $a_{12} \equiv 0$ in D then equation (4.31) is not of second order in D. If $a_{12} = 0$ at some points of D, then the region D of hyperbolicity is not a simply connected one, contrary to assumption. But if the identities $a_{11} \equiv a_{22} \equiv 0$ and the inequality $a_{12} \neq 0$ hold simultaneously everywhere in D, then the equation (4.31) has the following form in D:

$$\frac{\partial^2}{\partial x_1 \partial x_2} u + F = 0 \tag{4.40}$$

that is, it is in first canonical form, so that no further changes of variables are needed. Thus, assume that one of the coefficients a_{11}, a_{22}, say a_{11}, does not vanish everywhere in D.

Recall that the linear transformation

$$x_i = \alpha_i \xi_1 + \beta_i \xi_2, \quad i = 1, 2 \tag{4.41}$$

with α_1/α_2 and β_1/β_2 equal respectively to the roots r_1 and r_2 of the characteristic equation (4.34), reduces the hyperbolic quadratic form (4.32) to its first canonical form; that is, it nullifies the coefficients a_{11}^* and a_{22}^*. The hyperbolic partial differential equation corresponding to the characteristic equation (4.34) is a first-order nonlinear characteristic differential equation

$$a_{11} \left(\frac{\partial \xi}{\partial x_1} \right)^2 + 2a_{12} \frac{\partial \xi}{\partial x_1} \cdot \frac{\partial \xi}{\partial x_2} + a_{22} \left(\frac{\partial \xi}{\partial x_2} \right)^2 = 0 \tag{4.42}$$

or, in view of the assumption that $a_{11} \neq 0$,

$$\left[\frac{\frac{\partial \xi}{\partial x_1}}{\frac{\partial \xi}{\partial x_2}} \right]^2 + 2 \frac{a_{12}}{a_{11}} \frac{\frac{\partial \xi}{\partial x_1}}{\frac{\partial \xi}{\partial x_2}} + \frac{a_{22}}{a_{11}} = 0 \tag{4.43}$$

Let

$$y = \xi(x_1, x_2) \tag{4.44}$$

be a solution of this equation. Then the identity

$$y \equiv \text{const} \tag{4.45}$$

determines the differential equation

$$\frac{dx_2}{dx_1} = r(x_1, x_2) \tag{4.46}$$

where $r(x_1, x_2)$ is a root of the characteristic equation (4.34). Indeed, (4.45) yields

$$\frac{dx_2}{dx_1} = -\frac{\frac{\partial \xi}{\partial x_1}}{\frac{\partial \xi}{\partial x_2}} \tag{4.47}$$

Substitution of (4.47) into (4.43) implies

$$\left(\frac{dx_2}{dx_1} \right)^2 - 2 \frac{a_{12}}{a_{11}} \frac{dx_2}{dx_1} + \frac{a_{22}}{a_{11}} = 0 \tag{4.48}$$

which proves (4.46).

Conversely, let $\xi(x_1, x_2) = \text{const}$ be a solution of equation (4.46). Take

$$y \equiv \xi(x_1, x_2) \qquad (4.49)$$

as a new variable and insert it into equation (4.43). We obtain

$$\left[\dfrac{\frac{\partial y}{\partial x_1}}{\frac{\partial y}{\partial x_2}}\right]^2 + 2\dfrac{a_{12}}{a_{11}}\dfrac{\frac{\partial y}{\partial x_1}}{\frac{\partial y}{\partial x_2}} + \dfrac{a_{22}}{a_{11}} = 0 \qquad (4.50)$$

because of (4.45) and (4.46), so that y is the solution of the characteristic equation (4.42).

By assumption, the equation in question is hyperbolic in the entire region D. Hence the characteristic equation (4.34) has two different real roots, $r_1(x_1, x_2)$ and $r_2(x_1, x_2)$. Thus the characteristic differential equations

$$\frac{dx_2}{dx_1} = r_i(x_{1,2}), \quad i = 1, 2 \qquad (4.51)$$

determine two 1-parametric nontangent families of real integral curves

$$y_1(x_1, x_2) = \text{const}, \quad y_2(x_1, x_2) = \text{const} \qquad (4.52)$$

called the characteristics of the partial differential equation (4.31). Taking y_i as new coordinate lines, we reduce the hyperbolic equation (4.31) to its first canonical form.

It remains only to show that this change of variable is nondegenerate, that is, its Jacobian does not vanish everywhere in D. We have

$$\frac{D(y_1, y_2)}{D(x_1, x_2)} = \left(\frac{\partial y_2}{\partial x_2}\right)^2 \cdot \begin{bmatrix} \frac{\partial y_1}{\partial x_1} & \frac{\partial y_2}{\partial x_1} \\ \frac{\partial y_1}{\partial x_2} & \frac{\partial y_2}{\partial x_2} \end{bmatrix} = \left(\frac{\partial y_2}{\partial x_2}\right)^2 [r_2(x_1, x_2) - r_1(x_1, x_2)] \qquad (4.53)$$

Since $r_1 \neq r_2$ and $\partial y_2/\partial x_2 \neq 0$, $\partial y_2/\partial x_2 \neq \infty$, we see that

$$\frac{D(y_1, y_2)}{D(x_1, x_2)} \neq 0 \qquad (4.54)$$

Q.E.D.[5]

Now consider an *elliptic* equation. In this case the characteristic equation has complex conjugated roots:

$$r_1 = \alpha + i\beta, \quad r_2 = \alpha - i\beta \qquad (4.55)$$

Let us assume temporarily that all the coefficients of the differential form Φ^* and the function F are analytic functions of all their arguments. Then the solution of the equation

$$\Phi^* + F = 0 \qquad (4.56)$$

---

[5] The equality $dy_2/dx_2 = \infty$ would mean that $r_2 = r_1$, which is incompatible with our assumption that the equation is hyperbolic.

may be sought in the complex plane of $z = x_1 + ix_2$. The characteristics curves will be complex conjugate to one another:

$$\xi_k(x_1, x_2) = y_1(x_1, x_2) + (-1)^k iy_2(x_1, x_2), \quad k = 1, 2 \qquad (4.57)$$

Taking them as new independent variables, we obtain the first canonical form of the equation

$$\frac{\partial^2 u}{\partial \xi_1 \partial \xi_2} + F = 0 \qquad (4.58)$$

Passing to the second canonical form by means of the substitution

$$y_1 = \frac{\xi_1 + \xi_2}{2}, y_2 = \frac{\xi_1 - \xi_2}{2i} \qquad (4.59)$$

we have

$$\frac{\partial^2}{\partial y_1^2} u + \frac{d^2}{dy_2^2} u + F = 0 \qquad (4.60)$$

which returns the equation to the real plane. Thus, in order to reduce an equation of elliptic type to canonical form, one must determine solutions

$$y_1(x_1, x_2) = \text{const}, \quad y_2(x_1, x_2) = \text{const} \qquad (4.61)$$

of the differential equations

$$\frac{dx_2}{dx_1} = \mathfrak{R}(x_1, x_2), \quad \frac{dx_2}{dx_1} = \mathfrak{I}(x_1, x_2) \qquad (4.62)$$

where $\mathfrak{R}$ and $\mathfrak{I}$ are the real and imaginary parts of the roots of the characteristic equation (4.34) and take y_1 and y_2 as new independent variables.

Finally, consider the *parabolic* case. In this case there is only one real root $r(x_1, x_2)$ of the characteristic equation (more precisely, the equation has a double root). Therefore we may equate only one coefficient, say a_{22}^*, to zero, taking as a new independent variable the left hand side of a solution

$$\xi(x_1, x_2) = \text{const} \qquad (4.63)$$

of the equation

$$\frac{dx_2}{dx_1} = r(x_1, x_2) \qquad (4.64)$$

We claim that if one takes as a second new independent variable any smooth function of x_2,

$$\eta = \Psi(x_2) \qquad (4.65)$$

such that

$$\frac{d\Psi}{dx_2} \neq 0 \quad \forall (x_1, x_2) \in D \qquad (4.66)$$

then the equation is reduced to canonical form, not containing the mixed second derivative. Indeed, introducing ξ and η as new independent variables, we obtain

$$a_{12}^* = a_{11} \frac{\partial \xi}{\partial x_1} \frac{\partial \eta}{\partial x_1} + a_{12} \left(\frac{\partial \xi}{\partial x_1} \frac{\partial \eta}{\partial x_1} + \frac{\partial \xi}{\partial x_2} \frac{\partial \eta}{\partial x_2} \right) + a_{22} \frac{\partial \xi}{\partial x_2} \frac{\partial \eta}{\partial x_2}$$

$$\equiv \frac{d\Psi}{dx_2} \frac{\partial \xi}{x_2} \cdot [a_{22} - a_{12} \cdot r] = 0 \tag{4.67}$$

Q.E.D.

Note that if F^* is independent of $\partial u / \partial \eta$ then the equation degenerates into an ordinary differential equation, no longer dependent on η except as a parameter. If F^* depends linearly on $\partial u / \partial \eta$ then the equation turns out to be an equation of the form

$$\frac{\partial^2}{\partial x^2} u(x, t) + \lambda \frac{\partial}{\partial t} u + F(x, t, u) = 0 \tag{4.68}$$

that is, the heat-conduction equation with nonlinear (if $\partial F / \partial u \neq 0$) or linear source term and

$$\lambda = \pm 1 \tag{4.69}$$

If $\lambda = -1$ then (4.68) is the usual nonhomogeneous heat-conduction equation, whereas if $\lambda = 1$ it is the heat-conduction equation with reversed time.

3. Canonical form of linear partial differential equations with constant coefficients

Our reduction of an equation to canonical form was carried out using a change of the *independent* variables only. Linear equations with constant coefficients may be further simplified by introducing a new *dependent* variable. Thus, consider a second-order linear equation in canonical form:

$$Lu = \sum_{i=1}^{n} \lambda_i \frac{\partial^2}{\partial x_i^2} u + \sum_{i=1}^{n} b_i \frac{\partial}{\partial x_i} u + cu = f(x_1, \dots, x_n) \tag{4.70}$$

Introduce a new dependent variable v by setting

$$u = v \exp \left(\sum_{i=1}^{n} \alpha_i x_i \right) \tag{4.71}$$

This substitution transforms equation (4.70) into the equation

$$Lv = \sum_{i=1}^{n} \lambda_i \frac{\partial^2}{\partial x_i^2} v + \sum_{i=1}^{n} [2\alpha_i \lambda_i + b_i] \frac{\partial}{\partial x_i} v + \left(c + \sum_{i=1}^{n} \left[\lambda_i \alpha_i^2 + b_i \alpha_i \right] \right) v$$

$$= f(x_1, \dots, x_n) \exp \left(-\sum_{i=1}^{n} \alpha_i x_i \right) \tag{4.72}$$

Suppose equation (4.70) is elliptic. Then

$$\lambda_i = 1, \quad i = 1, 2, \ldots, n \tag{4.73}$$

Take

$$\alpha_i = -\frac{b_i}{2}, \quad i = 1, 2, \ldots, n \tag{4.74}$$

Then equation (4.72) becomes

$$Lv = \sum_{i=1}^{n} \frac{\partial^2}{\partial x_i^2} v + \left(c - \frac{1}{4} \sum_{i=1}^{n} b_i^2 \right) v = f(x_1, \ldots, x_n) \exp\left(\frac{1}{2} \sum_{i=1}^{n} b_i x_i \right) \tag{4.75}$$

If (4.70) is a hyperbolic equation then

$$\lambda_i = 1 \quad \text{if } i = 1, 2, \ldots, n-1, \qquad \lambda_n = -1 \tag{4.76}$$

Hence, by setting

$$\alpha_i = -\frac{b_i}{2}, \quad i = 1, 2, \ldots, n-1, \qquad \alpha_n = \frac{b_n}{2} \tag{4.77}$$

one reduces the equation to

$$Lv = \sum_{i=1}^{n-1} \frac{\partial^2}{\partial x_i^2} v - \frac{\partial^2}{\partial x_n^2} v + \left(c + \frac{1}{4} \sum_{i=1}^{n-1} b_i^2 + \frac{3}{4} b_n \right) v$$

$$= f(x_1, \ldots, x_n) \exp\left(\frac{1}{2} \sum_{i=1}^{n-1} b_i x_i - \frac{1}{2} b_n x_n \right) \tag{4.78}$$

Finally, for a parabolic equation not degenerating into an ordinary differential equation,

$$\lambda_i = 1 \quad \text{if } i = 1, 2, \ldots, n-1, \qquad \lambda_n = 0, \quad b_n \neq 0 \tag{4.79}$$

In this case,[6] setting

$$\alpha_i = \frac{b_i}{2}, \quad i = 1, 2, \ldots, n-1, \qquad \alpha_n = \left(c - \frac{1}{4} \sum_{i=1}^{n-1} b_i x_i^2 \right) \Big/ b_n \tag{4.80}$$

yields

$$Lv = \sum_{i=1}^{n-1} \frac{\partial^2}{\partial x_i^2} v \mp \frac{\partial}{\partial x_n} v = f(x_1, \ldots, x_n) \exp\left(\frac{1}{2} \sum_{i=1}^{n-1} b_i x_i - \alpha_n x_n \right) \tag{4.81}$$

Thus the canonical form of a linear second-order partial differential equation with constant coefficients in $\mathbb{R}_n$, or in $\mathbb{R}_n \times (-\infty < t < \infty)$, is:

- in the elliptic case,

$$\Delta u + cu + f(x) = 0 \tag{4.82}$$

---

[6] In the nondegenerate parabolic case, one can make b_n in (4.75) equal to ± 1 by setting $\tilde{x}_n = x_n/|b_n|$.

- in the hyperbolic case,

$$\Delta u - \frac{\partial^2}{\partial t^2} u + cu + f(x) = 0 \qquad (4.83)$$

- in the nondegenerate parabolic case,

$$\Delta u - \frac{\partial}{\partial t} u - f(x, t) = 0 \quad \text{or} \quad \Delta u + \frac{\partial}{\partial t} u - f(x, t) = 0 \qquad (4.84)$$

Here $x = (x_1, x_2, \ldots, x_n)$ is a point in $\mathbb{R}_n$ and Δ the Laplace operator in $\mathbb{R}_n$,

$$\Delta u = \sum_{i=1}^{n} \frac{\partial^2}{\partial x_i^2} u(x) \qquad (4.85)$$

This explains, from the purely mathematical point of view, why these specific equations are necessarily the most important objects of study in the theory of partial differential equations of mathematical physics. This conclusion is in full accordance with the basic physical facts that steady-state processes are described by elliptic equations, oscillatory processes by hyperbolic equations, and dissipative inertialess processes by nondegenerate parabolic equations.[7]

4. Cauchy problem for partial differential equations with linear principal part. Classification of equations

Another criterion for the classification of second-order partial differential equations with linear principal part is whether or not the Cauchy problem can be solved. We restrict ourselves to consideration of the problem in $\mathbb{R}_2$.[8]

Suppose we are given the equation

$$\sum_{i=1}^{2} \sum_{j=1}^{2} a_{ij} \frac{\partial^2}{\partial x_i \partial x_j} + f\left(x_1, x_2, u, \frac{\partial u}{\partial x_1}, \frac{\partial u}{\partial x_2}\right) = 0 \qquad (4.86)$$

The Cauchy problem may be formulated as follows: Given a smooth curve Γ in $\mathbb{R}_2$ without self-intersections,

$$\Gamma = \{x_1, x_2 : x_i = x_i(s), i = 1, 2; \alpha^* < \alpha \leq s \leq \beta < \beta^*\} \qquad (4.87)$$

---

[7]Equation (4.82), known as the nonhomogeneous Helmholtz equation, is the basic equation of the theory of diffraction. Equation (4.83) is the non–steady-state counterpart of the Helmholtz equation; it may be reduced to the latter by Fourier's method of separation of variables (see Chapter 19). The first of equations (4.84) is the classical equation of thermal conductivity or diffusion, whereas the second is the heat-conduction equation with reversed time. This equation appears naturally in the climatological context (see Chapter 2, Section 12). We restrict our presentation to a study of the Poisson, wave, and heat-conduction equations.

[8]The Cauchy problem in $\mathbb{R}_n$ is considered in Chapter 7, Section 1.

find $u(x_1, x_2)$ in the neighborhood of Γ such that

$$u\,|_\Gamma = u_0(s), \quad \frac{\partial}{\partial\nu}u\,|_\Gamma = u_1(s) \qquad (4.88)$$

Here ν is a vector not tangent to Γ, and $u_0(s)$, $u_1(s)$ are continuous functions of the parameter s.

Note that the data (4.88) are equivalent to data

$$u\,|_\Gamma = u_0(s), \quad \frac{\partial}{\partial x_i}u\,\bigg|_\Gamma = u_i(s), \quad i = 1,2 \qquad (4.89)$$

such that

$$du_0(s) = u_1(s)\,dx_1(s) + u_2(s)\,dx_2(s) \qquad (4.90)$$

Indeed,

$$\frac{\partial}{\partial\nu}u\,|_\Gamma = u_1(s)\cos(x_1\nu) + u_2(s)\cos(x_2\nu) \qquad (4.91)$$

$$\frac{\partial}{\partial s}u\,|_\Gamma = u_1(s)\cos(x_1s) + u_2(s)\cos(x_2s) \qquad (4.92)$$

Here $\cos(x_i\nu)$ and $\cos(x_is)$ are the direction cosines of the vectors ν and s, respectively. Since ν is not a tangent vector, the determinant of this system is not equal to zero:

$$\begin{vmatrix} \cos(x_1\nu) & \cos(x_2\nu) \\ \cos(x_1s) & \cos(x_2s) \end{vmatrix} \neq 0 \qquad (4.93)$$

so that $u_1(s)$ and $u_2(s)$ may be determined uniquely. Thus, we shall consider the Cauchy problem (4.86), (4.89), (4.90).

Let us assume temporarily that $u_i(s)$, $i = 0, 1, 2$, and moreover that all the coefficients of equation (4.86) and its free term are analytic functions of all their arguments. Then one may expect that $u(x_1, x_2)$ can be found by expanding in Taylor series in the neighborhood of the interval (α, β), provided that the Cauchy data (4.89) and equation (4.86) enable one to determine derivatives of u of all orders. We have

$$a_{11}\frac{\partial^2}{\partial x_1^2}u + 2a_{12}\frac{\partial^2}{\partial x_1\partial x_2}u + a_{22}\frac{\partial^2}{\partial x_2^2} = -f$$

$$\frac{dx_1}{ds}\frac{\partial^2}{\partial x_1^2}u + \frac{dx_2}{ds}\frac{\partial^2}{\partial x_1\partial x_2}u = \frac{d}{ds}u_1 \qquad (4.94)$$

$$\frac{dx_1}{ds}\frac{\partial^2}{\partial x_1\partial x_2}u + \frac{dx_2}{ds}\frac{\partial^2}{\partial x_2^2}u = \frac{d}{ds}u_2$$

This system uniquely determines all three derivatives $\partial^2 u/\partial x_1^2$, $\partial^2 u/\partial x_1\partial x_2$, $\partial^2 u/\partial x_2^2$ if its determinant does not vanish:

$$\Delta = \begin{vmatrix} a_{11} & 2a_{12} & a_{22} \\ \dfrac{dx_1}{ds} & \dfrac{dx_2}{ds} & 0 \\ 0 & \dfrac{dx_1}{ds} & \dfrac{dx_2}{ds} \end{vmatrix} \neq 0 \qquad (4.95)$$

It is easy to see that this inequality is a necessary and sufficient condition for all the derivatives of u to be determinable, provided (of course), that the assumption of analyticity holds. Indeed, suppose that all the derivatives

$$v \stackrel{\text{def}}{=} u_{mn} \stackrel{\text{def}}{=} \frac{\partial^{m+n}}{\partial x_1^m \partial x_2^n} u, \quad m + n = k \tag{4.96}$$

have been found, so that v, $\partial v/\partial x_1$ and $\partial v/\partial x_2$ are known along Γ:

$$v\big|_\Gamma = v_0(s), \quad \frac{\partial}{\partial x_1} v = v_1(s), \quad \frac{\partial}{\partial x_2} v = v_2(s) \tag{4.97}$$

Differentiation of equation (4.86) m times with respect to x_1 and n times with respect to x_2 yields

$$a_{11} \frac{\partial^2}{\partial x_1^2} v + 2a_{12} \frac{\partial^2}{\partial x_1 \partial x_2} v + a_{22} \frac{\partial^2}{\partial x_2^2} v = -\tilde{f} \tag{4.98}$$

At the same time, equalities (4.97) yield

$$\frac{dx_1}{ds} \frac{\partial^2}{\partial x_1^2} v + \frac{dx_2}{ds} \frac{\partial^2}{\partial x_1 \partial x_2} v = \frac{d}{ds} v_1(s) \tag{4.99}$$

$$\frac{dx_1}{ds} \frac{\partial^2}{\partial x_1 \partial x_2} v + \frac{dx_2}{ds} \frac{\partial^2}{\partial x_1^2} v = \frac{d}{ds} v_2(s) \tag{4.100}$$

Hence all three second derivatives of v are determined by a system with the same determinant as the system (4.33). **Q.E.D.**

Thus a solution of the Cauchy problem (4.86)–(4.90) does not exist or is not unique if

$$\Delta\big|_\Gamma = 0 \tag{4.101}$$

On the other hand, a unique solution of the Cauchy problem (4.86)–(4.90) exists formally if

$$\Delta\big|_\Gamma \neq 0 \tag{4.102}$$

Note that we may assume without loss of generality that

$$\frac{dx_1}{ds} \neq 0 \quad \forall s \in [\alpha^*, \beta^*] \tag{4.103}$$

Indeed, the equality

$$\left(\frac{dx_1}{ds} \right)^2 + \left(\frac{dx_2}{ds} \right)^2 = 0 \tag{4.104}$$

is impossible, since it would imply that the curve Γ degenerated to a point. Hence

$$\frac{dx_1}{ds} = 0 \Rightarrow \frac{dx_2}{ds} \neq 0 \tag{4.105}$$

so that, by renaming the arguments, we obtain (4.103).

Let

$$r = \frac{\frac{dx_2}{ds}}{\frac{dx_1}{ds}} \equiv \frac{dx_2}{dx_1} \tag{4.106}$$

Taking (4.103) into account, we find that

$$\Delta = \frac{dx_1}{ds} D \tag{4.107}$$

where

$$D = \begin{vmatrix} a_{11} & 2a_{12} & a_{22} \\ 1 & r & 0 \\ 0 & 1 & r \end{vmatrix} \tag{4.108}$$

D is called the *characteristic determinant* and the equation

$$D \equiv a_{11} r^2 - 2a_{12} r + a_{22} = 0 \tag{4.109}$$

or, what is the same,

$$D \equiv a_{11} \left(\frac{dx_2}{dx_1} \right)^2 - 2a_{12} \frac{dx_2}{dx_1} + a_{22} = 0 \tag{4.110}$$

is called the *characteristic equation* and its integral curves the *characteristics curves*. Thus the Cauchy problem cannot be solved, or has no unique solution, if the Cauchy data are prescribed on characteristic curves.

Note that the characteristic equation (4.110) is exactly the same as the characteristic equation (4.48) (up to division by a_{11} in (4.48)) that we derived by invariance arguments. Since the reduction of a second-order equation to canonical form is based only on taking the characteristic curves as coordinate lines in a transformed coordinate system, we see that our consideration of the Cauchy problem may serve as the starting step of such reduction, or as the basis for classification of equations.

5. Cauchy problem for system of two quasilinear first-order partial differential equations with two independent variables; concept of characteristics[9]

Recall that any partial differential equation of order n may be transformed into a system of n first-order partial differential equations and vice versa (see Chapter 2, Section 10). Hence it is natural to approach directly the problem of classification and reduction to a canonical form for a system of first-order equations, even more so since it may be more convenient to consider a boundary-value problem formulated for a system of equations than for a single higher-order equation. (See, e.g., the formulation of the problem

---

[9]Without essential changes one can consider a system of any finite number of equations. Attention is restricted to two equations for the sake of simplicity only.

on the dynamic of sorption in Chapter 2, Section 8 and its treatment in Chapter 6, Section 3.) Thus, consider the following Cauchy problem. Given a smooth curve

$$\Gamma = \{x = x(s), \ y = y(s); \ \alpha^* \le s \le \beta^*\} \in \mathbb{R}_2 \tag{4.111}$$

find $u(x,y)$ and $v(x,y)$ in a neighborhood of Γ such that

$$a_1 \frac{\partial}{\partial x} u + b_1 \frac{\partial}{\partial y} u + c_1 \frac{\partial}{\partial x} v + d_1 \frac{\partial}{\partial y} v = f_1(x, y, u, v) \tag{4.112}$$

$$a_2 \frac{\partial}{\partial x} u + b_1 \frac{\partial}{\partial y} u + c_2 \frac{\partial}{\partial x} v + d_2 \frac{\partial}{\partial y} v = f_2(x, y, u, v) \tag{4.113}$$

$$u\big|_\Gamma = u_0(s), \quad v\big|_\Gamma = v_0(s), \quad \alpha^* < \alpha \le s \le \beta < \beta^* \tag{4.114}$$

By analogy with the case of second-order equations, the curve Γ is called a characteristic curve if the solution of this Cauchy problem does not exist or exists but is not unique. This means that the characteristics of system (4.112)–(4.113) are integral curves of the characteristic equation

$$\Delta = \begin{vmatrix} a_1 & b_1 & c_1 & d_1 \\ a_2 & b_2 & c_2 & d_2 \\ \dfrac{dx}{ds} & \dfrac{dy}{ds} & 0 & 0 \\ 0 & 0 & \dfrac{dx}{ds} & \dfrac{dy}{ds} \end{vmatrix} = 0 \tag{4.115}$$

Assume that

$$\begin{vmatrix} a_1 & b_1 \\ a_2 & b_2 \end{vmatrix} \ne 0 \frac{dx}{ds} \ne 0 \tag{4.116}$$

Then, taking

$$\rho = \frac{dy}{dx} \tag{4.117}$$

we find that (4.115) is a quadratic equation

$$\begin{vmatrix} a_1 & a_2 \\ c_1 & c_2 \end{vmatrix} \rho^2 - \left(\begin{vmatrix} a_1 & a_2 \\ d_1 & d_2 \end{vmatrix} + \begin{vmatrix} b_1 & b_2 \\ c_1 & c_2 \end{vmatrix} \right) \rho + \begin{vmatrix} b_1 & b_2 \\ d_1 & d_2 \end{vmatrix} = 0 \tag{4.118}$$

If the discriminant

$$\delta = \left(\begin{vmatrix} a_1 & a_2 \\ d_1 & d_2 \end{vmatrix} + \begin{vmatrix} b_1 & b_2 \\ c_1 & c_2 \end{vmatrix} \right)^2 - 4 \begin{vmatrix} a_1 & a_2 \\ c_1 & c_2 \end{vmatrix} \cdot \begin{vmatrix} b_1 & b_2 \\ d_1 & d_2 \end{vmatrix} \tag{4.119}$$

of the characteristic equation (4.118) is positive, this equation determines two families of reciprocally nontangent integral curves, also called characteristic curves. In this case system (4.112)–(4.113) is said to be hyperbolic. If $\delta < 0$ there exist two families of imaginary characteristic lines, and then the system is said to be elliptic. Finally, if $\delta = 0$, the characteristic equation has a double root and the system is said to be parabolic. Just as in the case of second-order equations, system (4.112) and (4.113) (in the hyperbolic and elliptic case) may be transformed, by introducing the characteristics as new coordinate lines, to a form reducible through a linear transformation to the canonical form in which each equation contains derivatives with respect to only one independent variable.

Examples

1. Consider system of equations describing the dynamics of sorption in convective transfer of a sorptive through a porous adsorbent (see Chapter 2, Section 8):

$$\frac{\partial}{\partial t}a = \beta\left(c - \frac{a}{\Gamma}\right), \qquad m\frac{\partial}{\partial t}c + s\frac{\partial a}{\partial t} + v\frac{\partial}{\partial x}c = 0 \qquad (4.120)$$

Recall that here a is the sorption per unit of specific external surface of grains of a porous adsorbent, S the specific surface of the latter,[10] c concentration of sorptive per unit volume of percolating solution, v the percolation rate, m the porosity and Γ Henry's coefficient. The characteristic determinant is

$$\Delta = \begin{vmatrix} m & v & s & 0 \\ 0 & 0 & 1 & 0 \\ \dfrac{dt}{ds} & \dfrac{dx}{ds} & 0 & 0 \\ 0 & 0 & \dfrac{dt}{ds} & \dfrac{dx}{ds} \end{vmatrix} \qquad (4.121)$$

so the characteristic equation is

$$\frac{dx}{ds}\left(m\frac{dx}{ds} - v\frac{dt}{ds}\right) = 0 \qquad (4.122)$$

Hence there are two families of real characteristics:

$$x = \text{const}, \qquad t - \frac{m}{v}x = \text{const} \qquad (4.123)$$

and system (4.120) is the hyperbolic.

The change of independent variables

$$\xi = x, \qquad \tau = t - \frac{m}{v}x \qquad (4.124)$$

reduces system (4.120) to canonical form

$$\frac{\partial c}{\partial \xi} + \frac{s\beta}{v}\left(c - \frac{a}{\Gamma}\right) = 0, \qquad \frac{\partial a}{\partial \tau} = \beta\left(c - \frac{a}{\Gamma}\right) \qquad (4.125)$$

---

[10] Recall that "specific surface" is the surface area of grains per unit volume of a porous material.

Let us eliminate a from this system. We obtain

$$\frac{\partial^2 c}{\partial \xi \partial \tau} + \frac{s\beta}{v} \cdot \frac{\partial c}{\partial \tau} + \frac{\beta}{\Gamma} \cdot \frac{\partial c}{\partial \xi} = 0 \qquad (4.126)$$

that is, a second-order hyperbolic equation in its first canonical form. In this case there is an easy physical interpretation of our passage to characteristic variables as new independent variables. Indeed, consider a semi-infinite layer of porous adsorbent, with the porous space initially saturated by a pure solvent (e.g. water). Assume that at the end $x = 0$ the adsorbent is in contact with a solution of some sorptive whose adsorption follows the linear Henry isotherm. At time $t = 0$ this solution begins to be pumped into the porous space of the adsorbent at a constant percolation rate v. Then the linear velocity of the solution will be v/m, which means that the solution entering the layer of adsorbent at $x = 0$ will reach a point $x > 0$ at time $t = mx/v$, so that the "characteristic time" τ is reckoned at any point x from the time at which adsorption begins at x.

2. Consider Cauchy–Riemann equations, the basis of the theory of analytic functions:

$$\frac{\partial u}{\partial x} - \frac{\partial v}{\partial y} = 0, \quad \frac{\partial u}{\partial y} + \frac{\partial v}{\partial x} = 0 \qquad (4.127)$$

The characteristic determinant is

$$\Delta = \begin{vmatrix} 1 & 0 & 0 & -1 \\ 0 & 1 & 1 & 0 \\ \dfrac{dy}{ds} & \dfrac{dx}{ds} & 0 & 0 \\ 0 & 0 & \dfrac{dy}{ds} & \dfrac{dx}{ds} \end{vmatrix} \qquad (4.128)$$

so the characteristic equation is

$$\left(\frac{dx}{ds}\right)^2 + \left(\frac{dy}{ds}\right)^2 = 0 \qquad (4.129)$$

which has only imaginary roots. Hence Cauchy–Riemann equations form an elliptic system.

3. Consider the system of telegraphic equations, with the wire capacity neglected (see Chapter 2, Problem 2.7.1):

$$L\frac{\partial i}{\partial t} + \frac{\partial v}{\partial x} = Ri, \quad \frac{\partial i}{\partial x} = -Gv \qquad (4.130)$$

The characteristic determinant is

$$\Delta = \begin{vmatrix} L & 0 & 0 & 1 \\ 0 & 1 & 0 & 0 \\ \dfrac{dt}{ds} & \dfrac{dx}{ds} & 0 & 0 \\ 0 & 0 & \dfrac{dt}{ds} & \dfrac{dx}{ds} \end{vmatrix} \equiv \left(\frac{dt}{ds}\right)^2 \qquad (4.131)$$

so the characteristic equation $\Delta = 0$ admits a double root:

$$\frac{dt}{ds} = 0 \tag{4.132}$$

Hence the system of telegraphic equations, with wire capacity neglected, is parabolic.

6. Characteristics as curves of weak discontinuity of second or higher order

Our next approach to the theory of characteristics is very important from the physical point of view. Whereas the considerations of Section 4 were based on the assumption that the carrier of the Cauchy data and the equation are analytic in the neighborhood of the carrier, we now turn our attention to discontinuous solutions of second-order linear equations. Consider an equation

$$Lu = a_{11}\frac{\partial^2}{\partial x_1^2}u + 2a_{12}\frac{\partial^2}{\partial x_1 \partial x_2}u + a_{22}\frac{\partial^2}{\partial x_1^2}u$$

$$+ a_1\frac{\partial}{\partial x_1}u + a_2\frac{\partial}{\partial x_2}u + cu = f(x_1, x_2) \tag{4.133}$$

where a_{ij} and a_i, $i, j = 1, 2$, are given functions of x_1 and x_2. In what follows we use the same notations as in Chapter 2. Namely, let D be a given region bounded by a contour Γ and let L be a smooth curve, without self-intersections, dividing D into two parts D^+ and D^-, so that

$$L = \bar{D}^+ \cap \bar{D}^- \tag{4.134}$$

Let the normal $\mathbf{n}$ to L be directed into D^+. Let

$$p \in L, \quad p^+ \in D^+, \quad p^- \in D^- \tag{4.135}$$

Assume that p^+ and p^- tend to p along a nontangential path, and denote

$$\lim_{p^+ \to p} F(p^+) = F^+(p), \quad \lim_{p^- \to p} F(p^-) = F^-(p), \quad [F(p)] = F^+(p) - F^-(p) \tag{4.136}$$

whatever the function F.

Now consider the class $C_L^k(D)$, $k \geq 2$, of functions $\Psi(x_1, x_2)$ defined as follows: $\Psi \in C_L^k(D)$ if Ψ is defined in D as a function continuous together with all its derivatives up to order $k - 1$ and all its kth derivatives are continuous in the closures of D^+ and D^-. Let

$$x_i = x_i(s), \quad i = 1, 2 \tag{4.137}$$

be the equation of the arc L, $F \in C_L^k(D)$, and

$$\Psi_{mn}^k(x_1, x_2) = \frac{\partial^{k-1}F}{\partial x_1^m \partial x_2^n}, \quad m + n = k - 1 \tag{4.138}$$

Then for any $p = [x_1(s), x_2(s)]$ there exist continuous $d\Psi_{mn}^k(p)/ds$, so that

$$
\begin{aligned}
\frac{d}{ds}\Psi_{mn}^k(p) &= \left(\frac{\partial}{\partial x_1}\Psi_{mn}^k(p)\right)^+ \frac{dx_1}{ds} + \left(\frac{\partial}{\partial x_2}\Psi_{mn}^k(p)\right)^+ \frac{dx_2}{ds} \\
&= \left(\frac{\partial}{\partial x_1}\Psi_{mn}^k(p)\right)^- \frac{dx_1}{ds} + \left(\frac{\partial}{\partial x_2}\Psi_{mn}^k(p)\right)^- \frac{dx_2}{ds}
\end{aligned}
\tag{4.139}
$$

Definition 4.6.1. The curve L is known as a curve of weak discontinuity of the kth order of the function $F(x_2) \in C_k(L)$ if for at least one m, $0 \leq m \leq k-1$, at least one of the jumps $[\partial\Psi_{mn}^k(p)/\partial x_1]$, $[\partial\Psi_{mn}^k(p)/\partial x_2]$ is not equal to zero.

Let $C^k(D)$ be the class of all functions defined in D as continuous together with all their derivatives up to the kth order. Assume that all the coefficients of equation (4.133) and its free term f satisfy conditions

$$
a_{ij} \in C^k(D), \quad i, j = 1, 2, \qquad a_i \in C^{k-1}(D), \quad c, \ f \in C^{k-2}(D)
\tag{4.140}
$$

Theorem 4.6.1. *A curve L may be a curve of weak discontinuity of order k $(k \geq 2)$ of a solution $u(x,t)$ of equation (4.133) only if L is a characteristic curve of the equation.*

PROOF. Let L be a curve of weak discontinuity of order k $(k \geq 2)$ of equation (4.133), and denote

$$
v \stackrel{\text{def}}{=} \frac{\partial^{m+n}u}{\partial x_1^m \partial x_2^n}, \quad m+n = k-2
\tag{4.141}
$$

Differentiating u m times with respect to x_1 and n times with respect to x_2, we obtain

$$
a_{11}\frac{\partial^2}{\partial x_1^2}v + 2a_{12}\frac{\partial^2}{\partial x_1 \partial x_2}v + a_{22}\frac{\partial^2}{\partial x_1^2}v = \Phi(x_1, x_2)
\tag{4.142}
$$

where

$$
\Phi = \Phi\left(x_1, x_2 \mid u, \frac{\partial u}{\partial x_1}, \frac{\partial}{\partial x_u}, \ldots, v, \frac{\partial v}{\partial x_1}, \frac{\partial v}{\partial x_2}, f, \frac{\partial f}{\partial x_1}, \frac{\partial f}{\partial x_2}, \ldots, \varphi, \frac{\partial\varphi}{\partial x_1}, \frac{\partial\varphi}{\partial x_2}\right)
$$

$$
\varphi = \frac{\partial^{m+n}u}{\partial x_1^m \partial x_2^n}
\tag{4.143}
$$

Note that by assumption F, as well as all coefficients on the left-hand side of (4.142), are continuous in a neighborhood of L.

Let $p \in L$. Then by (4.142), (()4.135), and (()4.136),

$$
a_{11}\left[\frac{\partial^2}{\partial x_1^2}v(p)\right] + 2a_{12}\left[\frac{\partial^2}{\partial x_1 \partial x_2}v(p)\right] + a_{22}\left[\frac{\partial^2}{\partial x_1^2}v(p)\right] = 0
\tag{4.144}
$$

On the other hand, $\partial v/\partial x_1$ and $\partial v/\partial x_2$ are continuous in the neighborhood of L. Hence

$$\frac{d}{ds}\left[\frac{\partial}{\partial x_1}v(p)\right] = \left[\frac{\partial^2}{\partial x_1^2}v(p)\right]\frac{d}{ds}x_1 + \left[\frac{\partial^2}{\partial x_1\partial x_2}v(p)\right]\frac{dx_2}{ds} = 0 \qquad (4.145)$$

$$\frac{d}{ds}\left[\frac{\partial}{\partial x_2}v(p)\right] = \left[\frac{\partial^2}{\partial x_1\partial x_2}v(p)\right]\frac{d}{ds}x_1 + \left[\frac{\partial^2}{\partial x_2^2}v(p)\right]\frac{d}{ds}x_2 = 0 \qquad (4.146)$$

The determinant of system (4.144)–(4.146) is the characteristic determinant

$$\Delta = \begin{vmatrix} a_{11} & 2a_{12} & a_{22} \\ \dfrac{d}{ds}x_1 & \dfrac{d}{dx}x_2 & 0 \\ 0 & \dfrac{d}{ds}x_1 & \dfrac{d}{ds}x_2 \end{vmatrix} \qquad (4.147)$$

Since by assumption L is the line of discontinuity of u of order k, at least one of the jumps entering into system (4.144)–(4.146) must be different from zero. This means that

$$\Delta = 0 \qquad (4.148)$$

which means that L is a characteristic curve of equation (4.133). **Q.E.D.**

7. Riemann's formula. Characteristics as curves of weak discontinuity of first order or as curves of strong discontinuity

Let us use the definitions (4.133)–(4.137) with $k = 0$ or $k = 1$. Let $F(p)$, $p = (x_1, x_2)$, belong to $C_L^k(D)$, and let

$$F_i(p) \stackrel{\text{def}}{=} \frac{\partial}{\partial x_i}F(p), \quad p = (x_1, x_2) \in D^+ \cup D^-, \quad i = 1, 2 \qquad (4.149)$$

so that for every $p \in L$ there exist continuous $F^+(p)$, $F^-(p)$ if $k = 0$, or $F_i^+(p)$, $F_i^-(p)$ if $k = 1$. Then we use the following definitions.

Definition 4.7.1. L is a curve of *weak discontinuity* of F of first order if at least one of the jumps $[F_i(p)]$, $i = 1$, 2, is not equal to zero, but $F(p)$ remains continuous.

Definition 4.7.2. L is a curve of *strong discontinuity* of F if the jump $[F(p)]$ is not equal to zero.

Let $u(x_1, x_2)$ be a solution of the equation

$$Lu = a_{11}\frac{\partial^2}{\partial x_1^2}u + 2a_{12}\frac{\partial^2}{\partial x_1\partial x_2}u + a_{22}\frac{\partial^2}{\partial x_2^2}u + a_1\frac{\partial}{\partial x_1}u + a_2\frac{\partial}{\partial x_2}u + cu$$

$$= f(x_1, x_2) \quad \forall p \in D^+ \cup D^- \qquad (4.150)$$

and let L be a curve of weak discontinuity of u of first order or a curve of strong discontinuity. In this case one cannot apply the technique of

Section 6 to prove that L is a characteristic curve, since equation (4.150) is not defined at the points of L. In this connection we introduce the concept of a generalized solution of equation (4.150). With this in mind, we first derive the following identities.

Let $v(x_1, x_2)$, $u(x_1, x_2)$ be twice continuously differentiable, $a_{ij}(x_1, x_2)$, $i, j = 1, 2$, continuously differentiable, and $a_i(x_1, x_2)$, $c(x_1, x_2)$ continuous functions with respect to both their arguments x_1 and x_2, $i, j = 1, 2$. Then the following identities hold:

$$va_{11}\frac{\partial^2}{\partial x_1^2}u = \frac{\partial}{\partial x_1}\left(va_{11}\frac{\partial}{\partial x_1}u\right) - \frac{\partial}{\partial x_1}\left(u\frac{\partial}{\partial x_1}(a_{11}v)\right)$$
$$+ u\frac{\partial^2}{\partial x_1^2}(va_{11}) \tag{4.151}$$

$$va_{12}\frac{\partial^2}{\partial x_1 \partial x}u = \frac{\partial}{\partial x_1}\left(va_{12}\frac{\partial}{\partial x_2}u\right) - \frac{\partial}{\partial x_2}\left(u\frac{\partial}{\partial x_1}(a_{12}v)\right)$$
$$+ u\frac{\partial^2}{\partial x_1 \partial x_2}(va_{12}) \tag{4.152}$$

$$va_{12}\frac{\partial^2}{\partial x_1 \partial x_2}u = \frac{\partial}{\partial x_2}\left(va_{12}\frac{\partial}{\partial x_1}u\right) - \frac{\partial}{\partial x_1}\left(u\frac{\partial}{\partial x_2}(a_{12}v)\right)$$
$$+ u\frac{\partial^2}{\partial x_1 \partial x_2}(va_{12}) \tag{4.153}$$

$$va_{22}\frac{\partial^2}{\partial x_2^2}u = \frac{\partial}{\partial x_2}\left(va_{22}\frac{\partial}{\partial x_2}u\right) - \frac{\partial}{\partial x_2}\left(u\frac{\partial}{\partial x_2}(a_{22}v)\right)$$
$$+ u\frac{\partial^2}{\partial x_2^2}(va_{22}) \tag{4.154}$$

$$va_1\frac{\partial}{\partial x_1}u = \frac{\partial}{\partial x_1}(a_1 vu) - u\frac{\partial}{\partial x_1}(a_1 v) \tag{4.155}$$

$$va_2\frac{\partial}{\partial x_2}u = \frac{\partial}{\partial x_2}(a_2 vu) - u\frac{\partial}{\partial x_2}(a_2 v) \tag{4.156}$$

$$vcu = cvu \tag{4.157}$$

Summing identities (4.151)–(4.157) we obtain Riemann's identity,

$$vLu = uMv + \frac{\partial}{\partial x_2}H - \frac{\partial}{\partial x_1}K \tag{4.158}$$

where M is the operator adjoined to L:

$$Mv = \frac{\partial^2}{\partial x_1^2}(a_{11}v) + 2\frac{\partial^2}{\partial x_1 \partial x_2}(a_{12}v) + \frac{\partial^2}{\partial x_2^2}(a_{22}v)$$
$$- \frac{\partial}{\partial x_1}(a_1 v) - \frac{\partial}{\partial x_2}(a_2 v) + cv \quad \forall p \in D^+ \cup D^- \tag{4.159}$$

and H and K are bilinear forms

$$H(u,v) = v\left(a_{12}\frac{\partial u}{\partial x_1} + a_{22}\frac{\partial u}{\partial x_2} + a_2 u\right)$$
$$- u\left(\frac{\partial}{\partial x_1}(a_{12}v) + \frac{\partial}{\partial x_2}(a_{22}v)\right) \qquad (4.160)$$

$$K(u,v) = -v\left(a_{11}\frac{\partial u}{\partial x_1} + a_{12}\frac{\partial u}{\partial x_2} + a_1 u\right)$$
$$+ u\left(\frac{\partial}{\partial x_1}(a_{11}v) + \frac{\partial}{\partial x_2}(a_{12}v)\right) \qquad (4.161)$$

Let D^* be an arbitrary subregion of D bounded by a piecewise smooth contour Γ. Assume first that u and v are twice continuously differentiable functions. Then integration over D^* yields

$$\int_{D^*}[vLu - uMv]dx_1dx_2 = \int_{D^*}\left(\frac{\partial H}{\partial x_2} - \frac{\partial K}{\partial x_1}\right)dx_1dx_2 \qquad (4.162)$$

Hence, by Green's formula,

$$\int_{D^*}[vLu - uMv]dx_1dx_2 = \int_{\Gamma}Hdx_1 + Kdx_2 \qquad (4.163)$$

where the contour Γ is described in the positive sense (i.e., counterclockwise).

Riemann's formula (4.162) has been derived for any functions u and v that are twice continuously differentiable in the closure of the region D^*. This formula is the basis for our definition of a generalized solution of the equation

$$Lu(x_1, x_2) - f(x_1, x_2) = 0 \qquad (4.164)$$

Definition 4.7.3. A function $u(x_1, x_2)$ is a generalized solution of equation (4.164) in the region D if it has the following properties.

(a) u is continuous in D together with both its first partial derivatives, everywhere except for a finite number of piecewise smooth arcs L_k, $k = 1, 2, \ldots, m$, intersecting one another in at most finitely many points.

(b) u, $\partial u/\partial x_1$, and $\partial u/\partial x_2$ have continuous one-sided limits u^+, $(\partial u/\partial x_1)^+$, $(\partial u/\partial x_2)^+$, and u^-, $(\partial u/\partial x_1)^-$, $(\partial u/\partial x_2)^-$ at each arc L_k.

(c) Riemann's identity (4.162) is valid for any piecewise smooth contour $\Gamma \in D^* \in D$ and for any continuously differentiable function v in D^*, so that

$$\int_{D^*}[vLu - uMv]dx_1dx_2 = \int_{\Gamma}Hdx_1 + Kdx_2 \qquad (4.165)$$

Theorem 4.7.2. Let $u(x_1, x_2)$ be a generalized solution of equation (4.164) and L a curve of first-order weak or strong discontinuity of u. Then L is a characteristic curve of the equation.

PROOF. Let $D^* \subset D$ be a subregion intersected by one of the curves, say L, of first-order weak or strong discontinuity of u and let Γ be the piecewise smooth contour bounding D^*. Assume that D^* is so small that L is the only curve of discontinuity intersecting it, and let M and N be the points of intersection of L and Γ. By contracting Γ to L, we obtain

$$\int_{MN} \left(H \frac{\partial x_1}{\partial s} + K \frac{\partial x_2}{\partial s} \right)^+ ds + \int_{NM} \left(H \frac{\partial x_1}{\partial s} + K \frac{\partial x_2}{\partial s} \right)^- ds = 0 \quad (4.166)$$

which yields

$$\int_{MN} \left([H] \frac{\partial x_1}{\partial s} + [K] \frac{\partial x_2}{\partial s} \right) ds = 0 \quad (4.167)$$

By condition (b), the one-sided limits of u and its first-order derivatives are continuous along L. Since the arc MN is arbitrarily small, it follows from (4.167) that

$$[H] \frac{\partial x_1}{\partial s} + [K] \frac{\partial x_2}{\partial s} \bigg|_L \equiv 0 \quad (4.168)$$

Let us first suppose that L is a curve of weak discontinuity of u of first order, so that

$$[u] \big|_L = 0 \quad (4.169)$$

Take

$$v \equiv 1 \quad (4.170)$$

This choice of v turns the expressions of forms H and K into

$$H(u,v) = \left(a_{12} \frac{\partial u}{\partial x_1} + a_{22} \frac{\partial u}{\partial x_2} + a_2 u \right) - u \left(\frac{\partial}{\partial x_1}(a_{12}) + \frac{\partial}{\partial x_2}(a_{22}) \right) \quad (4.171)$$

$$K(u,v) = -\left(a_{11} \frac{\partial u}{\partial x_1} + a_{12} \frac{\partial u}{\partial x_2} + a_1 u \right) + u \left(\frac{\partial}{\partial x_1}(a_{11}) + \frac{\partial}{\partial x_2}(a_{12}) \right) \quad (4.172)$$

Since a_{ij}, i, $j = 1$, 2, are continuously differentiable by assumption, it follows from (4.171) and (4.172) that

$$[H] = a_{12} \left[\frac{\partial u}{\partial x_1} \right] + a_{22} \left[\frac{\partial u}{\partial x_2} \right], \quad [K] = -a_{11} \left[\frac{\partial u}{\partial x_1} \right] - a_{12} \left[\frac{\partial u}{\partial x_2} \right] \quad (4.173)$$

Hence (4.168) becomes

$$\left(a_{12} \frac{dx_1}{ds} - a_{11} \frac{dx_2}{ds} \right) \left[\frac{\partial u}{\partial x_1} \right] + \left(a_{22} \frac{dx_1}{ds} - a_{12} \frac{dx_2}{ds} \right) \left[\frac{\partial u}{\partial x_2} \right] = 0 \quad (4.174)$$

By assumption, u^+ and u^- are continuously differentiable along L and $[u] \big|_L = 0$. Hence $du^+/ds \big|_L = du^-/ds \big|_L$, which implies that

$$\left[\frac{\partial u}{\partial x_1} \right] \frac{dx_1}{ds} \bigg|_L + \left[\frac{\partial u}{\partial x_2} \right] \frac{dx_2}{ds} \bigg|_L = 0 \quad (4.175)$$

By assumption, at least one of the jumps $[\partial u/\partial x_1]$, $[\partial u/\partial x_2]$ is not zero. Hence the determinant of system (4.174), (4.175) is equal to zero:

$$-\Delta = \begin{vmatrix} a_{12}\dfrac{dx_1}{ds} - a_{11}\dfrac{dx_2}{ds} & a_{22}\dfrac{dx_1}{ds} - a_{12}\dfrac{dx_2}{ds} \\[2mm] \dfrac{dx_1}{ds} & \dfrac{dx_2}{ds} \end{vmatrix} = 0 \tag{4.176}$$

or, what is the same,

$$\Delta = \begin{vmatrix} a_{11} & 2a_{12} & a_{22} \\[2mm] \dfrac{dx_1}{ds} & \dfrac{dx_2}{ds} & 0 \\[2mm] 0 & \dfrac{dx_1}{ds} & \dfrac{dx_2}{ds} \end{vmatrix} = 0 \tag{4.177}$$

But this is the characteristic equation. Hence line L is the characteristic curve of equation (4.150).

Now let L be a curve of strong discontinuity of u. In this case we let v be a function, twice continuously differentiable in the neighborhood of L, identically equal to zero along L, and such that at least one of its derivatives does not vanish along L. Then, by the definitions of H and K,

$$H(u,v) = -u\left(a_{12}\frac{\partial v}{\partial x_1} + a_{22}\frac{\partial v}{\partial x_2}\right), \quad K(u,v) = u\left(a_{11}\frac{\partial v}{\partial x_1} + a_{12}\frac{\partial v}{\partial x_2}\right) \tag{4.178}$$

Hence

$$[H(u,v)] = -[u]\left(a_{12}\frac{\partial v}{\partial x_1} + a_{22}\frac{\partial v}{\partial x_2}\right), \quad [K(u,v)] = [u]\left(a_{11}\frac{\partial v}{\partial x_1} + a_{12}\frac{\partial v}{\partial x_2}\right) \tag{4.179}$$

This and (4.168) show that, along L,

$$\left(a_{11}\frac{dx_2}{ds} - a_{12}\frac{dx_1}{ds}\right)\frac{\partial v}{\partial x_1} + \left(a_{12}\frac{dx_2}{ds} - a_{22}\frac{dx_1}{ds}\right)\frac{\partial v}{\partial x_2} = 0 \tag{4.180}$$

since by assumption $[u] \neq 0$.

Because (by definition) $v \equiv 0$ along L, we also have

$$\frac{\partial v}{\partial x_1}\frac{dx_1}{ds} + \frac{\partial v}{\partial x_2}\frac{dx_2}{ds} = 0 \tag{4.181}$$

The determinant of the system (4.180) and (4.181) is

$$\Delta = \begin{vmatrix} a_{11}\dfrac{dx_2}{ds} - a_{12}\dfrac{dx_1}{ds} & a_{12}\dfrac{dx_2}{ds} - a_{22}\dfrac{dx_1}{ds} \\[2mm] \dfrac{dx_1}{ds} & \dfrac{dx_2}{ds} \end{vmatrix} \tag{4.182}$$

or, what is the same,

$$\Delta = \begin{vmatrix} a_{11} & 2a_{12} & a_{22} \\[2mm] \dfrac{dx_1}{ds} & \dfrac{dx_2}{ds} & 0 \\[2mm] 0 & \dfrac{dx_1}{ds} & \dfrac{dx_2}{ds} \end{vmatrix} \tag{4.183}$$

By assumption, at least one of the derivatives $\partial v/\partial x_i$, $i = 1$, 2, is not equal to zero, so that

$$\Delta = 0 \qquad (4.184)$$

which again shows that L is the characteristic curve of equation (4.150). Thus, if L is a curve of the weak discontinuity of first order or of strong discontinuity of a generalized solution $u(x_1, x_2)$ of equation (4.150), then it is a characteristic curve of that equation. **Q.E.D.**

Corollary 1. *Theorems 4.6.1 and 4.7.1 show that no solution of any linear elliptic equation can have discontinuities inside its open domain of definition.*

Recall that harmonic functions in $\mathbb{R}_2$ may be treated as the real or imaginary part of an analytic function, so that it can be expanded into a convergent double Taylor series in x_1 and x_2 variables. Hence 2-dimensional harmonic functions are analytic.[11]

Corollary 2. *A nondegenerate parabolic equation with linear principal part*

$$\frac{\partial^2 u}{\partial x^2} \pm \frac{\partial u}{\partial t} + f(x, t, u) = 0 \qquad (4.185)$$

has a double characteristic line

$$t = \text{const} \qquad (4.186)$$

Hence discontinuities of its solutions (weak or strong) cannot propagate in time.

Problems

P4.1.–2. Determine the types of the following equations, and reduce them to canonical form.

$$\frac{\partial^2}{\partial x^2} u + 2e^{x+y} \frac{\partial^2}{\partial x \partial y} u + e^{2y} \frac{\partial^2}{\partial y^2} u = 0$$

$$\frac{\partial^2}{\partial x^2} u + 2 \frac{\partial^2}{\partial x \partial y} u + 4 \frac{\partial^2}{\partial x \partial z} \cdot u + 5 \frac{\partial^2}{\partial z^2} u + \frac{\partial}{\partial x} u + 2 \frac{\partial}{\partial y} u = 0$$

P4.3. Determine the type of a system of telegraphic equations (see Problem 2.7.1) according as which of the terms can be ignored owing to the smallness of its coefficients. In the hyperbolic case, reduce the system to canonical form.

---

[11] The solution of any linear elliptic equation with analytic coefficients (and even of a very broad class of nonlinear systems of elliptic equations) is an analytic function [4];[48];[56].

P4.6. Suppose that the following boundary-value problem has been solved numerically: Find $u(x,t)$ satisfying the conditions

$$\frac{\partial^2}{\partial x^2}u + \sin x = \frac{\partial}{\partial t}u = 0 \quad \forall x > 0$$

$$u(x,0) = |x-1| \quad \forall x \geq 0, u(0,t) = 1 \quad \forall t > 0, \ |u| < \infty$$

The first-order derivative was computed by using central differences; as a result it was found that

$$\frac{\partial}{\partial x}u < 0.532 \quad \text{for } x < 0.75t$$

$$\frac{\partial}{\partial x}u > 0.533 \quad \text{for } x > 0.75t$$

Estimate the degree of accuracy of the computation.

P4.7. The following problem was solved:

$$\frac{\partial}{\partial t}a = (c-a); \quad \frac{\partial}{\partial t}c + \frac{\partial}{\partial t}a + \frac{\partial}{\partial x}c = 0, \quad x > 0, \ t > 0$$

$$u = a = 0 \quad \text{at } t = 0, \ x > 0, \quad u = 1 \quad \text{at } x = 0, \ t > 0$$

It was found that, for all $t > 0$,

$$u \equiv 0 \quad \text{for } 0 < x < 10t, \quad u > e^{-2t} \quad \text{for } x > 10t$$

Is this true? Prove the correctness of the answer; and explain its physical meaning.

5

Cauchy and mixed problems for the wave equation in $\mathbb{R}_1$. Method of traveling waves

1. Small vibrations of infinite string. Method of traveling waves

Let $\mathbf{x} = (x_1, x_2, \dots, x_n)$ and $\mathbf{c} = (c_1, c_2, \dots, c_n) = \text{const}$ be vectors in $\mathbb{R}_n$, t the time. Then, whatever the constant a and the dimension n of the space, the function

$$f = f(\mathbf{c} \cdot \mathbf{x} - at) \tag{5.1}$$

is called a *plane wave* traveling in direction $\mathbf{c}$ with speed $a/|\mathbf{c}|$. The characteristic property of plane-traveling waves is that they propagate without any distortion of their shape.

Consider the Cauchy problem

$$a^2 \frac{\partial^2}{\partial x^2} u - \frac{\partial^2}{\partial t^2} u = 0, \quad -\infty < x < \infty, \ t > 0 \tag{5.2}$$

$$u(x, 0) = f(x), \quad \frac{\partial}{\partial t} u(x, 0) = \varphi(x) \tag{5.3}$$

for the homogeneous wave equation in $\mathbb{R}_1$. Recall that the characteristics of this equation are

$$\xi = x - at, \quad \eta = x + at \tag{5.4}$$

so that the first canonical form of (5.2) is

$$\frac{\partial^2}{\partial \xi \partial \eta} u = 0 \tag{5.5}$$

Hence

$$u = \Phi(\eta) + F(\xi) \tag{5.6}$$

where F and Φ are arbitrary functions. Returning to the original coordinates, we have

$$u(x, t) = \Phi(x + at) + F(x - at) \tag{5.7}$$

144

The initial conditions (5.3) yield

$$\Phi(x) + F(x) = f(x), \qquad \frac{d}{dx}\Phi(x) - \frac{d}{dx}F(x) = \frac{1}{a}\varphi(x) \tag{5.8}$$

In view of the left side of equation 5.8,

$$\frac{d}{dx}\Phi(x) + \frac{d}{dx}F(x) = \frac{d}{dx}f(x) \tag{5.9}$$

Hence

$$\frac{d}{dx}\Phi(x) = \frac{1}{2}\left(\frac{d}{dx}f(x) + \frac{1}{a}\varphi(x)\right) \tag{5.10}$$

$$\frac{d}{dx}F(x) = \frac{1}{2}\left(\frac{d}{dx}f(x) - \frac{1}{a}\varphi(x)\right) \tag{5.11}$$

This implies

$$\Phi(x) = \frac{1}{2}\left(f(x) + \frac{1}{a}\int_{c_0}^{x}f(x)\right) + c_1 \tag{5.12}$$

$$F(x) = \frac{1}{2}\left(f(x) - \frac{1}{a}\int_{c_0}^{x}f(x)\right) + c_2 \tag{5.13}$$

where c_0, c_1, and c_2 are arbitrary constants of integration. Together with (5.7) this gives

$$u(x,t) = u_1(x,t) + u_2(x,t) \tag{5.14}$$

where

$$u_1(x,t) = \frac{f(x-at) + f(x+at)}{2} \tag{5.15}$$

and

$$u_2(x,t) = \frac{1}{2a}\int_{x-at}^{x+at}\varphi(s)ds \tag{5.16}$$

Obviously, u_1 and u_2 are solutions of equation (5.2) satisfying the conditions

$$u_1(x,0) = f(x), \qquad \frac{\partial}{\partial t}u_1(x,0) = 0 \tag{5.17}$$

$$u_2(x,0) = 0, \qquad \frac{\partial}{\partial t}u_2(x,0) = \varphi(x) \tag{5.18}$$

The solution (5.14) of problems (5.2) and (5.3) is named for D'Alembert who first obtained it.

Definition 5.1.1. The (x,t) plane, $(x,t) \in \mathcal{E}_{n+1} = \mathbb{R}_n \times (0 \le T < \infty)$, is called the *phase plane*.

Definition 5.1.2. Let $h(x,t)$ be defined everywhere in the phase space $\mathcal{E}_{n+1}$. The complement G to the entire space $\mathbb{R}_n$ of the maximal doubly-connected open region in which

$$h(x,t) \equiv 0 \tag{5.19}$$

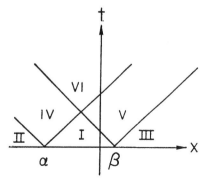

Figure 5.1.   Regions of influence of initial string disturbances located in an interval (α, β).

at the time $t = \text{const}$ is called the *locus of spatial localization* of $h(x, t)$ at time t. The *locus T of time localization* is defined for a fixed x in a similar manner. If the region $D = G \times T$ is not empty, then D is called the *region of space–time localization* of $h(x, t)$.

Let α and β be given constants, $\alpha < \beta$. The characteristics

$$\Gamma_\alpha : x - at = \alpha, \quad \Gamma_\beta : x + at = \beta \tag{5.20}$$

divide the phase plane (x, t) into subregions D_i, $i = 1, 2, \ldots, 6$, where (see Figure 5.1)

$$D_1 = \{x, t : \alpha + at \le x \le \beta - at, \ 0 \le t \le (\beta - \alpha)/2a\} \tag{5.21}$$

$$D_2 = \{x, t : x \le \alpha - at, \ 0 \le t < \infty\} \tag{5.22}$$

$$D_3 = \{x, t : x \ge \beta + at, \ 0 \le t < \infty\} \tag{5.23}$$

$$D_4 = \{x, t : \alpha - at \le x \le \min(\beta - \alpha t, \alpha + at), \ 0 \le t \le \infty\} \tag{5.24}$$

$$D_5 = \{x, t : \max(\alpha + at, \beta - at) \le x \le \beta + at), \ 0 \le t < \infty\} \tag{5.25}$$

$$D_6 = \{x, t : \beta - at \le x \le \alpha + at, \ t \ge (\beta - \alpha)/2\} \tag{5.26}$$

Theorem 5.1.1. *Let $f(x)$ and $\varphi(x)$ be localized in the segment $[\alpha, \beta]$. Then*

$$u_1(x, t) = \frac{f(x - at) + f(x + at)}{2}, \quad u_2(x, t) = \frac{1}{2a} \int_{x-at}^{x+at} \varphi(s) ds$$

$$\forall (x, t) \in D_1 \tag{5.27}$$

$$u_1(x, t) = u_2(x, t) \equiv 0 \quad \forall (x, t) \in D_2 \cup D_3 \tag{5.28}$$

$$u_1(x, t) = \frac{f(x + at)}{2}, \quad u_2(x, t) = \frac{1}{2a} \int_\alpha^{x+at} \varphi(s) ds \quad \forall (x, t) \in D_4 \tag{5.29}$$

$$u_1(x, t) = \frac{f(x - at)}{2}, \quad u_2(x, t) = \frac{1}{2a} \int_{x-at}^\beta \varphi(s) ds \quad \forall (x, t) \in D_5 \tag{5.30}$$

$$u_1(x, t) = 0, \quad u_2(x, t) = \frac{1}{2a} \int_\alpha^\beta \varphi(s) ds \quad \forall (x, t) \in D_5 \tag{5.31}$$

The validity of these equalities follows from the definitions (5.15) and (5.16) of u_1, u_2 and D_i, $i = 1, 2, \ldots, 6$. The assertions of Theorem 5.1.1 admit the following physical interpretation.

The segment $[\alpha, \beta]$ of perturbation of the equilibrium state of an elastic medium occupying the space $\mathbb{R}_1$ (see Chapter 2, Problem 2.6.1) is called the region of influence of initial perturbations or simply the *region of influence*. Within the region of influence, plane-traveling waves generated by the nonzero initial displacements (i.e., the wave $u_1(x,t)$) and those generated by the nonzero initial rate of displacement (i.e., the waves $u_2(x,t)$) begin to propagate. Each is the superposition of two waves: one traveling forward (to the right) and described by a function of the type $U(x - at)$; the other traveling backward and described by a function of the type $U(x + at)$. The region D_1 is that part of space at whose points both forward- and backward-traveling waves are present. D_2 and D_3 are the regions where neither forward nor backward waves generated in the region of influence are present. Everywhere in D_4 there are only backward waves, and in D_5 only forward waves. Finally, D_6 is the region from which both types of waves depart, so that all its points are in an equilibrium state. If the initial rate $\varphi(x)$ of the perturbations is equal to zero, then the equilibrium state of the medium in part D_6 of the phase plane coincides with its initial equilibrium state. But if $\varphi(x)$ does not vanish identically and if, moreover,

$$I = \int_\alpha^\beta \varphi(s)ds \neq 0 \tag{5.32}$$

then all points of D_6 are displaced by the constant value I. Thus, if the initial rate of displacement of the points of an elastic string vanished identically everywhere, then the perturbations of the equilibrium state are localized in space and in time. But if the initial rate does not vanish identically within the region of influence and the integral I does not vanish, then the perturbations of the equilibrium state of the string are not localized in time.

2. Small vibrations of semi-infinite and finite strings with rigidly fixed or free ends. Method of prolongation

We now consider small vibrations of a semi-infinite unloaded string with rigidly fixed end:

$$a^2 \frac{\partial^2}{\partial x^2} u = \frac{\partial^2}{\partial t^2} u, \quad 0 < x < \infty, \ t > 0 \tag{5.33}$$

$$u(x,0) = f(x), \quad \frac{\partial}{\partial t} u(x,0) = \varphi(x) \tag{5.34}$$

$$u(0,t) = 0 \tag{5.35}$$

The D'Alembert solution (5.14)–(5.16) of the problem of vibration of an infinite string may be used to obtain a solution to the problem under consideration by using the method of prolongation. This method proceeds as follows. Instead of dealing with a semi-infinite string, let us consider an infinitely long string, with initial data identical to those prescribed in the region $x > 0$, and defined in the region $x < 0$ in such a way that condition (5.35) is fulfilled. Since the function f is prescribed independently of φ and φ independently of f, we require that

$$u_1(0,t) = \frac{f(-at) + f(at)}{2} = 0 \tag{5.36}$$

and

$$u_2(0,t) = \frac{1}{2a} \int_{-at}^{at} \varphi(s)ds = 0 \tag{5.37}$$

Conditions (5.36) and (5.37) mean that f and φ must be continued into the left half-line $x < 0$ as odd functions:

$$f(x) = -f(-x), \qquad \varphi(x) = -\varphi(-x) \tag{5.38}$$

Assume now that the end $x = 0$ of the string is free (i.e., that there is a local mechanical equilibrium), so that instead of (5.35) we have

$$\frac{\partial}{\partial x}u(0,t) = 0 \tag{5.39}$$

Here again, the method of prolongation is also applicable, but $f(x)$ and $\varphi(x)$ must be continued into the left half-line as even functions. Indeed, both conditions

$$\frac{\partial}{\partial x}u_1(0,t) = 0, \quad \frac{\partial}{\partial x}u_2(0,t) = 0 \tag{5.40}$$

again must hold, in view of the mutual independence of f and φ. Thus we must demand that

$$\frac{\partial}{\partial x}u_1(0,t) = \frac{1}{2a}\left(\frac{\partial}{\partial t}f(at) - \frac{\partial}{\partial t}f(-at)\right) = 0 \tag{5.41}$$

and

$$\frac{\partial}{\partial x}u_2(0,t) = \varphi(at) - \varphi(-at) = 0 \tag{5.42}$$

Q.E.D.

We now consider a string of finite length L with rigidly fixed ends. In this case we must solve the boundary-value problem

$$a^2\frac{\partial^2}{\partial x^2}u - \frac{\partial^2}{\partial t^2}u = 0 \quad \forall x \in (0,L), \ \forall t > 0 \tag{5.43}$$

$$u(x,0) = f(x), \quad \frac{\partial}{\partial t}u(x,0) = \varphi(x) \tag{5.44}$$

$$u(0,t) = 0, \quad u(L,t) = 0 \quad \forall t > 0 \tag{5.45}$$

We can again apply the method of prolongation of the initial data from the segment [0,L]. Namely, the following statement is true.

Theorem 5.1.2. *The D'Alembert formula (5.14)–(5.16) represents a solution of problem (5.43)–(5.45) if f and φ are continued into the half-lines*

$x < 0$ and $x > L$ as functions odd with respect to the both ends $x = 0$ and $x = L$; thus f and φ must be extended to the entire real line $(-\infty < x < \infty)$ as functions odd with respect to the origin $x = 0$ and periodic with period $2L$.

3. Generalized solution of problem of vibration of loaded string with nonhomogeneous boundary conditions

Until now we have demonstrated only that D'Alembert's formulas satisfy the initial and homogeneous boundary conditions, without proving that it is actually a solution of the wave equation. This is certainly true if $f(x)$ is twice and $\varphi(x)$ once continuously differentiable. However, these requirements are obviously redundant from the physical point of view. Indeed, we cannot assume that the initial shape $f(x)$ of the string is discontinuous, since any discontinuity would mean that the string was broken. However, it is not natural to require that the initial shape of the string be smooth. For example, the best approximate description of the case where the string is stressed at $x = 0$ by a sharp knife is $f(x) = |x|$ in the neighborhood of $x = 0$. And if the taut string is subjected at the initial time to a blow by the sharp edge of a rigid hammer, the best approximation to the initial rate of displacement would be: $\varphi(x)$ is a piecewise continuous function with a jump at the point $x = 0$. In other words, we must allow the displacement $u(x, t)$ of the points of the string to have weak discontinuity of the first order.

Thus, consider the following problem: Find a function $u(x, t)$ in the region

$$D^* = \{x, t : \alpha < x < \beta, t > 0\}, \quad -\infty < \alpha < \beta < \infty \tag{5.46}$$

such that

$$\Box u \stackrel{\text{def}}{=} a^2 \frac{\partial^2}{\partial x^2} u(x, t) - \frac{\partial^2}{\partial t^2} u(x, t) = F(x, t) \quad \forall (x, t) \in D^* \tag{5.47}$$

$$u(\alpha, t) = \mu_1(t), \quad u(\beta, t) = \mu_2(t) \quad \forall t > 0 \tag{5.48}$$

$$u(x, 0) = f(x), \quad \frac{\partial}{\partial t} u(x, 0) = \varphi(x) \quad \forall x \in (\alpha, \beta) \tag{5.49}$$

If $\alpha = -\infty$ or $\beta = \infty$ then the appropriate one of conditions (5.48) must be omitted. Assume that the input data F, μ_1, μ_2, f, and φ satisfy the following conditions:

- F and φ are piecewise continuous functions everywhere in their regions of definition;
- f, μ_1, and μ_2 are piecewise smooth continuous functions.

By a *generalized solution* of the problem we mean a function $u(x, t)$ admitting at most finitely many curves of weak discontinuities of first order whose limits at $t = +0$, $x = \alpha + 0$, and $x = \beta - 0$ coincide with $f(x)$, $\mu_1(t)$, and $\mu_2(t)$, respectively, at every point of continuity of the latter.

Let $M(x, t) \in D^*$ be a fixed point and $(\xi, \tau) \in D^*$ a variable point. Let us apply Riemann's formula (4.162) with $v \equiv 1$ to some subregion $D \subset D^*$

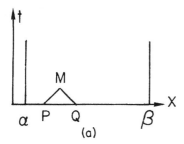

Figure 5.2a. Traveling waves'
reflection from walls, case (a).

whose definition depends on the location of M within D^* (see Figures 5.2).
In the case under consideration,

$$H = -\frac{\partial}{\partial \tau}u(\xi,\tau), \quad K = -a^2\frac{\partial}{\partial \xi}u(\xi,\tau) \tag{5.50}$$

so that

$$\int_\Gamma a^2\frac{\partial}{\partial \xi}u(\xi,\tau)d\tau + \frac{\partial}{\partial \tau}u(\xi,\tau)d\xi = \int_D F(\xi,\tau)d\xi d\tau \tag{5.51}$$

In the case shown in Figure 5.2a, the region D is bounded by characteristics
MP and MQ emanating from M in the direction of decreasing time t and
by the segment PQ of the axis $t=0$. This gives

$$\int_{MP} + \int_{PQ} + \int_{QM} = \int_D \tag{5.52}$$

where

$$M = (x,t), \quad P = (x-at,0), \quad Q = (x+at,0) \tag{5.53}$$

Along MP, $d\xi = ad\tau$, so that

$$\int_{MP} = -a\int_{MP} du = a[f(x-at) - u(x,t)] \tag{5.54}$$

Along PQ, $d\tau = 0$, so that

$$\int_{pq} = \int_{x-at}^{x+at} \varphi(s)ds \tag{5.55}$$

Along QM, $d\xi = -ad\tau$ so that

$$\int_{QM} = -a\int_{QM} du = a[f(x+at) - u(x,t)] \tag{5.56}$$

Finally,

$$\int_D = \int_0^t d\tau \int_{x-a(t-\tau)}^{x+a(t-\tau)} F(\xi,\tau)d\xi \tag{5.57}$$

Substitution of (5.54)–(5.57) into (5.51) yields

$$u(x,t) = u_1(x,t) + u_2(x,t) - \frac{1}{2a}\int_0^t d\tau \int_{x-a(t-\tau)}^{x+a(t-\tau)} F(\xi,\tau)d\xi \tag{5.58}$$

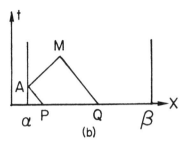

Figure 5.2b. Traveling waves'
reflection from walls, case (b).

where u_1 and u_2 are defined by (5.15) and (5.16). Therefore, in the case
of an unloaded string, the generalized solution coincides with that given
by D'Alembert's formula. Hence the latter in fact determines not only a
classical solution but also a generalized solution of the Cauchy problem
under consideration.

Now consider the case shown in Figure 5.2b. Here D is the region D
bounded by the characteristics MA, AP and QM and by the segment PQ
of the axis $t = 0$, so that

$$\int_\Gamma = \int_{MA} + \int_{AP} + \int_{PQ} + \int_{QM} \tag{5.59}$$

Similarly to (5.54)–(5.56), we now have

$$\int_{MA} = a[u(M) - u(A)], \int_{AP} = a[-u(A) - u(P)]$$

$$\int_{PQ} = \int_{x_p}^{x_q} \frac{\partial}{\partial t} u(\xi, 0)d\xi, \qquad \int_{QM} = a[u(M) - u(Q)] \tag{5.60}$$

The coordinates of the points A, P, and Q are clearly

$$A = (\alpha, t_a), \quad P = (x_p, 0), \quad Q = (x_q, 0) \tag{5.61}$$

where

$$t_a = t - (x - \alpha)/a, \quad x_p = 2\alpha - x + at, \quad x_q = x + at \tag{5.62}$$

Thus

$$u(M) = u(A) + \frac{1}{2}[u(Q) - u(P)] + \frac{1}{2a}\int_{x_p}^{x_q} \varphi(\xi)d\xi + \int_D F(\xi, \tau)d\xi d\tau \tag{5.63}$$

which may be rewritten

$$u(x,t) = u_\alpha(x,t) + u_q(x,t) + U(x,t) \tag{5.64}$$

where

$$u_\alpha(x,t) = \mu_1(t_a) - \frac{1}{2}f(x_p), \qquad u_q(x,t) = \frac{1}{2}f(x_q) + \frac{1}{2a}\int_{x_p}^{x_q} \varphi(\xi)d\xi$$

$$U(x,t) = \int_D F(\xi, \tau)d\xi d\tau \tag{5.65}$$

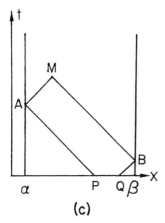

(c)

Figure 5.2c. Traveling waves'
reflection from walls, case (c).

Let us interpret the problem in terms of the propagation of sound through an elastic solid bounded by acoustically impermeable walls, represented by the planes $x = \alpha$ and $x = \beta$, and vibrating with given displacements $\mu_1(t)$ and $\mu_2(t)$, respectively. The sound is generated by an initial perturbation of the equilibrium state and the presence of constantly acting generators of sound, distributed continuously within the body with volume density $F(x,t)$. When the backward wave generated at the initial time reaches the left wall, it is reflected there, reaching the point x at time t as a forward propagating wave u_α. The wave u_q is the superposition of the initially generated waves that reach x at time t without being first reflected at the walls, and U is the influence function of the distributed sources of sound. An analogous picture is shown in Figure 5.2c, which corresponds to a wave reflected from the right wall $x = \beta$ and reaching x at time t. Finally, in Figure 5.2d, the point $M(x,t)$ of the phase plane is located so that the point x is reached at time t by a finite number of forward and backward waves reflected from both the walls. In this case Riemann's identity must be applied successively to a sequence of regions D_k, $k = 1, 2, 3, \dots, n+1$, such that the boundaries of D_k are composed of characteristics $M_{k-1}A_k$, A_kM_k, M_kB_k, and B_kM_{k-1} if $k = 1, 2, \dots, n$, and such that D_{n+1} is bounded by characteristics M_nA_{n+1}, $A_{n+1}P$, QB_{n+1}, $B_{n+1}M_n$, and the segment PQ of the axis $M_0 = M$. This sequence of operations determines the value of u at M_k in terms of the values of the wall displacements μ_1 and μ_2 and the unknown value of u at M_{k+1}. The coordinates of all the points A_k, B_k, M_k are determined by their predecessors, and the coordinates of P and Q are determined by those of A_{n+1} and B_{n+1}. Hence the sequence of computations allow us successively to determine values of u_k, $k = n, n-1, \dots, 1$, in terms of those of u_{k+1}, and those of μ_1 and μ_2 at the points A_{k+1} and B_{k+1}, respectively. It is easy to show that if the vibrating body is not loaded (i.e., if $F \equiv 0$), and if both the walls are rigidly fixed (i.e., if $\mu_1 = \mu_2 \equiv 0$),

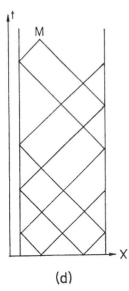

(d)

Figure 5.2d. Traveling waves'
reflection from walls, case (d).

then the generalized solution is precisely that given by the method of
prolongation.

Remark. It is obvious that this method of computation is easily applicable
within relatively small time intervals, depending on the size of the region
of influence. The larger the latter, the longer is the time interval that
can be covered in a relatively small number of steps. More convenient for
computation in large time intervals, provided that the spatial region is not
too large, is the Fourier method: expansion of the solution in a series of
eigenfunctions of the appropriate Sturm–Liouville's equation (see Chapter 20
and Appendix 3).

Problems

P5.1.1. Draw the profile of an infinite string at several different times if
a, the speed of wave motion, is equal to 0.01, 0.1, 1, and 10; the region of
the influence is $(-1,1)$; and initial data are

(1) $u(x,0) = 0, \partial u(x,0)/\partial t = 1$,
(2) $u(x,0) = 1, \partial u(x,0)/\partial t = 0$,
(3) $u(x,0) = 1, \partial u(x,0)/\partial t = 1$.

P5.1.2. Let $a^2 = 1.10^{-8}$ cm^2/sec^2; $f(x) = 3\exp(-x/3)\sin[(x+0.25)/5]\cos[(x-0.25)/2]$ and $\varphi(x) = 0.5\ln(3/2x)$ if $x \in (5,15)([x] = $ microns); $t = 20$ ($[t] = $
min); and $f, \varphi \equiv 0$ for all $x \notin (5,15)$. Determine $u(-25,4)$.

P5.2.1. Derive condition (5.39) for longitudinal and transversal vibrations
of a string (See Chapter 2, Section 10).

P5.2.2. Consider the boundary value problem

$$\frac{\partial^2}{\partial x^2}u = \frac{\partial^2}{\partial t^2}u \quad \forall x \in (0, 2\pi), \ \forall t > 0$$

$$u(0,t) = \sin(t), \qquad u(2\pi, t) = \cos(t)$$

$$u(x,0) = \frac{\partial}{\partial t}u(x,0) \equiv 0$$

Can this problem be solved by the method of traveling waves? If so, solve the problem and draw the solution at several instants of time. If not, explain why.

P5.2.3. Solve the problem

$$\frac{\partial^2}{\partial x^2}u = \frac{\partial^2}{\partial t^2}u \quad \forall x \in (-\infty, \infty), \ \forall t > 0$$

$$u(x,0) = |\tanh x|, \qquad \frac{\partial}{\partial t}u(x,0) = \begin{cases} 1 & \forall x < 0 \\ -1 & \forall x > 0 \end{cases}$$

P5.2.4. Derive rules of prolongation for a string of finite length for the following cases:

(a) both ends are rigidly fixed,
(b) both ends are free,
(c) the left end is rigidly fixed and the right end is free.

P5.3.1. Prove our assertion that the generalized solutions of homogeneous boundary-value problems for an unloaded string are identical to those obtained by the method of prolongation.

6

Cauchy and Goursat problems for a second-order linear hyperbolic equation with two independent variables. Riemann's method

1. Riemann's method

We now proceed to a study of the general Cauchy problem in $\mathbb{R}_2$ for a linear equation of hyperbolic type. Assume that the equation is given in first canonical form:

$$Lu = 2\frac{\partial^2 u}{\partial x \partial y} + a\frac{\partial u}{\partial x} + b\frac{\partial u}{\partial y} + cu = f(x,y) \qquad (6.1)$$

The reader will recall that the problem is to determine a function $u(x,y)$ satisfying equation (6.1) in the neighborhood of a smooth curve without self-intersections,

$$\Gamma = \{x, y : x = x(s), \ y = y(s); \ \alpha^* < \alpha \le s \le \beta < \beta^*\} \qquad (6.2)$$

on the assumption that no characteristic curve of equation (6.1) intersects Γ at more than one point, and such that along Γ,

$$u - u_0(s), \quad \frac{\partial u}{\partial x} - u_1(s), \quad \frac{\partial u}{\partial y} - u_2(s) \qquad (6.3)$$

where

$$\frac{du_0}{ds} = u_1(s)\frac{dx}{ds} + u_2(s)\frac{dy}{ds} \qquad (6.4)$$

Let $M = (x, y)$ be an arbitrary point, and let $P = (\xi, y) \in \Gamma$ and $Q = (x, \eta) \in \Gamma$ be points at which characteristics emanating from M intersect Γ. Let D be the region bounded by the contour $MPQM$ (see Figure 6.1). Further, let $q = (\xi, \eta)$ be a variable point of D and $v(M,q) = v(x, y, \xi, \eta)$ a solution of the equation

$$M(v) = 2\frac{\partial^2}{\partial \xi \partial \eta}v - \frac{\partial}{\partial \xi}(av) - \frac{\partial}{\partial \eta}(bv) + cv = 0 \qquad (6.5)$$

which is the adjoint of the equation $Lu(\xi, \eta) = f(\xi, \eta)$ (see Chapter 4,

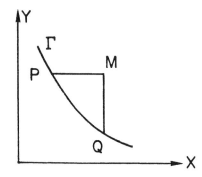

Figure 6.1.   Carrier Γ of Cauchy data and the region $MPQM$ of the influence.

Section 7). Then, applying Riemann's formula (4.165), we obtain

$$\int_{MPQM} H d\xi + K d\eta = \int_D v f d\xi d\eta \qquad (6.6)$$

Here (see (4.160) and (4.161))

$$H(u,v) = v\left(\frac{\partial}{\partial \xi}u + bu\right) - u\frac{\partial}{\partial \xi}v, \quad K(u,v) = -v\left(\frac{\partial}{\partial \eta}u + au\right) + u\frac{\partial}{\partial \eta}v \quad (6.7)$$

Along MP,

$$d\eta = 0, \quad H(u,v) = \frac{\partial}{\partial \xi}(uv) - 2u\left(\frac{\partial}{\partial \xi}v - \frac{1}{2}bv\right) \qquad (6.8)$$

Along QM,

$$d\xi = 0, K(u,v) = -\frac{\partial}{\partial \eta}(uv) + 2u\left(\frac{\partial}{\partial \eta}v - \frac{1}{2}av\right) \qquad (6.9)$$

Hence, by (6.6),

$$uv\Big|_M = \frac{1}{2}\left(uv\Big|_P + uv\Big|_Q - \int_{MP} 2u\left(\frac{\partial}{\partial \xi}v - \frac{1}{2}bv\right)d\xi\right.$$

$$+ \int_{QM} 2u\left(\frac{\partial}{\partial \eta}v - \frac{1}{2}av\right)d\eta$$

$$\left. + \int_{PQ} H(u,v)d\xi + K(u,v)d\eta - \int_D v f d\xi d\eta\right) \qquad (6.10)$$

This integral identity is valid for any function u such that all the partial derivatives occurring in the expression for L are continuous, where f is only an abbreviated notation for Lu. It does not determine a solution of the Cauchy problem, since the integrands of the integrals along the paths MP and QM involve the unknown values of u. However, the identity becomes a solution of the Cauchy problem (6.1) if v is chosen as a function of two pairs of variables (x,y) and (ξ,η),

$$v = g(x,y,\xi,\eta) \qquad (6.11)$$

with the properties

$$Mg(x, y, \xi, \eta) \equiv 0 \tag{6.12}$$

$$g(x, y, x, y) = 1 \tag{6.13}$$

$$\frac{\partial}{\partial \xi} g(x, y, \xi, y) - \frac{1}{2} b(\xi, y) g(x, y, \xi, y) = 0 \tag{6.14}$$

$$\frac{\partial}{\partial \eta} g(x, y, x, \eta) - \frac{1}{2} a(x, \eta) g(x, y, x, \eta) = 0 \tag{6.15}$$

Note that here x, y are considered as parameters and ξ, η as independent variables.

Indeed, if g satisfies these requirements then the identity (6.10) becomes[1]

$$u(x, y) = \frac{1}{2} \left(ug \, |_P + ug \, |_Q + \int_{PQ} H(u, g) d\xi + K(u, g) d\eta - \int_D g f d\xi d\eta \right) \tag{6.16}$$

where all functions in the right-hand side are known, since g is known and the Cauchy data are prescribed on the arc PQ. Riemann's method for the solution of the Cauchy problem consists in the introduction of a function $g(x, y, \xi, \eta)$ possessing properties (6.12)–(6.15). This function is therefore known as the Riemann function.

Note that integration of equations (6.14) and (6.15) implies

$$g(x, y, \xi, y) = \exp \left(-\frac{1}{2} \int_\xi^x b(s, y) ds \right) \tag{6.17}$$

$$g(x, y, x, \eta) = \exp \left(-\frac{1}{2} \int_\eta^y a(x, s) ds \right) \tag{6.18}$$

We have defined $g(x, y, \xi, \eta)$ as a function of the second pair of variables (ξ, η). Let us now consider its properties as a function of the first pair (x, y). With this in mind, consider the following problem: Find a function $u(x, y, \xi, \eta)$ in the region D bounded by the rectangle $MPNQM$, where

$$M = (x, y), \quad P = (\xi, y), \quad N = (\xi, \eta), \quad Q = (x, \eta) \tag{6.19}$$

(see Figure 6.2) which satisfies the equation

$$Lu = 2\frac{\partial^2 u}{\partial x \partial y} + a\frac{\partial u}{\partial x} + b\frac{\partial u}{\partial y} + cu = 0 \tag{6.20}$$

and the conditions

$$\frac{\partial u}{\partial y} + \frac{1}{2} au = 0 \quad \text{on } PN, \quad \frac{\partial u}{\partial x} + \frac{1}{2} bu = 0 \quad \text{on } NQ, \quad u = 1 \quad \text{at } N \tag{6.21}$$

---

[1]Two cases must be distinguished: (a) $\xi < x$ and $\eta < y$ or $\xi > x$ and $\eta > y$; in this case, the sense of description of Γ in equality (6.16) is counterclockwise. (b) $\xi > x$ and $\eta < y$ or $\xi < x$ and $\eta > y$; in this case, the sense of description in equality (6.16) is clockwise.

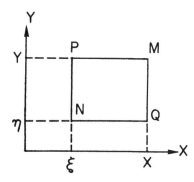

Figure 6.2. Carrier PNQ of Cauchy data and the region $MPNQM$.

Riemann's formula (6.16) now becomes

$$u(x,y,\xi,\eta) = \frac{1}{2}\left(ug\mid_P + ug\mid_Q + \int_{PN} K(u,g)ds + \int_{NQ} H(u,g)ds\right) \quad (6.22)$$

By (6.7), (6.14), (6.15), and (6.19),[2]

$$H[u(s,\eta), g(x,y,s,\eta)] = 2g(x,y,s,\eta)\left(\frac{\partial}{\partial s}u(s,\eta) + \frac{1}{2}b(s,\eta)u(s,\eta)\right)$$

$$- \frac{\partial}{\partial s}[u(s,\eta)g(x,y,s,\eta)]$$

$$\equiv -\frac{\partial}{\partial s}[u(s,\eta)g(x,y,s,\eta)] \quad \text{along } NQ \quad (6.23)$$

$$K[u(\xi,s)g(x,y,\xi,s)] = -2g(x,y,\xi,s)\left(\frac{\partial}{\partial s}u(\xi,s) + \frac{1}{2}a(\xi,s)u(\xi,s)\right)$$

$$+ \frac{\partial}{\partial s}[u(\xi,s)g(x,y,\xi,s)]$$

$$\equiv \frac{\partial}{\partial s}[u(\xi,s)g(x,y,\xi,s)] \quad \text{along } PN \quad (6.24)$$

Then it follows from (6.22) and (6.23) that

$$u(x,y,\xi,\eta) = g(x,y,\xi,\eta) \quad (6.25)$$

---

[2]Recall that u depends on x,y as independent variables and on ξ,η as fixed parameters. Thus $u \equiv u(x,y,\xi,\eta)$. Here and in the sequel, for abbreviation we write

$$u(\xi,s) \text{ instead of } u(x,s,\xi,\eta)\Big|_{x=\xi}, \quad u(s,\eta) \text{ instead of } u(s,y,\xi,\eta)\Big|_{y=\eta}$$

$$\frac{\partial}{\partial \xi}u(\xi,s) \text{ instead of } \frac{\partial}{\partial x}u(x,s,\xi,\eta)\Big|_{x=\xi}, \quad \frac{\partial}{\partial \eta}u(s,\eta) \text{ instead of } \frac{\partial}{\partial y}u(s,y,\xi,\eta)\Big|_{y=\eta}$$

Hence $g(x, y, \xi, \eta)$, as a function of the first pair (x, y) of its arguments, is the solution of the equation

$$Lg = 0 \tag{6.26}$$

satisfying the conditions

$$\frac{\partial}{\partial y}g(\xi, y, \xi, \eta) + \frac{1}{2}a(\xi, y)g(\xi, y, \xi, \eta) = 0 \quad \text{along } PN \tag{6.27}$$

$$\frac{\partial}{\partial x}g(x, \eta, \xi, \eta) + \frac{1}{2}b(x, \eta)g(x, \eta, \xi, \eta) = 0 \quad \text{along } NQ \tag{6.28}$$

$$g(\xi, \eta, \xi, \eta) = 1 \tag{6.29}$$

Integration of (6.27)–(6.29) yields equalities

$$g(x, \eta, \xi, \eta) = \exp\left(-\frac{1}{2}\int_{\xi}^{x} b(s, \eta)ds\right) \tag{6.30}$$

$$g(\xi, y, \xi, \eta) = \exp\left(-\frac{1}{2}\int_{\eta}^{y} a(\xi, s)ds\right) \tag{6.31}$$

which coincide with (6.17) and (6.18), respectively.

Note now that the operator Mv defined by (6.5) may be rewritten as

$$Mv = 2\frac{\partial^2 v}{\partial \xi \partial \eta} - a\frac{\partial v}{\partial \xi} - b\frac{\partial v}{\partial \eta} + \left(c - \frac{\partial a}{\partial \xi} - \frac{\partial b}{\partial \eta}\right) \tag{6.32}$$

so that the adjoint operator of Mv is Lu. Therefore let $G(\xi, \eta, x, y)$ be the Riemann function of the operator Mv, defined by conditions

$$LG(\xi, \eta, x, y) = 0 \tag{6.33}$$

$$G(\xi, y, x, y) = \exp\left(-\frac{1}{2}\int_{\xi}^{x} b(s, y)ds\right) \tag{6.34}$$

$$G(x, \eta, x, y) = \exp\left(-\frac{1}{2}\int_{\eta}^{y} a(x, s)ds\right) \tag{6.35}$$

Comparison of (6.33)–(6.35) and (6.17), (6.18) shows that

$$G(\xi, \eta, x, y) \equiv g(x, y, \xi, \eta) \tag{6.36}$$

Thus the Riemann function $g(x, y, \xi, \eta)$ of the operator L, considered as a function of the first pair (x, y) of its arguments, is the Riemann function of the adjoint operator M. This means that, formally speaking, *any Cauchy problem for equations*

$$Lu + f = 0, \qquad Mv + F = 0 \tag{6.37}$$

may be solved simultaneously if the Riemann function of these equations is known.

This statement needs some explanation. All we have proved is that if $g(x, y, \xi, \eta)$ is the Riemann function and if $u(x, y)$ is a solution of the

Cauchy problem (6.1)–(6.4), then Riemann's identity (6.16) gives an explicit expression of $u(x,y)$ in terms of the Cauchy data. But we cannot state a priori that if $u(x,y)$ is defined by (6.16) then it is a solution of the Cauchy problem (6.1)–(6.4). The proof of this assertion is given by the following theorem.

Theorem 6.1.1. *Let the Cauchy data $u_i(s)$, $i = 0, 1, 2$, and the source term $f(x,y)$ satisfy the following conditions:*

$$u_0(s) \in C^1(\Gamma), \quad u_i(s) \in C(\Gamma), \quad f(x,y) \in C(D) \tag{6.38}$$

$$\frac{d}{ds} u_0(s) = u_1(s) \frac{dx}{ds} + u_2(s) \frac{dy}{ds} \quad \text{along } \Gamma \tag{6.39}$$

Assume that the Riemann function $g(x, y, \xi, \eta)$ is such that[3]

$$\frac{\partial^2}{\partial x \partial \xi} g, \frac{\partial^2}{\partial y \partial \xi} g, \frac{\partial^3}{\partial x \partial y \partial \xi} g \in C(D) \tag{6.40}$$

and let $u(x,y)$ be defined by (6.16). Then $u(x,y)$ is the solution of the Cauchy problem (6.1)–(6.4).

PROOF. By assumption, Γ is a smooth curve without self-intersections and such that no characteristic crosses it at more than one point. Hence Γ can be represented by equations in which either x or y acts as the independent variable:

$$\Gamma = \{x, y : a < x < b, \ y = y(x)\} \equiv \{x, y : x = x(y), \ c < y < d\} \tag{6.41}$$

Therefore the coordinates of the points P and Q are

$$P = [x(y), y], \quad Q = [x, y(x)] \tag{6.42}$$

Equalities (6.7) imply

$$\int_{PQ} H d\xi + K d\eta = \int_{PQ} d(ug) - 2 \int_{PQ} u \left(\frac{\partial}{\partial \xi} - \frac{1}{2} b \right) g d\xi + g \left(\frac{\partial}{\partial \eta} + \frac{1}{2} a \right) u d\eta \tag{6.43}$$

This and (6.16) yield

$$u(x,y) = (ug) \big|_Q - \int_{PQ} u \left(\frac{\partial}{\partial \xi} - \frac{1}{2} b \right) g d\xi + g \left(\frac{\partial}{\partial \eta} + \frac{1}{2} a \right) u d\eta - \frac{1}{2} \int_D f g d\xi d\eta$$

$$\equiv v(x, y) + w(x, y) \tag{6.44}$$

where

$$v(x, y) = u_0[x, y(x)] \ g[x, y, x, y(x)]$$

$$- \int_{x(y)}^x u_0[\xi, y(\xi)] \left(\frac{\partial}{\partial \xi} - \frac{1}{2} b[\xi, y(\xi)] \right) g[x, y, \xi, y(\xi)] d\xi$$

$$- \int_y^{y(x)} g[x, y, x(\eta), \eta] \left(u_2[x(\eta), \eta] + \frac{1}{2} a[x(\eta), \eta] u_0[x(\eta), \eta] \right) d\eta \tag{6.45}$$

---

[3] Note that if g is a Riemann function, then the existence and continuity of $\partial^2 g / \partial x \partial \xi$ and $\partial^2 g / \partial y \partial \xi$ imply the existence and continuity of $\partial^3 g / \partial x \partial y \partial \xi$.

and

$$w(x,y) = -\frac{1}{2} \int_{x(y)}^{x} d\xi \int_{y(\xi)}^{y} f(\xi,\eta)g(x,y,\xi,\eta)d\xi d\eta \tag{6.46}$$

Note that the identities

$$x[y(x)] \equiv x, \quad y[x(y)] \equiv y, \quad g(x,y,x,y) \equiv 1 \tag{6.47}$$

imply

$$w[x,y(x)] = \frac{\partial}{\partial x}w[x,y(x)] = \frac{\partial}{\partial y}w[x(y),y] = 0 \tag{6.48}$$

$$v[x,y(x)] = u_0[x,y(x)] \tag{6.49}$$

By virtue of assumptions (6.40), we can differentiate the integrand in the right-hand side of (6.45) under the integral sign. This gives

$$\frac{\partial}{\partial y}v(x,y) = u_0[x,y(x)]\frac{\partial}{\partial y}g[x,y,x,y(x)]$$

$$+ \frac{dx}{dy}u_0[x(y),y]\left(\frac{\partial}{\partial\xi} - \frac{1}{2}b[\xi,y(\xi)]\right)g[x,y,\xi,y(\xi)]\mid_{\xi=x(y)}$$

$$+ g[x,y,x(y),y]\left[u_2[x(y),y] + \frac{1}{2}a[x(y),y]u_0[x(y),y]\right]$$

$$- \int_{x(y)}^{x} u_0[\xi,y(\xi)]\frac{\partial}{\partial y}\left(\frac{\partial}{\partial\xi} - \frac{1}{2}b[\xi,y(\xi)]\right)g[x,y,\xi,y(\xi)]d\xi$$

$$- \int_{y}^{y(x)} \frac{\partial}{\partial y}g[x,y,x(\eta),\eta]\left[u_2[x(\eta),\eta] + \frac{1}{2}a[x(\eta),\eta]u_0[x(\eta),\eta]\right]d\eta \tag{6.50}$$

Hence

$$\frac{\partial}{\partial y}v(x,y)\mid_{x=x(y)} = u_0[x,y(x)]\frac{\partial}{\partial y}g[x,y,x,y(x)]$$

$$+ \frac{d}{dy}x(y)u_0[x(y),y]\left(\frac{\partial}{\partial\xi} - \frac{1}{2}b[\xi,y(\xi)]\right)g[x(y),y,\xi,y(\xi)]\mid_{\xi=\tau(y)}$$

$$+ u_2[x(y),y] + \frac{1}{2}a[x(y),y]u_0[x(y),y] \tag{6.51}$$

since $g(x(y),y,x(y),y[x(y)]) \equiv 1$.

By (6.44), (6.48) and the identities (6.13), (6.14), and (6.27), this last equality yields

$$\frac{\partial}{\partial y}u(x,y)\mid_{x(y)} = u_2(x,y)\mid_{x(y)} \tag{6.52}$$

Moreover, since

$$\frac{d}{dy}u[x(y),y] = \frac{\partial}{\partial x}u[x(y),y]\frac{dx}{dy} + \frac{\partial}{\partial y}u[x(y),y] \tag{6.53}$$

$$\frac{d}{dy}u_0[x(y),y] = u_1[x(y),y]\frac{dx}{dy} + u_2[x(y),y] \tag{6.54}$$

and

$$\frac{dx}{dy} \neq 0 \tag{6.55}$$

it follows that

$$\frac{\partial}{\partial x} u[x(y), y] = u_1[x(y), y] \tag{6.56}$$

Equalities (6.52) and (6.56) mean that the Cauchy conditions (6.3) are satisfied.

Because $g(x, y, \xi, \eta)$, treated as a function of (x, y), is a solution of the equation

$$Lg = 0 \quad \forall (x, y) \in D \tag{6.57}$$

we obviously have

$$Lv = 0 \quad \forall (x, y) \in D \tag{6.58}$$

Hence it remains only to prove that

$$Lw + f = 0 \quad \forall (x, y) \in D \tag{6.59}$$

Assume that

$$Lw + F(x, y) = 0 \tag{6.60}$$

Since

$$w \mid_\Gamma = \frac{\partial}{\partial x} w \mid_\Gamma = \frac{\partial}{\partial y} w \mid_\Gamma = 0 \tag{6.61}$$

Riemann's formula (6.16) yields

$$w(x, y) = -\frac{1}{2} \int_{x(y)}^{x} d\xi \int_{y(\xi)}^{y} F(\xi, \eta) g(x, y, \xi, \eta) d\xi d\eta \tag{6.62}$$

At the same time, $w(x, y)$ is defined by (6.46); that is

$$w(x, y) = -\frac{1}{2} \int_{x(y)}^{x} d\xi \int_{y(\xi)}^{y} f(\xi, \eta) g(x, y, \xi, \eta) d\xi d\eta \tag{6.63}$$

Hence

$$J \overset{\text{def}}{=} \int_{x(y)}^{x} d\xi \int_{y(\xi)}^{y} \Phi(\xi, \eta) g(x, y, \xi, \eta) d\xi d\eta = 0 \tag{6.64}$$

where

$$\Phi = F - f \tag{6.65}$$

By virtue of (6.40) we may differentiate J twice under the integral sign. Hence it follows from (6.64) that

$$
\frac{\partial J}{\partial x} = \int_{y(x)}^{y} \Phi(x,\eta) g(x,y,x,\eta) d\eta
$$

$$
+ \int_{x(y)}^{x} d\xi \int_{y(\xi)}^{y} \Phi(\xi,\eta) \frac{\partial}{\partial x} g(x,y,\xi,\eta) d\eta \equiv 0 \tag{6.66}
$$

$$
\frac{\partial^2 J}{\partial x \partial y} = \Phi(x,y) + \int_{x(y)}^{x} \Phi(\xi,y) \frac{\partial}{\partial x} g(x,y,\xi,y) d\xi + \int_{y(x)}^{y} \Phi(x,\eta) \frac{\partial}{\partial y} g(x,y,x,\eta) d\eta
$$

$$
+ \int_{x(y)}^{x} d\xi \int_{y(\xi)}^{y} \Phi(\xi,\eta) \frac{\partial^2}{\partial x \partial y} g(x,y,\xi,\eta) d\eta \equiv 0 \tag{6.67}
$$

Denote

$$
V(x,y \mid \Phi) \stackrel{\text{def}}{=} \int_{x(y)}^{x} \Phi(\xi,y) \frac{\partial}{\partial x} g(x,y,\xi,y) d\xi + \int_{y(x)}^{y} \Phi(x,\eta) \frac{\partial}{\partial y} g(x,y,x,\eta) d\eta
$$

$$
+ \int_{x(y)}^{x} d\xi \int_{y(\xi)}^{y} \Phi(\xi,\eta) \frac{\partial^2}{\partial x \partial y} g(x,y,\xi,\eta) d\eta \tag{6.68}
$$

so that

$$
\Phi(x,y) + V(x,y \mid \Phi) = 0 \tag{6.69}
$$

But this is a homogeneous integral equation of the Volterra type, with a smooth operator V. Hence its solution exists and is unique,[4] so that

$$
\Phi(x,y) \equiv F(x,y) - f(x,y) \equiv 0 \tag{6.70}
$$

Together with (6.60), this implies that

$$
Lw + f(x,y) \equiv 0 \tag{6.71}
$$

Q.E.D.

Remark 6.1.1. Theorem 6.1.1 means that a solution of the Cauchy problem (6.1)–(6.4) exists and is unique if and only if there exists a unique Riemann function $g(x,y,\xi,\eta)$ satisfying conditions (6.40).

2. Goursat problem.[5] Existence and uniqueness of Riemann's function

Note, first of all, that the problem

$$
Lu = 2\frac{\partial^2 u}{\partial x \partial y} + a\frac{\partial u}{\partial x} + b\frac{\partial u}{\partial y} + cu = 0 \tag{6.72}
$$

$$
\frac{\partial u}{\partial \eta} + au = 0 \quad \text{along } PN, \qquad \frac{\partial u}{\partial \xi} + bu = 0 \quad \text{along } NQ \tag{6.73}
$$

$$
u = 1 \quad \text{at } N \tag{6.74}
$$

---

[4]Concerning Volterra integral equations, see Chapter 17, Section 1.

[5]See E. Goursat [28].

(see Figure 6.2) determining the Riemann function is not a Cauchy problem. Indeed, the curve PNQ is composed of segments PN and NQ of the characteristics $\xi \equiv x = \mathrm{const}$ and $\eta \equiv y = \mathrm{const}$, whereas the Cauchy problem is to determine a solution of a hyperbolic equation in a neighborhood of the carrier Γ of the Cauchy data, which is the curve not tangent to a characteristic at any point. The problem of determining a solution of a hyperbolic equation taking prescribed values on characteristics of two different families emanating from the same point is known as the *Goursat problem*. Thus the Riemann function, considered as the function of the first pair of variables (x, y) with the second pair (ξ, η) treated as fixed parameters, is a solution not of the Cauchy problem but of the Goursat problem.

The Goursat problem is thus as follows: Find a function $u(x, y, \xi, \eta)$ inside an angle $D^* \in \mathbb{R}_2$,

$$D^* = \{x, y : x > \xi, \; y > \eta\} \tag{6.75}$$

such that

$$Lu \stackrel{\mathrm{def}}{=} 2\frac{\partial^2}{\partial x \partial y}u + a(x,y)\frac{\partial}{\partial x}u + b(x,y)\frac{\partial}{\partial y}u + c(x,y)u + f(x,y) = 0 \tag{6.76}$$

$$u(\xi + 0, s, \xi, \eta) = \varphi(\xi, s, \xi, \eta) \quad \forall s \in [\eta, y] \tag{6.77}$$

$$u(s, \eta + 0, \xi, \eta) = \Psi(s, \eta, \xi, \eta) \quad \forall s \in [\xi, x] \tag{6.78}$$

where φ and ψ are continuously differentiable functions and also

$$\varphi(\xi, \eta, \xi, \eta) = \Psi(\xi, \eta, \xi, \eta) \tag{6.79}$$

Theorem 6.2.1. *If coefficients a, b, c of equation (6.76) and its free term $f(x, y)$ are bounded continuous functions in D^* and conditions (6.77)–(6.79) are satisfied, then a global solution of problem (6.76)–(6.79) exists and is unique.*

PROOF. First consider an auxiliary problem: Find $\tilde{v}(x, y, \xi, \eta)$ such that

$$L^*\tilde{v} \stackrel{\mathrm{def}}{=} 2\frac{\partial^2}{\partial x \partial y}\tilde{v} + F(x, y) = 0 \tag{6.80}$$

$$\tilde{v}(\xi + 0, y, \xi, \eta) = \varphi(\xi, y, \xi, \eta), \tilde{v}(x, \eta + 0, \xi, \eta) = \Psi(x, \eta, \xi, \eta) \tag{6.81}$$

Let $\tilde{v}(x, y, \xi, \eta)$ be the solution of this auxiliary problem.[6] Then integration yields

$$\int_\xi^x dx_1 \int_\eta^y \frac{\partial^2}{\partial x_1 \partial y_1}\tilde{v}(x_1, y_1, \xi, \eta)dy_1 = -\frac{1}{2}\int_\xi^x dx_1 \int_\eta^y F(x_1, y_1)dy_1 \tag{6.82}$$

---

[6]The Riemann function for the 1-dimensional wave operator obviously exists, is unique, and is identically equal to 1 (see Chapter 5, Section 3). Hence the solution of the auxiliary problem (6.80) and (6.81) exists and is unique. Assumption of differentiability of functions φ and ψ allows us to consider a classical solution of the auxiliary problem rather than a generalized one.

or, by virtue of (6.81),

$$\tilde{v}(x, y, \xi, \eta) = \varphi(\xi, y, \xi, \eta) + \Psi(x, \eta, \xi, \eta) - \Psi(\xi, \eta, \xi, \eta) - \frac{1}{2} \int_{\xi}^{x} dx_1 \int_{\eta}^{y} F(x_1, y_1) dy_1 \tag{6.83}$$

We now return to the original Goursat problem by setting

$$\tilde{v} = u, F = f + a\frac{\partial}{\partial x}u + b\frac{\partial}{\partial y}u + cu \tag{6.84}$$

Equation (6.83) now becomes an integrodifferential equation for u:

$$u(x, y, \xi, \eta) = \varphi(x, y, \xi, \eta)\big|_{x=\xi} + \Psi(x, y, \xi, \eta)\big|_{y=\eta} - \Psi(x, y, \xi, \eta)\big|_{x=\xi, y=\eta}$$

$$- \frac{1}{2} \int_{\xi}^{x} dx_1 \int_{\eta}^{y} \left(f(x_1, y_1) + a(x_1, y_1)\frac{\partial}{\partial x_1} u(x_1, y_1, \xi, \eta) \right.$$

$$\left. + b(x_1, y_1)\frac{\partial}{\partial y_1} u(x_1, y_1, \xi, \eta) + c(x_1, y_1) u(x_1, y_1, \xi, \eta,) \right) dy_1 \tag{6.85}$$

Denote

$$\Phi_0(x, y, \xi, \eta) \overset{\text{def}}{=} \varphi(\xi, y, \xi, \eta) + \Psi(x, \eta, \xi, \eta) - \Psi(\xi, \eta, \xi, \eta)$$

$$- \frac{1}{2} \int_{\xi}^{x} dx_1 \int_{\eta}^{y} f(x_1, y_1) dy_1 \tag{6.86}$$

$$\Phi_1(x, y, \xi, \eta) \overset{\text{def}}{=} \frac{\partial}{\partial x}\Phi_0(x, y, \xi, \eta) \tag{6.87}$$

$$\Phi_2(x, y, \xi, \eta) \overset{\text{def}}{=} \frac{\partial}{\partial y}\Phi_0(x, y, \xi, \eta) \tag{6.88}$$

and

$$v(x, y, \xi, \eta) = \frac{\partial}{\partial x}u(x, y, \xi, \eta), \qquad w(x, y, \xi, \eta) = \frac{\partial}{\partial y}u(x, y, \xi, \eta) \tag{6.89}$$

It is obvious that solution of the integrodifferential equation (6.83) is equivalent to solution of the system of integral equations

$$u = \Phi_0 + \lambda \int_{\xi}^{x} dx_1 \int_{\eta}^{y} (av + bw + cu) dy_1 \tag{6.90}$$

$$v = \Phi_1 + \lambda \int_{\eta}^{y} (av + bw + cu)\big|_{x_1=x} dy_1 \tag{6.91}$$

$$w = \Phi_2 + \lambda \int_{\xi}^{x} (av + bw + cu)\big|_{y_1=y} dx_1 \tag{6.92}$$

if

$$\lambda = -\frac{1}{2} \tag{6.93}$$

Let us seek a solution of system (6.90)–(6.92) as a series expansion:

$$u = \sum_{n=0}^{\infty} u_n \lambda^n, \quad v = \sum_{n=0}^{\infty} v_n \lambda^n, \quad w = \sum_{n=0}^{\infty} w_n \lambda^n \tag{6.94}$$

Assume that these series are convergent in $\bar{D}$,[7] uniformly in λ. Then, inserting them into equations (6.90)–(6.92), changing the order of summation and integration, and equating coefficients of like powers of λ, we find that

$$u_0 = \Phi_0, \quad v_0 = \Phi_1, \quad w_0 = \Phi_2 \tag{6.95}$$

$$u_{n+1} = \int_\xi^x dx_1 \int_\eta^y (av_n + bv_n + cu_n)\,dy_1 \tag{6.96}$$

$$v_{n+1} = \int_\eta^y (av_n + bw_n + cu_n)\,\big|_{x_1 = x}\, dy_1 \tag{6.97}$$

$$w_{n+1} = \int_\xi^x (av_n + bw_n + cu_n)\,\big|_{y_1 = y}\, dx_1 \tag{6.98}$$

Let $\alpha > 0$ be arbitrarily large, denote

$$D_\alpha = \{x, y : \xi < x < \xi + \alpha, \ \eta < y < \eta + \alpha\} \tag{6.99}$$

and assume that

$$|\Phi_i| < M, \quad i = 0, 1, 2, \qquad |a|, |b|, |c| < K \quad \forall (x, y) \in D \tag{6.100}$$

Then it follows from (6.95)–(6.98) that

$$|v_1| < 3MK(y - \eta), \quad |w_1| < 3MK(x - \xi), \quad |u_1| < 3MK(x - \xi)(y - \eta) \tag{6.101}$$

or, majorizing,

$$|v_1| < 3MK\frac{x - \xi + y - \eta}{1}, \qquad |w_1| < 3MK\frac{x - \xi + y - \eta}{1} \tag{6.102}$$

Moreover, noting that

$$(x - \xi)(y - \eta) < \frac{(x - \xi)^2 + (y - \eta)^2}{2} < \frac{(x - \xi + y - \eta)^2}{2} \tag{6.103}$$

we see that

$$|u_1| < 3MK\frac{(x - \xi + y - \eta)^2}{2} \tag{6.104}$$

Taking

$$N = 2K(1 + \alpha) \tag{6.105}$$

we shall prove by induction that, for every integer $n \geq 1$,

$$|v_n|, |w_n| < 3MKN^{n-1}\frac{(x - \xi + y - \eta)^n}{n!}, \ |u_n| < 3MKN^{n-1}\frac{(x - \xi + y - \eta)^{n+1}}{(n + 1)!} \tag{6.106}$$

---

[7]Recall that D is the interior of the rectangle $MPNQM$ and that $\bar{D}$ is the closure of D.

Suppose that (6.106) is true. Then, by (6.98)–(6.100),

$$|v_{n+1}| < \frac{3MKN^{n-1}}{n!} \int_0^{y-\eta} [2K(x-\xi+s)^n + \frac{K}{n+1}(x-\xi+s)^{n+1}]ds$$

$$< \frac{3MKN^n}{n!} \int_\eta^y (x-\xi+s)^n ds < \frac{3MKN^n}{n!} \int_0^{x-\xi+y-\eta} s^n \cdot ds$$

$$= \frac{3MKN^{n+1}}{(n+1)!}(x-\xi+y-\eta)^{n+1} \tag{6.107}$$

Clearly, the same estimate is true for $|w_{n+1}|$. Furthermore,

$$|u_{n+1}| < \frac{3MKN^{n-1}}{n!} \int_\xi^x dx_1 \int_\eta^y \left[2K(x_1-\xi+y_1-\eta)^n \right.$$

$$\left. + \frac{K}{n+1}(x_1-\xi+y_1-\eta)^{n+1} \right] dy_1$$

$$< \frac{3MKN^n}{n!} \int_\xi^x dx_1 \int_\eta^y (x_1-\xi+y_1-\eta)^n dy_1 \tag{6.108}$$

which yields

$$|u_{n+1}| < \frac{3MKN^n}{(n+2)!}(x-\xi+y-\eta)^{n+2} \tag{6.109}$$

Since (6.106) is true for $n=1$, our claim is proved. Hence the series (6.94) are majorized by the series

$$3MK \sum_{n=1}^n \frac{|\lambda|^n N^{n-1}(x-\xi+y-\eta)^{n+1}}{(n+1)!} \tag{6.110}$$

$$3MK \sum_{n=1}^n \frac{|\lambda|^n N^{n-1}(x-\xi+y-\eta)^n}{n!} \tag{6.111}$$

which have infinitely large radii of convergence. Hence the series (6.94) converge uniformly and absolutely inside D_α for any λ, in particular for $\lambda = -\frac{1}{2}$. Therefore, the Goursat problem indeed has a solution.[8]

It is easy to verify that the solution is unique. Indeed, suppose that there exist two solutions of this problem, (u,v,w) and (u',v',w'). Let

$$\tilde{u} = u - u', \quad \tilde{v} = v - v', \quad \tilde{w} = w - w' \tag{6.112}$$

Then $\tilde{u}, \tilde{v}, \tilde{w}$ satisfy system (6.90)–(6.92) of integral equations with

$$\Phi_i \equiv 0, \quad i = 0, 1, 2 \tag{6.113}$$

Obviously

$$u = v = w \equiv 0 \tag{6.114}$$

---

[8]This proof of the existence theorem is borrowed from Goursat's monograph [28].

in some sufficiently small neighborhood of the point (ξ, η). Indeed, let

$$\max(|u| + |v| + |w|) < K_0 \quad \forall x \in (\xi, \xi + \alpha), \ \forall y \in (\eta, \eta + \alpha) \tag{6.115}$$

Then it follows from (6.90)–(6.92) and (6.113) that

$$K_0 < 2KK_0\alpha + KK_0\alpha^2 \tag{6.116}$$

Hence, if

$$2\alpha + \alpha^2 < \frac{1}{K} \tag{6.117}$$

then (6.116) yields the contradictory inequality

$$K_0 < K_0 \tag{6.118}$$

Thus the solution of the Goursat problem (6.76)–(6.79) is locally (i.e., for sufficiently small α) unique. To prove that it is also globally unique, suppose the contrary. Note that there exists $\alpha > 0$ such that inequality (6.115) is valid. Consider a Dedekind cut (A, B) of the ray $\alpha > 0$ defined by

$$\alpha \in A \quad \text{if} \ u = v = w \equiv 0 \ \forall(x, y) \in D_\alpha = \{x, y : \xi < x < \xi + \alpha, \ \eta < y < \eta + \alpha\} \tag{6.119}$$

and

$$\alpha \in B \quad \text{if} \ \exists(x_0, y_0), \ x_0 > \xi + \alpha \ \text{or} \ y_0 > \eta + \alpha \tag{6.120}$$

such that

$$u^2(x_0, y_0) + v^2(x_0, y_0) + w^2(x_0, y_0) > 0 \tag{6.121}$$

Let $\tilde{\alpha}$ be the number defined by this cut:

$$\tilde{\alpha} = \bar{A} \cap \bar{B} \tag{6.122}$$

Then for all $\varepsilon > 0$ there exists $K_0 > 0$ such that

$$\xi + \tilde{\alpha} < x < \xi + \tilde{\alpha} + \varepsilon, \ \eta + \tilde{\alpha} < y < \eta + \tilde{\alpha} + \varepsilon$$
$$\Rightarrow \max(|u| + |v| + |w|) < K_0 \tag{6.123}$$

By (6.90)–(6.92) and (6.113), this implies

$$0 < K_0 < KK_0(2\varepsilon + \varepsilon^2) \tag{6.124}$$

But again this inequality is contradictory if

$$2\varepsilon + \varepsilon^2 < \frac{1}{K} \tag{6.125}$$

Q.E.D.

We now return to the properties of the Riemann function. Theorem 6.2.1 proves that it exists and is unique. However, this is not enough: We must also prove that the Riemann function satisfies the conditions[9]

$$\frac{\partial^2}{\partial x \partial \xi}g, \ \frac{\partial^2}{\partial y \partial \xi}g, \ \frac{\partial^3}{\partial x \partial y \partial \xi}g \in C(D) \tag{6.126}$$

---

[9]See note 3 on p.160.

Theorem 6.2.2. *Let $\varphi(x,y,\xi,\eta)$ and $\Psi(x,y,\xi,\eta)$ possess in D^* continuous and uniformly bounded derivatives*

$$\frac{\partial}{\partial x}\varphi(x,y,\xi,\eta)\big|_{x=s}, \quad \frac{\partial}{\partial x}\Psi(x,y,\xi,\eta)\big|_{x=s},$$

$$\frac{\partial^2}{\partial x \partial y}\varphi(x,y,\xi,\eta)\big|_{x=s} \quad and \quad \frac{\partial}{\partial \xi}\Psi(s,y,\xi,\eta)$$

Then the solution of the Goursat problem (6.75)–(6.79) possesses in D^* continuous and uniformly bounded derivatives $\partial^2 u/\partial x \partial\xi, \partial^2 u/\partial y \partial\xi, \partial^3 u/\partial x \partial y \partial\xi$.

Theorem 6.2.3. *If the coefficients a and b of the adjoint operator M of L are continuously differentiable, then the Riemann function $g(x,y,\xi,\eta)$ satisfies the conditions of Theorem 6.2.2, so that the derivatives $\partial^2 g/\partial x \partial\xi$, $\partial^2 g/\partial y \partial\xi$, $\partial^3 g/\partial x \partial y \partial\xi$ exist and are indeed continuous and uniformly bounded in D^*.*

Remark. In the definition of M it is tacitly assumed that the coefficients a and b are differentiable.

PROOF OF THEOREM 6.2.3. Assume that Theorem 6.2.2 has been proved. Equalities (6.30) and (6.31) yield

$$\frac{\partial}{\partial \xi}g(x,y,\xi,\eta)\big|_{y=\eta}=\frac{1}{2}b(\xi,\eta)\exp\left(-\frac{1}{2}\int_\xi^x b(s,\eta)ds\right) \tag{6.127}$$

$$\frac{\partial}{\partial \xi}g(x,y,\xi,\eta)\big|_{x=\xi}=-\frac{1}{2}\int_\eta^y \frac{\partial}{\partial \xi}a(\xi,s)ds\exp\left(-\frac{1}{2}\int_\eta^y a(\xi,s)ds\right) \tag{6.128}$$

Defining

$$\varphi(s,y,\xi,\eta)=g(s,y,\xi,\eta), \quad \Psi(x,s,\xi,\eta)=g(x,s,\xi,\eta) \tag{6.129}$$

we find that the Goursat data satisfy all the conditions of Theorem 6.2.2. **Q.E.D.**

PROOF OF THEOREM 6.2.2. Denote

$$\bar{u}(x,y,\xi,\eta)=\frac{\partial}{\partial \xi}u(x,y,\xi,\eta), \quad \bar{v}(x,y,\xi,\eta)=\frac{\partial}{\partial \xi}v(x,y,\xi,\eta)$$

$$\bar{w}(x,y,\xi,\eta)=\frac{\partial}{\partial \xi}w(x,y,\xi,\eta) \tag{6.130}$$

and

$$\bar{\Phi}_0(x,y,\xi,\eta)\overset{\text{def}}{=}\frac{\partial}{\partial \xi}\Phi_0(x,y,\xi,\eta)$$

$$-\frac{1}{2}\int_\eta^y (a(\xi,y_1)v(\xi,y_1,\xi,\eta)+b(\xi,y_1)w(\xi,y_1,\xi,\eta)$$

$$+c(\xi,y_1)u(\xi,y_1,\xi,\eta))\,dy_1 \tag{6.131}$$

$$\bar{\Phi}_1(x,y,\xi,\eta)\overset{\text{def}}{=}\frac{\partial}{\partial x}\bar{\Phi}_0(x,y,\xi,\eta), \bar{\Phi}_2(x,y,\xi,\eta)\overset{\text{def}}{=}\frac{\partial}{\partial y}\bar{\Phi}_0(x,y,\xi,\eta) \tag{6.132}$$

Where Γ_0 is defined by (6.86), it follows from (6.90)–(6.93) and (6.130)–(6.132) that

$$\bar{u} = \bar{\Phi}_0 + \lambda \int_\xi^x dx_1 \int_\eta^y (a\bar{v} + b\bar{w} + c\bar{u})dy_1 \tag{6.133}$$

$$\bar{v} = \bar{\Phi}_1 + \lambda \int_\eta^y (a\bar{v} + b\bar{w} + c\bar{u})\,|_{x_1=x}\, dy_1 \tag{6.134}$$

$$\bar{w} = \bar{\Phi}_2 + \lambda \int_\xi^x (a\bar{v} + b\bar{w} + c\bar{u})\,|_{y_1=y}\, dx_1 \tag{6.135}$$

Here $\bar{\Phi}_i(x,y,\xi,\eta)$, $i = 0,1,2$, may be treated as known, since the solution of system (6.90) and (6.91) may be assumed to be known owing to Theorem 6.2.1. Hence system (6.133)–(6.135) has a unique global solution for any λ, in particular for $\lambda = -1/2$. It remains only to prove that insertion of the solution of (6.133)–(6.135) into equalities (6.130)–(6.132) makes them identities. To verify this, note that if one considers $\bar{u}, \bar{v}, \bar{w}$ as given functions, then the values of their integrals are defined up to a constant of integration, which may be chosen arbitrarily. Taking this into account, define

$$\tilde{u}(x,y,\xi,\eta) = -\int_\xi^x \bar{u}(x,y,s,\eta)ds + u(x,y,x,\eta) \tag{6.136}$$

$$\tilde{v}(x,y,\xi,\eta) = -\int_\xi^x \bar{v}(x,y,s,\eta)ds + v(x,y,x,\eta) \tag{6.137}$$

$$\tilde{w}(x,y,\xi,\eta) = -\int_\xi^x \bar{w}(x,y,s,\eta)ds + w(x,y,x,\eta) \tag{6.138}$$

Then the identities (6.130) will follow from the identities

$$\delta u(x,y,\xi,\eta) \equiv 0, \quad \delta v(x,y,\xi,\eta) \equiv 0, \quad \delta u(x,y,\xi,\eta) \equiv 0 \tag{6.139}$$

where

$$\delta u \stackrel{\text{def}}{=} \tilde{u} - u, \delta v \stackrel{\text{def}}{=} \tilde{v} - v, \delta w \stackrel{\text{def}}{=} \tilde{w} - w \tag{6.140}$$

These identities may in turn be derived from the following.

Lemma 6.2.1. *The differences $\delta u, \delta v, \delta w$ are solutions of a homogeneous system of integral equations of Volterra type with continuous and uniformly bounded kernels; since the solution of such a system is unique, this implies the validity of identities (6.139).*

Assume that Lemma 6.2.1 has been proved. By (6.136)–(6.139), u, v, w are differentiable with respect to ξ, so that the definitions (6.136)–(6.138) imply (6.130), proving Theorem 6.2.2.

PROOF OF LEMMA 6.2.1. The definitions (6.131), (6.132), and (6.136)–(6.138), equations (6.133)–(6.135) with $\lambda = -1/2$, and (6.90) yield

$$\delta u(x,y,\xi,\eta) = u(x,y,x,\eta) - u(x,y,\xi,\eta) - \int_\xi^x \bar{u}(x,y,s,\eta)ds = J_1 + J_2 \tag{6.141}$$

where

$$J_1 = \frac{1}{2} \int_\xi^x ds \int_\eta^y \{a(s,y_1)[v(s,y_1,s,\eta) - v(s,y_1,\xi,\eta)] + b(s,y_1)[w(s,y_1,s,\eta)$$
$$- w(s,y_1,\xi,\eta)] + c(s,y_1)[u(s,y_1,s,\eta) - u(s,y_1,\xi,\eta)]\}dy_1 \qquad (6.142)$$

$$J_2 = \frac{1}{2} \int_\xi^x dx_1 \int_{x_1}^x ds \int_\eta^y [a(x_1,y_1)\bar{v}(x_1,y_1,\xi,\eta)$$
$$+ b(x_1,y_1)\bar{w}(x_1,y_1,\xi,\eta) + \bar{c}(x_1,y_1)u(x_1,y_1,\xi,\eta)]dy_1 \qquad (6.143)$$

Changing the order of integration and using definitions (6.136)–(6.138), we obtain

$$J_2 = -\frac{1}{2} \int_\xi^x ds \int_\eta^y [a(s,y_1)\delta v(s,y_1,\xi,\eta) + b(s,y_1)\delta w(s,y_1,\xi,\eta)$$
$$+ c(s,y_1)\delta u(s,y_1,\xi,\eta)]dy_1 - J_1 \qquad (6.144)$$

Hence, by (6.143),

$$\delta u(x,y,\xi,\eta) = -\frac{1}{2} \int_\xi^x ds \int_\eta^y [a(s,y_1)\delta v(s,y_1,\xi,\eta) + b(s,y_1)\delta w(s,y_1,\xi,\eta)$$
$$+ c(s,y_1)\delta u(s,y_1,\xi,\eta)]dy_1 \qquad (6.145)$$

Quite similarly, we obtain

$$\delta v(x,y,\xi,\eta) = -\int_\eta^y [a(x,y_1)\delta v(x,y_1,x,\eta) + b(x,y_1)\delta w(x,y_1,x,\eta)$$
$$+ c(x,y_1)\delta u(x,y_1,x,\eta)]dy_1 \qquad (6.146)$$

and

$$\delta w(x,y,\xi,\eta) = -\int_\zeta^x [a(x_1,y)\delta v(x_1,y,\xi,\eta) + b(x_1,y)\delta w(x_1,y,\xi,\eta)$$
$$+ c(x_1,y)\delta u(x_1,y,\xi,\eta)]dx_1 \qquad (6.147)$$

Thus $(\delta u, \delta v, \delta w)$ is indeed the solution of a homogeneous system of integral equations of the Volterra type. **Q.E.D.**

3. Dynamics of sorption from solution percolating through layer of porous adsorbent. Riemann function for a linear hyperbolic equation with constant coefficients

The dynamics of sorption by a porous adsorbent, from a solution percolating through the porous space at a constant velocity much greater than the rate of redistribution of concentration due to diffusion, is described by the following boundary value problem (see Chapter 2, Section 8 and Chapter 4,

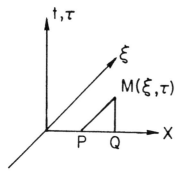

Figure 6.3.   Region of influence in coordinates (x,t) and (ξ,τ).

Section 5).

$$m\frac{\partial c}{\partial t} + s\frac{\partial a}{\partial t} + v\frac{\partial c}{\partial x} = 0, \quad 0 < x < \infty, \ t > 0 \tag{6.148}$$

$$\frac{\partial a}{\partial t} = \beta\left(c - \frac{a}{\Gamma}\right), \quad 0 < x < \infty, \ t > 0 \tag{6.149}$$

$$c(0,t) = c_0 > 0, \quad a(0,t) = 0, \quad t > 0 \tag{6.150}$$

$$c(x,0) = a(x,0) = 0, \quad 0 < x < \infty \tag{6.151}$$

The characteristics of system (6.148) and (6.149) are (see Figure 6.3)

$$\xi = x, \qquad \tau = t - \frac{m}{v}x \tag{6.152}$$

Introducing the characteristics as independent variables, we obtain

$$\frac{\partial \tilde{c}}{\partial \xi} + \frac{s}{v}\beta\left(\tilde{c} - \frac{\tilde{a}}{\Gamma}\right) = 0, \quad \xi > 0 \tag{6.153}$$

$$\frac{\partial \tilde{a}}{\partial \tau} = \beta\left(\tilde{c} - \frac{\tilde{a}}{\Gamma}\right), \quad \tau > -\frac{m}{v}\xi \tag{6.154}$$

and the boundary and initial conditions become

$$\tilde{c}(0,\tau) = c_0 > 0, \quad \tilde{a}(0,t) = 0, \quad \tau > 0 \tag{6.155}$$

$$\tilde{c}(\xi, -\frac{m}{v}\xi) = \tilde{a}(\xi, -\frac{m}{v}\xi) = 0, \quad 0 < \xi < \infty \tag{6.156}$$

where

$$\tilde{c}(\xi,\tau) \stackrel{\text{def}}{=} c(c,t), \tilde{a}(\xi,\tau) \stackrel{\text{def}}{=} a(x,t) \tag{6.157}$$

By scaling

$$u = \frac{\tilde{c}}{c_0}, \quad w = \frac{\tilde{a}}{\Gamma c_0}, \quad y = \frac{\xi}{L}, \quad z = \frac{\tau}{T}, \quad L = \frac{v}{s\beta}, \quad T = \frac{\Gamma}{\beta}, \quad \alpha = \frac{m}{s\Gamma} \tag{6.158}$$

system (6.152)–(6.157) can be reduced to the form

$$\frac{\partial u}{\partial y} + u - w = 0, \quad y > 0, \ z > -\alpha y \tag{6.159}$$

$$\frac{\partial w}{\partial z} = u - w, \quad y > 0, \ z > -\alpha y \tag{6.160}$$

$$u = 1 \quad \text{at } y = 0, \ z > 0 \tag{6.161}$$

$$u = w = 0 \quad \text{at } z = -\alpha y, \ y > 0 \tag{6.162}$$

Elimination of w yields

$$\frac{\partial^2 u}{\partial y \partial z} + \frac{\partial u}{\partial z} + \frac{\partial u}{\partial y} = 0 \tag{6.163}$$

$$u(0, z) = 1 \quad \forall z > 0 \tag{6.164}$$

$$u(y, -\alpha y) = 0, \quad \frac{\partial}{\partial y} u(y, z) \big|_{z=-\alpha y} = 0, \quad y > 0 \tag{6.165}$$

Note that $(0,0)$ is a point of discontinuity of u. This discontinuity propagates along the characteristic $z = 0$, which divides the region

$$D = \{y, z : y > 0, \ z > -y\} \tag{6.166}$$

in which u is to be determined, into two subregions

$$D_1 = \{y, z : y > 0, \ -y < z < 0\} \tag{6.167}$$
$$D_2 = \{y, z : y > 0, \ z > 0\} \tag{6.168}$$

Problem (6.163)–(6.165) is a Cauchy problem within D_1 with the line

$$\Gamma = \{y, z : y > 0, \ z = -y\} \tag{6.169}$$

as carrier of homogeneous Cauchy data

$$u = \frac{\partial u}{\partial y} = \frac{\partial u}{\partial z} \equiv 0 \quad \text{along } \Gamma \tag{6.170}$$

Hence

$$u(y, z) \equiv 0 \quad \forall (y, z) \in D_1 \tag{6.171}$$

so that

$$u(y, -0) \equiv 0 \tag{6.172}$$

We now determine $u(y, +0)$. To that end we note that $(0,0)$ is not a strong discontinuity point of w, because of conditions (6.149) and (6.151). Hence $w(y, z)$ is continuous along Γ:

$$w(y, -0) \equiv w(y, +0) \tag{6.173}$$

But by (6.159) and (6.171),

$$w(y, z) \equiv 0 \quad \forall (y, z) \in \bar{D}_1 \tag{6.174}$$

Hence by (6.173) and (6.174) and (6.159),

$$\frac{\partial u}{\partial y} + u \equiv 0 \quad \forall y > 0, \ z = +0 \tag{6.175}$$

Integration of (6.175), taking the boundary condition (6.161) into account, yields

$$u(y, +0) = e^{-y} \quad \forall y > 0 \tag{6.176}$$

Thus $u(y, z)$ is the solution in D_2 of the Goursat problem

$$\frac{\partial^2 u}{\partial y \partial z} + \frac{\partial u}{\partial z} + \frac{\partial u}{\partial y} = 0 \quad \forall (y, z) \in D_2 \tag{6.177}$$

$$u(0, z) = 1 \quad \forall z > 0, \quad u(y, 0) = \exp(-y) \quad \forall y > 0 \tag{6.178}$$

Note that the substitution

$$u(y, z) = \exp[-(y + z)]U(y, z) \tag{6.179}$$

reduces problem (6.177) and (6.178) to the problem

$$\frac{\partial^2 U}{\partial y \partial z} - U = 0 \quad \forall (y, z) \in D_2 \tag{6.180}$$

$$U(0, z) = \exp(z) \quad \forall z > 0 \tag{6.181}$$

$$U(y, 0) = 1 \quad \forall y > 0 \tag{6.182}$$

Except for a 1-dimensional wave equation, the equation

$$\frac{\partial^2}{\partial x \partial y} u + cu = 0 \tag{6.183}$$

is the simplest one for which Riemann's function may be constructed in an analytically closed, explicit form. Indeed, in this case the definition of the Riemann function gives

$$\frac{\partial^2}{\partial x \partial y} g + cg = 0, \quad g(x, y, x, \eta) \equiv 1, \quad g(x, y, \xi, y) \equiv 1 \tag{6.184}$$

It follows from (6.184) that we may introduce a similarity variable by setting

$$g(x, y, \xi, \eta) \equiv g(\lambda), \quad \lambda = 2[c(x - \xi)(y - \eta)]^{1/2} \tag{6.185}$$

Substitution of (6.185) into (6.183) yields

$$\frac{d^2}{d\lambda^2} g + \frac{1}{\lambda} \frac{d}{dx} g + g = 0, \quad g(0) = 1 \tag{6.186}$$

Thus (see Appendix 2)

$$g(x, y, \xi, \eta) = J_0 \left[2[c(x - \xi)(y - \eta)]^{1/2} \right] \tag{6.187}$$

where J_0 is the Bessel function of zero order.

In the case under consideration $c = -1$. Hence

$$g(y, z, \eta, \zeta) = I_0 \left[2[(y - \eta)(z - \zeta)]^{1/2} \right] \tag{6.188}$$

where I_0 is the Bessel function of zero order of an imaginary argument. Thus, by (6.22)–(6.24), (6.180)–(6.182), and (6.184), we have

$$U(y, z) = Ug \left|_{\eta=\zeta=0} + \int_0^z g \frac{\partial U}{\partial \zeta} \right|_{\eta=0} d\zeta + \int_0^y g \frac{\partial U}{\partial \eta} \left|_{\zeta=0} d\eta \right.$$

$$\Rightarrow U(y, z) = I_0[(2yz)^{1/2}] + \int_0^z \exp(\zeta) I_0 \left([\, z - \zeta)y]^{1/2} \right) d\zeta \tag{6.189}$$

Returning to the concentration $u(y, z)$, we obtain

$$u(y, z) = \exp[-(y + z)]I_0[(2yz)^{1/2}] + \int_0^z \exp[-(y + \zeta)]I_0[(2y\zeta)^{1/2}]d\zeta \quad (6.190)$$

Problems

P6.2.1. Proof of Lemma 6.2.1 was given without details of computations. Complete the proof.

P6.3.1. Can Riemann's method be used to describe the vibrations of an infinitely long, semi-infinite and finite, loaded or unloaded elastic string, instead of D'Alembert's method and the method of prolongation? If so, demonstrate; if not, explain.

P6.3.2. Propose a method to solve the following problem: Find a function $u(x, t)$ satisfying the conditions

$$\frac{\partial^2}{\partial x^2}u - \frac{\partial^2}{\partial t^2}u = 0 \quad \forall(x, t) \in \bar{D}, \quad u = f_i(t) \quad \forall(x, t) \in \Gamma_i, \quad i = 1, 2$$

$$u = u_0(x), \quad \frac{\partial}{\partial t}u = u_1(x) \quad \text{at } t = 0$$

where

$$D = \{x, t : 0 < t < \infty, \ t < x < t + a\},$$
$$\Gamma_1 = \{x, t : 0 < t < \infty, \ x = t\},$$
$$\Gamma_2 = \{x, t : 0 < t < \infty, \ x = t + a, \ a = \text{const} > 0\}$$

Is it possible to extend Riemann's method to this kind of problem? If so, demonstrate; if not, explain.

7

Cauchy problem for a 2-dimensional wave equation. The Volterra–D'Adhemar solution

1. Characteristic manifold of second-order linear hyperbolic equation with n independent variables

The formulation of the Cauchy problem for a linear hyperbolic equation of second order with n independent variables is quite similar to that in $\mathbb{R}_2$. Namely, consider a given smooth manifold

$$\Sigma = \{x : \Phi(x) = 0\}, \quad x = (x_1, x_2, \ldots, x_n) \in \mathbb{R}_n \tag{7.1}$$

without self-intersections, as well as the equation

$$U(x|u) \overset{\text{def}}{=} \sum_{k=1}^{n} \sum_{m=1}^{n} a_{km}(x) \frac{\partial^2}{\partial x_k \partial x_m} u(x) + F\left(x, u, \frac{\partial u}{\partial x_1}, \frac{\partial u}{\partial x_2}, \ldots, \frac{\partial u}{\partial x_n}\right) = 0 \tag{7.2}$$

Assume that the value of u and its derivative $\partial u / \partial \nu$ in the direction of a vector ν not tangent to Σ are prescribed on Σ:

$$u(x) = u_0(x), \quad \frac{\partial}{\partial \nu} u(x) = v(x) \quad \forall x \in \Sigma \tag{7.3}$$

The problem is: *Find a function $u(x)$ satisfying equation (7.2) in a neighborhood of Σ and satisfying conditions (7.3) on Σ.*

Let C_n be an n-dimensional circular cylinder of radius $\rho > 0$ with its axis directed along the normal $\mathbf{n}(x)$ to Σ at a point x and height h so small that $\sigma = \Sigma \cap C_n$ is a simple connected region. Assume that

$$\frac{\partial}{\partial x_n} \Phi(x) \neq 0 \quad \forall x \in \sigma \tag{7.4}$$

so that the equation of σ may be written as

$$\sigma = \{x : x_n = \Psi(x_1, x_2, \ldots, x_{n-1})\} \tag{7.5}$$

Hence the equations of $u\big|_\sigma$ and $\partial u/\partial \nu\big|_\sigma$ may be written as

$$u\big|_\sigma = u_0\,[x_1, x_2, \dots, x_{n-1}, \quad \Psi(x_1, x_2, \dots, x_{n-1})] \stackrel{\text{def}}{=} \Pi_0(x_1, x_2, \dots, x_{n-1}) \quad (7.6)$$

$$\Pi_\nu(x_1, x_2, \dots, x_{n-1}) \stackrel{\text{def}}{=} \frac{\partial}{\partial \nu} u\big|_\sigma = v\,[x_1, x_2, \dots, x_{n-1}, \Psi(x_1, x_2, \dots, x_{n-1})] \quad (7.7)$$

where, since Σ is assumed to be smooth, Π_0 is continuously differentiable.

We have

$$\Pi_k(x_1, x_2, \dots, x_{n-1}) \stackrel{\text{def}}{=} \frac{\partial}{\partial x_k}\Pi_0 = \frac{\partial}{\partial x_k} u\big|_\sigma + \frac{\partial}{\partial x_n} u\big|_\sigma \frac{\partial}{\partial x_k}\Psi, \quad k = 1, 2, \dots, n-1$$

$$(7.8)$$

and

$$\Pi_\nu = \sum_{k=1}^{n} \frac{\partial}{\partial x_k} u\big|_\sigma \cos(\widehat{x_k \nu}) \quad (7.9)$$

By assumption, ν is not tangent to Σ. Hence

$$\Delta = \begin{vmatrix} 1 & 0 & 0 & \Psi_1 \\ 0 & 1 & 0 & \Psi_2 \\ \vdots & \vdots & \cdots & \vdots & \vdots \\ 0 & 0 & 0 & \Psi_{n-1} \\ \cos(\widehat{x_1\nu}) & \cos(\widehat{x_2\nu}) & \cos(\widehat{x_{n-1}\nu}) & \cos(\widehat{x_n\nu}) \end{vmatrix} \neq 0 \quad (7.10)$$

so that system (7.7)–(7.9) determines all the partial derivatives of u on σ:

$$\frac{\partial}{\partial x_k} u\big|_\sigma = \Pi_k(x_1, x_2, \dots, x_{n-1}), \quad k = 1, 2, \dots, n \quad (7.11)$$

The quantities Π_k are connected by the equalities

$$\frac{\partial}{\partial x_k}\Pi_0 = \Pi_k + \Pi_n \frac{\partial}{\partial x_k}\Psi, \quad k = 1, 2, \dots, n-1 \quad (7.12)$$

Thus the Cauchy problem for equation (7.2) is to determine a function u in the neighborhood of Σ that satisfies this equation and conditions (7.11) and (7.12).

By analogy with the question considered in Chapter 3, we can now try to ascertain whether all these data allow us to compute all the second derivatives of u. We have

$$\frac{\partial}{\partial x_j}\Pi_i = \frac{\partial^2}{\partial x_i \partial x_j} u\big|_\sigma + \frac{\partial^2 u}{\partial x_i \partial x_n}\big|_\sigma \frac{\partial \Psi}{\partial x_j}, \quad i, j = 1, 2, \dots, n-1 \quad (7.13)$$

$$\frac{\partial}{\partial x_i}\Pi_n = \frac{\partial^2}{\partial x_i^2} u\big|_\sigma + \frac{\partial^2 u}{\partial x_i \partial x_n}\big|_\sigma \frac{\partial \Psi}{\partial x_i}, \quad i = 1, 2, \dots, n-1 \quad (7.14)$$

Hence

$$\frac{\partial^2}{\partial x_i \partial x_j} u\big|_\sigma = \frac{\partial}{\partial x_j}\Pi_i - \frac{\partial}{\partial x_i}\Pi_n \frac{\partial}{\partial x_j}\Psi + \frac{\partial^2}{\partial x_n^2} u\big|_\sigma \frac{\partial}{\partial x_i}\Psi \frac{\partial}{\partial x_j}\Psi,$$

$$i, j = 1, 2, \dots, n-1 \quad (7.15)$$

$$\frac{\partial^2}{\partial x_i \partial x_n} u\big|_\sigma = \frac{\partial}{\partial x_i}\Pi_n - \frac{\partial^2}{\partial x_n^2} u\big|_\sigma \frac{\partial}{\partial x_i}\Psi, \quad i = 1, 2, \dots, n-1 \quad (7.16)$$

Inserting (7.15) and (7.16) into (7.2) yields

$$\left(\sum_{i,j=1}^{n-1} a_{ij} \frac{\partial \Psi}{\partial x_i} \frac{\partial \Psi}{\partial x_j} - \sum_{i=1}^{n-1} a_{in} \frac{\partial \Psi}{\partial x_i} + a_{nn} \right) \frac{\partial^2 u}{\partial x_n^2} + F^* = 0 \qquad (7.17)$$

where

$$F^* = F + \sum_{j,j=1}^{n-1} a_{ij} \left(\frac{\partial}{\partial x_i} \Pi_i - \frac{\partial}{\partial x_i} \Pi_n \frac{\partial \Psi}{\partial x_j} \right) + \sum_{i=1}^{n-1} a_{in} \frac{\partial \Pi_n}{\partial x_i} \qquad (7.18)$$

At the points of the manifold σ, the function F^* and all the terms in parentheses in the right-hand side of (7.17) are known. Thus equation (7.18) determines $\partial^2 u/\partial x_n^2$ on σ, and together with it all the other second derivatives of u on σ, if and only if

$$\mathcal{F} \overset{\text{def}}{=} \sum_{i,j=1}^{n-1} a_{ij} \frac{\partial \Psi}{\partial x_i} \frac{\partial \Psi}{\partial x_j} - \sum_{i=1}^{n-1} a_{in} \frac{\partial \Psi}{\partial x_n} + a_{nn} \neq 0 \quad \forall x \in \sigma \qquad (7.19)$$

Obviously, there is no need to represent the equation of σ in the form resolved with respect to x_n. Indeed, we have at σ

$$x_n = \Psi(x_1, x_2, \dots, x_{n-1}), \quad \Phi(x_1, x_2, \dots, x_n) \equiv 0, \quad \frac{\partial}{\partial x_n} \Phi \neq 0 \Rightarrow \frac{\partial \Psi_n}{\partial x_i} = -\frac{\frac{\partial \Phi}{\partial x_i}}{\frac{\partial \Phi}{\partial x_n}} \qquad (7.20)$$

so that (7.19) takes the symmetric form

$$\mathcal{F} \frac{\partial \Phi}{\partial x_n} = \sum_{i,j=1}^{n} a_{ij} \frac{\partial \Phi}{\partial x_i} \frac{\partial \Phi}{\partial x_j} \neq 0 \quad \forall x \in \sigma \qquad (7.21)$$

where no one of the coordinates x_k, $k = 1, \dots, n$, plays a special role.
    Therefore (7.19) may be replaced by the inequality

$$\sum_{i,j=1}^{n} a_{ij} \frac{\partial \Phi}{\partial x_i} \cdot \frac{\partial \Phi}{\partial x_j} \neq 0 \quad \forall x \in \Sigma \qquad (7.22)$$

Thus, if the equation

$$\tilde{\Phi}(x) \overset{\text{def}}{=} \sum_{i,j=1}^{n} a_{ij} \frac{\partial \Phi}{\partial x_i} \cdot \frac{\partial \Phi}{\partial x_j} = 0 \qquad (7.23)$$

holds along the manifold Σ, then the Cauchy problem (7.1) and (7.2) is either unsolvable or not uniquely solvable.
    If $n = 2$ then equation (7.23) is a characteristic equation of the second-order partial differential equation $U(x_1, x_2 | u) = 0$, so that it is a hyperbolic equation in a neighborhood of Σ if the form

$$f = \sum_{i,j=1}^{2} a_{ij} \xi_i \xi_j \qquad (7.24)$$

has real nontrivial roots. In such a case the characteristic equation has real integrals determining the characteristic curves. Similar terminology and results are valid for any dimension n of the space. Equation (7.23) is called the *characteristic equation*, and its integral

$$\Phi(x) = \text{const} \tag{7.25}$$

is known as the *characteristic manifold*. If the form

$$f = \sum_{i,j=1}^{n} a_{ij}\xi_i\xi_j \tag{7.26}$$

has only real nontrivial roots then the equation $U(x|u) = 0$ is a hyperbolic equation.

Since the characteristic equation is a first-order nonlinear partial differential equation, the theory of characteristics of such equations (see Chapter 3, Section 2) can be applied to it. Hence the characteristic equation

$$\Psi \stackrel{\text{def}}{=} \sum_{i,j=1}^{n} a_{ij}P_iP_j = 0 \tag{7.27}$$

where

$$P_i(x) \stackrel{\text{def}}{=} \frac{\partial}{\partial x_i}\Phi, \quad i = 1, 2, \dots, n \tag{7.28}$$

is to be associated with the system of ordinary differential equations

$$\frac{dx_i}{\sum_{j=1}^{n} a_{ij}P_j} = -\frac{dP_i}{\sum_{k,m=1}^{n} \frac{\partial a_{km}}{\partial x_i}P_kP_m} = \frac{d\Phi}{0}, \quad i = 1, 2, \dots, n \tag{7.29}$$

System (7.29) is called the system of differential equations of *bicharacteristics*; the integral curves of these equations are also known as *bicharacteristics*.

As an example, let us consider the wave equation in $\mathbb{R}_2$:

$$\frac{\partial^2}{\partial x^2}u + \frac{\partial^2}{\partial y^2}u - \frac{\partial^2}{\partial t^2}u = 0, \tag{7.30}$$

assuming that the rate a of the wave propagation is made 1 by a suitable scaling. The characteristic equation is

$$2\Phi \equiv \left(\frac{\partial\Psi}{\partial x}\right)^2 + \left(\frac{\partial\Psi}{\partial y}\right)^2 - \left(\frac{\partial\Psi}{\partial t}\right)^2 = 0 \tag{7.31}$$

Hence the differential equations of bicharacteristics are

$$\frac{dx}{X} = \frac{dy}{Y} = -\frac{dt}{T} = \frac{dX}{0} = \frac{dY}{0} = \frac{dT}{0} = \frac{d\Psi}{0} = ds \tag{7.32}$$

where we have used the notation

$$X \stackrel{\text{def}}{=} \frac{\partial\Psi}{\partial x}, \quad Y \stackrel{\text{def}}{=} \frac{\partial\Psi}{\partial y}, \quad T \stackrel{\text{def}}{=} \frac{\partial\Psi}{\partial t} \tag{7.33}$$

The integrals of equation (7.32) are

$$X = \text{const}, \quad Y = \text{const}, \quad T = \text{const}, \quad \Psi = \text{const} \qquad (7.34)$$

and

$$x = \xi + Xs, \quad y = \eta + Ys, \quad t = \tau - Ts \qquad (7.35)$$

The arbitrary parameters ξ, η, τ vary along a prescribed manifold—the carrier of Cauchy data. The parameter s varies along the bicharacteristics. The equation of the characteristic manifold must be added (see Chapter 3, Section 2), so that

$$X^2 + Y^2 - T^2 = 0 \qquad (7.36)$$

Let

$$p = (x, y), \quad q = (\xi, \eta), \quad r_{pq} = [(x - \xi)^2 + (y - \eta)^2]^{1/2} \qquad (7.37)$$

Equalities (7.34)–(7.36) imply

$$r_{pq} - |t - \tau| = 0 \qquad (7.38)$$

This is the equation of a cone, known as the *characteristic cone*. In the 3-dimensional case,

$$p = (x, y, z), \quad q = (\xi, \eta, \zeta) \qquad (7.39)$$

The characteristic cone is a cone in a 4-dimensional space,

$$\mathcal{E}_4 = \mathbb{R}_3 \times (|t - \tau| \geq 0) \qquad (7.40)$$

Evidently, by means of a suitable rescaling we may return to an arbitrary rate of wave propagation. The results obtained admit the following obvious interpretation in wave optics.

The characteristic manifold of the wave equation

$$a^2 \Delta u - \frac{\partial^2}{\partial t^2} u = 0 \qquad (7.41)$$

is the wave front, and the bicharacteristics are the rays. The generators of the characteristic cone are rays. A light signal appearing at the point (q, τ) of the phase space $\mathcal{E}_4$ propagates at a rate a and reaches the point $(p, t) \in \mathcal{E}_4$ on the surface of the characteristic cone.

2. Cauchy problem for the 2-dimensional wave equation. Volterra–D'Adhemar solution [1;28]

Riemann's method for solution of the 1-dimensional second-order hyperbolic equation is based on the use of the Riemann identity (4.158). An analogous identity may be obtained in

$$\mathcal{E}_3 = \mathbb{R}_2 \times (0 \leq t < \infty) \qquad (7.42)$$

For the sake of simplicity, we shall restrict ourselves to the wave equation

$$\Box u \overset{\text{def}}{=} \Delta u - \frac{\partial^2}{\partial t^2} u = 0 \qquad (7.43)$$

Define vectors

$$\mathbf{A} = \mathbf{i}\frac{\partial u}{\partial x} + \mathbf{j}\frac{\partial u}{\partial y} - \boldsymbol{\tau}\frac{\partial u}{\partial t}, \quad \mathbf{B} = \mathbf{i}\frac{\partial v}{\partial x} + \mathbf{j}\frac{\partial v}{\partial y} - \boldsymbol{\tau}\frac{\partial v}{\partial t} \qquad (7.44)$$

where $\mathbf{i}, \mathbf{j}, \boldsymbol{\tau}$ are unit basis vectors in $\mathcal{E}_3$. Clearly

$$\Box u = \operatorname{div}\mathbf{A}, \qquad \Box v = \operatorname{div}\mathbf{B} \qquad (7.45)$$

Let u and v be functions twice continuously differentiable in $\mathcal{E}_3$ and D be a closed region in $\mathcal{E}_3$ with a boundary Σ to which the Gauss divergence theorem is applicable. Further, let

$$\mathbf{n}^0 = \alpha\mathbf{i} + \beta\mathbf{j} + \gamma\boldsymbol{\tau} \qquad (7.46)$$

be the unit vector of the outward normal to Σ, and let $\boldsymbol{\nu}^0$ be the unit vector

$$\boldsymbol{\nu}^0 = \alpha\mathbf{i} + \beta\mathbf{j} - \gamma\boldsymbol{\tau} \qquad (7.47)$$

symmetric to $\mathbf{n}^0$ with respect to a plane, normal to $\boldsymbol{\tau}$ (i.e., normal to the time axis) and named by D'Adhemar the unit vector of a *conormal* to Σ. The identity

$$vu - u\Box v \equiv \operatorname{div}(v\mathbf{A} - u\mathbf{B}) \qquad (7.48)$$

is obvious. Applying the Green's formula to the identity (7.48) we obtain

$$\int_D (v\Box u - u\Box v)d\omega = \int_\Sigma \left(v\frac{\partial u}{\partial \nu} - u\frac{\partial}{\partial \nu}\right)d\sigma \qquad (7.49)$$

This identity is the aforementioned analog to Riemann's identity.

Now consider the following Cauchy problem: Given a smooth surface

$$\Sigma_0 = \{x, y, t : t = T(x,y)\} \subset \mathcal{E}_3 \qquad (7.50)$$

without self-intersections, and given functions

$$u_i(x,y), \quad i = 0, 1, 2 \qquad (7.51)$$

defined on Σ_0 and satisfying the conditions

$$\frac{\partial}{\partial x}u_0 = u_1 + u_3\frac{\partial T}{\partial x}, \quad \frac{\partial}{\partial y}u_0 = u_2 + u_3\frac{\partial T}{\partial y} \qquad (7.52)$$

the problem is: *In a neighborhood U of Σ_0, find a solution u of the wave equation*

$$\Box u = 0, \quad (p,t) \in U \qquad (7.53)$$

satisfying conditions (7.52) on Σ_0.

Let $p = (x,y,t)$ and $q = (\xi,\eta,\tau)$ be a fixed point and a variable point, respectively. Let D_p be the region bounded by the characteristic cone S_p with apex at p and by the part Σ_{0p} of Σ_0 cut out by S_p (see Figure 7.1). The equation of the plane T_0 tangent to S_p at a point $q_0 = (\xi_0, \eta_0, \tau_0)$ is

$$(\xi - \xi_0)(x - \xi_0) + (\eta - \eta_0)(y - \eta_0) - (t - \tau_0)(\tau - \tau_0) = 0 \qquad (7.54)$$

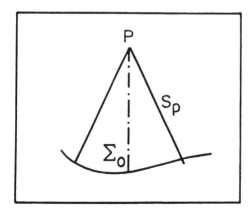

Figure 7.1. Sketch of a carrier of Cauchy data and the characteristic cone; the region D_p.

Hence

$$\alpha = \frac{x - \xi_0}{\mathfrak{M}}, \quad \beta = \frac{y - \eta_0}{\mathfrak{M}}, \quad \gamma = -\frac{t - \tau_0}{\mathfrak{M}} \tag{7.55}$$

where

$$\mathfrak{M} = [(x - \xi_0)^2 + (y - \eta_0)^2 + (t - \tau_0)^2]^{1/2} \tag{7.56}$$

On the other hand, the equation of the generatrix L of the characteristic cone S_p passing through the point q is

$$L = \left\{ \frac{\xi - \xi_0}{x - \xi_0} = \frac{\eta - \eta_0}{y - \eta_0} = \frac{\tau - \tau_0}{t - \tau_0} \right\} \tag{7.57}$$

Hence the direction cosines of L coincide with those of the conormal ν.

Now let the function

$$v(q, \tau) \stackrel{\text{def}}{=} g(p, t, q, \tau) \tag{7.58}$$

be a solution of the equation

$$\Box v = 0 \tag{7.59}$$

such that

$$v \Big|_{(q,\tau) \in S_p} = 0 \tag{7.60}$$

and

$$\frac{\partial}{\partial \nu} v \Big|_{(q,\tau) \in S_p} = 0 \tag{7.61}$$

Then, applying the integral identity (7.49) to the functions u and v, we obtain

$$\int_{D_p} (v \Box u - u \Box v) d\omega = 0 \tag{7.62}$$

In contrast to the 1-dimensional case (see Chapter 6, Section 1), this identity does not enable us to determine the value of $u(p, t)$ if $v(q, \tau) = g(p, t, q, \tau)$

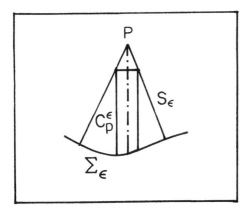

Figure 7.2.   Sketch of the Volterra–D'Adhemar solution of the Cauchy problem; the region D_M^ε.

does not possess some singularity that provides g with a "filtering" property. The Volterra–D'Adhemar method for solution of this Cauchy problem consists in the introduction of such a filtering function g. To this end, we define

$$v = v(z), \quad z = \frac{t - \tau}{r} \quad (r = r_{pq}) \tag{7.63}$$

and introduce polar coordinates (r, φ) with origin at the point p. Assuming $v(z)$ to be independent of φ, we find that

$$\Box v = \left(\frac{1}{r}\frac{\partial}{\partial r}\left[r\frac{\partial}{\partial r}v\right]\right) - \frac{\partial^2}{\partial t^2}v \equiv z\frac{d}{dz}\left(z\frac{d}{dz}v\right) - \frac{d^2}{dz^2}v = 0 \tag{7.64}$$

Recall that the equation of the characteristic cone is

$$s_p : r_{pq} - |t - \tau| = 0 \tag{7.65}$$

so that

$$z\Big|_{s_\mu} = 1 \tag{7.66}$$

Therefore the integral of equation (7.64) satisfying conditions (7.60) and (7.61) is

$$v = \ln[z + (z^2 - 1)^{1/2}] \tag{7.67}$$

This function has a logarithmic singularity along the axis $r = 0$ of the cone S, and this singularity gives v the desired filtering property. Indeed, take an arbitrarily small $\varepsilon > 0$ and consider the region (see Figure 7.2)

$$D_p^\varepsilon = D_p \setminus C_p^\varepsilon \tag{7.68}$$

where C^ε is the circular cylinder

$$C^\varepsilon = \{\tau : r = \varepsilon, \ 0 \le \varphi \le 2\pi, \ -\infty < \tau < \infty\} \tag{7.69}$$

Let $u(q, \tau)$ be a solution of the Cauchy problem under consideration, and apply the integral identity (7.49) to the region D_p^ε:

$$0 = \int_{D_p^\varepsilon} (v \,\square\, u - u \,\square\, v)d\omega = \int_{\partial D_p^\varepsilon} \left(v\frac{\partial u}{\partial \nu} - u\frac{\partial v}{\partial \nu}\right)d\sigma$$

$$- \left(\int_{\partial C_p^\varepsilon} + \int_{\Sigma_{0p}^\varepsilon}\left(u\frac{\partial v}{\partial \nu} - v\frac{\partial u}{\partial \nu}\right)d\sigma\right) \tag{7.70}$$

since

$$v\,\Big|_S = \frac{\partial v}{\partial \nu}\,\Big|_S \equiv 0 \tag{7.71}$$

In the chosen polar coordinate system,

$$\int_{\partial C_p^\varepsilon} = \varepsilon \int_0^{2\pi} d\varphi \int_{t(\varepsilon,\varphi)}^{t(\varepsilon)} \left(u\frac{\partial v}{\partial \nu} - v\frac{\partial u}{\partial \nu}\right)d\tau \tag{7.72}$$

Here

$$\lim_{\varepsilon \downarrow 0} t(\varepsilon) = t \tag{7.73}$$

and

$$\tau = t(\rho,\varphi) = T(\xi,\eta) \tag{7.74}$$

are the equations of Σ in the chosen polar and cartesian coordinates, respectively.

Note that

$$\lim_{\varepsilon \downarrow 0}(\varepsilon v) = \lim_{\varepsilon \downarrow 0}\left(\varepsilon \cdot \ln\left[\frac{(t-\tau) + [(t-\tau)^2 - \varepsilon^2]^{1/2}}{\varepsilon}\right]\right) = 0 \tag{7.75}$$

and that the direction of the conormal ν coincides here with inner normal to ∂C_p^ε. Therefore,

$$\lim_{\varepsilon \downarrow 0}\left(v\frac{\partial v}{\partial \nu}\right) = -\lim_{\varepsilon \downarrow 0}\left(\frac{(t-\tau)}{[(t-\tau)^2 - \varepsilon^2]^{1/2}}\right) = -1 \tag{7.76}$$

Hence

$$\lim_{\varepsilon \downarrow 0}\int_{\partial C_p^\varepsilon} = -2\pi\int_{T(x,y)}^t u(x,y,\tau)d\tau \tag{7.77}$$

since

$$u\,\Big|_{\rho=0} = u(x,u,\tau) \tag{7.78}$$

Thus

$$\int_{T(x,y)}^t u(x,y,\tau)d\tau = \frac{1}{2\pi}\int_{\Sigma_{0p}}\left(v\frac{\partial u}{\partial \nu} - u\frac{\partial v}{\partial \nu}\right)d\sigma \tag{7.79}$$

Differentiation of (7.79) yields Volterra's result:

$$u(x,y,t) = \frac{1}{2\pi}\frac{\partial}{\partial t}\int_{\Sigma_{0p}}\left(v\frac{\partial u}{\partial \nu} - u\frac{\partial v}{\partial \nu}\right)d\sigma \tag{7.80}$$

We have thus proved the following theorem.

Theorem 7.2.1 (Volterra [28]). *Let $u(x,y,t)$ be a solution of the following Cauchy problem: In a neighborhood $U \subset \mathcal{E}_3$ of a smooth surface Σ_0 without self-intersections,*

$$\Sigma_0 = \{x, y, t : t = T(x,y)\} \subset \mathcal{E}_3 \tag{7.81}$$

find a solution of the wave equation

$$\Box u \stackrel{\text{def}}{=} \Delta u - \frac{\partial^2}{\partial t^2} \cdot u = 0 \tag{7.82}$$

such that

$$u\big|_{\Sigma_0} = u_0, \quad \frac{\partial u}{\partial x}\bigg|_{\Sigma_0} = u_1, \quad \frac{\partial u}{\partial y}\bigg|_{\Sigma_0} = u_2, \quad \frac{\partial u}{\partial t}\bigg|_{\Sigma_0} = u_3 \tag{7.83}$$

where $u_i(x,y)$, $i = 0,1,2,3$, are functions prescribed on Σ_0;

$$\frac{\partial}{\partial x}u_0 = u_1 + u_3\frac{\partial T}{\partial x}, \quad \frac{\partial}{\partial y}u_0 = u_2 + u_3\frac{\partial T}{\partial y} \tag{7.84}$$

Then

$$u(x,y,t) = \frac{1}{2\pi}\frac{\partial}{\partial t}\int_{\Sigma_{0p}}\left(v\frac{\partial u}{\partial \nu} - u\frac{\partial v}{\partial \nu}\right)d\sigma \tag{7.85}$$

where Σ_{0p} is the part of Σ_0 cut out by the characteristic cone S_p with its apex at the point

$$(p,t) \in U, \quad p = (x,y) \in \mathbb{R}_2, (q,\tau) \in \Sigma_p, \quad q = (\xi,\eta) \tag{7.86}$$

and

$$v = v(z) \equiv \ln\left[z + \sqrt{(z^2 - 1)}\right], \quad z = (t-\tau)\left((x - \xi)^2 + (y - \eta)^2\right)^{-1/2} \tag{7.87}$$

We again emphasize that this result follows from the existence of a logarithmic singularity in the definition of the Volterra function $v(z)$ and that – just as in the case of Riemann's solution of the hyperbolic equation in 1-dimensional space – one cannot assert without further proof that the integral (7.85) furnishes a solution of the Cauchy problem for any functions u_i, $i = 0,1,2,3$, satisfying conditions (7.84). A proof was given by D'Adhemar [1]; it is omitted here. (See Problem 7.2.4.)

Problems

P7.2.1. Why must the surface Σ_0 in the formulation of the Cauchy problem be a surface without self-intersections?

P7.2.2. Find an explicit expression for the integral (7.85)–(7.87) if Σ_0 is a plane in $\mathbb{R}_2$.

P7.2.3. Is it possible to apply Volterra's result if Σ_0 is bounded by the coordinate angle $x > 0$, $y \geq 0$?

P7.2.4. Try to retrieve the omitted D'Adhemar's proof. Formulate conditions under which the equality (7.85)–(7.87) is indeed a solution to the Cauchy problem for given input data.

8

Cauchy problem for the wave equation in $\mathbb{R}_3$. Methods of averaging and descent. Huygens's principle

1. Method of averaging

Consider the Cauchy problem

$$\Box u(p,t) \overset{\text{def}}{=} a^2 \Delta u(p,t) - \frac{\partial^2}{\partial t^2} u(p,t) = 0, \quad (p,t) \in \mathcal{E}_4 = \mathbb{R}_3 \times (t \geq 0)$$

$$u(p,0) = \varphi(p), \quad \frac{\partial}{\partial t} u(p,o) = \psi(p) \tag{8.1}$$

Let $M = (x,y,z)$ and $q = (\xi, \eta, \zeta)$ be a fixed point and a variable point, respectively. Introduce polar coordinates

$$\xi = x + \alpha\rho, \quad \eta = y + \beta\rho, \quad \zeta = z + \gamma\rho \tag{8.2}$$

where

$$\alpha = \cos\varphi \sin\vartheta, \quad \beta = \sin\varphi \sin\vartheta, \quad \gamma = \cos\vartheta \tag{8.3}$$

are the direction cosines of the vector Mq. Let K_M^r be a sphere of the radius r centered at M, S_M^r its surface, and Ω a solid angle such that along S_M^r,

$$d\sigma_q = r^2 \sin\vartheta d\vartheta d\varphi = r^2 d\Omega \tag{8.4}$$

Let $v(r,t)$ be the average value of u over S_M^r:

$$v(r,t) = \frac{1}{4\pi r^2} \int_{S_M^r} u(q) d\sigma_q \tag{8.5}$$

Assume that $u(q,t)$ is a solution of the Cauchy problem (8.1). Then

$$a^2 \int_{K_M^r} \Delta u(q) d\omega_q = \int_{K_M^r} \frac{\partial^2}{\partial t^2} u(q) d\omega_q \tag{8.6}$$

Since

$$d\omega = \rho^2 d\Omega d\rho = d\sigma_q d\rho \tag{8.7}$$

this equality, together with (8.4), yields

$$\int_{K_M^r} \frac{\partial^2}{\partial t^2} u(q) d\omega_q = \int_0^r d\rho \int_{S_M^r} \frac{\partial^2}{\partial t^2} u(q) d\sigma_q \qquad (8.8)$$

By (8.5), this means that

$$\frac{\partial}{\partial r} \int_{K_M^r} \frac{\partial^2}{\partial t^2} u(q) d\omega_q = 4\pi r^2 \int_{S_M^r} \frac{\partial^2}{\partial t^2} u(q) d\sigma_q = 4\pi r^2 \frac{\partial^2}{\partial t^2} v(r,t) \qquad (8.9)$$

On the other hand, by the Gauss divergence theorem

$$\int_{K_M^r} \Delta u(q) d\omega_q = \int_{K_M^r} \text{div}[\text{grad}\, u(q)] d\omega_q$$

$$= \int_{S_M^r} \frac{\partial u}{\partial r} d\sigma_q = r^2 \int_{S_M^r} \frac{\partial u}{\partial r} d\Omega = r^2 \frac{\partial}{\partial r} \int_{S_M^r} u d\Omega \qquad (8.10)$$

Hence

$$\frac{\partial}{\partial r} \int_{K_M^r} \Delta u(q) d\omega_q = \frac{\partial}{\partial r} \left(r^2 \frac{\partial}{\partial r} \int_{S_M^r} u d\Omega \right) \qquad (8.11)$$

or, what is the same,

$$\frac{\partial}{\partial r} \int_{K_M^r} \Delta u(q) d\omega_q = 4\pi \frac{\partial}{\partial r} \left(r^2 \frac{\partial}{\partial r} \left(\frac{1}{4\pi r^2} \int_{S_M^r} u d\sigma_q \right) \right) = 4\pi \frac{\partial}{\partial r} \left(r^2 \frac{\partial v}{\partial r} \right) \qquad (8.12)$$

Comparison of (8.6), (8.9), and (8.12) yields

$$\frac{a^2 \partial}{r^2 \partial r} \left(r^2 \frac{\partial v}{\partial r} \right) = \frac{\partial^2 v}{\partial t^2} \qquad (8.13)$$

Thus $v(r,t)$ is a solution of the wave equation with spherical symmetry.

Let

$$w(r,t) = rv(r,t) \qquad (8.14)$$

It follows from (8.13) that

$$a^2 \frac{\partial^2}{\partial r^2} w = \frac{\partial^2}{\partial t^2} w \qquad (8.15)$$

By D'Alembert's method of traveling waves (see Chapter 5), equation (8.15) implies

$$w(r,t) = \Phi\left(t + \frac{r}{a} \right) + F\left(t - \frac{r}{a} \right) \qquad (8.16)$$

where Φ and F must be determined from the Cauchy data. To this end, we note that by the definition (8.5) of v and equalities (8.1)–(8.4),

$$v\big|_{r=0} = \frac{1}{4\pi} \int_{S_M^r} u(x + \alpha r, y + \beta r, z + \gamma r, t)\big|_{r=0} d\Omega = u(M,t) \qquad (8.17)$$

Hence

$$w\big|_{r=0} = 0, \quad \lim_{r\downarrow 0}\frac{w}{r} = \frac{\partial}{\partial r}w(r,t)\big|_{r=0} = u(M,t) \tag{8.18}$$

By (8.16) and (8.18),

$$\Phi(t) + F(t) = 0, \quad \frac{1}{a}\left(\frac{d}{dt}\Phi(t) - \frac{d}{dt}F(t)\right) = u(M,t) \tag{8.19}$$

This yields

$$u(M,t) = \frac{2}{a}\frac{d}{dt}\Phi(t) \tag{8.20}$$

On the other hand, differentiating (8.5) with respect to t and taking (8.14) into account, we obtain

$$\frac{\partial}{\partial t}w(r,0) = \frac{1}{4\pi r}\int_{S_M^r}\psi(q)d\sigma_q \tag{8.21}$$

Besides,

$$w(r,0) = \frac{1}{4\pi r}\int_{S_M^r}\varphi(q)d\sigma_q \tag{8.22}$$

Finally, we note that

$$w(r,t)\big|_{t=0} = \Phi\left(\frac{r}{a}\right) + F\left(-\frac{r}{a}\right) \tag{8.23}$$

This yields

$$\frac{\partial}{\partial r}w(r,t)\big|_{t=0} = \frac{1}{a}\frac{d}{dt}\left(\Phi(t)\big|_{t=r/a} - F(t)\big|_{t=-r/a}\right) \tag{8.24}$$

But by (8.16),

$$\frac{\partial}{\partial t}w(r,t)\big|_{t=0} = \frac{d}{dt}\left(\Phi(t)\big|_{t=r/a} + F(t)\big|_{t=-r/a}\right) \tag{8.25}$$

Hence it follows from (8.23)–(8.25) that

$$\frac{2}{a}\frac{d}{dt}\Phi(t)\big|_{t=r/a} = \frac{\partial}{\partial r}w(r,t)\big|_{t=0} + \frac{1}{a}\frac{\partial}{\partial t}w(r,t)\big|_{t=0} \tag{8.26}$$

Comparison of (8.26) with (8.21) and (8.22) yields

$$u(M,t) = \frac{1}{4\pi a}\left(\frac{\partial}{\partial t}\int_{S_M^{at}}\frac{\varphi(q)}{r}d\sigma_q + \int_{S_M^{at}}\frac{\psi(q)}{r}d\sigma_q\right) \tag{8.27}$$

Thus we have proved the following theorem.

Theorem 8.1.1 (Poisson). *Let $u(p,t)$ be a solution of the Cauchy problem*

$$\Box u(p,t) = 0, \quad (p,t) \in \mathcal{E}_4 = \mathbb{R}_3 \times (t \geq 0), \quad u(p,0) = \varphi(p), \quad \frac{\partial}{\partial t}u(p,0) = \psi(p) \tag{8.28}$$

Then $u(M,t)$ admits an integral representation

$$u(M,t) = \frac{1}{4\pi a}\left(\frac{\partial}{\partial t}\int_{S_M^{at}}\frac{\varphi(q)}{r}d\sigma_q + \int_{S_M^{at}}\frac{\psi(q)}{r}d\sigma_q\right) \qquad (8.29)$$

for any point $M \in \mathbb{R}_3$.

Conditions under which $u(M,t)$ represented by Poisson's formula (8.29) is indeed a solution of the Cauchy problem (8.1) are given by the following theorem.

Theorem 8.1.2. *Let $\varphi(q)$ and $\psi(q)$ be thrice and twice continuously differentiable functions, respectively, and define $u(M,t)$ by (8.29). Then $u(M,t)$ is a solution of problem (8.28) for any point $(M,t) \in \mathcal{E}_4$.*

PROOF. Let

$$u_1 = \frac{1}{4\pi a}\int_{S_M^{at}}\frac{\psi(q)}{r}d\sigma_q, \qquad u_2 = \frac{1}{4\pi a}\cdot\frac{\partial}{\partial t}\int_{S_M^{at}}\frac{\varphi(q)}{r}d\sigma_q \qquad (8.30)$$

Substituting the coordinates of the point q, we obtain

$$u_1 = \frac{t}{4\pi}\int_{S_M^{at}}\psi(x+\alpha at, y+\beta at, z+\gamma at)d\Omega \qquad (8.31)$$

$$u_2 = \frac{1}{4\pi}\frac{\partial}{\partial t}\left(t\int_{S_M^{at}}\varphi(x+\alpha at, y+\beta at, z+\gamma at)d\Omega\right) \qquad (8.32)$$

It is obvious from (8.31) that

$$\frac{\partial}{\partial t}u_1\Big|_{t=0} = \psi(M), \qquad u_2\Big|_{t=0} = \varphi(M) \qquad (8.33)$$

Further, by (8.32),

$$\frac{\partial}{\partial t}u_2\Big|_{t=0} = \frac{a}{2\pi}\left(\frac{\partial}{\partial x}\varphi(x,y,z)\int_{S_M^{at}}\alpha d\Omega\right.$$
$$\left. + \frac{\partial}{\partial y}\varphi(x,y,z)\int_{S_M^{at}}\beta d\Omega + \frac{\partial}{\partial z}\varphi(x,y,z)\int_{S_M^{at}}\gamma d\Omega\right) \qquad (8.34)$$

We have

$$\int_0^{2\pi}\cos\varphi d\varphi\int_0^{\pi}\sin^2\vartheta d\vartheta = 0,$$
$$\int_0^{2\pi}\sin\varphi d\varphi\int_0^{\pi}\sin^2\vartheta d\vartheta = 0, \quad \int_0^{2\pi}d\varphi\int_0^{\pi}\sin\vartheta\cos\vartheta d\vartheta = 0 \qquad (8.35)$$

so that

$$\frac{\partial}{\partial t}u_2\Big|_{t=0} = 0 \qquad (8.36)$$

Thus it remains only to demonstrate that u_1 and u_2 are solutions of the wave equation.

First consider u_1. We have

$$\frac{\partial}{\partial t}u_1 = \frac{u_1}{t} + \frac{at}{4\pi}\int_{S_M^{at}}\left(\alpha\frac{\partial}{\partial\xi} + \beta\frac{\partial}{\partial\eta} + \gamma\frac{\partial}{\partial\zeta}\right)\psi(q)d\Omega \qquad (8.37)$$

Since the radius vector Mq is directed along the outward normal to S_M^{at} we have

$$\alpha\frac{\partial}{\partial\xi} + \beta\frac{\partial}{\partial\eta} + \gamma\frac{\partial}{\partial\zeta} = \frac{\partial}{\partial n} \qquad (8.38)$$

Moreover, along S_M^{at}

$$atd\Omega = \frac{1}{at}d\sigma_q \qquad (8.39)$$

Hence

$$\frac{\partial}{\partial t}u_1 = \frac{u_1}{t} + \frac{1}{4\pi at}\int_{S_M^{at}}\frac{\partial}{\partial n}\psi(q)d\sigma_q \qquad (8.40)$$

Applying the Gauss divergence theorem, we obtain

$$\frac{\partial}{\partial t}u_1 = \frac{u_1}{t} + \frac{1}{4\pi at}\int_{K_M^{at}}\Delta\psi(q)d\omega_q = \frac{u_1}{t} + \frac{1}{4\pi at}\int_0^{at}d\rho\int_{S_M^{at}}\Delta\psi(q)d\sigma_q \quad (8.41)$$

Differentiation yields

$$\frac{\partial^2}{\partial t^2}u_1 = \frac{1}{4\pi t}\int_{S_M^{at}}\Delta\psi(q)d\sigma_q = \frac{a^2 t}{4\pi}\int_{S_M^{at}}\Delta\psi(\xi,\eta,\zeta)d\Omega \qquad (8.42)$$

At the same time, it follows from the definition (8.30) that

$$a^2\Delta u_1 = \frac{a^2 t}{4\pi}\int_{S_M^{at}}\Delta\psi(\xi,\eta,\zeta)d\Omega \qquad (8.43)$$

Hence indeed

$$a^2\Delta u_1 = \frac{\partial^2}{\partial t^2}u_1 \qquad (8.44)$$

Since the operators Δ and $\partial/\partial t$ are permutable, this result and the definition (8.30) imply that

$$a^2\Delta u_2 = \frac{\partial^2}{\partial t^2}u_2 \qquad (8.45)$$

Q.E.D.

Remark. Since Poisson's formula furnishes an integral representation for any solution of the Cauchy problem, and since conversely it defines a solution of the Cauchy problem, the solution will be unique, provided that the Cauchy data satisfy the conditions of Theorem 8.1.2.

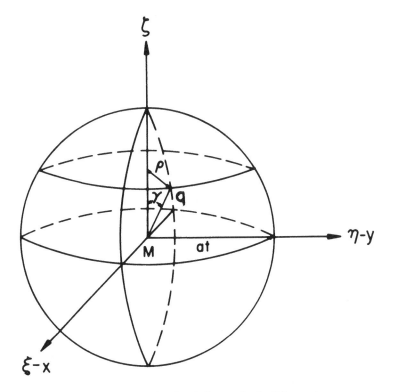

Figure 8.1.   Sketch of the method of descent.

2. Method of descent

Consider the following Cauchy problem in $\mathbb{R}_2$:

$$\Box u(x,y,t) = 0, -\infty < x,y < \infty, \ t > 0 \tag{8.46}$$

$$u\mid_{t=0} = \varphi(x,y), \quad \frac{\partial}{\partial t}u\mid_{t=0} = \psi(x,y) \tag{8.47}$$

To apply the method of descent, we consider u as a solution of the Cauchy problem in $\mathbb{R}_3$ with Cauchy data φ and ψ independent of z. Assume that M lies in the plane $z = 0$. This plane divides the sphere S_M^{at} into two parts $\widehat{S}_M^{at}$ and $\widecheck{S}_M^{at}$, lying above and below the plane, respectively. Since

$$d\sigma_q = \frac{d\xi d\eta}{\cos\gamma} \tag{8.48}$$

where γ is the angle between the outward normal $\mathbf{n}$ to S_M^{at} at a point q and the direction of the vector Mq (see Figure 8.1), Poisson's formula yields

$$u(x,y,t) = \frac{1}{2a\pi}\left(\frac{\partial}{\partial t}\int_{C_M^{at}}\frac{\varphi(\xi,\eta)d\xi d\eta}{r\cos\gamma} + \int_{C_M^{at}}\frac{\varphi(\xi,\eta)d\xi d\eta}{r\cos\gamma}\right) \tag{8.49}$$

Here C_M^{at} is the projection of the sphere K_M^{at} on the plane $z = 0$, so that the radius is $r = at$ and

$$\cos\gamma = \frac{\zeta}{at}, \quad \zeta = [a^2t^2 - (x-\xi)^2 - (y-\eta)^2]^{1/2} \qquad (8.50)$$

Thus we finally have

$$u(x,y,t) = \frac{1}{2a\pi}\left(\frac{\partial}{\partial t}\int_{C_M^{at}}\varphi(\xi,\eta)[a^2t^2 - (x-\xi)^2 - (y-\eta)^2]^{-1/2}d\xi d\eta\right.$$

$$\left. + \int_{C_M^{at}}\psi(\xi,\eta)[a^2t^2 - (x-\xi)^2 - (y-\eta)^2]^{-1/2}d\xi d\eta\right) \qquad (8.51)$$

The method of descent is clearly applicable to obtaining a solution of the Cauchy problem in $\mathbb{R}_1$ (i.e., of D'Alembert's formula). Indeed, assume that φ and ψ depend on only one coordinate, say z. Let (ρ,α,ζ) be cylindrical coordinates with pole at $M = (0,0,z)$, so that

$$d\sigma_q = r d\zeta d\alpha, \quad r = at, \quad -at < \zeta - z < at, \ 0 \le \varphi \le 2\pi \qquad (8.52)$$

Hence Poisson's formula may be rewritten as

$$u(z,t) = \frac{1}{4a\pi}\left(\frac{\partial}{\partial t}\int_{z-at}^{z+at}\varphi(\xi)d\zeta\int_0^{2\pi}d\alpha + \int_{z-at}^{z+at}\psi(\xi)d\zeta\int_0^{2\pi}d\alpha\right)$$

$$= \frac{\varphi(z+at)+\varphi(z-at)}{2} + \frac{1}{2a}\int_{z-at}^{z+at}\psi(\xi)d\zeta \qquad (8.53)$$

which is exactly D'Alembert's formula.

3. Huygens's principle

Let us return to Poisson's formula,

$$u(M,t) = \frac{1}{4\pi a}\left(\frac{\partial}{\partial t}\int_{S_M^{at}}\frac{\varphi(q)}{r}d\sigma_q + \int_{S_M^{at}}\frac{\psi(q)}{r}d\sigma_q\right) \qquad (8.54)$$

Assume that φ and ψ are localized in a bounded region D, so that

$$\varphi^2(q) + \psi^2(q) = 0 \quad \forall q \notin D, \quad \varphi^2(q) + \psi^2(q) \ne 0 \quad \forall q \in D \qquad (8.55)$$

Take some fixed point p, and denote

$$m = \min_{q\in\bar{D}}r_{pq}, \quad M = \max_{q\in\bar{D}}r_{pq} \qquad (8.56)$$

Then formula (8.54) implies that

$$u(p,t) = 0 \quad \text{if } t < \frac{m}{a} \text{ or } t > \frac{M}{a} \qquad (8.57)$$

since in these cases all points of the sphere K_p^{at} are located outside the region D in which the initial perturbations of the equilibrium state of the medium are localized (see Figure 8.2).

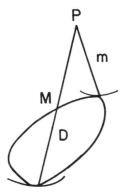

Figure 8.2. Sketch of Huygens's principle: localization of spatial disturbances in time.

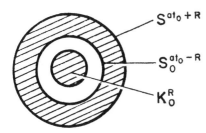

Figure 8.3. Sketch of Huygens's principle: localization of spatial disturbances in space, leading and trailing fronts.

Now let $t = t_0 > 0$ be fixed. Consider the locus W of all points at which $u(p,t)$ is not equal to zero:

$$u(p, t_0) \neq 0 \quad \forall p \in W \tag{8.58}$$

In order to determine D, note that

$$p \in W \quad \text{if } S_p^{at_0} \cap D \neq \emptyset \tag{8.59}$$

The boundary of W consists of surfaces enveloping the localization region D of the initial perturbation.[1] There exist two enveloping surfaces: the leading and trailing fronts. To illustrate, let D be the sphere K_O^R of radius R centered at the origin O of the cartesian coordinate system (see Figure 8.3). Then the leading front is the sphere $S_O^{R+at_0}$ and the trailing front is the sphere $S_O^{R-at_0}$. The above statements are combined in the Huygens's principle: *If the initial perturbations of the equilibrium state of a continuous medium occupying 3-dimensional space* $\mathbb{R}_3$ *are spatially localized then they are localized in space and in time.*

---

[1] For the theory of enveloping curves and surfaces, see textbooks on differential geometry (e.g., the classical book [27]).

Clearly Huygens's principle is not valid in $\mathbb{R}_2$ and $\mathbb{R}_1$. Indeed, in $\mathbb{R}_2$ the perturbations are localized neither spatially nor in time; the situation in $\mathbb{R}_1$ is similar if the initial rate of vibrations is not equal identically to zero (see Chapter 5, Section 1).

Problems

P8.3.1.   Does there exist in $\mathbb{R}_2$ a sharp leading or trailing front?

9

Basic properties of harmonic functions

1. Convex, linear, and concave functions in $\mathbb{R}_1$

Let us recall some simple properties of functions of one real variable. Generalization of these properties to n-dimensional space provides important information concerning the basic facts of the theory of elliptic boundary-value problems, which have counterparts in electrostatic, magnetostatics, Newtonian gravitation theory, and continuum mechanics.

Let $u(x)$ be a twice continuously differentiable function, defined in an interval (a, b) of finite length. Then $u(x)$ possesses the following properties:

(a) If $d^2u/dx^2 \leq 0$ ($d^2u/dx^2 \geq 0$) for all $x \in (a, b)$ (i.e., if $u(x)$ is convex (concave) in (a, b)) and is continuous in $[a, b]$, then its minimum (maximum) is achieved at the endpoint of this closed interval. This property is known as the *weak maximum principle*.

(b) If $u(x)$ is a twice continuously differentiable function that is convex (concave) in (a, b) and reaches its minimum (maximum) at some interior point of (a, b), then $u(x) \equiv$ const in (a, b). This property is known as the *strong maximum principle*.

(c) Let $u(x)$ be convex (concave) in (a, b), continuous in $[a, b]$ and not identically equal to a constant. Assume that $u(x)$ is differentiable at that end point of the interval (a, b) where it reaches its minimum (maximum). Then the inward derivative of u at this point is strictly positive (negative). This statement is known as *Hopf's lemma*.

(d) If $u(x)$ is linear and integrable in $[a, b]$, then

$$\frac{1}{b-a} \int_a^b u(x)dx = \frac{u(a) + u(b)}{2} \tag{9.1}$$

This property is known as the *mean value theorem*.

195

(e) If $u(x)$ is defined in some interval (A, B) and equality (9.1) holds in every closed subinterval $[a, b] \subset (A, B)$, then $u(x, t)$ is a linear function in (A, B). This assertion is known as the *inverse mean value theorem*.

(f) The flux of a linear function $u(x)$ through the boundary of the interval (a, b) in which it is linear is equal to zero; that is,

$$\frac{d}{dx}u(b) - \frac{d}{dx}u(a) = 0 \qquad (9.2)$$

(g) Let $u(x)$ be a linear bounded function, defined everywhere in $\mathbb{R}_1$. Then

$$u(x) \equiv \text{const} \qquad (9.3)$$

This assertion is known as *Liouville's theorem*.

(h) Any function $u(x)$, linear in (a, b), is *analytic*.

All these assertions and their respective names remain in force for analogs of convex, linear, and concave functions in multidimensional space $\mathbb{R}_n$. Their rigorous formulations and proofs (which are far from trivial) are given in this chapter.

2. Classes of twice continuously differentiable superharmonic, harmonic, and subharmonic functions in multidimensional regions

Let $p = (x_1, x_2, \ldots, x_n)$ be a point in $\mathbb{R}_n$, where $(x_1, x_2, \ldots, x_n)$ is a rectangular cartesian coordinate system, and let D be a region in $\mathbb{R}_n$.

Definition 9.1.1. A function $u(p) \in C^2(D)$ is called *superharmonic, harmonic,* or *subharmonic*, respectively, if

$$\Delta u(p) \leq 0, \quad \Delta u(p) = 0, \quad \Delta u(p) \geq 0 \qquad (9.4)$$

Here Δ is the Laplace operator

$$\Delta u = \text{div}[\text{grad}(u)] = \sum_{k=1}^{n} \frac{\partial^2}{\partial x_k^2} u \qquad (9.5)$$

In $\mathbb{R}_1$, $\Delta = \partial^2/\partial x^2$ and D is some interval $D = \{x : a < x < b\}$ so that

$$u(x) \text{ is } \begin{cases} \text{superharmonic} & \text{if } \dfrac{d^2}{dx^2}u \leq 0 \ \forall x \in (a, b) \\[2mm] \text{harmonic} & \text{if } \dfrac{d^2}{dx^2}u = 0 \ \forall x \in (a, b) \\[2mm] \text{subharmonic} & \text{if } \dfrac{d^2}{dx^2}u \geq 0 \ \forall x \in (a, b) \end{cases} \qquad (9.6)$$

Thus, in the 1-dimensional case, convex functions are superharmonic, linear functions are harmonic, and concave functions are subharmonic. We now proceed to the formulation and prove the n-dimensional generalization of assertions (a)–(h) of Section 1.

Theorem 9.2.1 (Weak maximum principle). *Let $u(p)$ be a function superharmonic (subharmonic) in D and continuous in the closure $\bar{D}$. Then $u(p)$ achieves its minimum (maximum) at some point of the boundary Σ of D.*

PROOF. Let $u(p)$ be superharmonic in D and continuous in $\bar{D}$. If

$$\Delta u < 0 \quad \forall p \in D \tag{9.7}$$

then the validity of the assertion is obvious. Indeed, let

$$\min_{p \in \Sigma} u(p) = M > \min_{p \in D} u(p) = m \tag{9.8}$$

By Weierstrass's theorem, any function continuous in the domain $\bar{D}$ achieves its minimum. Hence there exists $q \in D$ such that

$$u(q) = m \Rightarrow \frac{\partial^2}{\partial x_i^2} u(q) \geq 0 \ \forall i = 1, 2, \ldots, n \Rightarrow \Delta u(q) \geq 0 \tag{9.9}$$

which contradicts (9.4). Hence the theorem need be proved only if there are points inside D where $\Delta u = 0$. In such cases one can use the method of auxiliary functions. Indeed, as before, let

$$u(q) = m, \quad q \in D \tag{9.10}$$

Let $d > 0$ be the diameter of D; that is,

$$d = \max_{p \in D} \max_{s \in \Sigma} r_{ps} \tag{9.11}$$

Introduce the auxiliary function

$$v(p) = u(p) - \frac{M - m}{2d^2} r_{pq}^2 \tag{9.12}$$

Then by (9.8) and (9.11),

$$v(p) \geq \frac{M + m}{2} \quad \forall p \in \Sigma, \quad v(q) = m < \frac{M + m}{2} \tag{9.13}$$

Hence $v(p)$ achieves its minimum at some interior point p_0 of D. This means that

$$\Delta v(p_0) \geq 0 \tag{9.14}$$

On the other hand for every $p \in D$ we have

$$\Delta r_{pq}^2 = 2n \Rightarrow \Delta v(p) = \Delta u(p) - \frac{M - m}{d^2} n < 0 \tag{9.15}$$

by superharmonicity of u. Since (9.15) contradicts (9.14), the assumption (9.8) is contradictory. **Q.E.D.**

Corollary 1. *Harmonic functions achieve their maximum and minimum on the boundary of the region of harmonicity, since any harmonic function is simultaneously super- and subharmonic.*

Corollary 2 (Uniqueness theorem for solution of the Dirichlet problem for the Poisson equation in a bounded region). *Let $u(p) \in C^2(D) \cap C(\bar{D})$ and*

$$\Delta u(p) + f(p) = 0 \quad \forall p \in D, \quad u(p) = \varphi(p) \quad \forall p \in \Sigma \qquad (9.16)$$

Then $u(p)$, if it exists, is unique.

PROOF. Suppose that there exist two solutions $u_1(p)$ and $u_2(p)$ of problem (9.16). Let

$$u(p) \overset{\text{def}}{=} u_1(p) - u_2(p) \qquad (9.17)$$

Then

$$u(p) \in C^2(D) \cap C(\bar{D}) \wedge (\Delta u(p) = 0 \quad \forall p \in D) \wedge (u(p) = 0 \quad \forall p \in \Sigma) \quad (9.18)$$

so that, by Corollary 1,

$$\min u(p) = \max u(p) = 0 \Rightarrow u(p) \equiv 0 \qquad (9.19)$$

Q.E.D.

Theorem 9.2.2. *Let*

$$u(p) \in C^2(D) \cap \bar{C}(D) \wedge (\Delta u(p) + c(p)u(p) = 0 \quad \forall p \in D) \wedge (c(p) \leq 0 \quad \forall p \in D) \qquad (9.20)$$

Then $u(p)$ achieves a positive maximum (if it exists) and a negative minimum (if it exists) on Σ.

The proof is quite analogous to that of the Theorem 9.2.1, and we therefore omit it.

Remark. The condition that $c(p)$ be nonpositive is essential. For example, let

$$u(x) = \sin x \quad \forall x \in (-\pi, \pi) \qquad (9.21)$$

We have

$$\frac{d^2}{dx^2} u + u = 0 \quad \forall x \in (-\pi, \pi),$$

$$u(-\pi) = u(\pi) = 0, \quad u(\pi/2) = 1, \quad u(-\pi/2) = -1 \qquad (9.22)$$

so that u achieves a positive maximum and negative minimum not at an endpoint of the interval $(-\pi, \pi)$ but inside it. Thus, in this case the assertion of the theorem is false.

Now consider a quasilinear elliptic operator, given in divergent form:

$$A[u] \overset{\text{def}}{=} \text{div}[k(u, p) \, \text{grad} \, u] - \mathbf{v}(p) \, \text{grad} \, u \quad \forall p \in D \qquad (9.23)$$

where $\mathbf{v}(p)$ is a vector continuous in D. Assume that $k(u, p)$ satisfies the inequality

$$k(u, p) \geq k_0 > 0 \quad \forall p \in D, \; \forall u \qquad (9.24)$$

and that there exists $N = \text{const} > 0$ such that

$$|\operatorname{grad} k(u, p)| < N \quad \forall u(p) \in C^2(D), \ \forall p \in D \qquad (9.25)$$

Then the following theorem, analogous to the weak maximum principle for the Laplace operator, is valid.

Theorem 9.2.3. *If conditions (9.23)–(9.25) hold and*

$$m = \max_{p \in \Sigma} u(p), \ \ M = \max_{p \in \bar{D}} u(p), \quad \left[M = \min_{p \in \Sigma} u(p), \ \ m = \min_{p \in \bar{D}} u(p) \right] \qquad (9.26)$$

then

$$A[u(p)] \geq 0 \ (A[u(p)] \leq 0) \quad \forall p \in D \Rightarrow M = m \qquad (9.27)$$

PROOF. Assume that

$$A[u(p)] \geq 0 \quad \forall p \in D \qquad (9.28)$$

If

$$A[u(p)] > 0 \qquad (9.29)$$

everywhere in D, then it is obvious that (9.27) is valid. Indeed, suppose that

$$m < M \qquad (9.30)$$

Then there exists $q \in D$ such that

$$u(q) = M \qquad (9.31)$$

so that

$$\operatorname{grad} u(q) = 0, \quad \Delta u(q) \leq 0 \qquad (9.32)$$

which implies

$$A[u(q)] = k[u(q), q]\Delta u(q) + \operatorname{grad} u(q) \cdot \operatorname{grad} k[u(q), q] - \mathbf{v}(q) \cdot \operatorname{grad} u(q)$$
$$= k[u(q), q]\Delta u(q) \leq 0 \qquad (9.33)$$

Hence our assumption (9.30) is false, since (9.33) contradicts (9.29). Thus we must prove the theorem only if there exists $q \in D$ such that

$$A[u(q)] = 0 \wedge u(q) = M > m \qquad (9.34)$$

We again employ the method of auxiliary functions.

Consider the linear operator

$$B[w(p)] \overset{\text{def}}{=} \operatorname{div}[k(u, p) \operatorname{grad} w(p)] - \mathbf{v}(p) \cdot \operatorname{grad} w(p) \quad \forall p \in D \qquad (9.35)$$

and construct a nonnegative function $w_0(p)$ such that

$$w_0(p) \in C(\bar{D}) \cap C^2(D) \wedge (Bw_0(p) > 0 \ \forall p \in D) \wedge (w_0(p) \leq 1 \ \forall p \in D) \qquad (9.36)$$

Now define a second auxiliary function by

$$w(p) = u(p) + \frac{M-m}{3} w_0(p) \tag{9.37}$$

for all $p \in D$;

$$B[w(p)] = B[u(p)] + \frac{M-m}{3} B[w_0(p)] \equiv A[u(p)] + \frac{M-m}{3} B[w_0(p)] > 0 \tag{9.38}$$

and

$$w(p)\big|_{p\in\Sigma} \leq m + \frac{M-m}{3}, \quad w(q) = M + \frac{M-m}{3} w_0(q) > M - \frac{M-m}{3} \tag{9.39}$$

It follows from (9.39) that

$$w(q) - w(p) > \frac{M-m}{3} > 0 \quad \forall p \in \Sigma \tag{9.40}$$

Hence $w(p)$ achieves its maximum in some point $s \in D$. This yields

$$\Delta w(s) \leq 0, \quad \operatorname{grad} w(s) = 0 \tag{9.41}$$

and consequently

$$B[w(s)] = k(u,s)\Delta w(s) + \{\operatorname{grad}[k(u,s) - \mathbf{v}(s)]\} \cdot \operatorname{grad} w(s) = k(u,s)\Delta w(s) \leq 0 \tag{9.42}$$

Hence our assumption (9.30) is false if the required function $w_0(p)$ exists, since (9.42) contradicts (9.38).

It remains to prove the existence of $w_0(p)$. Let us assume that the origin O of the cartesian coordinate system with basis vectors $\mathbf{e}_k$, $k = 1, 2, \ldots, n$, is located outside the region D. Let

$$\mathbf{r} = \sum_{k=1}^{n} x_k \mathbf{e}_k \Rightarrow r^2 = \sum_{k=1}^{n} x_k^2 \tag{9.43}$$

be the radius vector of the point p, and define $w_0(p)$ as

$$w_0(p) = \exp(-\alpha r^2) \tag{9.44}$$

where $\alpha = \mathrm{const} > 0$ will be chosen later. We have

$$\frac{\partial}{\partial x_k} w_0 = -2\alpha x_k w_0, \quad \frac{\partial^2}{\partial x_k^2} w_0 = (4\alpha^2 x_k^2 - 2\alpha) w_0 \tag{9.45}$$

so that

$$\Delta w_0 = (4\alpha^2 r^2 - 2\alpha n) w_0, \quad \operatorname{grad} w_0 = -2\alpha w_0 \sum_{k=1}^{n} x_k \mathbf{e}_k \tag{9.46}$$

We have

$$Bw_0 = k(4\alpha^2 r^2 - 2\alpha n) w_0 + 2\alpha w_0 [\mathbf{v} - \operatorname{grad} k] \sum_{k=1}^{n} x_k \mathbf{e}_k$$

$$\geq k(4\alpha^2 r_0^2 - 2\alpha n) w_0 + 2\alpha w_0 [\mathbf{v} - \operatorname{grad} k] \sum_{k=1}^{n} x_k \mathbf{e}_k \tag{9.47}$$

where

$$r_0 = \min_{q \in D} r_{qO} > 0 \tag{9.48}$$

Since $|v|$ and $|\operatorname{grad} k|$ are bounded, it follows from (9.47) that $\alpha = \text{const} > 0$ may be chosen so large that

$$Bw_0(p) > 0 \quad \forall p \in D \tag{9.49}$$

Because the definition (9.44) implies

$$w_0(p) \in C(\bar{D}) \cap C^2(D) \wedge [w_0(p) < 1] \quad \forall p \in \bar{D} \tag{9.50}$$

we see that a function $w_0(p)$ satisfying conditions (9.36) indeed exists. **Q.E.D.**

Remark. The weak maximum principle is valid for a general operator strongly elliptic in D, not only for a quasilinear operator given in divergence form. Theorem 9.2.3 was stated and proved for the latter case solely for the sake of simplicity.[1]

3. Hopf's lemma and strong maximum principle [38]

The generalization of the (trivial) statement (c) of the Section 1, called Hopf's lemma, is as follows.

Lemma 9.3.1 (Hopf). *Let K be a sphere of radius R in $\mathbb{R}_n$ with its center at a point s, and let $u(p) \in C(\bar{K})$ be a subharmonic function in K. Assume that*

$$\max_{p \in \bar{D}} u(p) = u(P) \underset{P \in \Sigma}{\overset{\text{def}}{=}} M \wedge [u(p) < M \quad \forall p \in \bar{K} \setminus (P)] \tag{9.51}$$

so that P is the only maximum point of u in the closure of K and lies on the boundary Σ of K. Assume that $(\partial/\partial\nu)u(P)$ exists, where ν is the vector pointing from P into K. Then

$$\frac{\partial}{\partial\nu}u(P) < 0 \tag{9.52}$$

---

[1]The quasilinear operator

$$\sum_{k=1}^{1}\sum_{m=1}^{1} a_{km}[u(p), p]\frac{\partial^2 u(p)}{\partial x_k \partial x_m} + F\left[p, u(p), \frac{\partial}{\partial x_1}u(p), \dots, \frac{\partial}{\partial x_n}u(p)\right]$$

is said to be *strongly elliptic* at point p if there exists $\alpha(p) > 0$ such that, for all admissible $u(p)$,

$$\sum_{k=1}^{n}\xi_k^2 \geq \xi > 0 \Rightarrow \sum_{k=1}^{n}\sum_{m=1}^{n} a_{km}\xi_k\xi_m \geq \alpha(p)$$

If there exists α_0 such that $\alpha(p) \geq \alpha_0 > 0$ for all $p \in D$ then this operator is said to be strongly elliptic in D. Note that the divergence form of quasilinear operators is the one that appears usually in physical applications.

PROOF. Note, first of all, that if the strict inequality were weakened to

$$\frac{\partial}{\partial \nu} u(P) \leq 0 \tag{9.53}$$

then the assertion of the lemma would be trivial. However, the proof of the strict inequality (9.53) is by no means trivial.

Define the auxiliary function

$$w(r) = 1 - \exp[\alpha(R^2 - r^2)] \tag{9.54}$$

where $\mathbf{r}$ is the radius vector of the point $p \in K$ and $r = |\mathbf{r}|$. Let K_0 be a sphere of radius $R/2$ concentric with K, and let Σ_0 be its boundary. Consider $w(r)$ within the spherical cavity

$$D = K \setminus \bar{K}_0 \tag{9.55}$$

We have

$$\Delta w = -(4\alpha^2 r^2 - 2\alpha n) \exp[\alpha(R^2 - r^2)] < 0 \tag{9.56}$$

if $\alpha > 0$ is large enough.

Now let

$$v(p) = u(p) - M - w(p) \tag{9.57}$$

By assumption, u is subharmonic in K. Hence (9.57) and (9.56) imply

$$\Delta v(p) > 0 \quad \forall p \in D \tag{9.58}$$

Besides, the definition (9.54) and condition (9.51) yield

$$(v(P) = 0, \; w(p) < 0) \wedge (u(p) < M \quad \forall p \in [\Sigma \setminus (P)] \cup \Sigma_0) \tag{9.59}$$

Hence $v(p)$, as a subharmonic function, assumes its maximum at point P, so that

$$\frac{\partial}{\partial \nu} v(P) = \frac{\partial}{\partial \nu} u(P) - \frac{\partial}{\partial \nu} w(P) \leq 0 \tag{9.60}$$

We have

$$\frac{\partial}{\partial \nu} w(P) = 2\alpha r \exp[\alpha(R^2 - r^2)] \frac{dr}{d\nu} \bigg|_{r=R} = 2R \cos \varphi \tag{9.61}$$

where φ is the angle between $\mathbf{r}$ and $\boldsymbol{\nu}$. Since $\mathbf{r}$ is an outward and $\boldsymbol{\nu}$ an inward pointing vector,

$$\cos \varphi < 0 \tag{9.62}$$

which implies

$$\frac{\partial}{\partial \nu} w(P) < 0 \tag{9.63}$$

Hence, by (9.60),

$$\frac{\partial}{\partial \nu} u(P) \leq \frac{\partial}{\partial \nu} w(P) < 0 \tag{9.64}$$

Q.E.D.[2]

Theorem 9.3.1 (Strong maximum principle). *If a function $u(p)$, p 204 that is superharmonic (subharmonic) in a region $D \subset \mathbb{R}_n$ attains its minimum m (maximum M) at some interior point of D, then*

$$u(p) \equiv m, \quad [u(p) \equiv M], \quad p \in D \tag{9.65}$$

PROOF. The strong maximum principle is a corollary of Hopf's lemma. Indeed, assume that the superharmonic function $u(p)$ attains its minimum inside D, so that

$$\mathcal{M} = \{p \in D, \ u(p) = m\} \neq \emptyset \tag{9.66}$$

Since $u(p)$ is continuous, $\mathcal{M}$ is a closed set. If

$$\mathcal{M} = \bar{D} \tag{9.67}$$

then the assertion of the theorem is true. Suppose that

$$D \setminus \mathcal{M} \overset{\text{def}}{=} \mathcal{E} \neq \emptyset \tag{9.68}$$

Let K be sphere in $\mathcal{E}$ that touches $\mathcal{M}$ in some point $q \notin \Sigma$, where Σ is the boundary of D. We may assume that the radius ρ of K is so small that $\bar{K} \subset D$. Let $\mathbf{n}_q$ be the inward normal to the boundary of K at q. By Hopf's lemma,

$$\frac{\partial}{\partial n_q} u(q) > 0 \tag{9.69}$$

But this is impossible, since q is an interior minimum point of u, so that

$$\frac{\partial}{\partial \nu} u(q) = 0 \tag{9.70}$$

whatever the direction of the vector ν is. The contradiction obtained proves that $\mathcal{E}$ is an empty set. Q.E.D.[3]

Hopf's lemma is important not only as an auxiliary tool. It may be used to prove a number of uniqueness theorems where boundary conditions

---

[2]The assertion of Hopf's, and that of the strong maximum principle proved in what follows, is valid not only for super- and subharmonic operators, but in the general case of a strongly elliptic operator as well. (In this connection see the proof in Section 2 of the weak maximum principle for a quasilinear elliptic operator in divergent form).

[3]Rigorously speaking, the existence of such a sphere K must be proved. A rigorous proof of existence of K is required in Problem 9.3.4.

of the Neumann or irradiation type are prescribed on some parts of the boundary. As examples, let us consider the following problems.

A. Neumann problem for the Poisson equation

Let D be a region in $\mathbb{R}_n$ with boundary Σ such that, for every boundary point $p \in \Sigma$, there is a sphere $K_p \subset D$ touching Σ at p. Let $C_n^1(\bar{D})$ denote the set of functions defined and continuous in the closure $\bar{D}$ of D together with their first order derivatives with respect to the direction of the normal to Σ. The problem is: Find $u(q) \in C_n^1(\bar{D}) \cap C^2(D)$ such that

$$\Delta u + f(q) = 0 \quad \forall q \in D, \quad \frac{\partial}{\partial n_q} u = \varphi(q) \quad \forall q \in \Sigma = \bar{D} \setminus D \qquad (9.71)$$

The solution of this problem, if it exists, is unique up to an additive constant.

Indeed, suppose that there exist two different solutions u_1 and u_2, and let u denote the difference between them. We must prove that

$$u(q) \equiv \text{const} \qquad (9.72)$$

The function $u(q)$ is a solution of the homogeneous problem

$$u \in C_n^1(\bar{D}) \cap C^2(D) \wedge u \in \Delta u = 0 \quad \forall q \in D \wedge \left(\frac{\partial}{\partial n_q} u = 0 \quad \forall q \in \Sigma = \bar{D} \setminus D \right)$$

$$(9.73)$$

Hence, by the weak maximum principle, the extremum of u occurs at some boundary point q. If $u \neq \text{const}$ then by Hopf's lemma, which is assumed to be applicable,

$$\frac{\partial}{\partial n_q} u \neq 0 \qquad (9.74)$$

which contradicts (9.73). Hence (9.72) is true.   **Q.E.D.**

B. Radiation of a black body into vacuum

Consider a black body occupying a region $D \in \mathbb{R}_3$, bounded by a surface Σ of finite curvature, so that at every boundary point there is a sphere touching Σ from within D. The problem is: Find the temperature $u(p)$ of the body D such that

$$u(p) \in C_n^1(\bar{D}) \cap C^2(D) \ \wedge \ \Delta u(p) + f(p) = 0 \ \forall p \in D$$

$$\wedge \ \frac{\partial}{\partial n_q} u(q) = -\sigma \cdot u^4(q) \ \forall q \in \Sigma \qquad (9.75)$$

where $\sigma > 0$ is the Stefan constant and u is measured in degrees Kelvin.[4] If a solution $u(p)$ of this problem exists, it is unique up to an additive constant.

---

[4] Since the temperature is measured in degrees Kelvin, u_i, $(i = 1, 2)$, are assumed positive.

Suppose the contrary, and let $u(p)$ denote the difference between two solutions u_1 and u_2. Then u is the solution of the problem

$$\Delta u = 0 \quad \forall p \in \Sigma, u(p) \equiv \text{const} \quad \forall p \in D \tag{9.76}$$

$$\frac{\partial}{\partial n_q} u(q) \stackrel{\text{def}}{=} v(q) = -\sigma \left[u_1^4(q) - u_2^4(q) \right]$$

$$= -\sigma [u_1(q) - u_2(q)][u_1(q) + u_2(q)][u_1^2(q) + u_2^2(q)] \quad \forall q \in \Sigma \tag{9.77}$$

If

$$v(q) = 0 \quad \forall q \in \Sigma \tag{9.78}$$

then, by what has been proved in the previous example,

$$u(p) \equiv \text{const} \tag{9.79}$$

contrary to assumption. Thus, assume that

$$\max_{p \in \bar{D}} u(p) = M > \min_{p \in \bar{D}} u(p) = m \tag{9.80}$$

By the maximum principle, the values M and m are achieved at some points r and s of Σ. By Hopf's lemma,

$$v(s) < 0 \tag{9.81}$$

which implies, by definition (9.77) and since σ is positive, that

$$m > 0 \tag{9.82}$$

At the same time, Hopf's lemma yields

$$v(r) > 0 \tag{9.83}$$

which implies

$$M < 0 \tag{9.84}$$

Comparing (9.82) and (9.84), we see that

$$M < m \tag{9.85}$$

which contradicts (9.80). This contradiction proves that

$$u(p) \equiv \text{const} \tag{9.86}$$

Q.E.D.

4. Green's formulas. Flux of harmonic function through closed surface. Uniqueness theorems

Let D be a region in $\mathbb{R}_3$ and $\Sigma = \bar{D} \setminus D$ its boundary, such that the Gauss divergence theorem is applicable. Let

$$u(p), v(p) \in C^2(\bar{D}) \tag{9.87}$$

Integration of the identity

$$v\Delta u \equiv v \, \text{div}(\text{grad}\, u) = \text{div}[v \, \text{grad}\, u] - \text{grad}\, u \cdot \text{grad}\, v \tag{9.88}$$

yields

$$\int_D v\Delta u d\omega = \int_D \text{div}[v \, \text{grad} \, u] d\omega - \int_D \text{grad} \, u \cdot \text{grad} \, v d\omega \qquad (9.89)$$

By the Gauss divergence theorem,

$$\int_D \text{div}[v \, \text{grad} \, u] d\omega = \int_\Sigma v(q) \, \text{grad} \, u(q) \cdot \mathbf{d\sigma}_q \qquad (9.90)$$

where $\mathbf{d\sigma}_q$ is a vector emanating from the point $q \in \Sigma$ in the direction of the outward normal $\mathbf{n}_q$, so that

$$\mathbf{d\sigma} = d\sigma \mathbf{n}_q^0, \quad d\sigma = |\mathbf{d\sigma}|, \quad \mathbf{n}_q^0 = \frac{\mathbf{n}_q}{|\mathbf{n}_q|} \qquad (9.91)$$

Hence (9.90) may be rewritten as

$$\int_D \text{div}[v \, \text{grad} \, u] d\omega = \int_\Sigma v(q) \frac{\partial}{\partial n_q} u(q) d\sigma_q \qquad (9.92)$$

Inserting (9.92) into (9.90) we obtain

$$\int_D v\Delta u d\omega = \int_\Sigma v(q) \frac{\partial}{\partial n_q} u(q) d\sigma_q - \int_D \text{grad} \, u \cdot \text{grad} \, v d\omega \qquad (9.93)$$

This identity is usually called *Green's first formula.*[5] Interchanging v and u and subtracting the resulting identity from (9.93), we obtain

$$\int_D [v\Delta u - u\Delta v] d\omega = \int_\Sigma \left[v(q) \frac{\partial}{\partial n_q} u(q) - u(q) \frac{\partial}{\partial n_q} v(q) \right] d\sigma_q \qquad (9.94)$$

This identity is known as *Green's second formula.*

Remark. The requirement that u and v be twice continuously differentiable in the closure $\bar{D}$ and harmonic inside D may be weakened. It is sufficient

---

[5]Let $v \equiv u$. Then equality (9.93) becomes

$$G(u) = \int_\Sigma u(q) \frac{\partial}{\partial n_q} u(q) d\sigma_q - \int_D u(q) \Delta u(q) d\omega_q$$

where

$$G(u) = \int_D \text{grad}^2 u(q) d\omega_q$$

is known as the Dirichlet integral. Note that the Laplace equation $\Delta u = 0$ is the Euler equation for the extremum of $G(u)$. Thus, let $\varphi(p)$ be a given function continuous on Σ, and let $\mathfrak{M}(D)$ be a class of functions continuous in $\bar{D}$ possessing piecewise continuous derivatives in D, equal to $\varphi(p)$ on Σ and such that $u \in \mathfrak{M}(D) \Rightarrow G(u) < \infty$. Then $\min G(u)$ is the solution of the Dirichlet problem $\Delta u = 0$ in D, $u = \varphi$ on Σ. This assertion is known as *Dirichlet principle*. There is an extensive literature on the Dirichlet principle (see, e.g., [13]).

to require that u and v be harmonic inside D and differentiable inside $\bar{D}$. Indeed, consider a monotonic sequence of regions

$$D_1 \subset \bar{D}_1 \subset D_2 \subset \bar{D}_2 \subset \cdots \subset D_n \subset \bar{D}_n \subset \cdots \tag{9.95}$$

such that

$$\lim_{n \uparrow \infty} D_n = D \tag{9.96}$$

Green's first and second formulas are applicable to D_n. Since the second derivatives of u and v do not occur in the right-hand sides of (9.93) and (9.94), we may pass to the limit $n = \infty$, which proves that Green's first and second formulas are valid under the aforementioned weakened conditions.

Green's formulas provide an important source of information about the properties of harmonic functions, as the following theorems show.

Theorem 9.4.1. *The flux of an harmonic function through a closed surface is equal to zero:*

$$\int_\Sigma \frac{\partial}{\partial n_q} u(q) d\sigma_q = 0 \tag{9.97}$$

PROOF. Let

$$\Delta u \equiv 0, \quad v \equiv 1 \tag{9.98}$$

Then Green's first formula yields (9.97). **Q.E.D.**

Uniqueness theorems

1. Dirichlet Problem

We have to prove that if

$$u(p) \in C(\bar{D}), \quad \Delta u = 0 \quad \forall p \in D, \quad u(p) = 0 \quad \forall p \in \Sigma \tag{9.99}$$

then

$$u(p) = 0 \quad \forall p \in \bar{D} \tag{9.100}$$

PROOF. Let us take

$$v \equiv u \tag{9.101}$$

Then Green's first formula yields

$$\int_D \operatorname{grad}^2 u \, d\omega = 0 \tag{9.102}$$

Since $u \in C^1(\bar{D})$, it follows from (9.102) that

$$\operatorname{grad} u \equiv 0 \Rightarrow u \equiv \operatorname{const} \tag{9.103}$$

which, since $u \equiv 0$ for all $p \in \Sigma$, implies that

$$u \equiv 0 \quad p \in \bar{D} \tag{9.104}$$

Q.E.D.[6]

2. Neumann Problem

We have to prove that if

$$u(p) \in C^1(\bar{D}), \quad \Delta u = 0 \quad \forall p \in D, \quad \frac{\partial}{\partial n_q} u(q) = 0 \quad \forall q \in \Sigma \tag{9.105}$$

then

$$u(p) = \text{const} \quad \forall p \in \bar{D} \tag{9.106}$$

PROOF. In this case it follows from (9.102) that

$$\text{grad} \, u \equiv \text{const} \Rightarrow u \equiv \text{const} \tag{9.107}$$

Q.E.D.[7]

5. Integral identity. Mean value theorem. Inverse mean value theorem

Let us determine an important solution of the Laplace equation, one that possesses radial symmetry. Let (r, φ, ϑ) be the polar coordinate system with pole at a point m. Then the Laplace equation

$$Lv(p) \overset{\text{def}}{=} \left(\frac{\partial^2}{\partial r^2} + \frac{2}{r} \frac{\partial}{\partial r} + \frac{1}{r^2 \sin^2(\vartheta)} \frac{\partial^2}{\partial \varphi^2} \right.$$
$$\left. + \frac{1}{r^2 \sin(\vartheta)} \frac{\partial}{\partial \vartheta} \left(\sin(\vartheta) \frac{\partial}{\partial \vartheta} v(r, \varphi, \vartheta) \right) \right) v(p) = 0,$$
$$p = (r, \varphi, \vartheta), \quad r = r_{pm} \tag{9.108}$$

becomes

$$\left(\frac{\partial^2}{\partial r^2} + \frac{2}{r} \frac{\partial}{\partial r} \right) v \equiv \frac{1}{r} \frac{\partial^2}{\partial r^2} (rv) = 0 \tag{9.109}$$

the general solution of which is

$$v = a + \frac{b}{r} \tag{9.110}$$

Taking $a = 0$ and $b = 1/4\pi$, we obtain

$$v(p) = \frac{1}{4\pi r_{pm}} \tag{9.111}$$

---

[6] Compare this proof with that of Corollary 2 to Theorem 9.2.1.

[7] Compare with Section 3A on the use of Hopf's lemma, (9.71)–(9.74).

Thus $v(p)$ is the potential at point p in a vacuum, induced by a unit charge placed at point m, if rational Gauss electrostatic units are used. This potential is known as the fundamental solution of the Laplace equation in $\mathbb{R}_3$, or simply *the fundamental solution*.

The following integral identity, which is a corollary of Green's second formula, is extremely important, and yields many fundamental facts.

Theorem 9.5.1 (Fundamental identity). *Let* $u(p) \in C^1(\bar{D}) \cap C^2(D)$, *where* $D \subset \mathbb{R}_3$ *is a region with boundary* Σ *in which the Gauss divergence theorem is applicable. Then* $u(p)$ *satisfies the identity*

$$u(p) \equiv \frac{1}{4\pi} \int_\Sigma \left(\frac{1}{r_{pq}} \frac{\partial}{\partial n_q} u(q) - u(q) \frac{\partial}{\partial n_q} \frac{1}{r_{pq}} \right) d\sigma_q - \frac{1}{4\pi} \int_\Sigma \frac{1}{r_{pq}} \Delta u(q) d\omega_q$$

$$(9.112)$$

PROOF. Let m be a fixed point in D and K_m^ρ a sphere of radius ρ with center at m, where ρ is so small that

$$\bar{K}_m^\rho \subset D \qquad (9.113)$$

Let

$$D_\rho = D \setminus \bar{K}_m^\rho \qquad (9.114)$$

Take $v(p) = 1/4\pi r_{pm}$. Since $m \notin D_\rho$, Green's second formula is applicable to $u(p)$ and to $v(p)$ in D_ρ, where D_ρ is the doubly connected region with boundary

$$\Sigma_\rho = \Sigma \cup S_\rho \qquad (9.115)$$

S_ρ being the boundary of the sphere K_m^ρ. The normal to S_ρ at a point q, pointing out of D_ρ, has the same direction as the vector $\mathbf{r}_{qm}$, that is, the inner normal with respect to K_m^ρ. Green's second formula therefore becomes

$$\int_{D_\rho} \frac{1}{4\pi r_{pm}} \Delta u(q) d\omega_q = \frac{1}{4\pi} \int_\Sigma \left[\frac{1}{r_{qm}} \frac{\partial}{\partial n_q} u(q) - u(q) \frac{\partial}{\partial r_{qm}} \frac{1}{r_{qm}} \right] d\sigma_q$$

$$- \frac{1}{4\pi} \int_{S_\rho} \left[\frac{1}{\rho} \frac{\partial}{\partial n_q} u(q) - \rho^{-2} u(q) \right] d\sigma_q \qquad (9.116)$$

Using the mean value theorem, we obtain

$$\frac{1}{4\pi} \int_{S_\rho} \left[\frac{1}{\rho} \frac{\partial}{\partial n_q} u(q) - \rho^{-2} u(q) \right] d\sigma_q = \left[\frac{1}{\rho} \frac{\partial}{\partial n_q} u(q) - \rho^{-2} u(q) \right] \rho^2, \quad q \in S_\rho$$

$$(9.117)$$

Hence, letting $\rho \to 0$, we have by continuity of u and $\partial u / \partial n_q$ in D,

$$\lim_{\rho \downarrow 0} \frac{1}{4\pi} \int_{S_\rho} \left[\frac{1}{\rho} \frac{\partial}{\partial n_q} u(q) - \rho^{-2} u(q) \right] d\sigma_q = \lim_{\rho \downarrow 0} \left[\frac{1}{\rho} \frac{\partial}{\partial n_q} u(q) - \rho^{-2} u(q) \right] \rho^2$$

$$= -u(m) \qquad (9.118)$$

The first integral in the right-hand side of (9.116) is independent of ρ. Hence the existence of the limit of the second integral implies that of the integral on the left, so that

$$\lim_{\rho\downarrow 0}\int_{D_\rho}\frac{1}{4\pi r_{pm}}\Delta u(q)d\omega_q = \int_D \frac{1}{4\pi r_{pm}}\Delta u(q)d\omega_q \qquad (9.119)$$

Thus it follows from (9.116), (9.118), and (9.119) that

$$u(m) \equiv \frac{1}{4\pi}\int_\Sigma \left(\frac{1}{r_{qm}}\frac{\partial}{\partial n_q}u(q) - u(q)\frac{\partial}{\partial n_q}\frac{1}{r_{qm}}\right)d\sigma_q - \frac{1}{4\pi}\int_\Sigma \frac{1}{r_{qm}}\Delta u(q)d\omega_q$$

$$(9.120)$$

Since m is an arbitrary point of D, this proves (9.112). **Q.E.D.**

Remark. The fundamental solution of the Laplace equation in $\mathbb{R}_2$ is[8]

$$\frac{1}{2\pi}\ln\left(\frac{1}{r_{pm}}\right) \qquad (9.121)$$

and the fundamental identity becomes

$$u(p) \equiv \frac{1}{2\pi}\int_\Sigma \left(\ln\left(\frac{1}{r_{qp}}\right)\frac{\partial}{\partial n_q}u(q) - u(q)\frac{\partial}{\partial n_q}\ln\left(\frac{1}{r_{qp}}\right)\right)ds_q$$

$$-\frac{1}{2\pi}\int_\Sigma \ln\left(\frac{1}{r_{qp}}\right)\Delta u(q)d\sigma_q \qquad (9.122)$$

The mean value theorem, which is trivial in the 1-dimensional case, is an obvious corollary of the fundamental identity. It reads as follows.

Theorem 9.5.2 (Mean Value Theorem). *The value of an harmonic function u at the center of a sphere in its region of harmonicity equals the mean value of u over the surface of the sphere.*

PROOF. Let $K_M^R \subset \mathbb{R}_3$ be a sphere of radius R with center at M, and let $u(p) \in C^1\left(\bar{K}_M^R\right)\cap C^2\left(K_M^R\right)$ be an harmonic function within K_M^R. Then the fundamental identity is applicable. Since

$$\left.\frac{\partial}{\partial n_q}\frac{1}{r_{qm}}\right|_{q\in\Sigma} = -\frac{1}{R^2} \qquad (9.123)$$

this and Theorem 9.5.1 imply

$$u(M) = \frac{1}{4\pi}\int_\Sigma u(q)d\sigma \qquad (9.124)$$

Q.E.D.

The mean value theorem represents a characteristic property of harmonic functions. This is seen from the following theorem.

---

[8]For motivation see Chapter 12, Section 4.

Theorem 9.5.3 (Inverse mean value theorem).[9] *Let D be a region in $\mathbb{R}_n$ ($n = 2$ or 3) and let $u(p) \in C(D)$ be a function such that, for any sphere $K_M^R \subset \bar{K}_M^R \subset D$, $u(M)$ is equal to the mean value of $u(q)$ over $S_M^R = \bar{K}_M^R \setminus K_M^R$. Then u is harmonic in D.*

PROOF. To fix ideas, let $n = 3$, and let M be an arbitrary point of D. Take an arbitrary sphere $K_M^R \subset \bar{K}_M^R \subset D$. Assume first that $u \in C^2(D)$, so that the fundamental identity is applicable. Then

$$u(M) = \frac{1}{4\pi R^2} \int_{S_M^R} u(q) d\sigma_q \qquad (9.125)$$

and simultaneously

$$u(M) \equiv \frac{1}{4\pi R} \int_{S_M^R} \left(\frac{\partial}{\partial n_q} u(q) + \frac{u(q)}{R} \right) d\sigma_q - \frac{1}{4\pi} \int_{K_M^R} \frac{1}{r_{qM}} \Delta u(q) d\omega_q \qquad (9.126)$$

so that

$$\frac{1}{4\pi R} \int_{S_M^R} \frac{\partial}{\partial n_q} u(q) d\sigma_q = \frac{1}{4\pi} \int_{K_M^R} \frac{1}{r_{qM}} \Delta u(q) d\omega_q \qquad (9.127)$$

By the Gauss divergence theorem,

$$\frac{1}{4\pi R} \int_{S_M^R} \frac{\partial}{\partial n_q} u(q) d\sigma_q = \frac{1}{4\pi} \int_{K_M^R} \frac{1}{R} \Delta u(q) d\omega_q \qquad (9.128)$$

so that by (9.126) and (9.128),

$$\int_{K_M^R} \left(\frac{1}{R} - \frac{1}{r_{qM}} \right) \Delta u(q) d\omega_q = 0 \qquad (9.129)$$

By the mean value theorem of integral calculus, it follows from (9.129) that

$$\Delta u(\tilde{q}) \int_{K_M^R} \left(\frac{1}{R} - \frac{1}{r_{qM}} \right) d\omega_q = 0 \Rightarrow -\frac{2}{3}\pi R^2 \Delta u(\tilde{q}) = 0 \Rightarrow \Delta u(\tilde{q}) = 0 \qquad (9.130)$$

where $\tilde{q} \in K_M^R$. Here R is arbitrary, so that it is legitimate to let $R \to 0$. Since $u \in C^2(D)$, this yields

$$\lim_{R \downarrow 0} \Delta u(\tilde{q}) = \Delta u(M) = 0 \qquad (9.131)$$

Thus it remains only to demonstrate that the assumption that u is twice differentiable may be dropped. This results from the following lemma.

Lemma 9.5.1. *If u satisfies the conditions of Theorem 9.5.2 then it is infinitely differentiable.*

PROOF. Let (x, y, z) and (ξ, η, ζ) be the cartesian coordinates of fixed and variable points p and q, respectively. Introduce the auxiliary function

$$f(r^2) = \begin{cases} \exp[-(R^2 - r^2)^{-1}] & \text{if } r < R \\ 0 & \text{if } r \geq R \end{cases} \qquad (9.132)$$

---

[9] The proof is borrowed from [13].

The function $f(r^2)$ is continuously differentiable infinitely many times and identically equal to zero at $r \geq R$. By multiplying (9.125) (with R replaced by r) and integrating from zero to R, we obtain

$$u(M) \int_0^R r^2 f(r^2) dr = \frac{1}{4\pi} \int_0^R f(r^2) dr \int_{S_M^r} u(q) d\sigma_q \qquad (9.133)$$

or, what is the same,

$$Au(M) = \int_{K_M^R} f(r^2) u(q) d\omega_q \qquad (9.134)$$

where

$$A = 4\pi \int_0^R r^2 f(r^2) dr = \text{const} > 0 \qquad (9.135)$$

Introducing cartesian coordinates and taking into account that

$$f(r^2) \equiv 0 \quad \forall r \geq R \qquad (9.136)$$

we obtain

$$Au(M) = \int_{K_M^R} f[(x - \xi)^2 + (y - \eta)^2 + (z - \zeta)^2] u(\xi, \eta, \zeta) d\xi d\eta d\zeta$$

$$\equiv \int_{-\infty}^{\infty} d\xi \int_{-\infty}^{\infty} d\eta \int_{-\infty}^{\infty} f[(x - \xi)^2 + (y - \eta)^2$$

$$+ (z - \zeta)^2] u(\xi, \eta, \zeta) d\xi d\eta d\zeta \qquad (9.137)$$

Since f is infinitely differentiable and vanishes identically outside K_M^R, the integral in the right-hand side may be differentiated arbitrary many times under the integral sign. Hence u has derivatives of any order $(k + m + n)$:

$$\frac{\partial^{k+m+n} u}{\partial x^k \partial y^m \partial z^n} = \int_{-\infty}^{\infty} d\xi \int_{-\infty}^{\infty} d\eta \int_{-\infty}^{\infty} \frac{\partial^{k+m+n}}{\partial x^k \partial y^m \partial z^n} f[(x - \xi)^2 + (y - \eta)^2$$

$$+ (z - \zeta)^2] u(\xi, \eta, \zeta) d\xi d\eta d\zeta \qquad (9.138)$$

Because the only assumption used is that u satisfies the conditions of the theorem, it follows from (9.138) that Lemma 9.5.1 and thus Theorem 9.5.2 are true. **Q.E.D.**

Corollary. *Any function harmonic inside a region D is differentiable infinitely many times inside D.*

Remark. All the results of this section are valid in spaces of arbitrary many dimensions, provided that the fundamental solution of the n-dimensional Laplace equation is appropriately defined.

Problems

P9.2.1. Prove Theorem 9.2.2.

P9.2.2. Prove that there may exist no more than one function harmonic inside a region D that satisfies the Dirichlet condition on one part of the boundary Σ of D and the condition of Newtonian irradiation on the other part. (First formulate the problem rigorously.)

P9.2.3. Let $D \subset \mathbb{R}_3$ and $D_0 \subset \bar{D}_0 \subset D$, $\Sigma_0 = \bar{D}_0 \setminus D_0$, $\Sigma = \bar{D} \setminus D$. Assume that Σ and Σ_0 are surfaces of finite curvature. Prove that the solution of the following problem is unique:

$$\Delta u + f(p) = 0 \quad \forall p \in D \setminus \bar{D}_0, \ \forall p \in D_0$$

$$\frac{\partial}{\partial n_q} u(q) = \varphi(q) \quad \forall q \in \Sigma, \quad [u] = \left[k \frac{\partial}{\partial n_q} u\right] = 0 \quad \forall q \in \Sigma_0$$

where $k \geq k_0 > 0$ and $[k] \neq 0$. ($[\psi]$ denotes the jump of ψ for any ψ.)

P9.3.1. Prove the existence of the sphere K_0 in the proof of the strong maximum principle.

P9.4.1. (a) Is our proof of the uniqueness of the solution of the Dirichlet problem correct? If so, compare it with the corresponding proof based on the maximum principle; which proof is preferable? If not, correct the proof and carry out the same comparison. (b) Is our proof of the uniqueness of the solution of the Neumann problem correct? If so, compare it with the proof based on the use of Hopf's lemma. If not, correct the proof and perform the same comparison. Which method is preferable?

P9.5.1. Prove the strong maximum principle using the mean value theorem.

P9.5.2. Prove that any function harmonic in $D \subset \mathbb{R}_2$ is analytic.

P9.5.3. Define a fundamental solution of the Laplace equation in $\mathbb{R}_n$ for $n > 3$, and derive the appropriate fundamental identity.

10

Green's functions

1. Definitions. Main properties

In what follows we consider the Dirichlet and Neumann problems in $\mathbb{R}_3$.
  Let

$$D_i \subset \mathbb{R}_3, \quad \Sigma = \bar{D}_i \setminus D_i, \quad D_e = \mathbb{R}_3 \setminus \bar{D}_i \tag{10.1}$$

Let us assume in addition that D_e contains the point at infinity, and that Σ is a closed, convex, piecewise smooth surface so that Gauss's divergence theorem and consequently Green's second formula are applicable.
  Let

$$u(p) \in C^1(\bar{D}_i) \cap C^2(D_i), \quad v(p,q) \in C^1(\bar{D}_i) \quad \forall p = \text{const} \in D_i, \ \forall q \in \bar{D}_i \tag{10.2}$$

and[1]

$$\Delta_q v = 0 \quad \forall p = \text{const} \in D_i, \ \forall q \in D_i \tag{10.3}$$

Then the fundamental identity (9.123) and Green's second formula (9.94) yield, for all $p \in D_i$,

$$u(p) = \frac{1}{4\pi} \int_\Sigma \frac{1}{r_{pq}} \frac{\partial}{\partial n_q} u(q) d\sigma_q - \frac{1}{4\pi} \int_\Sigma u(q) \frac{\partial}{\partial n_q} \frac{1}{r_{pq}} d\sigma_q$$

$$- \frac{1}{4\pi} \int_{D_i} \Delta_q u(q) \frac{1}{r_{pq}} d\omega_q \tag{10.4}$$

$$0 = \int_\Sigma v(p,q) \frac{\partial}{\partial n_q} u(q) d\sigma_q - \int_\Sigma u(q) \frac{\partial}{\partial n_q} v(p,q) d\sigma_q$$

$$- \int_{D_i} \Delta u(q) v(p,q) d\omega_q \tag{10.5}$$

---

[1]We write $\Delta_p v(p,q)$ or $\Delta_q v(p,q)$ to indicate differentiation with respect to p or q, respectively.

Let us denote

$$g(p,q) = \frac{1}{4\pi r_{pq}} - v(p,q) \tag{10.6}$$

Subtracting (10.5) from (10.4), we obtain

$$u(p) = \int_{\Sigma} g(p,q) \frac{\partial}{\partial n_q} u(q) d\sigma_q - \int_{\Sigma} u(q) \frac{\partial}{\partial n_q} g(p,q) d\sigma_q - \int_{D_i} \Delta_q u(q) g(p,q) d\omega_q \tag{10.7}$$

Now let $u(p)$ be a solution of the Dirichlet problem

$$u(p) \in C^1(\bar{D}_i) \cap C^2(D_i), \quad \Delta u(p) + F(p) = 0 \quad \forall p \in D_i,$$
$$u(p) = f(p) \quad \forall p \in \Sigma \tag{10.8}$$

Then, by (10.7),

$$u(p) = \int_{\Sigma} g(p,q) \frac{\partial}{\partial n_q} u(q) d\sigma_q - \int_{\Sigma} f(q) \frac{\partial}{\partial n_q} g(p,q) d\sigma_q + \int_{D_i} F(q) g(p,q) d\omega_q \tag{10.9}$$

This integral identity may be treated as an integrodifferential equation, but not as an integral representation of $u(p)$, since the first integral in the right-hand side depends on the unknown boundary values of $\partial u(q)/\partial n_p \big|_{q \in \Sigma}$. Moreover, this function cannot be prescribed arbitrarily, since the identity (10.9) might then involve a contradiction. Indeed, the solution of the Dirichlet problem, if it exists, is unique. The proper formulation of the Dirichlet problem demands continuity of u, but not necessarily of its first derivatives, in the closure $\bar{D}_i$ of D_i. This and the basic existence theorem[2] mean that the first derivatives of u are not necessarily continuous in the closure of D and cannot be prescribed arbitrarily, even in a case where such continuity is guaranteed.

However, the function $v(p,q)$ can be chosen in such a way that the first integral on the right of (10.7) vanishes; this yields an integral representation of $u(p)$ in the proper formulation of the Dirichlet problem:

$$u(p) \in C(\bar{D}_i) \cap C^2(D_i), \quad \Delta u(p) + F(p) = 0 \quad \forall p \in D_i,$$
$$u(p) = f(p) \quad \forall p \in D_i \tag{10.10}$$

This implies the following theorem.

Theorem 10.1.1. *Let $g(p,q)$ be the Green's function of the Dirichlet problem on D_i, defined by the conditions*

$$g(p,q) = \frac{1}{4\pi r_{pq}} - v(p,q) \tag{10.11}$$

---

[2]See the basic Perron theorem and definition of a fundamental region in Chapter 11, Sections 3, and 4.

where

$$v(p,q) \in C(\bar{D}_i), \quad \Delta_q v = 0 \quad \forall p = \text{const} \in D_i,$$

$$v = \frac{1}{4\pi r_{pq}} \quad \forall p = \text{const} \in D_i, \quad \forall q \in \Sigma \tag{10.12}$$

Then the solution of the Dirichlet problem admits an integral representation

$$u(p) = -\int_{\Sigma} f(q) \frac{\partial}{\partial n_q} g(p,q) d\sigma_q + \int_{D_i} F(q) g(p,q) d\omega_q \tag{10.13}$$

dependent on the input data $f(p)$ and $F(p)$ and on the structure of the region D_i.

The converse is also true:

Theorem 10.1.2. Let D_i be a fundamental region for the Dirichlet problem and $f(p) \in C(\Sigma)$, $F(p) \in C(D_i)$. Then the solution of the Dirichlet problem exists and admits the integral representation (10.13).

This theorem is the corollary of a basic Perron theorem, which guarantees the existence of Green's function (See Chapter 11, Sections 3, 4 and 5).

Green's functions possess the following properties.

Theorem 10.1.3. Let $g(p,q)$ be the Green's function of the Dirichlet problem in D_i. Then

(a) for all $p = \text{const} \in D_i$, and all $q \in \bar{D}_i$,

$$0 \le g(p,q) < \frac{1}{4\pi r_{pq}} \tag{10.14}$$

(b) $g(p,q)$ is symmetric,

$$g(p,q) = g(q,p) \quad \forall p,q \in D_i \tag{10.15}$$

(c) for all $q = \text{const} \in D_i$ and all $p \in D_i$,

$$\Delta_p v(p,q) = 0 \tag{10.16}$$

so that v is an harmonic function of each of its arguments everywhere in D, and consequently g is an harmonic function of each of its arguments everywhere in D except for the point $p = q$.

(d) Let D_i be a bounded region and d its diameter. Then

$$\int_{D_i} g^2(p,q) d\omega_q < \frac{d}{12\pi} \quad \forall p \in \bar{D}_i \tag{10.17}$$

PROOF. (a) Let p be an arbitrarily fixed point inside D_i and K_p^ρ a sphere with center at p and radius ρ so small that $\bar{K}_p^\rho \subset D_i$. Since p is a fixed inner point of D_i and v is harmonic, there exists $M = M(p)$ such that

$$v(p,q) < M(p) \quad \forall q \in S_p^\rho = \bar{K}_p^\rho \setminus K_p^\rho \tag{10.18}$$

At the same time,

$$\lim_{q \to p} \frac{1}{r_{pq}} = \infty \tag{10.19}$$

and so there exists $\rho > 0$ so small that

$$\frac{1}{r_{pq}} > 4\pi M(p) \tag{10.20}$$

Since $1/r_{pq} > 0$ for all p, q, the definition of $v(p, q)$ and the maximum principle imply that

$$0 \leq v(p, q) < \frac{1}{4\pi r_{pq}} \ \forall q \in \bar{D}_i \backslash \bar{K}_p^\rho \Rightarrow 0 < g(p, q) < \frac{1}{4\pi r_{pq}} \ \forall q \in \bar{D}_i \backslash \bar{K}_p^\rho \tag{10.21}$$

Since p is an arbitrarily chosen point of D_i, this proves (a).

(b) Let p and m be two different arbitrary points of D_i, and let

$$2\rho < r_{pm} \tag{10.22}$$

Let

$$D_\rho = D_i \backslash (\bar{K}_p^\rho \cup \bar{K}_m^\rho) \tag{10.23}$$

so that the boundary S_ρ of D_ρ is

$$S_\rho = S_p^\rho \cup S_m^\rho \cup \Sigma \tag{10.24}$$

Inside D_ρ, $g(p, q)$ and $g(m, q)$ are harmonic functions without singularities. Since they vanish on Σ, Green's second formula (9.94) yields

$$\int_{S_p^\rho} \left(g(m, q) \frac{\partial g(p, q)}{\partial n_q} - g(p, q) \frac{\partial g(m, q)}{\partial n_q} \right) d\sigma_q$$

$$+ \int_{S_m^\rho} \left(g(m, q) \frac{\partial g(p, q)}{\partial n_q} - g(p, q) \frac{\partial g(m, q)}{\partial n_q} \right) d\sigma_q = 0 \tag{10.25}$$

Now let $\rho \to 0$. Since $v(p, q)$ and $v(m, q)$ are harmonic in D_i, the function v and its derivatives are bounded on S_m^ρ and S_p^ρ, respectively. The same is true tor $g(p, q)$ and $(\partial/\partial n_q) g(p, q)$ on S_m^ρ and for $g(m, q)$ and $(\partial/\partial n_q) g(m, q)$ on S_p^ρ. In addition, on S_p^ρ and S_m^ρ (respectively),

$$r_{pq} = \rho, \ \frac{\partial}{\partial n_q} \frac{1}{r_{pq}} = \frac{1}{\rho^2} \wedge r_{mq} = \rho, \ \frac{\partial}{\partial n_q} \frac{1}{r_{pq}} = \frac{1}{\rho^2} \tag{10.26}$$

Hence

$$\frac{1}{4\pi\rho} \int_{S_m^\rho} \frac{\partial g(p, q)}{\partial n_q} d\sigma_q - \frac{1}{4\pi\rho^2} \int_{S_m^\rho} g(p, q) d\sigma_q - \frac{1}{4\pi\rho} \int_{S_p^\rho} \frac{\partial g(m, q)}{\partial n_q} d\sigma_q$$

$$+ \frac{1}{4\pi\rho^2} \int_{S_p^\rho} g(m, q) d\sigma_q = 0 \tag{10.27}$$

or, by the mean value theorem of integral calculus,

$$\rho \frac{\partial g(p, \tilde{q}_m)}{\partial n_q} - g(p, \tilde{q}_m) - \rho \frac{\partial g(m, \tilde{q}_p)}{\partial n_q} + g(m, \tilde{q}_p) = 0 \tag{10.28}$$

where $\tilde{q}_m \in S_m^\rho$, $\tilde{q}_p \in S_p^\rho$. Since $g(p,q)$ and $g(m,q)$ are continuous in neighborhoods of m and p, respectively, passage to the limit $\rho \to 0$ yields

$$g(m,p) = g(p,m) \qquad (10.29)$$

Since p and m are arbitrary interior points of D_i, this proves (b).

(c) This assertion follows from the harmonicity of $g(p,q)$ as a function of q and its symmetry.

(d) Let p be a fixed interior point of D_i and K_p^d the sphere of the radius d with center at p. By (10.20) we have

$$\int_{D_i} g^2(p,q) d\omega_q < \frac{1}{16\pi^2} \int_{K_p^d} \frac{1}{r_{pq}^2} d\omega_q = \frac{d}{12\pi} \qquad (10.30)$$

Q.E.D.[3]

The definition and properties of the Green's function admit the following electrostatic interpretation.

Let D be a closed region in vacuum bounded by a metallic grounded envelope Σ. A unit positive charge placed at some interior point $p \in D$ induces a single layer of negative charges distributed with a certain density over Σ, to compensate the potential of Σ created by the charge at p. The Green's function is exactly this potential at point $q \in \bar{D}$ created by the unit positive charge situated at p within a grounded metal shell. In particular, the symmetry of the Green's function is simply a mathematical expression of the physical principle of reciprocity, which in electrostatic terminology may be stated as follows: The potential at a point q due to the charge located at the point p is equal to the potential at p due to a charge of the same magnitude placed at q. The positiveness of the Green's function simply expresses the fact that the resultant potential of the induced charges inside the region is smaller in absolute value than the potential created by the real charge that induces the appearance of the negative charges on Σ. The fact that all these charges are negative constitutes the physical contents of Hopf's lemma (see Chapter 9, Section 3), with the density of the induced charges given by the normal derivative of the Green's function at the points of Σ.

Finally, note that the very expression (10.13) establishes a remarkable and somewhat mysterious relation between the potential of a point charge in a grounded metal shell and the field of dipoles distributed over the inner surface Σ of this shell. To elaborate on this, assume that there are no space charges encapsulated by the shell: $F(q) \equiv 0$ (i.e., the second term in (10.13) vanishes; whenever it does not, its interpretation is obvious from the meaning of $g(p,q)$ and the principle of reciprocity. Indeed, the potential at p of a distributed space charge is just a superposition of the potentials of unit charges placed accordingly and smeared all over the volume encapsulated

---

[3]Bounds for the integral of the square of Green's and Neumann's functions are important for a rigorous justification of the use of their Fourier series expansions.

by Σ, to produce the given space charge density $F(q)$.) A solution of the Dirichlet problem may be viewed as a potential of an electric double layer distributed with a given surface density of dipoles over the inner surface of a grounded metal shell. The trouble is that these dipoles induce on the shell an electric charge of an unspecified surface density as reflected by (10.4). (Recall that $(1/4\pi)(\partial/\partial n_q)(1/r_{pq})$ is the potential induced at p by a unit dipole oriented in the direction of the normal at q, whereas $1/4\pi r_{pq}$ is the potential of a unit charge placed there.) The expression (10.13) miraculously resolves this difficulty by relating the resulting potential to that of a unit charge in a grounded shell in the absence of dipoles, that is, to $g(p,q)$. This likely has to do with the electrostatic contents of another reciprocity relation (10.5) (with $\Delta u \equiv 0$ for the purpose of this discussion). Recall that since there is no electric field within the bulk of the shell, the normal derivative of the potential at some point of the interior shell surface equals the surface density of the induced charges at this point. The bulk of the shell being grounded, the same is true concerning the potential at this point as related to the surface density of the dipoles.[4]

Now consider the Neumann problem:

$$u(p) \in C^1(\bar{D}_i) \cap C^2(D_i), \quad \Delta u(p) + F(p) = 0 \quad \forall p \in D_i, \frac{\partial u}{\partial n_p} = \varphi(p) \quad \forall p \in \Sigma$$
$$(10.31)$$

In this case the identity (10.9) yields

$$u(p) = \int_\Sigma g(p,q)\varphi(q)d\sigma_q - \int_\Sigma u(q)\frac{\partial}{\partial n_q}g(p,q)d\sigma_q + \int_{D_i} F(q)g(p,q)d\omega_q \quad (10.32)$$

Here the second integral on the right is not known in advance, since $u(q)$ is not prescribed. Hence this integral must be eliminated by special choice of $v(p,q)$. The option

$$\frac{\partial}{\partial n_q}g(p,q) = 0 \quad \forall q \in \Sigma \ \forall p \in D_i \quad (10.33)$$

is impossible. Indeed, the flux of harmonic function through the boundary of a bounded region vanishes, so that (10.33) would imply

$$\int_\Sigma \frac{\partial}{\partial n_q}v(p,q)d\sigma_q = \frac{1}{4\pi}\int_\Sigma \frac{\partial}{\partial n_q}\frac{1}{r_{pq}}d\sigma_q = 0 \quad \forall p \in D_i \quad (10.34)$$

But by Lemma 12.7.1, which is valid independently of considerations of this chapter,

$$\frac{1}{4\pi}\int_\Sigma \frac{\partial}{\partial n_q}\frac{1}{r_{pq}}d\sigma_q = 1 \quad \forall p \in D_i \quad (10.35)$$

This contradiction shows that (10.33) is impossible. Instead, let us define $v(p,q)$ by the boundary condition

$$\frac{\partial}{\partial n_q}v(p,q) = \frac{\partial}{\partial n_q}\frac{1}{r_{pq}} - \frac{1}{S} \quad (10.36)$$

---

[4]For a systematic presentation of the potential theory, see Chapter 12, Sections 4–7.

where S is the area of the surface Σ,

$$S = \int_\Sigma d\sigma_4 \qquad (10.37)$$

By (10.35), we obtain

$$\int_\Sigma \frac{\partial}{\partial n_q} v(p,q) d\sigma_q = 0 \qquad (10.38)$$

so that the necessary condition for the existence of an harmonic function is now satisfied.

The function $N(p,q)$, known as *Neumann's function*,[5] is defined by the conditions

$$N(p,q) = \frac{1}{4\pi r_{pq}} - w(p,q) \qquad (10.39)$$

$$w(p,q) \in C(\bar{D}_i), \quad \Delta w = 0 \quad \forall p = \text{const} \in D_i,$$

$$\frac{\partial}{\partial n_q} w = \frac{1}{4\pi} \frac{\partial}{\partial n_q} \frac{1}{r_{pq}} - \frac{1}{S} \quad \forall p = \text{const} \in D_i, \ \forall q \in \Sigma \qquad (10.40)$$

Note that (10.40) implies

$$\frac{\partial}{\partial n_q} N(p,q) = \frac{1}{S} \qquad (10.41)$$

The identity (10.32) now implies an integral representation for the solution of the Neumann problem

$$u(p) = \int_\Sigma \varphi(q) N(p,q) d\sigma_q + \int_{D_i} F(q) N(p,q) d\omega_q - \frac{1}{S} \int_\Sigma u(q) d\sigma_q \qquad (10.42)$$

so that, in accordance with the uniqueness theorem, the solution of the Neumann problem is defined uniquely up to an additive constant.

We have the following analog of Theorem 10.1.1.

Theorem 10.1.4. *Neumann's function possesses the following properties.*

(a) *There exists $c = \text{const} > 0$ such that*

$$|N(p,q)| < c \frac{1}{r_{pq}} \quad \forall p, q \in \bar{D}_i \qquad (10.43)$$

(b) *$N(p,q)$ is symmetric,*

$$N(p,q) = N(q,p) \quad \forall p, q \in D_i \qquad (10.44)$$

---

[5](a) One often says "Green's function of the Neumann problem" instead of "Neumann's function." (b) The sufficiency of this condition for the existence of Neumann's function, for a convex region bounded by surfaces of finite curvature, follows from the existence theorems of the theory of electrostatic potentials (see Chapter 12, Section 8).

so that $N(p,q)$ is an harmonic function of each of points p and q inside the region D_i if $p \neq q$.

(c) *Let D_i be a bounded region of diameter d. Then there exists a positive constant $A(d)$, independent of the location of point p, such that*

$$\int_{D_i} N(p,q)^2 d\omega_q < A(d) \tag{10.45}$$

The validity of part (a) is a far-from-trivial consequence of the maximum principle, like that of part (a) of Theorem 10.1.1. It can be proved by successive approximation of the Green's and Neumann functions and their normal derivatives as series of double and single layer potentials, as done by Tichonov [83] for the parabolic Dirichlet and Neumann problems. As this is outside the scope of our book, we omit the proof of assertion (a). The proof of symmetry is a literal repetition of the analogous part of Theorem 10.1.1. Finally, inequality (10.45) is a trivial corollary of (a).

Remark. A proof of (10.45) independent of Tichonov's theory may be found in Gunter [33]. This proof (though simple), is rather lengthy, and is also outside the scope of this book.

There is no natural electrostatic interpretation of the concept of Neumann's function. The most convenient interpretation is a diffusional one. Consider a region D of a continuum bounded by a hollow shell with the internal boundary Σ permeable to a substance diffusing from a pointlike unit source inside D toward the boundary. Let the hollow interior of the shell be constantly stirred, so that the diffusing substance is distributed spatially evenly all over it. Such a shell represents a diffusional instance of "concentrated capacity" introduced in Chapter 2, Section 10. Let the diffusing substance be continuously removed from the shell at a spatially constant rate with a unit total discharge, to compensate for the mass production by the source and thus to allow for the existence of a diffusional steady state in D (see also Chapter 17, Section 2). The Neumann function represents the steady-state field of concentration of the diffusing substance in D.

2. Sommerfeld's method of electrostatic images (method of superposition of sources and sinks)

An effective construction of Green's functions is furnished by Sommerfeld's method of electrostatic images (superposition of sources and sinks). Recall that to solve the problem of a vibrating string of finite length by the method of prolongation, one replaces the real, finite string by a fictitious, infinitely long one, introducing an initial displacement and rate of displacement of the prolonged (fictitious) string in such a way that the boundary conditions are satisfied. The basic idea is thus to extend the prescribed data to continuous and smooth functions. Sommerfeld's method of electrostatic images employs an analogous idea, except that it not only extends the given smooth function to smooth functions but also introduces fictitious positive and negative charges outside the region, distributed so that the resultant potential vanishes at the boundary of the region. The implementation of this idea is possible only if the region possesses some kind of symmetry. We illustrate by means of several characteristic examples.

A. Half-space

Let (x, y, z) be a rectangular cartesian coordinate system, and let

$$D = \{x, y, z : -\infty < x, y < \infty, \ z > 0\} \tag{10.46}$$

Let a unit positive charge be placed at a point $p_0 = (x, y, z) \in D$ and let $q = (\xi, \eta, \zeta)$ be a variable point. Place a fictitious unit negative charge at the point $p_1 = (x, y, -z)$. The resultant potential at q is

$$g(p, q) = \frac{1}{4\pi} \left(\frac{1}{r_{pq}} - \frac{1}{r_{p_1 q}} \right) \tag{10.47}$$

Since

$$q = (\xi, \eta, 0) \Rightarrow r_{pq} = r_{p_1 q} \Rightarrow g(p, q) = 0 \tag{10.48}$$

$g(p, q)$ is indeed the desired Green's function.

B. Strip

Let D be a strip

$$D = \{x, y, z : -\infty < x, y < \infty, \ 0 < z < h\} \tag{10.49}$$

bounded by planes

$$P_0 = \{x, y, z : -\infty < x, y < \infty, \ z = 0\}, P_h = \{x, y, z : -\infty < x, y < \infty, \ z = h\} \tag{10.50}$$

Let a unit positive charge be placed at a point $p_0 = p(x, y, z) \in D$. Place fictitious unit negative charges at the points p_1 and p_2 symmetric to p_0 with respect to P_0 and to P_h, respectively. The superposition of the potentials due to the charges at p_0 and p_1 is equal to zero at P_0, whereas that of the charges at points p_0 and p_2 is equal to zero at P_h. However, the potentials at P_0 and P_h due to the charges at p_2 and at p_1 (respectively), do not vanish. Therefore, in order to compensate for the potential at P_0, a positive charge must be placed at the point p_3 symmetric to p_2 with respect to P_0, as well as a positive charge at the point p_4 symmetric to p_1 with respect to P_h. These charges must be compensated by negative charges at the points p_5 and p_6 symmetric to p_4 and to p_3, and so on ad infinitum. Thus two sequences of positive charges must be placed at the points p_{4n} and p_{4n+3}, and two sequences of negative charges must be placed at the points p_{4n+1} and p_{4n+2}, $n = 0, 1, 2, \ldots$. The z coordinates of these points are

$$z_{4n+1} = -z_{4n}, \quad z_{4n+2} = 2h - z_{4n-1}, \quad z_{4n+3} = -z_{4n+2}, \tag{10.51}$$

so that

$$z_{4n} = 2nh + z, \quad z_{4n+1} = -(2nh + z),$$
$$z_{4n+2} = 2(n+1)h - z, \quad z_{4n+3} = -[2(n+1)h + z] \tag{10.52}$$

Let us denote

$$r_{kn} \overset{\text{def}}{=} r_{p_{4n+k}}, \quad n = 0, 1, 2, \ldots, \quad k = 0, 1, 2, 3 \tag{10.53}$$

Then

$$g(p, q) = g_0(p, q) - g_1(p, q) - g_2(p, q) \tag{10.54}$$

where

$$g_0 = \frac{1}{4\pi r_{pq}}, \quad g_1 = \frac{1}{4\pi} \sum_1^\infty \left(\frac{1}{r_{1n}} - \frac{1}{r_{3n}} \right), \quad g_2 = \frac{1}{4\pi} \sum_1^\infty \left(\frac{1}{r_{2n}} - \frac{1}{r_{4n}} \right) \tag{10.55}$$

Note that all the terms in the series g_1 and g_2 have no singularities. Moreover, these series converge uniformly for any $p = \text{const} \in D$ and $q \in \bar{D}$. Indeed,

$$\frac{1}{r_{1n}} - \frac{1}{r_{3n}} = \frac{r_{3n}^2 - r_{1n}^2}{r_{1n} r_{3n} (r_{3n} + r_{1n})} \tag{10.56}$$

We have

$$r_{3n}^2 - r_{1n}^2 = (2(n+1)h + z - \zeta)^2 - (2nh + z - \zeta)^2 = O(n) \tag{10.57}$$

and

$$r_{1n} r_{3n} (r_{1n} + r_{3n}) = O(n^3) \tag{10.58}$$

so that

$$\frac{1}{r_{1n}} - \frac{1}{r_{3O}} = O \left(\frac{1}{n^2} \right) \tag{10.59}$$

Hence the series g_1 converges absolutely and uniformly, and moreover the same is true for series obtained by differentiating g_1 term by term any number of times. This is obviously true for the series g_2 as well. Thus the function $g(p, q)$ defined by (10.54) is indeed the desired Green's function. Note that since $r_{2n}^2 = r_{3n}^2$ and $r_{0n}^2 = r_{1n}^2$,

$$g(p, q) = \sum_{-\infty}^\infty \left(\frac{1}{r_{0n}} - \frac{1}{r_{1n}} \right) \tag{10.60}$$

Remark 1. Obviously the function

$$N(p, q) = \frac{1}{4\pi} \left(\frac{1}{r_{pq}} + \frac{1}{r_{p_1 q}} \right) \tag{10.61}$$

is Neumann's function in the half-space $z > 0$.

Remark 2. If $D \subset \mathbb{R}_2$, then all the preceding results remain valid, provided the 3-dimensional fundamental solution $1/4\pi r_{pq}$ is replaced by the 2-dimensional solution $(1/2\pi) \ln(1/r_{pq})$. (See Problem 10.2.6.)

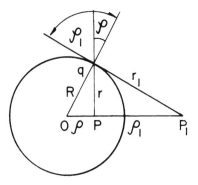

Figure 10.1. Symmetry with
respect to a sphere K_O^R; similarity
of triangles OPQ and OQP_1.

C. Sphere $K \overset{\text{def}}{=} \mathbf{K}_o^R$ of radius R

Recall that points p and p_1 are said to be *symmetric* with respect to a
sphere $S \overset{\text{def}}{=} S_o^R = \bar{K} \setminus K$ of radius R with center at the origin O of the
cartesian coordinate system, if p and p_1 lie on the same prolonged diameter
of K and if

$$\frac{\rho_1}{R} = \frac{R}{\rho} \tag{10.62}$$

where

$$\rho = r_{op}, \quad \rho_1 = r_{op_1} \tag{10.63}$$

Now let $q \in S$, p, and p_1 be symmetric points. Then the triangles $\triangle_{opq}$ and
$\triangle_{oqp_1}$ are similar (see Figure 10.1), so that

$$\frac{\rho}{R} = \frac{R}{\rho_1} = \frac{r}{r_1} \tag{10.64}$$

where

$$r = r_{pq}, \quad r_1 = r_{p_1q} \tag{10.65}$$

The Green's function $g(p, q)$ for K is

$$g(p, q) \overset{\text{def}}{=} \frac{1}{4\pi r_{pq}} - v(p, q) = \frac{1}{4\pi}\left(\frac{1}{r} - \frac{R}{\rho} \cdot \frac{1}{r_1}\right) \tag{10.66}$$

Indeed, if $p \in K$ and $q \in \bar{K}$ then $r_1 \neq 0$, so that $g(p, q)$ is a harmonic function
of q inside K. At the same time, if $q \in S$ then it follows from (10.64) and
(10.65) that $g(p, q) = 0$, so that $g(p, q)$ as defined by (10.66) is indeed the
desired Green's function.

Remark 3. The Green's function for a circle $C \in \mathbb{R}_2$ is

$$g(p, q) = \frac{1}{2\pi}\ln\left(\frac{\rho r_1}{Rr}\right) \tag{10.67}$$

3. Poisson integral

Let $u(p)$ be the solution of the Dirichlet problem

$$u(p) \in \bar{C}(K_o^R) \wedge \Delta u(p) = 0 \quad \forall p \in K_o^R \wedge u(p) = f(p) \quad \forall p \in S_o^R \qquad (10.68)$$

Then, by (10.12) and (10.13),

$$u(p) = -\int_{S_o^R} f(q) \frac{\partial}{\partial n_q} g(p,q) d\sigma_q \qquad (10.69)$$

where g is defined by (10.66). Let φ and φ_1 be the angles between the outward normal $\mathbf{n}_q$ to S_o^R at q and the vectors $\mathbf{r}_{pq}$ and $\mathbf{r}_{p_1 q}$ respectively. Then

$$\rho^2 = R^2 + r^2 - 2Rr \cos\varphi \qquad (10.70)$$

so that

$$\cos\varphi = \frac{R^2 + r^2 - \rho^2}{2Rr} \qquad (10.71)$$

and similarly

$$\cos\varphi_1 = \frac{R^2 + r_1^2 - \rho^2}{2Rr_1} \qquad (10.72)$$

We have

$$\frac{\partial}{\partial n_q} g(p,q) = -\frac{1}{4\pi} \left(\cos\varphi \frac{\partial}{\partial r} \frac{1}{r} - \frac{R}{\rho} \cos\varphi_1 \frac{\partial}{\partial r_1} \frac{1}{r_1} \right) \equiv \frac{1}{4\pi} \left(\frac{\cos\varphi}{r^2} - \frac{R}{\rho} \frac{\cos\varphi_1}{r_1^2} \right) \qquad (10.73)$$

Using (10.72) and (10.63), we may eliminate $\cos\varphi_1$ and r_1 from this equality, obtaining

$$-\frac{\partial}{\partial n_q} g(p,q) = \frac{R^2 - \rho^2}{4\pi R} \cdot \frac{1}{r^3} \qquad (10.74)$$

Thus we have proved the following theorem.

Theorem 10.3.1. *If $u(p)$ is the solution of the Dirichlet problem (10.68), then it admits an integral representation in the form of a Poisson integral*

$$u(p) = \frac{R^2 - \rho^2}{4\pi R} \int_{S_o^R} \frac{f(q) d\sigma_q}{r^3} \qquad (10.75)$$

Here

$$r^2 = R^2 + \rho^2 - 2R\rho \cos\theta \qquad (10.76)$$

where θ is the angle between the vectors $\mathbf{r}_{op}$ and $\mathbf{r}_{op_1}$. Now let $(\rho, \varphi, \vartheta)$ be a spherical coordinate system with origin at O and $(\rho, \varphi, \vartheta)$ and $(\rho_1, \varphi_1, \vartheta_1)$ the coordinates of the points p and p_1, respectively. Then

$$\cos\theta = \sin\vartheta \cdot \sin\vartheta_1 \cdot \cos(\varphi - \varphi_1) + \cos\vartheta \cdot \cos\vartheta_1 \qquad (10.77)$$

We now prove the converse.

Theorem 10.3.2. Let $f(p) \in C(S_o^R)$ and let $u(p)$ be represented by the Poisson integral (10.75). Then $u(p)$ is the solution of the Dirichlet problem (10.68).

PROOF. We have to prove that:

(a) $u(p)$ is harmonic in K_o^R; and
(b) $u(p)$ is continuous in $\bar{K}_o^R$ and $u(p) = f(p)$ for all $p \in S_o^R$.

(a) That $u(p)$ is harmonic follows from the fact that it is legitimate to change the order of differentiation and integration, so that

$$\Delta u(p) = -\Delta \int_{S_o^R} f(q) \frac{\partial}{\partial n_q} g(p,q) d\sigma_q = - \int_{S_o^R} f(q) \Delta \frac{\partial}{\partial n_q} g(p,q) d\sigma_q$$

$$= - \int_{S_o^R} f(q) \frac{\partial}{\partial n_q} \Delta g(p,q) d\sigma_q \equiv 0 \quad \forall p \in K_o^R \qquad (10.78)$$

(b) Let $m \in S_o^R$ be an arbitrarily fixed point and Γ a continuous curve cutting S_o^R at m nontangentially. We must prove that if $p \in K_o^R$ tends to m along Γ then

$$\lim_{p \to m} u(p) = f(m) \qquad (10.79)$$

Note, first of all, that since 1 is the solution of the Dirichlet problem (10.68) with $f(p) \equiv 1$, it may be represented by a Poisson integral:

$$1 = \frac{R^2 - \rho^2}{4\pi R} \int_{S_o^R} \frac{d\sigma_q}{r^3} \qquad (10.80)$$

Since m is fixed it follows from (10.75) and (10.80) that

$$u(p) - f(m) = \frac{R^2 - \rho^2}{4\pi R} \int_{S_o^R} \frac{f(q) - f(m)}{r^3} d\sigma_q \qquad (10.81)$$

Let $\varepsilon > 0$ be arbitrarily small. Since $f(p)$ is continuous, there exist $M = \text{const} > 0$ and $\delta = \text{const} > 0$ such that, for all $q, m \in S_o^R$,

$$|f(p)| < M \qquad (10.82)$$

and

$$r_{pm} < \delta \Rightarrow |f(p) - f(m)| < \varepsilon/2 \qquad (10.83)$$

Let

$$S_1 = S_o^R \cap K_m^{\delta/2}, \quad S_2 = S_o^R \setminus K_m^{\delta/2} \qquad (10.84)$$

Then

$$|u(p) - f(m)| \leq I_1 + I_2 \qquad (10.85)$$

where

$$I_1 = \frac{R^2 - \rho^2}{4\pi R} \int_{S_1} \frac{|f(q) - f(m)|}{r^3} d\sigma_q, \quad I_2 = \frac{R^2 - \rho^2}{4\pi R} \int_{S_2} \frac{|f(q) - f(m)|}{r^3} d\sigma_q$$

$$(10.86)$$

By (10.80), (10.83), and (10.84),

$$I_1 < \frac{R^2 - \rho^2}{4\pi R} \int_{S_o^R} \frac{|f(q) - f(m)|}{r^3} d\sigma_q < \frac{\varepsilon}{2} \qquad (10.87)$$

Note, further, that there exists a positive constant d such that

$$p \in K_m^{\delta/2} \cap \Gamma \wedge q \in S_2 \Rightarrow r^3 > d \qquad (10.88)$$

Thus it follows from (10.88) and (10.82) that

$$I_2 < \frac{2R^2}{3} \cdot \frac{M}{d} (R^2 - \rho^2) \qquad (10.89)$$

Hence there exists $\delta_1 \leq \delta$ so small that

$$R - \rho < \delta_1 \Rightarrow I_2 < \varepsilon/2 \qquad (10.90)$$

Thus, by (10.85), (10.87), and (10.90),

$$r_{pm} < \delta_1 \Rightarrow |u(p) - f(m)| < \varepsilon \qquad (10.91)$$

Since $\varepsilon > 0$ is arbitrarily small and m is an arbitrary point of S_o^R, this proves (b). **Q.E.D.**

Remark. Let $C_o^R \subset \mathbb{R}_2$ be a circle of radius R and circumference C. Then Theorems 10.3.1 and 10.3.2 remain true if the 3-dimensional Poisson integral is replaced by its 2-dimensional counterpart,

$$u(p) = \frac{R^2 - \rho^2}{4\pi R} \int_C \frac{f(q) ds_q}{r^2} \qquad (10.92)$$

Problems

P10.1.1. Give a hydrodynamic interpretation of the Green's function and Neumann's function.

P10.2.1. The expression for the Neumann's function for the semispace and strip has been given without confirming its correctness. Do so.

P10.2.2. Are the series (10.54), (10.55), and (10.60) equivalent? If so, prove your statement; if not, explain.

P10.2.3. Consider the divergent series

$$N(p, q) \stackrel{\text{def}}{=} \frac{1}{4\pi} \sum_{-\infty}^{\infty} \left(\frac{1}{r_{0n}} + \frac{1}{r_{1n}} \right)$$

Is it possible to make $N(p, q)$ meaningful? If so, do this and prove that the result represents the Neumann function in the strip $0 < z < h$. *Hint:* See Appendix 4, Section 5.

P10.2.4. Construct the Green's function for a channel

$$D_1 = \{x, y, z : 0 < x < a, 0 < y < b, -\infty < z < \infty\}$$

and for a cube

$$D_2 = \{x, y, z : 0 < x < a, 0 < y < b, 0 < x < c\}$$

P10.2.5. Construct the Green's function for the angle $0 < \varphi < \pi/a$. For what values of a can this be done using Sommerfeld's method of electrostatic images? Are there any restrictions on the value of a, and of what kind? Explain your answer.

P10.2.6. In the 2-dimensional case, series (10.55) determining g_1 and g_2 must be replaced by divergent series

$$g_1 = \frac{1}{2\pi} \sum_1^\infty \ln \frac{r_{3n}}{r_{1n}}, \qquad g_2 = \frac{1}{2\pi} \sum_1^\infty \ln \frac{r_{4n}}{r_{2n}}$$

Is it possible to justify the assertion of remark 2? In what sense does this remark hold? *Hint:* See Chapter 12, Section 5 and Appendix 4, Section 5.

11

Sequences of harmonic functions. Perron's theorem. Schwarz alternating method

1. Harnack's theorems

Because the real and imaginary parts of any analytic function are harmonic functions, the properties of analytic and harmonic functions exhibit parallel features. One of the main theorems in the theory of analytic functions is Weierstrass's theorem that the limit of any uniformly convergent sequence of analytic functions is again an analytic function. The harmonic functions analogy of this theorem is as follows.

Theorem 11.1.1 (Harnack's first theorem). *Let $D \subset \mathbb{R}_n$ be a region with boundary Σ, and let $\{u_n(p)\}\big|_{n=1}^{\infty}$ be a sequence of functions continuous in $\bar{D}$, harmonic inside D, and uniformly convergent on Σ. Then the sequence converges uniformly to a harmonic function in every region $D^* \subset \bar{D}^* \subset D$.*

PROOF. Let $\varepsilon > 0$ be arbitrarily small. Then, in view of the uniform convergence of $\{u_n(p)\}$ on Σ, there exists $N > 0$ so large that

$$|u_n(p) - u_m(p)| < \varepsilon \quad \forall p \in \Sigma, \ \forall m, n > N \tag{11.1}$$

By the maximum principle, this means that

$$|u_n(p) - u_m(p)| < \varepsilon \quad \forall p \in D \tag{11.2}$$

Hence $\{u_n(p)\}$ converges uniformly everywhere in D. Let $M \in D$ be an arbitrarily fixed point, and let K_M^R be a sphere of arbitrarily small radius $R > 0$ such that $\bar{K}_M^R \subset D$. By the mean value theorem,

$$u_n(M) = \frac{1}{4\pi} \int_{S_M^R} u_n(q) d\sigma_q \tag{11.3}$$

Let

$$u(p) = \lim_{n \uparrow \infty} u_n(p) \quad \forall p \in D \tag{11.4}$$

229

Because of the uniform convergence of $\{u_n\}$ on S_M^R, we may evaluate the limit under the integral sign in (11.3), obtaining

$$u(M) = \lim_{n\uparrow\infty} \frac{1}{4\pi} \int_{S_M^R} u_n(q) d\sigma_q$$

$$= \frac{1}{4\pi} \int_{S_M^R} \lim_{n\uparrow\infty} u_n(q) d\sigma_q = \frac{1}{4\pi} \int_{S_M^R} u(q) d\sigma_q \tag{11.5}$$

Since R is arbitrary, the inverse mean value theorem is applicable, so that (11.5) implies that $u(M)$ is harmonic and consequently, since M is an arbitrary point of D, that u is harmonic everywhere in D. **Q.E.D.**

Harnack's second theorem deals with monotonic sequences of harmonic functions and has no counterpart in the theory of analytic functions, since the set of complex numbers is unordered. The theorem reads as follows.

Theorem 11.1.2 (Harnack's second theorem). *Let $\{u_n\}$ be a monotonic sequence of functions, harmonic inside a region $D \in \mathbb{R}_n$, that converges at some point $M \in D$. Then it converges uniformly inside any region D^* contained in D together with its boundary, so that, by Harnack's first theorem,*

$$u(p) = \lim_{n\uparrow\infty} u_n(p) \tag{11.6}$$

is a function harmonic in D^*.

PROOF. The theorem is a corollary of the Heine–Borel lemma and the following.

Lemma 11.1.1 (Harnack's large inequality). *Consider a region $D \in \mathbb{R}_n$, a subregion $D_1 \subset \bar{D}_1 \subset D$, and a function $u \in C(\bar{D})$, harmonic in D. Then there exist positive constants c, C, dependent on D and $d = \min_{p\in\partial D, q\in\partial D} r_{pq}$ only, such that, for any fixed point $M \in D$ and any $p \in D_1$,*

$$cu(M) \le u(p) \le Cu(M) \tag{11.7}$$

Assume that the assertion of this lemma is true. Let $\{u_k\}$ be (say) a nondecreasing sequence, so that

$$u(p) \overset{\text{def}}{=} u_n(p) - u_m(p) \ge 0 \quad \forall n > m \ge 1 \tag{11.8}$$

We may assume that $M \in D_1$, since otherwise we may replace D_1 by a region D_2 such that

$$M \in D_2 \subset \bar{D}_2 \subset D \wedge D_1 \subset \bar{D}_1 \subset D_2 \tag{11.9}$$

Let $\varepsilon = \text{const} > 0$ be arbitrarily small. Then the convergence of the sequence $\{u_n(M)\}$, together with (11.9) and (11.7), implies

$$0 < u(M) < \varepsilon/C \Rightarrow u(p) < \varepsilon \quad \forall p \in D_1 \tag{11.10}$$

which proves the theorem on the assumption that the lemma is true.

Harnack's large inequality is in turn a corollary of the following lemma.

Lemma 11.1.2 (Harnack's small inequality). *Let $K_M^{R_1}$ be a sphere of radius R_1 with center at a point M, embedded in the sphere K_M^R:*

$$K_M^{R_1} \subset \bar{K}_M^{R_1} \subset K_M^R \tag{11.11}$$

and let $u(p)$ be a function continuous in $\bar{K}_M^R$, harmonic and nonnegative inside K_M^R. Then there exist positive constants α, β dependent only on $R - R_1$ (i.e., independent of u) such that

$$\alpha u(M) < u(p) < \beta u(M) \quad \forall p \in K_M^{R_1} \tag{11.12}$$

Assume that the assertion of this lemma is true, and let D_2 be a region such that

$$D_1 \subset \bar{D}_1 \subset D_2 \subset \bar{D}_2 \subset D \tag{11.13}$$

Let Σ and Σ_2 be the boundaries of D and D_2, respectively. Define δ by

$$\delta \overset{\text{def}}{=} \frac{1}{4} R_{\Sigma, \Sigma_2} > 0 \tag{11.14}$$

where

$$R_{\Sigma, \Sigma_2} = \min_{p \in \Sigma} \min_{q \in \Sigma_2} r_{pq} \tag{11.15}$$

Let p be an arbitrary point of D_1. Consider a covering of $\bar{D}_1$ by spheres of radius δ. By the Heine–Borel lemma, we can select a finite subcovering of $\bar{D}_1$ consisting of N spheres of radius δ with centers at points p_k, $k = 0, \dots, N$, numbered so that $p_0 = M$ and $p_N = p$. Let L be the polygonal line with vertices at the centers of the spheres in this finite covering of $\bar{D}_1$ and with endpoints p_0 and $p_n = p$, consisting of n segments, numbered so that $n \leq N$. The total length of the segments is less than 2δ, and hence Harnack's small inequality can be used to estimate $u(p_k)$. This gives

$$\alpha u(p_0) < u(u_1) < \beta u(p_0), \quad \alpha u(p_1) < u(u_2) < \beta u(p_1), \dots,$$
$$\alpha u(p_{n-1}) < u(p_n) < \beta u(p_{n-1}) \tag{11.10}$$

Consequently,

$$\alpha^n u(m) \prod_{k=1}^{n-1} u(p_k) < u(p) \prod_{k=1}^{n-1} u(p_k) < \beta^n u(m) \prod_{k=1}^{n-1} u(p_k)$$
$$\Rightarrow \alpha^n u(M) < u(p) < \beta^n u(M) \tag{11.17}$$

Since α and β depend only on δ which in turn depends only on $R - R_1$, and since p is an arbitrary point of D_1, inequality (11.17) proves Harnack's large inequality on the assumption that Harnack's small inequality is true.

It remains to prove the latter. Represent $u(p)$ by a Poisson integral

$$u(p) = \frac{R^2 - \rho^2}{4\pi R} \int_{S_M^R} \frac{u(q)}{r_{pq}^3} d\sigma_q \tag{11.18}$$

where $\rho = r_{pm}$. Since $u(q) \geq 0$ for any $q \in K_M^R$, the mean value theorem of integral calculus is applicable, so that

$$u(p) = \frac{R^2 - \rho^2}{4\pi r_{pq^*}^3} \int_{S_M^R} u(q) d\sigma_q = \frac{R^2 - \rho^2}{4\pi r_{pq^*}^3} Ru(M) \tag{11.19}$$

Here q^* is a point of S_M^R. If $p \in K_M^{R_1}$, then

$$0 \leq \rho \leq R_1, \quad R - R_1 \leq r_{pq^*} \leq R + R_1 \tag{11.20}$$

Therefore, taking

$$\alpha = \frac{R^2(R - R_1)}{(R + R_1)^2}, \quad \beta = \frac{R^2(R + R_1)}{(R - R_1)^3} \tag{11.21}$$

we obtain

$$\alpha u(M) \leq u(p) \leq \beta u(M) \tag{11.22}$$

$$\textbf{Q.E.D.}$$

2. Complete classes of (continuous) superharmonic and subharmonic functions

The purpose of this chapter is to prove the basic Perron theorem, determining the most general conditions for the existence of a classical solution to the Dirichlet problem. Harnack's second theorem is the main tool in its proof. However, we must first consider the complete classes of superharmonic and subharmonic functions. The reader will recall that we defined these functions (in Chapter 9, Section 2), as twice continuously differentiable functions; these, however, constitute only subsets of the complete classes of continuous super- and subharmonic functions, as the following definitions show.

Definition 11.2.1. Let D be a region in $\mathbb{R}_n$ and $K_M^R \subset D$ a sphere of radius R with center at the point M. To fix ideas, let $n = 3$. Let $u(p) \in C(D)$. Then $\mathfrak{M}(p \,|\, u, M, R)$ is the continuous function in D defined by the equalities

$$\mathfrak{M}(p \,|\, u, M, R) = \begin{cases} u(p) & \text{if } p \in D \setminus K_M^R \\ \Delta\mathfrak{M}(p \,|\,) = 0 & \text{if } p \in K_M^R \end{cases} \tag{11.23}$$

By the mean value theorem,

$$\mathfrak{M}(M \,|\, u, M, R) = \frac{1}{4\pi R^2} \int_{S_M^R} u(q) d\sigma_q \tag{11.24}$$

Definition 11.2.2. Let $u(p) \in C(D)$. Then $u(p)$ is said to be *superharmonic* in D if, for any $K_M^R \subset D$,

$$\mathfrak{M}(p \,|\, u, M, R) \leq u(p) \quad \forall p \in D \tag{11.25}$$

It is said to be *subharmonic* in D if, for any $K_M^R \subset D$,

$$\mathfrak{M}(p \,|\, u, M, R) \geq u(p) \quad \forall p \in D \tag{11.26}$$

If $u(p)$ is simultaneously super- and subharmonic, it is said to be *harmonic*.

The main properties of super- and subharmonic functions are as follows.

Lemma 11.2.1. *If $u(p)$ is superharmonic in D then $-u(p)$ is subharmonic and conversely.*

Lemma 11.2.2. *If $u(p)$ and $v(p)$ are superharmonic (subharmonic) in D then $u(p) + v(p)$ is superharmonic (subharmonic) in D.*

The validity of these lemmas is obvious.

Lemma 11.2.3. *Let $u(p) \in C^2(D)$ be superharmonic in D. Then*

$$\Delta u(p) \leq 0 \quad \forall p \in D \tag{11.27}$$

PROOF. Suppose there exists $M \in D$ such that

$$\Delta u(M) > 0 \tag{11.28}$$

Then by the continuity of Δu there exists $\rho > 0$ such that

$$\Delta u(p) > 0 \quad \forall p \in K_M^\rho \tag{11.29}$$

But by definition,

$$\Delta \mathfrak{M}(p \mid u, M, \rho) = 0 \quad \forall p \in K_M^\rho, \quad \mathfrak{M}(p \mid u, M, \rho) = u(p) \quad \forall p \in S_M^\rho \tag{11.30}$$

Hence

$$\begin{aligned}\Delta[u(p) - \mathfrak{M}(p \mid u, M, \rho)] > 0 \quad \forall p \in K_M^\rho \\ u(p) - \mathfrak{M}(p \mid u, M, \rho) = 0 \quad \forall p \in S_M^\rho\end{aligned} \tag{11.31}$$

so that, by the maximum principle,

$$u(p) - \mathfrak{M}(p \mid u, M, \rho) < 0 \quad \forall p \in K_M^\rho \tag{11.32}$$

which contradicts the assumption that $u(p)$ is superharmonic. Thus (11.28) cannot hold. **Q.E.D.**

Lemma 11.2.4. *Let*

$$\Delta u(p) \leq 0 \quad \forall p \in D \tag{11.33}$$

Then $u(p)$ is superharmonic in D.

PROOF. Let $K_M^R \subset D$. Then by the definition of $\mathfrak{M}(p \mid \ )$ and (11.33),

$$\begin{aligned}u(p) - \mathfrak{M}(p \mid u, M, R) = 0 \quad \forall p \in S_M^R \\ \Delta[u(p) - \mathfrak{M}(p \mid u, M, R)] \leq 0 \quad \forall p \in K_M^R\end{aligned} \tag{11.34}$$

Hence, by the maximum principle,

$$u(p) - \mathfrak{M}(p \mid u, M, R) \geq 0 \tag{11.35}$$

which means that u is a superharmonic function, since $K_M^R \subset D$ was chosen arbitrarily. **Q.E.D.**

Remark. Lemmas 11.2.3–4 confirm that our new definition of super-harmonicity in $C(D)$ is compatible with the definition in $C^2(D)$ from Chapter 9.

The following lemmas are essential for Perron theory.

Lemma 11.2.5 (Maximum principle). *Let $u(p) \in C(\bar{D})$ be a superharmonic (subharmonic) function inside D. Then it achieves its minimum (maximum) at the boundary Σ of D.*

PROOF. Let $u(p) \in C(\bar{D})$ be superharmonic in D. Suppose to the contrary that

$$M = \min_{p \in \Sigma} u(p) > \min_{p \in \bar{D}} u(p) = m \tag{11.36}$$

Define sets

$$F = \left\{ p \in \bar{D} \colon u(p) = m \right\}, \quad G = D \setminus F \tag{11.37}$$

F is a closed set and G is open, both are not empty, and moreover the distance between F and Σ is positive:

$$2R = \min_{q \in \Sigma} \min_{p \in F} > 0 \tag{11.38}$$

Clearly there exists $Q \in F$ such that

$$\min_{q \in \Sigma} r_{Qq} - 2R > 0 \tag{11.39}$$

Let K_Q^R be a sphere of radius R with center at Q and boundary S. The set F divides S into two parts,

$$S_1 = S \cap F, \quad S_2 = S \setminus F \tag{11.40}$$

Obviously

$$\mu_1 = \int_{S_1} d\sigma > 0, \int_{S_2} d\sigma = 4\pi R^2 - \mu_1 \overset{\text{def}}{=} \mu_2 < 4\pi R^2 \tag{11.41}$$

By Definition 11.2.1 and the mean value theorem,

$$\mathfrak{M}(Q \mid u, Q, R) = \frac{1}{4\pi R^2} \left(\int_{S_1} + \int_{S_2} \right) u(q) d\sigma_q \tag{11.42}$$

Hence it follows from (11.37), (11.40), and (11.41) that

$$\mathfrak{M}(Q \mid u, Q, R) > m\mu_1 + m\mu_2 = m = u(Q) \tag{11.43}$$

which contradicts the assumption that u is superharmonic. **Q.E.D.**

Lemma 11.2.6. *Let u_1 and u_2 be superharmonic (subharmonic) in D and define*

$$u(p) = \min[u_1(p), u_2(p)] \quad (u(p) = \max[u_1(p), u_2(p)]) \quad \forall p \in D \quad (11.44)$$

Then u is superharmonic (subharmonic) in D.

PROOF. Let $K_Q^R \subset D$ be an arbitrarily chosen sphere in D, and let S be its boundary. Let

$$S_1 = \{q \in S: u(q) = u_1(q)\}, S_2 = \{q \in S: u(q) = u_2(q)\} \quad (11.45)$$

Then

$$\mathfrak{M}(Q \mid u, Q, R) = \frac{1}{4\pi R^2} \left(\int_{S_1} u_1(q) d\sigma_q + \int_{S_2} u_2(q) d\sigma_q \right)$$

$$< \frac{1}{4\pi R^2} \int_S u_1(q) d\sigma_q = u_1(Q) \quad (11.46)$$

and similarly

$$\mathfrak{M}(Q \mid u, Q, R) < u_2(Q) \quad (11.47)$$

In view of (11.46) and (11.47),

$$\mathfrak{M}(Q \mid u, Q, R) \leq \min[u_1(Q), u_2(Q)] = u(Q) \quad (11.48)$$

Since Q is an arbitrary point of D, this completes the proof of the lemma. **Q.E.D.**

Lemma 11.2.7. *Let $u(p)$ be superharmonic (subharmonic) in D and let $K \overset{\text{def}}{=} K_Q^R \subset D$ be a sphere. Then*

$$v(p) \overset{\text{def}}{=} \mathfrak{M}(p \mid u, Q, R) \quad (11.49)$$

is a superharmonic (subharmonic) function in D.

PROOF. Let u be superharmonic in D. We must show that for any $\bar{K}_0 \overset{\text{def}}{=} \bar{K}_s^\rho \subset D$,

$$\mathfrak{M}(p \mid v, s, \rho) \leq v(p) \quad (11.50)$$

Let us consider successively all possible positions of K_0 relative to K (see Figure 11.1). These are

(a) $\bar{K}_0 \cap K = \emptyset$;
(b) $\bar{K}_0 \subset K$;
(c) $\bar{K} \subset K_0$;
(d) $D_{sQ} \overset{\text{def}}{=} (K_0 \cap K \neq \emptyset) \wedge (\text{not (b)}) \wedge (\text{not (c)})$.

We have

$$(a) \Rightarrow v(p) = u(p) \quad \forall p \in \bar{K}_0$$
$$\Rightarrow \mathfrak{M}(p \mid v, s, \rho) = \mathfrak{M}(p \mid u, s, \rho) \leq u(p) = v(p) \quad (11.51)$$
$$(b) \Rightarrow \mathfrak{M}(p \mid v, s, \rho) = v(p) \quad \forall p \in \bar{K}_0 \quad (11.52)$$

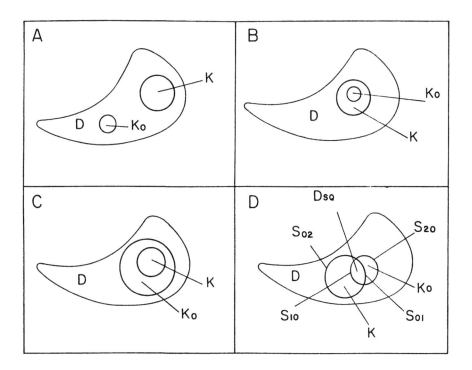

Figure 11.1.   Location of K_0 relative to K.

Indeed, $v(p)$ is harmonic inside K and consequently inside $\bar{K}_0$, so that the truth of (11.52) follows from the uniqueness of the solution to the Dirichlet problem;

$$(\text{c}) \Rightarrow \mathfrak{M}(p \mid v, s, \rho) = v(p) \equiv u(p) \;\; \forall p \in \bar{K}_0 \setminus K$$

$$\Rightarrow \mathfrak{M}(p \mid v, Q, R) = \mathfrak{M}(p \mid u, Q, R) = v(p) \quad \forall p \in K \qquad (11.53)$$

Consider (d). We have

$$(\text{d}) \Rightarrow \mathfrak{M}(p \mid v, s, \rho) \le v(p) \;\; \forall p \in K_0 \qquad (11.54)$$

Indeed let S, S_0, and Σ be the boundaries of K, K_0, and D_{sQ}, respectively. We have

$$\Sigma = S_{10} \cup S_{01}, \quad S = S_{02} \cup S_{01}, \quad S_0 = S_{10} \cup S_{20} \qquad (11.55)$$

where

$$S_{01} = \Sigma \cap K_0, \quad S_{10} = \Sigma \cap K \qquad (11.56)$$

and

$$S_{02} = S \setminus K_0, \quad S_{20} = S_0 \setminus K \qquad (11.57)$$

Further,

$$v(p) = u(p) \ \forall p \in D \setminus K \Rightarrow v(p) = u(p) \ \forall p \in S_{01} \tag{11.58}$$

and

$$v(p) = \mathfrak{M}(p \mid u, Q, R) \leq u(p) \quad \forall p \in S_{10} \tag{11.59}$$

Hence

$$v(p) \leq u(p) \quad \forall p \in (S_{01} \cup S_{102}) = \Sigma \tag{11.60}$$

By the maximum principle, it follows from (11.58) and (11.59) that

$$\mathfrak{M}(p \mid v, s, \rho) \leq \mathfrak{M}(p \mid u, s, \rho) \leq u(p) = v(p) \quad \forall p \in K_0 \setminus K \tag{11.61}$$

Since both $\mathfrak{M}(p \mid u, Q, R)$ and $\mathfrak{M}(p \mid u, s, \rho)$ are harmonic functions inside D_{sQ} and continuous in its closure $\bar{D}_{sQ}$, it follows from (11.58), by virtue of the maximum principle, that

$$\mathfrak{M}(p \mid v, s, \rho) \leq \mathfrak{M}(p \mid u, s, \rho) = v(p) \quad \forall p \in D_{sQ} \tag{11.62}$$

Thus, by (11.59) and (11.60),

$$\mathfrak{M}(p \mid v, s, \rho) \leq v(p) \quad \forall p \in K_0 \tag{11.63}$$

proving (11.54). Thus

$$\mathfrak{M}(p \mid v, s, \rho) \leq v(p) \quad \forall p \in D \tag{11.64}$$

in all possible cases. **Q.E.D.**

3. Basic Perron theorem [54]

Let $D \subset \mathbb{R}_n$ be region with boundary Σ (to fix ideas, we let $n = 3$). Let $f(p) \in C(\Sigma)$ be a given function. We introduce the following definitions.

Definition 11.3.1. $\mathcal{M}^+$ is the set of all functions $u(p) \in C(\bar{D})$ that are superharmonic inside D and such that

$$u(p) \in \mathcal{M}^+ \Rightarrow u(p) \geq f(p) \quad \forall p \in \Sigma \tag{11.65}$$

$\mathcal{M}^+$ is called the set of *upper functions*.

Definition 11.3.2. $\mathcal{M}^-$ is the set of all functions $u(p) \in C(\bar{D})$ that are subharmonic inside D and such that

$$u(p) \in \mathcal{M}^- \Rightarrow u(p) \leq f(p) \quad \forall p \in \Sigma \tag{11.66}$$

$\mathcal{M}^-$ is called the set of *lower functions*.

Lemma 11.3.1. *The sets of upper and lower functions are not empty.*

PROOF. Let

$$m = \min f(p), \quad M = \max f(p) \quad \forall p \in \Sigma \tag{11.67}$$

Then

$$M \in \mathcal{M}^+, \quad m \in \mathcal{M}^- \tag{11.68}$$

so that $\mathcal{M}^+$ and $\mathcal{M}^-$ are not empty.   **Q.E.D.**

Lemma 11.3.2. *The sets $\mathcal{M}^+$ and $\mathcal{M}^-$ are bounded from below and above, respectively.*

PROOF. Obviously, if $u(p) \in \mathcal{M}^+$ and $v(p) \in \mathcal{M}^-$ then

$$u(p) \geq v(p) \Rightarrow \inf \mathcal{M}^+ \geq m \wedge \sup \mathcal{M}^- \leq M \tag{11.69}$$
$$\textbf{Q.E.D.}$$

Definition 11.3.3. $u(p) \stackrel{\mathrm{def}}{=} \inf \mathcal{M}^+$ $(u(p) = \sup \mathcal{M}^-)$ is called a *Perron function.*

Our purpose is to prove the basic Perron theorem.[1]

Theorem 11.3.1 (Perron). *For any $f(p) \in C(\Sigma)$, the Perron function $u(p)$ is harmonic in D.*

PROOF. The theorem is a corollary of Harnack's second theorem and the following remark: All the lemmas of Section 2 of this chapter hold for upper and lower functions. This assertion is quite obvious and needs no proof.

We now proceed to the proof of Perron's theorem. It is sufficient to prove that $u(p)$ is harmonic inside any arbitrarily chosen sphere $K_M^R \subset \bar{K}_M^R \subset D$. To fix ideas, we assume that

$$u(M) = \inf \mathcal{M}^+ \tag{11.70}$$

Consider a monotonic sequence of numbers

$$\varepsilon_1 > \varepsilon_2 > \varepsilon_n > \cdots > 0, \quad \lim_{n \uparrow \infty} \varepsilon_n = 0 \tag{11.71}$$

By definition (11.70) there exists a sequence $\{u_n(p)\} \big|_{n=1}^{\infty}$ such that

$$0 \leq u_n(M) - u(M) < \varepsilon_n \quad \forall n = 1, 2, \ldots \tag{11.72}$$

Now define a sequence $\{v_n(p)\} \big|_{n=1}^{\infty}$ by

$$v_1(p) = \mathfrak{M}(p \mid u_1, R, M)$$
$$w_n(p) = \min[v_{n-1}(p), u_n(p)], \quad v_n(p) = \mathfrak{M}(p \mid w_n, R, M) \quad \forall n \geq 2 \tag{11.73}$$

By Lemmas 11.2.6 and 11.2.7,

$$v_n(p) \in \mathcal{M}^+ \quad \forall n \geq 1 \tag{11.74}$$

---

[1] The Perron function exists because the sets $\mathcal{M}^+$ and $\mathcal{M}^-$ are bounded from below and from above, respectively.

Moreover, by the choice of the sequence $\{u_n\}$,

$$v_n(p) \geq v_{n+1}(p) \quad \forall n \geq 1, \quad \lim_{n \uparrow \infty} v_n(M) = u(M) \tag{11.75}$$

Thus $\{v_n(p)\}$ is a monotonic sequence of functions continuous in $\bar{K}_M^R$ and harmonic inside K_M^R. Hence, by Harnack's second theorem,

$$\exists v(p) \overset{\text{def}}{=} \lim_{n \uparrow \infty} v_n(p) \quad \forall p \in K_M^{R^0}, \quad \forall R^0 < R \tag{11.76}$$

and moreover $v(p)$ is harmonic in $K_M^{R^0}$. Hence Perron's theorem will be proved if we can show that

$$v(p) \equiv u(p) \quad \forall p \in K_M^{R^0} \tag{11.77}$$

To prove this identity, suppose to the contrary that there exists an $s \in K_M^{R^0}$ such that

$$v(s) < u(s) \tag{11.78}$$

or

$$v(s) > u(s) \tag{11.79}$$

Inequality (11.78) is obviously false. Indeed, if it were true then there would exist $n > 0$ so large that

$$v_n(s) < u(s) \tag{11.80}$$

which is impossible since $v_n(p) \in \mathcal{M}^+$ and $u(p) = \inf \mathcal{M}^+$. Therefore, it remains to demonstrate that (11.79) is also false.

Suppose, then, that (11.79) is true. Let

$$v(s) - u(s) = 3\delta > 0 \tag{11.81}$$

Then

$$v_n(s) - u(s) > 3\delta > 0 \quad \forall n \geq 1 \tag{11.82}$$

Now let $w(p) \in \mathcal{M}^+$ be such that

$$w(s) < u(s) + \delta \tag{11.83}$$

This is possible because $u = \inf \mathcal{M}^+$. Then

$$v_n(s) - w(s) > 2\delta \geq 0 \quad \forall n \geq 1 \tag{11.84}$$

Now define $w_n(p)$ as

$$w_n(p) = \mathfrak{M} \{p \mid \min[w(p), v_n(p)], M, r\} \tag{11.85}$$

where

$$r = r_{M,s} \tag{11.86}$$

Since $\{v_n\}$ converges uniformly, it is uniformly bounded and equicontinuous in $K_M^{R^0}$. This means that there exists $\rho > 0$ so small that

$$K_s^\rho \in K_M^{R^0} \wedge [v_n - w(p) > \delta \; \forall p \in K_s^\rho] \tag{11.87}$$

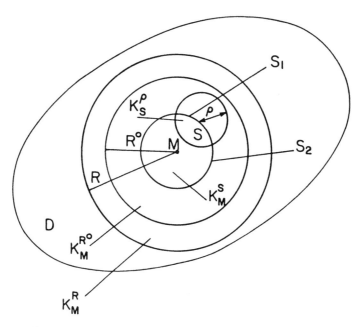

Figure 11.2. Sketch of the proof of the Perron theorem.

Let

$$S_1 = S_M^s \cap \bar{K}_s^\rho, S_2 = S_M^s \setminus K_s^\rho \tag{11.88}$$

(see Figure 11.2). We have

$$w_n(p) \le v_n(p) \quad \forall p \in S_2, w_n(p) < v_n(p) - \delta \quad \forall p \in S_1 \tag{11.89}$$

Hence

$$w_n(M) = \frac{1}{4\pi s^2} \left(\int_{S_1} + \int_{S_2} \right) w_n(q) d\sigma_q$$

$$< \frac{1}{4\pi s^2} \left(\int_{S_1} v_n(q) d\sigma_q + \int_{S_2} [v_n(q) - \delta] d\sigma_q \right)$$

$$< \frac{1}{4\pi s^2} \left(\int_{S_M^s} v_n(q) d\sigma_q - \delta \int_{S_2} d\sigma_q \right) = v_n(M) - \frac{\delta}{4\pi s^2} \int_{S_2} d\sigma_q \tag{11.90}$$

Now let $N > 0$ be so large that

$$\varepsilon_n < \frac{\delta}{4\pi s^2} \int_{S_2} d\sigma_q \quad \forall n > N \tag{11.91}$$

We obtain

$$w_n(M) < u(M) + \varepsilon_n - \frac{\delta}{4\pi s^2} \int_{S_2} d\sigma_q < u(M) \tag{11.92}$$

which contradicts the definition

$$u(p) = \inf \mathcal{M}^+ \tag{11.93}$$

since $w_n \in \mathcal{M}^+$. This contradiction proves that (11.79) is false. **Q.E.D.**

4. Existence theorem for Dirichlet problem. Barriers.

Let us now consider the Dirichlet problem

$$\Delta u(p) = 0 \quad \forall p \in D, u(p) = f(p) \quad \forall p \in \Sigma = \bar{D} \setminus D, \ u(p) \in C(\bar{D}) \tag{11.94}$$

To fix ideas, we assume as before that $D \subset \mathbb{R}_3$.

Definition 11.4.1. A point $P \in \Sigma$ is said to be a *regular boundary point* if the Perron function $u(p)$ is continuous in a neighborhood of P (whatever the admissible $f(p)$) and

$$u(P) = f(P) \tag{11.95}$$

Definition 11.4.2. A region **D** is said to be *fundamental with respect to the Dirichlet problem* (or simply *fundamental*) if all its boundary points are regular.

Since the Perron function is harmonic inside D, it is obvious – if the region under consideration is bounded – that this definition is equivalent to the following.

Definition 11.4.3. A region D is said to be fundamental with respect to the Dirichlet problem if the Dirichlet problem (11.94) is solvable whatever admissible $f(p)$.

Our next goal is to establish geometrical conditions for the regularity of boundary points. With this purpose in mind, we introduce the following definition.

Definition 11.4.4. A function $\Omega_Q(p)$ is called a *global barrier at a boundary point Q* if it possesses the following properties.

(a) $\Omega_Q(p)$ is continuous in $\bar{D}$;
(b) $\Omega_Q(p)$ is superharmonic inside D;
(c) $\Omega_Q(p) > 0$ for all $p \in \Sigma \setminus (Q)$;
(d) $\Omega_Q(Q) = 0$.

Theorem 11.4.1. *A boundary point Q is regular if and only if there exists a global barrier $\Omega_Q(p)$.*

PROOF. *Necessity.* Let Q be a regular point of Σ and let the boundary function be a continuous function $f(p)$ such that

$$f(p) \begin{cases} = 0 & \text{if } p = Q \\ > 0 & \text{if } p \neq Q \end{cases} \tag{11.96}$$

Then the corresponding Perron function $u(p)$ possesses all properties of a global barrier.

Sufficiency. Let ε be an arbitrarily small positive constant and let

$$v(p) = f(Q) + \varepsilon + C\Omega_Q(p) \tag{11.97}$$

where C is a positive constant to be chosen later.

Since by assumption $f(p)$ is continuous, there exist $\rho > 0$ and $M = \mathrm{const} > 0$ such that, if $p \in \Sigma$, then

$$|f(p) - f(Q)| < \varepsilon \text{ if } r_{pQ} < \rho, \quad |f(p)| < M \quad \forall p \in \Sigma \tag{11.98}$$

But by the definition of a global barrier, there exists $\delta > 0$ such that

$$\left(\Omega_Q(p) > \delta \; \forall p \in \Sigma\right) \wedge \left(r_{pQ} > \rho\right) \tag{11.99}$$

Hence $C > 0$ may be chosen so large that

$$\left(C\Omega_Q(p) > M \; \forall p \in \Sigma\right) \wedge \left(r_{pQ} > \rho\right) \tag{11.100}$$

which implies

$$v(p) > f(p) \; \forall p \in \Sigma \Rightarrow v(p) \in \mathcal{M}^+ \Rightarrow v(p) \geq u(p) \; \forall p \in D \tag{11.101}$$

Together with (11.97), this yields

$$\overline{\lim_{D \ni p \to Q}} u(p) \leq \lim_{D \ni p \to Q} v(p) = f(Q) + \varepsilon \tag{11.102}$$

Quite similarly, if

$$w(p) = f(Q) - \varepsilon - C\Omega_Q(p) \tag{11.103}$$

and C is chosen large enough, then

$$w(p) \in \mathcal{M}^- \Rightarrow w(p) < u(p) \tag{11.104}$$

which implies

$$\underline{\lim_{D \ni p \to Q}} u(p) \geq \lim_{D \ni p \to Q} w(p) = f(Q) - \varepsilon \tag{11.105}$$

Since $\varepsilon > 0$ is arbitrarily small, it follows from (11.102) and (11.105) that

$$\underline{\lim_{D \ni p \to Q}} u(p) = \overline{\lim_{D \ni p \to Q}} u(p) = f(Q) \tag{11.106}$$

so that indeed is Q a regular boundary point. **Q.E.D.**

Theorem 11.4.1 may create the wrong impression that being a regular boundary point is a global property. In fact, however, the regularity or irregularity of a boundary point Q depends only on the geometrical structure of Σ in an arbitrarily small neighborhood of Q. To prove this, we introduce the following definition.

Definition 11.4.2. Let U_Q be a relative neighborhood of the boundary point Q,

$$U_Q = K_Q^R \cap \bar{D} \tag{11.107}$$

where $R > 0$ is so small that Σ divides the boundary Σ_Q of U_Q into two connected parts

$$\Sigma_Q = \Sigma_{Q_1} + \Sigma_{Q_2} \tag{11.108}$$

where

$$\Sigma_{Q_1} = \Sigma_Q \cap \Sigma, \quad \Sigma_{Q_2} = \Sigma_Q \cap D \tag{11.109}$$

A function $w_Q(p)$ is called a *local barrier* if it is defined and continuous in $\bar{U}_Q$ and possesses the following properties:

(a) $w_Q(p)$ is superharmonic in U_Q;
(b) $u(p) > 0$ for all $p \in \Sigma_Q$, $p \neq Q$;
(c) $u(Q) = 0$.

Theorem 11.4.2. *There exists a global barrier $\Omega_Q(p)$ if and only if there exists a local barrier $w_Q(p)$.*

PROOF. Assume that $\Omega_Q(p)$ exists. Since

$$r_{Qp} > 0 \quad \forall p \in \Sigma_{Q_2} \tag{11.110}$$

and $\Omega_Q(p)$ is continuous in D, there exists $k > 0$ such that $\Omega_Q(p) \geq k$ for any $p \in \Sigma_{Q_2}$. Thus $\Omega_Q(p)$ possesses all the properties of a local barrier.
    Assume now that $w_Q(p)$ exists. Then there exists $k > 0$ such that

$$w_Q(p) > k \quad \forall p \in \Sigma_{Q_2} \tag{11.111}$$

Define $\Omega_Q(p)$ by

$$\Omega_Q(p) = \begin{cases} 1 & \text{if } p \in D \setminus \bar{U}_Q \\ \min\left(1, \dfrac{w_Q(p)}{k}\right) & \text{if } p \in U_Q \end{cases} \tag{11.112}$$

We claim that $\Omega_Q(p)$ is a global barrier. Indeed, it is continuous in $\bar{D}$, positive in Σ everywhere except at the point Q, and equal to zero at Q. Hence we need only show that the function $\Omega_Q(p)$ defined by (11.112) is superharmonic in D, in other words, that for any $K_M^R \subset \bar{K}_M^R \subset D$,

$$\mathfrak{M}(p \mid \Omega_Q, M, R) \leq \Omega_Q(p) \tag{11.113}$$

As in the proof of Lemma 11.2.7, we consider all possible positions of K_M^R relative to U_Q:

(a) $K_M^R \subset U_Q$;
(b) $K_M^R \subset D \setminus \bar{U}_Q$;
(c) $K \overset{\text{def}}{=} K_M^R \cap U_{Q_2} \neq \emptyset$.

(see Figure 11.3). Obviously, it follows from (11.112) that $\Omega_Q(p)$ is superharmonic in $D \setminus \bar{U}_Q$, so that in case (a) (11.112) holds. By Lemma 11.2.6, the same is true in case (b). Hence we need only prove (11.112) in case (c).

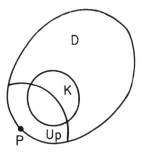

Figure 11.3.   Location of K_M^R
relative to U_Q; case (c).

Let S be the boundary of K_M^R. Then

$$S = S_1 + S_2 \tag{11.114}$$

where

$$S_1 \in \bar{U}_Q, \quad S_2 \in D \setminus \bar{U}_Q \tag{11.115}$$

We have

$$\mathfrak{M}(p \mid \Omega_Q, M, R) = \Omega_Q(p) \equiv 1 \quad \forall p \in S_2$$

$$\mathfrak{M}(p \mid \Omega_Q, M, R) = \Omega_Q(p) = \min\left(1, \frac{\omega_Q(p)}{k}\right) \quad \forall p \in S_1 \tag{11.116}$$

Hence, by the maximum principle,

$$\mathfrak{M}(p \mid \Omega_Q, M, R) \le \Omega_Q(p) \quad \forall p \in K_M^R \tag{11.117}$$

Thus inequality (11.113) holds in all three possible cases.   **Q.E.D.**

We can now establish geometrical conditions for the existence of a local barrier, and consequently for the regularity of a boundary point.

Theorem 11.4.3. *A boundary point q of a region D is regular if there exists a sphere K_O^R lying outside D that touches $\Sigma = \bar{D} \setminus D$ at q (see Figure 11.4).*

PROOF. Assume that such a sphere exists. Then for any $p \in \Sigma$, $p \ne q$,

$$r_{Op} > R \tag{11.118}$$

Let

$$\omega_q(p) = \frac{1}{R} - \frac{1}{r_{Op}} \tag{11.119}$$

This function is harmonic in D, positive on Σ everywhere except at q, and equal to zero at q. Hence it is a barrier and consequently q is a regular boundary point.   **Q.E.D.**

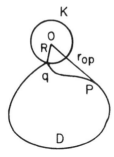

Figure 11.4.   The existence of a barrier; case (a).

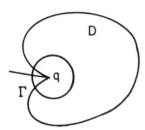

Figure 11.5.   The existence of a barrier; case (b).

The following condition is specific to the 2-dimensional case.

Theorem 11.4.4. *Let $D \in \mathbb{R}_2$ and let q be a boundary point*

$$q \in \Sigma = \bar{D} \setminus D \qquad (11.120)$$

such that there exists a segment Γ of a Jordan curve with endpoint at q lying outside D. Then q is a regular boundary point (see Figure 11.5).

PROOF. Let D be a region in the plane of the complex variable $z = \rho e^{i\varphi}$ and let C_q^r be the circle of radius r with center at q, where r is so small that C_q^r intersects Γ. Assume that Γ is parameterized in such a way that the parameter λ increases from 0 to some positive value λ_0 when Γ is described from q to the point of intersection q_0 of C_q^r with Γ, where q lies outside D. Let us now define the function

$$\omega_q(p) = -\Re \frac{1}{\ln(z)} = \frac{\ln(1/\rho)}{\ln^2(1/\rho) + \varphi^2} \qquad (11.121)$$

This function is univalent in the region $C_q^r \setminus \Gamma$, positive there except for the point q, equal to zero at q, and harmonic inside C_q^r. Hence it possesses all the properties of a local barrier, which proves the regularity of the point q.   **Q.E.D.**

We now formulate, without proof, a criterion for the regularity of a boundary point Q that is applicable in multidimensional cases. (For a proof, see e.g. [13; 57].)

Theorem 11.4.5. *Let $D \in \mathbb{R}_n$ be a region with boundary Σ. Then a boundary point Q is regular if there exists a sphere K_Q^r of radius $r > 0$ so small that there is a circular cone C_Q with apex at Q such that*

$$C_Q^* \stackrel{def}{=} (C_Q \setminus \bar{D}) \cap K_Q^r \neq \emptyset \qquad (11.122)$$

Of some theoretical interest are regions whose boundaries have conical points. There is an essential difference between cases (a) and (b) in which the cusp of the cone is directed outward and inward, respectively. In the case (a), such a boundary point is always regular since it satisfies the conditions of the Theorem 11.4.3. In the case (b), Theorem 11.4.4 shows us that in two dimensions such points are regular. In the 3-dimensional case, however, such points may be irregular. One example was given by Lebesgue, who considered a body of revolution generated by rotation about an axis of a plane curve that has a very high order of osculation relative to the axis (see e.g. [13]).

5. Schwarz alternating method

There are two kinds of existence theorems. Theorems of the first group, which may be called "pure" existence theorems, infer the existence of solutions to the problem without explicitly defining an algorithm (consisting of finitely or denumerably many steps) making it possible to construct the desired solution. For example, theorems based on the use of "degree of mapping" (Brouwer degree of a map) belong to this category. The second group of existence theorems, which we might call "constructive" existence theorems, infer the existence of a solution by actually describing an algorithm to construct the solution in a finite or denumerable sequence of operations. All theorems based on the Cacciopoli–Banach contraction mapping theorem, or on the solution of Fredholm or Volterra integral equations, belong to this second category of existence theorems.[2] Obviously the Perron theorem is a pure existence theorem, since it does not indicate any algorithm for constructing the infimum of the set of upper functions.

---

[2] We shall return to this issue in Chapter 18, in connection with application of the Schwarz alternating process to solution of the Dirichlet problem for the heat-conduction equation.

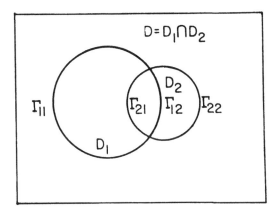

Figure 11.6. A composite region $D = D_1 \cup D_2$.

We now proceed to present the Schwarz alternating algorithm, which – after a very slight generalization – yields an existence theorem of the same generality as Perron's pure theorem.

The Schwarz alternating process enables us to construct a solution of the Dirichlet problem in a region that is the union of two fundamental regions having a common subregion. Since the Dirichlet problem is solvable in quadratures in (for example), a sphere or a half-space, the Schwarz method is actually applicable to any region that is the intersection of two spheres, two half-planes, or a half-plane and a sphere, or even for the intersection of a finite number of such regions.[3]

Theorem 11.5.1 (Schwarz alternating process). *Let D_1 and D_2 be fundamental regions with respect to the Dirichlet problem, and let*

$$D = D_1 \cup D_2, \Gamma = \bar{D} \setminus D, \Gamma_i = \bar{D}_i \setminus D_i, \quad i = 1, 2 \qquad (11.123)$$

Let $f(p) \in C(\Gamma)$ be a given function. Then the solution $u(p)$ of the Dirichlet problem

$$u(p) \in C(\bar{D}), \quad \Delta u(p) = 0 \quad \forall p \in D, u(p) = f(p) \quad \forall p \in \Gamma \qquad (11.124)$$

exists and can be constructed by means of the following alternating process: Let

$$\Gamma_{12} = \Gamma_1 \cap D_2, \Gamma_{11} = \Gamma_1 \setminus \Gamma_{12}, \qquad \Gamma_{21} = \Gamma_2 \cap D_1, \Gamma_{22} = \Gamma_2 \setminus \Gamma_{21} \qquad (11.125)$$

(see Figure 11.6). Define $f_{11}(p) \in C(\Gamma_1)$ by the condition

$$f_{11}(p) = \begin{cases} f(p) & \text{if } p \in \Gamma_{11} \\ \varphi(p) & \text{if } p \in \Gamma_{12} \end{cases} \qquad (11.126)$$

---

[3]Definition 11.4.3 is applicable not only to bounded regions but to unbounded regions as well.

where $\varphi(p)$ is an arbitrarily chosen continuous function such that

$$\max_{p \in \Gamma_{12}} |\varphi(p)| \leq \max_{p \in \Gamma} |f(p)| \leq M \qquad (11.127)$$

Let $u_1^1(p)$ be a solution of the Dirichlet problem

$$u_1^1(p) \in C(\bar{D}_1), \quad \Delta u_1^1(p) = 0 \quad \forall p \in D_1, \quad u_1^1(p) = f_{11}(p) \quad \forall p \in \Gamma_1 \quad (11.128)$$

and $u_1^2(p)$ a solution of the Dirichlet problem

$$u_1^2(p) \in C(\bar{D}_2), \quad \Delta u_1^2(p) = 0 \quad \forall p \in D_2,$$

$$u_1^2(p) = \begin{cases} f(p) & \text{if } p \in \Gamma_{22} \\ u_1^1(p) & \text{if } p \in \Gamma_{21} \end{cases} \qquad (11.129)$$

Assume that $u_k^1(p)$ and $u_k^2(p)$ are defined for $k = 1, 2, \ldots, m$, and define $u_{m+1}^1(p)$ and $u_{m+1}^2(p)$ by

$$u_{m+1}^1(p) \in C(\bar{D}_1), \quad \Delta u_{m+1}^1(p) = 0 \quad \forall p \in D_1$$

$$u_{m+1}^1(p) = \begin{cases} f(p) & \text{if } p \in \Gamma_{11} \\ u_m^2(p) & \text{if } p \in \Gamma_{12} \end{cases} \qquad (11.130)$$

$$u_{m+1}^2(p) \in C(\bar{D}_2), \quad \Delta u_{m+1}^2(p) = 0 \quad \forall p \in D_1$$

$$u_{m+1}^2(p) = \begin{cases} f(p) & \text{if } p \in \Gamma_{22} \\ u_{m+1}^1(p) & \text{if } p \in \Gamma_{21} \end{cases} \qquad (11.131)$$

Then both sequences $\{u_m^1(p)\}$ and $\{u_m^2(p)\}$ converge uniformly in D_1 and D_2, respectively, to functions

$$u_1(p) = \lim_{m \uparrow \infty} u_m^1(p), \quad u_2(p) = \lim_{m \uparrow \infty} u_m^2(p) \qquad (11.132)$$

and, moreover,

$$u_1(p) = u_2(p) \quad \forall p \in D_1 \cap D_2 \qquad (11.133)$$

so that

$$u(p) = \begin{cases} u_1(p) & \text{if } p \in D_1 \\ u_2(p) & \text{if } p \in D_2 \end{cases} \qquad (11.134)$$

is a solution of problem (11.124).[4]

PROOF. The theorem is a corollary of the following two lemmas.

Lemma 11.5.1 (Generalized maximum principle). *Let $D \subset \mathbb{R}_2$ be a region bounded by a closed curve Γ and let $\Gamma_0 \subset \Gamma$ be a set of points with zero linear measure. Let $f(p)$ be a bounded function defined on Γ and continuous*

---

[4]In what follows we restrict ourselves to the 2-dimensional case, in order to avoid complicated description of the geometry of intersecting 3-dimensional bodies.

on $\Gamma_1 = \Gamma \setminus \Gamma_0$. *Finally, let* $u(p)$ *be a function bounded in* $\bar{D}$, *continuous in* $\bar{D} \setminus \Gamma_0$, *equal to* $f(p)$ *at all its points of continuity, and harmonic in* D. *Then*

$$\sup_{p \in \bar{D}} u(p) = \sup_{p \in \Gamma} f(p), \quad \inf_{p \in \bar{D}} u(p) = \inf_{p \in \Gamma} f(p) \tag{11.135}$$

Lemma 11.5.2 (Schwarz).[5] *Let* L^* *be a smooth curve intersecting* Γ *at two noncorner points* A *and* B *and not tangent to* Γ *at either point. Let* L *be the arc of* L^* *with endpoints* A *and* B. *Let* Γ_1 *and* Γ_2 *be the open arcs of* Γ *into which* Γ *is divided by* L^*, *assuming that* Γ_2 *is described in the counterclockwise sense from* B *to* A *and* Γ_1 *from* A *to* B. *Finally, let* $u(p)$ *be a function, bounded in* D, *continuous in* $\bar{D}$ *except at* A *and* B, *harmonic inside* D, *and equal to 0 and to 1 on* Γ_1 *and* Γ_2, *respectively:*

$$|u| \le M \quad \forall p \in \bar{D}, \quad \Delta u = 0 \quad \forall p \in D, \quad u(p) = \begin{cases} 0, & p \in \Gamma_1 \\ 1, & p \in \Gamma_2 \end{cases} \tag{11.136}$$

Then there exists $q = \text{const}$, $0 < q < 1$, *such that*

$$0 \le u(p) \le q \quad \forall p \in L \tag{11.137}$$

We first prove Theorem 11.5.1 on the assumption that Lemmas 11.5.1–2 are true. Note, first of all, that all the functions u_k^1 and u_k^2, $k = 1, 2, \ldots$, may be considered as known, since the regions D_1 and D_2 are fundamental by assumption. Let

$$v_m = u_{m+1}^1 - u_m^1, \quad w_m = u_{m+1}^2 - u_m^2 \tag{11.138}$$

By the definitions (11.130) and (11.131),

$$v_m(p) = 0 \quad \forall p \in \Gamma_{11}, \quad v_m(p) = w_{m-1}(p) \quad \forall p \in \Gamma_{12}, \; \forall m \ge 2 \tag{11.139}$$

and

$$w_m(p) = 0 \quad \forall p \in \Gamma_{22}, \quad w_m(p) = v_m(p) \quad \forall p \in \Gamma_{21}, \; \forall m \ge 1 \tag{11.140}$$

By condition (11.127) and Lemmas 11.5.1–2,

$$|u_1^1(p)| \le M \; \forall p \in \bar{D}_1 \Rightarrow |v_1(p)| < qM \; \forall p \in \Gamma_{21} \tag{11.141}$$

Hence, by Lemmas 11.5.1–2,

$$|w_1(p)| < qM \; \forall p \in \bar{D}_2 \Rightarrow |w_1(p)| < q^2 M \; \forall p \in \Gamma_{12} \tag{11.142}$$

Let us assume that for all $k = 1, 2, \ldots, m$,

$$|v_k(p)| < q^{m+1} M \quad \forall p \in \Gamma_{12} \tag{11.143}$$

---

[5]The following proof, which is borrowed from [28], is applicable to the 2-dimensional problem only. For a proof applicable to the 3-dimensional Dirichlet problem, based on the use of potential theory (see Chapter 12, and [13]); for an application to the parabolic Dirichlet problem, see Chapter 18, Section 5.

and

$$|w_k(p)| < q^{m+2}M \quad \forall p \in \Gamma_{21} \tag{11.144}$$

Then, by Lemmas 11.5.1–2, these inequalities are valid for $k = m+1$. Because they are valid for $m = 1$, inequalities (11.143)–(11.145) hold by induction for all $m \geq 1$. Since $0 < q < 1$ this means that the series

$$u^1 \sim u_1^1 + \sum_{m=1}^{\infty} v_m^1, \quad u^2 \sim u_1^2 + \sum_{m=1}^{\infty} w_m^2 \tag{11.145}$$

converge uniformly in $\bar{D}_1$ and in $\bar{D}_2$, so that their sums are u^1 and u^2, respectively. Since all the functions v_m and w_m are harmonic, u^1 and u^2 are harmonic too, by Harnack's first theorem. Finally, since

$$v_{m+1} = w_m \quad \forall p \in \Gamma_{12}, \quad w_m = v_m \quad \forall p \in \Gamma_{21}, \ \forall m \geq 1 \tag{11.146}$$

and the boundary of the region

$$D_{12} \stackrel{\text{def}}{=} D_1 \cap D_2 \tag{11.147}$$

is

$$\Gamma^{12} \stackrel{\text{def}}{=} \Gamma_{12} \cup \Gamma_{21} \tag{11.148}$$

we conclude that

$$u^1 = u^2 \ \forall p \in \Gamma^{12} \Rightarrow u^1 = u^2 \ \forall p \in D_{12} \tag{11.149}$$

Since

$$u_m^1 = f \quad \forall p \in \Gamma_{11}, \quad u_m^2 = f \quad \forall p \in \Gamma_{22} \tag{11.150}$$

we have

$$u^1 = \lim_{m\uparrow\infty} u_m^1 = f \quad \forall p \in \Gamma_{11}, \quad u^2 = \lim_{m\uparrow\infty} u_m^2 = f \quad \forall p \in \Gamma_{22} \tag{11.151}$$

Thus we see that if

$$u(p) = \begin{cases} u^1(p) & \forall p \in D_1 \\ u^2(p) & \forall p \in D_2 \end{cases} \tag{11.152}$$

then

$$u(p) \in C(\bar{D}), \quad \Delta u(p) = 0 \quad \forall p \in D, \quad u(p) = f(p) \quad \forall p \in \Gamma \tag{11.153}$$

that is, $u(p)$ is a solution of problem (11.124). Thus Theorem 11.5.1 is proved, provided that we can prove Lemmas 11.5.1–2.

To prove Lemma 11.5.1, let us assume that

$$m = \inf_{p\in\Gamma} f(p), \quad M = \sup_{p\in\Gamma} f(p) \tag{11.154}$$

and that the conditions of the lemma are satisfied, that is,

$$u \in C(\bar{D} \setminus \Gamma_0), \quad \Delta u \in D, \quad |u| < K \quad \forall p \in \bar{D}, \quad u(p) = f(p) \in \Gamma_1 \tag{11.155}$$

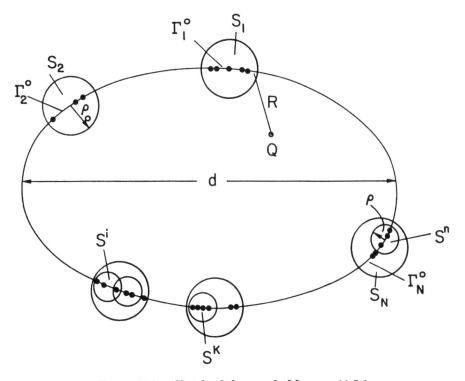

Figure 11.7. Sketch of the proof of Lemma 11.5.1.

We must prove that for any fixed point $Q \in D$,

$$m \leq u(Q) \leq M \tag{11.156}$$

Since u is bounded,

$$\lim_{D \ni p \to q \in \Gamma} \frac{u(p)}{\ln(1/r_{pq})} = 0 \tag{11.157}$$

Let

$$d = \dim D \stackrel{\text{def}}{=} \max_{p \in D} \max_{q \in D} r_{pq}, \quad R = \inf_{q \in \Gamma_0} r_{Qq} > 0 \tag{11.158}$$

(see Figure 11.7) Let $\rho_0 = \text{const} \in (0, R)$. Since Γ_0 is assumed to be a set of zero linear measure, it can be covered by a finite number N of circles of radius ρ_0. Let S be the union of all these circles and $\Gamma^0 = S \cap \Gamma$, so that $\Gamma_0 \subset \Gamma^0$. Let $\varepsilon > 0$ be arbitrarily small and $\eta > 0$ so small that

$$\eta \int_{\Gamma^0} \ln\left(\frac{d}{r_{qQ}}\right) ds_q < \varepsilon \tag{11.159}$$

Fix this η and take a positive $\rho < \rho_0$ so small that

$$|u(p)| < \eta \int_{\Gamma^0} \ln\left(\frac{d}{r_{qQ}}\right) ds_q \quad \text{if } r_{p\Gamma^0} < \rho \tag{11.160}$$

Now cover Γ_0 by circles of radius ρ and let $\mathcal{E}$ denote the set of all these circles. Let

$$D_\rho = D \setminus \mathcal{E}, \quad \Gamma_\rho = \bar{D}_\rho \setminus D_\rho \tag{11.161}$$

We have

$$\Gamma_\rho = \Gamma_\rho^1 \cup \Gamma_\rho^2 \tag{11.162}$$

where

$$\Gamma_\rho^1 = \Gamma_\rho \cap \Gamma, \quad \Gamma_\rho^2 = \Gamma_\rho \cap D \tag{11.163}$$

We introduce the auxiliary function

$$v(p) = M + \eta \int_{\Gamma^0} \ln\left(\frac{d}{r_{pq}}\right) ds_q \tag{11.164}$$

By (11.154) and (11.160),

$$v(p) > M, \quad u(p) < M \quad \forall p \in \Gamma_\rho^1, \quad v(p) \geq u(p) \quad \forall p \in \Gamma_\rho^2 \tag{11.165}$$

The maximum principle is applicable to $v(p)$ in D_ρ. Hence it follows from (11.165) that

$$u(p) < M + \eta \int_{\Gamma^0} \ln\left(\frac{d}{r_{pq}}\right) ds_q \quad \forall p \in D_\rho \tag{11.166}$$

By (11.159), it follows from (11.166) that

$$u(Q) < M + \varepsilon \tag{11.167}$$

Quite similarly,

$$u(Q) > m - \varepsilon \tag{11.168}$$

Since $\varepsilon > 0$ is arbitrarily small, these inequalities yield

$$m \leq u(Q) \leq M \tag{11.169}$$

Q.E.D.

We now prove Schwarz's Lemma 11.5.2. Consider first the following generalized Dirichlet problem.

Let $f(p)$ be a piecewise continuous function defined on a closed contour Γ bounding a fundamental region D in the plane $z = x + iy$, with discontinuities of the first kind at points $p_i = (x_i, y_i)$, $i = 1, 2, \ldots, n$, numbered in the positive (counterclockwise) sense; let

$$f(p_i - 0) = \alpha_i, \quad f(p_i + 0) = \beta_i \tag{11.170}$$

so that the jump of $f(p)$ at the point p_i is

$$[f(p_i)] = \beta_i - \alpha_i, \quad i - 1, 2, \ldots, n \tag{11.171}$$

Assume that p_i are corner points with angles equal to $\omega_i > 0$ measured in the positive sense (see Figure 11.8). Let $u(p)$ be a bounded solution of the

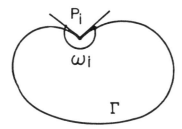

Figure 11.8. A contour Γ with a corner point.

generalized Dirichlet problem

$$u(p) \in C\left(\bar{D} \setminus \cup_1^n(p_i)\right), \quad |u(p)| < M \quad \forall p \in \bar{D}$$
$$\Delta u(p) = 0 \quad \forall p \in D, u(p) = f(p) \in \Gamma \setminus \cup_1^n(p_i) \tag{11.172}$$

Consider the function

$$z - z_i = \rho e^{i\varphi} \tag{11.173}$$

Its argument φ increases with passage through the point z_i, by the value ω_i. At the same time, the function

$$w_i(p) = \frac{\alpha_i - \beta_i}{\omega_i} \Im \ln(z - z_i) \tag{11.174}$$

is continuous on Γ everywhere except at $z = z_i$, where it experiences a jump equal to

$$[w_i(p_i)] = \beta_i - \alpha_i \tag{11.175}$$

Hence the function

$$w(p) = \sum_{i=1}^{n} \frac{\alpha_i - \beta_i}{\omega_i} \Im \ln(z - z_i) \tag{11.176}$$

has the same jumps at the points p_i as $f(p)$. But

$$w(p) \in C\left(\bar{D} \setminus \cup_1^n(p_i)\right) \wedge \Delta w(p) = 0 \quad \forall p \in D \tag{11.177}$$

Taking this into account, we define

$$\psi(p) = f(p) - w(p) \quad \forall p \in \Gamma \setminus \cup_1^n(p_i) \tag{11.178}$$

and solve the problem

$$v(p) \in C(\bar{D}), \quad \Delta v(p) = 0 \quad \forall p \in D, u(p) = \psi(p) \quad \forall p \in \Gamma \qquad (11.179)$$

Since by construction $\psi(p)$ is continuous on Γ and D is by assumption a fundamental region, problem (11.179) is solvable and its solution is unique. Setting

$$u(p) = v(p) + w(p) \quad \forall p \in \bar{D} \qquad (11.180)$$

we obtain a solution of the generalized Dirichlet problem. By Lemma 11.5.1, this solution is unique.[6]

We are now in a position to prove Lemma 11.5.2. The boundary function $f(p)$ is

$$f(p) = \begin{cases} 0 & \forall p \in \Gamma_1 \\ 1 & \forall p \in \Gamma_2 \end{cases} \qquad (11.181)$$

so that

$$\alpha_1 = 1, \quad \beta_1 = 0, \alpha_2 = 0, \quad \beta_2 = 1 \qquad (11.182)$$

and the angles ω_1 and ω_2 are equal to π by assumption. Hence the solution $u(p)$ of the generalized Dirichlet problem (11.136) may be represented as

$$u(p) = v(p) + \frac{1}{\pi} [\Im \ln(z - z_1) - \Im \ln(z - z_2)] \qquad (11.183)$$

Let us evaluate the limits of $u(p)$ at points p_1 and p_2 as p approaches them along Γ_1, Γ_2, and L. Let

$$z - z_1 = \rho e^{i\varphi} \quad \forall p \in L, \varphi \big|_{z = z_1} = \pi \qquad (11.184)$$

(see Figure 11.9) and

$$v(p) = v_1(p) + \frac{1}{\pi} \Im \ln(z - z_1) \qquad (11.185)$$

so that $v_1(p)$ is a continuous function at p_1. Then

$$\frac{1}{\pi} \Im \ln(z - z_1) \big|_{\Gamma_1 \ni p \to p_1} = 0, \quad \frac{1}{\pi} \Im \ln(z - z_1) \big|_{\Gamma_2 \ni p \to p_1} = 1 \qquad (11.186)$$

so that

$$\lim_{\Gamma_1 \ni p \to p_1} u(p) = 0, \quad \lim_{\Gamma_2 \ni p \to p_1} u(p) = 1 \qquad (11.187)$$

We now evaluate

$$u^1 = \lim_{L \ni p \to p_1} u(p), \quad u^2 = \lim_{L \ni p \to p_2} u(p) \qquad (11.188)$$

---

[6]The assumption in Lemma 11.5.1 that $u(p)$ is bounded is essential. For example, let $u(x, y) = k[1 - 2x/(x^2 + y^2)]$. This function is equal to zero at the boundary of the circle $(x - 1)^2 + y^2 < 1$, harmonic inside it everywhere except for the origin $x = y = 0$, but unbounded in the neighborhood of the origin.

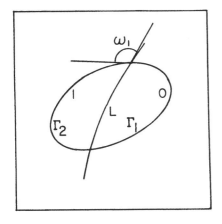

Figure 11.9. The section of D by L.

For each circuit around the contour $\Gamma_1 \cup L$, the function $(1/\pi)\Im\ln(z - z_1)$ increases by

$$\frac{1}{\pi}\Im\ln(z - z_1) \overset{\text{def}}{=} q_1 = 1 - \frac{\omega_1}{\pi} < 1 \qquad (11.189)$$

so that

$$u^1 = q_1, \quad 0 < q_1 < 1 \qquad (11.190)$$

Quite similarly, we find that

$$u^2 = q_2, \quad 0 < q_2 < 1 \qquad (11.191)$$

On the other hand,

$$\sup_{p\in\Gamma} u(p) = 1, \quad \inf_{p\in\Gamma} u(p) = 0 \qquad (11.192)$$

so that by Lemma 11.5.1

$$0 < u(p) < 1 \qquad (11.193)$$
$$\phantom{0 < u(}p\in L$$

Since by (11.190) and (11.191) this inequality holds at the ends p_1 and p_2 of L, there exists $q < 1$ such that

$$0 < u(p) \le q < 1 \quad \forall p \in \bar{L} \qquad (11.194)$$

Q.E.D.

The following theorem demonstrates that a slight generalization of the Schwarz alternating method makes it applicable to solution of the Dirichlet problem in any fundamental region.

Theorem 11.5.2. *Let D be a fundamental region. Then a solution of the Dirichlet problem in D for any continuous boundary function can be constructed as the limit of a sequence of solutions of Dirichlet problems, solved by applying the Schwarz alternating method to a monotonic sequence of regions that are contained in D together with their boundaries and that converge to D.*

PROOF. Consider a sequence of positive numbers

$$\rho_1 > \rho_2 > \rho_n > \cdots \qquad \lim_{n\uparrow\infty} \rho_n = 0 \qquad (11.195)$$

and cover the region $\bar{D}$ by spheres of radius ρ_1. By the Heine–Borel lemma, one can extract from a covering a finite number of these spheres. Let ρ_1 be so small that only some of the spheres cover Γ, whereas the others lie in the interior of D. Let D_1 denote the union of the latter, so that

$$D_1 \subset \bar{D}_1 \subset D \qquad (11.196)$$

Suppose now that the region D_n has been constructed. Cover $\bar{D}_n$ by spheres of radius ρ_n and construct a region D_{n+1} by the procedure used to construct D_1. Thus, by induction, we obtain a monotonic sequence of regions

$$D_1 \subset \bar{D}_1 \subset D_2 \subset \bar{D}_2 \cdots \subset D_n \subset \bar{D}_n \subset \cdots, \ \lim_{n\uparrow\infty} = D \qquad (11.197)$$

Let

$$g(p,q) = \frac{1}{2\pi} \ln\left(\frac{1}{r_{pq}}\right) - v(p,q) \qquad (11.198)$$

be the Green's function of the first boundary-value problem in D. Recall that by definition, for any constant $p \in D$,

$$\Delta v(p,q) = 0 \quad \forall q \in D, \quad v(p,q) = \frac{1}{2\pi} \ln\left(\frac{1}{r_{pq}}\right) \quad \forall q \in \Gamma, \ v(p,q) \in C(\bar{D}) \qquad (11.199)$$

Let

$$g_n(p,q) = \frac{1}{2\pi} \ln\left(\frac{1}{r_{pq}}\right) - v_n(p,q) \qquad (11.200)$$

be the Green's function of the first boundary-value problem in D_n, so that v_n is defined by restricting (11.199) to D_n. By the maximum principle it follows from (11.199) and (11.197) that, for any fixed $p \in D_n$ and any $q \in \bar{D}_n$,

$$v_n \geq v_{n+1} \quad \forall n \geq 1 \qquad (11.201)$$

Fix $m \geq 1$ and consider the sequence $\{v_n(p,q)\}\,|_{m+1}^{\infty}$. This sequence is monotonic and bounded. Hence, by Harnack's second theorem, it converges uniformly in D_m to a function

$$v(p,q) \overset{\text{def}}{=} \lim_{n\uparrow\infty} v_n(p,q) \quad \forall q, \ p = \text{const} \in D_m \qquad (11.202)$$

Now define

$$w_n(p,q) = \begin{cases} \dfrac{1}{2\pi} \ln\left(\dfrac{1}{r_{pq}}\right) & q \in D \setminus D_n \\[4mm] \min\left[\dfrac{1}{2\pi} \ln\left(\dfrac{1}{r_{pq}}\right), v_n(p,q)\right] & q \in \bar{D}_n \end{cases} \qquad (11.203)$$

Applying the same arguments as used in Section 3 in the construction of local barriers, we find that $w_n(p,q)$ belongs to the set $\mathcal{M}^+$ of upper functions for the boundary function $(1/2\pi)\ln(1/r_{pq})$, since for any fixed $m \geq 1$

$$w_n(p,q) = v_n(p,q) \quad \forall n > m \tag{11.204}$$

It follows from Perron's basic theorem that

$$v(p,q) = \inf \mathcal{M}^+ \tag{11.205}$$

But by assumption D is a fundamental region, so $v(p,q)$ satisfies the boundary condition

$$v(p,q) = \frac{1}{2\pi}\ln\left(\frac{1}{r_{pq}}\right) \quad \forall q \in \Gamma \tag{11.206}$$

which means that

$$g(p,q) = \frac{1}{2\pi}\ln\left(\frac{1}{r_{pq}}\right) - v(p,q) \tag{11.207}$$

is the Green's function of the first boundary-value problem in D. Since the existence of the Green's function implies the solvability of the Dirichlet problem for any continuous boundary function, we see that the assertion of Theorem 11.5.2 is true. **Q.E.D.**

Problems

P11.3.1. Is it possible to define a set of upper and lower functions for the Neumann problem? If so, formulate the definition and prove an analogue of Perron's theorem. If not, explain.

P11.4.1. Let D be a region in the complex plane $z = x + iy$:

$$D = \left\{ z \colon 0 < x^2/4 + y^2/16 < 1 \right\}$$

Is this region fundamental? If so, construct a global or local barrier. If not, explain.

P11.5.1. Let D be the region in the complex z plane, defined as

(a) $D = C_1 \cup C_2, C_1 = \{|z - 1| < 1\}, C_2 = \{|z + 1/2| < 2/3\}$
(b) $D = C_1 \cup C_2, C_1 = \{|z - 1| < 1\}, C_2 = \{|z + 1| < 1\}$

Is the Schwarz alternating method applicable? If so, estimate the rate of convergence of the alternating sequence. If not, explain.

12

Outer boundary-value problems. Elements
of potential theory

1. Isolated singular points of harmonic functions

In order to consider boundary-value problems in unbounded regions – or,
more precisely, to study outer problems, that is, problems posed in the
complement (relative to all of space) of bounded regions (or unbounded
regions, if their complements contain the point at infinity) – we need an
analog of the concept of removable singularities in the theory of analytic
functions. This is the motive for the following definition.

Definition 12.1.1. Let $u(p)$ be a function, harmonic in the closure of a
region D everywhere except for an interior point M. Then M is called an
isolated singular point. An isolated singularity is said to be *removable* if
$u(p)$ can be made harmonic everywhere in D by redefining it at the point
M.

Theorem 12.1.1. *Let $u(p)$ be harmonic in a region $D \subset \mathbb{R}_3$, everywhere
except for an isolated singular point M. Then the singularity of u at M is
removable if*

$$\lim_{p \to M} u(p) r_{PM} = 0 \tag{12.1}$$

PROOF. Let M be an isolated singular point and assume that condition
(12.1) holds. Let K_M^R be a sphere with center at M, of radius R so small
that M is the only singular point of $u(p)$ in K_M^R. Let Q be a fixed point
such that

$$Q \in K_M^R, \quad r_{MQ} = \rho > 0 \tag{12.2}$$

and $v(p)$ a solution of the Dirichlet problem

$$v(p) \in C\left(\bar{K}_M^R\right), \quad \Delta v(p) = 0 \quad \forall p \in K_M^R$$

$$v(p) = u(p) \quad \forall p \in S_M^R = \bar{K}_M^R \setminus K_M^R \tag{12.3}$$

Take some small $\tilde{\rho} = \text{const} > 0$ (whose exact value will be chosen later) and $\varepsilon = \text{const} > 0$ arbitrarily small; define a spherical cavity

$$G = K_M^R \setminus K_M^{\tilde{\rho}} \tag{12.4}$$

with boundary

$$\partial G = S_M^R \cup S_M^{\tilde{\rho}} \tag{12.5}$$

Let $w(p)$ be the auxiliary function defined by

$$w(p) = v(p) - u(p) + \frac{\eta}{r_{PM}} \tag{12.6}$$

where $\eta > 0$ is a fixed number so small that

$$\frac{\eta}{r_{QM}} < \varepsilon \tag{12.7}$$

Now take $\tilde{\rho}$ so small that

$$w(p) > 0 \quad \forall p \in S_M^{\tilde{\rho}} \tag{12.8}$$

This is possible because v is bounded and u satisfies condition (12.1).

The function $w(p)$ is continuous in $\bar{G}$, harmonic inside G, and positive on the boundary ∂G of G. Hence, by the maximum principle,

$$w(p) > 0 \quad \forall p \in G \tag{12.9}$$

This and (12.7) yield

$$u(Q) < v(Q) + \varepsilon \tag{12.10}$$

Quite similarly, we prove that

$$u(Q) > v(Q) - \varepsilon \tag{12.11}$$

Since $\varepsilon > 0$ is arbitrarily small, it follows from (12.10) and (12.11) that

$$u(Q) = v(Q) \tag{12.12}$$

Since Q is an arbitrary point of $K_M^R \setminus (M)$, we can define $u(M)$ to be equal to $v(M)$, making $u(p)$ a harmonic function everywhere in K_M^R. **Q.E.D.**

Remark. In the 2-dimensional case, condition (12.1) must be replaced by the condition

$$\lim_{p \to Q} u(p) \Big/ \ln\left(\frac{1}{r_{pQ}}\right) = 0 \tag{12.13}$$

2. Regularity of harmonic functions at infinity

We now proceed to consideration of harmonic functions in outer regions. Let us recall that functions harmonic in $\mathbb{R}_2$ are real (or imaginary) parts of functions analytic in the complex plane z. Hence the behavior at infinity of a function harmonic in $\mathbb{R}_2$ is determined by that of the corresponding analytic function.

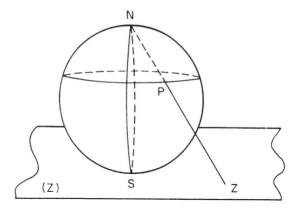

Figure 12.1. Spherical projection.

Elements at infinity are treated differently in different mathematical disciplines. Thus, in the theory of functions of one variable, one has two "improper" (infinitely distant) elements ($+\infty$ and $-\infty$). In the projective geometry of the Euclidean plane one assumes the existence of one straight line at infinity, whereas in $\mathbb{R}_3$ one needs a plane at infinity. In the theory of analytic functions, however, only one point at infinity is introduced. The reason for this must be sought in the theory of stereographic projections. The reader will recall that this kind of projection is defined as follows (see Figure 12.1). Let T be the complex z plane and K a sphere of unit radius (called Riemann's sphere or sphere of complex numbers, or complex sphere) touching T at the south pole S. Let L be the straight line connecting a point Q of the plane T and the north pole N, and P its point of intersection with K. Then there is a one-to-one correspondence between the spherical coordinates of P and the coordinate z of Q. If Q tends to infinity in an arbitrary direction, its image P on K tends to N. Clearly, to preserve a one-to-one correspondence between the complex sphere K and the extended (i.e., including points at infinity) complex plane T, one must assume that the latter contains only one point at infinity.

Let now $L_1 \subset T$ and $L_2 \subset T$ be two curves intersecting at a point $q \in T$; let Γ_1 and Γ_2 be their images on K and p the image of q; and let $\tilde{\Gamma}_1$ and $\tilde{\Gamma}_2$ be projections of Γ_1 and Γ_2 on the tangent plane T_K to K at p. The angle α between L_1 and L_2 at q is equal to the angle β between $\tilde{\Gamma}_1$ and $\tilde{\Gamma}_2$ at p; moreover, the positive sense of rotation from L_1 to L_2 coincides with that from $\tilde{\Gamma}_1$ to $\tilde{\Gamma}_2$. Thus, mapping of the complex plane T onto the complex sphere K is conformal and vice versa. Therefore the theory of functions, defined in the complex plane, may be built as the theory of their projection onto the complex sphere and vice versa. This means, in particular, that the regularity at infinity of an analytic function $f(z)$ defined in the complex plane must be defined as the regularity of its image analytic on K at the north pole N. Now note that any rotation of K as a solid body represents a conformal mapping of K into itself. In particular, the rotation that transforms the north pole N into the south pole S of K is conformal, so that a function analytic at N maps into a function analytic at S. The rotation of K, which transforms N into S, corresponds to the transformation of T into itself that maps the point at infinity into the origin $z = 0$, and the resultant function must be analytic there. A conformal mapping that transforms the point at infinity to the origin is $z = 1/\zeta$. This means that the analytic function $f(z)$ should be called regular at infinity if its image is analytic at $\zeta = 0$; more precisely, since $\zeta = 0$ is an isolated singular point of the function $z = 1/\zeta$, the singularity of the function $F(\zeta) = f(1/\zeta)$ at the origin must be removable.

Now note that the aforementioned rotation of K that transforms N into S is the result of two successively applied transformations of T: an inversion with respect to a unit circle about the origin, and mirror reflection with respect to the real axis. Each of these two transformations is the so-called conformal transformation of the second kind (i.e., in contrast to a conformal transformation that transforms the positive sense of description into a negative one), so that the real and imaginary parts of the corresponding function remain harmonic. Since any functions harmonic in $\mathbb{R}_2$ are real parts of analytic functions and since the Laplace equation is invariant under the two aforementioned transformations, we must define the regularity of a function $u(p)$ at the point at infinity as harmonicity of its image under inversion with respect to the unit circle about the origin, provided the singularity at that point is removable.

The invariance of the Laplace equation under inversion is easily established directly, without reference to the theory of analytic functions. Indeed, let (r, f) be polar coordinates and let

$$\rho = \frac{1}{r}, \quad \psi = \varphi, \quad v(\rho, \psi) = u(r, \varphi) \tag{12.14}$$

We have

$$\Delta u = \left(\frac{\partial^2}{\partial r^2} + \frac{1}{r}\frac{\partial}{\partial r} + \frac{1}{r}\frac{\partial^2}{\partial \varphi^2} \right) u \tag{12.15}$$

It follows from (12.14) that

$$\frac{\partial u}{\partial r} = \frac{\partial v}{\partial \rho}\frac{d\rho}{dr} = -\rho^2\frac{\partial v}{\partial \rho} \Rightarrow r\frac{\partial u}{\partial r} = -\rho\frac{\partial v}{\partial \rho}$$

$$\Rightarrow \Delta u = \rho^4 \left(\frac{\partial^2}{\partial \rho^2} + \frac{1}{\rho}\frac{\partial}{\partial \rho} + \frac{1}{\rho}\frac{\partial^2}{\partial \psi^2} \right) v \tag{12.16}$$

so that indeed

$$\Delta u = 0 \Rightarrow \Delta v = 0. \tag{12.17}$$

Generalization of these ideas to $\mathbb{R}_3$ proceeds as follows. Let Φ be some transformation of $\mathbb{R}_3$ into itself that maps the point at infinity into the origin and leaves the Laplace equation invariant. An harmonic function in $\mathbb{R}_3$ is said to be regular at *infinity* if there exists Φ such that under it the origin becomes a removable singular point.

The Laplace equation in $\mathbb{R}_3$ is not invariant under inversion with respect to a sphere. However the *Kelvin transformation*

$$\rho = \frac{1}{r}, \quad \varphi = \psi, \quad \theta = \vartheta, \quad u(r,\varphi,\vartheta) = \rho v(\rho,\psi,\theta) \tag{12.18}$$

where (r,φ,ϑ) and (ρ,ψ,θ) are polar coordinates, yields

$$\Delta u = \rho^5 \Delta v, \quad \text{i.e.,} \quad \Delta u = 0 \Rightarrow \Delta v = 0 \tag{12.19}$$

Using Theorem 12.1.1, we arrive at the following conclusions.

(a) In $\mathbb{R}_3$: If v is an harmonic function in the neighborhood of the origin $\rho = 0$ and the singularity of v at the origin is removable, then

$$\lim_{\rho \downarrow 0} \rho v = 0 \text{ and } v\big|_{\rho \downarrow 0} = O(1) \Rightarrow u\big|_{r \uparrow \infty} = \frac{O(1)}{r}\bigg|_{r \uparrow \infty}$$

$$\frac{\partial v}{\partial \rho}\bigg|_{\rho \downarrow 0} = O(1) \Rightarrow \frac{\partial u}{\partial r}\bigg|_{r \uparrow \infty} = \frac{O(1)}{r^2} \tag{12.20}$$

$$\frac{\partial^2 v}{\partial \rho^2}\bigg| = O(1) \Rightarrow \frac{\partial^2}{\partial r^2} u\bigg|_{r \uparrow \infty} = \frac{O(1)}{r^3}$$

(b) In $\mathbb{R}_2$: If v is an harmonic function in the neighborhood of the origin $\rho = 0$ and the singularity of v at the origin is removable, then

$$\lim_{\rho \downarrow 0} \frac{v}{\ln(1/\rho)} = 0 \Rightarrow v\bigg|_{\rho \downarrow 0} = O(1) \Rightarrow u\bigg|_{r \uparrow \infty} = O(1)\bigg|_{r \uparrow \infty}$$

$$\frac{\partial v}{\partial \rho}\bigg|_{\rho \downarrow 0} = O(1) \Rightarrow \frac{\partial u}{\partial r}\bigg|_{r \uparrow \infty} = \frac{O(1)}{r^2}, \tag{12.21}$$

$$\frac{\partial^2 v}{\partial \rho^2}\bigg|_{\rho \downarrow 0} = O(1) \Rightarrow \frac{\partial^2}{\partial r^2} u\bigg|_{r \uparrow \infty} = \frac{O(1)}{r^3}$$

Since the origin may be located at any arbitrarily chosen, fixed point M, we see that the equalities (12.20) and (12.21) justify the following definition.

Definition 12.2.1. A function $u(p)$ (not necessarily harmonic) is said to be *regular in the neighborhood of the point at infinity* if for any fixed point m there exist positive constants A and R, independent of p, such that if $r_{pm} > R$ then

(a) In $\mathbb{R}_3$:

$$|u(p)| < \frac{A}{r_{pm}}, \quad \left|\frac{\partial}{\partial r} u(p)\right| < \frac{A}{r_{pm}^2}, \quad \left|\frac{\partial^2}{\partial r^2} u(p)\right| < \frac{A}{r_{pm}^3} \tag{12.22}$$

(b) In $\mathbb{R}_2$:

$$|u(p)| < A, \quad \left|\frac{\partial}{\partial r} u(p)\right| < \frac{A}{r_{pm}^2}, \quad \left|\frac{\partial^2}{\partial r^2} u(p)\right| < \frac{A}{r_{pm}^3} \tag{12.23}$$

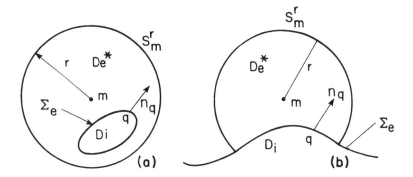

Figure 12.2.   The region D_e^*: (a) D_i bounded; (b) D_i unbounded.

3. Extension of the fundamental identity to unbounded regions. Liouville's theorem

In what follows, D_i denotes a given bounded or unbounded region whose complement D_e relative to all of space contains the point at infinity.

Theorem 12.3.1. *Let* $u(p) \in C^1(\bar{D}_e) \cap C^2(D_e)$ *be regular at infinity. Then*

$$D_e \subset \mathbb{R}_3 \Rightarrow u(p) = \frac{1}{4\pi} \int_{\Sigma_e} \left(u(q) \frac{\partial}{\partial n_q} \frac{1}{r_{pq}} - \frac{1}{r_{pq}} \frac{\partial}{\partial n_q} u(q) \right) d\sigma_q$$

$$- \frac{1}{4\pi} \int_{D_e} \frac{1}{r_{pq}} \Delta u(q) d\omega \qquad (12.24)$$

$$D_e \subset \mathbb{R}_2 \Rightarrow u(p) = \frac{1}{2\pi} \int_{\Sigma_e} \left(u(q) \frac{\partial}{\partial n_q} \ln \left[\frac{1}{r_{pq}} \right] - \ln \left[\frac{1}{r_{pq}} \right] \frac{\partial}{\partial n_q} u(q) \right) ds_q$$

$$- \frac{1}{2\pi} \int_{D_e} \Delta u(q) \ln \left(\frac{1}{r_{pq}} \right) d\sigma_q \qquad (12.25)$$

PROOF. Let $m \in D_e$ be a fixed point and K_m^r a sphere of radius r with center at m. Assume (to fix ideas) that $D_i \subset \mathbb{R}_3$ is bounded, and define

$$D_e^* = D_e \cap K_m^r, \quad \Sigma_e^* \stackrel{\text{def}}{=} \bar{D}_e^* \setminus D_e^* = S_m^r \cup \Sigma_e \qquad (12.26)$$

where $S_m^r = \bar{K}_m^r \setminus K_m^r$ (see Figure 12.2). Then, by the fundamental identity,

$$u(m) = u_1(m) + u_2(m) \qquad (12.27)$$

where

$$u_1(m) = \frac{1}{4\pi} \int_{\Sigma_e} \left(u(q) \frac{\partial}{\partial n_q} \frac{1}{r_{mq}} - \frac{1}{r_{mq}} \frac{\partial}{\partial n_q} u(q) \right) d\sigma_q$$

$$- \frac{1}{4\pi} \int_{D_e^*} \frac{1}{r_{mq}} \Delta u(q) d\omega_q \qquad (12.28)$$

$$u_2(m) = -\frac{1}{4\pi} \int_{s_m^r} \left(u(q) \frac{\partial}{\partial n_q} \frac{1}{r_{mq}} - \frac{1}{r_{mq}} \frac{\partial}{\partial n_q} u(q) \right) d\sigma_q$$

$$= \frac{1}{4\pi} \int_0^{2\pi} d\varphi \int_0^\pi r^2 \sin\vartheta \left(u(q) \frac{1}{r^2} - \frac{1}{r} \frac{\partial}{\partial n_q} u(q) \right) d\vartheta \qquad (12.29)$$

and[1] $q = (r, \varphi, \vartheta)$

By assumption, $u(p)$ is regular at infinity. Hence there exists $R > 0$ such that

$$|u(q)| < \frac{A}{r}, \quad \left| \frac{\partial}{\partial n_q} u(q) \right| < \frac{A}{r^2}, \quad |\Delta u(q)| < \frac{A}{r^3} \quad \forall r > R \qquad (12.30)$$

Hence

$$|u_2(m)| < \frac{A}{r} \Rightarrow \lim_{r \uparrow \infty} u_2(m) = 0 \qquad (12.31)$$

Since

$$\lim_{r \uparrow \infty} D_e^* = D_e \qquad (12.32)$$

and $u(m)$ is independent of r, we deduce from (12.27) and (12.31) that

$$\lim_{r \uparrow \infty} \frac{1}{4\pi} \int_{D_e^*} \frac{\Delta u(q)}{r_{mq}} d\omega_q = \frac{1}{4\pi} \int_{D_e} \frac{\Delta u(q)}{r_{mq}} d\omega_q \qquad (12.33)$$

so that (12.24) is valid if $p = m$. Since m is an arbitrary, the truth of (12.24) follows. Clearly the same is true if D_i is unbounded. It is also obvious that (12.25) holds. Thus the theorem is true in all possible cases. **Q.E.D.**

Corollary. *If the Dirichlet and Neumann problems are formulated for an unbounded region with the added condition that $u(p)$ is regular at infinity, then the integral representations of the solutions of these problems (see Chapter 9, Section 5) remain valid.*

We are now in a position to prove Liouville's theorem, which is trivial in one dimension.

Theorem 12.3.2 (Liouville). *If $u(p)$ is an harmonic function everywhere in space and regular at infinity, then*

(a) In $\mathbb{R}_3$:

$$u(p) \equiv 0 \quad \forall p \in \mathbb{R}_3 \qquad (12.34)$$

---

[1] As $\mathbf{n}_q$ in (12.29) is a vector directed toward the interior of D_e, the integral in question is evaluated with a minus sign.

(b) In $\mathbb{R}_2$:

$$u(p) \equiv \text{const} = \lim_{p \to \infty} u(p) \qquad (12.35)$$

PROOF. Let p be an arbitrarily fixed point and K_p^r a sphere of radius r with center at p. By the mean value theorem,

(a) In $\mathbb{R}_3$:

$$u(p) = \frac{1}{4\pi r^2} \int_{S_p^r} u(q) d\sigma_q \Rightarrow \lim_{r\uparrow\infty} u(\tilde{q}) \frac{1}{4\pi r^2} \int_{S_p^r} d\sigma_q = 0 \qquad (12.36)$$

(b) In $\mathbb{R}_2$:

$$u(p) = \frac{1}{2\pi r} \int_{S_p^r} u(q) ds_q \Rightarrow u(\tilde{q}) \frac{1}{2\pi r} \int_{S_p^r} ds_q = u(\tilde{q}) \qquad (12.37)$$

where $\tilde{q} \in S_p^r$. Since p is fixed (i.e., independent of r), the following limit exists:

$$\lim_{r\uparrow\infty} \frac{1}{2\pi r} \int_{S_p^r} u(q) ds_q = \lim_{r\uparrow\infty} u(\tilde{q}) = u(p) \qquad (12.38)$$

which proves (12.35). **Q.E.D.**

4. Electrostatic potentials

Let $D \subset \mathbb{R}_3$ be a region bounded by a surface Σ. According to the fundamental identity (9.112), any function $u(p) \in C^1(\bar{D}) \cap C^2(D)$ may be represented as the sum of three integrals of the following types.

(a) *Potential of a single layer with surface charge density* $f(q)$:

$$V[p \mid f(q)] = \frac{1}{4\pi} \int_\Sigma f(q) \frac{1}{r_{pq}} d\sigma_q \qquad (12.39)$$

(b) *Potential of a double layer with dipoles surface density* $F(q)$:

$$W[p \mid F(q)] = -\frac{1}{4\pi} \int_\Sigma F(q) \frac{\partial}{\partial n_q} \frac{1}{r_{pq}} d\sigma_q \qquad (12.40)$$

where $\mathbf{n}_q$ is the vector of the outer normal to Σ.[2]
(c) *Volume (Newtonian) potential with charge density* Φ:

$$U[p \mid \Phi(q)] = \frac{1}{4\pi} \int_D \Phi(q) \frac{1}{r_{pq}} dw_q \qquad (12.41)$$

Let us recall the electrostatic motivation for this terminology (see also Chapters 1, 2, and 10). Let us use the Gaussian rational electrostatic system of units. Suppose that a (negative) unit test charge is placed at a point

---

[2] If Σ is a closed surface, then $\mathbf{n}_q$ is the outward normal. Otherwise, the direction of the "outer" normal is chosen arbitrarily.

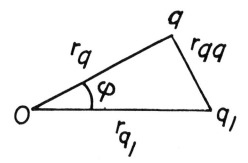

Figure 12.3. Sketch to the
definition of a dipole.

p in vacuum. Then, by Coulomb's law, the force with which it attracts a
positive charge $\Phi(q)d\omega_q$ placed at point q is

$$\mathbf{f} = -\frac{\Phi(q)}{4\pi r_{pq}^2}d\omega_q \mathbf{r}_{pq}^0 \qquad (12.42)$$

where $\mathbf{r}_{pq}^0$ is the unit vector in the direction from p to q. Hence the force
with which a unit test charge at p attracts charges distributed in D with
the density $\Phi(q)$ is equal to

$$\mathbf{f}^* = -\int_D \frac{\Phi(q)}{4\pi r_{pq}^2}\mathbf{r}_{pq}^0 d\omega_q \qquad (12.43)$$

and the potential of this force is precisely the potential $U[p \mid \Phi(q)]$ defined
by (12.41).[3]

Now suppose that charges are distributed with density $f(q)$ along a two-
sided surface Σ (not necessarily closed). Then the value of the electrostatic
potential at p is precisely the potential $V[p \mid f(q)]$ defined by (12.39).

We now recall the definition of the moment of an electrostatic dipole.
Suppose that positive and negative charges e^+ and e^- of magnitude e are
placed into points q and p, let $\ell = r_{qq_1}$, and let the vector $\boldsymbol{\ell} = \ell \mathbf{r}_{qq_1}^0$ be
directed from q to q_1 (see Figure 12.3). The dipole moment $\mathbf{N}$ of the pair
(e^+, e^-) is

$$\mathbf{N} \overset{\text{def}}{=} N\mathbf{r}_{qq_1}^0 = e\ell\mathbf{r}_{qq_1}^0 \qquad (12.44)$$

---

[3]Since the Newtonian law of mass attraction is formally identical with Coulomb's
electrostatic law, the potential $U(p \mid \Phi)$ may also be termed the Newtonian
potential.

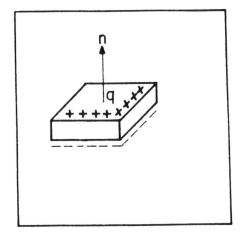

Figure 12.4. Sketch to the definition of a double layer.

The potential created at p by this pair is

$$w = \frac{1}{4\pi} \cdot \frac{N}{\ell}\left(\frac{1}{r_{pq_1}} - \frac{1}{r_{pq}}\right) \tag{12.45}$$

Now suppose that N remains constant as $\ell \downarrow 0$. Then

$$\lim_{\ell \downarrow 0} w = -\frac{N}{4\pi}\lim_{\ell \downarrow 0}\frac{r_{pq_1} - r_{pq}}{r_{pq_1}r_{pq}} = -\frac{N}{4\pi}\frac{\partial}{\partial \ell}\left(\frac{1}{r_{pq}}\right) \tag{12.46}$$

This is the electric potential of a limiting object known as a dipole and characterized by its dipole moment vector $\mathbf{N}$, defined by (12.44).

Let now Σ be a two-sided (not necessarily closed) surface, and choose the direction of its normal $\mathbf{n}$ (see footnote 2 on page 266). Suppose that dipoles are distributed over Σ with density $F(q)$, in such a way that the positive charges lie on the positive ("outer") side of Σ and the negative charges on the negative ("inner") side (see Figure 12.4). Then the potential created by these dipoles at point $p \notin \Sigma$ is precisely the potential $W[p \mid F(q)]$ defined by (12.40).

In 2-dimensional space $\mathbb{R}_2$, the potentials (12.39)–(12.41) must be replaced by the corresponding logarithmic potentials.

(a) *Logarithmic potential of a single layer with the linear charge density* $f(q)$:

$$V[p \mid f(q)] = \frac{1}{2\pi}\int_\Gamma f(q)\ln\left(\frac{1}{r_{pq}}\right)ds_q \tag{12.47}$$

(b) *Logarithmic potential of a double layer with linear dipole density* $F(q)$:

$$W[p \mid F(q)] = -\frac{1}{2\pi}\int_\Gamma F(q)\frac{\partial}{\partial n_q}\frac{1}{r_{pq}}ds_q \tag{12.48}$$

where $\mathbf{n}_q$ is the vector of the outer normal to Γ.

(c) *Volume (Newtonian) logarithmic potential with plane charge density* Φ:

$$U[p \mid \Phi(q)] = \frac{1}{2\pi} \int_D \Phi(q) \ln\left(\frac{1}{r_{pq}}\right) d\sigma_q \qquad (12.49)$$

where D is a plane region and Γ its boundary.

Once again, the electrostatic motivation of this terminology is as follows. Let (x, y, z) be the cartesian coordinates, and imagine a negative unit test charge placed at a point $(x, y, 0)$. Let positive charges be distributed with constant density f along the axis

$$L = \{\xi, \eta, \zeta: \xi = 0, \ \eta = 0, \ -\infty < \zeta < \infty\} \qquad (12.50)$$

Then the force of attraction between this axis and the negative test charge is

$$\mathbf{F}(p) = -\frac{f}{4\pi} \int_{-\infty}^{\infty} \frac{1}{r_{pq}^3 \mathbf{r}_{pq}} d\zeta, \quad q = (0, 0, \zeta) \qquad (12.51)$$

where $\mathbf{r}_{pq}$ is a unit vector in the direction from p to q. Let

$$\rho \overset{\text{def}}{=} \rho\rho^0 = \mathbf{r}_{po}, \quad o = (0, 0, 0) \qquad (12.52)$$

so that

$$\mathbf{r}_{pq} = \rho\rho^0 + \zeta\mathbf{k} \qquad (12.53)$$

where $(\mathbf{i}, \mathbf{j}, \mathbf{k})$ are the basis vectors of the axes x, y, z. Equalities (12.53) and (12.51) yield

$$\mathbf{F}(p) = R\rho^0 + Z\mathbf{k} \qquad (12.54)$$

where

$$Z = \frac{f}{4\pi} \int_{-\infty}^{\infty} \frac{\zeta d\zeta}{[\rho^2 + \zeta^2]^{3/2}} = 0 \qquad (12.55)$$

and

$$R = \frac{f}{4\pi} \int_{-\infty}^{\infty} \frac{\rho d\zeta}{[\rho^2 + \zeta^2]^{3/2}} = \frac{f}{4\pi\rho} \int_{-\infty}^{\infty} \frac{d\xi}{[1 + \xi^2]^{3/2}} = \frac{f}{2\pi\rho} \qquad (12.56)$$

Thus the force with which a negative test charge placed at a point $(x, y, 0)$ attracts the charged axis z is

$$\mathbf{F} = \frac{f}{2\pi\rho} \cdot \rho^0 \qquad (12.57)$$

and the potential of this force is equal to the logarithmic potential

$$w(p) = \frac{f}{2\pi} \ln\left(\frac{1}{r_{pq}}\right) \qquad (12.58)$$

that justified definitions (12.47)–(12.49).

5. Integrals with polar singularities

All the potentials introduced in Section 4 are integrals of the form

$$\int_D \Psi(p,q)d\omega_q \tag{12.59}$$

where D is a region in $\mathbb{R}_n$ ($n = 2,3$) or a surface (or a curve if D is a surface in $\mathbb{R}_3$, or $n = 2$) bounding a region D in $\mathbb{R}_n$. The integrands $\Psi(p,q)$ are functions of the form

$$\Psi(p,q) = \frac{\Phi(p,q)}{r_{pq}^\alpha} \quad \text{or} \quad \Psi(p,q) = \Phi(p,q)\ln\left(\frac{1}{r_{pq}}\right) \tag{12.60}$$

where $\Phi(p,q)$ is continuous everywhere in the domain of integration and

$$\Phi(p,p) \neq 0 \quad \forall p \in \bar{D} \tag{12.61}$$

Such integrals are known as *integrals of functions with polar singularities*.

Before studying the specific properties of these potentials, we recall the basic definition and properties of improper integrals in several dimensions.

Definition 12.5.1. Assume that $f(p,q)$ is continuous for all $p \in \bar{D}$, $q \in \bar{D}$, and $q \neq p$, and that

$$\lim_{q \to p} |f(p,q)| = \infty \tag{12.62}$$

Then the improper integral

$$I(p) \stackrel{\text{def}}{=} \int_D f(p,q)d\omega_q \tag{12.63}$$

converges at the point p and is equal to $I(p)$ if, for any neighborhood $U_p \subset D$ of p of diameter δ, the limit

$$\lim_{\delta \downarrow 0} \int_{D \setminus U_p} f(p,q)d\omega_q \tag{12.64}$$

exists and is equal to $I(p)$.

If there is a sequence of neighborhoods of p shrinking to p for which this limit does not exist, then the integral (12.63) is said to be *divergent at the point p*. If the integral (12.63) is divergent but there exists a sequence $\{\tilde{U}_p\}$ of neighborhoods of p shrinking to p for which the limit

$$\lim_{\delta \downarrow 0} \int_{D \setminus \tilde{U}_\delta} f(p,q)d\omega_q \tag{12.65}$$

exists and is equal to $I(p)$, then the improper integral is said to be *convergent to $I(p)$ in the sense of the Cauchy principal value*. Integrals that are convergent in the sense of the Cauchy principal value will be indicated by

$$\int_D^\bullet f(p,q)d\omega_q = I(p) \tag{12.66}$$

(This notation must be accompanied by a definition of the sequence of neighborhoods that determine the value of $I(p)$.)

Example.

$$\int_{-a}^{b} \frac{dx}{x} \tag{12.67}$$

is divergent. Indeed, for it to be convergent the limit

$$\lim_{\eta \downarrow 0} \int_{-a}^{-\eta} \frac{dx}{x} + \lim_{\varepsilon \downarrow 0} \int_{\varepsilon}^{b} \frac{dx}{x} \tag{12.68}$$

must exist as ε and η tend to zero independently. Since

$$\int_{-a}^{-\eta} \frac{dx}{x} + \int_{\varepsilon}^{b} \frac{dx}{x} = \ln\left(\frac{b}{a}\right) + \ln\left(\frac{\eta}{\varepsilon}\right) \tag{12.69}$$

we see that the limit (12.67) does not exist. However,

$$\eta = \varepsilon \Rightarrow \int_{-a}^{-\eta} \frac{dx}{x} + \int_{\varepsilon}^{b} \frac{dx}{x} = \ln\left(\frac{b}{a}\right) \tag{12.70}$$

so that

$$\int_{-a}^{\bullet b} \frac{dx}{x} = \ln\left(\frac{b}{a}\right) \tag{12.71}$$

Definition 12.5.2. The improper integral (12.63) is said to be *uniformly converge* at a point m if, for any $\varepsilon > 0$, there exists $\delta > 0$ such that if d is the diameter of U_m then

$$d < \delta \Rightarrow \left| \int_{U_m} f(p,q) d\omega_q \right| < \varepsilon \quad \text{for } \forall p \in U_m \tag{12.72}$$

If the integral (12.63) is uniformly convergent at every point $p \in D$, then it is convergent uniformly in **D**.

The following theorems are obvious.

Theorem 12.5.1. *If*

$$I(p) = \int_{D} f(p,q) d\omega_q \tag{12.73}$$

is uniformly convergent at a point p, then $I(p)$ is continuous at this point.

Theorem 12.5.2 (Weierstrass). *Suppose that for any $p, q \in D$,*

$$|f(p,q)| < F(p,q) \tag{12.74}$$

and

$$J(p) = \int_{D} F(p,q) d\omega_q \tag{12.75}$$

is uniformly convergent in D. Then

$$I(p) = \int_D f(p,q)d\omega_q \tag{12.76}$$

is also uniformly convergent in D.

Theorem 12.5.3. *Let $D \subset \mathbb{R}_n$, $n = 1, 2, 3$, be a smooth manifold, and let*

$$J(p) = \int_D \frac{f(p,q)}{r_{pq}^\alpha} d\omega_q \tag{12.77}$$

where

$$f(p,q) \in C(\bar{D}), \quad f(p,p) \neq 0 \quad \forall p \in D \tag{12.78}$$

Then

$$\begin{aligned} &J(p) \text{ is divergent if } \alpha \geq n \\ &J(p) \text{ is uniformly convergent if } \alpha < n \end{aligned} \tag{12.79}$$

PROOF. Because of conditions (12.78), it is sufficient to prove (12.79) for the integral

$$I(p) = \int_D \frac{1}{r_{pq}^\alpha} d\omega_q \tag{12.80}$$

Assume first that D is a 3-dimensional region. Then for an arbitrary neighborhood U_m of diameter $\delta < \rho/2$ and for $\alpha < 3$, if $p \in U_m$ then

$$\int_{U_m} \frac{1}{r_{pq}^\alpha} d\omega_q < 4\pi \int_0^\rho r^{2-\alpha} dr = \frac{4\pi}{3-\alpha} \rho^{3-\alpha} \Rightarrow \lim_{\rho \downarrow 0} \int_{U_m} \frac{1}{r_{pq}^\alpha} d\omega_q = 0 \tag{12.81}$$

so that $I(p)$ is indeed uniformly convergent if $\alpha < 3$.

Now let D be a smooth surface in $\mathbb{R}_3$ and let T be the tangent plane to D at a point m. Introduce a polar coordinate system (ρ, φ) in T, with pole at m. Let $\tilde{p}$ and $\tilde{q}$ be the projections of $p \in D \cap U_m$ and $q \in D$ on T, and let $\mathbf{r}_{pq}$ be the projection on T of the vector $\mathbf{r}_{pq}$. Let γ be the angle between the normals $\mathbf{n}_q$ and $\mathbf{n}_p$ to D at q and p, respectively. Since D is assumed to be smooth, there exists $R > 0$ so small that

$$r_{pq} < R \Rightarrow \cos \gamma > 1/2 \tag{12.82}$$

Let

$$R > \delta \tag{12.83}$$

where $\delta/2$ is the diameter of the neighborhood U_m of m. Then, if $r_{pq} \geq r_{\tilde{p}\tilde{q}}$ and $\alpha < 2$, we have

$$\left| \int_{U_p} \frac{1}{r_{pq}^\alpha} d\omega_q \right| < \int_0^{2\pi} d\varphi \int_0^R \frac{\rho \, d\rho}{r_{pq}^\alpha \cos \gamma} < 4\pi \int_0^R \rho^{1-\alpha} d\rho = \frac{4\pi R^{2-\alpha}}{2-\alpha}$$

$$\Rightarrow \lim_{R \downarrow 0} \left| \int_{U_p} \frac{1}{r_{mq}^\alpha} d\omega_q \right| = 0 \tag{12.84}$$

which again proves the uniform convergence of $I(p)$ if $\alpha < n$. Clearly, the same is true if D is a smooth curve.

That $I(p)$ is divergent in the case $\alpha \geq n$ is obvious. Hence the theorem is valid in all possible cases.   **Q.E.D.**

In connection with the reduction of the boundary-value problems to integral equations based on the use of electrostatic potentials (see Section 8), we shall also need improper integrals of the type

$$J_{\alpha\beta}(p, m) = \int_D \frac{\Phi(p, q, m)}{r_{pq}^\alpha r_{mq}^\beta} d\omega_q \tag{12.85}$$

where

$$\Phi(p, q, m) \in C(\bar{D}) \quad \forall p, q, m \in D$$
$$\Phi(p, p, m) \neq 0 \quad \text{for} \quad p \neq m \tag{12.86}$$
$$\Phi(p, m, m) \neq 0 \quad \text{for} \quad p \neq m$$

and

$$\alpha < n, \quad \beta < n \tag{12.87}$$

and, as above, $D \subset \mathbb{R}_n$ $(n = 1, 2, 3)$ is a smooth manifold.

Theorem 12.5.4.

$$J_{\alpha\beta}(p, m) = \begin{cases} \bar{\Phi}(p, m) r_{pm}^{2-\alpha-\beta} & \text{if } \alpha + \beta > n \\[2mm] \bar{\Phi}(p, m) \ln\left(\dfrac{1}{r_{pm}}\right) & \text{if } \alpha + \beta = n \\[2mm] \text{is uniformly convergent} & \text{if } \alpha + \beta < n \end{cases} \tag{12.88}$$

where

$$\bar{\Phi}(p, m) \in C(\bar{D}) \wedge \Phi(p, p) \neq 0 \quad \forall p \in \bar{D} \tag{12.89}$$

PROOF. By the mean value theorem,

$$\int_{U_p} \frac{\Phi(p, q, m)}{r_{pq}^\alpha r_{mq}^\beta} d\omega_q = \Phi(p, \tilde{q}, m) \int_{U_p} \frac{1}{r_{pq}^\alpha r_{mq}^\beta} d\omega_q \tag{12.90}$$

where $\tilde{q} \in D$. Since Φ is continuous and

$$\Phi(p, q, m) \neq 0 \quad \forall q \in U_p \tag{12.91}$$

it is sufficient to prove the theorem for $\Phi \equiv 1$.

Let us first assume that $D \subset \mathbb{R}_3$ is a plane. Let $r_{pm} = r > 0$ and let $r > 0$ be so small that

$$K_p^{2r} \subset D \tag{12.92}$$

Let

$$D_1 = D \setminus K_p^{2r}, \quad D_2 = K_p^{2r}, \quad J_{\alpha\beta} = J_{\alpha\beta}^o + J_{\alpha\beta}^{oo} \tag{12.93}$$

where

$$J^o_{\alpha\beta} = \int_{D_1} \frac{1}{r^\alpha_{pq} r^\beta_{mq}} d\omega_q, \quad J^{oo}_{\alpha\beta} = \int_{D_2} \frac{1}{r^\alpha_{pq} r^\beta_{mq}} d\omega_q \qquad (12.94)$$

We first consider $J^{oo}_{\alpha\beta}$. Introduce a polar coordinate system (ρ, φ) in D with pole at p, so that

$$m = (r, 0), \quad q = (\rho, \varphi), \quad r_{qm} = (r^2 + \rho^2 - 2r\rho \cos\varphi)^{1/2} \qquad (12.95)$$

and consider the similarity transformation

$$\rho = \frac{\tilde{\rho}}{r}, \quad \varphi = \psi \qquad (12.96)$$

Let $\tilde{p}$, $\tilde{q}$, and $\tilde{m}$ be the images of p, q, and m and let $K^2_{\tilde{p}}$ be the image of D_2. Then

$$J^{oo}_{\alpha\beta} = Cr^{2-\alpha-\beta} \qquad (12.97)$$

where

$$C \stackrel{\text{def}}{=} \int_{K^2_{\tilde{p}}} \frac{d\omega_q}{r^\alpha_{\tilde{p}\tilde{m}} r^\beta_{\tilde{q}\tilde{m}}} \qquad (12.98)$$

Since by assumption (12.87) $\alpha < 2$, $\beta < 2$ and $r_{\tilde{p}\tilde{m}} = 1$, the integral (12.98) is convergent by Theorem 12.5.3, so that $J^{oo}_{\alpha\beta}$ is uniformly convergent if $\alpha + \beta < 2$ and diverges if $\alpha + \beta \geq 2$.

Now consider $J^o_{\alpha\beta}$. Suppose that, for any $q \in D_1$,

$$D_{11} = \{q: r_{qm} > r_{pq}\}, \quad D_{12} = \{q: r_{qm} < r_{pq}\}$$

$$J_{\alpha\beta 1} = \int_{D_{11}} \frac{1}{r^\alpha_{\tilde{p}\tilde{m}} r^\beta_{\tilde{q}\tilde{m}}} d\omega_q, \quad J_{\alpha\beta 2} = \int_{D_{12}} \frac{1}{r^\alpha_{\tilde{p}\tilde{m}} r^\beta_{\tilde{q}\tilde{m}}} d\omega_q \qquad (12.99)$$

We first consider $J_{\alpha\beta 1}$. In polar coordinates (ρ, φ) with pole at p,

$$D_{11} = \{\rho, \varphi: 0 \leq \varphi \leq 2\pi, \ 2r < \rho < R(\varphi)\} \qquad (12.100)$$

where $R(\varphi)$ is separated from $\rho = 2r$, that is, there exists R_0 such that

$$0 < R_0 - \text{const} \leq R(\psi) - 2r, \quad R(\psi) < \infty \quad \forall \psi \in [0, 2\pi] \qquad (12.101)$$

It follows from (12.99)–(12.101) that

$$J_{\alpha\beta 1} < 2\pi \int_{2r}^\infty \rho^{1-\alpha-\beta} d\rho = \frac{2\pi(2r)^{2-\alpha-\beta}}{2-\alpha-\beta} \quad \text{if } \alpha + \beta < 2 \qquad (12.102)$$

and

$$J_{\alpha\beta 1} > \begin{cases} 2\pi \displaystyle\int_{2r}^{R_0} \rho^{1-\alpha-\beta} d\rho = \dfrac{2\pi}{2-\alpha-\beta} \rho^{2-\alpha-\beta} \Big|_{2\rho}^{R^0} & \text{if } \alpha + \beta > 2 \\[4mm] 2\pi \displaystyle\int_{2r}^{R_0} \frac{1}{\rho} d\rho = \ln\left(\dfrac{R_0}{2r}\right) & \text{if } \alpha + \beta = 2 \end{cases} \qquad (12.103)$$

Interchanging the roles of p and m, we conclude that estimates (12.102) and (12.103) are valid for $J_{\alpha\beta 2}$ too. Comparing (12.93)–(12.94) and (12.95)–(12.101), we convince ourselves that Theorem 12.5.4 is true if D is a plane.

If D is not a plane but a smooth convex surface, then we can project the neighborhood of $m \in D$ on the tangent plane to D at m. The same arguments as before will show the validity of the theorem in this case, too. Similarly we can consider the volume or contour integrals, which completes the proof of the theorem. **Q.E.D.**

6. Properties of electrostatic volume potential

Theorem 12.6.1. *Let $D_i \subset \mathbb{R}_3$ and let $\Phi(q)$ be the charge density of the electrostatic volume potential*

$$U[p \mid \Phi(q)] = \frac{1}{4\pi} \int_{D_i} \Phi(q) \frac{1}{r_{pq}} d\omega_q \qquad (12.104)$$

Then

(a) *$U(p \mid \Phi(q) \in C^\infty(D_e) \cup C^1(D_i)$ if Φ is integrable (bounded) in $\bar{D}_i$, so that, in particular,*

$$\Delta U[p \mid \Phi(q)] = 0 \quad \forall p \in D_e \qquad (12.105)$$

(b) *Let $\boldsymbol{\ell}$ be a vector of arbitrary direction and let $\Phi(p) \in C^1(\bar{D}_i)$. Then, for all $p \in D_i$,*

$$\frac{\partial^2}{\partial \ell^2} U[p \mid \Phi(q)] = -\frac{\Phi(p)}{3} + \frac{1}{4\pi} \int_D^{\bullet} \Phi(q) \frac{\partial^2}{\partial \ell_p^2} \left[\frac{1}{r_{pq}} \right] d\omega_q \qquad (12.106)$$

where the integral in the sense of the Cauchy principal value is determined by the system of spheres in D with centers at p.

(c) *If $\Phi(p) \in C^1(\bar{D}_i)$ then for all $p \in D_i$,*

$$\Delta U[p \mid \Phi(q)] = -\Phi(p) \qquad (12.107)$$

The statements (a), (b), and (c) are also valid for the logarithmic volume potential,

$$U[p \mid \Phi(q)] = \frac{1}{2\pi} \int_{D_i} \Phi(q) \ln \left[\frac{1}{r_{pq}} \right] d\omega_q \qquad (12.108)$$

PROOF. Part (a) is true since $r_{pq} \neq 0$ for all $p \in D_e$ and $q \in D_i$, so that $U(p \mid \Phi)$ can be differentiated under the integral sign infinitely many times and $\Delta(1/r_{pq}) = 0$ if $r_{pq} > 0$. Its truth for all $p \in D_i$ follows from Theorem 12.5.3.

To prove parts (b) and (c), let (x, y, z) be a cartesian coordinate system and let

$$p = (x, y, z), \quad q = (\xi, \eta, \zeta), \quad m = (x_0, y_0, z_0) \qquad (12.109)$$

where $p \in D_i$ and $m \in D_i$ is a fixed point. Let

$$K_m^r \subset \bar{K}_m^r \subset D_i, \quad S_m^r = \bar{K}_m^r \setminus K_m^r, \quad p \in K_m^r \qquad (12.110)$$

We have

$$U(p \mid \Phi) = U_1(p \mid \Phi) + U_2(p \mid \Phi) \qquad (12.111)$$

where

$$U_1(p \mid \Phi) = \frac{1}{4\pi} \int_{K_m^r} \Phi(q) \frac{1}{r_{pq}} d\omega_q, \quad U_2(p \mid \Phi) = \frac{1}{4\pi} \int_{D_i \backslash K_m^r} \Phi(q) \frac{1}{r_{pq}} d\omega_q$$

(12.112)

By part (a), $U_2(p \mid \Phi) \in C^\infty(D_i \backslash K_m^r)$, so that

$$\frac{\partial^2}{\partial \ell^2} U_2[p \mid \Phi(q)] = \frac{1}{4\pi} \int_{D_i \backslash K_m^r} \Phi(q) \frac{\partial^2}{\partial \ell^2} \left(\frac{1}{r_{pq}} \right) d\omega_q$$

(12.113)

whatever the direction of ℓ.

Consider $U_1(p \mid \Phi)$. By part (a) and the equality

$$r_{pq} = [(x - \xi)^2 + (y - \eta)^2 + (z - \zeta)^2]^{1/2}$$

(12.114)

we have

$$\frac{\partial}{\partial x} U_1(p \mid \Phi) = \frac{1}{4\pi} \int_{K_m^r} \Phi(q) \frac{\partial}{\partial x} \left(\frac{1}{r_{pq}} \right) d\omega_q = -\frac{1}{4\pi} \int_{K_m^r} \Phi(q) \frac{\partial}{\partial \xi} \left(\frac{1}{r_{pq}} \right) d\omega_q$$

$$= \frac{1}{4\pi} \int_{K_m^r} \Phi(q) \operatorname{div}_p \left(\frac{1}{r_{pq}} \mathbf{i} \right) d\omega_q = -\frac{1}{4\pi} \int_{K_m^r} \Phi(q) \operatorname{div}_q \left(\frac{1}{r_{pq}} \mathbf{i} \right) d\omega_q$$

$$= -\frac{1}{4\pi} \int_{K_m^r} \operatorname{div}_q \left(\Phi(q) \frac{1}{r_{pq}} \mathbf{i} \right) d\omega_q + \frac{1}{4\pi} \int_{K_m^r} \frac{1}{r_{pq}} \operatorname{grad}_q [\Phi(q)] \mathbf{i} d\omega_q$$

$$= -\frac{1}{4\pi} \int_{K_m^r} \operatorname{div}_q \left(\Phi(q) \frac{1}{r_{pq}} \mathbf{i} \right) d\omega_q + \frac{1}{4\pi} \int_{K_m^r} \frac{1}{r_{pq}} \frac{\partial}{\partial \xi} \Phi(q) d\omega_q \quad (12.115)$$

so that by the Gauss divergence theorem,

$$\frac{\partial}{\partial x} U_1(p \mid \Phi) = -\frac{1}{4\pi} \int_{S_m^r} \Phi(q) \frac{1}{r_{pq}} \alpha d\sigma_q + \frac{1}{4\pi} \int_{K_m^r} \frac{1}{r_{pq}} \frac{\partial}{\partial \xi} \Phi(q) d\omega_q \quad (12.116)$$

where α, β and γ are the direction cosines of $\mathbf{r}_{mq}$.

By part (a) the second integral in the right-hand side of (12.116) is differentiable for every $p \in K_m^r$, so that, in particular,

$$\frac{\partial^2}{\partial x^2} U_1(m \mid \Phi) = J_{11} + J_{12}$$

(12.117)

where

$$J_{11} = -\frac{1}{4\pi r^2} \int_{S_m^r} \Phi(q) \alpha^2 d\sigma_q$$

(12.118)

and

$$J_{12} = \frac{1}{4\pi} \int_{K_m^r} \frac{\alpha}{r_{mq}^2} \frac{\partial}{\partial \xi} \Phi(q) d\omega_q$$

(12.119)

Similarly,

$$\frac{\partial^2}{\partial y^2} U_1(m \mid \Phi) = J_{21} + J_{22}, \qquad \frac{\partial^2}{\partial z^2} U_1(m \mid \Phi) = J_{31} + J_{32} \quad (12.120)$$

where

$$J_{21} = -\frac{1}{4\pi r^2} \int_{S_m^r} \Phi(q)\beta^2 d\sigma_q, \qquad J_{31} = -\frac{1}{4\pi r^2} \int_{S_m^r} \Phi(q)\gamma^2 d\sigma_q \qquad (12.121)$$

and

$$J_{22} = \frac{1}{4\pi} \int_{K_m^r} \frac{\beta}{r_{mq}^2} \frac{\partial}{\partial \xi} \Phi(q) d\omega_q, \qquad J_{32} = \frac{1}{4\pi} \int_{K_m^r} \frac{\gamma}{r_{mq}^2} \frac{\partial}{\partial \xi} \Phi(q) d\omega_q \qquad (12.122)$$

Now note that by symmetry

$$J \overset{\text{def}}{=} \frac{1}{4\pi r^2} \int_{S_m^r} \alpha^2 d\omega_q = \frac{1}{4\pi r^2} \int_{S_m^r} \beta^2 d\omega_q = \frac{1}{4\pi r^2} \int_{S_m^r} \gamma^2 d\omega_q \qquad (12.123)$$

so that

$$\alpha^2 + \beta^2 + \gamma^2 = 1 \Rightarrow J = 1/3 \qquad (12.124)$$

By assumption, Φ is a continuous function. Therefore, using the mean value theorem of integral calculus and (12.124) we obtain

$$\lim_{r \downarrow 0} J_{i1} = -\frac{\Phi(m)}{3}, \quad i = 1, 2, 3, \qquad \lim_{r \downarrow 0} \sum_{i=1}^{3} J_{i1} = -\Phi(m) \qquad (12.125)$$

Note also that by Theorem 12.5.3 the continuity of Φ implies convergence of integral J_{i2}, $i = 1, 2, 3$, so that

$$\lim_{r \downarrow 0} J_{i2} = 0, \quad i = 1, 2, 3 \qquad (12.126)$$

Let us return to equality (12.111). We have proved that $U_1(p)$ is a twice continuously differentiable function in a neighborhood of m and that there exists the limit of $\partial^2 U_1/\partial \ell^2$ as $r \downarrow 0$ whatever the direction ℓ. Because $\partial^2 U(m)/\partial \ell^2$ is independent of r this means that there exist

$$\lim_{r \downarrow 0} \frac{1}{4\pi} \int_{D_i \backslash K_m^r} \Phi(q) \frac{\partial^2}{\partial \ell^2} \frac{1}{r_{pq}} d\omega_q \qquad (12.127)$$

Recall that m had been fixed arbitrarily, so that $U(p)$ is a function twice continuously differentiable along any direction everywhere it exists. By definition of an integral in the sense of the Cauchy principal value, it follows that

$$\frac{\partial^2}{\partial \ell^2} U[p \mid \Phi(q)] = -\frac{\Phi(p)}{3} + \frac{1}{4\pi} \int_{D_i}^{\bullet} \Phi(q) \frac{\partial^2}{\partial \ell^2} \frac{1}{r_{pq}} d\omega_q \qquad (12.128)$$

Thus part (b) is proved. Finally, (12.128) and

$$\Delta \frac{1}{r_{pq}} = 0 \quad \text{if } r_{pq} \neq 0 \qquad (12.129)$$

shows that

$$\Delta U[p \mid \Phi(q)] = -\Phi(p) \qquad (12.130)$$

so that the truth of part (c) and together with this the validity of all assertions of Theorem 12.6.1 are proved. **Q.E.D.**

7. Properties of electrostatic potentials of double and single layers

We shall need the following definitions.

Definition 12.7.1. Let Σ be the part of some two-sided smooth convex surface that is bounded by a closed Jordan curve Γ, and let m be a point outside Σ. Let C_m be a cone with apex m generated by straight lines, and let Γ be its directrix. Finally, let S be part of a spherical surface with center at m which is cut out by the rays going from m to Σ. It is assumed that each of these rays intersects Σ in one point only, and that the radius R of S is so small that the points of intersection of the rays and S lie between m and Σ. Then $C_{m\Sigma}$ is the closed surface bounded by Σ, S, and the cone C_m, and D_c is the region bounded by $C_{m\Sigma}$.

Definition 12.7.2. Let Σ and $C_{m\Sigma}$ be as in Definition 12.7.1. Assume that the positive direction of the normal $\mathbf{n}$ to Σ is chosen so that if Σ is complemented to a closed convex surface then $\mathbf{n}$ becomes an outward normals (see Figure 12.5). Then the solid angle $\Omega_{m\Sigma}$ that Σ subtends at m is defined as

$$
\Omega_{m\Sigma} = \begin{cases} R^{-2} \displaystyle\int_S d\sigma & \text{if all the rays going from } m \text{ to } \Sigma \\ & \text{intersect it on its negative side} \\[2mm] -R^{-2} \displaystyle\int_S d\sigma & \text{if all the rays going from } m \text{ to } \Sigma \\ & \text{intersect it on its positive side} \end{cases} \tag{12.131}
$$

Let Σ be a convex surface (bounded or not), smooth everywhere except for a finite set $\mathcal{M}$ of conic points and smooth convex edges $\mathcal{L}$. Let $W(p\,|\,\Sigma)$ be the double-layer electrostatic potential of dipoles distributed over Σ with unit moment density (see (12.40)).

Definition 12.7.3.

$$
W(p\,|\,\Sigma) = -\frac{1}{4\pi}\int_\Sigma \frac{\partial}{\partial n_q}\frac{1}{r_{pq}}\,d\sigma_q \quad \text{if } p \notin \mathcal{M}\cup\mathcal{L} \tag{12.132}
$$

$$
W(p\,|\,\Sigma) = -\frac{1}{4\pi}\int_\Sigma^{\bullet} \frac{\partial}{\partial n_q}\frac{1}{r_{pq}}\,d\sigma_q \quad \text{if } p \in \mathcal{M}\cup\mathcal{L} \tag{12.133}
$$

where the integral in the sense of the Cauchy principal value is defined by a sequence of spheres with centers at p. In what follows we shall omit the mark "$\bullet$" when considering double-layer potentials at points of $\mathcal{M}\cup\mathcal{L}$.

The following lemmas are basic in the theory of electrostatic potentials.

Lemma 12.7.1. *Let Σ and D_c be as defined in Definitions 12.7.1 and 12.7.2, and let $p \in D_c$. Let*

$$
W(p\,|\,1,\Sigma) = -\frac{1}{4\pi}\int_\Sigma \frac{\partial}{\partial n_q}\frac{1}{r_{pq}}\,d\sigma_q = \int_\Sigma \frac{\cos\varphi_{pq}}{4\pi r_{pq}^2}\,d\sigma_q \tag{12.134}
$$

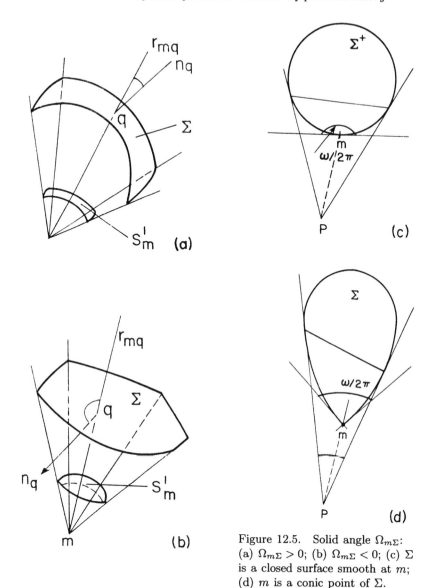

Figure 12.5. Solid angle $\Omega_{m\Sigma}$:
(a) $\Omega_{m\Sigma} > 0$; (b) $\Omega_{m\Sigma} < 0$; (c) Σ
is a closed surface smooth at m;
(d) m is a conic point of Σ.

be the electrostatic double-layer potential at the point p generated by dipoles
distributed with unit moment density along Σ. Then

$$W(p \mid 1, \Sigma) = \frac{\Omega_p}{4\pi} \qquad (12.135)$$

PROOF. Let $W(p \mid 1, C_{p\Sigma})$ be the electrostatic double-layer potential with unit dipole moment density, where $C_{p\Sigma}$ is as in Definition 12.7.1. Then

$$W(p \mid 1, C_{p\Sigma}) = -\frac{1}{4\pi} \int_{C_{p\Sigma}} \frac{\partial}{\partial n_q} \frac{1}{r_{pq}} d\sigma_q = 0 \qquad (12.136)$$

as the flux of a harmonic function through a closed surface.

We have

$$W(p \mid 1, C_{p\Sigma}) = I_1 + I_2 + I_3 \qquad (12.137)$$

where, by Definitions 12.7.1 and 12.7.2,[4]

$$I_1 = W(p \mid 1, \Sigma), I_2 = \int_{C_p} \frac{\cos \varphi_{pq}}{4\pi r_{pq}^2} d\sigma_q, I_3 = -\frac{\Omega_p}{4\pi} \qquad (12.138)$$

Recall that φ_{pq} is the angle between $\mathbf{r}_{pq}$ and $\mathbf{n}_q$, where $\mathbf{n}_q$ is the outward normal to the respective boundary at q. Since $\mathbf{n}_q$ for $q \in C_p$ is orthogonal to the generators C_p of the cone,

$$I_2 = 0 \qquad (12.139)$$

Hence, by (12.135), (12.137), and (12.138),

$$W(p \mid 1, \Sigma) = \frac{\Omega_p}{4\pi} \qquad (12.140)$$

Q.E.D.

Lemma 12.7.2 (Double-layer potential of unit dipole moment density). *Let Σ be a closed convex surface bounding a region $D_i \subset \mathbb{R}_3$, smooth everywhere except for a finite number of conical points m_i, $i = 1, 2, \ldots, k_m$, and smooth edges ℓ_i, $i = 1, 2, \ldots, k_l$. Let $\mathcal{M}$ and $\mathcal{L}$ denote the sets of all conic points m_i and all edges ℓ_i, respectively. Let $W(p \mid 1)$ be the electrostatic double-layer potential generated by dipoles, distributed with unit moment density over Σ. Then*

$$W(p \mid 1) = \begin{cases} 1 & \text{if } p \in D \\ 1/2 & \text{if } p \subset \Sigma \setminus (\mathcal{M} \cup \mathcal{L}) \\ \dfrac{\omega_p}{4\pi} & \text{if } p \in (\mathcal{M} \cup \mathcal{L}) \\ 0 & \text{if } p \notin \bar{D}_i \end{cases} \qquad (12.141)$$

where ω_p is the solid angle subtended by Σ at $p \in \mathcal{M} \cup \mathcal{L}$.

Remark 1. *If $D_i \subset \mathbb{R}_2$, then (12.141) remains true with 4π replaced by 2π.*

PROOF. First let $p \in D_i$ and let Π be a plane that divides D_i in two non-empty parts. Let $p \notin \Pi$. The solid angle Ω_p subtended by Σ at p is 4π. Let

---

[4]Ω_p occurs in I_3 with a minus sign, since the outward normal to K is the inner normal to $C_{p\Sigma}$.

$\Gamma = \Pi \cap \Sigma$ be the directrix of a cone C_p with solid angle $\Omega_p < 4\pi$. The cone divides Σ into two parts Σ_p^1 and Σ_p^2, which subtend at p a positive solid angle $\Omega_p^1 < 4\pi$ and positive solid angle $\Omega_p^2 = 4\pi - \Omega_1$. We have

$$W(p \mid 1) = W_1(p \mid 1) + W_2(p \mid 1) \tag{12.142}$$

where

$$W_i(p \mid 1) = -\frac{1}{4\pi} \int_{\Sigma_p^i} \frac{\partial}{\partial n_p} \frac{1}{r_{pq}} d\sigma_q, \quad i = 1, 2 \tag{12.143}$$

By Lemma 12.7.1,

$$W_1(p \mid 1) = \frac{\Omega_p^1}{4\pi}, \quad W_2(p \mid 1) = \frac{\Omega_p^2}{4\pi} = \frac{4\pi - \Omega_p^1}{4\pi} \tag{12.144}$$

Hence, by (12.142),

$$W(p \mid 1) = 1 \tag{12.145}$$

so that the first of assertions (12.141) is true.

Now let $P \notin \bar{D}_i$ and $C_{p\Sigma}$ be a cone with apex at p and generators tangent to Σ along a directrix Γ. This time Γ divides Σ in two parts Σ_p^1 and Σ_p^2 that subtend at p a positive solid angle Ω_p^1 and a negative solid angle Ω_p^2, equal to Ω_p^1 by its absolute value. Thus

$$W_1(p \mid 1) = \frac{\Omega_p^1}{4\pi}, \quad W_2(p \mid 1) = -\frac{\Omega_p^1}{4\pi} \tag{12.146}$$

Equality (12.142) remains valid, so that

$$W(p \mid 1) = 0 \tag{12.147}$$

which proves the fourth of assertions (12.141).

Now let $p \in \Sigma$. Take a point $m \in D_e$ at distance ρ from p. Repeating the first part of the proof (replacing p by m), we obtain

$$W^1(m \mid 1) = \frac{1}{4\pi} \Omega_m^1 \tag{12.148}$$

Now let ρ tend to zero. The generators of the cone C_m will tend to the straight lines tangent to Σ at p, so that

$$\lim_{\rho \downarrow 0} \Omega_m^1 = \omega_p \tag{12.149}$$

At the same time, if ρ tends to zero then the part Σ_m^2 vanishes. Thus

$$W(p \mid 1) = \lim_{\rho \downarrow 0} W(m \mid 1) = \lim_{\rho \downarrow 0} W^1(m \mid 1) = \frac{1}{4\pi} \Omega_p^1 = \omega_p \tag{12.150}$$

If Σ is smooth at p then

$$\omega_p = 2\pi \tag{12.151}$$

Otherwise, if $p \in \mathcal{M} \cup \mathcal{L}$,

$$\omega_p < 2\pi \tag{12.152}$$

since Σ is a convex surface by assumption. Thus equalities (12.151) and (12.152) prove the second and third assertions of (12.141), which completes the proof of Lemma 12.7.2. **Q.E.D.**

In what follows we consider the region D_i as in Definition 12.7.3 and Lemma 12.7.2. The following theorems are corollaries of Lemmas 12.7.1–2. Let

$$\Sigma^* = \Sigma \setminus (\mathcal{M} \cup \mathcal{L}) \tag{12.153}$$

so that Σ^* is an open manifold consisting of k pieces of convex surfaces of finite curvature Σ_i^* with boundaries $\mathcal{L}_i$, $i = 1, 2, \ldots, k$, and $\mathcal{M} \cup \mathcal{L}$ is its boundary:

$$\Sigma^* = \bigcup_{i=1}^{k} \Sigma_i^*, \quad \mathcal{L}_i = \bar{\Sigma}_i^* \setminus \Sigma_i^*, \quad \mathcal{M} \cup \mathcal{L} = \bigcup_{i=1}^{k} \mathcal{L}_i \tag{12.154}$$

Let $C(\Sigma^*)$ be a space of functions continuous on a closure of each Σ_i^* so that

$$f(p) \in C(\Sigma^*) \Rightarrow f(p) \in C(\Sigma_i^*), \quad i = 1, 2, \ldots, k \tag{12.155}$$

Let Γ_m be a curve intersecting Σ at m. Σ divides Γ_m in two parts Γ_i and Γ_e belonging to D_i and D_e, respectively.

Theorem 12.7.1.

(a) *Let $F(p)$ be bounded for all $p \in \Sigma$. Then*

$$W[p \mid F(q)] \in C^\infty(\mathbb{R}_3 \setminus \Sigma) \cup C(\Sigma^*) \tag{12.156}$$

which implies

$$\Delta W(p \mid F) = 0 \quad \forall p \notin \Sigma \tag{12.157}$$

(b) *Let $F(q) \in C(\Sigma^*)$. Then for any $m \in \Sigma^*$,*

$$W_i(m \mid F) \overset{\text{def}}{=} \lim_{\Gamma_i \ni p \to m} W(p \mid F) = \frac{1}{2}F(m) + W(m \mid F) \tag{12.158}$$

$$W_e(m \mid F) \overset{\text{def}}{=} \lim_{\Gamma_e \ni p \to m} W(p \mid F) = -\frac{1}{2}F(m) + W(m \mid F) \tag{12.159}$$

(c) *Let Γ intersect Σ at a point m belonging to one of L_i, $i = 1, 2, \ldots, k$, and let ω_m be the solid angle at m. Then*

$$W_i(m \mid F) \overset{\text{def}}{=} \lim_{\Gamma_i \ni p \to m} W(p \mid F) = \frac{4\pi - \omega_m}{4\pi}F(m) + W(m \mid F) \tag{12.160}$$

$$W_e(m \mid F) \overset{\text{def}}{=} \lim_{\Gamma_e \ni p \to m} W(p \mid F) = -\frac{\omega_m}{4\pi}F(m) + W(m \mid F) \tag{12.161}$$

Remark 2. If $D_i \subset \mathbb{R}_2$, S its boundary and $\{m_i\}\big|_{i=1}^{\ell}$ the set of all corner points of $\mathbf{S}$, then parts (a), (b), and (c) remain valid if equalities (12.160) and (12.161) are replaced by

$$W_i(m \mid F) \overset{\text{def}}{=} \lim_{\Gamma_i \ni p \to m} W(p \mid F) = \frac{2\pi - \alpha_m}{2\pi} F(m) + W(m \mid F) \quad (12.162)$$

$$W_e(m \mid F) \overset{\text{def}}{=} \lim_{\Gamma_e \ni p \to m} W(p \mid F) = -\frac{\alpha_m}{2\pi} F(m) + W(m \mid F) \quad (12.163)$$

where α_m is the angle between links of S at m.

Theorem 12.7.2. (Discontinuities of the normal derivatives of a single-layer electrostatic potential.)

(a) *If $f(p)$ is bounded then*

$$V[p \mid f(q)] \in C(\mathbb{R}_3) \cap C^\infty(\mathbb{R}_3 \setminus \Sigma) \Rightarrow \Delta V[p \mid f(q)] = 0 \quad \forall p \in \mathbb{R}_3 \setminus \Sigma \quad (12.164)$$

(b) *If $f(p) \in C(\Sigma)$ and $m \in \Sigma_i^*$, $i = 1, 2, \dots, k$, then*

$$\frac{\partial}{\partial n_p} V_i(m \mid f) \overset{\text{def}}{=} \lim_{\Gamma_i \ni p \to m} \frac{\partial}{\partial n_p} V(p \mid f) = \frac{1}{2} f(m) + \frac{\partial}{\partial n_p} V(p \mid f) \quad (12.165)$$

$$\frac{\partial}{\partial n_p} V_e(m \mid f) \overset{\text{def}}{=} \lim_{\Gamma_e \ni p \to m} \frac{\partial}{\partial n_p} V(p \mid f) = -\frac{1}{2} f(m) + \frac{\partial}{\partial n_p} V(p \mid f) \quad (12.166)$$

PROOF OF THEOREM 12.7.1. Part (a) is a corollary of Theorem 12.5.1. Indeed, let $p \notin \Sigma$ so that, since F is bounded (say, $|F| < M$) and $1/r_{pq}$ is harmonic, the integrand in the definition of the double-layer potential is an harmonic function. Therefore,

$$W(p \mid F) \in C^\infty(\mathbb{R}_3 \setminus \Sigma) \wedge \Delta W = 0 \quad \forall p \in \mathbb{R}_3 \setminus \Sigma \quad (12.167)$$

Now let $p \in \Sigma^*$ and $q \in \Sigma^*$, and let r_{mq} be the distance between m and q. We can assume that m and q belong to the same Σ_i^*, say Σ_1^*. Let T be a tangent plane to Σ_1^* at q. Introduce a cartesian orthogonal coordinate system (x, y, z) with origin at m and the z axis directed along an outward normal $\mathbf{n}_m$ to Σ^*. Let $q = (\xi, \eta, \zeta)$ and r_{qm} be the distance of q from m. Let the equation of Σ in the neighborhood of m be

$$\zeta = Z(\xi, \eta) \quad (12.168)$$

so that the normal equation of the tangent plane T_p to Σ at p is

$$\frac{-\frac{\partial}{\partial \xi} Z(\xi, \eta)(X - \xi) - \frac{\partial}{\partial \eta} Z(\xi, \eta)(Y - \eta) + Z - Z(\xi, \eta)}{\left[\left(\frac{\partial}{\partial \xi} Z(\xi, \eta) \right)^2 + \left(\frac{\partial}{\partial \eta} Z(\xi, \eta) \right)^2 + 1 \right]^{1/2}} = 0 \quad (12.169)$$

where X, Y, Z are coordinates of the variable point at T_q. Since Σ is a surface of a finite curvature in the part under consideration, $Z(0, 0) = 0$, and T is the tangent plane to Σ at m, it follows from Lagrange's formula of finite increments that

$$Z(\xi, \eta) = \frac{1}{2} \left(\frac{\partial^2}{\partial x^2} Z(\tilde{\xi}, \tilde{\eta}) \xi^2 + 2 \frac{\partial^2}{\partial x \partial y} Z(\tilde{\xi}, \tilde{\eta}) \xi \eta + \frac{\partial^2}{\partial y^2} Z(\tilde{\xi}, \tilde{\eta}) \eta^2 \right) \quad (12.170)$$

and

$$\frac{\partial}{\partial\xi}Z(\xi,\eta) = \frac{\partial^2}{\partial\xi^2}Z(s)\xi + \frac{\partial^2}{\partial\xi\partial\eta}Z(s)\eta, \quad \frac{\partial}{\partial\eta}Z(\xi,\eta) = \frac{\partial^2}{\partial\xi\partial\eta}Z(s)\xi + \frac{\partial^2}{\partial\eta^2}Z(s)\eta$$

(12.171)

where s is some point of T_q lying between m and q. Moreover, there exists $N = \text{const} \geq 0$ such that

$$\left|\frac{\partial^2}{\partial\xi^2}Z(s)\right|, \quad \left|\frac{\partial^2}{\partial\xi\partial\eta}Z(s)\right|, \quad \left|\frac{\partial^2}{\partial\eta^2}Z(s)\right| < N$$

(12.172)

Hence by (12.169), (12.171), and (12.172),

$$|\alpha|, |\beta|, |\gamma| < Nr_{qm}$$

(12.173)

where α, β, γ are direction cosines of $\mathbf{r}_{qm}$. Hence the integrand in the double-layer potential

$$W[m \mid F(q)] = \frac{1}{4\pi}\int_{\Sigma_k} F(q)\frac{\cos\varphi}{r_{qm}^2}d\omega_q$$

(12.174)

satisfies the inequality

$$|F(q)||\frac{\cos\varphi}{r_{qm}^2}| < \frac{3NM}{r_{qm}} \quad \forall q \in \Sigma_i, \ i = 1, 2, \dots, k$$

(12.175)

so that by Theorem 12.5.1 this integral converges uniformly and as a result.

$$W[m \mid F(q)] \in C(\Sigma^*)$$

(12.176)

which completes the proof of part (a).

Parts (b) and (c) are corollaries of Lemmas 12.7.1–2. Indeed, by assumption $F(p) \in C(\Sigma^*)$. Let m be the interior or boundary point of $\Sigma_j^* (1 \leq j \leq k)$. For any $p \in \Gamma_i$, by Lemma 12.7.2 we have

$$\begin{aligned}
J(p) &\stackrel{\text{def}}{=} W[p \mid F(q)] - W[m \mid F(q)] - cF(m) \\
&= W[p \mid F(q)] - W[m \mid F(q)] - W[m \mid F(m)] \\
&= W[p \mid F(q)] - W[p \mid F(m)] + W[m \mid F(m)] - W[m \mid F(q)] \\
&= W[p \mid F(q) - F(m)] - W[m \mid F(q) - F(m)]
\end{aligned}$$

(12.177)

where $W[s \mid F(m)]$ is the potential of a double layer with constant dipole moment density $F(m)$ at point s, and $c = 1/2$ if m is an interior point of Σ_j or $c = (4\pi - \omega_m)/4\pi$ if m belongs to the boundary of Σ_j. Thus

$$J(p) = \frac{1}{4\pi}\int_{\Sigma_k} [F(q) - F(m)]\left(\frac{\cos\varphi}{r_{pq}^2} - \frac{\cos\psi}{r_{mq}^2}\right)d\sigma_q$$

(12.178)

where φ is the angle between $\mathbf{r}_{pq}$ and $\mathbf{n}_q$ and ψ is the angle between $\mathbf{r}_{mq}$ and $\mathbf{n}_q$.[5]

---

[5]Recall that

$$\frac{\partial}{\partial n_q}\frac{1}{r_{pq}} = -r_{pq}^{-2}\frac{\partial}{\partial n_q}r_{pq} = \frac{\cos\varphi}{r_{pq}^2}$$

where φ is the angle between vectors $\mathbf{r}_{pq}$ and $\mathbf{n}_q$.

It is easy to see that $J(p)$ is continuous at the point $p = m$. Indeed, since Σ_j is a convex surface, $\cos\varphi$ and $\cos\psi$ have constant signs. Hence, by the mean value theorem,

$$J(p) < |[F(q) - F(m)]|\frac{1}{4\pi}\left(\left|\int_{\Sigma_j}\frac{\cos\varphi}{r_{pq}^2}d\sigma_q\right| + \left|\int_{\Sigma_j}\frac{\cos\psi}{r_{mq}^2}d\sigma_q\right|\right) \qquad (12.179)$$

so that, by Lemma 12.7.1, there exists $N = \text{const} > 0$ such that

$$J(p) < N|[F(q) - F(m)]| \qquad (12.180)$$

Let $\varepsilon > 0$ be arbitrarily small. By the continuity of F there exists $\delta = \delta(\varepsilon) > 0$ so small that

$$r_{qm} < \delta \Rightarrow |J(p)| < \varepsilon \qquad (12.181)$$

Since $\varepsilon > 0$ is arbitrarily small, this proves the continuity of J and consequently the validity of (12.158) and (12.160).

To prove (12.159) and (12.161) we can use an analogous construction. Instead of (12.177) we need only define $J(p)$ as

$$J(p) \overset{\text{def}}{=} W[p \mid F(q)] - W[m \mid F(q)] + cF(m) \qquad (12.182)$$

and repeat all the manipulations. Thus parts (b) and (c) are valid.   **Q.E.D.**

PROOF OF THEOREM 12.7.2. Theorem 12.7.2 is a corollary to Theorem 12.7.1, since the normal derivatives of single-layer potentials can be represented in terms of a sum of a double-layer potential and a single-layer potential. Indeed (see Figure 12.6), let ψ be the angle between $\mathbf{r}_{pq}$ and $\mathbf{n}_m$. Then we have[6]

$$\frac{\partial}{\partial n_m}\frac{1}{r_{pq}} = -\frac{\cos\psi}{r_{pq}^2} \qquad (12.183)$$

Let (x, y, z) be a cartesian coordinate system with origin at m and the z axis parallel to $\mathbf{n}_q$, and let (r, ψ, ϑ) be the respective polar coordinates. Let $\mathbf{i}, \mathbf{j}, \mathbf{k}$ be the basis vectors of this system. Let ω be the angle between $\mathbf{n}_m$ and $\mathbf{n}_q$, and let ϑ and θ the angles between $\mathbf{n}_m$ and the x-axis and between $\mathbf{r}_{pq}$ and the x-axis, respectively. The superscript "0" will denote the unit vectors in any direction. We have

$$\mathbf{n}_m^0 = \cos\vartheta\sin\omega\mathbf{i} + \sin\vartheta\sin\omega\mathbf{j} + \cos\vartheta\mathbf{k},$$
$$\mathbf{r}_{pq}^0 = \cos\theta\sin\varphi\mathbf{i} + \sin\theta\sin\varphi\mathbf{j} + \cos\vartheta\mathbf{k} \qquad (12.184)$$

so that

$$\cos\psi = \mathbf{n}_m^0\mathbf{r}_{pq}^0 = \cos(\vartheta - \theta)\sin\omega\sin\varphi + \cos\varphi\cos\omega \qquad (12.185)$$

---

[6]As above, we assume that p and q belong to the same Σ_i^*.

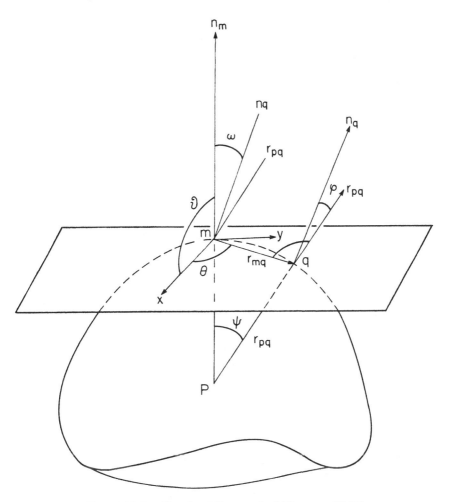

Figure 12.6.   Sketch of the proof of Theorem 12.7.2.

Hence

$$\frac{\partial}{\partial n_m} V[p \mid f(q)] = W[p \mid H(p,q,m)] + V[p \mid h(p,q,m)] \qquad (12.186)$$

where

$$H(p,q,m) = f(q)\cos\omega, \quad h(p,q,m) = [\cos(\vartheta - \theta)\sin\omega\sin\varphi]r_{pq}^{-1} \qquad (12.187)$$

We have

$$\omega = \frac{\pi}{2} - \varphi_m \qquad (12.188)$$

where φ_m is the angle between $\mathbf{r}_{qm}$ (i.e., the vector directed from q to m) and $\mathbf{n}_q$. We have

$$|\sin\omega| = |\cos\varphi_m| < |\alpha_m\alpha_q + \beta_m\beta_q| \qquad (12.189)$$

where $\alpha_m, \beta_m, \gamma_m$ are the direction cosines of $\mathbf{r}_{mq}$ and $\alpha_q, \beta_q, \gamma_q$ those of $\mathbf{n}_q$. Obviously there exists $N = \text{const} > 0$ such that

$$|\sin\omega| < 2Nr_{mq} \tag{12.190}$$

which, together with (12.187), implies that the density $h(p,q,m)$ of the single-layer potential $V[p \mid h(p,q,m)]$ is bounded. Hence, by Theorems 12.5.1 and 12.5.3, it is also continuous. Thus

$$V_i[m \mid h(m,q,m)] \overset{\text{def}}{=} \lim_{\Gamma_i \ni p \to m} V[p \mid h(p,q,m)]$$

$$V_e[m \mid h(m,q,m)] \overset{\text{def}}{=} \lim_{\Gamma_e \ni p \to m} V[p \mid h(p,q,m)] \tag{12.191}$$

But

$$H(p,q,m) \in C(\Sigma^*) \tag{12.192}$$

so that Theorem 12.7.1 is applicable to $W[p \mid H(p,q,m)]$. Hence

$$W_i[m \mid H(m,q,m)] \overset{\text{def}}{=} \lim_{\Gamma_i \ni p \to m} W[p \mid H(p,q,m)]$$

$$= \frac{1}{2}H(m,q,m) + W[m \mid H(m,q,m)] \tag{12.193}$$

$$W_e[m \mid m,q,m)] \overset{\text{def}}{=} \lim_{\Gamma_e \ni p \to m} W[p \mid H(p,q,m)]$$

$$= -\frac{1}{2}H(m,q,m) + W[m \mid H(m,q,m)] \tag{12.194}$$

Since

$$H(m,q,m) = F(m) \tag{12.195}$$

the truth of (12.165) and (12.166) follows from (12.186) and (12.195).   **Q.E.D.**

8. Dirichlet and Neumann boundary-value problems. Reduction to integral equations. Existence theorems

We now turn to the application of potential theory to the solution of the inner and outer Dirichlet and Neumann problems in convex regions. Thus, let $D_i \subset \mathbb{R}_3$ be a convex region bounded by a closed surface Σ of finite curvature, with D_e its complement relative to the whole space $\mathbb{R}_3$. Let $u_{1i}(p)$ and $u_{1e}(p)$ be solutions of the inner and outer Dirichlet problems, and let $u_{2i}(p)$ and $u_{2e}(p)$ be solutions of the inner and outer Neumann problems, so that

$$u_{1i}(p) \in C(\bar{D}_i) \cap C^2(D_i) \wedge \Delta u_{1i}(p) = 0 \quad \forall p \in D_i$$

$$u_{1i}(p) = f_{1i}(p) \quad \forall p \in \Sigma \tag{12.196$_i$}$$

$u_{1e}(p) \in C(\bar{D}_e) \cap C^2(D_e)$ and $u_{1e}(p)$ is regular at infinity, and

$$\Delta u_{1e}(p) = 0 \quad \forall p \in D_e, \qquad u_{1e}(p) = f_{1e}(p) \quad \forall p \in \Sigma \tag{12.196$_e$}$$

$$u_{2i}(p) \in C^1(\bar{D}_i) \cap C^2(D_i) \wedge \Delta u_{2i}(p) = 0 \quad \forall p \in D_i$$

$$\frac{\partial}{\partial n}u_{2i}(p) = F_{2i}(p) \quad \forall p \in \Sigma \tag{12.197$_i$}$$

Figure 12.7. The angles φ and ψ between $\mathbf{r}_{pq}$ and $\mathbf{n}_q$, and between $\mathbf{r}_{qp}$ and $\mathbf{n}_p$.

$u_{2e}(p) \in C^1(\bar{D}_e) \cap C^2(D_e)$ and $u_{2e}(p)$ is regular at infinity, and

$$\Delta u_{2e}(p) = 0 \quad \forall p \in D_e, \qquad \frac{\partial}{\partial n_p} u_{2e}(p) = F_{2e}(p) \quad \forall p \in \Sigma \qquad (12.197_e)$$

assuming that the flux of u_{2i} and u_{2e} through Σ is equal to zero:

$$\int_\Sigma F_{2i}(p)d\sigma_p = 0, \int_\Sigma F_{2e}(p)d\sigma_p = 0 \qquad (12.197_{ie})$$

In all cases we assume that the vector $\mathbf{n}_p$ is directed out of D_i.

Let us seek solutions of the Dirichlet problems (12.196_i) and (12.196_e) in the form of the double-layer potentials, and solutions of the Neumann problems (12.197_i) and (12.197_e) as single-layer potentials:

$$u_{1s}(p) = W[p \mid \mu_s(q)] = \frac{1}{4\pi} \int_\Sigma \mu_s(q) \frac{\cos\varphi}{r_{pq}^2} d\sigma_q, \quad s = i, e \qquad (12.198)$$

$$u_{2s}(p) = V[p \mid \nu_s(q)] = \frac{1}{4\pi} \int_\Sigma \nu_s(q) \frac{1}{r_{pq}} d\sigma_q, \quad s = i, e \qquad (12.199)$$

Assume that the boundary conditions hold. By Theorem 12.7.1,

$$\mu_i(p) = 2f_{1i}(p) - \frac{1}{2\pi} \int_\Sigma \mu_i(q) \frac{\cos\varphi}{r_{pq}^2} d\sigma_q \qquad (12.200)$$

$$\mu_e(p) = -2f_{1e}(p) + \frac{1}{2\pi} \int_\Sigma \mu_e(q) \frac{\cos\varphi}{r_{pq}^2} d\sigma_q \qquad (12.201)$$

By Theorem 12.7.2,

$$\nu_i(p) = 2F_{2i}(p) + \frac{1}{2\pi} \int_\Sigma \nu_i(q) \frac{\cos\psi}{r_{pq}^2} d\sigma_q \qquad (12.202)$$

$$\nu_e(p) = 2F_{2e}(p) - \frac{1}{2\pi} \int_\Sigma \nu_e(q) \frac{\cos\psi}{r_{pq}^2} d\sigma_q \qquad (12.203)$$

Recall that φ and ψ are the angles between $\mathbf{r}_{pq}$ and $\mathbf{n}_q$ and between $\mathbf{r}_{pq}$ and $\mathbf{n}_p$, respectively (see Figure 12.7). Let ω be the angle between $\mathbf{r}_{qp}$ and $\mathbf{n}_p$, so that

$$\omega = \pi - \psi \Rightarrow \cos\psi = -\cos\omega \qquad (12.204)$$

Equations (12.200)–(12.203) are all Fredholm integral equations of the second kind, that is, equations of the type

$$h(p) = \Phi(p) + \lambda \int_\Sigma h(p)\tilde{K}(p,q)d\sigma_q \qquad (12.205)$$

where

$$\tilde{K}(p,q) \stackrel{\text{def}}{=} K(p,q) = \frac{\cos\varphi}{r_{pq}^2} \qquad (12.206)$$

for the inner and outer Dirichlet problems and

$$\tilde{K}(p,q) = K(q,p) = \frac{\cos\psi}{r_{pq}^2} \qquad (12.207)$$

for the inner and outer Neumann problems, whereas

$$\lambda = -1 \qquad (12.208)$$

for the inner Dirichlet and the outer Neumann problems and

$$\lambda = 1 \qquad (12.209)$$

for the outer Dirichlet and the inner Neumann problems.

Thus the solvability of these integral equations is a question of the general theory of solvability of Fredholm integral equations of the second kind with kernels possessing polar singularities. For the reader's convenience, we recall formulation of Fredholm theorems and the relevant definitions.

A. Fredholm integral equations of the second kind with bounded kernels

Definition 12.8.1. Consider a given Fredholm integral equation of the second kind with bounded kernel $K(p,q)$ and continuous free term $f(p)$,

$$u(p) = f(p) + \lambda \int_{\Sigma} u(q)K(p,q)d\sigma_q \qquad (12.210)$$

where $\Sigma \in \mathbb{R}_n$ is a closed smooth manifold. If the corresponding homogeneous equation has a nontrivial solution $u(p\,|\,\lambda)$ (i.e., if $f(p) \equiv 0$ and $u((p\,|\,\lambda) \not\equiv 0)$ for some λ), then λ is called an *eigenvalue* and the nontrivial solution $u(p\,|\,\lambda)$ is called an *eigenfunction*, belonging to the eigenvalue λ.

Definition 12.8.2. The kernel $\tilde{K}(p,q) \equiv K(q,p)$ is known as the kernel *adjoint* to $K(p,q)$. The equation

$$v(p) = F(p) + \lambda \int_{\Sigma} v(q)K(q,p)d\sigma_q \qquad (12.211)$$

is called the adjoint equation of the equation with kernel $K(p,q)$, or simply the *adjoint equation*.

Definition 12.8.3. Let

$$K_0(p,q) \equiv K(p,q)$$

$$K_n(p,q) = \int_{\Sigma} K_0(p,r)K_{n-1}(r,q)d\sigma_r \quad \forall n = 1,2,\dots \qquad (12.212)$$

Then $K_n(p,q)$ and the sum of the series

$$R(p,q,\lambda) = \sum_{n=0}^{\infty} \lambda^n \cdot K_n(p,q) \qquad (12.213)$$

are known as the nth *iterated kernel* and the *resolvent*, respectively.

Remark. The nth iterated kernel $K_n(q,p)$ is adjoint to $K_n(p,q)$, and as the resolvent $R(q,p,\lambda)$ is adjoint to $R(p,q,\lambda)$. In the theory of Fredholm equations it is proved that $R(p,q,\lambda)$ and $R(q,p,\lambda)$ are meromorphic functions of λ, whose sequence of poles coincide with the eigenvalues of the respective equations.

If λ is a pole of the resolvent of multiplicity m, then m is called the *multiplicity* of the eigenvalue λ.

Theorem 12.8.1 (Fredholm's first theorem). *All the eigenvalues of the given Fredholm equation of the second kind are eigenvalues of the adjoint equation. If λ is not the eigenvalue, then the nonhomogeneous equations have unique solutions represented by*

$$
\begin{aligned}
u(p) &= f(p) + \lambda \int_{\Sigma} f(q) R(p,q,\lambda) d\sigma_q \\
v(p) &= F(p) + \lambda \int_{\Sigma} F(q) R(q,p,\lambda) d\sigma_q
\end{aligned}
\qquad (12.214)
$$

Theorem 12.8.2 (Fredholm's second theorem). *The number of linearly independent eigenfunctions belonging to the eigenvalue λ of multiplicity* **n** *is n for both original and adjoint equations.*

Theorem 12.8.3 (Fredholm's third theorem). *If λ is an eigenvalue of multiplicity m, then solutions of the nonhomogeneous equations (original and adjoint) exist and are unique if and only if their free terms are orthogonal to all the eigenfunctions of the adjoint and original equations, respectively, belonging to λ.*[7]

B. Fredholm integral equations of the second kind with kernels, possessing polar singularities

Let $K(p,q)$ be a kernel satisfying conditions of Theorem 12.5.4, that is, such that

$$K(p,q) = \frac{\tilde{K}(p,p)}{r_{pq}^{\alpha}}, \quad \tilde{K}(p,p) \neq 0, \quad \tilde{K}(p,q) \in C(\Sigma) \quad \forall (p,q) \in \Sigma, \ \alpha < 2 \qquad (12.215)$$

---

[7]Proofs of Fredholm's theorems may be found in all textbooks on the theory of integral equations. We recommend, first of all, the classical Goursat book [28].

The Fredholm theory, which was developed for bounded kernels only, is applicable to the equation obtained after a finite number of iterations. However, the following assertions are true.

Theorem 12.8.4. *Let $u(p)$ be a solution of the equation*

$$u(p) = f(p) + \lambda \int_\Sigma u(q)K(p,q)d\sigma_q \qquad (12.216)$$

Then (a) $u(p)$ is the solution of the equation

$$u(p) = f_n(p) + \lambda^n \int_\Sigma u(q)K_n(p,q)d\sigma_q \qquad (12.217)$$

where

$$f_n(p) = f(p) + \sum_{k=0}^{n-1} \lambda^k \int_\Sigma f(q)K_k(p,q)d\sigma_q \qquad (12.218)$$

and (b) the Fredholm theory is applicable to equation (12.217).

PROOF. (a) If $n=1$, the validity of (12.218) follows from (12.216). Assume that (12.217) is valid for $n=m$. Then, by (12.216) and (12.218),

$$u(p) = f(p) + \lambda \int_\Sigma K(p,q) \left(f_n(q) + \lambda^n \int_\Sigma K_n(q,s)u(s)d\sigma_s \right) ds_q \qquad (12.219)$$

Changing the order of integration, we obtain

$$\lambda^{n+1} \int_\Sigma K(p,q)d\sigma_q \int_\Sigma K_n(q,s)u(s)d\sigma_s = \lambda^{n+1} \int_\Sigma u(s)d\sigma_s \int_\Sigma K(q,s)K_n(p,q)d\sigma_q$$

$$= \lambda^{n+1} \int_\Sigma u(s)K_{n+1}(p,s)ds \qquad (12.220)$$

Similarly, Definition 12.8.3 and (12.218) yield

$$f(p) + \lambda \int_\Sigma K(p,q)f_n(q)d\sigma_q = f_{n+1}(p) \qquad (12.221)$$

so that the assertion of the theorem is true by induction.   **Q.E.D.**

(b) By Theorem 12.5.4, the nth iterated kernel is bounded if

$$n > \frac{\alpha}{2-\alpha} \qquad (12.222)$$

Since at the same time, by Theorem 12.5.3 all integrals figuring in the definition of f_n are uniformly convergent, the assertion follows.   **Q.E.D.**

Theorem 12.8.5. *Let $v(p) \overset{\text{def}}{=} u_n(p)$ be a solution of the nonhomogeneous equation (12.217) and let λ^n be different from all the eigenvalues of the equation. Then $v(p)$ is a solution of equation (12.216).*

PROOF. Let

$$v_0(p) \overset{\text{def}}{=} v(p), \quad v_k(p) = \lambda \int_\Sigma K(p,q)v_{k-1}(q)d\sigma_q, \quad k=1,2,\ldots,n \qquad (12.223)$$

which, by the definition of iterated kernels, implies that

$$v_n(p) = f_n(p) + \lambda^n \int_\Sigma K_n(p,q)v(q)d\sigma_q \Rightarrow v_n(p) = v(p) \qquad (12.224)$$

Now let

$$u(p) = \frac{1}{n} \sum_{k=1}^{n} v_k(p) \qquad (12.225)$$

Then by (12.223) and (12.224)

$$u(p) = f(p) + \lambda \int_\Sigma \frac{1}{n}\left(v(q) + \sum_{k=1}^{n-1} v_k(q)\right) K(p,q)d\sigma_q$$

$$= f(p) + \lambda \int_\Sigma \left(\frac{1}{n}\sum_{k=1}^{n} v_k(q)\right) K(p,q)d\sigma_q$$

$$= f(p) + \lambda \int_\Sigma u(q)K(p,q)d\sigma_q \qquad (12.226)$$

Hence the function $u(p)$ defined by (12.225) is a solution of the integral equation (12.216). Note that if λ^n is not an eigenvalue of equation (12.217) then

$$u(p) \equiv v(p) \qquad (12.227)$$

Indeed, in this case the solution of the nonhomogeneous equation (12.217) is unique, which implies the identities

$$v_k(p) \equiv v(p) \Rightarrow u(p) \equiv v(p) \qquad (12.228)$$

$$\text{Q.E.D.}^8$$

We can now return to consideration of the Dirichlet and Neumann problems, that is, to consideration of the integral equations (12.200)–(12.203).

Theorem 12.8.6. (a) *The parameter* $\lambda = -1$ *is not an eigenvalue of the integral equations* (12.200) *and* (12.202) *corresponding to the inner Dirichlet and outer Neumann problems. Hence, by Fredholm's first theorem, these equations are uniquely solvable.*

(b) $\lambda = 1$ *is an eigenvalue of multiplicity* 1 *of the integral equations* (12.201) *and* (12.203) *corresponding to the outer Dirichlet and inner Neumann problems. The eigenfunction of equation* (12.201) *belonging to the eigenvalue* $\lambda = 1$ *is a constant. Hence, by Fredholm's third theorem, equation* (12.203) *is uniquely solvable if and only if the free term* $2F_{2i}(p)$ *is orthogonal to a constant, that is, if and only if*

$$\int_\Sigma F_{2i}(p)d\sigma_p = 0 \qquad (12.229)$$

---

[8]The proof of Theorem 12.8.5 is borrowed from Privalov [61].

Hence, the integral equation corresponding to the inner Neumann problem is uniquely solvable if the prescribed flux satisfies the necessary condition for the Neumann problem to be solvable. The integral equation (12.201) corresponding to the outer Dirichlet problem is solvable if the free term $2f_{1e}(p)$ (i.e., the prescribed potential at the boundary Σ) is orthogonal to a corresponding eigenfunction of equation (12.202).

PROOF. (a) We first prove that $\lambda = -1$ is not an eigenvalue. Indeed, suppose the contrary. Then the integral equation

$$\mu(p) + \int_{\Sigma} \mu(q)K(p,q)d\sigma_q = 0 \tag{12.230}$$

has a nontrivial solution. As a continuous function, it achieves its maximum at some point $p \in \Sigma$. Since by Lemma 12.7.1

$$\int_{\Sigma} K(p,q)d\sigma_q \overset{\text{def}}{=} \frac{1}{2\pi}\int_{\Sigma}\frac{\cos\varphi}{r_{pq}^2}d\sigma_q = 1 \tag{12.231}$$

it follows from (12.230) that

$$\int_{\Sigma}[\mu(p) + \mu(q)]K(p,q)d\sigma_q = 0 \tag{12.232}$$

However, this contradicts the assumption that Σ is convex, which implies $\cos\varphi > 0$. Hence

$$\mu(p) \equiv 0 \tag{12.233}$$

(b) The proof of the assertion that $\lambda = 1$ is an eigenvalue is quite analogous. Indeed, in this case we have

$$\mu(p) - \int_{\Sigma}\mu(q)K(p,q)d\sigma_q = 0 \tag{12.234}$$

instead of (12.230), so that at a maximum point p of μ we have

$$\int_{\Sigma}[\mu(p) - \mu(q)]K(p,q)d\sigma_q = 0 \tag{12.235}$$

Hence

$$\mu(q) \equiv \mu(p) \tag{12.236}$$

Therefore $\lambda = 1$ is indeed an eigenvalue of multiplicity 1, since all possible eigenfunctions belonging to $\lambda = 1$ are constants; that is, there do not exist two linearly independent eigenfunctions belonging to $\lambda = 1$.

Thus parts (a) and (b) of the theorem are valid. **Q.E.D.**

Remark 1. An eigenfunction $\nu(p)$ of the integral equation corresponding to the inner Neumann problem has the following physical interpretation: Suppose that the region D_i is occupied by a conductor charged up to potential $v(p)$. It is known that the electric field of a charged conductor is equal to the constant generated by charges distributed over the boundary Σ of D_i with some density ν (see Chapter 1, Section 3), so that

$$v(p) \stackrel{\text{def}}{=} V(p \mid \nu) \equiv \text{const} \tag{12.237}$$

Thus $\nu(p)$ is a solution of the homogeneous equation

$$\nu(p) = \frac{1}{2\pi} \int_{\Sigma} \nu(q) \frac{\cos \psi}{r_{pq}^2} d\sigma_q \tag{12.238}$$

Hence $\nu(p)$ is a eigenfunction of the integral equation (12.203).

Remark 2. The statement of Theorem 12.8.6, according to which the integral equation (12.201) has a solution only if $f_{1e}(p)$ is orthogonal to the eigenfunction $\nu(p)$ of the adjoint equation (12.202), clearly contradicts the assumption that the double-layer potential $W(p \mid \mu)$ represents a harmonic function regular at infinity. Indeed, application of the Kelvin transformation reduces the outer Dirichlet problem to an inner one with a removable singularity at the origin. Hence a solution of the problem exists when $f_{1e}(p)$ is not orthogonal to $\nu(p)$.

The source of this contradiction is quite clear. Indeed, regularity of a function $\Phi(p)$ at infinity means that $\Phi(p) = O(1/r)$ in the neighborhood of point at infinity, whereas the double-layer potential with a bounded moment density is a function of the order $O(1/r^2)$. Hence the very attempt to find a solution of the outer Dirichlet problem in the form of a double-layer potential is contradictory. Instead, let us seek a solution in the form

$$u_e(p) = \frac{A}{r_{pO}} + W[p \mid \mu_{1e}(q)] \tag{12.239}$$

where O is some fixed point inside D_i and A is an undetermined coefficient (electrostatically, the total charge in $\bar{D}$). Then, instead of the integral equation (12.201), one obtains the equation

$$\mu_{1e}(p) = \frac{2A}{r_{pO}} - 2f_{2e}(p) + \frac{1}{2\pi} \int_{\Sigma} \mu_{1e}(q) \frac{\cos \varphi}{r_{pq}^2} d\sigma_q \tag{12.240}$$

so that a solution of this equation exists and is unique if and only if

$$\int_{\Sigma} \left(\frac{2A}{r_{pO}} - 2f_{2e}(p) \right) \nu(p) d\sigma_p = 0 \Rightarrow A = \frac{\int_{\Sigma} f_{2e}(p) \nu(p) d\sigma_p}{\int_{\Sigma} \frac{\nu(p)}{r_{pO}} d\sigma_p} \tag{12.241}$$

Since

$$\int_{\Sigma} \frac{\nu(p)}{r_{pO}} d\sigma_p = \text{const} \neq 0 \tag{12.242}$$

the desired constant A is determined, which eliminates the contradiction.

Remark 3. All these results remain in force in the case of two dimensions except for that relating to the outer Neumann problem, which requires some additional consideration in $\mathbb{R}_2$. Indeed, if $\nu_e(p)$ is bounded for every $p \in \Gamma$, where Γ is the boundary of a 2-dimensional region D_i, then the single-layer potential

$$V[p \mid \nu_e(q)] = \frac{1}{2\pi} \int_\Gamma \nu_e(q) \ln\left(\frac{1}{r_{pq}}\right) ds_q \qquad (12.243)$$

is not a function regular at infinity because

$$\lim_{r_p\Gamma\uparrow\infty} \left| \int_\Gamma \nu_e(q) \ln\left(\frac{1}{r_{pq}}\right) ds_q \right| = \infty \qquad (12.244)$$

unless some special assumption is made concerning the properties of $F_{2e}(p)$. However, the following assertion is valid.

Lemma 12.8.1. *Let $\nu_{2e}(p)$ be a solution of the integral equation*

$$\nu_e(p) = 2F_{2e}(p) - \frac{1}{\pi} \int_\Gamma \nu_e(q) \frac{\cos\psi}{r_{pq}} \cdot ds_q \qquad (12.245)$$

which corresponds in the 2-dimensional case to equation (12.203). Then the single-layer logarithmic potential (12.243) is regular at infinity if and only if

$$\int_\Gamma F_{2e}(q) ds_q = 0 \qquad (12.246)$$

PROOF. Indeed, by (12.245),

$$\int_\Gamma \nu_e(p) ds_p = 2 \int_\Gamma F_{2e}(p) ds_p - \frac{1}{\pi} \int_\Gamma ds_p \int_\Gamma \nu_e(q) \frac{\cos\psi}{r_{pq}} ds_q$$

$$\Rightarrow \int_\Gamma F_{2e}(p) ds_p = \frac{1}{2} \int_\Gamma \nu_e(p) ds_p$$

$$- \int_\Gamma \nu_e(q) ds_q \frac{1}{2\pi} \int_\Gamma \frac{\cos\omega}{r_{pq}} \cdot ds_p \qquad (12.247)$$

The inner integral in a double integral on the right is a double-layer logarithmic potential of unit moment density at $q \in \Gamma$. Hence, by Lemma 12.5.1, this integral is equal to $1/2$, and so

$$\int_\Gamma F_{2e}(p) ds_p = 0 \qquad (12.248)$$

Thus

$$\int_\Gamma \nu_e(p) ds_p = 0 \qquad (12.249)$$

if and only if (12.246) is true. At the same time, it is obvious that $V[p \mid \nu_e(q)]$ remains bounded at infinity if and only if (12.247) holds. Hence the assertion of Lemma 12.8.1 is true.   **Q.E.D.**[9]

---

[9]The obvious electrostatic interpretation of this lemma relates to the fact that the fundamental solution in $\mathbb{R}_2$ (the logarithmic potential), in contrast to that in $\mathbb{R}_3$, is unbounded at infinity.

Problems

P12.1.1. Prove Theorem 12.1.1 in the 2-dimensional case.

P12.3.1. Functions harmonic in $\mathbb{R}_1$ are linear. Give an electrostatic interpretation to this fact.

P12.5.1. Prove Theorem 12.5.1 and Theorem 12.5.2.

P12.5.2. Formulate and prove a theorem justifying the differentiation of improper integrals under the integral sign.

P12.5.3. Prove Theorems 12.5.3 and 12.5.4 in the 2-dimensional case.

P12.6.1. Prove Theorems 12.6.1 and 12.6.2 for outer problems.

P12.6.2. Prove equalities (12.162) and (12.163).

P12.7.2. Formulate and prove the theorems concerning the tangential derivatives of single- and double-layer electrostatic potentials.

P12.8.1. How must the statement of Theorem 12.8.6 be changed in order to make it true in the theory of logarithmic potential?

P12.8.2. Prove that the integral equations (12.202) and (12.203) have no complex eigenvalues and have no real eigenvalues smaller than $\lambda = 1$.

P12.8.3. Reduce the problem

$$\Delta u(p) = 0 \quad \forall p \in D_1$$

$$\frac{\partial}{\partial n_p} u(p) - hu(p) = f(p) \quad \forall p \in \Sigma$$

to a Fredholm integral equation, and prove the existence and uniqueness theorem on the assumption that h is a positive constant.

P12.8.4. There is a difference in formulation of a meaningful electrostatic outer Dirichlet problem in $\mathbb{R}_3$ and $\mathbb{R}_2$. Thus, a typical problem in $\mathbb{R}_3$ reads: Find the potential $\varphi(p)$ regular at infinity, harmonic in D_e, continuous in $\bar{D}_e$, and equal to a given value φ_0 on Σ. Yet, an electrostatically meaningful problem in $\mathbb{R}_2$ reads: Find the potential $\varphi(p)$, harmonic in D_e, equal to a given value φ_0 on Σ, and such that the total charge carried by D_i is q. (Recall that $D_e = \mathbb{R}_n \setminus \bar{D}_i$, $n = 3, 2$.) Is this $\mathbb{R}_2$ formulation accurate? Explain the difference in the formulation in $\mathbb{R}_3$ and $\mathbb{R}_2$. How do these formulations relate to the standard formulation of the outer Dirichlet problem given in the main text?

P12.8.5. What feature of electrostatic potentials is behind the St. Elmo light – electric discharge at sharp edges of prominent objects of ships in stormy weather? *Hint*: Consider the intensity of the electric field near the corner of an edge in the interior Dirichlet problem in the plane at different edge apertures.

13

Cauchy problem for heat-conduction equation

1. Fundamental solution of Fourier equation. Heaviside unit function and Dirac δ function

We begin our study of boundary-value problems of parabolic type with the construction of the fundamental solution of the Fourier (heat-conduction) equation. With this purpose in mind, we consider the following simple physical problem.

Suppose we are given two semi-infinite thermally conductive slabs of the same material, in contact with each other. Assume that these slabs are perfectly thermally insulated along their lateral boundaries; let the initial temperature of one slab be 1 and that of the other 0. Our purpose is to determine the redistribution of temperature within the system over time. Thus, we are considering the following problem: Find a function $\tilde{u}(x,t)$ satisfying the conditions

$$a^2 \frac{\partial^2}{\partial x^2} \tilde{u}(x,t) = \frac{\partial}{\partial t} \tilde{u}(x,t) \quad \forall x \in (-\infty, \infty), \ \forall t > 0 \tag{13.1}$$

$$\tilde{u}(x,0) = \begin{cases} 0 & \forall x < 0 \\ \tilde{u}_0 = \text{const} & \forall x > 0 \end{cases} \tag{13.2}$$

where a is a constant independent of x and t.

Because of the great conceptual importance of the construction that follows, we shall go into detail to a degree unusual even for our generally detailed presentation. The variables $\tilde{u}, x, t$ and parameters $a, \tilde{u}_0$ in equations (13.1) and (13.2) are dimensional, with respective dimensions,

$$[\tilde{u}] = T(^\circ \text{K}), \quad [x] = L \text{ (cm)}, \quad [t] = \tau \text{ (sec)}$$

$$[a^2] = \frac{L^2}{\tau} \left(\frac{\text{cm}^2}{\text{sec}} \right), \quad [\tilde{u}_0] = T(^\circ \text{K}) \tag{13.3}$$

The only parameters in (13.1) and (13.2) with the dimensions of space or time are x and t. Moreover, no parameters other than x and t with these

296

dimensions may be formed as a combination of the dimensional parameters in (13.1) and (13.2). Thus there are no a priori space and time scales in this formulation. (x varies over an infinite interval and the initial temperature is constant in both half-spaces $x > 0$ and $x < 0$, so that no typical length scale may be inferred from the initial condition, either). We emphasize at this point that the entire construction to follow is based on the belief that, in principle, *all information about the solution is already present in the formulation of the problem*. Thus it is believed, in particular, that no new typical time or space scales may show up in the solution in addition to those already present, as represented by the dimensional parameters in the formulation of the problem or dimensional combinations of them. With all this in mind, we introduce a dimensionless dependent variable u:

$$u(x,t) = \tilde{u}(x,t)/\tilde{u}_0 \qquad (13.4)$$

The dependence of the *dimensionless* variable u on the *dimensional* variables x and t should be such that x and t enter into u only in some dimensionless combination with other dimensional parameters involved in the formulation of the problem. Indeed, by Weierstrass's theorem, $u(x,t)$, as a continuous function, can be uniformly approximated by series in powers of x and t. Each term in these series should be dimensionless, since u is. The presence in $u(x,t \mid \tilde{u}_0, a^2)$ of any of the dimensional parameters $(x, t, \tilde{u}_0, a^2)$, such as x, other than in a dimensionless combination with the others, would imply the need for an additional dimensional parameter (of dimension L in this case) to nondimensionalize the powers of x in the series. On the other hand, the existence of such an additional parameter not included in the original formulation of the problem contradicts our just-mentioned fundamental belief. Hence u may depend only on various mutually independent dimensionless combinations of the original, dimensional parameters of the problem.

Note that in the absence of an explicit space scale in the formulation of (13.1) and (13.2), the only such combination is $x/a\sqrt{t}$ (or some power or any other dimensionless function of this combination). Thus, we define a dimensionless independent variable

$$z = x/(2a\sqrt{t}) \qquad (13.5)$$

(the factor 2 has been introduced for convenience in subsequent computations). Define

$$w(z) = u(x,t) \qquad (13.6)$$

We emphasize once again that the reduction in the number of essential variables from two (x,t) to one (z) is due exclusively to the absence of a priori space or time scales x_0 or t_0, respectively (the existence of one would imply that of the other, since $t_0 = x_0^2/a^2$). Thus the dependence of the dimensionless temperature $u(x/x_0, t/t_0)$ on two independent dimensionless variables is necessarily reduced to dependence on a single *similarity* (self-similar) variable z. Accordingly, the entire original situation described by

(13.1) and (13.2), which permits this reduction in the number of essential independent variables via arguments of dimensional symmetry (the absence of a priori scales), is known as a similarity or *self-similar* situation.[1]

In terms of the similarity variables $w(z)$ and z, the partial differential equation (13.1) reduces to an ordinary differential equation:

$$\frac{d^2}{dz^2}w + 2z\frac{d}{dz}w = 0; \qquad -\infty < z < \infty \qquad (13.7)$$

and the initial conditions (13.2) reduce to boundary conditions

$$w(-\infty) = 0; \qquad w(+\infty) = 1 \qquad (13.8)$$

Note that, uniformly for all $\delta = \text{const} > 0$ and x such that $|x| \geq \delta$,

$$\lim_{t\downarrow 0} z = +\infty \quad \forall x = \text{const} > 0, \quad \lim_{t\downarrow 0} z = -\infty \quad \forall x = \text{const} < 0 \qquad (13.9)$$

The boundary conditions (13.8) are obvious corollaries of (13.9) and of the initial conditions (13.2).[2]

Integration of (13.7) yields

$$w(z) = \alpha + \beta\,\text{erf}(z) \qquad (13.10)$$

where

$$\text{erf}(z) = 2\pi^{-1/2}\int_0^z e^{-\lambda^2}d\lambda \qquad (13.11)$$

---

[1]We have permitted ourselves this lengthy exposition (essentially an example of the application of the π theorem of the theory of similarity) because of its evident physical depth, as well as the didactic merit of presenting in great detail one very simple, prototypical example of a broad class of similarity situations. Of course, the result could have been arrived at in a few lines, via the following "continuous transformation group" [5] reasoning: Introduce new variables $\xi = x/X$, $\tau = t/T$, and $v(\xi, \tau) = u(x, t)$. In these variables equation (13.1) transforms into

$$(a^2/X^2)v_{\xi\xi} = (1/T)v_\tau \qquad (\alpha)$$

and the initial conditions (13.2) become

$$v(\xi, 0) = 0 \text{ for } \xi < 0 \quad \text{and} \quad v(\xi, 0) = 1 \text{ for } \xi > 0 \qquad (\beta)$$

If $X^2 = a^2/4T$, then (α) becomes

$$v_{\xi\xi} = v_\tau \qquad (\gamma)$$

and the initial conditions (β) remain unchanged. Since the Cauchy problem (γ), (β) is independent of T, its solution is invariant with respect to change in T. Fix arbitrary x and t and take $T = t$. Then we obtain

$$u(x, t) = v(x/2a\sqrt{t}, 1) \equiv v(z) \qquad (\delta)$$

As x and t were fixed arbitrarily, this implies that $u(x, t)$ is a function of the similarity argument z only.

[2]Compare this with Gelfand's heuristic treatment of the problem of decay of an arbitrary initial discontinuity in the Cauchy problem for a quasilinear first-order partial differential equation. (See Chapter 3, Section 4.)

is the error function (sometimes called the Cramp function). The constants of integration α and β must be determined by the boundary conditions (13.8). Using the equalities

$$\lim_{z \uparrow \infty} \operatorname{erf}(z) = 1, \qquad \lim_{z \downarrow -\infty} \operatorname{erf}(z) = -1 \tag{13.12}$$

we find that $\alpha = \beta = {}^1\!/_2$, and so

$$w(z) = {}^1\!/_2 (1 + \operatorname{erf}(z)) \tag{13.13}$$

Thus, returning to the original variables, one has

$$u(x, t) = {}^1\!/_2 \left(1 + \operatorname{erf}\left(x/2at^{1/2} \right) \right) \tag{13.14}$$

Note that by (13.14),

$$u(0, t) = {}^1\!/_2 \quad \text{if } t > 0 \tag{13.15}$$

and

$$u(x, 0) = \begin{cases} 1 & \text{if } x > 0 \\ 0 & \text{if } x < 0 \end{cases} \tag{13.16}$$

The discontinuous function defined by

$$\eta(x) = \begin{cases} 1 & \text{if } x > 0 \\ {}^1\!/_2 & \text{if } x = 0 \\ 0 & \text{if } x < 0 \end{cases} \tag{13.17}$$

is known as the *Heaviside unit function*. Thus, if one sets

$$u(0, 0) = {}^1\!/_2 \tag{13.18}$$

then equation (13.16) can be rewritten as

$$u(x, 0) = \eta(x) \tag{13.19}$$

Note next that if $u(x, t)$ is a solution of the heat-conduction equation then, for $t > 0$, any of its derivatives is also a solution. In particular, so is

$$\frac{\partial}{\partial x} u(x, t) \overset{\text{def}}{=} E(x, a^2 t) \equiv \frac{\exp(-x^2/4a^2 t)}{2a\sqrt{\pi t}} \tag{13.20}$$

This is the so-called *fundamental solution of the Fourier equation*, which plays an extremely important role in the theory of heat conduction and diffusion, like that of the fundamental solution $1/4\pi r$ in electrostatics and gravimetry. Figure 13.1 shows $E(x, a^2 t)$ for a sequence of time $t_1 < t_2 < t_3 < t_4$.

Note that

$$E(x, 0) = 0 \quad \forall x \neq 0 \tag{13.21}$$

and

$$E(0, t) = (2\sqrt{\pi t})^{-1} \tag{13.22}$$

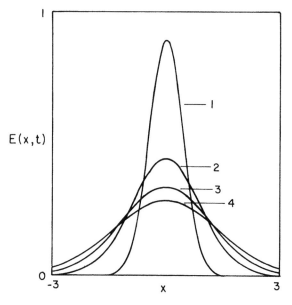

Figure 13.1. Fundamental solution $E(x,t)$ of the
heat-conduction equation in $\mathbb{R}_1$. 1: $t = 0.1$; 2:
$t = 0.4$; 3: $t = 0.7$; 4: $t = 1.0$.

so that

$$\lim_{t\downarrow 0} E(0,t) = \infty \tag{13.23}$$

Thus $E(x, a^2 t)$ satisfies the initial conditions

$$E(x,0) = \begin{cases} 0, & x \neq 0 \\ \infty, & x = 0 \end{cases} \tag{13.24}$$

Moreover,

$$\int_{-\infty}^{\infty} E(x, a^2 t)dx = 1 \quad \forall t > 0 \tag{13.25}$$

and this means that

$$\lim_{t\downarrow 0} \int_{-\infty}^{\infty} E(x, a^2 t)dx = 1 \tag{13.26}$$

If it were permissible to change the order of passage to the limit and
integration, we would have

$$\int_{-\infty}^{\infty} \lim_{t\downarrow 0} E(x, a^2 t)dx = \int_{-\infty}^{\infty} E(x,0)dx = 1 \tag{13.27}$$

Note that $E(x,0)$, as defined by (13.24) and (13.27), is not a function in
the classical Dirichlet sense. Rather, it is a convenient new object, which
Paul Dirac first introduced in the quantum–mechanical context [16] and
called the δ *function*. A few years after Dirac had first defined it, δ

functions and similar mathematical objects became the subject of a new branch of mathematics known today as the theory of generalized functions or distributions [72; 75].

Using Dirac's notation, we write

$$E(x,0) = \delta(x) \tag{13.28}$$

Once more, if it were admissible to change the order of passage to the limit and differentiation, we would get

$$\lim_{t \downarrow 0} \frac{\partial}{\partial x} \operatorname{erf}\left(\frac{x}{2a\sqrt{t}}\right) = \frac{\partial}{\partial x}\left(\lim_{t \downarrow 0} \operatorname{erf}\left(\frac{x}{2a\sqrt{t}}\right)\right) \tag{13.29}$$

so that by the definitions (13.19) and (13.28) and by (13.14) and (13.20) we could write

$$\delta(x) = \frac{d}{dx}\eta(x) \tag{13.30}$$

Naturally, this equality should not be understood in the sense of the classical differential calculus, but rather as a definition of the "differentiation" of an η function.

Note that formal application of properties (13.24) and (13.26), that is,

$$\delta(x) = \begin{cases} 0, & x \neq 0 \\ \infty, & x = 0 \end{cases}; \qquad \int_{-\infty}^{\infty} \delta(x)dx = 1 \tag{13.31}$$

implies that *the Dirac δ function possesses the following filtering property. For any $f(x) \in C^0$ (i.e., for any $f(x)$ continuous in $(-\infty, \infty)$) and for any $\varepsilon > 0$,*

$$\int_{x-\varepsilon}^{x+\varepsilon} f(\xi)\delta(x-\xi)d\xi = f(x) \tag{13.32}$$

Indeed, by formal application of the mean value theorem,

$$\int_{x-\varepsilon}^{x+\varepsilon} f(\xi)\delta(x-\xi)d\xi = f(\xi^0) \int_{x-\varepsilon}^{x+\varepsilon} \delta(x-\xi)d\xi$$

$$= f(\xi_0) \int_{-\infty}^{\infty} \delta(x-\xi)d\xi = f(\xi_0) \tag{13.33}$$

where ξ_0 is some point of the interval $(x-\varepsilon, x+\varepsilon)$. Since $\varepsilon > 0$ is arbitrarily small and $f(x)$ is assumed to be continuous, this means that for any $f(x)$ continuous in a neighborhood of the point $x \in (-\infty, \infty)$,

$$\int_{-\infty}^{\infty} f(\xi)\delta(x-\xi)d\xi = f(x) \tag{13.34}$$

Clearly, properties (13.31) are corollaries of the identity (13.34) if the latter is treated as the definition of the δ function. This is in fact Dirac's original definition.[3]

---

[3] Calculations based on using the δ function always lead to correct results, but for a long time mathematicians regarded them as unrigorous. Later it was understood

2. Solution of Cauchy problem for 1-dimensional Fourier equation. Poisson integral

While (13.34) is at most a *definition* of the δ function, the equality

$$\lim_{t\downarrow 0}\int_{\infty}^{\infty} f(\xi)E(x-\xi,a^2t)d\xi = f(x) \qquad (13.35)$$

for any function f continuous in some neighborhood of x can be proved rigorously by the methods of classical calculus, as we shall now see.

Theorem 13.2.1. *Let $f(x)$ be a function continuous in the left and right half-neighborhoods of any point x_0. Assume that there exist positive constants A, α such that*

$$|f(x)| < Ae^{\alpha x^2} \quad \forall |x| \qquad (13.36)$$

Then the Poisson integral

$$u(x,t) = \int_{-\infty}^{\infty} f(\xi)E(x-\xi,a^2t)d\xi \qquad (13.37)$$

is a solution of the equation

$$L_a u \stackrel{\text{def}}{=} a^2 \frac{\partial^2}{\partial x^2} u(x,t) - \frac{\partial}{\partial t} u(x,t) = 0 \quad \forall t > 0 \qquad (13.38)$$

such that if the limits

$$\lim_{\xi\to x_0-0} f(\xi) \quad and \quad \lim_{\xi\to x_0+0} f(\xi) \qquad (13.39)$$

exist then

$$\lim_{t\downarrow 0} u(x_0,t) = \frac{1}{2}\left(f(x_0-0)+f(x_0+0)\right) \qquad (13.40)$$

PROOF. Obviously, there exists $c = \text{const} > 0$ such that

$$L_a u(x,t) = 0 \quad \forall x, \ \forall t \in (0,c) \qquad (13.41)$$

---

that integrals of type (13.34) may be treated as Stieltjes integrals with the Heaviside unit function η as the integrating function:

$$\int_{-\infty}^{\infty} f(\xi)d\eta(x-\xi)$$

or as limits of weakly convergent sequences. After the development of the theory of distributions, it was understood that δ function is a distribution rather than a function in Dirichlet's sense. Nevertheless, such classical mathematicians as Cauchy, Poisson, and Weierstrass in fact used δ functions when they treated the integrals (13.34) as limits of integrals with a δ-shaped integrand. According to Weierstrass,

$$\int_{-\infty}^{\infty} f(\xi)\delta(x-\xi)d\xi = \lim_{t\downarrow 0}\int_{-\infty}^{\infty} f(\xi)E(x-\xi,a^2t)d\xi = f(x)$$

which is true for every function $f(x)$ continuous at the point x and increasing to infinity no more rapidly than $A\exp(x^2)$ (see Section 2 of this chapter).

Indeed, given x, take $N > 0$ such that

$$-N < x < N \tag{13.42}$$

We have

$$u(x,t) = \sum_{i=1}^{3} u_i(x,t) \tag{13.43}$$

where

$$u_1 = \int_{-\infty}^{-2N} , \qquad u_2 = \int_{-2N}^{2N} , \qquad u_3 = \int_{2N}^{\infty} \tag{13.44}$$

By (13.36),

$$|u_1(x,t)| \leq A \int_{-\infty}^{-2N} \frac{\exp(-(x-\xi)^2/4a^2t + \alpha\xi^2)}{2a(\pi t)^{1/2}} d\xi \tag{13.45}$$

Take

$$t < c = \frac{1}{4a^2\alpha} \tag{13.46}$$

Then

$$|u_1(x,t)| < A \int_{-\infty}^{-2N} \frac{\exp(-(x^2 - 2x\xi + \beta\xi^2)/4a^2t)}{4a(\pi t)^{1/2}} d\xi = I_1(x,t) \tag{13.47}$$

where $\beta = \text{const} > 0$. Hence $u_1(x,t)$ is majorized by a uniformly convergent integral. Moreover, all the integrals

$$\int_{\infty}^{2N} \frac{\partial^{k+m}}{\partial x^k \partial t^m} \frac{\exp(-(x^2 - 2x\xi + \beta\xi^2)/4a^2t)}{4a(\pi t)^{1/2}} d\xi \quad \forall(k,m) \geq 0, \ \forall t \in [\varepsilon, c) \tag{13.48}$$

converge uniformly and absolutely for any $\varepsilon > 0$. Hence $u_1(x,t)$ may be differentiated arbitrarily many times under the integral sign with respect to x and t, from which it follows in particular that

$$L_a u_1(x,t) = \int_{-\infty}^{-2N} f(\xi) L_a E(x - \xi, a^2 t) d\xi = 0 \tag{13.49}$$

Clearly, the same is true for $u_3(x,t)$. Concerning $u_2(x,t)$, it is enough to point out that $u_2(x,t)$ is an integral with no singularities for any $t \geq \varepsilon > 0$, so that u_2 may also be differentiated arbitrarily many times under the integral sign. Thus

$$L_a u(x,t) \equiv 0 \quad \forall x, \ \forall t \in (0 < \varepsilon \leq t < c) \tag{13.50}$$

The restriction $t < c$ may clearly be omitted. Indeed, take

$$t_0 = c - \varepsilon > c/2 \tag{13.51}$$

and consider t_0 as a new initial time. Then

$$u(x,t) = \int_{-\infty}^{\infty} u(\xi, t_0) E(x - \xi, t - t_0) d\xi \tag{13.52}$$

where

$$|u(x,t_0)| < Ae^{\alpha x^2} \tag{13.53}$$

Hence we see that

$$L_a(x,t) = 0 \quad \forall x, \ \forall t \in (0, 2c - \varepsilon) \tag{13.54}$$

Now take the sequence

$$\varepsilon_0 = \varepsilon, \quad \varepsilon_n = \frac{\varepsilon_{n-1}}{2}, \quad n = 1, 2, \ldots \tag{13.55}$$

We have

$$\lim_{n\uparrow\infty} \left(nc - \sum_{m=1}^{n} \varepsilon_{m-1} \right) = \lim_{n\uparrow\infty} \left(nc - \sum_{m=1}^{n} \varepsilon \cdot 2^{-m} \right) > \lim_{n\uparrow\infty} nc - \varepsilon = \infty \tag{13.56}$$

Hence, repeating the previous argument infinitely often, we have

$$L_a u(x,t) = 0 \quad \forall x, \ \forall t > 0 \tag{13.57}$$

This proves the first part of the theorem. The second part is a corollary of the following lemma.

Lemma 13.2.1. *Let $f(x)$ be continuous in the interval $[\alpha, \beta]$, and let*

$$J_{\alpha,\beta} = \int_{\alpha}^{\beta} f(\xi) E(x - \xi, a^2 t) d\xi \tag{13.58}$$

Then

$$\lim_{t\downarrow 0} J_{\alpha\beta}(x,t) = \begin{cases} f(\beta)/2 & \text{if } x = \beta \\ f(x) & \text{if } x \in (\alpha, \beta) \\ f(\alpha)/2 & \text{if } x = \alpha \\ 0 & \text{if } x \notin [\alpha, \beta] \end{cases} \tag{13.59}$$

Indeed, suppose first that $x \notin [\alpha, \beta]$. Then $E(x - \xi, a^2 t)$ has no singularities in the integration interval, so that indeed

$$\lim_{t\downarrow 0} J_{\alpha\beta}(x,t) = 0 \quad \text{if } x \notin [\alpha, \beta] \tag{13.60}$$

Now let $x = \alpha$. By assumption, $f(x)$ is continuous from the right at $x = \alpha$. Hence, for any $\varepsilon > 0$ there exists $\rho > 0$ such that

$$|f(x) - f(\alpha)| < \varepsilon \quad \text{if } \alpha < x < \alpha + \rho < \beta \tag{13.61}$$

Fix some such ρ. Then

$$J_{\alpha\beta}(\alpha,t) = J^* + J^{**} = \int_{\alpha}^{\alpha+\rho} + \int_{\alpha+\rho}^{\beta} f(\xi) E(\alpha - \xi, a^2 t) d\xi \tag{13.62}$$

Note that by (13.60),

$$\lim_{t\downarrow 0} J^{**}(\alpha, t) = 0 \tag{13.63}$$

Using the mean value theorem, we now see that

$$J^*(\alpha, t) = f(s) \int_\alpha^{\alpha+\rho} E(\alpha - \xi, a^2 t) d\xi \equiv \frac{f(s)}{\pi^{1/2}} \int_0^{\rho/(2a\sqrt{t})} e^{-z^2} dz \qquad (13.64)$$

Hence

$$\overline{\lim_{t \downarrow 0}} J^*(\alpha, t) = f(\tilde{s})/2, \quad \tilde{s} \in (\alpha, \alpha + \rho) \qquad (13.65)$$

Together with (13.61) this yields

$$\overline{\lim_{t \downarrow 0}} J^*(\alpha, t) < f(\alpha)/2 + \varepsilon \qquad (13.66)$$

Quite similarly,

$$\lim_{t \downarrow 0} J^*(\alpha, t) > f(\alpha)/2 - \varepsilon \qquad (13.67)$$

Since $\varepsilon > 0$ is arbitrarily small, this, together with (13.62) and (13.63), yields

$$\varliminf_{t \downarrow 0} J_{\alpha\beta}(\alpha, t) = \varlimsup_{t \downarrow 0} J_{\alpha\beta}(\alpha, t) = \lim_{t \downarrow 0} J_{\alpha\beta}(\alpha, t) = f(\alpha)/2 \qquad (13.68)$$

Quite similarly, one proves that

$$\lim_{t \downarrow 0} J_{\alpha\beta}(\beta, t) = f(\beta)/2 \qquad (13.69)$$

Using (13.68) and (13.69), we also conclude that

$$\lim_{t \downarrow 0} J_{\alpha\beta}(x, t) = {}^1\!/\!{}_2[f(x - 0) + f(x + 0)] \qquad (13.70)$$

Q.E.D.

Theorem 13.2.1 is an obvious corollary of Lemma 13.2.1 and the absolute and uniform convergence of the integrals $u_1(x, t)$ and $u_3(x, t)$. Indeed, N may be chosen so large that

$$|u_1(x, t)| + |u_3(x, t)| < \varepsilon \qquad (13.71)$$

so that by Lemma 13.2.1 and (13.43),

$${}^1\!/\!{}_2[f(x - 0) + f(x + 0)] \quad c < \varliminf_{t \downarrow 0} u(x, t) < \varlimsup_{t \downarrow 0} u(x, t)$$

$$< {}^1\!/\!{}_2[f(x - 0) + f(x + 0)] + \varepsilon \qquad (13.72)$$

Since $\varepsilon > 0$ is arbitrarily small, this means that

$$\lim_{t \downarrow 0} u(x, t) = {}^1\!/\!{}_2[f(x - 0) + f(x + 0)] \qquad (13.73)$$

Q.E.D.

Theorem 13.2.2. *If there exist $A = \text{const} > 0$ and $\alpha > 0$ such that*

$$|f(x)| < A e^{\alpha x^2} \quad \forall x \qquad (13.74)$$

then the solution of the Cauchy problem is unique. Otherwise, that is, if there exist positive constants σ, α, A such that

$$|f(x)| > A e^{\alpha x^{2+\sigma}} \quad \forall x \qquad (13.75)$$

the solution of the problem may not be unique.

For a proof of this theorem the reader is referred to the special literature.[4]

The Poisson integral easily yields a physical interpretation of the fundamental solution $E(x - \xi, a^2 t)$ of the heat-conduction equation. Indeed, let us return to our example of an infinitely long thermally conducting slab that is perfectly thermally insulated along its lateral boundary. Assume that the initial temperature $f(x)$ is a positive continuous function that vanishes everywhere except for a small interval $(x_0 - \delta, x_0 + \delta)$. Hence the temperature distribution of the slab at $t > 0$ is defined by the integral

$$u(x,t) = \int_{x_0 - \delta}^{x_0 + \delta} f(\xi) E(x - \xi, a^2 t) d\xi \qquad (13.76)$$

Let $c = \text{const}$ be the specific heat capacity and $\rho = \text{const}$ the linear density of the material of the slab. Then the amount of heat necessary to heat the slab to temperature $f(x)$ is

$$Q = c\rho \int_{-\infty}^{\infty} f(\xi) d\xi = c\rho \int_{x_0 - \delta}^{x_0 + \delta} f(\xi) d\xi \qquad (13.77)$$

Using the mean value theorem and taking into account that $f(x)$ is positive, we obtain from (13.76) that

$$u(x,t) = E(x - \tilde{x}, a^2 t) \int_{x_0 - \delta}^{x_0 + \delta} f(\xi) d\xi = \frac{Q}{c\rho} E(x - \tilde{x}, a^2 t) \qquad (13.78)$$

where $\tilde{x}$ is some point of the interval $(x_0 - \delta, x_0 + \delta)$.

Assume now that

$$Q = c\rho \qquad (13.79)$$

and that Q remains unchanged as $\delta \downarrow 0$. Then one says that an instantaneous unit source of heat is placed at time $t = 0$ at the point $x = x_0$ of the slab. Thus, if an instantaneous unit source of heat were placed at time $t = 0$ at $x = x_0$, then by (13.78)

$$u(x,t) = E(x - x_0, a^2 t) \qquad (13.80)$$

---

[4]The problem of uniqueness of the solution to the Cauchy problem for the heat-conduction equation was first studied by A. N. Tichonov [79], who proved Theorem 13.2.2 in 1935 and provided an example of nonuniqueness. His result is reproduced in many modern books (e.g., in [9; 62; 91]). Note that if $|f(x)| \downarrow 0$ as $|x| \uparrow \infty$, the uniqueness of the solution of this Cauchy problem is a simple corollary of the maximum principle (see Chapter 14); but if there exist positive constants A, α such that $|f(x)| < A \exp(\alpha |x|)$ for all x, uniqueness follows from the uniqueness of the Fourier integral representation of a function that has exponential growth at infinity (see Appendix 3).

In other words, the fundamental solution of the 1-dimensional heat-conduction equation $E[x - \xi, a^2(t - \tau)]$ is the temperature at time t at the point x of the slab (with thermal diffusivity coefficient $a^2 = \text{const}$), with zero temperature at times $t < \tau$ and with an instantaneous unit source of heat switched on at time τ at the point ξ. The fundamental solution $E[x - \xi, a^2(t - \tau)]$ is often referred to as the *influence function* of an instantaneous source of heat placed at the point ξ at time τ.

Note now that

$$E[x - \xi, a^2(t - \tau)] > 0 \quad \forall x, \ \forall (t - \tau) > 0 \qquad (13.81)$$

This means that the rate of redistribution of the temperature disturbance is infinite. This paradoxical result is obviously a consequence of the Fourier law

$$q = -k\frac{\partial}{\partial x}u \qquad (13.82)$$

determining the heat flux. The Fourier equation (13.1) (heat-conduction equation) is a corollary of this law, and so are all the properties of its solutions.

The physical source of this paradox is the fact that the Fourier law neglects the finiteness of the local relaxation time. Recall that relaxation is the process whereby a system returns to the state of equilibrium after all external disturbing factors have been removed. When dealing with heat conduction or diffusion we must distinguish between two relaxation time scales – short and long, corresponding to fast and slow relaxation processes. Fast relaxation occurs in volumes whose linear dimensions are comparable with the free path of molecules in gases (or the amplitudes of molecular vibrations in solids), whereas slow relaxation occurs in volumes of macroscopic dimensions. *A macroscopic system needs the fast relaxation time to arrive at local equilibrium.* In statistical–mechanical terms this important notion represents the process whereby the molecular velocity distribution acquires its local equilibrium Maxwellian form.[5] Essentially, this means that the microscopic molecular components of the system "forget" their inertia; simultaneously, the local intensive characteristics of the system (such as temperature or equilibrium pressure, related to density via the equation of state) become significant.

An example of a fast relaxation process in a macroscopic system whose microscopic components are themselves macroscopic objects is provided by Brownian particles in a viscous fluid that forget their inertia (see

---

[5]This means that the probability $dW(\mathbf{v})$ of the velocity of a macroscopic particle p belonging (in the state of statistical equilibrium) to the interval $(\mathbf{v}, \mathbf{v} + d\mathbf{v})$ is $dW(v) = (m/2\pi kT)^{3/2} \exp[-mv^2/2kT]$, where m is the mass of the particle, T the temperature in degrees Kelvin K, and k is the Boltzmann constant $k = 1.38 \times 10^{-23}$ JK^{-1} (J denotes Joules). Other physical facts and objects cited in this section were introduced in Chapters 1 and 2.

Chapter 2, Section 9). Consider again a collection of identical Brownian particles, heavier than the surrounding fluid (i.e., of higher density) settling under gravity. Recall that the time necessary for these particles to forget their initial velocity and reach their equilibrium Stokes settling velocity $v_e = {}^2/_9 \cdot (ar^2 g/\eta)(\rho_p - \rho_0)$ is approximately $\tau = (m/6\pi\eta r)$ where η is the dynamic viscosity of the fluid, r the radius of a particle, ρ_p the density of a particle, ρ_0 the density of the fluid, g the acceleration of free fall, and m the mass of the particle. Accordingly, only after a time $\approx \tau$ after the beginning of the evolution will the spatial–temporal distribution of a system of Brownian particles be governed by the diffusion equation in the gravity force field (an analog of the Fourier equation (13.1) for heat conduction), instead of the more complicated Fokker–Planck equation, which describes the redistribution of particles in the 6-dimensional velocity coordinate space and is valid for times $< \tau$.

To summarize, assuming an instantaneous local (fast) relaxation implies that one is neglecting microscopic inertia. By doing so, one naturally permits the system to have infinite acceleration, resulting in the aforementioned infinite rate of signal propagation in the solution of the Fourier equation. We reiterate that the Fourier equation views the fast relaxation process as instantaneous $(\tau = 0)$, and is therefore fit only for the treatment of sufficiently slow (macroscopic) relaxation processes. Moreover, some caution must be used regarding physical interpretation of the results based on equation (13.1) when considering the initial stage of a fast dissipation process, at which time one expects inertia to be significant.

As a purely phenomenological way of generalizing the Fourier law (13.82) to account for fast relaxation, one might replace it by the following hyperbolic expression for the heat flux q (the use of the term "hyperbolic" will be explained shortly):

$$q = -k\frac{\partial}{\partial x}u - \tau\frac{\partial}{\partial t}q, \quad k, \tau = \text{const} > 0 \tag{13.83}$$

Indeed, replacing equation (13.82) by equation (13.83) amounts to replacing the parabolic heat-conduction equation (13.1) by a hyperbolic equation. (Compare with Chapter 2: the hyperbolic diffusion equation (2.240), equation (2.137), and Problem 2.7.1.) Thus, consider the 1-dimensional process of heat transfer in direction x. Recall that the local energy conservation law reads

$$c\rho\frac{\partial}{\partial t}u = -\frac{\partial}{\partial x}q \tag{13.84}$$

On the other hand, differentiation of (13.83) with respect to x yields

$$\frac{\partial}{\partial x}q = -k\frac{\partial^2}{\partial x^2}u - \tau\frac{\partial^2}{\partial x\partial t}q \tag{13.85}$$

whereas by differentiating (13.84) with respect to t we obtain

$$c\rho\frac{\partial^2}{\partial t^2}u = -\frac{\partial^2}{\partial x\partial t}q \tag{13.86}$$

Substituting (13.84) and (13.86) into (13.85), we obtain the following hyperbolic equation for $u(x,t)$:

$$a^2 \frac{\partial^2}{\partial x^2} u = \tau \frac{\partial^2}{\partial t^2} u + \frac{\partial}{\partial t} u \qquad (13.87)$$

where as usual $a^2 = k/c\rho$.

Thus, inclusion of the finite fast relaxation time τ in the phenomeno-logical law for heat flux is indeed equivalent to replacing the parabolic heat-conduction equation (13.1) by the hyperbolic equation (13.87) with $a\tau^{-1/2}$ as the rate of signal propagation.

In rarefied gases, the value of τ may occasionally be sufficiently large to warrant the consideration of equation (13.87) rather than equation (13.1). In solids, however, τ is of the order of 10^{-11}–10^{-12} sec, so that in most macroscopic, slow-transport situations it is not expedient to replace equation (13.1) by equation (13.87).[6]

3. Moments of solution of Cauchy problem.[7] Asymptotic behavior of the Poisson integral as $t \uparrow \infty$

Let us begin with a consideration of the moments of the solution to the Cauchy problem studied in the previous section. An nth moment of the solution $u(x,t)$ to the Cauchy problem is defined by

$$M_n(t) = \int_{-\infty}^{\infty} x^n u(x,t)\,dx, \quad n = 1, 2, 3, \ldots \qquad (13.88)$$

Theorem 13.3.1. *Let $f(x)$ be a function that vanishes at infinity at a rate that is at least exponential, so that*

$$M_n(0) = \int_{-\infty}^{\infty} |x^n f(x)|\,dx \text{ is convergent for all } n \geq 0 \qquad (13.89)$$

Then the moments $M_n(t)$ satisfy the recurrence relations

$$M_0(t) \equiv M_0(0), \quad M_1(t) \equiv M_1(0), \qquad (13.90)$$

$$\frac{d}{dt} M_n(t) = n(n-1) M_{n-2}(t) \quad \forall n \geq 2 \qquad (13.91)$$

PROOF. We may assume without loss of generality, via appropriate scaling, that the coefficient of thermal diffusivity is unity ($a^2 = 1$). Then by (13.37)

---

[6]The asymptotic transition from the hyperbolic heat-conduction equation to the parabolic equation as $\tau \downarrow 0$ is treated in Chapter 21, Section 1.

[7]We shall consider the 1-dimensional case only. It can easily be shown that all properties of the solution of the Cauchy problem considered in this section remain valid in the multidimensional case as well. We shall not repeat the relevant arguments in our discussion of the multidimensional Cauchy problem in Section 5.

and (13.88),

$$M_0(t) = \int_{-\infty}^{\infty} dx \int_{-\infty}^{\infty} f(\xi)E(x - \xi, t)d\xi \tag{13.92}$$

Since this integral is uniformly and absolutely convergent for any x and any $t > 0$, as is the integral

$$\int_{-\infty}^{\infty} f(\xi)d\xi \int_{-\infty}^{\infty} \frac{\partial}{\partial t} E(x - \xi, t)dx \tag{13.93}$$

for any positive t separated from zero, the order of differentiation and integration may be changed. Thus

$$\frac{d}{dt}M_0(t) = \int_{-\infty}^{\infty} dx \int_{-\infty}^{\infty} f(\xi)\frac{\partial}{\partial t}E(x - \xi, t)d\xi$$

$$\equiv \int_{-\infty}^{\infty} dx \int_{-\infty}^{\infty} f(\xi)\frac{\partial^2}{\partial x^2}E(x - \xi, t)d\xi \tag{13.94}$$

Changing the order of integration, we obtain

$$\frac{d}{dt}M_0(t) = \int_{-\infty}^{\infty} f(\xi)d\xi \int_{-\infty}^{\infty} \frac{\partial^2}{\partial x^2}E(x - \xi, t)dx \equiv 0 \tag{13.95}$$

Similarly

$$\frac{d}{dt}M_1(t) \equiv 0 \tag{13.96}$$

Furthermore, we have

$$\frac{d}{dt}M_n(t) = \frac{d}{dt}\int_{-\infty}^{\infty} x^n dx \int_{-\infty}^{\infty} f(\xi)E(x - \xi, t)d\xi \tag{13.97}$$

Changing the order of integration, we obtain

$$\frac{d}{dt}M_n(t) = \int_{-\infty}^{\infty} f(\xi)d\xi \int_{-\infty}^{\infty} x^n \frac{\partial^2}{\partial x^2}E(x - \xi, t)dx \tag{13.98}$$

Finally, integrating by parts, we infer that

$$\frac{d}{dt}M_n(t) = n(n - 1)\int_{-\infty}^{\infty} x^{n-2}dx \int_{-\infty}^{\infty} f(\xi)E(x - \xi, t)d\xi$$

$$= n(n - 1)M_{n-2}(t) \tag{13.99}$$

Q.E.D.

Remark 1. If the conditions of Theorem 13.3.1 are satisfied, so that the moments M_0 and M_1 are finite, then by appropriate scaling of x and t and a suitable displacement of the origin $x = 0$, we may assume without loss of generality that

$$M_0 = 1, \qquad M_1 = 0 \tag{13.100}$$

To discuss the physical contents of this theorem it is somewhat more convenient to use the diffusional interpretation. Indeed, the equality

$M_0(t) = $ const merely states that, lacking a source of mass, the total mass of particles in an infinitely large domain is conserved (remains constant over time). Hence it is immediate that the physical meaning of the second equality in (13.90) is: During the diffusion process the position of the center of mass of the system remains unchanged over time.

Now choose an arbitrary $b = $ const, and consider the moment of inertia of the system about the point $x = b$ as defined by the integral

$$I(b,t) = \int_{-\infty}^{\infty} (x-b)^2 u(x,t)dx \qquad (13.101)$$

Equations (13.90), (13.92), and (13.96) yield

$$\frac{\partial}{\partial t} I(b,t) = \frac{d}{dt} M_2(t) - \frac{d}{dt}[2bM_1(t) - b^2 M_0(t)] = 2M_0(t) \equiv 2M(0) \quad (13.102)$$

Thus the moment of inertia of a system of particles diffusing on a line increases linearly over time. Note that

$$\min_b \int_{-\infty}^{\infty} (x-b)^2 u(x,t)dx = \int_{-\infty}^{\infty} \left(x - \frac{M_1(t)}{M_0(t)}\right)^2 u(x,t)dx$$

$$= \int_{-\infty}^{\infty} \left(x - \frac{M_1(0)}{M_0(0)}\right)^2 u(x,t)dx \qquad (13.103)$$

In other words, at any instant of time the minimum moment of inertia is the moment of inertia about the immobile center of mass of the system; this is exactly the definition of the principal momentum of inertia in mechanics.

We shall now study the asymptotic behavior of $u(x,t)$ for $t \uparrow \infty$.

Theorem 13.3.2. *Let $u(x,t)$ be a solution of the Cauchy problem given by (13.37) with $f(x)\exp(|x|)$ bounded (positive). Then, asymptotically as $t \uparrow \infty$,*

$$u(x,t) \asymp E(x,t)\left(M_0 + \frac{xM_1}{2t} - \frac{M_2(0)}{4t}\right.$$

$$+ \frac{1}{2!}(4t)^{-2}[4x^2 M_2(0) - 4xM_3(0) + M_4(0)]$$

$$+ \frac{1}{3!}(4t)^{-3}[8x^3 M_3(0) - 12x^2 M_4(0) + 6xM_5(0) - M_6(0)]$$

$$\left. + O(t^{-4})\right) \qquad (13.104)$$

Remark 2. The assumption that $f(x)$ is positive may occasionally be required to accord with the physical content of the Cauchy problem. Indeed, in the diffusion interpretation, the concentration of particles cannot be negative; the same is true of the temperature measured on the absolute (Kelvin) scale in the thermal interpretation.

PROOF. By equations (13.52) and (13.20),

$$u(x,t) = \int_{-\infty}^{\infty} f(\xi) \frac{\exp(-(x-\xi)^2/4t)}{2(\pi t)^{1/2}} d\xi \qquad (13.105)$$

so that

$$u(x,t)\exp\left(\frac{x^2}{4t}\right) = \int_{-\infty}^{\infty} f(\xi)\frac{\exp(-(\xi^2 - 2x\xi)/4t)}{2(\pi t)^{1/2}}d\xi \qquad (13.106)$$

Expanding the exponential function in the integrand in Taylor series, we obtain

$$u(x,t)\exp\left(\frac{(x)^2}{4t}\right)2(\pi t)^{1/2} = \int_{-\infty}^{\infty} f(\xi)\left(1 + \left(\frac{2x\xi - \xi^2}{4t}\right) + \frac{1}{2}\left(\frac{2x\xi - \xi^2}{4t}\right)^2\right.$$

$$\left. + \frac{1}{3!}\left(\frac{2x\xi - \xi^2}{4t}\right)^3 + \cdots\right)d\xi \qquad (13.107)$$

which may be rewritten as

$$u(x,t)\exp\left(\frac{(x)^2}{4t}\right)2(\pi t)^{1/2}$$

$$= \left(M_0 + \frac{x \cdot M_1}{2t} - \frac{M_2(0)}{4t}\right.$$

$$+ \frac{1}{2!}(4t)^{-2}[4x^2 M_2(0) - 4x M_3(0) + M_4(0)]$$

$$+ \frac{1}{3!}(4t)^{-3}[8x^3 M_3(0) - 12x^2 M_4(0) + 6x M_5(0) - M_6(0)]$$

$$\left. + O(t^{-4})\right) \qquad (13.108)$$

This proves (13.104). **Q.E.D.**

4. Prigogine principle, Glansdorf–Prigogine criterion, and solution of Cauchy problem for heat-conduction equation [26]

We recall the *Dirichlet Principle*: Let D be a bounded region in $\mathbb{R}_3$. The solution of the Laplace equation

$$\Delta u(p) = 0, \quad p \in D \qquad (13.109)$$

with appropriate (linear) boundary conditions, minimizes the Dirichlet integral

$$I(u) = \int\int_D\int \text{grad}^2 u(q)d\omega_q \qquad (13.110)$$

over the class of "admissible" functions satisfying the given boundary conditions (see Chapter 9, note 5). For diffusion or thermally conductive systems with small variation of concentration or temperature, the Dirichlet integral determines the rate of production of entropy due to diffusion (either mass or thermal) in the open domain D. The physical counterpart of the Dirichlet principle is the *Prigogine principle* of minimal production of entropy: The rate of production of entropy in a linear diffusion system in the steady state is minimal over the entire class of states compatible with the given boundary conditions.

The heat-conduction equation is not the Euler–Lagrange equation for any functional subject to minimization, nor (in particular) for the Dirichlet integral. Hence the nonstationary solutions of the heat equation do not minimize this integral. At any fixed time, however, the solutions satisfy the boundary conditions and thus belong to the class of admissible functions over which the minimum of the Dirichlet integral may be sought. This in turn suggests another physical principle, known as the *Glansdorf–Prigogine evolution criterion*: The rate of entropy production by a linear diffusion system decreases monotonically during the evolution of the system.

Let us trace the origin of the Glansdorf–Prigogine evolution criterion in terms of the Dirichlet integral. Consider a diffusion system governed by the heat-conduction equation:

$$\Delta u(p,t) = \frac{\partial}{\partial t} u(p,t) \quad \forall p \in D, \ \forall t > 0 \tag{13.111}$$

with boundary condition

$$\frac{\partial}{\partial n} u(p,t) = 0 \quad \vee \quad u(p,t) = f(p) \quad \forall p \in \partial D \tag{13.112}$$

Equation (13.111) yields

$$\int_D \frac{\partial}{\partial t} u(q,t) \Delta u(q,t) d\omega_q = \int_D [\Delta u(q,t)]^2 d\omega_q \tag{13.113}$$

We have

$$\int_D \frac{\partial}{\partial t} u(q,t) \Delta u(q,t) d\omega_q = \int_D \frac{\partial}{\partial t} u(q,t) \operatorname{div}[\operatorname{grad} u(q,t)] d\omega_q$$

$$\equiv \int_D \operatorname{div} \left[\frac{\partial}{\partial t} u(q,t) \operatorname{grad} u(q,t) \right] d\omega_q$$

$$- \frac{1}{2} \frac{\partial}{\partial t} \int_D [\operatorname{grad} u(q,t)]^2 d\omega_q \tag{13.114}$$

By the Gauss divergence theorem

$$\int_D \operatorname{div} \left[\frac{\partial}{\partial t} u(q,t) \operatorname{grad} u(q,t) \right] d\omega_q = \int_{\partial D} \frac{\partial}{\partial t} u(q,t) \frac{\partial}{\partial n_q} u(q,t) d\sigma_q \tag{13.115}$$

so that, in view of the boundary condition (13.112),

$$\int_D \operatorname{div} \left[\frac{\partial}{\partial t} u(q,t) \operatorname{grad} u(q,t) \right] d\omega_q \equiv 0 \tag{13.116}$$

Hence (13.113) and (13.114) imply

$$\frac{\partial}{\partial t} \int_D [\operatorname{grad} u(q,t)]^2 d\omega_q = -2 \int_D [\Delta u(q,t)]^2 d\omega_q < 0 \tag{13.117}$$

The left-hand side of (13.117) is the time derivative of the Dirichlet integral, that is, the rate of bulk entropy production in the system. Thus the rate of bulk entropy production in a linear, diffusion system, with time-independent boundary condition of the first kind or a homogeneous condition of the second kind, decreases monotonically during the evolution of the system until it reaches a minimum in equilibrium or in steady state.

The following remark is of a terminological nature. Consider the ordinary differential equation

$$\frac{dx}{dt} = f(x,t) \tag{13.118}$$

Let $x(t\,|\,x_0)$ be the trajectory passing through the point $x(0) = x_0$. Then if

$$\lim_{t\uparrow\infty} x(t\,|\,x_0) = x^* \tag{13.119}$$

the point x^* termed the *attractor* of the trajectory $x(t\,|\,x_0)$. This terminology may be applied to any evolutionary system. Thus the equilibrium or steady state of a linear diffusion system of finite size approached in time by the solution of a nonstationary problem with the same time-independent boundary conditions is the attractor for the latter. By the Prigogine principle and the Glansdorf–Prigogine evolution criterion, this attractor is characterized by a value for the rate of entropy production which is minimal over all admissible states and toward which the appropriate nonstationary rate decreases monotonically throughout the evolution of the system. The question naturally arises as to whether this also applies to nonstationary attractors, such as the fundamental or similarity solution. (It has been shown in the previous section that the fundamental solution of the heat-conduction equation is the attractor for the solution of any Cauchy problem with the initial distribution satisfying the conditions of Theorem 13.3.1. A similar result for the similarity solution (13.14) will be proved below (see Theorem 13.4.2).) More precisely, does the excess rate of entropy production of the solution of the Cauchy problem, compared with that of its attractor (i.e., of the corresponding similarity solution), decrease monotonically as $t\uparrow\infty$? Some answers to this question follow from the long-time asymptotic formulas of the previous section and are provided by the following two theorems.

Theorem 13.4.1. *Let $u(x,t)$ be a solution of the Cauchy problem*

$$\frac{\partial^2}{\partial x^2}u = \frac{\partial}{\partial t}u, \quad u(x,0) = f(x), \quad -\infty < x < \infty, \ t > 0 \tag{13.120}$$

with $f(x)$ satisfying the conditions of Theorem 13.3.1. Then the excess rate of entropy production by $u(x,t)$, as compared with the attractor $E(x,t)$, increases monotonically as $t\uparrow\infty$ for $M_2(0) > 0$ and decreases monotonically for $M_2(0) < 0$.

PROOF. In view of Remark 1 following the proof of Theorem 13.3.1, we may assume that

$$M_0 = 1, \quad M_1 = 0 \tag{13.121}$$

Then (13.104) becomes

$$u(x,t) \asymp E(x,t)\left(1 - \frac{M_2(0)}{4t}\right.$$

$$+ \frac{1}{2!} \cdot (4t)^{-2}[4x^2 M_2(0) - 4x M_3(0) + M_4(0)]$$

$$+ \frac{1}{3!}(4t)^{-3}[8x^3 M_3(0) - 12x^2 M_4(0) + 6x M_5(0) - M_6(0)]$$

$$\left. + O(t^{-4})\right) \tag{13.122}$$

Let us compute the excess rate of entropy production by $u(x,t)$ compared with its attractor $E(x,t)$. Denote the appropriate rates by

$$I(t \mid u) \quad \text{and} \quad I(t \mid E) \tag{13.123}$$

and the excess of $I(t \mid u)$ over $I(t \mid E)$ by

$$\delta I(t \mid u, E) = I(t \mid u) - I(t \mid E) \tag{13.124}$$

We find

$$I(t \mid u) = \int_{-\infty}^{\infty}\left(\frac{\partial}{\partial x}u(x,t)\right)^2 dx, \qquad I(t \mid E) = \int_{-\infty}^{\infty}\left(\frac{\partial}{\partial x}E(x,t)\right)^2 dx \tag{13.125}$$

Integration by parts yields

$$I(t \mid u) = -\int_{-\infty}^{\infty} u(x,t)\frac{\partial^2}{\partial x^2}u(x,t)dx$$

$$= -\int_{-\infty}^{\infty} u(x,t)\frac{\partial}{\partial t}u(x,t)dx = -\frac{1}{2}\frac{\partial}{\partial t}\int_{-\infty}^{\infty} u^2(x,t)dx \tag{13.126}$$

Similarly

$$I(t \mid E) = -\frac{1}{2}\frac{\partial}{\partial t}\int_{-\infty}^{\infty} E^2(x,t)dx \tag{13.127}$$

Using (13.122) and definition (13.124), we obtain

$$\delta I(t \mid u, E) = -\frac{1}{2}\frac{\partial}{\partial t}\int_{-\infty}^{\infty} E^2(x,t)\left\{1 - \left(1 - M_2(0)(4t)^{-1}\right.\right.$$

$$+ \frac{1}{2!}(4t)^{-2}[4x^2 M_2(0) - 4x M_3(0) + M_4(0)]$$

$$+ \frac{1}{3!}(4t)^{-3}[8x^3 M_3(0) - 12x^2 M_4(0) + 6x M_5(0) - M_6(0)]$$

$$\left.\left. + O(t^{-4})\right)^2\right\}dx \tag{13.128}$$

Denote

$$z = \frac{x}{2\sqrt{t}} \tag{13.129}$$

so that

$$E^2(x,t) = \exp(-2z^2)/(4\pi t) \qquad (13.130)$$

and

$$\delta I(t \mid u, E) = \frac{\partial}{\partial t} \int_{-\infty}^{\infty} E^2(z,t) \Big\{ 1 - \Big(1 - M_2(0)(4t)^{-1} $$

$$+ \frac{1}{2!}[z^2 M_2(0)t^{-1} - (zM_3(0)t^{-3/2}/2 + M_4(0)(4t)^{-2})]$$

$$+ \frac{1}{3!}[z^3 M_3(0)t^{-3/2} - \frac{3}{4}z^2 M_4(0)t^{-2} + \frac{3}{16}zM_5(0)t^{-3} - M_6(0)(4t)^{-3}]$$

$$+ O(t^{-4}) \Big)^2 \Big\} t^{1/2} dz \qquad (13.131)$$

Taking into account that all the integrals proportional to the odd moments vanish, we obtain[8]

$$\delta I(t \mid u, E) = -\frac{3M_2(0)}{2^{9/2}\pi^{1/2}} t^{-5/2} + o(t^{-5/2}) \qquad (13.132)$$

Thus

$$\operatorname{sign} \delta I(t \mid u, E) = -\operatorname{sign} M_2(0), \qquad \lim_{t \uparrow \infty} \delta I(t \mid u, E) = 0 \qquad (13.133)$$

Q.E.D.

Theorem 13.4.2. *Let $u(x,t)$ be a solution of the Cauchy problem*

$$\frac{\partial^2}{\partial x^2} u(x,t) = \frac{\partial}{\partial t} u(x,t), \qquad u(x,0) = f(x) \qquad (13.134)$$

where $f(x)$ is a nonintegrable (infinite mass) function such that

$$f(x) = \eta(x) + \varphi(x) \qquad (13.135)$$

where $\eta(x)$ is the Heaviside unit function and $\varphi(x)$ satisfies the conditions of Theorem 13.3.1, so that $\varphi(x)e^{|x|}$ is bounded at infinity. Then the rate of production of entropy by the attractor of $u(x,t)$ is minimal (compared to that by $u(x,t)$ itself) and is being approached monotonically as $t \uparrow \infty$.

PROOF. Let M_k, $k = 0, 1, 2, \ldots$, be the moments of $\varphi(x)$. Without loss of generality we may assume that the center of mass of $\varphi(x)$ is at the origin $x = 0$, so that

$$M_1 = 0 \qquad (13.136)$$

We have

$$u(x,t) = v(z) + w(x,t) \qquad (13.137)$$

---

[8]In general, it is not legitimate to differentiate asymptotic equalities. However, differentiation is justified in our case (see Chapter 21, Section 2)

where

$$z = \frac{x}{2\sqrt{t}}, \quad v(z) = \pi^{-1/2} \int_{-\infty}^{z} e^{-s^2} ds, \quad w(x,t) = \int_{-\infty}^{\infty} \varphi(\xi) E(x-\xi,t) d\xi$$

$$(13.138)$$

so that, asymptotically as $t \uparrow \infty$,

$$w(x,t) \asymp E(x,t) \left(M_0 - \frac{M_2(0)}{4t} + O(t^{-2}) \right) \qquad (13.139)$$

Introducing z and t as independent variables, using the notation

$$W(z,t) = w(x,t) \qquad (13.140)$$

and noting that

$$E(x,t) = \frac{dv}{dz} \Big/ 2\sqrt{t} \qquad (13.141)$$

we obtain

$$W(z,t) = \frac{1}{2\sqrt{t}} \frac{dv}{dz} \left(M_0 - \frac{M_2(0)}{4t} + O(t^{-2}) \right) \qquad (13.142)$$

Now $v(z)$ is the attractor of $u(x,t)$. Indeed,

$$\lim_{t \uparrow \infty} v(z) = 1 \wedge E(x,t) = O(t^{-1/2}) \quad \text{for } t \uparrow \infty$$

$$\Rightarrow u(x,t) = v(z) + O(t^{-1/2}) \quad \text{for } t \uparrow \infty \qquad (13.143)$$

Hence the excess rate of production of entropy by u as compared to that by its attractor v is equal to

$$\delta I(t \mid u, v) = \int_{-\infty}^{\infty} \left\{ \left(\frac{\partial}{\partial x} u(x,t) \right)^2 - \left(\frac{\partial}{\partial x} v(z) \right)^2 \right\} dx \qquad (13.144)$$

Repeating the computations in the proof of Theorem 13.4.1, we obtain

$$\delta I(t \mid u, v) = -\frac{1}{2} \frac{\partial}{\partial t} \left(\int_{-\infty}^{\infty} [u(x,t)^2 - v(z)^2] dx \right)$$

$$= -\frac{1}{2} \cdot \frac{\partial}{\partial t} \left(2\sqrt{t} \int_{-\infty}^{\infty} \left\{ 2v(z)W(z,t) + W(z,t)^2 \right\} dz \right) \qquad (13.145)$$

Hence, asymptotically up to the leading term in (13.142),

$$\delta I(t \mid u, v) = -\frac{1}{2} \frac{\partial}{\partial t} \left(2t^{1/2} \int_{-\infty}^{\infty} W(z,t)^2 dz \right)$$

$$= -\frac{M_0^2}{4\pi} \left(\frac{\pi}{2} \right)^{1/2} \frac{\partial}{\partial t} t^{-1/2} = \frac{M_0^2}{8\sqrt{2\pi}} t^{-3/2} > 0 \qquad (13.146)$$

Thus

$$\delta I(t \mid u, v) > 0, \qquad \lim_{t \uparrow \infty} \delta I(t \mid u, v) = 0 \qquad (13.147)$$

Q.E.D.

5. Fundamental solution of multidimensional heat-conduction equation

Prior to constructing the fundamental solution of the multidimensional heat-conduction equation, we offer the following remark. Denote

$$p = (x_1, x_2, \ldots, x_n), \quad q = (\xi_1, \xi_2, \ldots, \xi_n), \quad p, q \in \mathbb{R}_n \tag{13.148}$$

Let $u_i(x_i, t)$ be a solution of the heat-conduction equation in some bounded or unbounded region of variation of x_i:

$$a^2 \frac{\partial^2}{\partial x_i^2} u_i(x_i, t) = \frac{\partial}{\partial t} u_i(x_i, t), \quad \alpha_i < x_i < \beta_i \tag{13.149}$$

and

$$u(p, t) = \prod_{i=1}^{n} u_i(x_i, t) \tag{13.150}$$

Then

$$\Delta u(p, t) = \frac{\partial}{\partial t} u(p, t) \quad \forall p \in D_n \tag{13.151}$$

Here D_n is the topological product of the intervals (α_i, β_i):

$$D_n = \prod_{i=1}^{n} (\alpha_i, \beta_i) \subset \mathbb{R}_n \tag{13.152}$$

Define

$$E_n(p, q, a_2 t) = \prod_{i=1}^{n} E(x_i - \xi_i, a^2 t) \tag{13.153}$$

so that E_n is the solution of equation (13.151) in D_n for all $t > 0$. It is nearly evident that E_n possess the filtering property, so that the following theorem is true.

Theorem 13.5.1. *Let the boundary Σ_n of D_n posses a tangent plane at each point. If $f(p)$ is a function continuous in the closure $\bar{D}_n$ of D_n, then*

$$\lim_{t \downarrow 0} \int_{D_n} f(q) E_n(p, q, t) d\omega_q = \begin{cases} f(p) & \text{if } p \in D_n \\ {}^1/_2 f(p) & \text{if } p \in \Sigma_n = \bar{D}_n \setminus D_n \\ 0 & \text{if } p \notin \bar{D}_n \end{cases} \tag{13.154}$$

PROOF. Denote

$$J_n(p, t) \stackrel{\text{def}}{=} \int_{D_n} f(q) E_n(p, q, t) d\omega_q \tag{13.155}$$

It is obvious that

$$\lim_{t \downarrow 0} J_n(p, t) = 0 \quad \forall p \notin \bar{D} \tag{13.156}$$

since in this case the integrand in (13.155) has no singularities, so that it is legitimate to evaluate the limit under the integral sign.

Now let p be some interior point of D_n, and let Q_p be a cube in D_n with its center at p:

$$Q_p = \prod_{i=1}^{n}(x_i - \alpha_i, x_i + \alpha_i) \subset \bar{Q}_p \subset D_n \qquad (13.157)$$

We have

$$J_n(p,t) = J_n^*(p,t) + J_n^{**}(p,t) \qquad (13.158)$$

where

$$J_n^*(p,t) = \int_{Q_p} f(q)E_n(p,q,t)d\omega_q \qquad (13.159)$$

and

$$J_n^{**}(p,t) = \int_{D_n \backslash Q_p} f(q)E_n(p,q,t)d\omega_q \qquad (13.160)$$

By (13.156),

$$\lim_{t \downarrow 0} J_n^{**}(p,t) = 0 \qquad (13.161)$$

On the other hand, by the mean value theorem and (13.153),

$$J_n^*(p,t) = f(\check{p}) \prod_{i=1}^{n} \int_{x_i - \alpha}^{x_i + \alpha} E(x_i - \xi_i, t)d\xi_i, \quad \check{p} \in Q_p \qquad (13.162)$$

Hence

$$\varliminf_{t \downarrow 0} J_n^*(p,t) = f(\check{p}), \qquad \varlimsup_{t \downarrow 0} J_n^*(p,t) = f(\hat{p}) \qquad (13.163)$$

where $\check{p}$ and $\hat{p}$ are some intermediate points of Q_p. This, together with (13.161) and (13.158), yields

$$\varliminf_{t \downarrow 0} J_n(p,t) = f(\check{p}), \qquad \varlimsup_{t \downarrow 0} J_n(p,t) = f(\hat{p}) \qquad (13.164)$$

Now let $\alpha \downarrow 0$. This yields

$$\lim_{\alpha \downarrow 0} \check{p} = \lim_{\alpha \downarrow 0} \hat{p} = p \qquad (13.165)$$

Since $J_n(p,t)$ is independent of α, it follows from (13.165) and (13.164) that

$$\lim_{t \downarrow 0} J_n(p,t) = f(p) \qquad (13.166)$$

Assume now that

$$p \in \Sigma_n \qquad (13.167)$$

To fix ideas, let $n = 3$. Let $\varepsilon = \text{const} > 0$ be arbitrarily small. Introduce polar coordinates $(\rho, \varphi, \vartheta)$ with origin at p and the z axis directed along the

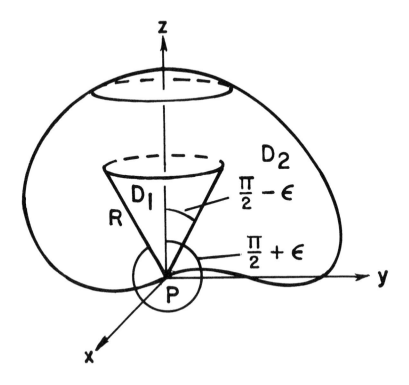

Figure 13.2. Sketch of the proof of a filtering property of Poisson's integral in $\mathcal{E}_{2+1}$.

inward normal to Σ_3 at p (see Figure 13.2). Since by assumption Σ_3 has a tangent plane at p, there exists $R > 0$ so small that

$$\forall q \in \Sigma_3, \quad r_{pq} < R \Rightarrow \frac{\pi}{2} - \varepsilon < \vartheta_q < \frac{\pi}{2} + \varepsilon \qquad (13.168)$$

where ϑ_q is the azimuth of the point q. Let K_p^R be the sphere of radius R with center at p. We have

$$D_3 = D_1 + D_2; \qquad \bar{D}_1 = D_3 \cap K_p^R; \qquad D_2 = \bar{D}_3 \setminus D_1 \qquad (13.169)$$

and

$$J_3(p,t) = J^1(p,t) + J^2(p,t) \qquad (13.170)$$

where

$$J^1(p,t) = \int_{D_1} f(q) E_n(p,q,t)\,d\omega_q, \quad J^2(p,t) = \int_{D_2} f(q) E_n(p,q,t)\,d\omega_q$$

$$(13.171)$$

Since $p \notin D_2$,

$$\lim_{t\downarrow 0} J^2(p,t) = 0 \qquad (13.172)$$

Consider $J^1(p,t)$. By the mean value theorem,

$$J^1(p,t) = f(\tilde{p})I(p,t) \qquad (13.173)$$

where $\tilde{p}$ is some intermediate point of D_1 and

$$I(p,t) = \int_{D_1} E_n(p,q,t)d\omega_q \qquad (13.174)$$

Consider a circular cone C_p with its vertex at p and azimuthal angle $\vartheta_p \in (\pi/2 - \varepsilon, \pi/2 + \varepsilon)$; let

$$K_p^1 = \bar{D}_1 \cap \bar{C}_p; \qquad K_p^2 = D_1 \setminus K_p^1 \qquad (13.175)$$

(see Figure 13.2) and set

$$I^1(p,t) = \int_{K_p^1} E_3(p,q,t)d\omega_q \qquad (13.176)$$

$$I^2(p,t) = \int_{K_p^2} E_3(p,q,t)d\omega_q \qquad (13.177)$$

so that

$$I = I^1 + I^2 \qquad (13.178)$$

Using polar coordinates (r, φ, ψ), we obtain

$$I^1(p,t) = \int_0^{2\pi} d\varphi \int_0^{\pi/2-\varepsilon} \sin(\vartheta)d\vartheta \int_0^R \frac{\exp[-\rho^2/4a^2t]}{8a^3\pi^{3/2}t^{3/2}}\rho^2 d\rho$$

$$= \pi^{-1/2}[1 - \sin(\varepsilon)] \int_0^{R^2/4a^2t} \sqrt{z}e^{-z}dz \qquad (13.179)$$

and consequently

$$\lim_{t\downarrow 0} I^1(p,t) = {}^1\!/{}_2[1 - \sin(\varepsilon)] \qquad (13.180)$$

Further, majorizing I^2, we find

$$I^2(p,t) \le \int_0^{2\pi} d\varphi \int_{\pi/2-\varepsilon}^{\pi/2+\varepsilon} \sin(\vartheta)d\vartheta \int_0^R \frac{\exp[-\rho^2/4a^2t]}{8a^3\pi^{3/2}t^{3/2}}\rho^2 d\rho \qquad (13.181)$$

which yields

$$\overline{\lim_{t\downarrow 0}} I^2(p,t) \le \sin(\varepsilon) \qquad (13.182)$$

Quite similarly,

$$\underline{\lim_{t\downarrow 0}} I^2(p,t) \ge -\sin(\varepsilon) \qquad (13.183)$$

Hence

$$\varlimsup_{t\downarrow 0} I(p,t) = \frac{1}{2}[1 - \sin(\varepsilon)] + O(\varepsilon) \qquad (13.184)$$

and

$$\varliminf_{t\downarrow 0} I(p,t) = \frac{1}{2}[1 - \sin(\varepsilon)] + O(\varepsilon) \qquad (13.185)$$

By (13.170), (13.172), and (13.173), this implies that

$$\varlimsup_{t\downarrow 0} J_3(p,t) = \frac{1}{2} f(\hat{p})[1 - \sin(\varepsilon)] + O(\varepsilon) \qquad (13.186)$$

and

$$\varliminf_{t\downarrow 0} J_3(p,t) = \frac{1}{2} f(\breve{p})[1 - \sin(\varepsilon)] + O(\varepsilon) \qquad (13.187)$$

where

$$\hat{p}, \breve{p} \in D_1 \qquad (13.188)$$

that is,

$$\varepsilon \downarrow 0 \Rightarrow \hat{p}, \breve{p} \to p \qquad (13.189)$$

Since $J_3(p,t)$ is independent of ε, it follows from (13.186) and (13.187) that

$$\lim_{t\downarrow 0} J_3(p,t) = \frac{1}{2} f(p) \quad \text{if } p \in \Sigma_3 \qquad (13.190)$$

Q.E.D.[9]

Let (x,y,z) and (ρ,φ,z) be rectangular cartesian and cylindrical coordinate systems, respectively, so that

$$x = \rho \cos \varphi, \quad y = \rho \sin \varphi \qquad (13.191)$$

Let P and Q be points in $\mathbb{R}_3$ and p and q their projections on the plane $z = 0$; let p_0 and q_0 be the projections of P and Q on the x axis, so that

$$P = (x,y,z), \quad Q = (\xi,\eta,\zeta)$$
$$p = (x,y,0), \quad q = (\xi,\eta,0), \quad p_0 = (x,0,0), \quad q_0 = (\xi,0,0) \qquad (13.192)$$

and

$$E_3(P,Q,a^2 t) = E_2(p,q,a^2 t) E_1(p_0,q_0,a^2 t) \qquad (13.193)$$

where

$$E_1(p_0,q_0,a^2 t) \equiv E(z - \zeta, a^2 t) \qquad (13.194)$$

We have

$$\int_{-\infty}^{\infty} E_3(P,Q,a^2 t) d\zeta = E_2(p,q,a^2 t)$$
$$\int_{-\infty}^{\infty} d\eta \int_{-\infty}^{\infty} E_3(P,Q,a^2 t) d\zeta = E_1(p_0,q_0,a^2 t) \qquad (13.195)$$

---

[9]This result clearly remains true for any $n \geq 1$. Our only purpose in taking $n = 3$ was to avoid the complicated computations when $n > 3$.

These equalities may be interpreted as follows.

E_3 is the influence function of an instantaneous unit source of heat placed at time $t = 0$ at the point Q of $\mathbb{R}_3$. Therefore, E_2 is the influence function of instantaneous unit sources of heat distributed uniformly along the axis $z = \zeta$ with unit linear density. Finally, E_1 is the influence function of instantaneous unit sources of heat distributed uniformly over the plane $x = \xi$ with unit surface density. Accordingly, E_3 is called the *influence function of an instantaneous point source of heat*, E_2 the *influence function of an instantaneous linear source of heat*, and E_1 the *influence function of an instantaneous planar source of heat*.

In problems with cylindrical symmetry it is convenient to introduce the influence function of a cylindrical source of heat. Let (r, φ, z) be cylindrical coordinates and $Q(\rho, \psi, \zeta)$ the points of an infinitely long circular cylinder C of radius ρ whose axis is the ζ axis of the cartesian coordinate system; let $P(r, \varphi, z)$ be some point of $\mathbb{R}_3$. Let p and q be the projections of P and Q on the plane $z = 0$. Then the influence function $E_C(p, q, a^2t)$ of unit cylindrical sources of heat uniformly distributed along C is defined as

$$
\begin{aligned}
E_C(p, q, a^2t) &= \int_C E_3(P, Q, a^2t) d\sigma_Q \\
&= \int_0^{2\pi} d\psi \int_{-\infty}^{\infty} \rho \frac{\exp[-r_{pq}^2/4a^2t]}{4a^2\pi t} \\
&\quad \cdot \frac{\exp[-(z-\xi)^2/4a^2t]}{2a(\pi t)^{1/2}} d\zeta
\end{aligned}
\tag{13.196}
$$

Since

$$
r_{pq}^2 = r^2 + \rho^2 - 2r\rho\cos(\varphi - \psi)
\tag{13.197}
$$

we obtain

$$
E_C(p, q, a^2t) = \int_0^{2\pi} \frac{\exp[-(r^2 + \rho^2 - 2r\rho\cos(\varphi - \psi))/4a^2t]}{4a^2\pi t} \rho d\psi
\tag{13.198}
$$

We have

$$
\frac{1}{2} \int_0^{2\pi} \exp\left[\frac{-2r\rho\cos(\varphi - \psi)}{4a^2t}\right] \rho d\psi = I_0\left(\frac{r\rho}{2a^2t}\right)
\tag{13.199}
$$

where $I_0(z)$ is the Bessel function of an imaginary argument of order zero (see Appendix 2, Sections 3, 6, and 8). Hence

$$
E_C(p, q, a^2t) = \exp\left[\frac{-(r^2 + \rho^2)}{4a^2t}\right] I_0\left(\frac{r\rho}{2a^2t}\right) \frac{1}{2a^2\pi t}
\tag{13.200}
$$

If the sources of heat are distributed over the surface of C with ζ-dependent surface density, the influence function becomes

$$
\begin{aligned}
E_c(r, z, \rho, \zeta, a_t) &= E(z - \zeta, a^2t) E_C(p, q, a^2t) \\
&= \frac{1}{4a^3(\pi t)^{3/2}} \cdot \exp\left[-\frac{r^2 + \rho^2 + (z - \zeta)^2}{4a^2t}\right] I_0\left(\frac{r\rho}{2a^2t}\right)
\end{aligned}
\tag{13.201}
$$

This influence function is applicable to problems with cylindrical symmetry evolving in $\mathbb{R}_3$, whereas E_C is used in planar problems.

Remark. All the results of Sections 3, and 4 are easily generalized to multidimensional Cauchy problems.

Problems

P13.2.1. Compute the integral

$$U(x,t) = \int_{-\infty}^{\infty} E(y - \xi, \lambda - \tau) E(x - y, t - \lambda) dy$$

Is $U(x,t)$ the solution of any Cauchy problem? If not, explain why; if so, formulate the problem.

P13.3.1. Formulate and prove a 2-dimensional analog of Theorem 13.3.2 for the case of a Cauchy problem with cylindrical symmetry.

P13.3.2. Is it possible to remove the requirement that $f(x)$ vanishes exponentially at infinity in the formulation and proof of Theorem 13.3.1, if all the integrals involved are interpreted in the sense of Abel summability? (See Appendix 4, Section 5.)

P13.3.3. Formulate and prove a 2-dimensional analog of Theorem 13.3.1 for the case of a Cauchy problem with cylindrical symmetry.

P13.5.1. Solve Problem 13.2.1 for a multidimensional Cauchy problem.

14

Maximum principle for parabolic equations

1. Notation

As mentioned in Chapter 1, there is a close relationship between the properties of solutions of equations of different types, in particular between those of elliptic and parabolic equations. One of our main purposes is precisely to trace these relationships. This relates in particular to the actual identity of the maximum principles for elliptic and parabolic operators and their applications (see Chapter 15, Sections 2 and 3; compare the uniqueness theorems of Section 5 with those of Chapter 9). This actual identity follows from the fact that a solution of any (Dirichlet) elliptic boundary-value problem is the long time asymptotic of the respective parabolic boundary-value problem (see Theorem 15.4.1 and Chapter 17, Section 2).

We first introduce a number of new definitions and notations.

- $\mathbb{R}_n = n$-dimensional Euclidean space
- $\mathcal{E}_{n+1} = \mathbb{R}_n \times (-\infty < t < \infty)$ – the topological product of $\mathbb{R}_n$ and the time axis
- $x = (x_1, x_2, \ldots, x_n) \in \mathbb{R}_n$, $(x, t) \in \mathcal{E}_{n+1}$ – points in rectangular cartesian coordinates
- $P = (x, t)$ or $P = (p, t)$ – point in $\mathcal{E}_{n+1}$
- $F(P) = F(p, t) = F(x, t)$ – any function in $\mathcal{E}_{n+1}$
- $D \subset \mathcal{E}_{n+1}$ – a region and $\bar{D}$ – the closure of D
- $D_\tau = D \cap (t = \tau) \in \mathbb{R}_n$ – the section of D by the plane $t = \tau$

Definition 14.1.1. A set

$$D_{t_0}^T = \bigcup_{t_0 < \tau \leq T} D_\tau$$

is called the *parabolic region*.

If D_τ is independent of τ for any $\tau \geq t_0$, so that $D_{t_0}^T$ is a cylindrical region with axis parallel to the time axis, we write simply D and Σ and

325

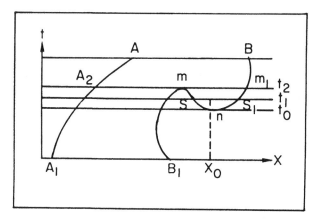

Figure 14.1. Sketch of a parabolic region.

consider them as the subsets of $\mathbb{R}_n$. If $t_0 = 0$ we write D^T, omitting the subscript in t_0 if this cannot cause any misunderstanding.

Definition 14.1.2. Let Σ_τ be the boundary of D_τ. Then the set

$$\Sigma_{t_0}^T = \bigcup_{t_0 < \tau \leq T} \Sigma_\tau$$

is called the *lateral boundary* of $D_{t_0}^T$ and $\Gamma_{t_0}^T = \Sigma_{t_0}^T + D_{t_0}$ the *parabolic boundary* of $D_{t_0}^T$.

Remark 1. Definitions 14.1.1 and 14.1.2 mean that a "parabolic" topology, different from that of the Euclidean space–time topology, is introduced. Indeed, in the latter the manifold D_T must be considered part of the boundary of $D_{t_0}^T$, rather than a subset of its interior. This distinction has an obvious physical meaning. Indeed, in physical problems – for example, in studying the redistribution of temperature inside some body – T may be interpreted as the current time of observation, which is not the time of the last observation. Hence T must be treated as a variable, and therefore the point (p, T) of the space–time region $D_{t_0}^T$ must be treated as an interior point.

Definition 14.1.3. $D_{t_0}^T$ is called a *normal parabolic region* if any pair of arbitrarily chosen points $(p_i, t_i) \in D_{t_0}^T$, $i = 1, 2$, may be connected by a continuous curve along which t varies monotonically.

Definition 14.1.4. $C(p, t)$ denotes the set of all points $(q, \tau) \in D_{t_0}^T$ which may be connected to $(p, t) \in D_{t_0}^T$ by the path along which τ increases. Such a path is termed *admissible*.

Remark 2. The physical motivation for the concept of a normal parabolic region is obvious. Indeed, consider the region D_0^t shown in Figure 14.1. The characteristics $\tau = t_0$ and $\tau = t_2$ are tangent to the lateral boundary of D_0^t at the points n and m, respectively. The body remains simply connected in the time interval $(0, t_0)$. At the time t_0 it splits into two separate parts, which are reconnected at time $\tau = t_2$. Then D_0^t is a normal parabolic region for any $t \in (0, t_0)$; the region $\tilde{D}_{t_0}^{t_2}$ with lateral boundary mnm_1 is also a normal parabolic region, as is the region $D_{t_2}^t$ for any $t > t_2$. However the whole region D_0^t for all $t \in (0 < t < \infty)$ is not normal, because parts $D_1 = A_1B_1mA_2A_1$ and $D_2 = mnm_1m$ of the whole region $D_{t_0}^{t_2}$ are not connected. The distribution of temperature within that part of the body represented by the space–time region D_2 is determined exclusively by the temperature of its lateral boundary, which is independent of the temperature distribution within D_1 – provided, of course, that there is no thermal interaction between these bodies by radiation. Therefore, *in what follows we shall consider only normal parabolic regions*, usually without mentioning this explicitly.

Definition 14.1.5. The operators

$$L_a u \stackrel{\text{def}}{=} a^2 \Delta u - \frac{\partial}{\partial t} u \tag{14.1}$$

$$L u \stackrel{\text{def}}{=} \Delta u - \frac{\partial}{\partial t} u \tag{14.2}$$

$$L_d u = \operatorname{div}[k(p,\tau)\operatorname{grad}(u)] + \sum_{i=1}^{n} b_i(p,\tau)\frac{\partial}{\partial x_i}u - c(p,\tau)\frac{\partial}{\partial \tau}u \tag{14.3}$$

are called *a-parabolic*, *parabolic*, and *divergent-parabolic*, respectively. Hence a parabolic operator is a-parabolic if $a = 1$.

With regard to divergent-parabolic operators, we shall assume that

$$c \geq c_0 = \text{const} > 0, \quad k \geq k_0 = \text{const} > 0 \tag{14.4}$$

Remark 3. The definition of L_a incorporates the value of the coefficient of diffusivity. Usually this is not necessary, since by appropriate scaling one can make its dimensionless value equal to 1. However, if one must compare solutions of boundary-value problems for equations with different coefficients of diffusivity, these different values must be emphasized. This may also be desirable in cases when the interaction of materials of different thermal diffusivity is considered.

Definition 14.1.6. Let $C^{2,1}(D^t)$ be the class of functions $f(p,\tau)$ twice continuously differentiable with respect to the spatial coordinates and continuously differentiable with respect to time within D^t. Then $u(p,t) \in D^t$ is said to be *a-superparabolic* if $u \in C^{2,1}(D^t)$ and $L_a u \leq 0$ for all $(p,\tau) \in D^t$. A function $u(p,t) \in D^t$ is called *d-subparabolic* if $u \in C^{2,1}(D^t)$ and $L_d u \geq 0$ for all $(p,\tau) \in D^t$, and analogously for a-subparabolic and d-superparabolic functions.[1] The adjectives "superparabolic" and "subparabolic" refer to the operator L rather than to L_d or L_a.

2. Weak maximum principle

Theorem 14.2.1 (Weak maximum principle). *Let $D^t \subset \mathcal{E}_{n+1}$ be a parabolic region, and let $u(p, \tau)$ be continuous in $\bar{D}^t$ and d-superparabolic (d-subparabolic) in D^t. Then its minimum (maximum) is on Γ^t.*

PROOF. Let $u(p, \tau)$ be a d-superparabolic function and

$$\min_{(p,\tau)\in\Gamma} u(p,\tau) = M, \quad \min_{(p,\tau)\in\bar{D}^t} u(p,\tau) = m \tag{14.5}$$

Assume that

$$m < M \tag{14.6}$$

Then there exists $(q, t_0) \in D^t$ such that

$$u(q, t_0) = m \tag{14.7}$$

(q, t_0) is the interior point of D^t in the parabolic topology. This means that

$$\text{grad}^2\, u(q, t_0) = 0, \quad \text{div}[k\,\text{grad}\,u] = k\Delta u \geq 0, \quad \frac{\partial}{\partial\tau}u(q, t_0) \leq 0$$

$$\Rightarrow L_d u(q, t_0) \geq 0 \tag{14.8}$$

$(\partial u(q, t_0)/\partial\tau = 0$ if $t_0 < t$ and $\partial u(q, t_0)/\partial\tau \leq 0$ if $t_0 = t$. Recall that in the parabolic topology (q, t) is the interior point of D^t, whereas in the usual topology of Euclidean space it is a boundary point.) If

$$L_d u(p, \tau) < 0 \quad \forall(p, \tau) \in D^t \tag{14.9}$$

then the contradiction proves the theorem. However, it is possible that

$$L_d u(q, t_0) = 0 \tag{14.10}$$

in which case we need an additional construction. Let $\varepsilon > 0$ be arbitrarily small, and introduce the auxiliary function

$$v(p, \tau) = u(p, \tau) + \frac{M - m}{2t_0}(\tau - t_0) \tag{14.11}$$

For all $\tau \leq t_0$,

$$v(q, t_0) = m, \quad v(p, \tau)\,\big|_{\Gamma^{t_0}} \geq M - \frac{M - m}{2} \geq \frac{M + m}{2} > m \tag{14.12}$$

Hence $v(p, \tau)$ achieves its minimum at some point $(\tilde{p}, \tilde{\tau}) \in D^{t_0}$ and therefore

$$L_d v(\tilde{p}, \tilde{\tau}) = L_d u(\tilde{p}, \tilde{\tau}) - \frac{M - m}{2t_0} \geq 0 \Rightarrow L_d u(\tilde{p}, \tilde{\tau}) > 0 \tag{14.13}$$

---

[1]The class of continuous (not necessarily differentiable) super- and subparabolic functions is defined in Chapter 18, Section 2.

which contradicts the assumption that u is d-superparabolic. Hence

$$\min u(p,\tau)\big|_{\Gamma^t} = \min u(p,\tau)\big|_{\bar{D}} \qquad (14.14)$$

Q.E.D.

Theorem 14.2.2. *Let* $u(p,\tau) \in C^{2,1}(D^t)$, $u(p,\tau) \in C(\bar{D}^t)$, *and*

$$L_d u + \delta u \leq 0 \ (L_d u + \delta u \geq 0) \quad \forall(p,\tau) \in D^t \qquad (14.15)$$

where

$$\delta < 0 \quad \forall(p,\tau) \in D^t \qquad (14.16)$$

Then u *achieves a nonpositive minimum (nonnegative maximum) on* Γ^t.

The proof duplicates that of Theorem 14.2.1.

Remark 4. Theorems 14.2.1–2 are proved for a general linear operator of a parabolic type presented in divergent form. This assumption is not necessary. One must assume only that the coefficients of the operator

$$\mathcal{L} = \sum_{i=1}^{n}\sum_{j=1}^{n} a_{ij}\frac{\partial^2}{\partial x_i \partial x_j}u + \sum_{i=1}^{n} a_i\frac{\partial}{\partial x_i}u + \delta u - \frac{\partial}{\partial \tau}\cdot u \qquad (14.17)$$

satisfy the *strong parabolicity* condition: There exists $a > 0$ such that

$$\sum_{i=1}^{n}\sum_{j=1}^{n} a_{ij}^2 \geq \alpha \quad \forall(p,\tau) \in D^t \qquad (14.18)$$

The assumption that u is continuous in D^t may be weakened as follows.

Theorem 14.2.3. *Let* D^t *be a bounded parabolic region with parabolic boundary* Γ^t, *lateral boundary* Σ^t, *and lower base* D. *Let* $u(p,t)$ *be a function uniformly bounded in* D^t, *continuous in* $\Gamma^t \setminus \Sigma_0$, *and subparabolic (superparabolic) in* D^t. *Then* $u(p,t)$ *achieves its supremum (infimum) on* Γ^t.

Theorem 14.2.4. *Let* $u(p,t)$ *be bounded from below (from above), superparabolic (subparabolic) in* $D^t \subset \mathcal{E}_{2+1}$, *and continuous in* $\bar{D}^t$ *everywhere except for a finite number* n *of smooth curves*

$$L_k = \{x_i, \tau: x_i = x_i(\tau),\ 0 \leq \tau \leq t,\ i = 1,2\} \subset \Sigma^t, \quad k = 1,2,\ldots,n \quad (14.19)$$

along which τ *varies monotonically (see Figure 14.2). Then*

$$\inf_{(p,\tau)\in\Gamma^t} u(p,\tau) = m \qquad \left(\sup_{(p,\tau)\in\Gamma^t} u(p,\tau) = M\right) \qquad (14.20)$$

implies

$$\inf_{(p,\tau)\in D^t} u(p,\tau) = m \qquad \left(\sup_{(p,\tau)\in D^t} u(p,\tau) = M\right) \qquad (14.21)$$

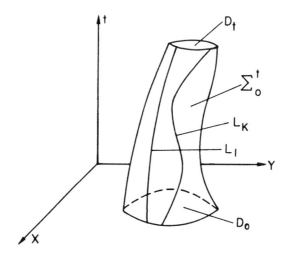

Figure 14.2.   Lines of discontinuity at Σ_0^t.

Remark 5. Theorem 14.2.4 has been formulated here for functions defined in $\mathcal{E}_{2+1}$ only for the sake of simplicity. A more complicated definition of the manifolds along which u experiences discontinuities would permit us to consider u in $\mathcal{E}_{n+1}$ for any dimension n.

PROOF OF THEOREM 14.2.3. Let

$$\sup_{(p,t)\in\Gamma^t} u(p,t) = m \tag{14.22}$$

Assume, contrary to the theorem's assertion, that

$$\sup_{(p,t)\in D^t} u(p,t) = M > m \tag{14.23}$$

Without loss of generality we may assume that

$$m = 0 \tag{14.24}$$

since otherwise we could consider $v = u - m$ instead of u. Thus we assume

$$M > 0 \tag{14.25}$$

By (14.23) there exists a point $(p, t_0) \in D^t$ such that

$$u(p, t_0) = M \tag{14.26}$$

Denote by $\mu(G)$ the measure of any set G in $\mathbb{R}_n$. Define two parabolic cylindrical regions

$$C^t = C \times (0 < \tau \le t), \quad C_0^t = C_0 \times (0 < \tau \le t) \subset C^t \tag{14.27}$$

such that there exists $\eta_0 < t_0$ with

$$C_0^{\eta_0} \subset D^{\eta_0} \subset C^{\eta_0} \tag{14.28}$$

and

$$\mu(C \setminus C_0) = \mu_0 \tag{14.29}$$

where μ_0 will be determined later.

We now take a function $\psi(q)$ in $\mathbb{R}_n$ such that

$$\psi(q) \equiv 0 \quad \forall q \in \mathbb{R}_n \setminus C, \quad \psi(q) = M \quad \forall q \in C \setminus C_0, \quad \psi(q) = \varepsilon \quad \forall q \in C_0 \tag{14.30}$$

where ε will be determined later. Let us take some $\eta < \eta_0$ and consider a Poisson integral

$$\Pi(p, t \mid \psi) = \int_{\mathbb{R}_n} \psi(q) E_n(p, q, t - \eta) d\omega_q \tag{14.31}$$

where

$$E(p, q, t) \stackrel{\text{def}}{=} \frac{\exp(-r_{pq}^2/4t)}{(4\pi t)^{n/2}} \tag{14.32}$$

is a fundamental solution of Fourier equation in $\mathbb{R}_n$.

By (14.30) and (14.32),

$$\Pi(p, t \mid \psi) < \frac{2M\mu(C \setminus C_0) + \varepsilon\mu(C_0)}{4\pi(t - \eta)^{n/2}} \tag{14.33}$$

Hence η_0, ε, and μ_0 can be chosen so that

$$\Pi(p, t_0 \mid \psi) < M \tag{14.34}$$

At the same time,

$$\Pi(p, t \mid \psi) > 0 > u(p, t) \quad \forall (p, t) \in \Sigma^t \tag{14.35}$$

and, by the choice of ψ and η,

$$\Pi(p, \eta \mid \psi) > u(p, \eta) \quad \forall p \in D \tag{14.36}$$

The weak maximum principle is applicable to $u(p, t)$ and $\Pi(p, t \mid \psi)$ in D_η^t, because both these functions are continuous there and their difference is there a superparabolic function. Hence $u(p, t) < \Pi(p, t \mid \psi)$ in D_η^t and, in particular,

$$u(p, t_0) < \Pi(p, t_0 \mid \psi) < M \tag{14.37}$$

But (14.37) contradicts (14.26). Thus assumption (14.23) is wrong.

Q.E.D.

Remark 6. Assume that $u(p, \tau)$ is defined in an infinite region and tends uniformly to zero at infinity. Then Theorem 14.2.3 may also be applied to estimate $u(p, t)$. Indeed, let Q_R be a cylinder of radius R in $\mathcal{E}_{n+1}$:

$$Q_R = \left(x \colon \sum_{i=1}^{n} x_i^2 \leq R^2 \right) \times (-\infty < \tau < \infty) \tag{14.38}$$

and let $\tilde{D}^t$ be its intersection with D^t,

$$\tilde{D}^t = Q_R \cap D^t \tag{14.39}$$

Let $\tilde{\Gamma}^t$ be the parabolic boundary of $\tilde{D}^t$ and $S = \tilde{\Gamma}^t \setminus \Gamma^t$. Then for every $\varepsilon > 0$ there exists $R > 0$ so large that

$$|u(p,\tau)| < \varepsilon \quad \forall (p,\tau) \in S \tag{14.40}$$

Repeating the proof of Theorem 14.2.3 and taking (14.39) into account, we obtain

$$|u(p,\tau)| < \varepsilon \quad \forall (p,\tau) \in \tilde{D}^t \tag{14.41}$$

which means that

$$u(p,\tau) < \varepsilon \quad \forall (p,\tau) \in D^t \tag{14.42}$$

since for any fixed (p,τ) there exists $R > 0$ so large that

$$(p,\tau) \in D^t \Rightarrow (p,\tau) \in \tilde{D}^t \tag{14.43}$$

Since $\varepsilon > 0$ is arbitrarily small and m is equated to zero, this means that

$$\sup_{(p,\tau)\in D^t} u(p,\tau) = \sup_{(p,\tau)\in\Gamma^t} u(p,\tau) \tag{14.44}$$

Q.E.D.[2]

PROOF OF THEOREM 14.2.4. Assume that u is superparabolic in D^t. Let

$$\inf_{(p,\tau)\in\Gamma^t} u(p,\tau) = m, \quad \inf_{(p,\tau)\in D^t} u(p,\tau) = M_0 < m \tag{14.45}$$

We use the abbreviated notation

$$q_i = q_i(\tau), \quad (q_i,\tau) \in L_i, \quad \tilde{q}_i = q_i(t), \quad i = 1,2,\ldots,n \tag{14.46}$$

for all $\tau \in [0,t]$. Consider the functions

$$W_i(p,t) = \int_0^t \frac{\exp[-r_{pq_i}^2/4(t-\tau)]}{4\pi^{1/2}(t-\tau)}d\tau, \quad i = 1,2,\ldots,n \tag{14.47}$$

By the triangle inequality,

$$r_{pq_i} \leq r_{p\tilde{q}_i} + r_{q_i\tilde{q}_i} \tag{14.48}$$

In addition, for each $i = 1,2,\ldots,n$ there exists $K_i = \text{const} > 0$ such that

$$r_{q_i\tilde{q}_i} \leq 2K_i(t-\tau) \tag{14.49}$$

since L_i is a smooth curve by assumption. By (14.47) and (14.48),

$$\exp\left[-\frac{r_{pq_i}^2}{4(t-\tau)}\right] \geq \exp\left[-\frac{r_{p\tilde{q}_i}^2}{4(t-\tau)}\right]\exp(-K_i^2(t-\tau) - K_i) \tag{14.50}$$

---

[2]Note that this kind of argument enables one to apply the weak maximum principle to infinite regions.

Hence

$$W_i(p,t) \geq \exp[-K_i^2 T - K_i] \int_0^t \frac{\exp[-r_{pq_i}^2/4(t-\tau)]}{4\pi^{1/2}(t-\tau)} d\tau$$

$$= \alpha_i \int_{\rho_i}^\infty e^{-z} \frac{dz}{z} \overset{\text{def}}{=} J(\rho_i), \quad i = 1, 2, \dots, n \tag{14.51}$$

where

$$\rho_i = r_{p\tilde{q}}^2/4t, \quad 0 \leq t \leq T \tag{14.52}$$

and

$$\alpha_i = \exp[-K_i^2 T - K_i] \tag{14.53}$$

We have

$$\lim_{\rho \downarrow 0} \int_\rho^\infty e^{-z} \frac{dz}{z} = \infty \tag{14.54}$$

Hence, for all $N = \text{const} > 0$, there exists $\rho > 0$ such that

$$r_{p\tilde{q}_i}^2 \leq 2\rho t^{1/2} \Rightarrow J(\rho_i) > N/n, \quad i = 1, 2, \dots, n \tag{14.55}$$

Let

$$W(p,t) \overset{\text{def}}{=} \sum_{i=1}^n W_i(p,t) \tag{14.56}$$

Then (14.55) yields

$$W(p,t) > N \quad \forall (p,t) \in D^T \tag{14.57}$$

Note now that $W(p,t)$ is a parabolic function everywhere, except for the set of all curves L_i, $i = 1, 2, \dots, n$. Let (p_0, t_0) be an arbitrarily fixed point of D^t. Take $\varepsilon > 0$ arbitrarily small and choose $\eta > 0$ so small that

$$\eta W(p_0, t_0) < \varepsilon \tag{14.58}$$

Fix this η and take $\rho > 0$ so small that for all $\tau \in (0, t_0]$,

$$\eta W(p, \tau) > m - M_0 \quad \text{if } r_{pq_i} < 2\rho \tau^{1/2} \tag{14.59}$$

With ρ chosen in this way, consider the tubes

$$\tilde{K}_i^t = \left\{ p, \tau \colon r_{pq_i} < 2\rho \tau^{1/2}, \ 0 < \tau \leq t \right\} \tag{14.60}$$

enclosing the lines L_i, and the regions

$$\tilde{D}^t = D^t \setminus \bigcup_{i=1}^n \tilde{K}_i^t \tag{14.61}$$

with parabolic boundary

$$\tilde{\Gamma}^t = \tilde{D}_0 \cup \tilde{\Sigma}^t \tag{14.62}$$

where

$$\tilde{D}_0 = D_0 \tag{14.63}$$

and

$$\tilde{\Sigma}^t = \tilde{\Sigma}_i^t + \tilde{\Sigma}_e^t \tag{14.64}$$

so that the part $\tilde{\Sigma}_e^t$ of the boundary $\tilde{\Gamma}^t$ lies outside and the part $\tilde{\Sigma}_i^t$ inside the region D^t. Definitions (14.45), (14.47), (14.56) and inequality (14.59) show that $W(p,\tau)$ is a bounded parabolic function within $\tilde{D}^t$, and also

$$\eta W(p,\tau) > 0 \quad \forall (p,\tau) \in \tilde{\Sigma}_e^t, \eta W(p,\tau) > M_0 \quad \forall (p,\tau) \in \tilde{\Sigma}_i^t$$
$$\eta W(p,\tau) = 0 \quad \forall p \in D_0 \tag{14.65}$$

Now define the auxiliary function

$$v(p,\tau) = u(p,\tau) + \eta W(p,\tau) - m \tag{14.66}$$

This function is defined within the region

$$D_{\tilde{k}}^t \overset{\text{def}}{=} D^t \setminus \tilde{K}^t \quad \left(\tilde{K} = \bigcup_{i=1}^n \tilde{K}_i^t \right) \tag{14.67}$$

It is bounded, superparabolic, continuous inside $\bar{D}_{\tilde{k}}^t$ except along the edge $\Sigma_{\tilde{k}}^t \cap \bar{D}_0$ of the parabolic boundary, and satisfies the boundary inequality

$$v(p,\tau) \geq 0 \quad \forall (p,\tau) \in \Gamma_{\tilde{k}}^t \tag{14.68}$$

Hence, by Theorem 14.2.3,

$$v(p,\tau) \geq 0 \quad \forall (p,\tau) \in D_{\tilde{k}}^t \tag{14.69}$$

so that

$$u(p,\tau) \geq m - \eta W(p,\tau) \quad \forall (p,\tau) \in D_{\tilde{k}}^t \tag{14.70}$$

Hence, by (14.40),

$$u(p_0,t_0) > m - \varepsilon \tag{14.71}$$

Since $u(p_0,t_0)$ is independent of ε and ε is arbitrarily small, this means that

$$u(p_0,t_0) \geq m \tag{14.72}$$

But (p_0,t_0) is an arbitrarily fixed point of D^t. Hence

$$u(p,\tau) \geq m \quad \forall (p,\tau) \in D^t \tag{14.73}$$

Q.E.D.

Remark 7. The requirement that $u(p,t)$ be uniformly bounded in D^t is essential for the truth of Theorems 14.2.3–4. Indeed, consider for example

$$u(p,t) = -\frac{\partial}{\partial x} E(x,t) \equiv x \cdot \frac{\exp(-x^2/4t)}{4\pi^{1/2}t^{3/2}} \tag{14.74}$$

This function is parabolic and positive in $(0 < x < \infty) \times (0 < t < \infty)$, vanishes at $x = 0$, $t < 0$ and at $x > 0$, $t = 0$, but tends to infinity along any path $x = kt^{1/2}$ for any $k \neq 0$ when $t \downarrow 0$, so that the assertions of the theorems are not valid.

3. Nirenberg's strong maximum principle [49]

We shall now show that the strong maximum principle, an analog of that of the theory of subharmonic and superharmonic functions, is valid for subparabolic and superparabolic functions.

Theorem 14.3.1 (Nirenberg's strong maximum principle). *Let $u(p,t)$ be a function continuous in the closure of a normal parabolic region D^t and subparabolic (superparabolic) in D^t. Let[3]*

$$M = \max u(p,\tau) > 0 \ \forall (p,\tau) \in \bar{D}^t \quad (m = \min u(p,\tau) \ \forall (p,\tau) \in \bar{D}^t) \quad (14.75)$$

and assume there exist p_0, t such that

$$u(p_0, t) = M, \ (p_0, t) \in D^t \quad (u(p_0, t) = m, \ (p_0, t) \in D^t) \quad (14.76)$$

Then

$$u(p,t) \equiv M \ (u(p,t) \equiv m) \quad \forall (p,\tau) \in \bar{D}^t \quad (14.77)$$

PROOF. The proof of Nirenberg's theorem is based on the following lemma.

Lemma 14.3.1. *Let $u(p,\tau)$ be subparabolic within D^t and achieve its maximum M at some point (p_0, t_0),*

$$u(p_0, t_0) = M, \quad (p_0, t_0) \in D^t, \ t_0 < t \quad (14.78)$$

Then

$$u(p, t_0) \equiv M \quad (14.79)$$

within any simple connected subset of D_{t_0}.

Assume that Lemma 14.3.1 is true. By assumption,

$$Lu = \Delta u - \frac{\partial}{\partial \tau} u \geq 0 \quad \forall (p,\tau) \in D^t \quad (14.80)$$

and there exists $(p_0, t) \in D^t$ such that

$$u(p_0, t) = M \quad (14.81)$$

Take some point[4]

$$(q_0, \tau_0) \in C(p_0, t), \quad \tau_0 < t, \quad (14.82)$$

where $C(p_0, t)$ is an admissible path between (q_0, τ_0) and (p_0, t).

Assume, contrary to the assertion, that

$$u(q_0, \tau_0) < M \quad (14.83)$$

Then there is a point $(q_1, \tau_1) \in C(p_0, t)$ nearest to (p_0, t) such that

$$u(q_1, \tau_1) = M, \quad \tau_0 < \tau_1 \leq t \quad (14.84)$$

---

[3]The assumption $M > 0$ clearly involves no loss of generality; compare with (14.24).

[4]See Definition 14.1.4.

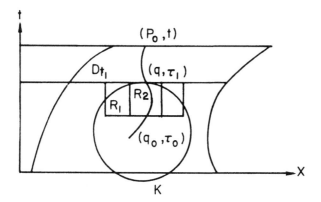

Figure 14.3.   Nirenberg's construction;   sketch of the proof of
Lemma 14.3.1.

Since $\tau_0 < \tau_1 \le t$, we can construct two cubes $Q_2 \subset Q_1 \subset D_{\tau_1}$ with their centers at (q_1, τ_1). Let R_i, $i = 1, 2$, be parallelepipeds such that $\bar{R}_2 \subset R_1 \subset \bar{R}_1 \subset D_{\tau_2}^{\tau_1}$, their upper faces coinciding with Q_2 and Q_1 (respectively) and their lower faces lying at the plane $\tau = \tau_2 > \tau_0$. We assume that $C(p_o, t) \cap [\tau_2, \tau_1] \subset R_2$ and that R_2 is chosen possibly small (see Figure 14.3). Obviously $u(p, \tau) < M$ for all $(p, \tau) \in R_1$, since if there exists $(\tilde{q}, \tilde{\tau}) \in R_1$ such that $u(\tilde{q}, \tilde{\tau}) = M$ then, by Lemma 14.3.1, $(\tilde{q}_1, \tilde{\tau}) \in C(p_0, t) \Rightarrow u(\tilde{q}_1, \tilde{\tau}) = M$, which contradicts the choice of (q_1, τ_1) as the point of $C(p_0, t)$ nearest (p_0, t) where $u - M$. Now let $K \in \mathcal{E}_{n+1}$ be a sphere of radius r, tangent to the plane $\tau = \tau_1$ at the point (q_1, τ_1), such that $(q_0, \tau_0) \in K$. Let (p^*, τ^*) be the center of K. Define

$$h(p, \tau) = r^2 - r_{p^* p}^2 - (\tau - \tau^*)^2 \qquad (14.85)$$

We have $p = (x_1, x_2, \ldots, x_n)$ and $p^* = (x_1^*, x_2^*, \ldots, x_n^*)$, so that

$$r_{p^* p}^2 = \sum_{i=1}^{n} (x_i - x_i^*)^2 \qquad (14.86)$$

and

$$r = \tau_1 - \tau^*, \quad Lh \equiv \Delta h - \frac{\partial h}{\partial \tau} = 2(\tau - \tau^* - n) \qquad (14.87)$$

Hence the radius r of K may be chosen so large that

$$Lh > 0 \quad \forall (Q, \tau) \in \bar{R}_1 \qquad (14.88)$$

Define the auxiliary function

$$v(p, \tau) = u(p, \tau) + \varepsilon h(p, \tau) \qquad (14.89)$$

where the constant $\varepsilon > 0$ will be chosen later. Let Γ be the part of the boundary of R_2 belonging to K, and S the part of the boundary of K belonging to R_2. All points of Γ are interior points of K, so that

$$u(p, \tau) < M \quad \forall (p, \tau) \in \Gamma \tag{14.90}$$

Clearly, there exists $\eta > 0$ such that

$$u(p, \tau) < M - \eta \quad \forall (p, \tau) \in \Gamma \tag{14.91}$$

since the distance r_0 between (q_1, τ_1) and Γ is positive:

$$r_0 \stackrel{\text{def}}{=} r[(q_1, \tau_1), \Gamma] > 0 \tag{14.92}$$

Hence $\varepsilon > 0$ may be chosen so small that

$$v(p, \tau) < M - \eta + \varepsilon h(p, \tau) < M \quad \forall (p, \tau) \in \Gamma \tag{14.93}$$

On the other hand,

$$h(p, \tau) = 0 \quad \forall (p, \tau) \in S \tag{14.94}$$

Consequently

$$v(p, \tau) = u(p, \tau) < M \quad \forall (p, \tau) \in S \setminus (q_1, \tau_1) \tag{14.95}$$

and

$$v(q_1, \tau_1) = u(q_1, \tau_1) = M \tag{14.96}$$

Hence

$$v(p, \tau) \le M \quad \forall (p, \tau) \in \bar{K} \setminus K \tag{14.97}$$

By the definition (14.89), the inequality (14.88), and the subparabolicity of u,

$$Lv(p, \tau) = Lu(p, \tau) + \varepsilon Lh(p, \tau) > 0 \tag{14.98}$$

Hence, by virtue of the weak maximum principle,

$$v(p, \tau) \le M \quad \forall (p, \tau) \in \bar{K} \tag{14.99}$$

This and (14.97) imply that

$$\frac{\partial}{\partial \tau} v(q_1, \tau_1) \ge 0 \tag{14.100}$$

Because

$$\frac{\partial}{\partial \tau} v(q_1, \tau_1) = \frac{\partial}{\partial \tau} u(q_1, \tau_1) - 2\varepsilon(\tau_1 - \tau^*) \tag{14.101}$$

it follows from (14.90) that

$$\frac{\partial}{\partial \tau} u(q_1, \tau_1) \ge 2\varepsilon(\tau_1 - \tau^*) > 0 \tag{14.102}$$

Now, since (q_1, τ_1) is an interior maximum point of u in $Q_1 \subset \mathbb{R}_n$, we have

$$\Delta u(q_1, \tau_1) \le 0 \tag{14.103}$$

and consequently

$$Lu(q_1, \tau_1) = \Delta u(q_1, \tau_1) - \frac{\partial}{\partial t} u(q_1, \tau_1) < 0 \qquad (14.104)$$

which contradicts the assumption that $u(p, \tau)$ is subparabolic. This contradiction proves that Nirenberg's theorem is valid provided Lemma 14.3.1 is true.

Thus we need only prove Lemma 14.3.1. This will be a consequence of the following lemma.

Lemma 14.3.2. *Let*

$$Lu(p, \tau) \geq 0 \quad \forall (p, \tau) \in D^t \qquad (14.105)$$

and

$$\max_{(p,\tau) \in D^t} u(p, \tau) = u(p_0, \tau_0) = M \qquad (14.106)$$

where (p_0, τ_0) belongs to the boundary S of the sphere $K \subset \bar{K} \subset D^t$. Assume that

$$u(p, \tau) < M \quad \forall (p, \tau) \in \bar{K} \setminus (p_0, \tau_0) \qquad (14.107)$$

Then the coordinates of p_0 are the spatial coordinates of the center O of K,

$$O = (x_{10}, x_{20}, \dots, x_{n0}) \Rightarrow p = (x_{10}, x_{20}, \dots, x_{n0}) \qquad (14.108)$$

Supposing that Lemma 14.3.2 is true, let us prove Lemma 14.3.1. Assume that conditions (14.107) and (14.108) are fulfilled and that

$$\tau_0 \leq t \qquad (14.109)$$

Let $\mathcal{M}$ be the set of all points of D_{τ_0} where $u = M$:

$$\mathcal{M} = \{\tilde{p}, \tau_0 \colon u(\tilde{p}, \tau_0) = M\} \qquad (14.110)$$

By the continuity of u, $\mathcal{M}$ is the closed set. Assume that the set of points

$$\mathcal{CM} = D_{\tau_0} \setminus \mathcal{M} \neq \emptyset \qquad (14.111)$$

is not empty. Clearly, there is a point $(\tilde{p}_0, \tau_0) \in \mathcal{CM}$ whose distance ρ_0 from $\mathcal{M}$ is less than its distance from the boundary of D_{τ_0}.[5] Since $u(\tilde{p}, \tau_0) < M$ there exists such a neighborhood U of $(\tilde{p}, \tau_0)$ lying in D^t in which $u < M$:

$$u(p, \tau) < M \quad \forall (p, \tau) \in U \subset D^t \qquad (14.112)$$

(see Figure 14.4). Let $\Gamma \subset U$ be a segment of the straight line parallel to the t axis with center at $(\tilde{p}, \tau_0)$, and construct a family $\{A^\rho\}$ of ellipsoids of rotation around Γ, the length ρ of whose semiaxes orthogonal to Γ varies from 0 to ρ_0. Since A^0 is Γ, we have

$$u(p, \tau) < M \quad \forall (p, \tau) \in A^0 \qquad (14.113)$$

---

[5]See the proof of the strong maximum principle in the theory of subharmonic functions (Chapter 8, Section 3).

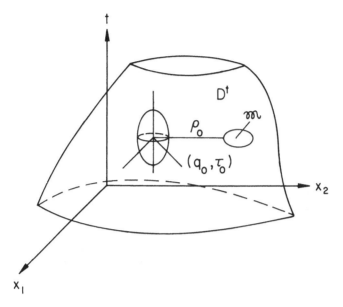

Figure 14.4. Nirenberg's construction; sketch of the proof of Lemma 14.3.2.

Let S^ρ be the boundary of A^ρ. Obviously, there is a point[6] $P \in S^{\rho_0}$ such that

$$u(P) = M \tag{14.114}$$

and, for all $Q \in A^{\rho_0}$,

$$u(Q) < M \tag{14.115}$$

Hence there exists $\lambda \in (0, \rho_0]$ such that

$$u(Q) < M \quad \forall Q \in A^\lambda \tag{14.116}$$

and $\tilde{Q} \in S^\lambda$ such that

$$u(\tilde{Q}) = M \tag{14.117}$$

Now let K^λ be a sphere lying in A^λ and touching S^λ at $\tilde{Q}$. Then, by Lemma 14.3.2, the x coordinates of the center of K^λ coincide with those of $\tilde{Q}$. But this is only possible if $\tilde{Q}$ is one of the vertices of A^λ, that is, if $\tilde{Q} \in \Gamma$, which contradicts the definition of U. This contradiction proves Lemma 14.3.1, provided that Lemma 14.3.2 is true. Thus it is only necessary to prove the latter, which we now proceed to do.

---

[6]Recall that $P = (p, \tau)$ means the point of $\mathcal{E}_{n+1}$.

PROOF OF LEMMA 14.3.2. Let

$$Lu \geq 0 \quad \forall (p, \tau) \in D^t \tag{14.118}$$

$$\max_{(p,\tau) \in D^t} u(p, \tau) = u(p_0, \tau_0) = M \tag{14.119}$$

where

$$(p_0, \tau_0) \in S = \bar{K} \setminus K, \quad K \in D^t \tag{14.120}$$

Suppose, contrary to the statement of the lemma, that the x coordinates of the center of K do not coincide with those of the maximum point (p_0, τ_0) of u. Suppose also that (p_0, τ_0) is the only point of $\bar{K}$ at which u achieves a maximum:

$$u(p, \tau) < M \quad \forall (p, \tau) \in \bar{K} \setminus (p_0, \tau_0) \tag{14.121}$$

This involves no loss of generality, since we may always replace K by a sphere of smaller radius if the assumption (14.121) fails to hold.

Assume that the origin of the coordinate system is at the center of K. Let us write

$$\rho^2 = \sum_{i=1}^{n} x_i^2, \quad r^2 = \rho^2 + \tau^2 \quad \forall (p, \tau) = (x_1, x_2, \dots, x_n, \tau) \tag{14.122}$$

so that r is the distance of the point (p, τ) from the center of K and ρ is the distance of the projection of (p, τ) on the plane $\tau = 0$ from this center.

Assume, contrary to the statement of the lemma, that

$$\rho_0 \neq 0 \tag{14.123}$$

Then $\tau_0 < t$, since $\bar{K} \subset D^t$. Hence there is a sphere K_1 with its center at (p_0, τ_0), of radius $r_1 > 0$ so small that $\bar{K}_1 \subset D^t$. The boundary S of K divides the boundary S_1 of $\bar{K}_1$ into two parts S_1^1 and S_1^2, where

$$S_1^1 = S_1 \cap K, \quad S_1^2 = S_1 \setminus \bar{K} \tag{14.124}$$

Obviously, there exists $\eta > 0$ so small that

$$u \leq M - \eta \quad \forall (p, \tau) \in S_1^1 \tag{14.125}$$

since by assumption (p_0, τ_0) is the only point of $\bar{K}_1$ where $u = M$. At the same time,

$$u(p, \tau) < M \quad \forall (p, \tau) \in S_1^2 \tag{14.126}$$

Now we define an auxiliary function

$$h(p, \tau) = \exp[-\alpha(\rho^2 + \tau^2)] - \exp(-\alpha r_0^2) \tag{14.127}$$

where r_0 is the radius of K and $\alpha = \text{const} > 0$ will be chosen later. We have

$$h(p, \tau) > 0 \quad \forall (p, \tau) \in K \tag{14.128}$$
$$h(p, \tau) = 0 \quad \forall (p, \tau) \in S \tag{14.129}$$
$$h(p, \tau) < 0 \quad \forall (p, \tau) \in D^t \setminus \bar{K} \tag{14.130}$$

Furthermore,

$$Lh = 2\exp[-\alpha(\rho^2 + \tau^2)](2\alpha^2\rho^2 - n\alpha + \alpha\tau) \tag{14.131}$$

By assumption $r_1 < \rho_0$, so that $\rho > 0$ for all $(p,\tau) \in \bar{K}_1$. Hence $\alpha > 0$ may be chosen so large that

$$Lh(p,\tau) > 0 \quad \forall (p,\tau) \in \bar{K}_1 \tag{14.132}$$

Now let

$$v(p,\tau) = u(p,\tau) + \varepsilon h(p,\tau) \tag{14.133}$$

where $\varepsilon = \text{const} > 0$ is arbitrarily small and, by virtue of (14.105) and (14.106),

$$Lv > 0 \quad \forall (P,\tau) \in K_1 \tag{14.134}$$

According to the weak maximum principle, v must attain a maximum at some point of S_1. But by (14.125), (14.126), and (14.130),

$$v \leq M - \eta + \varepsilon h \quad \forall (p,\tau) \in S_1^1, \quad v \leq u + \varepsilon h < M \quad \forall (p,\tau) \in S_1^2 \tag{14.135}$$

where $\eta > 0$ is fixed. Hence ε may be taken so small that

$$v < M \ \forall (p,\tau) \in S_1 \Rightarrow v < M \ \forall (p,\tau) \in K_1 \tag{14.136}$$

This means, in particular, that

$$v(p_0, \tau_0) = u(p_0, \tau_0) < M \tag{14.137}$$

since $(p_0, \tau_0) \in S$. But this contradicts the assumption that

$$u(p_0, \tau_0) = M \tag{14.138}$$

Thus the assumption that the x coordinates of the center of the sphere K do not coincide with those of the maximum point (p_0, τ_0) of u leads to a contradiction. This proves Lemma 14.3.2, and together with it Nirenberg's strong maximum principle.   **Q.E.D.**[7]

4. Vyborny–Friedman analog of Hopf's lemma [23; 89]

The following theorem is the parabolic analog of Hopf's Lemma from the theory of elliptic operators.

Theorem 14.4.1. *Let $u(p,\tau) \not\equiv \text{const}$ be subparabolic (superparabolic) within the region D^t, continuous in $\bar{D}^t$, and achieve a maximum (minimum) at some point $P \in \Gamma^t$. Assume that there is a sphere $K^t \subset \mathcal{E}_{n+1}$ such that its center belongs to D^t, its surface touches Γ^t at P, and the x coordinates of*

---

[7]Nirenberg's original theorem was proved for a strongly parabolic operator. The proof provided here reproduces Nirenberg's original theorem with a minor simplification, due to considering the heat-conduction operator $Lu = \Delta u - \partial u/\partial t$ instead of the general parabolic one.

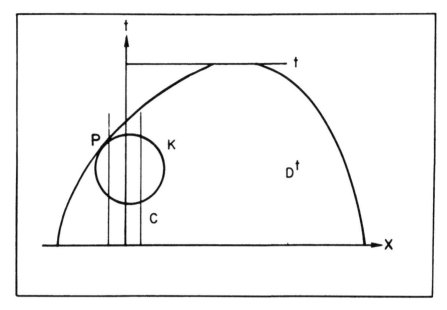

Figure 14.5.   Sketch of the proof of the Vyborny–Friedman theorem.

its center do not coincide with those of P. Let ν be a vector with its origin at P pointing into K^t. Then, if $\partial u(P)/\partial \nu$ exists,

$$\frac{\partial}{\partial \nu}u(P) < 0 \quad \left(\frac{\partial}{\partial \nu}u(P) > 0\right) \tag{14.139}$$

PROOF. Let u be subparabolic in D^t. Without loss of generality, we may assume that P is the only point of $S = \bar{K}^t \cap \bar{D}^t$ where u achieves it maximum, since otherwise we can consider a sphere $\bar{K}^t$ of smaller radius that has no points in common with Γ^t except P.

Supposing that the origin of the coordinate system is at the center of K^t, let us construct a cylinder C, coaxial with the τ axis, of radius r smaller than the distance d between P and this axis (see Figure 14.5). This means that

$$P \in \bar{G} = \bar{K}^t \setminus C \tag{14.140}$$

Let

$$S_g = \bar{G} \setminus G, \quad S^t = \bar{K}^t \setminus K^t \tag{14.141}$$

be the boundaries of G and K^t, respectively. Assume that

$$S_g^1 = S_g \cap S^t, \qquad S_g^2 = S_g \cap \bar{C} \tag{14.142}$$

and consider a function

$$h = h^* - \exp(-\alpha R^2), \quad h^* \stackrel{\text{def}}{=} \exp[-\alpha(\rho^2 + \tau^2)] \tag{14.143}$$

where ρ is defined by (14.122), R is the radius of K^t, and $\alpha = \text{const} > 0$ is chosen so great that for all $\rho > r$ and $\tau \geq 0$,

$$Lh = 2ah^*(2\alpha\rho^2 - n - \tau) > 0 \qquad (14.144)$$

Note that

$$h > 0 \quad \forall(p, \tau) \in K^t \qquad (14.145)$$

Now let

$$v = u + \varepsilon h \qquad (14.146)$$

where $\varepsilon > 0$ will be chosen later. Since u is subparabolic in D^t by assumption, it follows from (14.144) that

$$Lv > 0 \quad \forall(p, \tau) \in G \qquad (14.147)$$

Furthermore,

$$v = u < M \quad \forall(p, \tau) \in S_g^1 \setminus (P), \quad v(P) = M \qquad (14.148)$$

Moreover there exists $\eta = \text{const} > 0$ such that

$$u < M - \eta \quad \forall(p, \tau) \in S_g^2 \qquad (14.149)$$

since the distance $r_{PS_g^2}$ between P and S_g^2 is strictly positive. Hence $\varepsilon > 0$ may be chosen so small that

$$v(p, \tau) < M \quad \forall(p, \tau) \in S_g \setminus (P), \quad v(P) = M, \quad Lv > 0 \quad \forall(p, \tau) \in G \quad (14.150)$$

Hence v is a subparabolic function that is not a constant within G. Thus the strong maximum principle is applicable to v, so

$$v(p, \tau) < M \quad \forall(p, \tau) \in G \qquad (14.151)$$

Since $v(P) = M$ and $\boldsymbol{\nu}$ points into K^t, it follows from (14.151) that

$$\frac{\partial}{\partial\nu} v(P) \leq 0 \qquad (14.152)$$

Hence, by the definition (14.146) of v,

$$\frac{\partial}{\partial\nu} u(P) \leq -\varepsilon\frac{\partial}{\partial\nu} h(P) \qquad (14.153)$$

The definition (14.143) of h yields

$$\frac{\partial}{\partial\nu} h(P) = -\alpha R \exp(-\alpha R^2)\cos(\varphi) \qquad (14.154)$$

where φ is the angle between the radius vector $\mathbf{r}$ of P and the vector $\boldsymbol{\nu}$. Since $\boldsymbol{\nu}$ and $\mathbf{r}$ are directed inward to and outward from K, respectively,

$$\pi/2 < \varphi \leq \pi \Rightarrow \cos(\varphi) < 0 \qquad (14.155)$$

so that by (14.147)–(14.149)

$$\frac{\partial}{\partial\nu} u(P) < 0 \qquad (14.156)$$

Q.E.D.

Remark. The condition that the x coordinates of the center of K^t be distinct from those of P is essential. Consider, for example, the function

$$u \equiv -t^2, \quad t > 0 \tag{14.157}$$

Let K^t be the sphere

$$K^t = \left\{ p, \tau : r_{pO}^2 + \tau^2 = R^2, \ O = (0, 0, \dots, t_0); \ t_0 > 0 \right\} \tag{14.158}$$

where O is the center of K^t. The normal to the plane $\tau = 0$ at any point is parallel to the t axis. At the same time,

$$Lu = 2t > 0 \quad \forall t > 0 \tag{14.159}$$

$$\max_{t \geq 0} u(t) = u(0) = 0, \frac{\partial}{\partial t} u(0) = 0 \tag{14.160}$$

so that the assertion of the theorem is not valid. Clearly the assertion of the Vyborny–Friedman theorem fails to hold at extremum points located in the lower base D_0 of D^t because the condition in question is not fulfilled there.

5. Uniqueness theorems. Tichonov's comparison theorem

The maximum principle and the Vyborny–Friedman theorem enable us to prove a number of uniqueness theorems.

I. Dirichlet problem (first boundary-value problem)

Find a function $u(p, t)$, uniformly bounded within the parabolic region D^t, that satisfies the nonhomogeneous Fourier equation

$$Lu + f(p, t) = 0 \quad \forall (p, \tau) \in D^t \tag{14.161}$$

and the boundary and initial conditions

$$u(p, t) = \varphi(p, \tau) \quad \forall (p, \tau) \in \Sigma^t, \quad u(p, 0) = \psi(p) \quad \forall p \in D_0 \tag{14.162}$$

where φ and ψ are assumed to satisfy the conditions of Theorems 14.2.3–4 and f is assumed to be continuous within D^t.

Theorem 14.5.1. *If a solution of the problem (14.161) and (14.162) exists then it is unique.*

PROOF. Assume that there are two solutions u_1 and u_2 of problem (14.162); set

$$v(p, \tau) = u_1(p, t) - u_2(p, \tau) \tag{14.163}$$

Then $v(p, t)$ is uniformly bounded in D^t,

$$\sup u(p, \tau) = \inf u(p, \tau) = 0 \quad \forall (p, \tau) \in \Gamma^t \tag{14.164}$$

and parabolic within D^t. Hence, by Theorems 14.2.3–4,

$$\sup u(p, \tau) = \inf u(p, \tau) = 0 \quad \forall (p, \tau) \in D^t \tag{14.165}$$

Q.E.D.

II. Neumann problem (Second boundary-value problem)

Let $u(p, \tau)$ be a function continuous inside the closure of a parabolic region D^t, having a derivative $\partial u/\partial \nu$ in the direction of a normal ν to the section Σ_τ of the lateral boundary Σ^t by any plane $\tau = \text{const} \in (0, t]$, and satisfying the initial and boundary conditions

$$\frac{\partial}{\partial \nu} u(P)\big|_{P \in \Sigma^t} = \psi(P), \quad u(P)\big|_{P \in D^0} = \varphi(p) \tag{14.166}$$

and the nonhomogeneous equation

$$Lu + f(P) = 0 \quad \forall P \in D^t \tag{14.167}$$

Theorem 14.5.2. *If any point of the lateral boundary Σ^t of the region D^t may be touched from within D^t by a sphere K^t (i.e., if the lateral boundary Σ^t satisfies the conditions of the Vyborny–Friedman theorem at all its points), then the solution of the problem (14.166) and (14.167), if it exists, is unique.*

PROOF. Let u_1 and u_2 be two solutions satisfying all conditions of the theorem. Set

$$u(p, t) = u_1(p, t) - u_2(p, t) \tag{14.168}$$

Then

$$Lu = 0 \quad \forall P \in D^t, \quad u(P) = 0 \quad \forall P \in D_0, \quad \frac{\partial}{\partial \nu} u(P) = 0 \quad \forall P \in \Sigma \tag{14.169}$$

By the maximum principle, the maximum and the minimum of u occur at the parabolic boundary Γ^t of D^t. If extrema of u lie on the lateral boundary then, by the Vyborny–Friedman theorem, $\partial u/\partial \nu \neq 0$ at the extremum points. Hence, by the rightmost equation of (14.169), the extrema of u are achieved on the base D_0 of D^t. But this means, by the middle equation of (14.169), that

$$\max u = \min u = 0 \Rightarrow u \equiv 0 \tag{14.170}$$

Q.E.D.

III. Conjugation problem

Let $\bar{D}^t$ be the closure of two nonintersecting parabolic regions D_1^t and D_2^t whose lateral boundaries Σ_1^t and Σ_2^t have the common part

$$\Sigma_{12}^t = \Sigma_1^t \cap \Sigma_2^t, \quad \Sigma_i^t = \Sigma_i^t \setminus \Sigma_{12}^t, \quad i = 1, 2 \tag{14.171}$$

(see Figure 14.6), where Σ_{12}^t is a smooth manifold. Consider the conjugation problem: Find a function $u(P)$ continuous in $\bar{D}^t$, parabolic in D_1^t, and a-parabolic in D_2^t, that satisfies the boundary conditions

$$\frac{\partial}{\partial \nu} u(P) = -h[u(P) - f_1(P)] \quad \forall P \in \Sigma_1^t, \quad u(P) = f_2(p) \quad \forall P \in \Sigma_2^t \tag{14.172}$$

initial conditions

$$u(P) = \varphi(P) \quad \forall P \in D_{10} \cup D_{20} \tag{14.173}$$

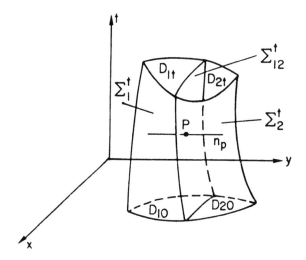

Figure 14.6. Sketch of the matching conditions; regions D_i^t and surface Σ_i^t, $i = 1, 2$.

and conjugation conditions

$$[u(P)] = 0, \quad \left[\lambda \frac{\partial}{\partial \nu} u(P)\right] = 0 \quad \forall P \in \Sigma_{12}^t \tag{14.174}$$

Here (as before) $[F]$ means the jump of F across the surface of discontinuity, whatever the function F;

$$\lambda = \begin{cases} k = \text{const} > 0 & P \in D_1^t \\ 1 & P \in D_2^t \end{cases} \tag{14.175}$$

ν is the normal to Σ_{12}^t or Σ_1^t directed outward with respect to D_1^t; and

$$h = h(P) \geq 0 \tag{14.176}$$

Theorem 14.5.3. Let Σ_{12}^t be a manifold every point of which can be touched by spheres from within both regions of D_i^t, $i = 1, 2$. Then the solution of the problem of conjugation, if it exists, is unique.

PROOF. Let u be the difference between two solutions u_1 and u_2 of the problem. Then

$$Lu = 0 \quad \forall P \in D_1^t, L_a u = 0 \quad \forall P \in D_2^t \tag{14.177}$$

$$\frac{\partial}{\partial \nu} u(P) = -h(P)u(P) \quad \forall P \in \Sigma_1^t \tag{14.178}$$

$$u(p) = 0 \quad \forall P \in \Sigma_2^t \tag{14.179}$$

$$[u(P)] = 0, \quad \left[\lambda \frac{\partial}{\partial \nu} u(P)\right] = 0 \quad \forall P \in \Sigma_{12}^t \tag{14.180}$$

$$u(p) = 0 \quad \forall P \in D_{10} \cup D_{20} \tag{14.181}$$

We must prove that $u \equiv 0$. Assume the contrary; for example, assume that

$$\max_{P \in \bar{D}^t} u(P) = u(Q) > 0 \qquad (14.182)$$

Then

$$Q \in (\Sigma_{12}^t \cup \Sigma_1^t) \qquad (14.183)$$

But $Q \in \Sigma_{12}^t$ is impossible since, by the Vyborny–Friedman theorem, in such a case $\partial u(P)/\partial \nu$ must change its sign across Σ_{12}^t at Q, which contradicts the second of conditions (14.180). Q cannot also lie at Σ_1^t because then, by the same theorem,

$$\frac{\partial}{\partial \nu} u(Q) > 0 \qquad (14.184)$$

which contradicts (14.178) and (14.182). Thus the assumption (14.182) is false.   **Q.E.D.**

Definition 14.5.1. A parabolic region D^t is said to be *fundamental* if the first boundary-value problem posed for the heat-conduction equation is solvable in it for any continuous boundary and initial data.[8]
   In what follows we shall assume (for the sake of simplicity) that $n \leq 3$, n being the dimension of the space R_n in which the heat-conduction equation is being considered. Our purpose is to prove the following.

Theorem 14.5.4 (Tichonov Comparison Theorem [81]). *Let D^t be a bounded fundamental parabolic region in $\mathcal{E}_{n+1}$, where n is taken equal to 3. Let $P = (x, \tau)$, $x = \{x_1, x_2, x_3\}$, be an arbitrarily chosen point of the lateral boundary Σ^t of D^t. Assume that Σ^t is defined by the equations*

$$x_i = x_i(u, v, \tau), \quad 0 \leq u, v \leq 1, \ 0 \leq \tau \leq t \qquad (14.185)$$

where x_i, $i = 1, 2, 3$, are continuously differentiable functions of all their arguments everywhere in their regions of definition.
   Let $U_i(P)$, $i = 1, 2$, be a_i parabolic functions in D^t, continuous in the closure $\bar{D}^t$ of D^t and such that

$$U_i(P) = f(P) \quad \forall P \in \Sigma^t \qquad (14.186)$$
$$U_i(p, 0) = \varphi(p) \quad \forall p \in D_0 \qquad (14.187)$$

where f and φ satisfy the following conditions:

$$P = [x(u, v, \tau), \tau] \in \Sigma^t, \quad P_0 = [x(u, v, 0), 0] \in \Sigma_0 \qquad (14.188)$$
$$f(P) \in C^1(\bar{D}^t), \quad \varphi(p) \in C^2(\bar{D}_0) \qquad (14.189)$$
$$a_1^2 \Delta \varphi(P_0) = \frac{\partial}{\partial \tau} f(P_0) - \sum_{i=1}^{3} \frac{\partial}{\partial x_i} f(P_0) \frac{\partial}{\partial \tau} x_i(u, v, 0) \qquad (14.190)$$

---

[8]Concerning conditions for a parabolic region to be fundamental, see Chapter 18, Sections 3 and 4.

$$\Delta\varphi(p) \le 0 \quad \forall p \in D_0 \tag{14.191}$$

$$\frac{\partial}{\partial\tau}f(P) - \sum_{i=1}^{3}\frac{\partial}{\partial x_i}f(P)\frac{\partial}{\partial\tau}x_i(u,v,\tau) \le 0 \quad \forall P \in \bar{\Sigma}^t \tag{14.192}$$

Then

$$a_1^2 > a_2^2 \Rightarrow U_1(P) < U_2(P) \quad \forall P \in D^t \tag{14.193}$$

PROOF. This theorem is an obvious corollary of the following lemma.

Lemma 14.5.1. *Under the assumptions of Theorem14.5.4, let*

$$V(P) \overset{\text{def}}{=} \frac{\partial}{\partial\tau}U_1(P) \tag{14.194}$$

Then $V(P)$ is a solution of the following boundary-value problem:

$$V(P) \in C(\bar{D}^t) \tag{14.195}$$

$$a_1^2\Delta V(P) = \frac{\partial}{\partial\tau}V(P) \quad \forall P \in D^t \tag{14.196}$$

$$V(p,0) = a_1^2\Delta\varphi(p) \quad \forall p \in D_0 \tag{14.197}$$

$$V(P) = \frac{\partial}{\partial\tau}f(P) - \sum_{i=1}^{3}\frac{\partial}{\partial x}f(P)\frac{\partial}{\partial\tau}x_i(u,v,\tau) \quad P \in \bar{\Sigma}^t \tag{14.198}$$

Assume that Lemma 14.5.1 is true. By virtue of the maximum principle, the inequalities (14.191) and (14.192) imply

$$\Delta U_1(P) \le 0 \quad \forall P \in \bar{D}^t \tag{14.199}$$

Now, let

$$U = U_1 - U_2 \tag{14.200}$$

Then the equalities

$$a_i^2 U_i(P) - \frac{\partial}{\partial\tau}U_i = 0 \quad \forall P \in D^t, \quad a^2 = a_1^2 - a_2^2 > 0 \tag{14.201}$$

imply

$$L_{a_2}U = -a^2\Delta U_1 > 0 \quad \forall P \in \bar{D}^t \tag{14.202}$$

This means, by (14.201), that U is a_2-subparabolic in D^t, continuous in $\bar{D}^t$, and vanishes identically on Γ^t. Hence, by virtue of the maximum principle,

$$U(P) < 0 \quad \forall P \in D^t \tag{14.203}$$

Q.E.D.

Thus it remains only to prove Lemma 14.5.1.

PROOF OF LEMMA 14.5.1. Let $\varphi^*(p)$ be the continuation of $\varphi(p)$ to a neighborhood D_0^* of D_0 as a function that is convex everywhere in D_0^* and twice continuously differentiable in $\bar{D}_0^*$. Let $\Pi(p,\tau)$ be the Poisson integral

$$\Pi(p,\tau) = \int_{D_0^*}\varphi^*(q)E_n(p,q,a_1^2\tau)d\omega_q \tag{14.204}$$

We have

$$\frac{\partial}{\partial \tau}\Pi(p,\tau) = a_1^2 \int_{D_0^*} \varphi^*(q)\Delta E_n(p,q,a_1^2\tau)d\omega_q \tag{14.205}$$

Hence, by Green's second formula,

$$\frac{\partial}{\partial \tau}\Pi(p,\tau) = a_1^2 \int_{D_0^*} E_n(p,q,a_1^2\tau)\Delta\varphi^*(q)d\omega_q$$

$$- a_1^2 \int_{\Sigma_0^*} \left(\varphi^*(q)\frac{\partial}{\partial n_q}E_n(p,q,a_1^2\tau) \right.$$

$$\left. - E_n(p,q,a_1^2\tau)\frac{\partial}{\partial n_q}\varphi^*(q) \right)d\omega_q \tag{14.206}$$

Using the filtering property of E_n, we find that

$$\lim_{\tau\downarrow 0}\frac{\partial}{\partial \tau}\Pi(p,\tau) = \Delta\varphi(p) \quad \forall p \in \bar{D}_0 \tag{14.207}$$

Now define $W_1(P)$ by the equality

$$W_1(P) = U_1(P) - \Pi(P) \tag{14.208}$$

Since $\Pi(P)$ is a_1-parabolic in D^t and

$$\Pi(p,0) = \varphi(p) \quad \forall p \in \bar{D}_0 \tag{14.209}$$

we see that

$$L_{a_1}W_1(P) = 0 \quad \forall P \in D^t \tag{14.210}$$
$$W_1(p,0) = 0 \quad \forall p \in D_0 \tag{14.211}$$

$$W_1(P) \overset{\text{def}}{=} F(P) = f(p,0) - \Pi(p,0) = 0 \quad \forall p \in \Sigma_0 \tag{14.212}$$

Moreover, by (14.189), (14.190), and (14.207),

$$\frac{\partial}{\partial \tau}F(p,0) = 0 \quad \forall p \in \Sigma_0 \tag{14.213}$$

Let $\bar{D}_{t_0}^t$ be an extension of $\bar{D}^t$ as a fundamental region, where $t_0 = \text{const} < 0$. Define a function W_1^* by the conditions

$$L_{a_1}W_1^* = 0 \quad \forall P \in D_{t_0}^t \tag{14.214}$$
$$W_1^*(p,t_0) = 0 \quad \forall p \in D_{t_0}^t \tag{14.215}$$

$$W_1^*(P) = \begin{cases} 0 & \forall P \in \bar{\Sigma}_t^0 \\ \dfrac{\partial}{\partial \tau}F(P) & \forall P \in \bar{\Sigma}_0^t \end{cases} \tag{14.216}$$

By (14.213), the function $W_1^*(P)$ is continuous on $\Sigma_{t_0}^t$. By construction, $D_{t_0}^t$ is a fundamental region, which means that W_1^* exists. Now let

$$\tilde{W}_1(P) = \int_{t_0}^t W_1^*(p,\tau)d\tau \tag{14.217}$$

Conditions (14.214) and (14.215) imply

$$L_{a_1}\tilde{W}_1(p,t) = \int_{t_0}^t a_1^2 \Delta W_1^*(p,\tau)d\tau - W_1^*(p,t)$$

$$= \int_{t_0}^t \left(a_1^2 \Delta W_1^*(p,\tau)d\tau - \frac{\partial}{\partial \tau}W_1^*(p,t)\right)d\tau$$

$$= \int_{t_0}^t L_{a_1}W_1^*(p,\tau)d\tau = 0 \tag{14.218}$$

Since, by the uniqueness theorem,

$$W_1^*(P) \equiv 0 \quad \forall P \in \bar{D}_{t_0}^0 \tag{14.219}$$

it follows from the definition (14.217) that

$$\tilde{W}_1(P) = \int_0^t W_1^*(p,\tau)d\tau \quad \forall t \geq 0, \ \forall P \in D_0^t \tag{14.220}$$

and consequently

$$\tilde{W}_1(P) = F(P) \quad \forall P \in \Sigma_0^t \tag{14.221}$$

Comparing equalities (14.210)–(14.212) and (14.218), (14.219), we conclude, by virtue of the uniqueness theorem, that

$$\tilde{W}_1(P) = W_1(P) \quad \forall P \in D^t \tag{14.222}$$

so that

$$\frac{\partial}{\partial \tau}W_1(P) = W_1^*(P) \quad \forall P \in D^t \tag{14.223}$$

Together with (14.208), this shows that

$$V(P) \stackrel{\text{def}}{=} \frac{\partial}{\partial \tau}U_1(P) = W_1^*(P) + \frac{\partial}{\partial \tau}\Pi(P) \tag{14.224}$$

Hence $V(P)$ satisfies all conditions (14.195)–(14.198). **Q.E.D.**

6. Remarks on time irreversibility in parabolic equations

Recall that the boundary-value problems posed for heat-conduction equations with reversed time are ill-posed in the Hadamard sense (see Chapter 2, Section 12, particularly, the problem of historical climatology), whereas those with forward time are well-posed owing to the validity of the maximum principle. We now show that the ill-posedness of heat-conduction equations with reversed time results from a specific irreversibility in time of operators of normal parabolic type.

Regarding their properties with respect to time reversal, the following types of evolutionary operators can be distinguished.

A. Operators with reversible time

These operators are invariant with respect to replacement of the time t by $-t$. Such differential operators contain partial derivatives with respect to time of the even order only, as in (for example) the wave operator

$$\Box(p,t\mid u) \stackrel{\text{def}}{=} \Delta u - \frac{\partial^2}{\partial t^2}u \qquad (14.225)$$

It is obvious that for an operator with reversible time, evolutionary boundary-value problems with increasing and decreasing time are solvable in the same functional spaces.

B. Operators with irreversible time

These operators are *not* invariant with respect to the time transformation $t \to -t$. Time derivatives of an odd order necessarily occur in this type of differential operator as in (for example) the hyperbolic diffusion operator and the heat-conduction operator:

$$H(p,t\mid u) \stackrel{\text{def}}{=} \Delta u - \tau\frac{\partial^2}{\partial t^2}u - \frac{\partial}{\partial t}u, \quad L(p,t\mid u) = \Delta u - \frac{\partial}{\partial t}u \qquad (14.226)$$

Operators with irreversible time must be divided into two groups:

(b$_1$) operators with parabolically irreversible time; and
(b$_2$) operators with hyperbolically reversible time.

Operators with *parabolically irreversible* time are those for which the initial–boundary-value problems with increasing time are solvable in an appropriate functional space, whereas the same boundary-value problem with reversed time is, generally speaking, unsolvable in the same functional space. Operators with *hyperbolically reversible* time are those where the initial–boundary-value problems with increasing and decreasing time are simultaneously solvable.

In order to understand the essence of parabolic time irreversibility, consider the following problem. Let $D_{T_o}^T$, $T_o < t \leq T$, $T_o < 0$, be a parabolic region fundamental with respect to a parabolic Dirichlet problem, $f(p,t)$ a function continuous on $\Sigma_{T_o}^T$, and $\varphi(p)$ a function defined in D_0 as continuous everywhere except at a point q of discontinuity of the first kind. Let $u(p,t)$ be a solution of the Dirichlet problem

$$\Delta u(p,t) - \frac{\partial}{\partial t}u(p,t) = 0 \quad \forall (p,t) \in D_0^T \qquad (14.227)$$

$$u(p,t) = f(p,t) \quad \forall (p,t) \in \Sigma_0^T \qquad (14.228)$$

$$u(p,0) = \varphi(p) \quad \forall p \in D_0 \qquad (14.229)$$

Assume that this problem is with hyperbolically reversible time. Then the problem

$$\Delta \tilde{u}(p,t) = \frac{\partial}{\partial t}\tilde{u}(p,t) \quad \forall (p,t) \in D_{T_o}^T \qquad (14.230)$$

$$\tilde{u}(p,t) = f(p,t) \quad \forall (p,t) \in \Sigma_{T_o}^T \tag{14.231}$$

$$\tilde{u}(p,T) = u(p,T) \quad \forall p \in D_T \tag{14.232}$$

is solvable and, moreover, by the uniqueness theorem,

$$\tilde{u}(p,t) \equiv u(p,t) \quad \forall (p,t) \in D_0^T \tag{14.233}$$

so that in particular

$$\tilde{u}(p,-0) = \varphi(p) \quad \forall p \in D_0 \tag{14.234}$$

Let now $\tilde{U}(p,t)$ be a solution of the problem

$$\Delta \tilde{U}(p,t) = \frac{\partial}{\partial t}\tilde{U}(p,t) \quad \forall (p,t) \in D_{T_o}^T \tag{14.235}$$

$$\tilde{U}(p,t) = \tilde{u}(p,t) \quad \forall (p,t) \in \Sigma_{T_o}^T \tag{14.236}$$

$$\tilde{U}(p,T_o) \equiv \tilde{u}(p,T_o) \quad \forall p \in D_{T_o} \tag{14.237}$$

so that, by the uniqueness theorem,

$$\tilde{U}(p,t) \equiv \tilde{u}(p,t) \Rightarrow \tilde{U}(p,0) = \varphi(p) \quad \forall p \in D_0 \tag{14.238}$$

But $\tilde{U}(p,t)$ is continuous everywhere in $D_{T_o}^T$, since discontinuities of the solution of a parabolic equation cannot propagate in time (see Chapter 4, Sections 6 and 7). Hence (14.234) is impossible. Thus the assumption of the hyperbolic time reversibility in heat-conduction equation is false. **Q.E.D.**

Obviously, the parabolic time irreversibility is the property not only of the heat-conduction equation but of any normal parabolic equation as well.

The time irreversibility in heat-conduction equations is of course the result of neglecting the fast relaxation time (i.e., inertia terms) in the hyperbolic heat-conduction equation, or (what is the same) in the equation of oscillations of a spring with distributed mass that is immersed in an immovable liquid resisting oscillations (see Chapter 2, Section 6). The time in this equation is hyperbolically reversible, as follows from the properties of solutions of Cauchy and Goursat problems considered in Chapter 6 and from the theory of Sturm–Liouville operators (see Appendix 3). Let us emphasize that in this case the mechanical energy dissipates and the spring tends asymptotically to equilibrium as $t \uparrow \infty$, but the amplitude of the displacements of the spring points is not a monotonic function of time.[9] Thus not only the dissipation is responsible for the parabolic time irreversibility in the heat-conduction equation. Dissipation entails the existence in an evolutionary system of a functional of the state variables (intensive variables) – for example, mechanical energy or entropy – that varies strictly monotonically in time; this is just an instance of the second

---

[9]The dissipation of mechanical energy in the hyperbolic heat-conduction equation can be traced quite similarly to the analysis of the heat-conduction equation in moving liquids (see Chapter 2, Section 4; see also Problem 14.6.1).

law of thermodynamics. On the other hand, parabolic time irreversibility is a very extreme realization of this law, typical for inertialess dissipative systems.

An insight into the physical nature if this irreversibility can be inferred from consideration of the second moment of the Cauchy problem solution for the diffusion (or heat-conduction) equation. Indeed, this moment $M_2(t)$ – which is the moment of inertia of a system of particles diffusing on a line – increases linearly in time (see Chapter 13, Section 3). Hence, while tracing $M_2(t)$ back in time, it will decrease with a constant rate. Since $M_2(t)$ is a positively determined form it cannot become negative, so that the solution of the problem cannot be traced backward in time farther than some particular moment ($t = 0$, depending on the problem's formulation).

The oscillatory character of solutions of hyperbolic equations (where the hyperbolicity is the result of the presence of inertia terms) is essentially different from the monotonicity in time of solutions of parabolic equations (with monotonic boundary data), which is a corollary of the maximum principle. The latter is the most important property of parabolic (i.e., inertialess) systems, and the absence of the maximum principle for hyperbolic operators is the fundamental distinction between these two types of operators.

Problems

P14.1.1. Consider the parabolic operator with reversed time:

$$L^- u = \Delta u + \frac{\partial}{\partial t} u$$

Give a suitable definition of a parabolic region for this operator.

P14.2–3.1. Is it possible to apply the weak and strong maximum principles to the operator L^-? If not, explain how the formulation of these principles should be changed in order to make them applicable. If so, propose an appropriately formulated proof of the weak maximum principle.

P14.2–3.2. Does there exist an analogue of the maximum principle for the hyperbolic heat-conduction equation? Verify your answer rigorously.

P14.4.1. Let $D^t = \{x, t: (x - t)^2 = 2t, \ 0 < t < \infty\}$. Let the function $u(x, t)$ be continuous in $\bar{D}^t$, subparabolic within D^t, not identically equal to a constant, and achieving a single maximum at the point $t = x = 0$. Is the Vyborny–Friedman theorem applicable to this function?

P14.5.1. Is it possible to remove the convexity condition imposed on the initial value of u in the formulation of Tichonov's theorem 14.5.4?

P14.6.1. Derive the total mechanical energy equation for the longitudinal string vibrations with friction (see Chapter 2, Section 6) (hyperbolic diffusion equation, see Chapter 2, Section 9). Consider an isolated system of diffusing particles whose behavior is described by the hyperbolic diffusion equation, and prove that this system tends asymptotically to the equilibrium as $t \uparrow \infty$.

15

Application of Green's formulas. Fundamental identity. Green's functions for Fourier equation

1. Fundamental identity

We now derive an identity, called the *fundamental identity*, that plays a role identical to that of the fundamental identity in the theory of harmonic functions.

Let $D_{t_0}^t$ be a normal parabolic region in $\mathcal{E}_{3+1}$ with lateral boundary $\Sigma_{t_0}^t$ and the lower base D_{t_0}. Assume that for all $\tau \in [t_0, t]$, Σ_τ may be defined by an equation

$$\Sigma_\tau = \{x, y, z : F(x, y, z \mid \tau) = 0\} \tag{15.1}$$

where (x, y, z) are orthogonal cartesian coordinates in $\mathbb{R}_3$ and F is a continuously differentiable function of all its arguments. Let

$$u(q, \tau), v(q, \tau) \in C^{2,1}(D_{t_0}^t) \cap C^{1,0}(\bar{D}_{t_0}^t) \tag{15.2}$$

Consider the adjoint operators

$$L_a u = a^2 \Delta u - \frac{\partial}{\partial \tau} u, \quad M_a v = a^2 \Delta v + \frac{\partial}{\partial \tau} v \tag{15.3}$$

Integration over $D_{t_0}^t$ yields

$$\int_{t_0}^t d\tau \int_{D_\tau} [v(q, \tau) L_a u(q, \tau) - u(q, \tau) M_a v(q, \tau)] d\omega_q$$

$$= a^2 \int_{t_0}^t d\tau \int_{\Sigma_\tau} [v(q, \tau) \Delta u(q, \tau) - u(q, \tau) \Delta v(q, \tau)] d\omega_q$$

$$- \int_{t_0}^t d\tau \int_{D_\tau} \frac{\partial}{\partial \tau} [v(q, \tau) u(q, \tau)] d\omega_q \tag{15.4}$$

354

By the second Green's formula,

$$a^2 \int_{t_0}^{t} d\tau \int_{D_\tau} [v(q,\tau)\Delta u(q,\tau) - u(q,\tau)\Delta u(q,\tau)] d\omega_q$$

$$= a^2 \int_{t_0}^{t} d\tau \int_{\Sigma_\tau} [v(q,\tau)\frac{\partial}{\partial n_q} u(q,\tau) - u(q,\tau)\frac{\partial}{\partial n_q} v(q,\tau)] d\omega_q \qquad (15.5)$$

where **n** is the outward normal to Σ_τ at $(q,\tau) \in \Sigma_\tau$. Now, by the rule for differentiation of time-dependent integrals (see (1.140)),

$$\int_{D_\tau} \frac{\partial}{\partial \tau}[v(q,\tau)u(q,\tau)] d\omega_q = \frac{\partial}{\partial t} \int_{D_\tau} [v(q,\tau)u(q,\tau)] d\omega_q$$

$$- \int_{\Sigma_\tau} [v(q,\tau)u(q,\tau)] \frac{\partial}{\partial \tau} n_q d\sigma_q \qquad (15.6)$$

where $\partial n_q/\partial \tau$ is the rate of displacement of Σ_τ in the instant direction of its normal $\mathbf{n}_q$ at the point (q,τ), that is,

$$\frac{\partial}{\partial \tau} n_q = -\frac{\frac{\partial F}{\partial \tau}}{\left[\left(\frac{\partial F}{\partial x}\right)^2 + \left(\frac{\partial F}{\partial y}\right)^2 + \left(\frac{\partial F}{\partial x}\right)^2\right]^{1/2}} = -\frac{\frac{\partial F}{\partial \tau}}{|\operatorname{grad} F|} \qquad (15.7)$$

Thus

$$\int_{t_0}^{t} d\tau \int_{D_\tau} \frac{\partial}{\partial \tau}[v(q,\tau)u(q,\tau)] d\omega_q$$

$$= \int_{D_t} [v(q,t)u(q,t)] d\omega_q - \int_{D_{t_0}} [v(q,t_0)u(q,t_0)] d\omega_q$$

$$- \int_{t_0}^{t} d\tau \int_{\Sigma_\tau} [v(q,\tau)u(q,\tau)] \frac{\partial}{\partial \tau} n_q d\sigma_q \qquad (15.8)$$

Inserting (15.8) into (15.4), we obtain

$$\int_{t_0}^{t} d\tau \int_{D_\tau} [v(q,\tau)L_a u(q,\tau) - u(q,\tau)M_a u(q,\tau)] d\omega_q$$

$$= a^2 \int_{t_0}^{t} d\tau \int_{\Sigma_\tau} \left[v(q,\tau)\frac{\partial}{\partial n_q} u(q,\tau) \right.$$

$$- u(q,\tau)\left(\frac{\partial}{\partial n_q} v(q,\tau) - a^{-2} v(q,\tau)\frac{\partial}{\partial \tau} n_q \right) \right] d\sigma_q$$

$$- \int_{D_t} v(q,t)u(q,t) d\omega_q + \int_{D_{t_0}} v(q,t_0)u(q,t_0) d\omega_q \qquad (15.9)$$

Assume now that

$$M_a v \equiv 0 \quad \forall Q \in D^t \qquad (15.10)$$

This yields

$$\int_{D_t} [v(q,t)u(q,t)]d\omega_q$$

$$= \int_{D_{t_0}} [v(q,t_0)u(q,t_0)]d\omega_q$$

$$+ a^2 \int_{t_0}^t d\tau \int_{\Sigma_\tau} \left[v(q,\tau) \frac{\partial}{\partial n_q} u(q,\tau) \right.$$

$$\left. - u(q,\tau) \left(\frac{\partial}{\partial n_q} v(q,\tau) - a^{-2} v(q,\tau) \frac{\partial}{\partial \tau} n_q \right) \right] d\sigma_q$$

$$- \int_{t_0}^t d\tau \int_{D_\tau} v(q,\tau) L_a u(q,\tau) d\omega_q \qquad (15.11)$$

This identity does not convey much information. But if v possesses filtering properties, it becomes the desired fundamental identity. To obtain it, set

$$v(q,\tau) = E[p,q,a^2(t-\tau+\varepsilon)] \qquad (15.12)$$

where ε is an arbitrarily small positive number and E is a fundamental solution of the Fourier equation in $\mathcal{E}_{3+1}$, that is, the solution of the equation $M_a E = 0$ considered as a function of (q,τ). Substituting (15.12) into (15.11), we obtain

$$\int_{D_t} u(q,t)E(p,q,a^2\varepsilon)d\omega_q$$

$$= \int_{D_{t_0}} u(q,t_0)E[p,q,a^2(t-t_0+\varepsilon)]d\omega_q$$

$$+ a^2 \int_{t_0}^t d\tau \int_{\Sigma_\tau} \left[E[p,q,a^2(t-t_0+\varepsilon)] \frac{\partial}{\partial n_q} u(q,\tau) \right.$$

$$\left. - u(q,\tau) \left(\frac{\partial}{\partial n_q} - a^{-2} \cdot \frac{\partial}{\partial \tau} n_q \right) E[p,q,a^2(t-t_0+\varepsilon)] \right] d\sigma_q$$

$$- \int_{t_0}^t d\tau \int_{D_\tau} E[p,q,a^2(t-t_0+\varepsilon)] L_a u(q,\tau) d\omega_q \qquad (15.13)$$

By the filtering property of $E(p,q,a^2\varepsilon)$ (see Theorem 13.5.1), passage to the limit $\varepsilon \downarrow 0$ yields

$$\nu(p,t)u(p,t) = \int_{D_{t_0}} u(q,t_0)E[p,q,a^2(t-t_0)]d\omega_q$$

$$+ a^2 \int_{t_0}^t d\tau \int_{\Sigma_\tau} \left[E[p,q,a^2(t-t_0)] \frac{\partial}{\partial n} u(q,\tau) \right.$$

$$\left. - u(q,\tau) \left(\frac{\partial}{\partial n_q} - a^{-2} \frac{\partial}{\partial \tau} n_q \right) E[p,q,a^2(t-t_0)] \right] d\sigma_q$$

$$- \int_{t_0}^t d\tau \int_{D_\tau} E[p,q,a^2(t-t_0)] L_a u(q,\tau) d\omega_q \qquad (15.14)$$

where

$$\nu(p,t) = \begin{cases} 1 & \text{if } (p,t) \in D^t \\ 1/2 & \text{if } (p,t) \in \Sigma^t \\ 0 & \text{if } (p,t) \notin \bar{D}^t \end{cases} \tag{15.15}$$

It should be emphasized that this fundamental identity is applicable to any function

$$u(p,t) \in C^{2,1}(D_{t0}^t) \cap C^{1,0}(\bar{D}_{t0}^t \setminus D_{t0}) \tag{15.16}$$

Quite similarly to what is true for the Laplace operator, the fundamental identity (15.14) admits a simple but very important generalization. Namely, the fundamental solution $E(p,q,a^2t)$ may be replaced by any function $g(p,t,q,\tau)$ such that (see (4.159))

$$M_a g(p,t,q,\tau) = 0 \quad \forall (p,t) = \text{const} \in D^t, \ \forall (q,\tau) \in \bar{D}^t \tag{15.17}$$

which possesses the same filtering properties as $E[p,q,a^2(t-\tau)]$. Taking this into account, let us put

$$g(p,a^2t,q,a^2\tau) = E[p,q,a^2(t-\tau)] - g^*(p,a^2t,q,a^2\tau) \tag{15.18}$$

where (p,t) is assumed to be a fixed and (q,τ) a variable point of D_{t0}^t, and

$$M_a g^* = 0 \quad \forall (q,\tau) \in D^t \tag{15.19}$$

Replacing E by g, we obtain

$$\nu(p,t)u(p,t)$$

$$= a^2 \int_{t_0}^t d\tau \int_{\Sigma_\tau} \left[g[p,q,a^2(t-\tau)] \frac{\partial}{\partial n_q} u(q,\tau) \right.$$

$$\left. - u(q,\tau) \left(\frac{\partial}{\partial n_q} - a^{-2} \frac{\partial}{\partial \tau} n_q \right) g(p,a^2t,q,a^2\tau) \right] d\sigma_q$$

$$+ \int_{D_{t0}} u(y,t_0) y[\nu, u^2 l, y, u^2 t_0] d\omega_q$$

$$- \int_{t_0}^t d\tau \int_{D_\tau} g(p,a^2t,q,a^2\tau) L_a u(q,\tau) d\omega_q \tag{15.20}$$

Note that in cylindrical regions $\partial n_q / \partial \nu \equiv 0$; in the 1-dimensional region

$$D^t = \{x,t: X_1(t) < x < X_2(t), \ 0 < t\} \tag{15.21}$$

equation (15.19) becomes

$$\nu(x,t)u(x,t)$$

$$= a^2 \int_0^t \left[g(x,a^2t,\xi,a^2\tau) \frac{\partial}{\partial \xi} u(\xi,\tau) \right.$$

$$\left. - u(\xi,\tau) \left(\frac{\partial}{\partial \xi} - a^{-2} \frac{\partial}{\partial \tau} \xi \right) g(x,a^2t,\xi,a^2\tau) \right] \Bigg|_{\xi=X_1(\tau)}^{X_2(\tau)} d\tau$$

$$+ \int_D u(\xi, 0) g(x, a^2 t, \xi, 0) d\xi$$

$$- \int_0^t d\tau \int_{X_1(\tau)}^{X_2(\tau)} g(x, a^2 t, \xi, a^2 \tau) L_a u(q, \tau) d\xi \qquad (15.22)$$

The region $X_1(t) < x < X_2(t)$, incidentally, may be semi-infinite or infinite. In the latter case the boundary-value problem becomes a Cauchy problem. The same is true of the multidimensional case if the region $D^t = \mathbb{R}_n \times (0 < \tau \le t)$.

2. Application of first Green's formula and uniqueness theorems

Apply now not the second but the first Green's formula:

$$\int_{D_\tau} v(q, \tau) \Delta u(q, \tau) d\omega_q = \int_{\Sigma_\tau} v(q, \tau) \frac{\partial}{\partial n_q} u(q, \tau) d\omega_q$$

$$- \int_{D_\tau} \operatorname{grad} v(q, \tau) \cdot \operatorname{grad} u(q, \tau) d\omega_q \qquad (15.23)$$

which yields

$$\int_0^t d\tau \int_{D_\tau} v(q, \tau) L u(q, \tau) d\omega_q = \int_0^t d\tau \int_{\Sigma_\tau} v(q, \tau) \frac{\partial}{\partial n_q} u(q, \tau) d\omega_q$$

$$- \int_0^t d\tau \int_{D_\tau} \operatorname{grad} v(q, \tau) \cdot \operatorname{grad} u(q, \tau) d\omega_q$$

$$- \int_0^t d\tau \int_{D_\tau} v(q, \tau) \frac{\partial}{\partial \tau} u(q, \tau) d\omega_q \qquad (15.24)$$

This identity may be used to prove uniqueness theorems, just as the first Green's formula is used to prove uniqueness theorems for the first, second, and third boundary-value problems for the Poisson equation. To fix ideas, let us consider the following boundary-value problem: Find a function

$$u(p, t) \in C^{2,1}(D^t) \cap C(\bar{D}^t) \qquad (15.25)$$

such that

$$Lu(p, t) + F(p, t) = 0 \quad \forall (p, t) \in D^t, u(p, 0) = \varphi(p) \quad \forall p \in D_0 \qquad (15.26)$$

$$u(p, t) = f_1(p, t) \quad \forall (p, t) \in \bar{\Sigma}_1^t, \frac{\partial}{\partial n_q} u(p, t) = f_2(p, t) \quad \forall (p, t) \in \Sigma_2^t \qquad (15.27)$$

where

$$\bar{\Sigma}_1^t \cup \Sigma_2^t = \Sigma^t \qquad (15.28)$$

$$\frac{\partial}{\partial n_p} u(p, t) = \lim_{q \to p} \frac{\partial}{\partial n_p} u(q, t) \in C(\Sigma_2^t), \quad (q, t) \in D^t \cap \mathbf{n}_p \qquad (15.29)$$

Theorem 15.2.1. *If*

$$\frac{\partial}{\partial t} n_p \le 0 \quad \forall (p, t) \in \Sigma_2^t \qquad (15.30)$$

then the solution of problem (15.25)–(15.29), *if it exists, is unique.*

Remark 1. Applying the thermal interpretation, one might explain condition (15.30) as meaning that the surface Σ^t bounding the thermally conductive body does not expand in that part of it where there is a prescribed influx of heat.

PROOF. If u_1 and u_2 are two solutions of the problem, then their difference

$$u = u_1 - u_2 \tag{15.31}$$

is the solution of the corresponding homogeneous problem, so that, taking $v \equiv u$, we obtain from (15.24)

$$\int_0^t d\tau \int_{D_\tau} [\operatorname{grad} u(q,\tau)]^2 d\omega_q = -\frac{1}{2} \int_0^t d\tau \int_{D_\tau} \frac{\partial}{\partial \tau} [u(q,\tau)]^2 d\omega_q$$

$$\equiv -\frac{1}{2} \int_{D_t} [u(q,t)]^2 d\omega_q$$

$$+ \frac{1}{2} \int_0^t d\tau \int_{\Sigma_2^\tau} u(q,\tau)^2 \frac{\partial}{\partial \tau} n_q d\sigma_q \tag{15.32}$$

By assumption (15.30), this yields

$$\frac{\partial}{\partial \tau} [u(q,\tau)] \equiv 0 \quad \forall (q,\tau) \in D^t \tag{15.33}$$

since the left- and right-hand sides of (15.32) are either different in sign or identically zero. But $u \equiv 0$ at $t = 0$ and (15.33) is true, so we have

$$u(q,\tau) \equiv 0 \quad \forall (q,\tau) \in D^t \tag{15.34}$$

Q.E.D.

Remark 2. If the subregion Σ_2^t is empty, then Theorem 15.2.1 coincides with Theorem 14.5.1. If the subregion Σ_1^t is empty, the Theorem 15.2.1 coincides with Theorem 14.5.2 except for the requirement (15.30) on the one hand and the assumption on applicability of the Vyborny–Friedman theorem on the other hand. This means, in particular, that condition (15.30) is redundant. This only emphasizes the shortcoming of the arguments used to prove Theorem 15.2.1 when compared with direct references to the maximum principle and its corollaries.

3. Green's functions

Based on the generalized fundamental identity (15.19), we can introduce Green's functions in the case of linear boundary-value problems for the Fourier equation. Thus, consider the boundary-value problem

$$L_a u(p,t) + F(p,t) = 0, \quad \forall (p,t) \in D^t, \ \forall t > 0 \tag{15.35}$$

$$\alpha(p,t) \frac{\partial}{\partial n} u(p,t) + \beta(p,t) u(p,t) = f(p,t) \quad \forall (p,t) \in \Sigma^t$$

$$u(p,0) = \varphi(p) \quad \forall p \in D \tag{15.36}$$

where α, β, f, φ, and F are prescribed functions such that

$$\alpha^2(p,t) + \beta^2(p,t) > 0 \quad \forall(P) \in \bar{\Sigma}_{t_0}^t \tag{15.37}$$

$$\alpha(p,t) \equiv 0 \Rightarrow \beta(p,t) \equiv 1, \beta(p,t) \equiv 0 \Rightarrow \alpha(p,t) \equiv 1 \tag{15.38}$$

$$\alpha(p,t) \neq 0, \ \beta(p,t) \neq 0 \ \forall(p,t) \in \Sigma_{t_0}^t \tag{15.39}$$

$$\Rightarrow \alpha \equiv 1, \ \beta \stackrel{\text{def}}{=} h(p,t), \ f(p,t) \stackrel{\text{def}}{=} h(p,t)\psi(p,t)$$

Thus conditions (15.38) and (15.39) correspond to the first, second, and third boundary-value problem, respectively.

Let $u(p,t)$ be a solution to the problem. Introduce all the input data into the generalized fundamental identity (15.19), which may be rewritten as

$$u(p,t) = \int_D \varphi(q)g(p, a^2t, q, 0)d\omega_q + I(p,t \mid f, \alpha, \beta)$$

$$+ \int_0^t d\tau \int_{D_\tau} F(q,\tau)g[p, a^2t, q, a^2\tau]d\omega_q \tag{15.40}$$

where

$$I(p,t \mid \alpha, \beta) = -a^2 \int_0^t d\tau \int_{\Sigma_\tau} f(q,\tau)\left(\frac{\partial}{\partial n_q} - a^{-2}\frac{\partial}{\partial \tau}n_q\right)g(p, a^2t, q, a^2\tau)d\sigma_q$$

$$+ a^2 \int_0^t d\tau \int_{\Sigma_\tau} g[p, a^2t, q, a^2\tau]\frac{\partial}{\partial n_q}u(q,\tau)d\sigma_q \tag{15.41}$$

for the first boundary-value problem,

$$I(p,t \mid \alpha, \beta) = a^2 \int_0^t d\tau \int_{\Sigma_\tau} f(q,\tau)g(p, a^2t, q, a^2\tau)d\sigma_q$$

$$- a^2 \int_0^t d\tau \int_{\Sigma_\tau} u(q,\tau)\left(\frac{\partial}{\partial n_q} - a^{-2}\frac{\partial}{\partial \tau}n_q\right)g(p, a^2t, q, a^2\tau)d\sigma_q \tag{15.42}$$

for the second boundary-value problem, and

$$I(p,t \mid \alpha, \beta) = a^2 \int_0^t d\tau \int_{\Sigma_\tau} h(q,\tau)\psi(q,\tau)g(p, a^2t, q, a^2\tau)d\sigma_q$$

$$- a^2 \int_0^t d\tau \int_{\Sigma_\tau} u(q,\tau)\left[\frac{\partial}{\partial n_q} + \left(h(q,\tau) - a^{-2}\frac{\partial}{\partial \tau}n_q\right)\right.$$

$$\left. \times g(p, a^2t, q, a^2\tau)\right]d\sigma_q \tag{15.43}$$

for the third boundary-value problem.

The second integrals on the right of (15.41)–(15.43) are unknown, since the integrands include factors dependent on $u(q,\tau)$ that are not prescribed in the formulations of the problem. However, as in the case of the Dirichlet problem in the theory of elliptic boundary-value problems, these unknown integrals may be eliminated by a special choice of the function g. Let us use the following terminology.

Definition 15.3.1. Let u and v be solutions of the adjoint equations

$$L_a u + F = 0, \quad M_a v = 0 \quad \forall (p,t) \in D^t \qquad (15.44)$$

Then the boundary conditions

$$u = f \quad \text{and} \quad v = \varphi \qquad (15.45)$$

$$\frac{\partial}{\partial n_q} u = f \quad \text{and} \quad \left(\frac{\partial}{\partial n_q} - a^{-2} \frac{\partial}{\partial \tau} n_q \right) v = \varphi \qquad (15.46)$$

$$\frac{\partial}{\partial n_q} u + h u = f \quad \text{and} \quad \frac{\partial}{\partial n_q} + \left(h(q,\tau) - a^{-2} \frac{\partial}{\partial \tau} n_q \right) v = \varphi \qquad (15.47)$$

are said to be *adjoint*. Operators and boundary conditions that equal their adjoints are said to be *self-adjoint*. If the operators and boundary conditions are self-adjoint then the corresponding boundary-value problems are said to be self-adjoint. In view of this terminology, we see that the parabolic operator is not self-adjoint. Indeed, the adjoint operator M_a of the parabolic operator L_a is a parabolic operator with reversed time. The boundary conditions corresponding to the second and third boundary-value problems are self-adjoint only in cylindrical regions, whereas in time-dependent regions they are non–self-adjoint. In contrast to this, the boundary conditions of the Dirichlet problem are self-adjoint in both cylindrical and noncylindrical regions.

We now define $g_i(p, a^2 t, q, a^2 \tau)$ for all $(p,t) \in D^t$ and $(q,\tau) \in \bar{D}^t$, $i = 1, 2, 3$, by

$$g_i(p, a^2 t, q, a^2 \tau) = E[p, q, a^2(t - \tau)] - g_i^*(p, a^2 t, q, a^2 \tau), \qquad (15.48)$$
$$i = 1, 2, 3$$

$$M_a g_i^*(p, a^2 t, q, a^2 \tau) = 0 \quad \forall (p,t) = \text{const} \in D^t, \ \forall (q,\tau) \in D^t, \qquad (15.49)$$
$$i = 1, 2, 3$$

$$g_i^*(p, a^2 t, a^2 q, t) = 0 \quad \forall (q,t) \in D^t, \quad i = 1, 2, 3 \qquad (15.50)$$

$$g_1^*(p, a^2 t, q, a^2 \tau) = E[p, q, a^2(t - \tau)] \quad \forall (q,\tau) \in \Sigma^t \qquad (15.51)$$

$$\left(\frac{\partial}{\partial n_q} - a^{-2} \frac{\partial}{\partial \tau} n_q \right) g_2^* = \left(\frac{\partial}{\partial n_q} - a^{-2} \frac{\partial}{\partial \tau} n_q \right) E \quad \forall (q,\tau) \in \Sigma^t \qquad (15.52)$$

$$\left[\frac{\partial}{\partial n_q} + \left(h(q,\tau) - a^{-2} \frac{\partial}{\partial \tau} n_q \right) \right] g_3^*$$
$$= \left[\frac{\partial}{\partial n_q} + \left(h(q,\tau) - a^{-2} \frac{\partial}{\partial \tau} n_q \right) \right] E \quad \forall (q,\tau) \in \Sigma^t \qquad (15.53)$$

so that g_i, $i = 1, 2, 3$, satisfy homogeneous boundary conditions adjoint to those of the first, second, and third boundary-value problem, respectively. Thus, if $u_i(p, t)$ is a solution of the ith boundary-value problem $(i = 1, 2, 3)$, then for all $(p, t) \in D^t$ and for any $t > 0$,

$$u_i(p, t) = \int_D \varphi(q) g_i(p, a^2 t, q, 0) d\omega_q + I_i(p, t)$$

$$+ \int_0^t d\tau \int_{D_\tau} F(q, \tau) g_i(p, a^2 t, q, a^2 \tau) d\omega_q \qquad (15.54)$$

where

$$I_1(p, t) = -a^2 \int_0^t d\tau \int_{\Sigma_\tau} f(q, \tau) \frac{\partial}{\partial n_q} g_1(p, a^2 t, q, a^2 \tau) d\sigma_q \qquad (15.55)$$

$$I_2(p, t) = a^2 \int_0^t d\tau \int_{\Sigma_\tau} \varphi(q, \tau) g_2(p, a^2 t, q, a^2 \tau) d\sigma_q \qquad (15.56)$$

$$I_3(p, t) = a^2 \int_0^t d\tau \int_{\Sigma_t} h(q, \tau) \psi(q, \tau) \frac{\partial}{\partial n_q} g_3(p, a^2 t, q, a^2 \tau) d\sigma_q \qquad (15.57)$$

The Green's functions $g_i(p, a^2 t, q, a^2 \tau)$, $i = 1, 2, 3$, are defined as functions of variable (q, τ) and fixed (p, t). Now consider them as functions of variable (p, t) and fixed (q, τ).

Theorem 15.3.1. *Let (p, t) be considered as a variable point and $(q, \tau) \in D^t$ as a fixed one. Then*

$$L_a g_i^*(p, a^2 t, q, a^2 \tau) = 0 \quad \forall (p, t) \in D_\tau^t, \quad i = 1, 2, 3 \qquad (15.58)$$

$$g_i^*(p, a^2 \tau, q, a^2 \tau) = 0, \quad i = 1, 2, 3 \qquad (15.59)$$

$$g_1^*(p, a^2 t, q, a^2 \tau) = E[p, q, a^2(t - \tau)] \quad \forall (p, t) \in \Sigma^t \qquad (15.60)$$

$$\frac{\partial}{\partial n_p} g_2^*(p, a^2 t, q, a^2 \tau) = \frac{\partial}{\partial n_p} E[p, q, a^2(t - \tau)] \quad \forall (p, t) \in \Sigma^t \qquad (15.61)$$

$$\left(\frac{\partial}{\partial n_p} + h(p, t) \right) g_3^*(p, a^2 t, q, a^2 \tau)$$

$$\qquad (15.62)$$

$$= \left(\frac{\partial}{\partial n_p} + h(p, t) \right) E[p, q, a^2(t - \tau)] \quad \forall (p, t) \in \Sigma^t$$

so that $g_i(p, a^2 t, q, a^2 \tau)$, being considered as a function of the first pair of variables, is a solution of the homogeneous heat-conduction equation satisfying boundary conditions of the first, second, or third boundary-value problem, $i = 1, 2$, and 3, respectively.

PROOF. Let

$$G_i(\rho, a^2 s, q, a^2 \tau) = E[\rho, q, a^2(s - \tau)] - G_i^*(\rho, a^2 s, q, a^2 \tau) \qquad (15.63)$$

where

$$L_a G_i^*(\rho, a^2 s, q, a^2 \tau) = 0 \quad \forall (\rho, s) \in D_\tau^t \qquad (15.64)$$

$$G_1^*(\rho, a^2 s, q, a^2 \tau) = E[\rho, q, a^2(s - \tau)] \quad \forall (\rho, s) \in \Sigma_\tau^S \qquad (15.65)$$

$$\frac{\partial}{\partial n_\rho}G_2^*(\rho, a^2s, q, a^2\tau) = \frac{\partial}{\partial n_\rho}E[\rho, q, a^2(s-\tau)] \quad \forall(\rho, s) \in \Sigma_\tau^s \qquad (15.66)$$

$$\left(\frac{\partial}{\partial n_\rho} - h(\rho, s)\right) G_3^*(\rho, a^2s, q, a^2\tau)$$

$$= \left(\frac{\partial}{\partial n_\rho} - h(\rho, s)\right) E[\rho, q, a^2(s-\tau)] \quad \forall(\rho, s) \in \Sigma_\tau^S \qquad (15.67)$$

$$G_i^*(\rho, a^2\tau, q, a^2\tau) = 0 \quad \forall \rho \in D_\tau, \ i = 1, 2, 3 \qquad (15.68)$$

The theorem states that

$$G_i(\rho, a^2s, q, a^2\tau) \equiv g_i(\rho, a^2s, q, a^2\tau), \quad i = 1, 2, 3 \qquad (15.69)$$

Taking arbitrarily small $\varepsilon > 0$, $\eta > 0$ and applying the identity (15.9) to the functions $G_i[p, a^2s, q, a^2(\tau - \eta)]$ and $g_i[p, a^2(t+\varepsilon), q, a^2s)]$ and to the region D_τ^t, we obtain

$$\int_\tau^t ds \int_{D_s} (g_i L_a G_i - G_i M_a g_i) d\omega_p$$

$$= a^2 \int_\tau^t ds \int_{\Sigma_s} \left[g_i \frac{\partial}{\partial n_\rho}G_i - G_i\left(\frac{\partial}{\partial n_\rho} - a^{-2}\frac{\partial}{\partial \tau}n_\rho\right)g_i\right]d\sigma_\rho$$

$$- \int_{D_t} g_i G_i d\omega_\rho + \int_{D_\tau} g_i G_i d\omega_\rho \qquad (15.70)$$

The integrands of all the integrals entering into (15.70) have no singularities, so that the left-hand side vanishes. Since g_i and G_i satisfy adjoint boundary conditions, both surface integrals on the right of (15.70) vanish. Hence,

$$\int_{D_t} G_i[\rho, a^2t, q, a^2(\tau - \eta)]g_i[p, a^2(t+\varepsilon), \rho, a^2t]d\omega_\rho$$

$$= \int_{D_\tau} G_i[\rho, a^2\tau, q, a^2(\tau - \eta)]g_i[p, a^2(t+\varepsilon), \rho, a^2\tau]d\omega_\rho \qquad (15.71)$$

Now let $\varepsilon \downarrow 0$. By the filtering property of g_i,

$$G_i[p, a^2t, q, a^2(\tau-\eta)] = \int_{D_\tau} G_i[\rho, a^2\tau, q, a^2(\tau-\eta)]g_i(p, a^2t, \rho, a^2\tau)d\omega_\rho \qquad (15.72)$$

Hence, by letting $\eta \downarrow 0$ and using the filtering property of G_i, we obtain

$$G_i(p, a^2t, q, a^2\tau) = g_i(p, a^2t, q, a^2\tau), \quad i = 1, 2, 3 \qquad (15.73)$$

Q.E.D.

As before, let $p = (x, y, z)$, $q = (\xi, \eta, \zeta)$, and also

$$r = j+k+m+n, \quad D_{pr} = \frac{\partial^r}{\partial x^j \partial y^k \partial z^m \partial t^n}, \quad D_{qr} = \frac{\partial^r}{\partial \xi^j \partial \eta^k \partial \zeta^m \partial \tau^n} \qquad (15.74)$$

The following theorems are corollaries of the maximum principle.

Theorem 15.3.2. *For all $r \geq 0$, $(p,t) \in D^t$, and $(q,\tau) \in D^t$, there exists $M_{ir} = \text{const} > 0$ such that*[1]

$$|D_{pr}g_i(p,t,q,\tau)| \leq M_{ir}|D_{pr}E(p,q,t-\tau)|$$

$$|D_{qr}g_i(p,t,q,\tau)| \leq M_{ir}|D_{qr}E(p,q,t-\tau)|, \quad i = 1,2,3 \qquad (15.75)$$

if the lateral boundary Σ^t of D^t possesses the relevant smoothness.

The proof of this theorem is beyond the scope of this book, despite its importance.[2]

Theorem 15.3.3. *The Green's function of the first boundary-value problem is nonnegative.*

PROOF. Take arbitrarily small $r > 0$ and compare $E(p,q,t-\tau)$ and $g^*(p,t,q,\tau)$ in the region

$$\tilde{D}^t = D^t \setminus K^t \qquad (15.76)$$

where K^t is the parabolic ball of radius r with center at (q,τ). Since g_1^* has no singularities inside D_r^t, it is bounded on $\tilde{\Sigma}^t$, so that there exists $M = \text{const} > 0$ such that

$$|g^*(p,t,q,\tau)| < M \quad \forall (p,t) \in \tilde{\Sigma}_k^t \qquad (15.77)$$

where $\tilde{\Sigma}_k^t$ is the parabolic boundary of K_r^t.
    At the same time,

$$\lim_{(p,t) \to (q,\tau)} E(p,q,t-\tau) = \infty \qquad (15.78)$$

so that there exists $r > 0$ so small that

$$E(p,q,t-\tau) > M \quad \forall (p,t) \in \tilde{\Sigma}_k^t \qquad (15.79)$$

Furthermore,

$$g^*(p,t,q,\tau) = E(p,q,t-\tau) \quad \forall (p,t) \in \Sigma_r^t, \ \forall t \geq \tau \qquad (15.80)$$

---

[1]The existence of derivatives of g_i of any order inside region D^t is an obvious corollary of the existence theorems for solutions of the boundary-value problems to be proved in Chapter 17.

[2]In the case of a 1-dimensional cylindrical region this assertion is trivial, since in this case the Green's functions g_i have explicit analytically closed expressions in terms of $E(x,t)$. In 1-dimensional noncylindrical regions, the validity of this theorem can be proved easily by consideration of integral equations of the Volterra type to which the boundary-value problem can be reduced (see Chapter 17, Section 1). For 3-dimensional cylindrical regions, the theorem is a corollary of Tichonov's analytically closed expressions for the Green's functions of the first and second boundary-value problems [83]. For noncylindrical 3-dimensional regions the proof of the theorem may be also obtained by using Tichonov's method, although this requires rather more complicated computations (see also [71]).

and

$$g^*(p, \tau, q, \tau) = 0, \quad E(p, q, 0) > 0 \tag{15.81}$$

Thus

$$|g^*| \leq E \quad \forall (p, t) \in \tilde{\Sigma}^t \tag{15.82}$$

so that, by the maximum principle,

$$|g^*| \leq E \quad \forall (p, t) \in \tilde{D}^t \tag{15.83}$$

Since $r > 0$ is arbitrarily small and (q, τ) is an arbitrary point of D^t, this means that, everywhere in $\bar{D}^t$,

$$g(p, t, q, \tau) \geq 0 \tag{15.84}$$

Q.E.D.

Remark. In the cylindrical case the Green's functions g_i, $i = 1, 2$, depend on $t - \tau$. The same is true for g_3 if $h(p, t)$ is independent of t. Indeed, the equation

$$Lu = 0 \tag{15.85}$$

is invariant with respect to the substitution

$$s = t - \tau \tag{15.86}$$

Let $G_i^*(p, q, s)$ be solutions of the problems

$$\Delta G_i^* - \frac{\partial}{\partial s} G_i^* = 0 \quad \forall p \in D, \ t > 0, \ q = \text{const} \in D \tag{15.87}$$

$$G_i^*(p, q, 0) = 0 \quad \forall p \in D \tag{15.88}$$

$$G_1^*(p, q, s) = E(p, q, s) \tag{15.89}$$

$$\frac{\partial}{\partial n_p} G_2^*(p, q, s) = \frac{\partial}{\partial n_p} E(p, q, s) \tag{15.90}$$

$$\frac{\partial}{\partial n_p} G_3^*(p, q, s) + h(p) G_3^*(p, q, s) = \frac{\partial}{\partial n_p} E(p, q, s) + h(p) E(p, q, s) \tag{15.91}$$

Thus $G_i^*(p, q, t - \tau)$ satisfy the same conditions as $g_i^*(p, t, q, \tau)$ which, by virtue of the uniqueness theorem, proves the validity of our Remark.

4. Relationship between Green's functions of Dirichlet problem in $\mathbb{R}_3$, corresponding to Laplace and Fourier operators (Tichonov's theorem) [83]

Consider the cylindrical region $D^t = D \times (0 \leq t < \infty)$. Assume that this region is fundamental with respect to the Dirichlet problem for the heat-conduction equation, so that the latter is solvable for any continuous boundary data; this means, in particular, that there exists a Green's function

$$g(p, q, t - \tau) = E(p, q, t - \tau) - g^*(p, q, t - \tau) \tag{15.92}$$

of the Dirichlet problem.

Theorem 15.4.1 (Tichonov). *Let*

$$G(p,q) = \int_0^\infty g(p,q,t)dt \qquad (15.93)$$

Then $G(p,q)$ is the Green's function of the electrostatic Dirichlet problem.

Remark. The existence of the Green's function implies the existence of a solution to the Dirichlet problem for any continuous boundary function. Tichonov's theorem may therefore be formulated as follows.

Theorem 15.4.1*. *Any cylindrical region in $\mathcal{E}_{3+1}$ that is fundamental with respect to the thermal Dirichlet problem is also fundamental with respect to the electrostatic Dirichlet problem.*

PROOF OF THEOREM 15.4.1. Note, first of all, that

$$\int_\tau^\infty E(p,q,t-\tau)dt = \frac{1}{4\pi r_{pq}} \int_0^\infty r_{pq} \frac{\exp(-r_{pq}^2/4(t-\tau))}{2\pi^{3/2}(t-\tau)^{3/2}}dt = \frac{1}{4\pi r_{pq}}J(p,q) \qquad (15.94)$$

where

$$J(p,q) = (2\sqrt{\pi})^{-1} \int_0^\infty r_{pq} t^{-3/2} \exp\left[\frac{-r_{pq}^2}{4t}\right] dt = 2\pi^{-1/2} \int_0^\infty \exp(-\lambda^2)d\lambda = 1 \qquad (15.95)$$

which means that

$$\int_\tau^\infty E(p,q,t-\tau)dt = \frac{1}{4\pi r_{pq}} \qquad (15.96)$$

Thus a fundamental solution of the Laplace equation in $\mathbb{R}_3$ is the integral with respect to time from τ to infinity of that of a fundamental solution of the Fourier equation.

We have

$$g(p,q,t) = E(p,q,t) - g^*(p,q,t) \qquad (15.97)$$

$$G(p,q) = \frac{1}{4\pi r_{pq}} - G^*(p,q) \qquad (15.98)$$

Hence the theorem will be true if

$$\int_0^\infty g^*(p,q,t)dt = G^*(p,q) \qquad (15.99)$$

We must therefore prove (15.99). Let

$$\tilde{G}(p,q \mid T) = \int_0^T g^*(p,q,t)dt \qquad (15.100)$$

For all $p, q \in \bar{D}$, $p \neq q$, and $t \in [0, \tau]$, $\Delta g^*(p, q, t)$ is a continuous function. Hence (15.100) yields

$$\Delta \tilde{G}(p, q \mid T) = \Delta \int_0^T g^*(p, q, t) dt = \int_0^T \Delta g^*(p, q, t) dt \qquad (15.101)$$

Since

$$\Delta g^*(p, q, t) = \frac{\partial}{\partial t} g^*(p, q, t) \qquad (15.102)$$

it follows from (15.101) that

$$\begin{aligned} \Delta \tilde{G}(p, q \mid T) &= \int_0^T \frac{\partial}{\partial t} g^*(p, q, t) dt \\ &= g^*(p, q, T) - g^*(p, q, 0) = g^*(p, q, T) \end{aligned} \qquad (15.103)$$

Since, further,

$$0 \leq g^* \leq E, \lim_{T \uparrow \infty} E(p, q, T) = 0 \qquad (15.104)$$

it follows from (15.103) that

$$\lim_{T \uparrow \infty} \Delta \tilde{G}(p, q \mid T) = 0 \qquad (15.105)$$

Taking into account the continuity of $\Delta \tilde{G}(p, q \mid T)$, we see that

$$\Delta \lim_{T \uparrow \infty} \tilde{G}(p, q \mid T) = \Delta \int_0^\infty g^*(p, q, t) dt = \Delta G^*(p, q) = 0 \qquad (15.106)$$

Finally,

$$g^* = E \quad \forall p \in \Sigma, \ \forall q \in D, \ \forall t > 0 \qquad (15.107)$$

and by (15.96),

$$G^*(p, q) = \frac{1}{4\pi r_{pq}} \quad \forall p \in \Sigma, \ \forall q \in D \qquad (15.108)$$

Thus $G^*(p, q)$ satisfies all the conditions defining the electrostatic Green's function, and since the solution of the electrostatic Dirichlet problem is unique, this proves (15.99). **Q.E.D.**

5. Examples of Green's functions

Sommerfeld's method of electrostatic images (see Chapter 10, Section 2) deals with the superposition of point sources and sinks. Clearly, the very possibility of using the method is independent of the nature of these sources (sinks). One can deal equally successfully with sources and sinks of heat or particles. Hence the expressions for the Green's function, derived for Dirichlet and Neumann problems in the elliptic case, can be automatically rewritten as follows for the corresponding problems of heat conduction and diffusion.

1. Green's function of Dirichlet problem in half-line $x \geq 0$

We have

$$g_1(x,\xi,t-\tau) = E(x-\xi,t-\tau) - E(x+\xi,t-\tau) \qquad (15.109)$$

Here the heat sink of the unit density is placed into the point $x_1 = -x$ symmetric with the x with respect to the boundary $x = 0$. The functions

$$g_2(x,y,\xi,\eta,t) = E(y-\eta,t)g_1(x,\xi,t) \qquad (15.110)$$

and

$$g_3(x,y,z,\xi,\eta,\zeta,t) = E(y-\eta,t)E(z-\zeta,t)g_1(x,\xi,t) \qquad (15.111)$$

are Green's functions of the thermal Dirichlet problem in the half-plane $x > 0$, $-\infty < y < \infty$ and in the half-space $x > 0$, $-\infty < y, z < \infty$, respectively.[3]

2. Green's function of Neumann problem in half-line $x \geq 0$

We have

$$G_1(x,\xi,t-\tau) = E(x-\xi,t-\tau) + E(x+\xi,t-\tau) \qquad (15.112)$$

Here the source of heat is placed into the point symmetrical to x with respect to the boundary $x = 0$. Analogously, the Neumann functions in the half-plane and the half-space $x > 0$ are

$$G_2(x,y,\xi,\eta,t) = E(y-\eta,t)G_1(x,\xi,t) \qquad (15.113)$$

and

$$G_3(x,y,z,\xi,\eta,\zeta,t) = E(y-\eta,t)E(z-\zeta,t)G_1(x,\xi,t) \qquad (15.114)$$

3. Green's function of Dirichlet problem in interval $0 \leq x \leq \ell$

As in the electrostatic Dirichlet problem in a strip, we now consider a denumerable number of reflections of sources and sinks with respect to the boundaries $x = 0$ and $x = \ell$. Thus

$$g_1(x,\xi,t-\tau) = \sum_{-\infty}^{\infty}[E(x-\xi+2n\ell,t-\tau) - E(x-\xi+2n\ell,t-\tau)] \qquad (15.115)$$

4. Green's function of the Neumann problem in an interval $0 \leq x \leq \ell$

In this case the denumerable sequence of reflections of heat sources must be performed:

$$G_1(x,\xi,t-\tau) = \sum_{-\infty}^{\infty}[E(x-\xi+2n\ell,t-\tau) + E(x+\xi+2n\ell,t-\tau)] \qquad (15.116)$$

---

[3]Recall that a fundamental solution of the heat-conduction equation in multidimensional cases is equal to the product of 1-dimensional fundamental solutions of this equation (see Chapter 13, Section 5). Thus it is easy to check that g_i and G_i, $i = 2, 3$, are indeed the required Green's functions.

Remark. Series (15.115) and (15.116) converge very quickly when $t - \tau$ is small. For example, if

$$4(t - \tau) < \ell \tag{15.117}$$

then the series

$$\sum_{n=1}^{\infty} [E(x \pm \xi \pm 2n\ell, t - \tau)] \tag{15.118}$$

is majorized by the rapidly convergent series

$$\frac{\exp[-(x - \xi)^2/4(t - \tau)]}{2\pi^{1/2}(t - \tau)^{1/2}} \sum_{n=1}^{\infty} \exp\left(-\frac{n^2\ell^2 \mp 2n\ell \cdot (x \mp \xi)}{t - \tau}\right) \tag{15.119}$$

But if

$$4(t - \tau) \gg \ell \tag{15.120}$$

then the series

$$\sum_{-\infty}^{\infty} [E(x - \xi + 2n\ell, t - \tau) \pm E(x + \xi + 2n\ell, t - \tau)] \tag{15.121}$$

converges very slowly. Therefore, along with the expansions (15.115) and (15.116), one should also use the Fourier expansion of g_1 and G_1:

$$g_1(x, \xi, t - \tau) = \frac{2}{\ell} \sum_{n=1}^{\infty} \exp\left[-\left(\frac{n\pi}{\ell}\right)^2 (t - \tau)\right] \sin\frac{n\pi x}{\ell} \cdot \sin\frac{n\pi \xi}{\ell} \tag{15.122}$$

and

$$G_i(x, \xi, tr\tau) = \frac{1}{\ell} + \frac{2}{\ell}\sum_{n=1}^{1} \exp\left[-\left(\frac{n\pi}{\ell}\right)^2 (t - \tau)\right] \cos\frac{n\pi x}{\ell} \cos\frac{n\pi \xi}{\ell} \tag{15.123}$$

These expansions are derived in Chapter 19, Section 2. Here we show only that they indeed provide the Fourier expansions of the Green's functions g_i and G_i. Note, first of all, that for all $t > \tau$ each term of these series is a parabolic function:

$$L\left(\exp\left[-\left(\frac{n\pi}{\ell}\right)^2 (t - \tau)\right] \sin\frac{n\pi x}{\ell}\right) = L\left(\exp\left[-\left(\frac{n\pi}{\ell}\right)^2 (t - \tau)\right] \cos\frac{n\pi x}{\ell}\right)$$

$$= 0 \tag{15.124}$$

and that the series obtained by term-by-term differentiation an arbitrary number of times are uniformly convergent in the interval $[0, \ell]$. Hence the functions g_1, G_1 represented by (15.122) and (15.123) are parabolic functions satisfying the conditions

$$g_1 = 0, \quad \frac{\partial}{\partial x}G_1 = 0 \quad \text{at } x = 0, x = \ell \quad \forall(t - \tau) > 0 \tag{15.125}$$

If $t = \tau$ then the series (15.122) and (15.123) are divergent. By the rules of operating with the Dirac δ function,[4]

$$\int_0^\ell \delta(x-\xi) \sin\frac{n\pi\xi}{\ell} d\xi = \sin\frac{n\pi x}{\ell}, \quad \int_0^\ell \delta(x-\xi) \cos\frac{n\pi\xi}{\ell} d\xi = \cos\frac{n\pi\xi}{\ell} \quad (15.126)$$

and hence at $t = \tau$ the series (15.122) and (15.123) are (respectively) the sine and cosine Fourier expansions of $\delta(x-\xi)$. Recall that according to (13.28),

$$E(x-\xi,0) = \delta(x-\xi) \quad (15.127)$$

Hence, if $t = \tau$ then

$$g_1(x,\xi,0) = G_1(x,\xi,0) = E(x-\xi,0) \quad (15.128)$$

Thus g_1 and G_1, as defined by (15.122) and (15.123), possess all the properties of the Green's functions of the parabolic Dirichlet and Neumann problems, respectively. Thus, since the Green's functions are unique, they are indeed the required functions.   **Q.E.D.**[5]

5. Green's function of Dirichlet problem in half-line $x > \alpha t$.

In this case there is no direct analogy with the stationary electrostatic problem, since now we are considering not a cylindrical region but rather a region with moving boundary. We must therefore provide all the necessary computations. Thus, we have to determine a function $g_1^*(x,a^2t,\xi,a^2\tau)$, satisfying the conditions

$$a^2 \frac{\partial^2}{\partial\xi^2} g_1^* + \frac{\partial}{\partial\tau} g_1^* = 0 \quad \forall\xi > \alpha\tau, \ \forall x = \text{const} > \alpha t, \ \forall\tau < t = \text{const} \quad (15.129)$$

$$g_1^*(x,l,\alpha\tau,\tau) = E[x-\alpha\tau,a^2(t-\tau)], \quad g_1^*(x,t,\xi,t) = 0 \quad (15.130)$$

Let us seek $g_1^*(x,a^2t,\xi,a^2\tau)$ as a function of (ξ,τ) in the form

$$g_1^*(x,a^2t,\xi,a^2\tau) = \exp[\varphi(x,t)]E[x_1-\xi,a^2(t-\tau)] \quad (15.131)$$

where

$$x_1 < \alpha t \quad (15.132)$$

Any such function satisfies equation (15.129) for every $\xi > \alpha\tau$ and $\tau < t$, as well as the condition on the right side of (15.130). Thus we need only worry about the boundary condition on the left side of (15.130), that is,

$$\varphi(x,t) - \frac{(x_1-\alpha t)^2}{4a^2(t-\tau)} = -\frac{(x-\alpha\tau)^2}{4a^2(t-\tau)} \quad (15.133)$$

---

[4] See Chapter 13, Section 1.

[5] A rigorous proof that g_1 and G_1 are indeed the desired Green's functions is given in Chapter 19, Section 2, without reference to the rules for operating with the δ function.

This yields

$$4a^2(t - \tau)\varphi(x,t) - x_1^2 + 2\alpha x_1 \tau = -x^2 + 2\alpha x\tau \tag{15.134}$$

where $\varphi(x,t)$ and x_1 are independent of ξ and τ. Hence

$$-4a^2\varphi(x,t) + 2\alpha x_1 = 2\alpha x, \; 4a^2 t\varphi(x,t) - x_1^2 + x^2 = 0 \tag{15.135}$$

whence

$$x_1 = 2\alpha t - x, \varphi(x,t) = \alpha\frac{\alpha t - x}{a^2} \tag{15.136}$$

It follows from (15.131), (15.136), and

$$g_1(x, a^2 t, \xi, a^2\tau) = E[x - \xi, a^2(t - \tau)] - g_1^*(x, a^2 t, \xi, a^2\tau) \tag{15.137}$$

that

$$g_1(x, a^2 t, \xi, a^2\tau) = \frac{\exp[-(x - \xi)^2/4a^2(t - \tau)]}{2a\pi^{1/2}(t - \tau)^{1/2}}$$

$$- \frac{\exp[-(2\alpha t - x - \xi)^2/4a^2(t - \tau)]}{2a\pi^{1/2}(t - \tau)^{1/2}}$$

$$\cdot \exp\left[-\frac{\alpha(\alpha t - x)}{a^2}\right] \tag{15.138}$$

or, what is the same,

$$g_1(x, a^2 t, \xi, a^2\tau) = E[x - \xi, a^2(t - \tau)]\left[1 - \exp\left(-\frac{(x - \alpha t)\cdot(\xi - \alpha\tau)}{a^2(t - \tau)}\right)\right] \tag{15.139}$$

6. Green's function $g_3(x, \xi, t - \tau)$ of problem of Newtonian irradiation in half-line $x > 0$.

The desired Green's function must satisfy the conditions

$$\left(\frac{\partial^2}{\partial x^2} - \frac{\partial}{\partial t}\right)g_3(x, \xi, t) = 0 \quad \forall x > 0, \; \forall t > 0, \; \forall \xi = \text{const} > 0 \tag{15.140}$$

$$\left(\frac{\partial}{\partial x} - h\right)g_3(0, \xi, t) = 0 \quad \forall \xi > 0, \; \forall t > 0 \tag{15.141}$$

$$g_3(x, \xi, 0) = E(x - \xi, 0) \equiv \delta(x - \xi) \quad \forall(x, \xi) > 0 \tag{15.142}$$

We assume that $h = \text{const} > 0$.

Let us seek g_3 in the form

$$g_3(x, \xi, t) = G_2(x, \xi, t) - v(x, \xi, t) \tag{15.143}$$

where G_2 is the Green's function of the Neumann problem in a half-line, that is,

$$G_2(x, \xi, t) = E(x - \xi, t) + E(x + \xi, t) \tag{15.144}$$

and $v(x, \xi, t)$ is subject to determination. Obviously, v must be a solution of the problem

$$\frac{\partial^2}{\partial x^2} v = \frac{\partial}{\partial t} v \quad \forall x > 0, \quad \forall t > 0, \quad v(x, \xi, 0) = 0, \quad \frac{\partial}{\partial x} v(0, \xi, t) = \varphi(\xi, t)$$
$$(15.145)$$

where, for all $\xi > 0$,

$$\varphi = h(v - G_2) \big|_{x=0} \qquad (15.146)$$

Assume that φ is known. Then v may be considered as a solution of the Neumann problem, so that

$$v(x, \xi, t) = \int_0^t \varphi(\xi, \tau) G_2(x, 0, t - \tau) d\tau \qquad (15.147)$$

Substituting φ from (15.146) into this expression, we see that φ must be a solution of the integral equation

$$\varphi(\xi, t) = -hG_2(0, \xi, t) - h \int_0^t \varphi(\xi, \tau) G_2(0, 0, t - \tau) d\tau \qquad (15.148)$$

so that, by the definition (15.144),

$$\varphi(\xi, t) = -h(\pi t)^{-1/2} \exp\left[-\frac{\xi^2}{4t}\right] - h\pi^{-1/2} \int_0^t \varphi(\xi, \tau)(t - \tau)^{-1/2} d\tau \qquad (15.149)$$

The simplest method for solving this generalized Abel integral equation is to apply the Laplace–Carson integral transform (see Chapter 20, Section 5). However, many problems of mathematical physics (in particular, nonlinear problems such as some versions of the Stefan problem [70]) involve the solution of this kind of integral equation when the Laplace–Carson transform is not applicable. Bearing this in mind, we shall solve this equation by reducing it to an ordinary differential equation.

Equation (15.149) implies that

$$\int_0^t \varphi(\xi, s)(t - s)^{-1/2} ds = -h\pi^{-1/2}[I_1(\xi, t) + I_2(\xi, t)] \qquad (15.150)$$

where

$$I_1(\xi, t) = \int_0^t \frac{\exp(-\xi^2/4s)}{s^{1/2}(t - s)^{1/2}} ds \qquad (15.151)$$

and

$$I_2(\xi, t) \stackrel{\text{def}}{=} \int_0^t (t - s)^{-1/2} ds \int_0^s \varphi(\xi, \tau)(s - \tau)^{-1/2} d\tau \qquad (15.152)$$

By (15.151),

$$I_1(\xi, t) = \int_0^t \frac{ds}{(t - s)^{1/2}} \int_\xi^\infty \lambda \frac{\exp(-\lambda^2/4s)}{2s^{3/2}} d\lambda$$

$$= \int_\xi^\infty d\lambda \int_0^t \lambda \frac{\exp(-\lambda^2/4s)}{2s^{3/2}} \frac{ds}{(t - s)^{1/2}} \qquad (15.153)$$

The substitution

$$z = \left(\frac{\lambda^2}{4s} - \frac{\lambda^2}{4t} \right)^{1/2} \tag{15.154}$$

yields

$$I_1(\xi, t) = 2t^{-1/2} \int_\xi^\infty \exp\left(\frac{-\lambda^2}{4t} \right) d\lambda \int_0^\infty \exp(-z^2) dz = 2\pi \int_\xi^\infty E(\lambda, t) d\lambda \tag{15.155}$$

Further,

$$I_2(\xi, t) = \int_0^t \varphi(\xi, \tau) d\tau \int_\tau^t [(t-s)(s-\tau)]^{-1/2} ds$$

$$= \int_0^t \varphi(\xi, \tau) d\tau \int_0^1 [\lambda(1-\lambda)]^{-1/2} d\lambda = \pi \int_0^t \varphi(\xi, \tau) d\tau \tag{15.156}$$

Hence it follows from (15.150) and (15.155) and (15.156) that

$$h^2 \int_0^t \varphi(\xi, \tau) d\tau = -h\pi^{-1/2} \int_0^t \varphi(\xi, s)(t-s)^{-1/2} ds - 2h^2 \int_\xi^\infty E(\lambda, t) d\lambda \tag{15.157}$$

Comparison of (15.149) and (15.157) implies

$$\varphi(\xi, t) = h^2 \int_0^t \varphi(\xi, \tau) d\tau - 2hE(\xi, t) + 2h^2 \int_\xi^\infty E(\lambda, t) d\lambda \tag{15.158}$$

But for all $\xi > 0$ it follows from (15.158) that

$$\frac{\partial}{\partial t} \varphi(\xi, t) = h^2 \varphi(\xi, t) + \frac{\partial}{\partial t} \left(2h^2 \int_\xi^\infty E(\lambda, t) d\lambda - 2h \cdot E(\xi, t) \right), \tag{15.159}$$

$$\lim_{t \downarrow 0} \varphi(\xi, t) = 0$$

This implies

$$\varphi(\xi, t) = \psi(\xi, t) + h^2 \int_0^t \psi(\xi, s) \exp[h^2(t-s)] ds \tag{15.160}$$

where

$$\psi(\xi, t) = 2h^2 \int_\xi^\infty E(\lambda, t) d\lambda - 2h \cdot E(\xi, t) \tag{15.161}$$

Inserting (15.160) and (15.161) into (15.147), we obtain an analytically closed representation of $v(x, \xi, t)$, as well as an expression for the Green's function. Omitting the rather cumbersome computations, we exhibit the final result:[6]

$$g_3(x, \xi, t) = G_2(x, \xi, t) - 2h \int_0^\infty e^{-hs} E(x + \xi + s, t) ds \tag{15.162}$$

---

[6]See Chapter 20, Problem 20.5.5.

7. Green's function for boundary-value problems in multidimensional regions

Theorem 15.5.1. *Let the parabolic region $D_\tau^t \subset \mathcal{E}_{k+m+1}$ be the topological product of disjoint parabolic regions $D_\tau^{kt} \in \mathcal{E}_{k+1}$ and $D_\tau^{mt} \in \mathcal{E}_{m+1}$,*

$$D_\tau^t = D_\tau^{kt} \times D_\tau^{mt} \tag{15.163}$$

and let Σ_τ^t be its lateral boundary,

$$\Sigma_\tau^t = \left(\Sigma_\tau^{kt} \cap \bar{D}_{\tilde{\tau}}^{mt} \right) \cup \left(\Sigma_\tau^{mt} \cap \bar{D}_\tau^{kt} \right) \tag{15.164}$$

so that

$$p = (x_1, x_2, \dots, x_k, x_{k+1}, x_{k+2}, \dots, x_{k+m}) \in \bar{D}_t \tag{15.165}$$

if

$$p_k = (x_1, x_2, \dots, x_k) \in \bar{D}_t^k, \quad p_m = (x_{k+1}, x_{k+2}, \dots, x_{k+m}) \in \bar{D}_t^m \tag{15.166}$$

Let $g^k(p_k, t, q_k, \tau)$ and $g^m(p_m, t, q_m, \tau)$ be the Green's functions of certain (first, second, or third) boundary-value problems (not necessarily the same) for a parabolic equation in D^{kt} and D^{mt}, respectively. Then

$$g^n(p, t, q, \tau) = g^k(p_k, t, q_k, \tau) g^m(p_m, t, q_m, \tau), \quad n = k + m \tag{15.167}$$

is the Green's function for D^t.

PROOF. By definition,

$$g^i(p_i, t, q_i, \tau) = E_i(p_i, q_i, t - \tau) - \tilde{g}^i(p_i, t, q_i, \tau), \quad i = k, m, n \tag{15.168}$$

$$L^i \tilde{g}^i(p_i, t, q_i, \tau) = 0 \; \forall p_i \in D_t^i, \; \forall q_i = \text{const} \in D_\tau^i, \; \forall \tau = \text{const} \leq t \tag{15.169}$$

$$\tilde{g}^i(p_i, \tau, q_i, \tau) = 0 \tag{15.170}$$

and

$$\tilde{g}^i(p_i, t, q_i, \tau) = E_i(p_i, q_i, t - \tau) \quad \forall (p_i, t) \in \Sigma_\tau^{it} \tag{15.171}$$

if g^i is the Green's function of the first boundary-value problem; or

$$\frac{\partial}{\partial n_{p\,i}} \tilde{g}^i(p_i, t, q_i, \tau) = \frac{\partial}{\partial n_{p_i}} E_i(p_i, q_i, t - \tau) \quad \forall (p_i, t) \in \Sigma_\tau^{it} \tag{15.172}$$

if g^i is the Green's function of the second boundary-value problem; or

$$\left(\frac{\partial}{\partial n_{p_i}} + h_i \right) \tilde{g}^i(p_i, t, q_i, \tau) = \left(\frac{\partial}{\partial n_{p_i}} + h_i \right) E_i(p_i, q_i, t - \tau) \quad \forall (p_i, t) \in \Sigma_\tau^{it} \tag{15.173}$$

if g^i is the Green function of the third boundary-value problem.

    Defining g^n and $\tilde{g}^n$ by (15.167) and (15.168), respectively, we shall prove that they satisfy all the requirements (15.168)–(15.173). Obviously $\tilde{g}^n(p, t, q, \tau)$ satisfies condition (15.169), since it is satisfied by $\tilde{g}^k(p, t, q, \tau)$ and $\tilde{g}^m(p, t, q, \tau)$. Thus we need only check the validity of boundary conditions (15.171)–(15.173) and the initial condition (15.170).

By (15.167), (15.168), and (15.48),

$$\tilde{g}^n(p,t,q,\tau) = -\tilde{g}^k(p_k,t,q_k,\tau)\tilde{g}^m(p_m,t,q_m,\tau)$$
$$+ E_k(p_k,t,q_k,\tau)\tilde{g}^m(p_m,t,q_m,\tau)$$
$$+ E_m(p_m,t,q_m,\tau)\tilde{g}^k(p_k,t,q_k,\tau) \tag{15.174}$$

In the case of the first boundary-value problem, it follows from (15.171) and (15.174) that

$$\tilde{g}^n(p,t,q,q,\tau)\big|_{p_k\in\Sigma_\tau^{kt}} = E_m(p_m,t,q_m,\tau)\tilde{g}^k(p_k,t,q_k,\tau)$$
$$= E_m(p_m,t,q_m,\tau)E_k(p_k,t,q_k,\tau)$$
$$= E_n(p,t,q,q,\tau)\big|_{p\in\Sigma_\tau^{kt}} \tag{15.175}$$

and similarly for $p_m \in \Sigma_\tau^{mt}$. Hence the boundary condition of the first kind is satisfied.

In the case of a boundary condition of the second kind, on Σ_τ^{kt} it follows from (15.174) that

$$\frac{\partial}{\partial n_{p_k}}\tilde{g}^n(p,t,q,\tau) = E_m(p_m,t,q_m,\tau)\frac{\partial}{\partial n_{p_k}}\tilde{g}^k(p_k,t,q_k,\tau)$$

$$= \frac{\partial}{\partial n_{p_k}}\Big(E_m(p_m,t,q_m,\tau)\tilde{g}^k(p_k,t,q_k,\tau)\Big)$$

$$= \frac{\partial}{\partial n_{p_k}}E_n(p,t,q,\tau) \tag{15.176}$$

so that the boundary condition of the second kind is also satisfied. Quite similarly, we verify condition (15.173). Thus it remains only to prove that g^n satisfies the initial condition (15.170), namely, that

$$\lim_{D_\tau^t \ni (p,t) \to (p^0,\tau)\in D_\tau} \tilde{g}^n(p,t) = 0 \tag{15.177}$$

If $(p^0,\tau) \in D_\tau^{it}$, $i = k,m$, then (15.177) follows from (15.170) and (15.167), since by assumption the regions D_τ^{kt} and D_τ^{mt} have no common points. Hence (15.177) needs proof only when (p^0,τ) lies on one of the lateral boundaries Σ_τ^{it}, say on Σ_τ^{mt}. But then (15.177) is a corollary of the following lemma.

Lemma 15.5.1. *Let s be the dimension of parabolic region $D_\tau^{st} \subset \mathcal{E}_{n+1}$. Then, for any $\delta = \mathrm{const} > 0$ and $t \geq \tau$, there exist $\alpha = \alpha(\delta) = \mathrm{const} > 0$ and $K(\delta) = \mathrm{const} > 0$ such that*

$$r_{p\Sigma_t^s} \geq \delta \wedge r_{q\Sigma_t^s} \geq \delta \Rightarrow |\tilde{g}^s(p,t,q,\tau)| < K(\delta)\exp\left[-\frac{\alpha(\delta)}{t-\tau}\right)$$

$$\left|\frac{\partial}{\partial n_p}\tilde{g}^s(p,t,q,\tau)\right| < K(\delta)\exp\left(-\frac{\alpha(\delta)}{t-\tau}\right) \tag{15.178}$$

Assume first that the lemma is true. Then the definition

$$E_s(p, q, t - \tau) = \frac{\exp(-r_{pq}^2/4(t-\tau))}{2^s \pi^{s/2}(t-\tau)^{s/2}}, \quad s = 1, 2, \ldots \quad (15.179)$$

and (15.178) imply

$$E_k(p_k, t, q_k, \tau)\tilde{g}^m(p_m, t, q_m, \tau) < K(\delta)\frac{\exp[-\alpha(\delta)/(t-\tau)]}{2^k \pi^{k/2}(t-\tau)^{k/2}} \quad (15.180)$$

which yields

$$\lim_{D_\tau^t \ni (p,t) \to (p_o, \tau) \in \Sigma_\tau^m} \tilde{g}^n(p, t) = \lim_{D_\tau^t \ni (p,t) \to (p_o, \tau) \in \Sigma_\tau^m} [E_k(p_k t, q_k, \tau)\tilde{g}^m(p_m, t, q_m, \tau)] = 0 \quad (15.181)$$

which completes the proof of the theorem. Hence it remains to prove Lemma 15.5.1.

PROOF OF LEMMA 15.5.1. Fix a point $(q, \tau) \in D_\tau^{st}$. Then there exists $N = N[r_{p\Sigma_t^s}] = \text{const}$ such that

$$|\tilde{g}^s|, \left|\frac{\partial}{\partial n_p}\tilde{g}^s\right| \leq N \quad (15.182)$$

This follows from the parabolicity of $\tilde{g}^s$, considered as a function of $(p, t) \in D_t$ with fixed (ξ, τ), since $\tilde{g}^n$ has no singularities when $r_{p\Sigma_t} > 0$. Recall that $\tilde{g}^s$ is defined by the fundamental identity

$$\tilde{g}^s(p, t, q, \tau) = \int_{-\tau}^t d\lambda \int_{\Sigma_\lambda} E_s(p, o, t - \lambda)\frac{\partial}{\partial n_o}\tilde{g}^s(o, \lambda, q, \tau)d\sigma_o$$
$$- \int_{-\tau}^t d\lambda \int_{\Sigma_\lambda} \tilde{g}^s(o, \lambda, q, \tau)\left(\frac{\partial}{\partial n_o} - \frac{\partial}{\partial \lambda}n_o\right)E_s(p, o, t - \lambda)d\sigma_o \quad (15.183)$$

We have

$$E_s(p, o, t - \lambda) = \frac{\exp[-r_{po}^2/4(t - \lambda)]}{2^s \pi^{s/2}(t - \lambda)^{s/2}}$$

$$\frac{\partial}{\partial n_o}E_s(p, o, t - \lambda) = -\frac{r_{po}}{t - \lambda}\frac{\exp[-r_{po}^2/4(t - \lambda)]}{2^s \pi^{s/2}(t - \lambda)^{s/2}}\cos(\varphi) \quad (15.184)$$

where φ is the angle between the vectors $\mathbf{n}_o$ and $\mathbf{r}_{po}$. By assumption $\partial n_o/\partial\lambda$ is bounded:

$$\left|\frac{\partial}{\partial\lambda}n_o\right| < A = \text{const} > 0 \quad (15.185)$$

Hence it follows from (15.182)–(15.184) that

$$|\tilde{g}^s(p, t, q, \tau)| < (1 + A)N \int_\tau^t d\lambda \int_{\Sigma_\lambda} \frac{\exp[-r_{po}^2/4(t - \lambda)]}{2^s \pi^{s/2}(t - \lambda)^{s/2}}d\sigma_o$$

$$+ N \int_\tau^t d\lambda \int_{\Sigma_\lambda} \frac{r_{po}}{t - \lambda} \cdot \frac{\exp[-r_{po}^2/4(t - \lambda)]}{2^s \pi^{s/2}(t - \lambda)^{s/2}}d\sigma_o \quad (15.186)$$

Let

$$\beta_s = \max |z^s \exp(-z^2/2)|, \quad H = \max_{p \in D_t} \max_{o \in D_\lambda} r_{po} \qquad (15.187)$$

If $r_{po} \geq \delta > 0$ then for all $\lambda \in [\tau, t]$,

$$\frac{\exp[-r_{po}^2/4(t-\lambda)]}{2^s \pi^{s/2}(t-\lambda)^{s/2}} \leq r_{po}^s \frac{\exp[-r_{po}^2/8(t-\lambda)]}{2^s \pi^{s/2}(t-\lambda)^{s/2}} \cdot \frac{\exp[-r_{po}^2/8(t-\lambda)]}{r_{po}^s}$$

$$\leq \frac{\beta_s}{\delta^s \pi^{s/2}} \exp\left(-\frac{r_{po}^2}{8(t-\tau)}\right) \qquad (15.188)$$

and similarly

$$\frac{r_{po}}{t-\lambda} \cdot \frac{\exp[-r_{po}^2/4(t-\lambda)]}{2^s \pi^{s/2}(t-\lambda)^{s/2}} \leq \frac{H}{\pi^{s/2}} \cdot \exp\left[-\frac{r_{po}^2}{8(t-\tau)}\right] \left(\frac{\beta_{s+2}}{\delta^{s+2}}\right) \qquad (15.189)$$

Majorizing inequality (15.186) by means of (15.188) and (15.189), we obtain

$$|\tilde{g}^s(p,t,q,\tau)| < N(1+A) \frac{\beta_s}{\delta^s \pi^{s/2}} \exp(-r_{po}^2/8(t-\tau)) \left(1 + A + \frac{\beta^2}{\delta^2}\right) \qquad (15.190)$$

Define

$$K(\delta) = N(1+A) \frac{\beta_s}{\delta^s \pi^{s/2}} \left(1 + A + \frac{\beta^2}{\delta^2}\right), \alpha(\delta) = \min_{o \in \Sigma_\tau} \left(\frac{r_{po}}{8}\right) \qquad (15.191)$$

This gives, by (15.190),

$$\left|\frac{\partial}{\partial n_p} \tilde{g}^s(p,t,q,\tau)\right| < K(\delta) \exp\left(-\frac{\alpha(\delta)}{t-\tau}\right) \qquad (15.192)$$

Q.E.D.

For example, let us construct the Green's function of the Dirichlet heat-conduction problem for a rectangular parallelepiped,

$$Q = \{x_i : 0 < x_i < \ell_i, \ i = 1, 2, 3\} \qquad (15.193)$$

By (15.115), the 1-dimensional Green's functions are

$$g_i(x_i, \xi_i, t-\tau) = \sum_{-\infty}^{\infty} [E(x_i - \xi_i + 2n\ell_i, t-\tau) - E(x_i + \xi_i + 2n\ell_i, t-\tau)], \quad i = 1, 2, 3 \qquad (15.194)$$

Hence, by Theorem 15.5.1,

$$g(p, q, t-\tau) = \prod_{i=1}^{3} g_i(x_i, \xi_i, t-\tau) \qquad (15.195)$$

Let us denote

$$q_{kmn}^1 = (\xi_1 - 2k\ell_1, \xi_2 - 2m\ell_2, \xi_3 - 2n\ell_3)$$
$$q_{kmn}^2 = (\xi_1 - 2k\ell_1, \xi_2 - 2m\ell_2, -\xi_3 - 2n\ell_3)$$
$$q_{kmn}^3 = (\xi_1 - 2k\ell_1, -\xi_2 - 2m\ell_2, -\xi_3 - 2n\ell_3)$$
$$q_{kmn}^4 = (\xi_1 - 2k\ell_1, -\xi_2 - 2m\ell_2, \xi_3 - 2n\ell_3)$$
$$q_{kmn}^5 = (-\xi_1 - 2k\ell_1, \xi_2 - 2m\ell_2, -\xi_3 - 2n\ell_3) \qquad (15.196)$$
$$q_{kmn}^6 = (-\xi_1 - 2k\ell_1, -\xi_2 - 2m\ell_2, \xi_3 - 2n\ell_3)$$
$$q_{kmn}^7 = (-\xi_1 - 2k\ell_1, -\xi_2 - 2m\ell_2, \xi_3 - 2n\ell_3)$$
$$q_{kmn}^8 = (-\xi_1 - 2k\ell_1, -\xi_2 - 2m\ell_2, -\xi_3 - 2n\ell_3)$$

Inserting (15.194) into (15.195) and using the notation (15.196), we obtain

$$g(p, q, t - \tau) = \sum_{k,m,n=-\infty}^{\infty} \cdot \sum_{i=1}^{3} (-1)^{i+1} E_3(p, q_{kmn}^i, t - \tau) \qquad (15.197)$$

Recall that

$$E_3(p, q, t) = \frac{\exp[-r_{pq}^2/4(t-\tau)]}{8\pi^{3/2}(t-\tau)^{3/2}} \qquad (15.198)$$

We now apply Tichonov's Theorem 15.4.1. The Green's function of the Dirichlet problem for the Laplace equation in the parallelepiped Q is

$$G(p, q) = \int_0^\infty g(p, q, t) dt = \sum_{k,m,n=-\infty}^{\infty} \cdot \sum_{i=1}^{3} \frac{(-1)^{i+1}}{4\pi r_{pq_{kmn}^i}} \qquad (15.199)$$

This representation of $G(p, q)$ can also be obtained directly by using Sommerfeld's superposition of electrostatic images. However, this would require a direct proof that the function $G(p, q)$ is indeed the desired Green's function, which is rather complicated [13].

One can also use the representation of g_i by its sine-series expansion. This yields

$$G(p, q) = \int_0^t \frac{8\pi^3}{\ell_2\ell_2\ell_3} \sum_{k,m,n=0}^{\infty} \exp\left(-\left[\frac{k^2}{\ell_1^2} + \frac{m^2}{\ell_2^2} + \frac{n^2}{\ell_3^2}\right]\pi^2 t\right)$$
$$\times \sin\frac{k\pi x_1}{\ell_1} \sin\frac{k\pi\xi_1}{\ell_1} \sin\frac{m\pi x_2}{\ell_2} \sin\frac{m\pi\xi_2}{\ell_2} \sin\frac{n\pi x_3}{\ell_3} \sin\frac{n\pi\xi_3}{\ell_3} \qquad (15.200)$$

Changing the order of integration and summation yields

$$G(p, q) = \frac{8\pi^{-2}}{\ell_2\ell_2\ell_3} \sum_{k,m,n=1}^{\infty} \sin\frac{k\pi x_1}{\ell_1} \sin\frac{k\pi\xi_1}{\ell_1} \sin\frac{m\pi x_2}{\ell_2} \sin\frac{m\pi\xi_2}{\ell_2}$$
$$\times \sin\frac{n\pi x_3}{\ell_3} \sin\frac{n\pi\xi_3}{\ell_3} \cdot \left[\frac{k^2}{\ell_1^2} + \frac{m^2}{\ell_2^2} + \frac{n^2}{\ell_3^2}\right]^{-1} \qquad (15.201)$$

However, this procedure is inadmissible because it is unknown whether the triple series converges or not [13]. Note that the double series

$$G(p,q) = \frac{1}{\ell_1\ell_2\pi^2} \sum_{k,m=1}^{\infty} \sin\tfrac{k\pi x_1}{\ell_1} \sin\frac{k\pi\xi_1}{\ell_1} \sin\frac{m\pi x_2}{\ell_2} \sin\frac{m\pi\xi_2}{\ell_2} \left[\frac{k^2}{\ell_1^2} + \frac{m^2}{\ell_2^2}\right]^{-1}$$
$$(15.202)$$

represents the Green's function of the Dirichlet problem for the Laplace equation in a rectangle (See Chapter 9, Section 2).[7]

Problems

P15.4.1. What is the physical meaning of Tichonov's Theorem 15.4.1? Is it possible to prove the analog of Tichonov's theorem for the Dirichlet problem in $\mathbb{R}_2$? If so, prove it; otherwise, explain. Is it possible to reformulate the theorem so as to ensure its validity in the 2-dimensional case as well? If so, do this; if not, explain.

P15.4.2. Is it possible to formulate an analog to Tichonov's Theorem 15.4.1 for the Neumman problem in $\mathbb{R}_3$? If so, prove the respective theorem; otherwise, explain.

P15.5.1. Is it expedient to refer to Theorem 15.5.1 for a rigorous proof of the equalities (15.109), (15.110), (15.113), and (15.114)?

P15.5.2. Prove the identity

$$\sum_{-\infty}^{\infty}[E(x - \xi + 2n\ell, t - \tau) - E(x + \xi + 2n\ell, t - \tau)]$$

$$\equiv \frac{2}{\ell} \sum_{n=1}^{\infty} \exp\left[-\left(\frac{n\pi}{\ell}\right)^2 (t - \tau)\right] \sin\frac{n\pi x}{\ell} \cdot \sin\frac{n\pi\xi}{\ell}$$

---

[7]For further examples of Green's functions, see Chapter 19.

16

Heat potentials

1. Volume heat potential

Let $u(p,t)$ be twice continuously differentiable with respect to the spatial coordinates in a parabolic region $D^t \subset \mathcal{E}_{3+1}$, once continuously differentiable in D^t with respect to time, and continuous in the closure of D^t together with its first-order derivative along the outward normals to the lateral boundary Σ_τ for all $\tau \in [0,t]$:

$$u(p,t) \in C^{2,1}(D^t) \cap C(\bar{D}^t) \tag{16.1}$$

so that

$$\lim_{n_p \ni q \to p \in \Sigma^t} \frac{\partial}{\partial n_p} u(q,t) = \frac{\partial}{\partial n_p} u(p,t), \qquad (q \in D_t), \tag{16.2}$$

The integral identity (15.14) represents any function $u(p,t)$ satisfying these conditions as the sum of a Poisson integral

$$\Pi[p,t \mid f(q)] = \int_{D_0} f(q) E(p,q,t) d\sigma_q \tag{16.3}$$

and the three integrals

$$U[p,t \mid F(q,\tau)] = \int_0^t d\tau \int_{D_\tau} F(q,\tau) E(p,q,t-\tau) d\omega_q \tag{16.4}$$

$$V[p,t \mid \varphi(q,\tau)] = \int_0^t d\tau \int_{\Sigma_\tau} \varphi(q,\tau) E(p,q,t-\tau) d\omega_q \tag{16.5}$$

$$W[p,t \mid \psi(q,\tau)] = \int_0^t d\tau \int_{\Sigma_\tau} \psi(q,\tau) \frac{\partial}{\partial n_q} E(p,q,t-\tau) d\omega_q \tag{16.6}$$

where

$$f(q) = u(q,0) \quad \forall q \in D_0 \tag{16.7}$$

$$F(q,\tau) = -Lu(q,\tau) \quad \forall (q,\tau) \in D^t \tag{16.8}$$

$$\varphi(q,\tau) = u(q,\tau) \quad \forall (q,\tau) \in \Sigma^t \tag{16.9}$$

$$\psi(q,\tau) = \frac{\partial}{\partial n_q} u(q,\tau) \quad \forall (q,\tau) \in \Sigma^t \tag{16.10}$$

Recall that the parabolic operator L was defined as

$$Lu \overset{\text{def}}{=} \Delta u - \frac{\partial}{\partial t} u \tag{16.11}$$

By analogy with electrostatic potentials, U, V, and W are called, respectively,

- $U(p,t \mid F)$: a volume heat potential
- $V(p,t \mid \varphi)$: the heat potential of a single layer
- $W(p,t \mid \psi)$: the heat potential of a double layer

The functions F, φ, ψ and F are called the *densities* of the respective integrals.[1] We begin our study of the properties of heat potentials with $U(p,t \mid F)$.

Theorem 16.1.1.

(a) *If the density $F(q,\tau)$ is bounded within a bounded parabolic region D^t, then $U(p,t \mid F)$ is continuous everywhere in $\mathcal{E}_{3+1}$, together with all its spatial derivatives of the first order, and parabolic outside $\bar{D}^t$.*

(b) *If $F(q,t)$ is continuous together with its first derivatives with respect to the spatial coordinates within D^t, then*

$$LU(p,t \mid F) = -F(p,t) \quad \forall (p,t) \in D^t \tag{16.12}$$

(c) *If $F(q,t)$ is continuous together with its first-order derivative with respect to time, then also*

$$LU(p,t \mid F) = -F(p,t) \quad \forall (p,t) \in D^t \tag{16.13}$$

Remark. These properties are analogous to those of the Newtonian volume potential, their verification is almost the same as in the latter case.

PROOF. Suppose that for all $t > 0$,

$$|F| < M = \text{const} > 0, \quad \int_{D_t} d\omega_q < N = \text{const} \tag{16.14}$$

In what follows, any inessential constant positive factor occurring in the estimates being proved will be denoted by c. According to Theorem 12.5.3,

---

[1]This is an abbreviation of terminology. F is the volume density of distributed sources of heat, φ the surface density of the single layer of sources of heat, and ψ the surface density of moments of heat dipoles. The definition of the latter is similar to that of the electrostatic dipoles. We use similar abbreviations dealing with electrostatic potentials.

for every $\alpha = \text{const} \in [0,3)$, there exists $N(\alpha)$ dependent on the diameter of D_t such that, for all $p \in \mathbb{R}_3$,

$$\int_{D_t} \frac{1}{r_{pq}^\alpha} d\omega_q < N(\alpha) < \infty \tag{16.15}$$

We have

$$|U(p,t \mid F)| < M \int_0^t d\tau \int_{D_\tau} \frac{\exp\left[-r_{pq}^2/4(t-\tau)\right]}{8\pi^{3/2}(t-\tau)^{3/2}} d\omega_q$$

$$\leq c\beta_2 M \int_0^t (t-\tau)^{-1/2} d\tau \int_{D_\tau} \frac{d\omega_q}{r_{pq}^2}$$

$$< c\beta_2 N(2) M \sqrt{t} \quad \left(\beta_k = \max|z^k| \exp(-z^2)\right) \tag{16.16}$$

Hence $U(p,t \mid F)$ is majorized by an integral that is uniformly convergent everywhere in $\mathcal{E}_{3+1}$. Therefore $U(p,t \mid F)$ is indeed continuous everywhere in $\mathcal{E}_{3+1}$.

Now let

$$p = (x_1, x_2, x_3) \tag{16.17}$$

and let α_i, $i = 1,2,3$, be the direction cosines of the vector $\mathbf{r}_{pq}$. Then

$$\left|\frac{\partial}{\partial x_i} U(p,t \mid F)\right| < M \int_0^t (t-\tau)^{-3/4} d\tau \int_{D_\tau} \frac{r_{pq}^{7/2} |\alpha_i| \exp\left[-r_{pq}^2/4(t-\tau)\right]}{2r_{pq}^{5/2} 8\pi^{3/2}(t-\tau)^{7/4}} d\omega_q$$

$$\leq \frac{M\beta_{7/2}}{16\pi} \int_0^t (t-\tau)^{-3/4} d\tau \int_{D_\tau} \frac{d\omega_q}{r_{pq}^{5/2}}$$

$$< cMN\left(\frac{7}{2}\right) \beta_{7/2} t^{1/4} \tag{16.18}$$

so that $\partial U(p,t \mid F)/\partial x_i$ is also majorized by a uniformly convergent integral. Since, moreover, $E(p,q,t-\tau)$ is a parabolic function without singularities for any $(q,\tau) \in D^t$ and any $(p,t) \notin \bar{D}^t$, and vanishes at $\tau = t$ and $p \neq q$, we have

$$LU(p,t \mid F) = L \int_0^t d\tau \int_{D_\tau} F(q,\tau) E(p,q,t-\tau) d\omega_q$$

$$= \int_0^t d\tau \int_{D_\tau} F(q,\tau) LE(p,q,t-\tau) d\omega_q = 0 \quad \forall p \notin D_t \tag{16.19}$$

This completes the proof of part (a).

We now prove part (b). Given $m \in D_t$, take $\alpha > 0$ and $\rho > 0$ so small that, if K is a sphere of the radius ρ with center at m, then

$$K \in D_\tau \quad \forall \tau \in (t-\alpha, t) \tag{16.20}$$

Fix $t_0 \in (t-\alpha, t)$ and

$$D_\tau^* = D_\tau \setminus K \tag{16.21}$$

Then, for every $p \in K$,

$$U(p, t \mid F) = \sum_{i=1}^{3} U_i(p, t \mid F) \tag{16.22}$$

where

$$U_1 = \int_{t_0}^{t} d\tau \int_{K} F(q, \tau) E(p, q, t - \tau) d\omega_q \tag{16.23}$$

$$U_2 = \int_{0}^{t_0} d\tau \int_{K} F(q, \tau) E(p, q, t - \tau) d\omega_q \tag{16.24}$$

$$U_3 = \int_{0}^{t} d\tau \int_{D_\tau^*} F(q, \tau) E(p, q, t - \tau) d\omega_q \tag{16.25}$$

By (16.19),

$$LU_3(p, t) = 0 \tag{16.26}$$

since $p \notin D_\tau^*$. We also have

$$LU_2(p, t) = 0 \tag{16.27}$$

since $E(p, q, t - \tau)$ has no singularities for any $\tau \in [0, t_0]$. Hence it remains only to consider U_1.

By part (a) of this theorem,

$$\frac{\partial}{\partial x_i} U_1 = \int_{t_0}^{t} d\tau \int_{K} F(q, \tau) \frac{\partial}{\partial x_i} E(p, q, t - \tau) d\omega_q \tag{16.28}$$

Since $p = (x_1, x_2, x_3)$, $q = (\xi_1, \xi_2, \xi_3)$, and $r_{pq}^2 = \sum_{1}^{3}(x_i - \xi_i)^2$, we have

$$\frac{\partial}{\partial x_i} E(p, q, t - \tau) = -\frac{\partial}{\partial \xi_i} E(p, q, t - \tau) \tag{16.29}$$

This yields

$$\frac{\partial}{\partial x_i} U_1 = -\int_{t_0}^{t} d\tau \int_{K} F(q, \tau) \frac{\partial}{\partial \xi_i} E(p, q, t - \tau) d\omega_q$$

$$= -\int_{t_0}^{t} d\tau \int_{K} \operatorname{div}[F(q, \tau) E(p, q, t - \tau) \mathbf{e}_i] d\omega_q$$

$$+ \int_{t_0}^{t} \tau \int_{K} E(p, q, t - \tau) \frac{\partial}{\partial \xi_i} F(q, \tau) d\omega_q \tag{16.30}$$

where $\mathbf{e}_i$, $i = 1, 2, 3$, are the basis vectors. By the Gauss divergence theorem,

$$\int_{t_0}^{t} d\tau \int_{K} \operatorname{div}[F(q, \tau) E(p, q, t - \tau) \mathbf{e}_i] d\omega_q = \int_{t_0}^{t} d\tau \int_{S} F(q, \tau) E(p, q, t - \tau) \alpha_i d\sigma_q \tag{16.31}$$

where $S = \bar{K} \setminus K$ is the boundary and α_i is the direction cosine of the inner normal to $D^t \setminus K$ at $q \in S = \partial K$. Hence (16.30) and (16.31) imply

$$\frac{\partial^2}{\partial x_i^2} U_1(p,t \mid F) = -\int_{t_0}^t d\tau \int_S \alpha_i F(q,\tau) \frac{\partial}{\partial x_i} E(p,q,t-\tau) d\sigma_q$$

$$+ \int_{t_0}^t d\tau \int_K \frac{\partial}{\partial x_i} E(p,q,t-\tau) \frac{\partial}{\partial \xi_i} F(q,\tau) d\omega_q \qquad (16.32)$$

Thus, at the center m of K,

$$\Delta U_1(m,t) = -\int_{t_0}^t d\tau \int_S F(q,\tau) \frac{\partial}{\partial r_{mq}} E(m,q,t-\tau) d\sigma_q$$

$$+ \int_{t_0}^t d\tau \int_K \operatorname{grad} E(m,q,t-\tau) \cdot \operatorname{grad} F(q,\tau) d\omega_q \qquad (16.33)$$

Let us fix the radius ρ of K and let $t_0 \uparrow t$. Since both integrals in the right-hand side of (16.33) are uniformly convergent, we find that

$$\lim_{t_0 \uparrow t} \Delta U_1(m,t \mid F) = 0 \qquad (16.34)$$

Now consider $\partial U_1/\partial t$. We have

$$\tilde{E}(r,t-\tau) \overset{\text{def}}{=} \int_0^r s^2 \frac{\exp\left[-s^2/4(t-\tau)\right]}{8\pi^{3/2}(t-\tau)^{3/2}} ds = \frac{1}{2\pi^{3/2}} \int_0^{r^2/4(t-\tau)} z^{1/2} e^{-z} dz$$

$$(16.35)$$

where $r = r_{mq}$. Hence

$$E(r,t-\tau) = r^{-2} \frac{\partial}{\partial r} \tilde{E}(r,t-\tau) \qquad (16.36)$$

This yields

$$U_1(m,t-\tau) = \int_{t_0}^t d\tau \int_K \frac{F(q,\tau)}{r^2} \frac{\partial}{\partial r} \tilde{E}(r,t-\tau) d\omega_q$$

$$= \int_{t_0}^t d\tau \int_0^\rho dr \int_S F(q,\tau) \frac{\partial}{\partial r} \tilde{E}(r,t-\tau) d\Omega_q \qquad (16.37)$$

where Ω_q is the solid angle. Integration by parts yields

$$U_1(m,t-\tau) = \int_{t_0}^t d\tau \int_S F(q,\tau) \tilde{E}(\rho,t-\tau) d\Omega_q$$

$$- \int_{t_0}^t d\tau \int_0^\rho dr \int_S \tilde{E}(r,t-\tau) \frac{\partial}{\partial r} F(q,\tau) d\Omega_q$$

$$= \int_{t_0}^t d\tau \int_S F(q,\tau) \tilde{E}(\rho,t-\tau) d\Omega_q$$

$$- \frac{1}{4\pi} \int_{t_0}^t d\tau \int_K \frac{1}{r^2} \tilde{E}(r,t-\tau) \frac{\partial}{\partial r} F(q,\tau) d\omega_q \qquad (16.38)$$

Note now that

$$\lim_{\tau \uparrow t} \tilde{E}(\rho, t - \tau) = \frac{1}{2\pi^{3/2}} \int_0^\infty z^{1/2} e^{-z} dz = \frac{1}{4\pi} \qquad (16.39)$$

Hence

$$\frac{\partial}{\partial t} U_1(m, t \mid F) = \frac{1}{4\pi} \int_S F(q, t) d\Omega_q$$

$$+ \int_{t_0}^t d\tau \int_S F(q, \tau) \frac{\partial}{\partial t} \tilde{E}(\rho, t - \tau) d\Omega_q$$

$$- \frac{1}{4\pi} \int_K \frac{1}{r^2} \frac{\partial}{\partial r} F(q, \tau) d\omega_q$$

$$+ \frac{1}{4\pi} \int_{t_0}^t d\tau \int_K \frac{1}{r^2} \frac{\partial}{\partial r} \tilde{E}(r, t - \tau) \frac{\partial}{\partial r} F(q, \tau) d\omega_q \qquad (16.40)$$

Since

$$\frac{1}{4\pi} \int_S F(q, t) d\Omega_q - \frac{1}{4\pi} \int_K \frac{1}{r^2} \frac{\partial}{\partial r} F(q, \tau) d\omega_q$$

$$= \frac{1}{4\pi} \left(\int_S F(q, t) d\Omega_q - \int_0^\rho \frac{\partial}{\partial r} \left[\int_S F(q, t) d\Omega_q \right] dr \right)$$

$$= \frac{1}{4\pi} F(m, t) \int_S d\Omega_q = F(m, t) \qquad (16.41)$$

it follows from (16.40) that

$$\frac{\partial}{\partial t} U_1(m, t \mid F) = F(m, t) + J_1(m, t) + J_2(m, t) \qquad (16.42)$$

where

$$J_1(m, t) = \frac{1}{4\pi} \int_{t_0}^t d\tau \int_S F(q, \tau) \frac{\partial}{\partial t} \tilde{E}(\rho, t - \tau) d\Omega_q \qquad (16.43)$$

and

$$J_2(m, t) = \int_{t_0}^t d\tau \int_K \frac{1}{r^2} \frac{\partial}{\partial \tau} \tilde{E}(r, t - \tau) \frac{\partial}{\partial r} F(q, \tau) d\omega_q \qquad (16.44)$$

The integrand in J_1 has no singularities. At the same time, the integral J_2 is absolutely and uniformly convergent. Indeed, it follows from (16.35) that

$$\frac{\partial}{\partial t} \tilde{E}(rt - \tau) = -\frac{r^3}{16\pi^{3/2}(t - \tau)^{5/2}} \exp\left(-\frac{r_{pq}^2}{4(t - \tau)}\right) \qquad (16.45)$$

Assuming that

$$\left| \frac{\partial}{\partial r} F \right| < N \qquad (16.46)$$

let us define

$$z = \frac{r^2}{4(t-\tau)} \quad \Rightarrow \quad \frac{d\tau}{t-\tau} = -\frac{dz}{z} \tag{16.47}$$

By (16.35),

$$\frac{\partial}{\partial \tau}\tilde{E}(r, t-\tau)d\tau = \frac{r^3}{16\pi^{3/2}(t-\tau)^{5/2}}\exp\left(-\frac{r_{pq}^2}{4(t-\tau)}\right) = \frac{z^{1/2}e^{-z}}{2\pi^{3/2}}dz \tag{16.48}$$

Hence, by (16.44) and (16.46),

$$|J_2(m,t)| < \frac{N}{2\pi^{3/2}}\int_0^\rho dr \int_S d\Omega \int_{\rho_2/4(t-\tau)}^\infty z^{1/2}e^{-z}dz < \frac{2N\rho}{\sqrt{\pi}}\int_0^\infty z^{1/2}e^{-z}dz = N\rho \tag{16.49}$$

which confirms our assertion. Hence our calculations are not only formal but also, under our assumptions, quite rigorous.

The uniform convergence of the integrals J_1 and J_2 implies

$$\lim_{t_0 \uparrow t} J_i(p,t) = 0, \quad i = 1, 2 \tag{16.50}$$

Hence it follows from (16.22), (16.26), (16.27), (16.34), and (16.45) that

$$LU(m,t \mid F) = -F(m,t) \tag{16.51}$$

which proves the validity of part (b), since m is an arbitrarily fixed point of D_t.

Remark 1. This result is clearly independent of the choice of the sphere K (i.e., of the fixed ρ) since both sides of (16.51) are independent of ρ.

The proof of part (c) is similar to that of part (b) and therefore we omit it. (See Problem 16.1.2.)

Remark 2. The definition of the operator LU includes a tacit assumption of the existence of Δu and $\partial u/\partial t$. Let us introduce another definition:

$$LU(p,t \mid F) \stackrel{\text{def}}{=} \lim_{\varepsilon \downarrow 0} LU_\varepsilon(p,t) \tag{16.52}$$

where

$$U_\varepsilon(p,t \mid F) = \int_0^{t-\varepsilon} d\tau \int_{D_\tau} F(q,\tau)E(p,q,t-\tau)d\omega_q \tag{16.53}$$

Then the equality

$$LU(p,t) + F(p,t) = 0 \tag{16.54}$$

is a trivial consequence of the filtering property of the Poisson integral. Indeed,

$$LU_\varepsilon = -\int_{D_\varepsilon} F(q,\tau)E(p,q,t-\tau)d\omega_q$$

$$+ \int_0^{t-\varepsilon} d\tau \int_{D_\tau} F(q,\tau) LE(p,q,t-\tau) d\omega_q$$

$$= - \int_{D_\varepsilon} F(q,\tau) E(p,q,t-\tau) d\omega_q \qquad (16.55)$$

since $LE \equiv 0$ for all $t - \tau \geq \varepsilon$. Using the filtering property of E, we obtain

$$LU(p,t) = \lim_{u \downarrow 0} LU_\varepsilon(p,t \mid F) = -F(p,t) \qquad (16.56)$$

$$\text{Q.E.D.}^2$$

2. Heat potentials of double and single layers

Heat potentials possess properties analogous to those of electrostatic potentials. If the region D under consideration is cylindrical then this follows trivially from the fact that the heat potentials of double and single layers can be represented as the corresponding electrostatic potentials, by means of division and multiplication of the integrands by an appropriately chosen power of r_{pq}, with a subsequent change in the order of integration. As a result, one obtains the electrostatic surface potentials with time-dependent densities, which possess the necessary smoothness properties if appropriate assumptions are made concerning the densities of the thermal potentials. The case of a noncylindrical parabolic region is technically (although not in principle) much more complicated. In what follows we shall present the computations in full detail. Before stating the analogs of the theorems about discontinuity of the electrostatic potential of a double layer and of the normal derivatives of the electrostatic potential of a single layer, we make some basic assumptions.

Let D^t be a parabolic region possessing the following properties.

(a) For any $t \geq 0$, $D^t = \cup_0^t D_\tau$ where D_τ is convex for each $\tau \in [0,t]$.
(b) Let m be an arbitrarily chosen point of the lateral boundary Σ_t. Introduce a rectangular cartesian coordinate system (x_1, x_2, x_3) with origin at m and x_3 axis directed along the outward normal $\mathbf{n}_m$ to Σ_t at m. Let C be a cylinder of radius $R \ll 1$ and height $2h$, coaxial with $\mathbf{n}_m$ and with bases lying in the planes $x_3 = \pm h$. Then there exist $\alpha > 0$, $\rho > 0$, $h > 0$, and $K = \text{const} > 0$ such that for any $\tau \in [t - \alpha, t]$, C cuts out from Σ_τ a simply connected surface Σ_τ^* whose equation may be written as

$$\Sigma_\tau^* = \left\{ \xi_1, \xi_2, \xi_3, \tau \colon \xi_3 = X_3(\xi_1, \xi_2, \tau); \ \xi_1^2 + \xi_2^2 = \rho^2 \leq R^2 \right\} \qquad (16.57)$$

---

[2] The requirement that F be differentiable with respect to the spatial coordinates or time may be weakened: It is sufficient if F is Hölder continuous with exponent $\lambda > 0$ with respect to the spatial coordinates or time.

where the function X_3 is twice continuously differentiable with respect to ξ_1 and ξ_2, continuously differentiable with respect to τ,

$$\left|\frac{\partial X_3}{\partial \tau}\right|, \left|\frac{\partial^2 X_3}{\partial \xi_i \partial \xi_j}\right| < K, \quad i, j = 1, 2, \ \forall q = (\xi_1, \xi_2) \in \Sigma_\tau \cap C \tag{16.58}$$

and for any $\tau_1, \tau_2 \in [t - \alpha, t]$ and $\xi_1, \xi_2 \in \Sigma_\tau^*$,

$$\left|\frac{\partial}{\partial x_i} X_3(\xi_1, \xi_2, \tau_1) - \frac{\partial}{\partial x_i} X_3(\xi_1, \xi_2, \tau_2)\right| < K|\tau_1 - \tau_2|^{1/2} \tag{16.59}$$

Any parabolic region satisfying these conditions will be called *admissible*. We introduce some convenient notations. As before, let

$$P = (p, t), \quad p = (x_1, x_2, x_3), \quad Q = (q, \tau), \quad q = (\xi_1, \xi_2, \xi_3) \tag{16.60}$$

and let $\Gamma_{pt} \subset \mathcal{E}_{3+1}$ be a smooth curve, not tangent to Σ^t, that cuts Σ^t at (p, t) and is divided by Σ^t into external and internal parts

$$\Gamma_{pt}^e = \Gamma_{pt} \setminus \bar{D}^t, \quad \Gamma_{pt}^i = \Gamma_{pt} \cap D^t \tag{16.61}$$

Then for any function $f(p, t)$, we define

$$f_e(p, t) = \lim_{\Gamma_t^e \ni Q \to P \in \Sigma^t} f(Q), \quad f_i(p, t) = \lim_{\Gamma_t^i \ni Q \to P \in \Sigma^t} f(Q) \tag{16.62}$$

Theorem 16.2.1 (Discontinuity of the heat potential of a double layer). *Let D^t be an admissible parabolic region. Then the double-layer heat potential*[3]

$$W[p, t \mid \varphi(q, \tau)] = -\int_0^t d\tau \int_{\Sigma_\tau} \varphi(q, \tau) \frac{\partial}{\partial n_q} E(p, q, t - \tau) d\tau \tag{16.63}$$

possesses the following properties.

(α) *If the density $\varphi(q, \tau)$ is bounded,*

$$|\varphi(q, \tau)| < N \quad \forall Q \in D^t \tag{16.64}$$

then $W(p, t \mid \varphi)$ is continuous everywhere in $\mathcal{E}_{3+1} \setminus \Sigma^t$ and in Σ^t and parabolic outside Σ^t:

$$LW(p, t \mid \varphi) \stackrel{\text{def}}{=} \Delta W - \frac{\partial}{\partial t} W = 0 \quad \forall P \in \mathcal{E}_{3+1} \setminus \Sigma^t \tag{16.65}$$

(β) *If $\varphi(q, \tau)$ is continuous in Σ^t then there exist $W_e(p, t \mid \varphi)$ and $W_i(p, t \mid \varphi)$ such that, for any $P = (p, t) \in \Sigma^t$,*

$$W_i(p, t \mid \varphi) = \frac{1}{2}\varphi(p, t) + W(p, t \mid \varphi) \tag{16.66}$$

$$W_e(p, t \mid \varphi) = -\frac{1}{2}\varphi(p, t) + W(p, t \mid \varphi) \tag{16.67}$$

---

[3]$\partial E(p, q, t - \tau)/\partial n_q$ in the definition of $W(p, t \mid \psi)$ is understood as $\lim_{q_i \to q} \frac{\partial E}{\partial n_q}$, where $\mathbf{n}_q$ is a vector normal to Σ_τ at q and $q_i \in n_q$ is an interior point of D_τ.

Theorem 16.2.2 (Discontinuities of the normal derivatives of the heat potential of a single layer). *Let D^t be an admissible region and assume that the curve Γ_{pt} in (16.61) coincides with the normal $\mathbf{n}_p$ to Σ_t. Then the heat potential*

$$V(p,t \mid \psi(q,\tau)) = \int_0^t d\tau \int_{\Sigma_\tau} \psi(q,\tau)E(p,q,t-\tau)d\sigma_q \qquad (16.68)$$

possesses the following properties.

(α) *If the density $\psi(q,\tau)$ is bounded,*

$$\psi(q,\tau) < N \quad \forall Q \in D^t \qquad (16.69)$$

then $V(p,t \mid \psi)$ is continuous everywhere in $\mathcal{E}_{3+1}$ and parabolic outside of Σ^t:

$$LV(p,t \mid \psi) = 0 \quad \forall P \in \mathcal{E}_{3+1} \setminus \Sigma^t \qquad (16.70)$$

(β) *If $\psi(q,\tau)$ is continuous for any $Q \in D^t$, then there exist $\partial V(p,t \mid \psi)_e/\partial n_p$ and $\partial V(p,t \mid \psi)_i/\partial n_p$ such that*

$$\frac{\partial}{\partial n_p}V(p,t \mid \psi)_e = -\frac{1}{2}\psi(p,t) + \frac{\partial}{\partial n}V(p,t \mid \psi) \quad \forall P \in \Sigma^t \qquad (16.71)$$

$$\frac{\partial}{\partial n_p}V(p,t \mid \psi)_i = \frac{1}{2}\psi(p,t) + \frac{\partial}{\partial n_p}V(p,t \mid \psi) \quad \forall P \in \Sigma^t \qquad (16.72)$$

PROOF OF THEOREM 16.2.1. Note first that the statement that W is parabolic everywhere outside Σ^t is obviously true, since for every $P \notin \Sigma^t$ the integrand in the expression for W is a parabolic function without singularities that vanishes at $t = 0$. Hence, to prove part (a) it is only necessary to prove that $W(p,t \mid \varphi)$ is continuous along Σ^t, i.e., that $W(p,t \mid \varphi)$ is convergent uniformly at $p = m$, m being an arbitrary point of Σ_t.

We have

$$W(p,t \mid \varphi) = W_1(p,t \mid \varphi) + W_2(p,t \mid \varphi) \qquad (16.73)$$

where

$$W_1(p,t \mid \varphi) = -\int_{t-\alpha}^t d\tau \int_{\Sigma_\tau \cap C} \varphi(q,\tau)\frac{\partial}{\partial n_q}E(p,q,t-\tau)d\sigma_q \qquad (16.74)$$

$$W_2(p,t \mid \varphi) = -\left(\int_0^{t-\alpha} d\tau \int_{\Sigma_\tau} + \int_0^t d\tau \int_{\Sigma_\tau \setminus C}\right)\varphi(q,\tau)$$

$$\times \frac{\partial}{\partial n_q}E(p,q,t-\tau)d\sigma_q \qquad (16.75)$$

where C is the cylinder defined in condition (b) at the beginning of this section.

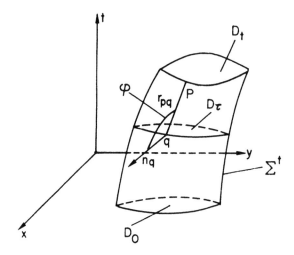

Figure 16.1. Sketch of a parabolic D^t and its
lateral boundary Σ^t; sections D_t and D_τ, and the
angle φ between $\mathbf{n}_q$ and $\mathbf{r}_{pq}$.

W_2 is continuous, since its integrand has no singularities. Consider W_1.
Introducing polar coordinates (ρ, ψ) in the plane T_p tangent to Σ_t at m
with pole at m, we find that

$$W_1(m, t \mid \varphi) = \int_{t-\alpha}^{t} d\tau \int_{\Sigma_\tau \cap C} \varphi(q, \tau) r_{mq} \cos \varphi \frac{\exp\left[-r_{mq}^2/4(t-\tau)\right]}{16\pi^{3/2}(t-\tau)^{3/2}} \cdot \frac{\rho \, d\rho}{\cos \gamma} \tag{16.76}$$

where γ is the angle between the x_3 axis and the normal $\mathbf{n}_q$ to Σ_τ at q,
and φ is the angle between vectors $\mathbf{r}_{pq}$ and the normal $\mathbf{n}_q$ so that

$$\cos \varphi = \frac{\mathbf{r}_{mq} \cdot \mathbf{n}_q^0}{r_{mq}}, \quad |\mathbf{n}_q^0| = 1 \tag{16.77}$$

(see Figure 16.1). Let $\mathbf{e}_1, \mathbf{e}_2, \mathbf{e}_3$ be the basis vectors of the cartesian
coordinate system. Then the equation of the plane T_q tangent to Σ_τ at
q is

$$-\frac{\partial}{\partial \xi_1} X_3(\xi_1, \xi_2, \tau)(\tilde{x}_1 - \xi_1) - \frac{\partial}{\partial \xi_2} X_3(\xi_1, \xi_2, \tau)(\tilde{x}_2 - \xi_2) + \tilde{x}_3 - \xi_3 = 0 \tag{16.78}$$

where $\tilde{p} = (\tilde{x}_1, \tilde{x}_2, \tilde{x}_3) \in T_p$ is a variable point. The radius vector $\mathbf{r}_{mq}$ is

$$\mathbf{r}_{mq} = \Sigma_{i=1}^{3} \xi_i \mathbf{e}_i \tag{16.79}$$

Hence

$$\cos \varphi = \frac{-\frac{\partial}{\partial \xi_1} X_3(\xi_1, \xi_2, \tau)\xi_1 - \frac{\partial}{\partial \xi_2} X_3(\xi_1, \xi_2, \tau)\xi_2 + 1}{r_{mq} \sqrt{1 + \left(\frac{\partial}{\partial \xi_1} X_3\right)^2 + \left(\frac{\partial}{\partial \xi_2} X_3\right)^2}} \tag{16.80}$$

At the same time,

$$\cos\gamma = \left(1 + \left[\frac{\partial}{\partial\xi_1}X_3\right]^2 + \left[\frac{\partial}{\partial\xi_2}X_3\right]^2\right)^{-1/2} \tag{16.81}$$

Finally,

$$r_{mq}^2 = \xi_1^2 + \xi_2^2 + X_3^2(\xi_1,\xi_2,\tau) \geq \rho^2 \tag{16.82}$$

Using the Lagrange formula of finite increments and taking into account that the plane T_m is tangent to Σ_t at the point $m = (0,0,0)$, we find that

$$\frac{\partial}{\partial\xi_i}X_3(\xi_1,\xi_2,\tau) = \frac{\partial}{\partial\xi_i}X_3(\xi_1,\xi_2,t) + \left(\frac{\partial}{\partial\xi_i}X_3(\xi_1,\xi_2,\tau) - \frac{\partial}{\partial\xi_i}X_3(\xi_1,\xi_2,t)\right)$$

$$= \frac{\partial^2}{\partial\xi_i\partial\xi_1}X_3(\tilde{\xi}_1,\tilde{\xi}_2,t)\xi_1 + \frac{\partial^2}{\partial\xi_i\partial\xi_2}X_3(\tilde{\xi}_1,\tilde{\xi}_2,t)\xi_2$$

$$+ \left(\frac{\partial}{\partial\xi_i}X_3(\xi_1,\xi_2,\tau) - \frac{\partial}{\partial\xi_i}X_3(\xi_1,\xi_2,t)\right) \tag{16.83}$$

where $\tilde{\xi}_i \in (0,\xi_i)$, $i = 1,2$. Hence, by (16.57)–(16.59), we have

$$\left|\frac{\partial}{\partial\xi_i}X_3(\xi_1,\xi_2,\tau)\right| < K(\sqrt{t-\tau} + 2\rho) \tag{16.84}$$

Similarly,

$$X_3(\xi_1,\xi_2,\tau) = \left\{\frac{\partial}{\partial\xi_1}X_3(\tilde{\xi}_1,\tilde{\xi}_2,t)\xi_1 + \frac{\partial}{\partial\xi_2}X_3(\tilde{\xi}_1,\tilde{\xi}_2,t)\xi_2\right\}$$

$$+ X_3(\xi_1,\xi_2,\tau) - X_3(\xi_1,\xi_2,t) \tag{16.85}$$

so that

$$|X_3(\xi_1,\xi_2,\tau)| < K[t - \tau + 2\rho\sqrt{t-\tau} + 4\rho^2] \tag{16.86}$$

Equalities (16.80) and (16.81) and estimates (16.82) yield

$$0 \leq \frac{\cos\varphi}{\cos\gamma} \leq 2K[t - \tau + 2\rho\sqrt{t-\tau} + 4\rho^2] \tag{16.87}$$

Let

$$\beta_i = \sqrt{2/\pi}\max[z^i\exp(-z^2)], \quad i \geq 0, \ 0 \leq z < \infty \tag{16.88}$$

Since $r_{mq} \geq \rho$, we have

$$\frac{\exp\left[-r_{mq}^2/4(t-\tau)\right]}{r_{mq}\ 2\pi^{1/2}(t-\tau)^{1/2}} \leq \beta_1\exp\left[-\frac{\rho^2}{8(t-\tau)}\right] \tag{16.89}$$

Hence it follows from (16.64), (16.87), and (16.89) that

$$W_1(m,t \mid \varphi) < C_0 \int_{t-\alpha}^{t} d\tau \int_0^R \exp\left[-\frac{\rho^2}{8(t-\tau)}\right]\frac{\rho\sqrt{t-\tau} + 2\rho^2}{(t-\tau)^2}\rho d\rho$$

$$< C_1 \int_{t-\alpha}^{t} (t-\tau)^{-1/2}d\tau \tag{16.90}$$

where C_0 and C_1 are constants dependent only on β_1, β_2, β_3, N, and K. Thus W_1 is majorized by a uniformly convergent integral and is therefore continuous along Σ^t. This proves part (a) of the theorem.

We now prove part (b). Let $\varphi(p,t)$ be continuous. We have

$$W[p,t \mid \varphi(q,\tau)] = W^1[p,t \mid \varphi(q,t)] + W^2[p,t \mid, \varphi(q,\tau)] \tag{16.91}$$

where

$$W^1[p,t \mid \varphi(q,t)] = -\int_0^t d\tau \int_{\Sigma_t} \varphi(q,t)\frac{\partial}{\partial n_q} E(p,q,t-\tau)d\tau$$

$$W^2[p,t \mid \varphi(q,t),\varphi(q,\tau)] = \int_0^t d\tau \int_{\Sigma_t} \varphi(q,t)\frac{\partial}{\partial n_q} E(p,q,t-\tau)d\tau \tag{16.92}$$

$$-\int_0^t d\tau \int_{\Sigma_\tau} \varphi(q,\tau)\frac{\partial}{\partial n_q} E(p,q,t-\tau)d\tau$$

Changing the order of integration, we get

$$W^1[p,t \mid \varphi(q,t)] = -\int_{\Sigma_t} \Phi(p,q,t)\frac{\cos\varphi}{4\pi r_{pq}^2}d\sigma_q \tag{16.93}$$

where

$$\Phi(p,q,t) \stackrel{\text{def}}{=} \int_0^t r_{pq}^3 \frac{\exp\left[-r_{pq}^2/4(t-\tau)\right]}{4\pi^{1/2}(t-\tau)^{5/2}}\varphi(q,t)d\tau$$

$$= \varphi(q,t)2\pi^{-1/2}\int_{r_{pq}^2/4t}^{\infty} \sqrt{z}\exp(-z)dz \tag{16.94}$$

Hence $\Phi(p,q,t)$ is a continuous function and also

$$\Phi(p,p,t) = \varphi(p,t) \tag{16.95}$$

Thus $W^1[p,t \mid \varphi(q,t)]$ is the electrostatic potential of a double layer with continuous density, so that, by the theorem about the discontinuity of such potentials,

$$W_i^1(m,t \mid \varphi) = {}^1\!/{}_2\varphi(m,t) + W^1(m,t \mid \varphi)$$
$$W_e^1(m,t \mid \varphi) = -{}^1\!/{}_2\varphi(m,t) + W^1(m,t \mid \varphi) \tag{16.96}$$

Hence it remains only to prove the continuity of $W^2[m,t \mid \varphi(q,\tau)]$. Again we have

$$W^2(p,t \mid \varphi) = W_0^2(p,t \mid \varphi) + W_1^2(p,t \mid \varphi) \tag{16.97}$$

where

$$W_0^2(p,t \mid \varphi) = -\int_{t-\alpha}^t d\tau \int_{\Sigma_\tau \cap C} \varphi(q,\tau)\frac{\partial}{\partial n_q} E(p,q,t-\tau)d\sigma_q$$

$$+\int_{t-\alpha}^t d\tau \int_{\Sigma_t \cap C} \varphi(q,t)\frac{\partial}{\partial n_q} E(p,q,t-\tau)d\sigma_q \tag{16.98}$$

and $W_1^2(p, t \mid \varphi)$ is a sum of integrals whose integrands have no singularities. Introducing polar coordinates (ρ, ψ) in the plane T_p, we obtain

$$W_0^2(p, t \mid \varphi) = I_1 - I_2 - I_3 \tag{16.99}$$

$$I_1 = \int_{t-\alpha}^t d\tau \int_0^{2\pi} d\psi \int_0^R [\varphi(q, \tau) - \varphi(q^1, \tau)]$$

$$\times \frac{r_{pq^1} \cos \varphi^1}{\cos \gamma^1} \cdot \frac{\exp\left[-r_{pq^1}^2 / 4(t-\tau)\right]}{16\pi^{3/2}(t-\tau)^{5/2}} \rho \, d\rho \tag{16.100}$$

$$I_2 = \int_{t-\alpha}^t d\tau \int_0^{2\pi} d\psi \int_0^R \varphi(q, \tau)$$

$$\times \left(\frac{r_{pq^1} \cos \varphi^1}{\cos \gamma^1} - \frac{r_{pq} \cos \varphi}{\cos \gamma}\right) \frac{\exp[-r_{pq^1}^2 / 4(t-\tau)]}{16\pi^{3/2}(t-\tau)^{5/2}} \rho \, d\rho \tag{16.101}$$

$$I_3 = \int_{t-\alpha}^t d\tau \int_0^{R\pi} d\psi \int_0^R \varphi(q, \tau)$$

$$\times \frac{r_{pq} \cos \varphi^1}{\cos \gamma} \left(\frac{\exp[-r_{pq^1}^2 / 4(t-\tau)]}{16\pi^{3/2}(t-\tau)^{5/2}} - \frac{\exp\left[-r_{pq}^2 / 4(t-\tau)\right]}{16\pi^{3/2}(t-\tau)^{5/2}}\right) \rho \, d\rho \tag{16.102}$$

where q^1 is the projection of $q \in \Sigma_\tau$ on Σ_t, so that

$$q^1 \in \Sigma_t, \qquad q \in \Sigma_\tau \tag{16.103}$$

$$\cos \varphi^1 = \frac{\mathbf{r}_{pq^1} \cdot n_{q^1}^0}{r_{pq^1}}, \qquad \cos \varphi = \frac{\mathbf{r}_{pq} \cdot n_q^0}{r_{pq}} \tag{16.104}$$

$$\cos \gamma^1 = \mathbf{n}_{q^1}^0 \cdot n_m^0, \qquad \cos \gamma = \mathbf{n}_q^0 \cdot n_m^0 \tag{16.105}$$

Since by assumption Σ_τ is convex for every $\tau \geq 0$, we have

$$\cos \varphi \geq 0, \quad \cos \varphi^1 \geq 0, \quad \cos \gamma > 0, \quad \cos \gamma^1 > 0 \tag{16.106}$$

Hence, by the mean value theorem,

$$|I_1| < \sup |\varphi(q, \tau) - \varphi(q^1, \tau)|$$

$$\int_{t-\alpha}^t d\tau \int_0^{2\pi} d\psi \int_0^R \frac{r_{pq^1} \cos \varphi^1}{\cos \gamma^1} \cdot \frac{\exp[-r_{pq^1}^2 / 4(t-\tau)]}{16\pi^{3/2}(t-\tau)^{5/2}} \rho \, d\rho$$

$$\leq \sup |\varphi(q, \tau) - \varphi(q^1, \tau)| \cdot \left|-\int_0^t d\tau \int_{\Sigma_\tau} \frac{\partial}{\partial n_q} E(p, q^1, t-\tau) d\tau\right|$$

$$\leq \sup |\varphi(q, \tau) - \varphi(q^1, \tau)| \tag{16.107}$$

We now have

$$J = \frac{r_{pq^1} \cos \varphi^1}{\cos \gamma^1} - \frac{r_{pq} \cos \varphi}{\cos \gamma}$$

$$= (r_{pq^1} - r_{pq}) \frac{\cos \varphi^1}{\cos \gamma^1} + r_{pq} \left(\frac{\cos \varphi^1}{\cos \gamma^1} - \frac{\cos \varphi}{\cos \gamma}\right) \tag{16.108}$$

If $R > 0$ and $\alpha > 0$ are sufficiently small, then

$$0 \leq \frac{\cos \varphi^1}{\cos \gamma^1} < 2 \qquad (16.109)$$

Besides, by the triangle inequality

$$|r_{pq^1} - r_{pq}| \leq r_{qq^1} \leq K(t - \tau) \qquad (16.110)$$

Furthermore,

$$q = [\xi_1, \xi_2, X_3(\xi_1, \xi_2, \tau)], \quad q^1 = [\xi_1, \xi_2, X_3(\xi_1, \xi_2, t)], \quad p = (0, 0, x_3) \quad (16.111)$$

Hence

$$
\begin{aligned}
J^* &= \frac{\cos \varphi^1}{\cos \gamma^1} - \frac{\cos \varphi}{\cos \gamma} \\
&= \left(X_3(\xi_1, \xi_2, t) - x_3 - \xi_1 \frac{\partial}{\partial \xi_1} X_3(\xi_1, \xi_2, t) - \xi_2 \frac{\partial}{\partial \xi_2} X_3(\xi_1, \xi_2, t) \right) \frac{1}{r_{pq^1}} \\
&\quad - \left(X_3(\xi_1, \xi_2, \tau) - x_3 - \xi_1 \frac{\partial}{\partial \xi_1} X_3(\xi_1, \xi_2, \tau) \right. \\
&\quad \left. - \xi_2 \frac{\partial}{\partial \xi_2} X_3(\xi_1, \xi_2, \tau) \right) \frac{1}{r_{pq}}
\end{aligned}
\qquad (16.112)
$$

or, what is the same,

$$
\begin{aligned}
J^* &= \frac{1}{r_{pq^1}} \left\{ X_3(\xi_1, \xi_2, t) - X_3(\xi_1, \xi_2, \tau) \right. \\
&\quad \left. - \sum_{i-1}^{2} \xi_i \frac{\partial}{\partial \xi_i} [X_3(\xi_1, \xi_2, t) - X_3(\xi_1, \xi_2, \tau)] \right\} \\
&\quad + \frac{r_{pq} - r_{pq^1}}{r_{pq^1} \cdot r_{pq}} \cdot \left(X_3(\xi_1, \xi_2, t) - X_3(\xi_1, \xi_2, \tau) \right. \\
&\quad \left. - \sum_{i=1}^{2} \xi_i \frac{\partial}{\partial \xi_i} X_3(\xi_1, \xi_2, t) \right)
\end{aligned}
\qquad (16.113)
$$

We have

$$r_{pq^1} > \rho, \quad r_{pq} > \rho, \quad |r_{pq^1} - r_{pq}| \leq r_{qq^1} < K(t - \tau) \qquad (16.114)$$

$$|X_3(\xi_1, \xi_2, t) - X_3(\xi_1, \xi_2, \tau)| < K(t - \tau) \qquad (16.115)$$

$$\left| \sum_{i=1}^{2} \xi_i \frac{\partial}{\partial \xi_i} X_3(\xi_1, \xi_2, t) \right| < 2K\rho^2 \qquad (16.116)$$

$$\sum_{i=1}^{2} \left| \xi_i \frac{\partial}{\partial \xi_i} [X_3(\xi_1, \xi_2, t) - X_3(\xi_1, \xi_2, \tau)] \right| < 2K\rho\sqrt{t - \tau} \qquad (16.117)$$

$$|X_3(\xi_1, \xi_2, t) - x_3| < r_{pq^1}, \quad \frac{r_{pq}}{r_{pq^1}} \leq 1 + \frac{r_{qq^1}}{r_{pq^1}} < 1 + \frac{K(t - \tau)}{r_{pq^1}} \qquad (16.118)$$

Inserting (16.109), (16.110), (16.113), and (16.114)–(16.118) into (16.108), we obtain

$$|J| < 2K(t-\tau) + \left(1 + \frac{K(t-\tau)}{r_{pq^1}}\right)[K(t-\tau) + 2K\rho\sqrt{t-\tau}]$$

$$+ \frac{K(t-\tau)}{r_{pq^1}}(r_{pq^1} + 2K\rho^2) \tag{16.119}$$

Inserting (16.108) and (16.119) into (16.101), we obtain

$$|I_2| = \sum_{i=1}^{5} I_{2i} \tag{16.120}$$

where

$$0 < I_{21} = NK \int_{t-\alpha}^{t} d\tau \int_{0}^{R} \frac{\exp\left[-\rho^2/4(t-\tau)\right]}{2\pi^{1/2}(t-\tau)^{3/2}}\rho d\rho$$

$$= NK \int_{0}^{R} d\rho \frac{2}{\sqrt{\pi}} \int_{\rho/2\sqrt{t-\tau}}^{\infty} \exp(-z^2)dz \le NKR \tag{16.121}$$

$$0 < I_{22} = NK^2 \int_{t-\alpha}^{t} d\tau \int_{0}^{R} \frac{\exp\left[-\rho^2/4(t-\tau)\right]}{8\pi^{3/2}(t-\tau)^{1/2}}\rho d\rho$$

$$< \frac{NK^2}{8\pi^{3/2}} \int_{t-\alpha}^{t} d\rho \int_{0}^{\infty} \exp(-z^2)dz = \frac{NK^2}{8\pi}\alpha \tag{16.122}$$

$$0 < I_{23} = NK \int_{t-\alpha}^{t} d\tau \int_{0}^{R} \rho^2 \frac{\exp\left[-r_{pq}^2/4(t-\tau)\right]}{4\pi^{3/2}(t-\tau)^2}d\rho$$

$$< \frac{4NK}{\pi^{3/2}} \int_{0}^{R} d\rho \int_{\rho^2/4t}^{\infty} z\exp(-z)dz < \frac{4NK}{\pi^{3/2}}R \tag{16.123}$$

$$0 < I_{24} = NK^? \int_{t-\alpha}^{t} d\tau \int_{0}^{R} \mu\frac{\exp\left[-r_{pq}^2/4(t-\tau)\right]}{2\pi^{3/2}(t-\tau)}d\mu$$

$$< \frac{NK^2}{\pi^{3/2}} \int_{t-\alpha}^{t} d\tau \int_{0}^{R^2/4(t-\tau)} \exp(-z)dz \le \frac{NK^2}{\pi^{3/2}}\alpha \tag{16.124}$$

$$0 < I_{25} \le NK^2 \int_{t-\alpha}^{t} d\tau \int_{0}^{R} \rho^2 \frac{\exp\left[-r_{pq}^2/4(t-\tau)\right]}{4\pi^{3/2}(t-\tau)^{3/2}}d\rho$$

$$< \frac{2NK^2}{\pi^{3/2}} \int_{t-\alpha}^{t} d\tau \int_{0}^{R^2/4(t-\tau)} z^2\exp(-z^2)dz \le NK^2\alpha \tag{16.125}$$

Hence I_2 is majorized by a sum of the uniformly convergent integrals, and also

$$|I_2| < C(N, K)(\alpha + R) \tag{16.126}$$

Finally, let us consider I_3. By the mean value theorem,

$$\tilde{J} = \exp\left(-\frac{r_{pq}^2}{4(t-\tau)}\right) - \exp\left(-\frac{r_{pq^1}^2}{4(t-\tau)}\right)$$

$$= \int_{r_{pq}^2/4(t-\tau)}^{r_{pq^1}^2/4(t-\tau)} \exp(-z)dz = -\exp\left(-\frac{r_{ps}^2}{4(t-\tau)}\right)\frac{r_{pq}^2 - r_{pq^1}^2}{4(t-\tau)} \quad (16.127)$$

where r_{ps} lies between r_{pq} and r_{pq^1}, so that

$$r_{ps} = r_{pq} + (r_{ps} - r_{pq}) \geq r_{pq} - r_{qq^1} > 0 \quad (16.128)$$

Recall that

$$q \in \Sigma_\tau, \; q^1 \in \Sigma_t \Rightarrow r_{qq^1} < K(t-\tau) \quad (16.129)$$

Hence

$$r_{pq}^2 - r_{pq^1}^2 < r_{qq^1}(r_{pq} + r_{q^1}) \leq K(t-\tau)(r_{pq} + K(t-\tau)) \quad (16.130)$$

and

$$r_{ps}^2 \geq r_{pq}^2 - 2RK(t-\tau) > r_{pq}^2 - K(t-\tau) \quad (16.131)$$

so that

$$|\tilde{J}| < \frac{\exp\left[-r_{pq}^2/4(t-\tau) + K/4\right]}{4(t-\tau)} r_{qq^1}(r_{pq} + r_{qq^1})$$

$$\leq \frac{K}{4}\exp\left[-\frac{r_{pq}^2}{4(t-\tau)} + \frac{K}{4}\right][r_{pq} + K(t-\tau)] \quad (16.132)$$

We now insert (16.132) into (16.102), taking into account inequalities (16.109) and (16.114). This yields

$$|I_3| < NK\exp\left(\frac{K}{4}\right)\int_{t-\alpha}^{t} dt \int_0^R r_{pq}\frac{\exp\left[-r_{pq}^2/4(t-\tau)\right]}{16\pi^{3/2}(t-\tau)^{3/2}}\cdot[r_{pq} + K(t-\tau)]\rho d\rho \quad (16.133)$$

Note that

$$r_{pq}^2\frac{\exp\left[-r_{pq}^2/4(t-\tau)\right]}{16\pi^{3/2}(t-\tau)} \leq \beta_2\exp\left(-\frac{r_{pq}^2}{8(t-\tau)}\right)$$

$$\leq \beta_2\exp\left(-\frac{\rho^2}{8(t-\tau)}\right) \quad (16.134)$$

$$r_{pq}\frac{\exp\left(-r_{pq}^2/4(t-\tau)\right)}{4(2\pi)^{3/2}(t-\tau)^{1/2}} \leq \beta_1\exp\left(-\frac{r_{pq}^2}{8(t-\tau)}\right)$$

$$\leq \beta_1\exp\left(-\frac{\rho^2}{8(t-\tau)}\right) \quad (16.135)$$

where β_1 and β_2 are defined by (16.88). Hence

$$|I_3| < NK \exp\left(\frac{K}{4}\right) \int_{t-\alpha}^{t} dt \int_0^R \frac{\exp\left[-\rho^2/8(t-\tau)\right]}{\pi^{1/2}(t-\tau)^{1/2}} [\beta_2 + 2K\beta_1] \cdot \rho \cdot d\rho$$

$$= C(N,K) \int_0^R d\rho \int_{\rho/2\sqrt{t-\tau}}^{\infty} \exp(-z^2) dz < C(NK)R \qquad (16.136)$$

Inserting (16.107), (16.126), and (16.136) into (16.99), we obtain

$$|W_0^2| < \sup_{(q,\tau) \in C \cap \Sigma^t, t-\alpha \leq \tau \leq t} |\varphi(q,\tau) - \varphi(q^1,\tau)| + C(N,K)(R+\alpha) \qquad (16.137)$$

Take $\varepsilon > 0$ arbitrarily small. Since φ is continuous, we can find $R > 0$ and $\alpha > 0$ so small that, by (16.137),

$$|W_0^2| < \varepsilon \qquad (16.138)$$

Hence, by the continuity of $W_1^2(m,t)$, it follows from (16.97), that

$$\varliminf_{p \to m \in \Sigma_t} W^2(p,t \mid \varphi) \leq W^1(m,t \mid \varphi) + \varepsilon$$

$$\varliminf_{p \to \min \Sigma_t} W^2(p,t \mid \varphi) \geq W^1(m,t \mid \varphi) - \varepsilon \qquad (16.139)$$

so that

$$\varlimsup_{p \to m \in \Sigma_t} W^2(p,t \mid \varphi) - \varliminf_{p \to m \in \Sigma_t} W^2(p,t \mid \varphi) \leq 2\varepsilon \qquad (16.140)$$

Since $\varepsilon > 0$ is arbitrarily small and $W^2(p,t \mid \varphi)$ is independent of ε, this means that

$$\varlimsup_{p \to m \in \Sigma_t} W^2(p,t \mid \varphi) - \varliminf_{p \to m \in \Sigma_t} W^2(p,t \mid \varphi) = 0 \qquad (16.141)$$

Thus $W^2(p,t \mid \varphi)$ is continuous at $p = m$. **Q.E.D.**

Looking back at this proof of Theorem 16.2.1, we see that its crucial point is the representation (16.91) of W as the sum of W^1 and W^2. Indeed, $W^1(p,T \mid \varphi)$ behaves like the heat potential of a double layer taken over a cylindrical region. This is precisely why we can treat it as an electrostatic potential of a double layer with continuous density, and consequently use the theorem on the discontinuity of the latter. On the other hand, $W^2 = W - W^1$ is continuous, and the proof that this is indeed the case involves only technical difficulties.

Remark. If p is a conic point of Σ_t with solid angle ω, then equalities (16.66) and (16.67) must be replaced by

$$W_i(p,t \mid \varphi) = \frac{4\pi - \omega}{4\pi} \varphi(p,t) + W(p,t \mid \varphi) \qquad (16.142)$$

$$W_e(p,t \mid \varphi) = -\frac{\omega}{4\pi} \varphi(p,t) + W(p,t \mid \varphi) \qquad (16.143)$$

exactly as in the case of the electrostatic potential of a double layer (see Theorem 12.7.1).

PROOF OF THEOREM 16.2.2. Let $D^t \subset \mathcal{E}_{3+1}$ so that $D \subset \mathbb{R}_3$ is the region of volume M, and also

$$\Sigma_\tau \subset K \quad \forall \tau \in [0, \infty) \tag{16.144}$$

where $K \subset \mathbb{R}_3$ is some sphere. Then, by the properties of integrals with polar singularities,[4] inequality (16.69) shows that $|V(p, t \mid \psi(q, \tau))|$ is majorized by a uniformly convergent integral, so that

$$|V(p, t \mid \psi(q, \tau))| < \int_D \frac{N}{r_{pq}} d\omega_q \int_0^t r_{pq} \frac{\exp\left[-r_{pq}^2 / 4(t - \tau)\right]}{8\pi^{3/2}(t - \tau)^{3/2}} d\tau$$

$$< \frac{N}{4\pi} \int_D \frac{1}{r_{pq}} d\omega_q \frac{2}{\sqrt{\pi}} \int_{r_{pq}/2\sqrt{t-\tau}}^{\infty} \exp(-z^2) dz \tag{16.145}$$

$V(p, t \mid \psi(q, \tau))$ is continuous everywhere in $\mathcal{E}_{3+1}$. Since the integrand of V has no singularities outside Σ_τ for any $\tau \in [0, t]$, it is also parabolic everywhere in $\mathcal{E}_{3+1} \setminus \Sigma_\tau$ for any $\tau \in [0, t]$. Thus we have to prove only part (β). To this end, note that expressions for the normal derivatives of the heat potential of a single layer may be obtained by replacing the density $\varphi(q, \tau)$ of the heat potential of a double layer by the density $\psi(q, \tau)$ of the heat potential of a single layer and replacing $\cos \varphi$ by $\cos \psi$, where ψ is the angle between vectors $\mathbf{r}_{pq}$ and the normal $\mathbf{n}_m$. This is precisely the situation in the theory of electrostatic potentials. In other words, the proof of part (β) is identical to that of part (b) of Theorem 16.2.1. We may therefore omit the continuation of the proof of Theorem 16.2.2.

Problems

P16.1.1. Prove Theorem 16.1.1 in the 1-dimensional case, without reference to the proof for dimension 3.

P16.1.2. Prove part (c) of Theorem 16.1.1.

P16.2.1. Theorems 16.2.1–2 were proved by referring to the corresponding theorems of the theory of electrostatic potentials. In dimension 1, however, heat potentials were defined although the respective electrostatic potentials had not been discussed. Is it possible to apply the method of descent to prove the theorems in dimension 1? If so, explain how; if not, provide direct proof of the theorems.

P16.2.2. Prove Theorem 16.2.2 for a cylindrical region in $\mathcal{E}_{2+1}$.

P16.2.3. Is it possible to generalize Theorems 12.7.2 and 16.2.2 for the case where the boundaries of the regions under consideration have conical points? If so, do this; otherwise, explain why not.

---

[4]See Chapter 12, Section 5.

17

Volterra integral equations and their application to solution of boundary-value problems in heat-conduction theory

1. Reduction of first, second, and third boundary-value problems for Fourier equation to Volterra integral equations. Existence theorems

It was shown in Chapter 12 how the properties of electrostatic potentials could be used to reduce the first and second inner and outer boundary-value problems for the Poisson equation to Fredholm integral equations of the second kind. Quite similarly, discontinuity theorems for the heat potential of a double layer and for the normal derivatives of the heat potential of a single layer may be used to reduce the first, second, and third boundary-value problems for the nonhomogeneous Fourier equation to integral equations of mixed type: Fredholm type with respect to the spatial variables and Volterra type with respect to time. The Volterra character of these equations dominates, and globally they behave as purely Volterra integral equations.

As shown in Section 1 of Chapter 16, the boundary-value problems for the nonhomogeneous heat-conduction equation with nonhomogeneous initial conditions can be reduced to the analogous problems with homogeneous initial data by using the representation:

$$u(p,t) = v(p,t) + U(p,t \mid f) + \Pi(p,y \mid \varphi) \qquad (17.1)$$

where $u(p,t)$ is the solution of the original problem, U the heat volume potential with density f of inner sources of heat, and Π the Poisson integral; φ is the initial temperature. We shall therefore restrict ourselves to problems for the homogeneous heat-conduction equation with homogeneous initial data. As to the region D^t in which the problem is being considered, we shall assume that it satisfies the conditions formulated in Section 2 of Chapter 16, so that Theorems 16.2.1 and 16.2.2 are applicable. We shall further assume that the boundary of D^t is bounded for all $t \geq 0$. Finally, we postulate that there exist positive numbers A and B such that the ratio of the areas $d\sigma(q,\tau)$ and $d\sigma(q,t)$ of corresponding elements of the surfaces Σ_τ and Σ_t

lies in the interval (A, B). This means the following. The surface Σ_τ may be defined parametrically as

$$\Sigma_\tau = \{x = x_i(u, v, \tau), \ i = 1, 2, 3; \ 0 < u, v < \ell\} \tag{17.2}$$

where $x_i(u, v, \tau)$ are functions twice continuously differentiable with respect to u and v and once continuously differentiable with respect to τ. Let $\mathfrak{E}$, $\mathfrak{G}$, and $\mathfrak{F}$ be the coefficients of the first quadratic form

$$ds^2 = \mathfrak{E}du^2 + 2\mathfrak{F}dudv + \mathfrak{G}dv^2 \tag{17.3}$$

so that

$$\mathfrak{E} = \sum_{i=1}^{3} \left(\frac{\partial x_i}{\partial u}\right)^2; \quad \mathfrak{G} = \sum_{i=1}^{3} \left(\frac{\partial x_i}{\partial v}\right)^2; \quad \mathfrak{F} = \sum_{i=1}^{3} \left(\frac{\partial x_i}{\partial u} \cdot \frac{\partial x_i}{\partial v}\right) \tag{17.4}$$

Then

$$d\sigma(q, \lambda) = \sqrt{\mathfrak{E}(u, v, \lambda) \cdot \mathfrak{G}(u, v, \lambda) - \mathfrak{F}(u, v, \lambda)^2}\, dudv \tag{17.5}$$

Thus we are postulating that for any τ, $0 \le \tau \le t < \infty$,

$$0 < A \le \frac{\sqrt{\mathfrak{E}(u, v, \tau) \cdot \mathfrak{G}(u, v, \tau) - \mathfrak{F}(u, v, \tau)^2}}{\sqrt{\mathfrak{E}(u, v, t) \cdot \mathfrak{G}(u, v, t) - \mathfrak{F}(u, v, t)^2}} \le B \tag{17.6}$$

We wish to find a solution of the heat-conduction equation

$$Lu \overset{\text{def}}{=} \Delta u(p, t) - \frac{\partial u}{\partial t}(p, t) = 0, \quad (p, t) \in D^t \tag{17.7}$$

satisfying the initial conditions

$$u(p, 0) = 0, \quad p \in D_0 \tag{17.8}$$

and one of the following boundary conditions:

$$u(p, t) = f(p, t) \quad \text{at } (p, t) \in \Sigma_t \tag{17.9}$$

for the first (Dirichlet) boundary-value problem;

$$\frac{\partial u}{\partial n_p} = f(p, t) \quad \text{at } (p, t) \in \Sigma_t \tag{17.10}$$

for the second (Neumann) boundary-value-problem; or

$$\frac{\partial u}{\partial n_p} + \alpha(p, t)u(p, t) = \alpha(p, t)f(p, t) \quad \text{at } (p, t) \in \Sigma_t \tag{17.11}$$

for the third (Newtonian irradiation) problem.[1]

---

[1] Recall that $\partial F(p, t)/\partial n_p$ denotes the derivative of a function $F(p, t)$ in a neighborhood of a point (p, t) of Σ^t in the direction of the outward normal to Σ_t. Recall also that D_t denotes the intersection of the region D^T by the plane $\tau = t = \text{const}$ where $t \le T$, and Σ_t is the intersection of the lateral boundary Σ^T of D^T with the plane $\tau = t$. In physical problems in regions that change shape while the process is evolving, these notational distinctions between superscripts and subscripts become essential.

Let us seek a solution of the first boundary-value problem as the heat potential of a double layer:

$$u(p,t) = W[p,t \mid \mu(q,\tau)]$$
$$\equiv \frac{1}{16\pi^{3/2}} \int_0^t d\tau \int_{\Sigma_\tau} \mu(q,\tau) \frac{r_{pq} \cos(\varphi_{pq}) \exp[-r_{pq}^2/4(t-\tau)]}{(t-\tau)^{5/2}} d\sigma_q \quad (17.12)$$

and a solution of the second and the third boundary-value problems as heat potentials of the single layer

$$u(p,t) = V[p,t \mid \mu(q,\tau)] \equiv \frac{1}{8\pi^{3/2}} \int_0^t d\tau \int_{\Sigma_\tau} \mu(q,\tau) \frac{\exp[-r_{pq}^2/4(t-\tau)]}{(t-\tau)^{3/2}} d\sigma_q$$
$$(17.13)$$

The densities $\mu(p,t)$ must be determined on the basis of the discontinuities of $W[p,t \mid \mu(q,\tau)]$ in the first case and of $\partial V[p,t \mid \mu(q,\tau)]/\partial n_q$ in the second. Thus, we obtain:

$$f(p,t) = \pm \frac{\mu(p,t)}{2} + W[p,t \mid \mu(q,\tau)] \quad (17.14)$$

for the first boundary-value problem,

$$f(p,t) = \pm \frac{\mu(p,t)}{2} + \frac{\partial}{\partial n_p} V[p,t \mid \mu(q,\tau)] \quad (17.15)$$

for the second boundary-value problem and

$$\alpha(p,t)f(p,t) = \pm \mu(p,t)/2 + \frac{\partial}{\partial n_p} V[p,t \mid \mu(q,\tau)] + \alpha(p,t)V[p,t \mid \mu(q,\tau)] \quad (17.16)$$

for the third. The sign "+" in equations (17.14)–(17.16) corresponds to the inner boundary-value problems and "−" to the outer ones. Recall that

$$\frac{\partial}{\partial n_p} V[p,t \mid \mu(q,\tau)] = \frac{1}{16\pi^{3/2}} \int_0^t d\tau \int_{\Sigma_\tau} \mu(q,\tau) \frac{r_{pq} \cos(\psi_{pq}) \exp[-r_{pq}^2/4(t-\tau)]}{(t-\tau)^{5/2}}$$
$$(17.17)$$

and that

$$\cos(\varphi_{pq}) = \frac{\mathbf{r}_{pq} \cdot \mathbf{n}_q^0}{r_{pq}}, \qquad \cos(\psi_{pq}) = \frac{\mathbf{r}_{pq} \cdot \mathbf{n}_p^0}{r_{pq}} \quad (17.18)$$

All three equations (17.14)–(17.16) may be written

$$\mu(p,t) = F(p,t) + \lambda \int_0^t d\tau \int_{\Sigma_\tau} \mu(q,\tau)K(p,q,t,\tau)d\sigma_q \quad (17.19)$$

where $\lambda = -1$ for the inner problems and $\lambda = +1$ for the outer problems. The modulus of the kernel $K(p,q,t,\tau)$ may be majorized by a function $\hat{K}(p,q,t-\tau)$ of the form

$$\hat{K}(p,q,t-\tau) = N \frac{\exp[-\delta r_{pq}^2/(t-\tau)]}{(t-\tau)^{3/2}} \quad (17.20)$$

so that

$$|K(p,q,t,\tau)| < N \frac{\exp[-\delta r_{pq}^2/(t-\tau)]}{(t-\tau)^{3/2}} \qquad (17.21)$$

with suitably chosen constants $N > 0$ and $\delta > 0$. Indeed, there exists $N_0 > 0$ such that

$$0 < \cos\varphi_{pq}, \quad \cos\psi_{pq} \le N_0[r_{pq} + \sqrt{t-\tau}] \quad \forall(p,q) \in \Sigma_\tau \qquad (17.22)$$

(see Chapter 16, Section 2). Hence the kernel $K(p,q,t,\tau)$ may indeed be majorized by a kernel $\hat{K}$ with δ equal to (say) $1/8$.

We now show that the integral equation (17.19) has a unique solution in any finite time interval $[0, T]$ if $F(p,t)$ is uniformly bounded in Σ^t for every $t \ge 0$. A formal solution of this integral equation may be sought as a power series in λ:

$$\mu(p,t) = \sum_{n=0}^{\infty} \mu_n(p,t)\lambda^n \qquad (17.23)$$

Assuming that the series (17.23) is uniformly convergent, we insert it into (17.19) and change the order of integration and summation. Comparing coefficients of like powers of λ, we obtain

$$\mu_0(p,t) = F(p,t) \qquad (17.24)$$

$$\mu_n(p,t) = \int_0^t d\tau \int_{\Sigma_\tau} \mu_{n-1}(q,\tau)K(p,q,t,\tau)d\sigma_q \quad \forall n \ge 1 \qquad (17.25)$$

To estimate μ_n, let

$$h = \sup(r_{pq}), \quad p \in \Sigma_t, \ q \in \Sigma_\tau, \quad 0 \le \tau < t < \infty \qquad (17.26)$$

$$p = x(\tilde{u}, \tilde{v}, t), \quad q = x(u, v, \tau), \quad \tilde{q} = x(u, v, t) \qquad (17.27)$$

We have[2]

$$r_{pq}^2 = \sum_{i=1}^{3} [x_i(\tilde{u}, \tilde{v}, t) - x_i(u, v, \tau)]^2$$

$$= \sum_{i=1}^{3} [x_i(\tilde{u}, \tilde{v}, t) - x_i(u, v, t)]^2 + \sum_{i=1}^{3} [x_i(u, v, t) - x_i(u, v, \tau)]^2$$

$$+ 2\sum_{i=1}^{3} [x_i(\tilde{u}, \tilde{v}, t) - x_i(u, v, \tau)][x_i(u, v, t) - x_i(u, v, \tau)]$$

$$= r_{pq}^2 + r_{q\tilde{q}}^2 + 2\sum_{i=1}^{3} [x_i(u, v, t) - x_i(u, v, \tau)][x_i(\tilde{u}, \tilde{v}, t) - x_i(u, v, \tau)] \quad (17.28)$$

---

[2] Here $x = (x_1, x_2, x_3)$, where x_1, x_2, x_3 are the cartesian coordinates of a point in $\mathbb{R}_3$.

Using the Cauchy inequality

$$\left(\sum_{i=1}^{n}|a_i b_i|\right)^2 \leq \sum_{i=1}^{n}a_i^2 \sum_{i=1}^{n}b_i^2 \tag{17.29}$$

we obtain

$$r_{pq}^2 \geq r_{p\tilde{q}}^2 + r_{q\tilde{q}}^2 - 2r_{p\tilde{q}}r_{q\tilde{q}} \tag{17.30}$$

where, by (17.26),

$$r_{p\tilde{q}} \leq h \tag{17.31}$$

By the assumption made in Chapter 16, Section 2 (see (16.10)),

$$r_{q\tilde{q}} \leq K(t-\tau) \tag{17.32}$$

where K may evidently be identified with the constant N_0 in (17.21).
We now apply inequality (17.6). We have

$$\left|\int_{\Sigma_\tau} \mu(q,\tau)K(p,q,t,\tau)d\sigma_q\right|$$

$$= \left|\int_0^\ell du \int_0^\ell \mu(q,\tau)K(p,q,t,\tau)\right.$$

$$\left.\times \sqrt{\mathfrak{E}(u,v,\tau)\cdot\mathfrak{G}(u,v,\tau) - \mathfrak{F}(u,v,\tau)^2}dv\right| \tag{17.33}$$

Hence

$$\left|\int_{\Sigma_\tau} \mu(q,\tau)K(p,q,t,\tau)d\sigma_q\right|$$

$$= \left|\int_0^\ell du \int_0^\ell \mu(q,\tau)K(p,q,t,\tau)\frac{\sqrt{\mathfrak{E}(u,v,\tau)\cdot\mathfrak{G}(u,v,\tau) - \mathfrak{F}(u,v,\tau)^2}}{\sqrt{\mathfrak{E}(u,v,t)\cdot\mathfrak{G}(u,v,t) - \mathfrak{F}(u,v,t)^2}}\right.$$

$$\left.\times \sqrt{\mathfrak{E}(u,v,t)\cdot\mathfrak{G}(u,v,t) - \mathfrak{F}(u,v,t)^2}dv\right| \tag{17.34}$$

Thus, by (17.6),

$$\left|\int_{\Sigma_\tau} \mu(q,\tau)K(p,q,t,\tau)d\sigma_q\right|$$

$$\leq B\int_0^\ell du \int_0^\ell |\mu(q,\tau)K(p,q,t,\tau)|\sqrt{\mathfrak{E}(u,v,t)\cdot\mathfrak{G}(u,v,t) - \mathfrak{F}(u,v,t)^2}dv$$

$$= B\int_{\Sigma_\tau} |\mu(q,\tau)K(p,q,t,\tau)|d\sigma_q \tag{17.35}$$

It follows from (17.21) and (17.25) that

$$|\mu_n| < BN \int_0^t d\tau \int_{\Sigma_\tau} |\mu_{n-1}| \frac{\exp[-\delta r_{pq}^2/(t-\tau) - 2N_0 h]}{(t-\tau)^{3/2}} d\sigma_q \qquad (17.36)$$

Now let T_p be the tangent plane to Σ_t at a point $p \in \Sigma_t$, and let γ be the acute angle between $\mathbf{r}_{p\bar{q}}$ and $\mathbf{n}_p$. By assumption, Σ_t is convex. Therefore there exist $R > 0$, $h > 0$, and $\eta > 0$, independent of p, such that the cylinder C of radius R and height h, whose axis coincides with the normal $\mathbf{n}_p$ to Σ_t at p cuts from Σ_t a connected part Σ_t^1 containing p such that

$$\cos(\gamma) > 1/2 \quad \forall q \in \Sigma_t^1, \ r_{pq} > \eta > 0 \quad \forall q \in \Sigma_t^2 = \Sigma_t \setminus \Sigma_t^1 \qquad (17.37)$$

We have

$$|\mu_n| < BN \exp(2N_0 h)(f_n^1 + f_n^2) \qquad (17.38)$$

where

$$f_n^i = \int_0^t d\tau \int_{\Sigma_t^i} |\mu_{n-1}(q,\tau)| \exp\left[-\frac{r_{pq}^2}{4(\delta-\tau)}\right] (t-\tau)^{-3/2} d\sigma_q; \quad i = 1, 2 \quad (17.39)$$

Assume that there exists Φ_{n-1} such that for any $\tau > 0$,

$$|\mu_{n-1}(q,\tau)| < \Phi_{n-1}(\tau) \qquad (17.40)$$

Let ρ be the length of the projection of $\mathbf{r}_{pq}$ onto the projection S_t^1 of Σ_t^1 onto T_p. Then

$$0 < f_n^1(p,t) \le 2 \int_0^t \Phi_{n-1}(\tau) d\tau \int_0^{2\pi} d\varphi \int_0^\rho \exp\left[-\frac{r_{pq}^2}{\delta(t-\tau)}\right] (t-\tau)^{-3/2} \rho \, d\rho \qquad (17.41)$$

This yields

$$0 < f_n^1(p,t) < 2\pi \delta^{-1} \int_0^t \Phi_{n-1}(\tau) d\tau \int_0^\infty \exp(-z) dz$$

$$= 2\pi \delta^{-1} \int_0^t \Phi_{n-1}(\tau) d\tau \qquad (17.42)$$

On the other hand,

$$0 < f_n^2(p,t) \le \int_0^t \Phi_{n-1}(\tau) d\tau \int_{\Sigma_t^2} \delta r_{pq}^2 (\delta\eta^2)^{-1} \exp\left[-\frac{r_{pq}^2}{4(\delta-\tau)}\right] (t-\tau)^{-3/2} d\sigma_q$$

$$< S_t (\delta\eta^2)^{-1} \int_0^t \Phi_{n-1}(\tau)(t-\tau)^{-1/2} d\tau \qquad (17.43)$$

where

$$S_t = \int_{\Sigma_t} d\sigma_q \qquad (17.44)$$

Let

$$\alpha = \max_t \left(S_t/\delta\eta^2 + 2\pi/\delta \right) \tag{17.45}$$

Then, inserting (17.40)–(17.42) and (17.45) into (17.38), we obtain

$$|\mu_n(p,t)| < L \int_0^t \Phi_{n-1}(\tau)(\tau)(t-\tau)^{-1/2}d\tau \tag{17.46}$$

where

$$L = \alpha BN \exp(2N_0 h) \tag{17.47}$$

There exists $c = \text{const} > 0$ such that

$$|\mu_0(p,t)| \equiv |F(p,t)| < c \quad \forall p \in \Sigma_t, \ \forall t \geq 0 \tag{17.48}$$

Hence

$$|\mu_1(p,t)| \leq 2cLt^{1/2} \tag{17.49}$$

Thus, we can set

$$\Phi_1(t) = 2cLt^{1/2} \tag{17.50}$$

Together with (17.46), this gives

$$|\mu_2(p,t)| < 2cL^2 \int_0^t \tau^{1/2}(t-\tau)^{-1/2}d\tau$$

$$= 2cL^2 B(3/2,1/2)t \equiv 2cL^2\Gamma(3/2)/\Gamma(1/2)\Gamma^{-1}(2)t \equiv \Phi_2(t) \tag{17.51}$$

where $B(p,q)$ and $\Gamma(p)$ are the Euler beta and gamma functions.

We claim that for every $n \geq 1$,

$$\Phi_n(t) = cL^n\Gamma^{n/2}(1/2)\Gamma(n/2+1)^{-1}t^{n/2} \tag{17.52}$$

This is certainly true for $n = 1$ and $n = 2$. Inserting (17.52) into (17.38) and assuming that (17.52) is true, we obtain, replacing $c\Gamma(3/2)$ by c,

$$|\mu_{n+1}| < cL^{n+1}\Gamma^{n/2}(1/2)\Gamma(n/2+1)^{-1}\frac{1}{\sqrt{\pi}}\int_0^t \tau^{n/2}(t-\tau)^{-1/2}d\tau$$

$$= cL^{n+1}\Gamma^{n/2}(1/2)\Gamma(n/2+1)^{-1}B(n/2+1,1/2)t^{\frac{n+1}{2}}$$

$$= cL^{n+1}\Gamma^{\frac{n+1}{2}}(1/2)\Gamma[(n+1)/2+1]^{-1}t^{\frac{n+1}{2}} = \Phi_{n+1}(t) \tag{17.53}$$

The truth of our claim now follows by induction.

Thus the series (17.23) is majorized by the series

$$c\sum_{n=0}^{\infty} \lambda^n \frac{L^n\Gamma^{n/2}(1/2)}{\Gamma[n/2+1]t^{n/2}} \tag{17.54}$$

The radius of the convergence of this series is infinity, as may be proved by the D'Alembert test. Hence the integral equations corresponding to all three boundary-value problems are indeed globally solvable (i.e., in any arbitrarily large interval on the time axis). Inserting the expression for $\mu(p,t)$ into the integral representations of the solutions of the three boundary-value problems and using the properties of heat potentials, we see that these problems are globally solvable, provided the preceding assumptions are valid.[3]

2. Asymptotic behavior of solution of first boundary-value problem and respective integral equations

The existence of global solutions of the integral equations was proved in Section 1 simultaneously for the first, second, and third boundary-value problems. However, there is a very essential difference between these equations.

Consider first the Neumann problem in a convex cylindrical region D, a parabolic region $D^t = D \times (0 < t < \infty,\ D \subset \mathbb{R}_n,\ n = 1, 2, 3)$:

$$Lu(p,t) \equiv \left(\Delta - \frac{\partial}{\partial t}\right) u(p,t) = 0 \quad \forall p \in D,\ \forall t > 0 \tag{17.55}$$

$$u(p,0) \equiv 0 \quad \forall p \in D \tag{17.56}$$

$$\frac{\partial}{\partial n_p} u(p,t) = f(p,t) \leq \delta < 0 \quad \forall p \in \Sigma,\ \forall t \geq 0 \tag{17.57}$$

Recall that $\mathbf{n}_p$ is the outward normal to Σ, so the meaning of (17.57) is that there is an influx into D of an unbounded amount of heat. Since the mass of the thermally conductive material occupying D remains unchanged, as does its volume heat capacity, this means that the temperature within D must increase infinitely.

It seems quite obvious that this unrestricted increase of temperature within D, as implied by (17.57), must be accompanied by an unrestricted increase of the density $\mu(p,t)$ of the heat potential of a single layer[4] representing the temperature within D:

$$f(p,t) \leq \delta < 0 \quad \forall p \in \Sigma,\ \forall t \geq 0$$
$$\Rightarrow \lim_{t\uparrow\infty} \mu(p,t) = \infty \tag{17.58}$$

---

[3](a) The proof that this process of successive approximations is convergent simultaneously proves that the solution of the integral equation (17.19) is both unique and stable with respect to variation of boundary data and physical parameters in the formulation of the problems under consideration. (b) The preceding existence theorems are not applicable if the lower base D_0 of the parabolic region D_0^t degenerates. Let then $D_{tn}^t \subset D_0^t$ be a monotonic sequence of subregions that converges to D_0^t and let $u_n(p,t)$ be the solution of the corresponding boundary-value problem posed in D_{tn}^t, with initial data convergent to the initial data of the original problem. Then it is possible to prove that the desired $u(p,t)$ is the limit of $u_n(p,t)$ as $n \uparrow \infty$.

[4]Here and in what follows we use the same abbreviated terminology as in Chapter 16 (see Chapter 16, note 1).

However, the validity of this last equality cannot be taken for granted; it must be proved rigorously. We shall do so, assuming for the sake of simplicity that

$$\max \left(\frac{1}{2\pi} \int_\Sigma \cos(\psi_{pq}) r_{pq}^{-2} d\sigma_q \right) \overset{\text{def}}{=} -m < -1 \tag{17.59}$$

PROOF. By assumption, D is convex. Hence the angle ψ_{pq} defined by (17.18) is obtuse, so that

$$\cos(\psi_{pq}) < 0 \quad \forall q \neq p \tag{17.60}$$

But in the case of the inner Neumann problem, the parameter λ in equation (17.19) is equal to -1, implying that all terms of the series (17.23) are positive.

Let $\mu^n(p,t)$ be the nth partial sum of the series (17.19),

$$\mu^n(p,t) = \sum_{k=0}^n \mu_k(p,t) \tag{17.61}$$

so that

$$\mu^n(p,t) = -\int_0^t d\tau \int_\Sigma \mu^{n-1}(q,\tau) \cos(\psi_{pq}) r_{pq} \frac{\exp\left[-r_{pq}^2/4(t-\tau)\right]}{16\pi^{3/2}(t-\tau)^{5/2}} d\sigma_q$$

$$= -\frac{1}{\pi^{3/2}} \int_\Sigma \cos(\psi_{pq}) r_{pq}^{-2} d\sigma_q$$

$$\times \int_0^t \mu^{n-1}(q,\tau) r_{pq}^3 \frac{\exp\left[-r_{pq}^2/4(t-\tau)\right]}{16(t-\tau)^{5/2}} d\tau \tag{17.62}$$

Since

$$F(p,t) = -2f(p,t) > 2\delta > 0 \tag{17.63}$$

we have

$$\mu^1(p,t) > -\frac{\delta}{\pi^{3/2}} \int_\Sigma \cos(\psi_{pq}) r_{pq}^{-2} d\sigma_q \int_0^t r_{pq}^3 \frac{\exp\left[-r_{pq}^2/4(t-\tau)\right]}{8(t-\tau)^{5/2}} d\tau$$

$$= -\frac{2\delta}{\pi^{3/2}} \int_\Sigma \cos(\psi_{pq}) r_{pq}^{-2} d\sigma_q \int_{r_{pq}^2/4t}^\infty \lambda^{3/2} e^{-\lambda} d\lambda \tag{17.64}$$

The integral in the right-hand side of (17.64) is uniformly convergent, so that

$$\lim_{t \uparrow \infty} \mu^1(p,q) \geq -\delta \max_{r \in \Sigma} \left(\frac{1}{2\pi} \int_\Sigma \cos(\psi_{pq}) r_{pq}^{-2} d\sigma_q \right) \overset{\text{def}}{=} \delta m \tag{17.65}$$

Assume that for all $k \leq n - 1$,

$$\lim_{t \uparrow \infty} \mu^k(p, t) = \lim_{t \uparrow \infty} \frac{1}{8\pi^{3/2}} \int_0^t d\tau \int_\Sigma \mu^{k-1}(q, \tau) \frac{\exp[-r_{pq}^2/4(t - \tau)]}{(t - \tau)^{3/2}} d\sigma_q > \delta m^k$$

(17.66)

then

$$\lim_{t \uparrow \infty} \mu^n(p, t) = \lim_{t \uparrow \infty} \frac{1}{8\pi^{3/2}} \int_0^t d\tau$$

$$\times \int_\Sigma \mu^{n-1}(q, \tau) \frac{\exp[-r_{pq}^2/4(t - \tau)]}{(t - \tau)^{3/2}} \cos(\psi_{pq}) r_{pq}^{-2} d\sigma_q$$

$$> \lim_{t \uparrow \infty} -\frac{2\delta}{\pi^{3/2}} \int_\Sigma \cos(\psi_{pq}) r_{pq}^{-2} d\sigma_q$$

$$\times \int_0^t \mu^{n-1}(q, \tau) r_{pq} \frac{\exp\left[-r_{pq}^2/4(t - \tau)\right]}{8(t - \tau)^{5/2}} d\tau$$

$$> m^{n-1} \lim_{t \uparrow \infty} -\frac{2\delta}{\pi^{3/2}} \int_\Sigma \cos(\psi_{pq}) r_{pq}^{-2} d\sigma_q \int_{r_{pq}^2/4t}^\infty \lambda^{1/2} e^{-\lambda} d\lambda > \delta m^n \quad (17.67)$$

Since this inequality is true for $n = 1$, it follows by induction that it is true for all n. Thus, by (17.59) and (17.60),

$$\lim_{t \uparrow \infty} \mu(p, t) = \lim_{n \uparrow \infty} \lim_{t \uparrow \infty} \mu(p, t) \geq \delta \lim_{n \uparrow \infty} m^n = \infty \quad (17.68)$$

Q.E.D.

It clearly follows from (17.68) and (17.13) that

$$\lim_{t \uparrow \infty} u(p, t) = \infty \quad (17.69)$$

Hence we have established that $u(p, t)$ increases unboundedly, irrespectively of the physical sense of the Neumann problem under consideration.

The situation with regard to the Dirichlet problem is quite different, as the following theorems show.

Theorem 17.2.1. *Let $D_\tau \equiv D$ be a cylindrical region and let*

$$\exists \lim_{t \uparrow \infty} f(p, t) = F(p) \quad \forall p \in \Sigma = \partial D \quad (17.70)$$

Then the solution of the first boundary-value problem

$$Lu(p, t) \equiv \left(\Delta - \frac{\partial}{\partial t}\right) u(p, t) = 0 \quad \forall p \in D, \ \forall t > 0 \quad (17.71)$$

$$u(p, 0) \equiv 0 \quad \forall p \in D, \quad u(p, t) = f(p, t) \quad \forall p \in \Sigma, \ \forall t \geq 0 \quad (17.72)$$

tends asymptotically to the solution of the electrostatic Dirichlet problem

$$\Delta \tilde{u}(p) = 0 \quad \forall p \in D, \qquad \tilde{u}(p) = F(p) \quad \forall p \in \Sigma \quad (17.73)$$

so that

$$\lim_{t \uparrow \infty} u(p, t) = \tilde{u}(p) \quad \forall p \in D \quad (17.74)$$

Theorem 17.2.2. *Under the assumptions of Theorem 17.2.1, the solution $\mu(p,t)$ of the integral equation (17.23),*

$$\mu(p,t) = 2f(p,t) - \int_\Sigma \frac{\cos(\varphi_{pq})}{2\pi \cdot r_{pq}^2} d\sigma_q \int_0^t \mu(q,\tau)r_{pq}^3 \frac{\exp\left[-r_{pq}^2/4(t-\tau)\right]}{4\pi^{1/2}(t-\tau)^{5/2}} d\tau$$

(17.75)

corresponding to the first boundary-value problem in heat-conduction theory, tends to the density $\tilde{\mu}(p)$ of the electrostatic potential $W[p \mid \mu_\infty(q)]$ representing the solution of the electrostatic Dirichlet problem

$$\Delta\tilde{u}(p) = 0 \quad \forall p \in D, \quad \tilde{u}(p) = F(p) \quad \forall p \in \Sigma \tag{17.76}$$

$$\exists \lim_{t\uparrow\infty} \mu(p,t) = \tilde{\mu}(p) \quad \forall p \in \Sigma \tag{17.77}$$

PROOF OF THEOREM 17.2.1. The assertion of Theorem 17.2.1 is an almost obvious corollary of the maximum principle. Indeed, let

$$\Phi(p,t) = 2[f(p,t) - F(p)]; \qquad v(p,t) = u(p,t) - \tilde{u}(p) \tag{17.78}$$

Take an arbitrarily small $\varepsilon > 0$. There exists $T = \text{const} > 0$ so large that

$$|\Phi(p,t)| < \varepsilon \quad \forall p \in \Sigma, \ \forall t \geq T \tag{17.79}$$

The function $v(p,t)$ is a solution of the boundary-value problem

$$\Delta v - \partial v/\partial t = 0 \quad \forall p \in D, \ \forall t \geq T \tag{17.80}$$

$$v(p,T) = u(p,T) - \tilde{u}(p) \quad \forall p \in D \tag{17.81}$$

$$v(p,t) = \Phi(p,t) \quad \forall p \in \Sigma, \ \forall t \geq T \tag{17.82}$$

Let

$$M = \max|u(p,t) - \tilde{u}(p)| \tag{17.83}$$

Define an auxiliary function

$$\widehat{W}(p,t) = \varepsilon + \Pi(p,t \mid M) - v(p,t) \tag{17.84}$$

where $\Pi(p,t \mid M)$ is the Poisson integral

$$\Pi(p,t \mid M) = M \int_D \frac{\exp\left[-r_{pq}^2/4(t-T)\right]}{8\pi^{3/2}(t-T)^{3/2}} d\omega_q \tag{17.85}$$

$\widehat{W}(p,t)$ is parabolic within D for every $t > T$, continuous in $\bar{D}$, positive inside D at $t = T$ and on Σ for every $t \geq T$. Hence, by the maximum principle,

$$\widehat{W}(p,t) > 0 \quad \forall p \in \bar{D}, \ \forall t \geq T \tag{17.86}$$

But

$$\lim_{t\uparrow\infty} \Pi(p,t \mid M) \equiv 0 \tag{17.87}$$

Hence

$$\overline{\lim_{t\uparrow\infty}} \, v(p,t) \leq \varepsilon \tag{17.88}$$

Similarly

$$\breve{W}(p,t) = -\varepsilon - \Pi(p,t \mid M) - v(p,t) < 0 \qquad (17.89)$$

so that

$$\lim_{t\uparrow\infty} v(p,t) \geq -\varepsilon \qquad (17.90)$$

Thus

$$-\varepsilon \leq \underline{\lim_{t\uparrow\infty}} v(p,t) \leq \overline{\lim_{t\uparrow\infty}} v(p,t) \leq \varepsilon \qquad (17.91)$$

Since $\varepsilon > 0$ is arbitrarily small, this means that

$$\lim_{t\uparrow\infty} v(p,t) = 0 \quad \forall p \in D \qquad (17.92)$$

Q.E.D.

PROOF OF THEOREM 17.2.2. Let

$$\nu(p,t) = \mu(p,t) - \tilde{\mu}(p), \qquad \Phi(p,t) = 2[f(p,t) - F(p)] \qquad (17.93)$$

We have

$$\mu(p,t) = 2f(p,t) - \int_0^t d\tau \int_\Sigma \mu(q,\tau) r_{pq} \cos(\varphi_{pq}) \frac{\exp\left[-r_{pq}^2/4(t-\tau)\right]}{8\pi^{3/2}(t-\tau)^{5/2}} d\sigma_q$$

$$\equiv 2f(p,t) - \int_\Sigma \frac{\cos(\varphi_{pq})}{2\pi r_{pq}^2} d\sigma_q$$

$$\times \int_0^t \mu(q,\tau) r_{pq}^3 \frac{\exp\left[-r_{pq}^2/4(t-\tau)\right]}{4\pi^{1/2}(t-\tau)^{5/2}} d\tau \qquad (17.94)$$

$$\tilde{\mu}(p) = 2F(p) - \int_\Sigma \frac{\tilde{\mu}(q)\cos(\varphi_{pq})}{2\pi r_{pq}^2} d\sigma_q \qquad (17.95)$$

Definitions (17.93) yield

$$\nu(p,t) = \Phi(p,t) - \int_\Sigma \frac{\cos(\varphi_{pq})}{2\pi r_{pq}^2} d\sigma_q \int_0^t \mu(q,\tau) r_{pq}^3 \frac{\exp\left[-r_{pq}^2/4(t-\tau)\right]}{4\pi^{1/2}(t-\tau)^{5/2}} d\tau$$

$$+ \int_\Sigma \frac{\tilde{\mu}(q)\cos(\varphi_{pq})}{2\pi r_{pq}^2} d\sigma_q \qquad (17.96)$$

which may be rewritten as

$$\nu(p,t) = \Phi(p,t) - \int_\Sigma \frac{\cos(\varphi_{pq})}{2\pi r_{pq}^2} d\sigma_q \int_0^t \nu(q,\tau) r_{pq}^3 \frac{\exp\left[-r_{pq}^2/4(t-\tau)\right]}{4\pi^{1/2}(t-\tau)^{5/2}} d\tau$$

$$+ \int_\Sigma \tilde{\mu}(q) \frac{\cos(\varphi_{pq})}{2\pi r_{pq}^2} d\sigma_q$$

$$\times \left(1 - \int_0^t r_{pq}^3 \frac{\exp\left[-r_{pq}^2/4(t-\tau)\right]}{4\pi^{1/2}(t-\tau)^{5/2}} d\tau\right) \qquad (17.97)$$

Note that

$$\lim_{t\uparrow\infty}\Phi(p,t)=0$$

(17.98)

$$\lim_{t\uparrow\infty}\int_0^T r_{pq}^3\frac{\exp\left[-r_{pq}^2/4(t-\tau)\right]}{4\pi^{1/2}(t-\tau)^{5/2}}d\tau=0 \quad \forall T=\text{const}>0$$

and

$$1-\int_0^t r_{pq}^3\frac{\exp\left[-r_{pq}^2/4(t-\tau)\right]}{4\pi^{1/2}(t-\tau)^{5/2}}d\tau=1-2\pi^{1/2}\int_{r_{pq}/4t}^\infty z^{1/2}e^{-z}dz$$

$$\equiv 2\pi^{1/2}\int_0^{r_{pq}^2/4t} z^{1/2}e^{-z}dz\downarrow 0$$

$$\text{when } t\uparrow\infty \qquad (17.99)$$

Let

$$\Phi_1(p,t)=\Phi(p,t)-\int_\Sigma\frac{\cos(\varphi_{pq})}{2\pi r_{pq}^2}d\sigma_q\int_0^T\nu(q,\tau)r_{pq}^3\frac{\exp\left[-r_{pq}^2/4(t-\tau)\right]}{4\pi^{1/2}(t-\tau)^{5/2}}d\tau$$

(17.100)

Then by (17.98), uniformly in $p\in\Sigma$,

$$\lim_{t\uparrow\infty}\Phi_1(p,t)=0$$

(17.101)

Hence, by the mean value theorem of integral calculus,

$$\overline{\lim_{t\uparrow\infty}}\,\nu(p,t)=\overline{\lim_{t\uparrow\infty}}\left\{-\int_\Sigma\nu(q,\tau^*)\frac{\cos(\varphi_{pq})}{2\pi r_{pq}^2}d\sigma_q\int_T^t r_{pq}^3\frac{\exp\left[-r_{pq}^2/4(t-\tau)\right]}{4\pi^{1/2}(t-\tau)^{5/2}}d\tau\right\}$$

(17.102)

where $\tau^*\in(T,t)$ and depends on q.

The relation (17.102) is equivalent to

$$\lim_{t\uparrow\infty}\nu(p,t)=\int_\Sigma\lim_{t\uparrow\infty}[-\nu(q,\tau^*)]\frac{\cos(\varphi_{pq})}{2\pi\cdot r_{pq}^2}d\sigma_q\lim_{t\uparrow\infty}\int_T^t r_{pq}^3\frac{\exp\left[-r_{pq}^2/4(t-\tau)\right]}{4\pi^{1/2}(t-\tau)^{5/2}}d\tau$$

$$=\int_\Sigma\lim_{t\uparrow\infty}[-\nu(q,\tau^*)]\frac{\cos(\varphi_{pq})}{2\pi\cdot r_{pq}^2}d\sigma_q$$

$$\times\overline{\lim_{t\uparrow\infty}}\frac{2}{\pi^{1/2}}\int_{r_{pq}^2/4(t-T)}^\infty \lambda^{1/2}e^{-\lambda}d\lambda$$

(17.103)

The inner integral in the right-hand side of (17.103) converges uniformly in t to 1 for every $p\neq q$. Thus

$$\overline{\lim_{t\uparrow\infty}}\,\nu(p,t)=\int_\Sigma\lim_{t\uparrow\infty}[-\nu(q,\tau^*)]\frac{\cos(\varphi_{pq})}{2\pi r_{pq}^2}d\sigma_q$$

$$=-\int_\Sigma[\nu(q,\tilde\tau^*)]\frac{\cos(\varphi_{pq})}{2\pi r_{pq}^2}d\sigma_q$$

(17.104)

where $\tilde{\tau}^* \in [T, \infty)$. Now let $T \uparrow \infty$. Then

$$\overline{\lim_{t \uparrow \infty}} \, \nu(p,t) = \int_{\Sigma} \overline{\lim_{t \uparrow \infty}} [-\nu(q,t)] \frac{\cos(\varphi_{pq})}{2\pi \cdot r_{pq}^2} d\sigma_q \qquad (17.105)$$

Quite similarly, we find that

$$\lim_{t \uparrow \infty} \nu(p,t) = \int_{\Sigma} \overline{\lim_{t \uparrow \infty}} [-\nu(q,t)] \frac{\cos(\varphi_{pq})}{2\pi \cdot r_{pq}^2} d\sigma_q \qquad (17.106)$$

By (17.105) and (17.106),

$$\overline{\lim_{t \uparrow \infty}} \, \nu(p,t) + \lim_{t \uparrow \infty} \nu(p,t)$$

$$= \int_{\Sigma} \left(\overline{\lim_{t \uparrow \infty}} \, \nu(q,t) + \lim_{t \uparrow \infty} \nu(q,t) \right) \frac{\cos(\varphi_{pq})}{2\pi r_{pq}^2} d\sigma_q \qquad (17.107)$$

$$\overline{\lim_{t \uparrow \infty}} \, \nu(p,t) - \lim_{t \uparrow \infty} \nu(p,t)$$

$$= \int_{\Sigma} \left(\overline{\lim_{t \uparrow \infty}} \, \nu(q,t) - \lim_{t \to \infty} \nu(q,t) \right) \frac{\cos(\varphi_{pq})}{2\pi r_{pq}^2} d\sigma_q \qquad (17.108)$$

Arguments identical to those of Chapter 12 (see (12.232), (12.233), (12.235), and (12.236)), show that

$$\overline{\lim_{t \uparrow \infty}} \, \nu(p,t) + \lim_{t \to \infty} \nu(p,t) \equiv 0 \qquad (17.109)$$

$$\overline{\lim_{t \uparrow \infty}} \, \nu(p,t) - \lim_{t \to \infty} \nu(p,t) = \text{const} \overset{\text{def}}{=} 2c_0 \qquad (17.110)$$

Hence

$$\overline{\lim_{t \uparrow \infty}} \, \nu(p,t) = - \lim_{t \to \infty} \nu(p,t) = c_0 \qquad (17.111)$$

Thus we must prove that

$$c_0 = 0 \qquad (17.112)$$

But this is a corollary of Theorem 17.2.1. Indeed, we have

$$u(p,t) = W(p,t \mid \mu), \qquad \tilde{u}(p) = W(p \mid \tilde{\mu}) \qquad (17.113)$$

where W and $\hat{W}$ are heat and electrostatic potentials of double layers. Hence by Theorem 17.2.1,

$$\lim_{t \uparrow \infty} \overline{W} = \lim_{t \uparrow \infty} \underline{W} = \lim_{t \uparrow \infty} W, \qquad \lim_{t \uparrow \infty} W(p,t \mid \mu) - \lim_{t \uparrow \infty} \hat{W}(p \mid \tilde{\mu}) \equiv 0 \quad (17.114)$$

But it follows from (17.111) that

$$\lim_{t \uparrow \infty} W(p,t \mid \mu) - \hat{W}(p \mid \tilde{\mu}) \equiv c_0 \qquad (17.115)$$

so that, by (17.114),

$$c_0 = 0 \qquad (17.116)$$

Q.E.D.

3. Solution of quasilinear Cauchy problem

Let us now consider the application of heat potentials to the solution of nonlinear boundary-value problems. We begin with the quasilinear Cauchy problem

$$\frac{\partial^2 u}{\partial x^2} + \varphi(u)\frac{\partial u}{\partial x} = \frac{\partial u}{\partial t}, \quad -\infty < x < \infty, \ t > 0 \tag{17.117}$$

$$u(x,0) = f(x), \quad -\infty < x < \infty \tag{17.118}$$

mentioned (in Chapter 3, Section 4) in connection with the decay of an arbitrary discontinuity. Recall that this problem is closely connected with the Buckley–Leverett theory of oil production.

Let us assume that $\varphi(u)$ is a continuously differentiable function throughout the real axis, and that there exists $K = \text{const} > 0$ such that

$$|\varphi(u)|, |\partial\varphi/\partial u| \le K \quad \forall u \in (-\infty, \infty) \tag{17.119}$$

We further assume that $f(x)$ is a piecewise continuous, piecewise smooth function bounded together with its first derivative. For the sake of simplicity we shall assume that f and df/dx have only one point of discontinuity, which is located at the origin. Thus, let

$$|f(x)|, \left|\frac{df}{dx}\right| < M, \quad f(+0) - f(-0) = \alpha > 0 \tag{17.120}$$

Obviously, we may assume without loss of generality that

$$8\alpha K^2 < 1 \tag{17.121}$$

(this inequality will be used later). Indeed, the change of variables

$$\xi = x/\ell, \quad \tau = t/\ell^2 \tag{17.122}$$

transforms equation (17.117) into

$$\frac{\partial^2 u}{\partial \xi^2} + \ell\varphi(u)\frac{\partial u}{\partial \xi} = \frac{\partial u}{\partial t} \tag{17.123}$$

so that, by choosing $\ell > 0$ sufficiently small, we can guarantee the validity of inequality (17.120).

Let $u(x,t)$ be a solution of the problem. Apply the fundamental identity (15.20), assuming that $g = E(x - \xi, t - \tau)$. It follows that $u(x,t)$ must be a solution of the integrodifferential equation

$$u(x,t) = \int_0^t d\tau \int_{-\infty}^{\infty} \varphi[u(\xi,\tau)]v(\xi,\tau)E(x - \xi, t - \tau)d\xi + \int_{-\infty}^{\infty} f(\xi)E(x - \xi, t)d\xi \tag{17.124}$$

where

$$v(x,t) = \frac{\partial}{\partial x}u(x,t) \tag{17.125}$$

Differentiation of (17.124) with respect to x yields

$$v(x,t) = \int_0^t d\tau \int_{-\infty}^{\infty} \varphi[u(\xi,\tau)]v(\xi,\tau)\frac{\partial}{\partial x}E(x-\xi,t-\tau)d\xi$$

$$+ \int_{-\infty}^{\infty} f(\xi)\frac{\partial}{\partial x}E(x-\xi,t)d\xi \qquad (17.126)$$

By (17.120),

$$\int_{-\infty}^{\infty} f\frac{\partial}{\partial x}E d\xi = -\left(\int_{\infty}^0 + \int_0^{\infty}\right)f\frac{\partial}{\partial \xi}E d\xi$$

$$= \alpha E(x,t) + \int_{-\infty}^{\infty} \frac{d}{d\xi}f(\xi)E(x-\xi,t)d\xi \qquad (17.127)$$

so that

$$v(x,t) = \alpha E(x,t) + \int_0^t d\tau \int_{-\infty}^{\infty} \varphi[u(\xi,\tau)]v(\xi,\tau)\frac{\partial}{\partial x}E(x-\xi,t-\tau)d\xi$$

$$+ \int_{-\infty}^{\infty} \frac{d}{d\xi}f(\xi)E(x-\xi,t)d\xi \qquad (17.128)$$

Let us seek $v(x,t)$ as

$$v(x,t) = \alpha E(x,t) + w(x,t) \qquad (17.129)$$

This reduces system (17.124) and (17.128) to the system

$$u(x,t) = \alpha \int_0^t d\tau \int_{-\infty}^{\infty} \varphi[u(\xi,\tau)]E(\xi,\tau)E(x-\xi,t-\tau)d\xi$$

$$+ \int_{-\infty}^{\infty} f(\xi)E(x-\xi,t)d\xi$$

$$+ \int_0^t d\tau \int_{-\infty}^{\infty} \varphi[u(\xi,\tau)]w(\xi,\tau)E(x-\xi,t-\tau)d\xi$$

$$\overset{\text{def}}{=} U(x,t\mid u,w) = \sum_{k=1}^{3} J_k(x,t\mid u,w) \qquad (17.130)$$

$$w(x,t) = \int_{-\infty}^{\infty} \frac{d}{d\xi}f(\xi)E(x-\xi,t)d\xi$$

$$+ \alpha \int_0^t d\tau \int_{-\infty}^{\infty} \varphi[u(\xi,\tau)]E(\xi,\tau)\frac{\partial}{\partial x}E(x-\xi,t-\tau)d\xi$$

$$+ \int_0^t d\tau \int_{-\infty}^{\infty} \varphi[u(\xi,\tau)]w(\xi,\tau)\frac{\partial}{\partial x}E(x-\xi,t-\tau)d\xi$$

$$\overset{\text{def}}{=} W(x,t\mid u,w) = \sum_{k=4}^{6} J_k(x,t\mid u,w) \qquad (17.131)$$

Thus, the integrodifferential equation (17.128) is reduced to the system of two integral equations (17.130) and (17.131). A local solution of this system may be constructed by Picard's method of successive approximations.[5]

Set

$$u_n(x,t) = U(x,t \mid u_{n-1}, w_{n-1}), \quad w_n(x,t) = W(x,t \mid u_{n-1}, w_{n-1}), \quad n = 1, 2, \ldots \tag{17.132}$$

and let $u_0(x,t)$ and $w_0(x,t)$ be arbitrarily chosen functions, defined for all $x \in (-\infty, \infty)$ and $t \geq 0$, continuous with respect to x and t for every $x \in (-\infty, 0] \vee [0, \infty)$ and $t \geq 0$, and bounded by some constant N_0:

$$|u_0|, |w_0| < N_0 \quad \forall x \in (-\infty, \infty), \ \forall t \geq 0 \tag{17.133}$$

Lemma 17.3.1. *There exist $T_0 = \text{const} > 0$ so small and $N_1 > 0$ so large that the sequences $\{u_n(x,t)\}\big|_{n=1}^{\infty}$, $\{w_n(x,t)\}\big|_{n=1}^{\infty}$ are uniformly bounded in $-\infty < x < \infty$, $0 \leq t \leq T_0$, by N_1:*

$$|u_n(x,t)|, |w_n(x,t)| < N_1 \quad \forall n \geq 0, \ \forall x \in (-\infty, \infty), \ \forall t \in [0, T_0] \tag{17.134}$$

Lemma 17.3.2. *There exists $T_1 = \text{const} > 0$ such that, for every $\eta > 0$, sequences $\{u_n\}\big|_{n_1}^{\infty}$, $\{w_n\}\big|_{n_0}^{\infty}$ are equicontinuous in $((-\infty < x \leq 0) \wedge (0 \leq x < \infty)) \times (0 \leq t \leq T_1)$ and in $(-\infty < x < \infty) \times (\eta \leq t \leq T_1)$; moreover, the series*

$$\sum_{n=1}^{\infty} |u_n(x,t) - u_{n-1}(x,t)|, \quad \sum_{n=1}^{\infty} |w_n(x,t) - w_{n-1}(x,t)| \tag{17.135}$$

are majorized there by a convergent geometrical progression.

PROOF OF LEMMA 17.3.1. We first estimate the integrals in (17.130) and (17.131). By definition,

$$E(x - \xi, t - \tau) = \frac{\exp\left[-(x - \xi)^2 / 4(t - \tau)\right]}{2\pi^{1/2}(t - \tau)^{1/2}} \tag{17.136}$$

By the mean value theorem and (17.119),

$$|J_1| = \left| \int_0^t d\tau \int_{-\infty}^{\infty} \varphi[u(\xi, \tau)] E(\xi, \tau) E(x - \xi, \tau - \tau) d\xi \right|$$

$$\leq \alpha K \left| \int_0^t d\tau \int_{-\infty}^{\infty} E(\xi, \tau) E(x - \xi, \tau - \tau) d\xi \right| \tag{17.137}$$

---

[5]Naturally, we could appeal to the Cacciopoli–Banach contracting mapping principle instead of Picard's method of successive approximation. However, in some cases the contracting mapping principle is not directly applicable to proving the existence and uniqueness of a global solution of the problem, whereas the direct application of Picard iterations does imply the existence and uniqueness of a global solution. One example was presented in Section 1 of this chapter. For another example, see Problem 17.1.1.

The integral

$$I^1 = \int_{-\infty}^{\infty} E(\xi, \tau) E(x - \xi, t - \tau) d\xi \qquad (17.138)$$

is a solution of the Cauchy problem

$$\frac{\partial^2}{\partial x^2} I^1 = \frac{\partial}{\partial t} I^1, \quad -\infty < x < \infty, \ t > \tau, \qquad I^1(x, \tau) = E(x, \tau) \qquad (17.139)$$

Hence, by the uniqueness theorem,

$$I^1 \equiv E(x, t) \qquad (17.140)$$

Inserting (17.140) into (17.137) and taking (17.119) into account, we obtain

$$|J_1| < \alpha K \int_0^t E(x, t) d\tau = \alpha K \exp\left[-\frac{x^2}{4t}\right] \sqrt{t/4\pi} \qquad (17.141)$$

It now follows from (17.121) that

$$|J_2| < M \int_{-\infty}^{\infty} E(x - \xi, t) d\xi = M \qquad (17.142)$$

Assume that

$$|w| < N \qquad (17.143)$$

Then we find, again using (17.119), that

$$|J_3| = \left| \int_0^t d\tau \int_{-\infty}^{\infty} \varphi(u) w E d\xi \right| < KNt \qquad (17.144)$$

The estimates (17.141), (17.142), and (17.144) show, by virtue of (17.130), that

$$|U(x, t \mid u, w)| < M + \alpha K \exp\left(-\frac{x^2}{4t}\right) \sqrt{t/4\pi} + KNt \qquad (17.145)$$

Further, it follows from (17.120) that

$$|J_4| < \left| \int_{-\infty}^{\infty} \frac{df}{d\xi} E d\xi \right| < M \qquad (17.146)$$

and

$$|J_5| = \left| \alpha \int_0^t d\tau \int_{-\infty}^{\infty} \varphi[u(\xi, \tau)] E(\xi, \tau) \frac{\partial}{\partial x} E(x - \xi, t - \tau) d\xi \right| \leq \alpha K I^2 \qquad (17.147)$$

where

$$I^2 = \int_0^t d\tau \int_{-\infty}^{\infty} E(\xi, \tau) \left| \frac{\partial}{\partial x} E(x - \xi, t - \tau) \right| d\xi$$

$$\leq \int_0^t (4\pi\tau)^{-1/2} d\tau \int_{-\infty}^{\infty} \left| \frac{\partial}{\partial x} E(x - \xi, t - \tau) \right| d\xi = I_2^* \qquad (17.148)$$

Note that

$$\text{sign}\,\frac{\partial}{\partial\xi}E(x-\xi,t-\tau) = -\,\text{sign}\,\frac{\partial}{\partial x}E(x-\xi,t-\tau) = \text{sign}(\xi-x) \qquad (17.149)$$

Hence

$$I_2^* = \int_0^t (4\pi\tau)^{-1/2}d\tau \left(\int_{-\infty}^x \frac{\partial}{\partial\xi}E(x-\xi,t-\tau)d\xi - \int_x^\infty \frac{\partial}{\partial\xi}E(x-\xi,t-\tau)d\xi \right)$$

$$= \int_0^t (\pi\tau)^{-1/2}E(0,t-\tau)d\tau = \frac{1}{2\pi}\int_0^t \frac{d\tau}{[\tau(t-\tau)]^{1/2}} = \frac{1}{2} \qquad (17.150)$$

Thus it follows from (17.147) that

$$|J_5| < \alpha K/2 \qquad (17.151)$$

Finally, by virtue of (17.119) and (17.143),

$$|J_6| = \left| \int_0^t d\tau \int_{-\infty}^\infty W(\xi,\tau)\varphi[u(\xi,\tau)]\frac{\partial}{\partial x}E(x-\xi,t-\tau)d\xi \right|$$

$$\le KN \int_0^t d\tau \int_0^\infty z\frac{\exp\left[-z^2/4(t-\tau)\right]}{2\pi^{1/2}(t-\tau)^{3/2}}dz$$

$$= KN \int_0^t (t-\tau)^{-1/2}d\tau \int_0^\infty e^{-\zeta}d\zeta = 2KN\sqrt{t/\pi} \qquad (17.152)$$

Combining all of these estimates, we conclude that

$$|W(x,t\mid u,w)| < M + \alpha K + 2KN\sqrt{t/\pi} \qquad (17.153)$$

Until now, the constant N has remained undefined. Let us stipulate that

$$N > 2(M+\alpha K) \qquad (17.154)$$

and take

$$T = \min\left(\frac{\pi}{16K^2}, \frac{1}{4K}, \frac{\pi}{4} \right) \qquad (17.155)$$

Then it follows from inequalities (17.121), (17.143), (17.145), and (17.153) that

$$|U(x,t\mid u,w)|, |W(x,t\mid u,w)| < N \quad \forall x \in (-\infty,\infty), \ \forall t \in [0,T] \qquad (17.156)$$

Hence, taking N_0 in (17.133) equal to N and T_0 in (17.134) equal to T, with N and T defined (respectively) by (17.154) and (17.155), we find

$$|u_n(x,t)|, |w_n(x,t)| < N \quad \forall x \in (-\infty,\infty), \ \forall t \in [0,T], \ \forall n \ge 0 \qquad (17.157)$$
$$\textbf{Q.E.D.}$$

PROOF OF LEMMA 17.3.2. Let (u,w) and $(\tilde{u},\tilde{w})$ be two pairs of functions satisfying the conditions of Lemma 17.3.1. Denote

$$\delta U(x,t) \overset{\text{def}}{=} U(x,t\mid u,w) - U(x,t\mid \tilde{u},\tilde{w}) \qquad (17.158)$$

$$\delta W(x,t) \overset{\text{def}}{=} W(x,t\mid u,w) - W(x,t\mid \tilde{u},\tilde{w}) \qquad (17.159)$$

We have

$$\delta U = I_1 + I_2; \qquad \delta W = I_3 + I_4 \tag{17.160}$$

where

$$I_1 = \int_0^t d\tau \int_{-\infty}^\infty \varphi[u(\xi,\tau)] - \varphi[\tilde{u}(\xi,\tau)]E(\xi,\tau)$$
$$\times E(x-\xi,t-\tau)d\xi \tag{17.161}$$

$$I_2 = \int_0^t d\tau \int_{-\infty}^\infty \{\varphi[u(\xi,\tau)]w(\xi,\tau) - \varphi[\tilde{u}(\xi,\tau)]\tilde{w}(\xi,\tau)\}$$
$$\times E(x-\xi,t-\tau)d\xi \tag{17.162}$$

$$I_3 = \alpha \int_0^t d\tau \int_{-\infty}^\infty \{\varphi[u(\xi,\tau)] - \varphi[\tilde{u}(\xi,\tau)]\}$$
$$\times E(\xi,\tau)\frac{\partial}{\partial x}E(x-\xi,t-\tau)d\xi \tag{17.163}$$

$$I_4 = \int_0^t d\tau \int_{-\infty}^\infty \{\varphi[u(\xi,\tau)]w(\xi,\tau) - \varphi[\tilde{u}(\xi,\tau)]\tilde{w}(\xi,\tau)\}$$
$$\times \frac{\partial}{\partial x}E(x-\xi,t-\tau)d\xi \tag{17.164}$$

Comparing definitions (17.131) of J_K and (17.161)–(17.164) of I_K and using inequalities (17.141), (17.144), (17.151), and (17.152), we obtain

$$|I_1| < \sup |\varphi(u) - \varphi(\tilde{u})|2\alpha\sqrt{t/\pi} \tag{17.165}$$
$$|I_2| < \sup |w\varphi(u) - \tilde{w}\varphi(\tilde{u})|t \tag{17.166}$$
$$|I_3| < \sup |\varphi(u) - \varphi(\tilde{u})|\alpha K \tag{17.167}$$
$$|I_4| < \sup |w\varphi(u) - \tilde{w}\varphi(\tilde{u})|2\sqrt{t/\pi} \tag{17.168}$$

According to (17.119),

$$|\varphi(u) - \varphi(\tilde{u})| < K|u - \tilde{u}| \tag{17.169}$$

so that, by (17.157),

$$|w\varphi(u) - \tilde{w}\varphi(\tilde{u})| \le |\varphi(u)||w - \tilde{w}| + |\tilde{w}||\varphi(u) - \varphi(\tilde{u})|$$
$$< KN|u - \tilde{u}| + K|w - \tilde{w}| \tag{17.170}$$

Hence

$$|\delta U| < K\left([2\alpha\sqrt{t/\pi} + Nt]\sup|u - \tilde{u}| + t\sup|w - \tilde{w}|\right) \tag{17.171}$$

$$|\delta W| < K\left([\alpha K + 2N\sqrt{t/\pi}]\sup|u - \tilde{u}| + 2N\sqrt{t/\pi}\sup|w - \tilde{w}|\right) \tag{17.172}$$

In view of (17.121), there exist $\rho = \text{const} \in (0,1)$ and $T_1 \in (0,T]$ such that

$$|\delta U|, |\delta W| < (\rho/2)[\sup|u - \tilde{u}| + \sup|w - \tilde{w}|] \tag{17.173}$$
$$\sup|u - \tilde{u}| < 2N, \qquad \sup|w - \tilde{w}| < 2N \tag{17.174}$$

We now apply these estimates to the sequences of successive approximations. By Lemma 17.3.1,

$$\sup|u_1 - u_0| < 2N, \qquad \sup|w_1 - w_0| < 2N \tag{17.175}$$

Hence

$$\sup|u_2 - u_1|, \ \sup|w_2 - w_1| < 2N\rho \tag{17.176}$$

and by induction,

$$\sup|u_n - u_{n-1}|, \ \sup|w_n - w_{n-1}| < 2N\rho^{n-1} \quad \forall n \geq 1 \tag{17.177}$$

Hence the series

$$\sum_{n=1}^{\infty} |u_n - u_{n-1}|, \ \sum_{n=1}^{\infty} |w_n - w_{n-1}| \tag{17.178}$$

are absolutely and uniformly convergent for all x and $t \in [0, T_1]$, since they are majorized by the convergent geometrical progression

$$2N\sum_{n=0}^{\infty} \rho^n \tag{17.179}$$

Finally, we note that the representation of all $u_n(x, t)$ and $w_n(x, t)$ by heat potentials means that the sequences $\{u_n(x, t)\}$, $\{w_n(x, t)\}$ are equicontinuous in the regions indicated in the statement of this lemma. Thus the proof is complete. **Q.E.D.**

Lemmas 17.3.1 and 17.3.2 imply the following theorem.

Theorem 17.3.1. *Under the assumptions of Lemma 17.3.1, there exists $T = \text{const} > 0$ such that a solution to the quasilinear Cauchy problem (17.117), (17.118) exists, is unique in the interval $[0, T]$, and can be constructed there by Picard's method of successive approximations.*

In Section 1 we saw that Picard's method sometimes makes it possible to prove the existence of not only local but also global solutions. In most nonlinear problems, however, the method is applicable only locally, and in order to prove the existence of global solutions one must use the method of *prolongation*, which we now proceed to describe.

Assume that successive approximations yield the solution $u(x, t)$ of some time-dependent boundary-value problem over a time interval $0 \leq t \leq T_1$, where $T_1 > 0$ is a monotonically decreasing function of the norm of the input data. Assume, to fix ideas, that this norm M_0 is defined as the maximum maximorum of all the derivatives of the input data up to some order. Assume, moreover, that the solution of the problem has all these derivatives. Take an arbitrarily small $\varepsilon > 0$ and consider the time $t = T_1 - \varepsilon = T_1^0$ as a new starting time.[6] Denote the norm of these new input data by M_1. Then the length $\delta T_2 = T_2 - T_1^0$ of the extension of the interval in which the solution is determined is a monotonically decreasing function of M_1. This process of prolonging the existence interval of the solution may obviously go on for a denumerable many times. Let δT_n be the length of the nth extension. There exist two possibilities:

(i) the series $\displaystyle\sum_{n=1}^{\infty} \delta T_n$ diverges;   (ii) the series $\displaystyle\sum_{n=1}^{\infty} \delta T_n$ converges

$$(17.180)$$

In the first case, the interval in which a solution can be constructed is infinitely long; this means that there exist a global solution. In the second case, however, the interval $(0, T)$ in which a solution of the problem can be determined is finite, so that the solution $u(x, T)$ at time $t = T$ can be determined only asymptotically as $t \uparrow T$. What happened at time T can be described as a *heat explosion* if $\lim_{t \uparrow T} |u(x, t)| = \infty$ or a *collapse* if $\overline{\lim}_{t \uparrow T} |u(x, t)| < \infty$.

A decision as to whether a heat explosion or collapse will occur, based on the use of the method of prolongation, requires an analysis of the convergence of the series (17.180). As an analysis of this kind may be very complicated or even unfeasible, the method of a priori estimates, which is based on methods of functional analysis, has become a very useful tool of the theory of partial differential equations; but that is beyond the scope of this book [45].

Returning to the quasilinear problem studied in this section, we note that the existence of its global solution is obvious. Indeed, let a local solution be constructed in the time interval $[0, T]$. Take $T_0 \in (0, T)$ as a new starting time, and consider the problem

$$\frac{\partial^2 \tilde{u}}{\partial x^2} + \varphi(\tilde{u})\frac{\partial \tilde{u}}{\partial x} = \frac{\partial \tilde{u}}{\partial t}, \quad -\infty < x < \infty, \ t > T_0 \qquad (17.181)$$

$$\tilde{u}(x, T_0) = u(x, T_0) \qquad (17.182)$$

Let

$$\tilde{v}(x, t) \overset{\text{def}}{=} \frac{\partial}{\partial x}\tilde{u}(x, t) \qquad (17.183)$$

Similarly to (17.128), we conclude that $(\tilde{u}, \tilde{v})$ must be a solution of the system of integral equations

$$\tilde{u}(x, t) = \int_{-\infty}^{\infty} u(\xi, T_0) E(x - \xi, t - T_0) d\xi$$

$$+ \int_{T_0}^{t} d\tau \int_{-\infty}^{\infty} \varphi[u(\xi, \tau)]\tilde{v}(\xi, \tau) E(x - \xi, t - \tau) d\xi \qquad (17.184)$$

---

[6]We take $T_1^0 = T_1 - \varepsilon$ rather than $T_1^0 = T_1$, because generally one cannot guarantee that the solution constructed will possess all the necessary derivatives at time T_1, whereas this can be done for T_1^0.

$$v(x,t) = \int_{-\infty}^{\infty} v(\xi, T_0)E(x - \xi, t - T_0)d\xi$$

$$+ \alpha \int_{T_0}^{t} d\tau \int_{-\infty}^{\infty} \varphi[\tilde{u}(\xi, \tau)]\tilde{v}(\xi, \tau)\frac{\partial}{\partial x}E(x - \xi, t - \tau)d\sigma \qquad (17.185)$$

By Theorem 17.1.1, this system has a global solution (see Problem 17.1.1). Let

$$U(x,t) = \begin{cases} u(x,t) & \forall x \in (-\infty, \infty), \ \forall t \in [0, T_0] \\ \tilde{u}(x,t) & \forall x \in (-\infty, \infty), \ \forall t > T_0 \end{cases} \qquad (17.186)$$

Since, by the uniqueness theorem,

$$u(x,t) \equiv \tilde{u}(x,t) \quad \forall x \in (-\infty, \infty), \ \forall t \in [T_0, T] \qquad (17.187)$$

$U(x,t)$ is a global solution of the original problem (17.117), (17.118).

4. One-dimensional one-phase Stefan problem with ablation

As a second example of applying the theory of heat potentials to the solution of nonlinear problems, let us consider the following version of the 1-dimensional Stefan problem (see Chapter 2, Section 11).

Let the half-space $-\infty < x < y(t)$ be filled with a solid body melting under the influence of irradiation of its boundary $x = y(t)$. It is assumed that the thermal flux $h(t)$ through the boundary is prescribed as a function of time and that ablation takes place, so that the product of the melting is removed instantly. The initial temperature of the body is prescribed and the melting point is assumed to be constant. The problem is to determine the temperature $T(x,t)$ of the body and its boundary $x = y(t)$, subject to the following conditions:

$$a^2 \frac{\partial^2}{\partial x^2}T = \frac{\partial}{\partial t}T, \quad -\infty < x < y(t), \ t > 0 \qquad (17.188)$$

$$T[y(t), t] = T_m, \quad t > 0 \qquad (17.189)$$

$$T(x,0) = T_0(x), \quad -\infty < x < 0 \qquad (17.190)$$

$$-\gamma\rho\frac{d}{dt}y = h(t) - k\frac{\partial}{\partial x}T[y(t), t], \quad t > 0 \qquad (17.191)$$

$$y(0) = 0 \qquad (17.192)$$

The specific latent heat γ and density ρ are assumed constant.

We first switch to dimensionless variables by scaling:

$$x_1 = \frac{x}{L}, \quad t_1 = \frac{a^2}{L^2}, \quad \alpha = \frac{kT_m}{\gamma\rho a^2} \qquad (17.193)$$

$$u(x_1, t_1) = \frac{T(x,t) - T_m}{T_m}, \quad y_1(t_1) = \frac{y(t)}{L}, \quad \varphi(t_1) = \frac{h(t)L}{\gamma\rho a^2} \qquad (17.194)$$

Omitting the subscript 1 in x_1, t_1, and $y_1(t_1)$, we obtain

$$\frac{\partial^2}{\partial x^2}u = \frac{\partial}{\partial t}u, \quad -\infty < x < y(t), \ t > 0 \tag{17.195}$$

$$u(x,0) = f(x), \quad -\infty < x < 0, \quad u[y(t),t] = 0, \quad t > 0 \tag{17.196}$$

$$\frac{d}{dt}y = -\varphi(t) + \alpha v(t), \quad t > 0 \tag{17.197}$$

where

$$v(t) \stackrel{\text{def}}{=} \frac{\partial}{\partial x}u[y(t),t] \tag{17.198}$$

In what follows we assume that $f(x)$ is twice continuously differentiable, and that

$$f(0) = 0, \ f(x) < 0 \quad \forall x < 0 \tag{17.199}$$

$$|f|, |df/dx|, |d^2 f/dx^2| < M \quad \forall x < 0, |\varphi| < M \quad \forall t \geq 0 \tag{17.200}$$

Let $\big(u(x,t),y(t)\big)$ be a solution of the problem (17.195)–(17.198). Then, by the fundamental identity of Chapter 15,

$$u(x,t) = \int_\infty^0 f(\xi)E(x-\xi,t)d\xi + \int_0^t v(\tau)E[x-y(\tau),t-\tau]d\tau \tag{17.201}$$

Here the second condition on the right side of (17.196) has been used, and therefore the heat potential of the double layer occurring in the fundamental identity is dropped.

Differentiation of (17.201) with respect to x yields

$$\frac{\partial}{\partial x}u(x,t) = -\int_{-\infty}^0 f(\xi)\frac{\partial}{\partial \xi}E(x-\xi,t)d\xi + \int_0^t v(\tau)\frac{\partial}{\partial x}E[x,y(\tau),t-\tau]d\tau \tag{17.202}$$

or, after integration by parts with allowance for (17.199),

$$\frac{\partial}{\partial x}u(x,t) = \int_{-\infty}^0 \frac{d}{d\xi}f(\xi)E(x-\xi,t)d\xi + \int_0^t v(\tau)\frac{\partial}{\partial x}E[x,y(\tau),t-\tau]d\tau \tag{17.203}$$

Letting $x \to y(t) - 0$, appealing to the theorem on discontinuity of the heat potential of a double layer, and recalling the notation (17.198), we obtain

$$v(t) = 2\int_{-\infty}^0 \frac{d}{d\xi}f(\xi)E\big(y(t)-\xi,t\big)d\xi + 2\int_0^t v(\tau)\frac{\partial}{\partial x}E[y(t),y(\tau),t-\tau]d\tau \tag{17.204}$$

To this we add condition (17.197), rewritten in integral form:

$$y(t) = \alpha\int_0^t v(\tau)d\tau - \psi(t) \tag{17.205}$$

where

$$\psi(t) = \int_0^t \varphi(\tau)d\tau \tag{17.206}$$

Thus the problem is reduced to the solution of a system of two nonlinear integral equations of Volterra type. For its solution we again use successive approximations, with

$$v_n(t) = 2 \int_{-\infty}^{0} \frac{d}{d\xi} f(\xi) E\big(y_{n-1}(t) - \xi, t\big) d\xi$$

$$+ 2 \int_{0}^{t} v_{n-1}(\tau) \frac{\partial}{\partial x} E[y_{n-1}(t), y_{n-1}(\tau), t - \tau] d\tau \tag{17.207}$$

$$y_n(t) = \alpha \int_{0}^{t} v_n(\tau) d\tau - \psi(t) \quad \forall n \geq 1 \tag{17.208}$$

where $v_0(t)$ is an arbitrary bounded function,

$$|v_0(t)| < N = \text{const} > 0 \quad \forall t > 0 \tag{17.209}$$

satisfying the initial condition

$$v_0(0) = \frac{d}{dx} f(0) \tag{17.210}$$

Lemma 17.4.1. *There exist $T > 0$ so small and $N > 0$ so large that*

$$|v_n(t)| < N \quad \forall t \in [0, T], \ \forall n > 0 \tag{17.211}$$

Lemma 17.4.2. *There exist $T^* > 0$ so small that for all $n \geq 1$ and $t \in [0, T^*]$,*

$$|v_{n+1}(t) - v_n(t)| \leq \rho |v_n(t) - v_{n-1}(t)| \tag{17.212}$$

where

$$0 < \rho = \text{const} < 1 \tag{17.213}$$

PROOF OF LEMMA 17.4.1. Indeed,

$$E(x - \xi, t - \tau) = \frac{\exp\big[-(x - \xi)^2/4(t - \tau)\big]}{2\pi^{1/2}(t - \tau)^{1/2}}$$

$$\frac{\partial}{\partial x} E(x - \zeta, t - r) = -(x - \zeta)\frac{\exp\big[-(x - \xi)^2/4(t - \tau)\big]}{4\pi^{1/2}(t - \tau)^{3/2}} \tag{17.214}$$

Hence, by the left side of (17.200) and the inequality

$$|v_m(t)| < N \quad \forall t \in [0, T] \tag{17.215}$$

we have

$$\int_{-\infty}^{y_m(t)} \left| \frac{d}{d\xi} f(\xi) E\big(y_m(t) - \xi, t\big) \right| d\xi < \frac{M}{\sqrt{\pi}} \int_{-\infty}^{0} e^{-z^2} dz = \frac{M}{2} \tag{17.216}$$

$$\int_{0}^{t} \left| v_m(\tau)[y_m(t) - y_m(\tau)] \frac{\exp\left(-\frac{[y_m(t) - y_m(\tau)]^2}{4(t - \tau)}\right)}{4\pi^{1/2}(t - \tau)^{3/2}} \right| d\tau$$

$$< N \frac{M + \alpha N}{4\sqrt{\pi}} \int_{0}^{t} (t - \tau)^{-1/2} d\tau = N \frac{M + \alpha N}{2\sqrt{\pi}} \sqrt{t} \tag{17.217}$$

Assuming now that

$$N > \frac{2M\sqrt{\pi}}{\alpha}, \qquad T < \frac{M\sqrt{\pi}}{2\alpha N^2} \qquad (17.218)$$

we find that

$$|v_m(t)| < N \Rightarrow |v_{m+1}(t)| < N \qquad (17.219)$$

Since inequality (17.219) is valid for $m = 0$, we see by induction that (17.211)–(17.213) is true. **Q.E.D.**

PROOF OF LEMMA 17.4.2. By Lemma 17.4.1,

$$|v_1(t) - v_0(t)| < 2N, \quad 0 \le t \le T \qquad (17.220)$$

Further, it follows from (17.207) that

$$v_{n+1}(t) - v_n(t) = 2\int_{-\infty}^0 \frac{d}{d\xi} f d\xi \int_{y_n(t)}^{y_{n-1}(t)} \frac{\partial}{\partial \xi} E(z - \xi, t) dz$$

$$+ 2\int_0^t \left(v_n(\tau)\frac{\partial}{\partial x} E[y_n(t) - y_n(\tau), t - \tau] \right.$$

$$\left. - v_{n-1}(\tau)\frac{\partial}{\partial x} E[y_{n-1}(t) - y_{n-1}(\tau), t - \tau] \right) d\tau$$

$$= I_1 + I_2 \qquad (17.221)$$

By changing the order of integration and integrating by parts, we obtain

$$|I_1| = 2\left| \int_{y_n(t)}^{y_{n-1}(t)} dz \int_{-\infty}^0 \frac{d}{d\xi} f(\xi)\frac{\partial}{\partial \xi} E(z - \xi, t) d\xi \right|$$

$$= 2\left| \int_{y_n(t)}^{y_{n-1}(t)} \frac{d}{d\xi} f(0) E(z, t) dz \right.$$

$$\left. - \int_{y_n(t)}^{y_{n-1}(t)} dz \int_{-\infty}^0 \frac{d^2}{d\xi^2} f(\xi) E(z - \xi, t) d\xi \right| \qquad (17.222)$$

Hence, by the left side of (17.200)

$$|I_1| < M(\pi t)^{-1/2}|y_n(t) - y_{n-1}(t)| + 2M|y_n(t) - y_{n-1}(t)| \qquad (17.223)$$

Thus, by (17.205),

$$|I_1| < \alpha M\pi^{-1/2}|\sqrt{t} + 2\sqrt{\pi}t \,| \,\delta v_n \qquad (17.224)$$

where

$$\delta v_n(t) = \max_{0 \le t \le T} |v_n(t) - v_{n-1}(t)| \qquad (17.225)$$

Consider I_2. We have

$$I_2 = I_2^* + I_2^{**} \qquad (17.226)$$

where

$$I_2^* = 2 \int_0^t [v_n(\tau) - v_{n-1}(\tau)] \frac{\partial}{\partial x} E(y_n(t) - y_n(\tau), t - \tau) d\tau \qquad (17.227)$$

$$I_2^{**} = 2 \int_0^t [v_{n-1}(\tau)] \left(\frac{\partial}{\partial x} E(y_n(t) - y_n(\tau), t - \tau) \right.$$
$$\left. - \frac{\partial}{\partial x} E(y_{n-1}(t) - y_{n-1}(\tau), t - \tau) \right) d\tau \qquad (17.228)$$

Using (17.205), (17.225), and the definition of $E(x,t)$, we obtain

$$|I_2^*| < \delta v_n(t) \int_0^t \frac{|y_n(t) - y_n(\tau)|}{2(t-\tau)} \cdot \frac{\exp\left[-\frac{(y_n(t) - y_n(\tau))^2}{4(t-\tau)} \right]}{2\pi^{1/2}(t-\tau)^{1/2}} d\tau$$
$$< (M + \alpha N)\sqrt{t/4\pi} \cdot \delta v_n \qquad (17.229)$$

Further, we have

$$I_2^{**} = 2 \int_0^t v_{n-1}(\tau) d\tau \int_{y_{n-1}(t) - y_{n-1}(\tau)}^{y_n(t) - y_n(\tau)} \frac{\partial^2}{\partial z^2} E(z, t - \tau) dz \qquad (17.230)$$

But

$$\frac{\partial^2}{\partial z^2} E(z, t - \tau) = -\frac{\exp\left[-z^2/4(t-\tau) \right]}{4\pi^{1/2}(t-\tau)^{3/2}} \left(1 - \frac{z^2}{2(t-\tau)} \right) \qquad (17.231)$$

and

$$\frac{z^2}{2(t-\tau)} < \frac{(M + \alpha N)^2}{2}(t - \tau) \quad \forall z \in [y_{n-1}(t) - y_{n-1}(\tau), y_n(t) - y_n(\tau)] \qquad (17.232)$$

so that

$$\left| \frac{\partial^2}{\partial z^2} E(z, t - \tau) \right| < \left(1 + \frac{(M + \alpha N)^2}{2} T \right) [4\sqrt{\pi}(t-\tau)^{3/2}]^{-1} \qquad (17.233)$$

This gives

$$|I_2^{**}| < 2NA \int_0^t |y_n(t) - y_n(\tau) - y_{n-1}(t) + y_{n-1}(\tau)|(t-\tau)^{-3/2} d\tau$$
$$= 4NA(N)t^{1/2}\delta v_n \qquad (17.234)$$

where

$$A(N) = (4\sqrt{\pi})^{-1} \left(1 + \frac{(M + \alpha N)^2}{2} T \right) \qquad (17.235)$$

Combining all the estimates, we get

$$\delta v_{n+1} < B(N,T)t^{1/2}\delta v_n \qquad (17.236)$$

where $B(N,T)$ is an increasing function of N and T, equal to $[4\pi^{1/2}]^{-1}$ at $T = 0$. Take now $T_1 = \text{const} > 0$ so small that

$$B(N,T)\sqrt{T_1} = \rho < 1 \qquad (17.237)$$

Then

$$\delta v_{n+1} \leq \rho \delta v_n \quad \forall t \in [0, T^*], \quad T^* = \min(T, T_1) \tag{17.238}$$

This yields

$$\delta v_{n+1} \prod_{m=1}^{n} \delta v_m < \delta v_1 \prod_{m=1}^{n} \rho \delta v_m \tag{17.239}$$

so that, by (17.220)

$$\delta v_{n+1} < 2N\rho^n, \quad \delta y_{n+1} < 2NT^*\rho^n \quad \forall n \geq 0 \tag{17.240}$$

and the assertion of Lemma 17.4.2 is true.   **Q.E.D.**

The convergence of Picard's iterations guarantees the existence and uniqueness of the solution of the preceding system of integral equations. Hence we have the following theorem.

Theorem 17.4.1. *Under the assumptions of Lemma 17.4.1, system (17.204) and (17.205) has a unique local solution that can be constructed by successive approximations.*

A priori, however, one cannot assert that the solution of this system of integral equations yields a solution of the original boundary-value problem. Indeed, all the conditions of the original problem are explicitly satisfied except the boundary condition

$$u\big(y(t), t\big) = 0 \tag{17.241}$$

although it was taken into account in the derivation of the integral equations, because in the fundamental identity the term that was dependent on the heat potential of the double layer was dropped. Clearly, a rigorous proof of the existence theorem requires direct proof that this guarantees the validity of condition (17.241). Taking this into account, we prove the following *equivalence* theorem.

Theorem 17.4.2. *Let $v(t)$ and $y(t)$ be the solution of the system of integral equations (17.204) and (17.205). Let $u(x,t)$ be defined by (17.201). Then the function $u(x,t)$ is twice continuously differentiable with respect to x and once continuously differentiable with respect to t in $(-\infty, y(t))$, for all $t \in (0, T]$. Moreover, it satisfies all conditions of the problem; in particular,*

$$\frac{\partial}{\partial x} u[y(t), t] = v(t), \qquad u[y(t), t] = 0 \tag{17.242}$$

PROOF. Note, first of all, that $v(t)$ is continuous, since it is represented by the heat potential of a double layer with bounded density along the differentiable curve $x = y(t)$. Hence $u(x,t)$ is twice differentiable with respect to x, once differentiable with respect to t inside the region

$$D = (-\infty, y(t)) \times (0, T] \tag{17.243}$$

and continuous in the closure $\bar{D}$ of D. This means that the fundamental identity of the theory of heat potentials (see Chapter 15) is applicable to $u(x,t)$. Hence

$$u(x,t) = \int_0^t v(\tau)E\big(x,y(\tau),t-\tau\big)d\tau$$

$$- \int_0^t u^*(\tau)\left(\frac{\partial}{\partial x} - \frac{d}{d\tau}y(\tau)\right)E\big(x-y(\tau),t-\tau\big)d\tau$$

$$+ \int_{-\infty}^0 f(\xi)E(x-\xi,t)d\xi \qquad (17.244)$$

where

$$u^*(t) = u[y(t),t] \qquad (17.245)$$

On the other hand, by (17.201) we have

$$u(x,t) = \int_0^t v(\tau)E\big(x-y(\tau),t-\tau\big)d\tau + \int_{-\infty}^0 f(\xi)E(x-\xi,t)d\xi \qquad (17.246)$$

Comparing the two integral representations of $u(x,t)$, we find that

$$\int_0^t u^*(\tau)\left(\frac{\partial}{\partial x} - \frac{d}{d\tau}y(\tau)\right)E\big(x-y(\tau),t-\tau\big)d\tau \equiv 0 \qquad (17.247)$$

Letting $x \to y(t)-0$ we find, again appealing to the theorem on discontinuity of the heat potential of a double layer, that

$$u^*(t) - 2\int_0^t u^*(\tau)\left(\frac{\partial}{\partial x} - \frac{d}{d\tau}y(\tau)\right)E\big(y(t) - y(\tau),t-\tau\big)d\tau = 0 \qquad (17.248)$$

Thus $u^*(t)$ is the solution of a homogeneous Volterra integral equation, so that by the uniqueness theorem

$$u^*(t) \equiv 0 \qquad (17.249)$$
$$\textbf{Q.E.D.}$$

Remark. Very often, when a physical problem is formulated as a boundary-value problem for partial differential equations, some essential features of the problem may appear to be omitted. This is just what happened in our formulation of the Stefan ablation problem. Indeed, from the point of view of the physical contents of the problem, one must first, assume, that the influx of heat into the melting body is positive (i.e., that $q(t) > 0$ for all $t > 0$) if the ablation process continues without a break. Second, one must assume that the initial temperature of the body everywhere is below the melting point, since otherwise the solid will be superheated locally or new phase interfaces will appear. Only the second of these requirements was taken into account in our formulation of the problem. Moreover, while constructing the solution we did not prove that the front $x = y(t)$ is monotonic, which is very natural because no such proof is possible without requiring $q(t)$ to be positive. In view of these remarks, we leave aside the question of the existence of a global solution.[7]

5. Determination of temperature of half-space $z > 0$ radiating heat according to Stefan–Boltzmann law

Finally, as an example of a 3-dimensional problem, let us consider radiation of heat into vacuum according to the Stefan–Boltzmann law:
Find a function $u(p,t)$ satisfying the conditions

$$\Delta u = \frac{\partial}{\partial t} u, \quad p \in D, \ t > 0, \tag{17.250}$$

$$u(p,0) = f(p), \quad p \in D, \quad \frac{\partial}{\partial z} u(p,t) = F[u(p,t)], \quad p \in \Sigma \tag{17.251}$$

where

$$D = \{x,y,z\colon x,y \in (-\infty,\infty), \ z \in (0,\infty)\} \tag{17.252}$$

$$\Sigma = \partial D = \{x,y,z\colon x,y \in (-\infty,\infty), \ z \equiv 0\} \tag{17.253}$$

$u(p,t)$ is the temperature in degrees Kelvin,

$$F(u) = \alpha u^4, \qquad \alpha = \sigma/k \tag{17.254}$$

k is the coefficient of thermal conductivity of the body, and σ is the Stefan constant. It is assumed that, by scaling, the coefficient of thermal diffusivity has been made equal to 1. Here and in what follows we use the notations

$$P = (x,y,z) \in D, \qquad Q = (\xi,\eta,\zeta) \in D \tag{17.255}$$

$$p = (x,y,0) \in \Sigma, \qquad q = (\xi,\eta,0) \in \Sigma \tag{17.256}$$

$$f(Q) \equiv f(q,\zeta), \quad u(Q,\tau) = u(q,\zeta,\tau), \quad u(q,0,\tau) = u(q,\tau) \tag{17.257}$$

Let

$$G(P,Q,t-\tau) = E(p,q,t-\tau)g(z,\zeta,t-\tau) \tag{17.258}$$

be the Green's function of the second boundary-value problem in D, so that

$$g(z,\zeta,t-\tau) = E(z-\zeta,t-\tau) + E(z+\zeta,t-\tau) \tag{17.259}$$

$$E(z,t-\tau) = \frac{\exp\left[-z^2/4(t-\tau)\right]}{2\pi^{1/2}(t-\tau)^{1/2}},$$

$$E(p,q,t-\tau) = \frac{\exp\left[-r_{pq}^2/4(t-\tau)\right]}{4\pi(t-\tau)} \tag{17.260}$$

If $u(P,t)$ is a solution of the problem, it has the integral representation

$$u(P,t) = \int_D f(Q)G(P,Q,t)d\omega_Q - \int_0^t d\tau \int_\Sigma F[u(q,\tau)]G(P,q,t-\tau)d\sigma_q \tag{17.261}$$

---

[7]The task of proving the existence of a global solution under the assumption that the front is indeed monotonic is included in the list of problems at the end of this chapter.

Letting $P \to p \in \Sigma$ we obtain, owing to the continuity of the heat potential of a single layer, an integral equation for determining $u(P, t)$:

$$u(p, t) = \int_D f(Q)G(p, Q, t)d\omega_Q - \int_0^t d\tau \int_\Sigma F[u(q, \tau)]G(p, q, t - \tau)d\sigma_q$$

$$(17.262)$$

which may be rewritten as

$$u(p, t) = \int_\Sigma E(p, q, t)d\sigma_q \int_0^\infty f(q, \zeta)g(z, \zeta, t) \, d\zeta$$

$$- \int_0^t [\pi(t - \tau)^{-1/2}]d\tau \int_\Sigma F[u(Q, \tau)]E(p, q, t - \tau)d\sigma_q$$

$$\overset{\text{def}}{=} J(p, t) - \Phi[p, t \mid u(q, \tau)] \overset{\text{def}}{=} \Psi(p, t \mid u) \qquad (17.263)$$

For a proof of the existence and uniqueness of a local solution of this equation, we appeal to the Cacciopoli–Banach *contracting mapping principle*: Let A be the contraction mapping of a compact subset of a complete normed space into itself; then A has one and only one fixed point.[8]

Thus, let M be the space of continuous functions defined in $\Sigma \times [0, T]$. Let $M(\Sigma, T, N, \omega)$ be the subset of M consisting of the uniformly bounded functions with common modulus of continuity ω, so that

$$u(p, t) \in M(\Sigma, T, N, \omega) \Rightarrow |u(p, t)| \le N \quad \forall (p, t) \in \Sigma \times [0, T] \qquad (17.264)$$

$$r_{pq} + |t - \tau| < \omega \Rightarrow |u(p, t) - u(q, \tau)| < \varepsilon \qquad (17.265)$$

Thus $M(\Sigma, T, N, \omega)$ is a set of equicontinuous and uniformly bounded functions, so that by the Arzela–Ascoli theorem $M(\Sigma, T, N, \omega)$ is a compact subset of M.

Now let A be the set of all function defined on $\Sigma \times [0, T]$ and bounded by a constant N. Define the set

$$\mathfrak{M} = \{v = \Psi(p, t \mid u); \ u \in A\} \qquad (17.266)$$

and assume that there exist functions $K(N)$ and $L(N)$ such that

$$|F(u)| < K(N), \ |F(u) - |F(v)| < L(N) \quad \text{if } |u|, |v| < N \qquad (17.267)$$

It is easy to see that there exist $T > 0$, $N > 0$, and $\omega(\varepsilon)$ such that

$$\mathfrak{M} \subset M(\Sigma, T, N, \omega) \qquad (17.268)$$

---

[8] We recall the formulation of the Cacciopoli–Banach principle without reference to the concept of compactness: Let $\mathfrak{M}$ be a set of bounded functions $f \colon |f| \le M_f = $ const (M_f may depend of f), such that any uniformly convergent sequence $\{f_n\}$ of functions belonging to $\mathfrak{M}$ has a limit belonging to $\mathfrak{M}$. Let A be an operator such that $f \in \mathfrak{M} \Rightarrow A(f) \in \mathfrak{M}$ and there exists $m = $ const, $0 < m < 1$, such that $f \in \mathfrak{M}$ and $\varphi \in \mathfrak{M} \Rightarrow |A(f) - A(\varphi)| < m \cdot \sup |f - \varphi|$ (supremum in $\mathfrak{M}$). Then the equation $\varphi = A(\varphi)$ has one and only one solution in $\mathfrak{M}$ (see [58]).

Indeed,

$$\int_\Sigma E(p,q,t)d\sigma_q = 1, \qquad 2\int_0^\infty E(z,\zeta,t)d\zeta = 1 \qquad (17.269)$$

since the first integral in the left-hand side is a solution of the 2-dimensional Cauchy problem and the second a solution of the 1-dimensional Cauchy problem, with initial temperature equal to 1 in both cases, and since the solutions to these problems are unique. Hence, by the mean value theorem, it follows from (17.264) that

$$|J(p,t)| \leq 2\int_\Sigma E(p,q,t)d\sigma_q \int_0^\infty |f(q,\zeta)|E(z,\zeta,t)d\omega_q \leq M \qquad (17.270)$$

and

$$|\Phi[p,t \mid u(q,\tau)]| \leq \int_0^t [\pi(t-\tau)]^{-1/2}d\tau \int_\Sigma |F[u(Q,\tau)]|E(p,q,t-\tau)d\sigma_q$$

$$\leq 2K(N)\pi^{-1/2}t^{1/2} \qquad (17.271)$$

Hence

$$|\Psi(p,t \mid u)| = |J - \Phi| \leq M + 2K(N)\pi^{-1/2}t^{1/2} \qquad (17.272)$$

if

$$0 \leq t \leq T \quad \text{and} \quad u \in \mathfrak{M} \qquad (17.273)$$

It is also clear that the functions $v \in \mathfrak{M}$ have a common modulus of continuity. Indeed, $J(p,t)$ is continuous in M and independent of $u \in A$. Let $\omega_1(\varepsilon)$ be its modulus of continuity. Now $\Phi(p,t \mid u(q,\tau))$ is the heat potential of a single layer with density F, bounded by a constant $K(N)$ independent of u for every $u \in A$. Hence $\Phi(p,t \mid u(q,\tau))$ is continuous in A with modulus of continuity ω_2 dependent only on the properties of the fundamental solution $E(p,q,t-\tau)$, the constant N, and the length T of the time interval under consideration:

$$\omega_2 = \omega_2(\varepsilon, N, T) \qquad (17.274)$$

Hence the modulus ω of continuity of $\Psi(p,t \mid u(q,t))$

$$\omega = \omega_1(\varepsilon) + \omega_2(\varepsilon, N, T) \qquad (17.275)$$

is independent of $u \in A$.

Now take

$$N > 2M, \ T_1 < \frac{\pi}{4K^2(N)} \qquad (17.276)$$

Then, by (17.272),

$$|\Psi(p,t \mid u)| < N \quad \forall u \in \mathfrak{M} \qquad (17.277)$$

which shows, by (17.275) and the definition (17.268) of $\mathfrak{M}$, that

$$\mathfrak{M} \subset M(\Sigma, T, N, \omega) \qquad (17.278)$$

Thus

$$v = \Psi(p, t \mid u) \qquad (17.279)$$

maps $\mathfrak{M}$ into itself.

We claim that this is also a contraction mapping if $T > 0$ is small enough. Indeed, let $u, v \in \mathfrak{M}$. Then (17.263) and (17.271) imply

$$|\Psi(p, t \mid u) - \Psi(p, t \mid v)| \le \rho|u - v| \quad \forall p \in \Sigma, \; \forall t \in [0, T] \qquad (17.280)$$

where

$$\rho = 2L(N)(t/\pi)^{1/2} \qquad (17.281)$$

Hence, taking

$$T_2 < \frac{\pi}{4L^2(N)} \qquad (17.282)$$

we obtain

$$0 < \rho < 1 \qquad (17.283)$$

so that (17.279) is indeed a contraction mapping if $t \in [0, T_2]$. Take, therefore,

$$T = \min(T_1, T_2) \qquad (17.284)$$

Then the mapping

$$v(p, t) = \Psi(p, t \mid u) \qquad (17.285)$$

is a contraction mapping of the subset $\mathfrak{M}$ of M into itself.   **Q.E.D.**[9]

Whenever the contraction mapping principle is applicable to the solution of a problem, this means that a local solution exists and is unique.

Problems

P17.1.1.   Let D be a bounded convex region in $\mathbb{R}_3$ with smooth boundary ∂D. Let $K(p, t, q, \tau)$ be a kernel such that

$$|K(p, t, q, \tau)| < L[t - \tau]^{-\lambda} \quad \forall p, q \in D, \; \forall t \ge 0, \; \forall \tau \in [0, t]$$

where $L = \mathrm{const} > 0$ and $0 \le \lambda < 1$. Consider the integral equation

$$u(p, t) = f(p, t) + \int_0^t d\tau \int_D \Phi[q, \tau, u(q, \tau)] K(p, t, q, \tau) d\omega_q$$

where f and Φ are bounded functions in $\bar{D} \times (0 \le t < \infty)$, Lipschitz-continuous with respect to u:

$$|\Phi(p, t, u) - \Phi(p, t, v)| < A|u - v|, \quad A = \mathrm{const} > 0,$$

---

[9] For the solution of the Stefan–Boltzmann irradiation problem in astrophysical context see [82].

for all uniformly bounded u, v defined in $\bar{D}$. Prove the existence, uniqueness, and stability of a global solution to this integral equation.

P17.1.2. Let $V(t \mid f)$ be a Volterra functional operator defined on the class of functions $f(t)$ continuous in $[0, \infty)$; that is, the value of $V(t \mid f)$ is defined by the values of $f(\tau)$ for all $\tau \in [0, t]$, and $V(0 \mid f)$ is independent of f. Assume that the following conditions hold:

(1) If the sequence $f_1(t), f_2(t), \dots, f_n(t), \dots$ is uniformly convergent in $[0, T]$ to a function $f(t)$, then

$$\lim_{n \uparrow \infty} V(t \mid f_n) = V(t \mid f) \quad \forall t \in [0, T]$$

(2) Let $\mathfrak{F}$ be the class of functions $\tilde{\varepsilon}(\delta, M, T)$ such that $\tilde{\varepsilon}(\delta, M, T)$ is an increasing function of δ continuous at $\delta = 0$, and let

$$\tilde{\varepsilon}(0, M, T) = 0$$

Then there exists $\tilde{\varepsilon}(\delta, M, T) \in \mathfrak{F}$ such that, for any f continuous in $[0, T]$ and bounded by M,

$$|f(t)| < M \in [0, T]$$

we have

$$|V(t_1 \mid f) - V(t_2 \mid f)| < \varepsilon(\delta, \mu, \tau) \quad \forall t_1, t_2 \in [0, t], \ |t_1 - t_2| < \delta$$

Prove the following theorem [82]. *If $V(t \mid f)$ is an operator satisfying conditions (1) and (2) then a solution of the equation*

$$\varphi(t) = v(t \mid \varphi)$$

may be constructed for sufficiently small $T > 0$ by the following method of approximations: Let $\{h_n\}\big|_{n=0}^{\infty}$ be a monotone sequence of positive numbers, convergent to zero as $n \uparrow \infty$, and let $\{t_n^m\}\big|_{n=0}^{\infty}$ be the sequence defined by

$$t_0^m = 0, \quad t_n^m = t_{n-1}^m + h^m \quad \forall m = 0, 1, 2, \dots, \ \forall n \geq 1$$

Then there exists $T \geq 0$ such that the sequence of functions

$$\varphi^m(t_n^m) = V(t_{n-1}^m \mid \varphi^m), \qquad \varphi^m(t) = \varphi^m(t_{n-1}^m) + [\varphi^m(t_n^m) - \varphi^m(t_{n-1}^m)]/h^m$$

converges uniformly in $[0, T]$ as $m \uparrow \infty$ to a solution $\varphi(t)$ of the equation $f = V(t \mid f)$.

Hint: Proceed as follows. (1) Prove that all the functions $\varphi^m(t)$ exist. (2) Prove that the sequence $\{\varphi^m(t)\}$ is uniformly bounded and equicontinuous, so that the Arzela–Ascoli theorem is applicable.

P17.1.3. Is the solution constructed by the method of Problem 17.1.2 unique? If so, prove the uniqueness theorem. If not, find a counterexample.

P17.2.1. Prove equality (17.69).

P17.2.2. What can be said about asymptotic behavior as $t \uparrow \infty$ of the solution of the third boundary-value problem (the problem of Newtonian heat irradiation) for a cylindrical region? Compare with Dirichlet and Neumann problems.

P17.2.3. Consider the quasilinear problem

$$\frac{\partial^2}{\partial x^2}u + \varphi(u) = \frac{\partial}{\partial t}u; \quad t < x < \infty, \ t > 0$$

$$u(x,0) = f(x), \quad 0 < x < \infty$$

assuming that $\varphi(u)$ is a continuously differentiable, bounded function of a constant sign, tending to zero for $|u| \uparrow \infty$, and that f is a twice continuously differentiable function tending to zero as $|x| \uparrow \infty$. Prove the existence, uniqueness, and stability of the solution to this problem.

Hint: Use the Green's function of the Dirichlet problem for the heat conduction equation for the region $t < x < \infty$.

P17.2.4. Solve the problem posed by the equations

$$\frac{\partial^2}{\partial x^2}u - \frac{\partial}{\partial x}u + \varphi(u) = \frac{\partial}{\partial t}u \quad 0 < x < \infty, \ t > 0, u(x,0) = f(x)$$

Compare this problem with Problem 17.2.3.

P17.3.1. What is the connection between the quasilinear problem studied in Section 3 and Gelfand's heuristic theory of the decay of discontinuities in the Cauchy problem posed for the quasilinear first-order partial differential equation?

P17.4.1. Prove the existence of a global solution to the problem of Section 4, assuming that the heat influx $q(t)$ is a constant. Generalize the result for $q(t)$ positive, bounded, continuous, and continuously differentiable for all $t > 0$.

P17.4.2. The Stefan–Neumann problem

$$\frac{\partial^2}{\partial x^2}u - \frac{\partial}{\partial t}u = 0, \quad 0 < x < y(t), \ t > 0$$

$$u(0,t) - -1, \quad t > 0, \qquad u(x,0) - 0, \quad x > 0$$

$$\frac{d}{dt}y(t) = \frac{\partial}{\partial x}u[y(t),t], \quad t > 0, \qquad y(0) = 0$$

has the solution

$$u(x,t) = -\operatorname{erfc}(x/2\sqrt{t}), \qquad y(t) = 2\gamma\sqrt{t}$$

where γ is the root of the transcendental algebraic equation

$$\gamma = e^{-\gamma^2}/\sqrt{\pi}$$

Construct this solution by reduction to integral equations, and (using the same method) prove that the solution is unique. Give another uniqueness proof using the maximum principle.

P17.5.1. Formulate the problem of radiation of heat from the surface of an infinitely long cylinder with constant boundary and initial temperature, assuming that the process obeys the Stefan–Boltzmann law. Prove an existence and uniqueness theorem.

Hint: Use the expression for the Green's function of the first cylindrical boundary-value problem. Also use estimates for the remainder term and its derivatives in the asymptotic representation of the Bessel function of an imaginary argument (see Appendix 2).

Sequences of parabolic functions

1. Parabolic analogs of Harnack's theorems

To complete our tracing of the analogies between the properties of harmonic and parabolic functions, we first demonstrate that there exist parabolic analogues of Harnack's first and second theorems and of their corollaries.

Theorem 18.1.1 (Parabolic analog of Harnack's first theorem). *Let D^t be a normal parabolic region in $\mathcal{E}_{n+1}$ and $\{u_m(p,t)\}$ a sequence of parabolic functions, continuous in $\bar{D}^t$ and uniformly convergent at the parabolic boundary Γ^t of D^t. Then the sequence $\{u_n\}$ converges uniformly everywhere within D^t to a parabolic function*

$$\exists u(p,t) = \lim_{n\uparrow\infty} u_n(p,t), \quad Lu(p,t) = 0 \quad \forall P \in D^t \tag{18.1}$$

(Recall that $L = \Delta - \partial/\partial t$).

PROOF. For every $\varepsilon > 0$ there exists $N > 0$ such that for all $n, m \geq N$ and $P \in \Gamma^t$,

$$|u_n(P) - u_m(P)| < \varepsilon \tag{18.2}$$

Since $u_n - u_m$ is parabolic in D^t and continuous in $\bar{D}^t$, the maximum principle is applicable, so that (18.2) implies

$$|u_n(P) - u_m(P)| < \varepsilon \quad \forall P \in D^t \tag{18.3}$$

Since $\varepsilon > 0$ is arbitrarily small, (18.3) means that $\{u_n\}$ is uniformly convergent within D^t.

Let M be an arbitrarily chosen point of D_t. Take a parabolic region

$$\tilde{D}^t_{t^o} = C \times (t^o, t] \tag{18.4}$$

where $C \subset \mathbb{R}_n$ is a cube with its center at M and $t - t^o > 0$ such that $\tilde{D}_{t^o}^t \subset D^t$. Then for all $m \geq 0$ and $p \in C$,

$$u_m(p,t) = \int_C u_m(q,t^o)g(p,q,t-t^o)d\omega_q$$

$$+ \int_{t^o}^t d\tau \int_{\Sigma_\tau} u_m(q,\tau)\frac{\partial}{\partial n_q}g(p,q,t-\tau)d\sigma_q \qquad (18.5)$$

where g is the Green's function of the first boundary-value problem in $\tilde{D}^t$.

For all $t > t^o$, both integrals in the right-hand side of (18.5) have no singularities, so that passage to the limit $m \uparrow \infty$ under the integral sign is admissible. Hence

$$u(p,t) = \lim_{m\uparrow\infty} u_m(p,t)$$

$$\stackrel{\text{def}}{=} \int_C \lim_{m\uparrow\infty} u_m(q,0)g(p,q,t-t^o)d\omega_q$$

$$+ \int_{t^o}^t d\tau \int_{\Sigma_\tau} \lim_{m\uparrow\infty} u_m(q,\tau)\frac{\partial}{\partial n_q}g(p,q,t-\tau)d\sigma_q$$

$$= \int_C u(q,t^o)g(p,q,t-t^o)d\omega_q$$

$$+ \int_{t^o}^t d\tau \int_{\Sigma_\tau} u(q,\tau)\frac{\partial}{\partial n_q}g(p,q,t-\tau)d\sigma_q \qquad (18.6)$$

which proves that u is parabolic at (M,t), since the right-hand side of (18.6) is a parabolic function in a neighborhood of this point. Since M is an arbitrary point of D_t, this proves that $u(p,t)$ is parabolic everywhere in D^t. **Q.E.D.**

Theorem 18.1.2 (Pini–Hadamard's parabolic analog of Harnack's second theorem [35; 59]). *Let D^t be a normal parabolic region in $\mathcal{E}_{n+1}$ and $\{u_m(P)\}$ a monotone sequence of functions continuous in the closure $\bar{D}^t$ of D^t and parabolic within D^t. Let*

$$0 < t^o < t_1 < t, \quad D_{1t^o}^{t_1} \subset \bar{D}_{1t^o}^{t_1} \subset D^t \qquad (18.7)$$

be a region belonging to D^t together with its boundary $\Gamma_{1t^o}^{t_1}$, such that the distance δ from $\bar{D}_{1t^o}^{t_1}$ to Γ^t is positive:

$$\max(r_{PQ}) = \delta > 0, \quad P \in \bar{D}_{1t_o}^{t_1}, \; Q \in \Gamma^t \qquad (18.8)$$

Then, if $\{u_m(P)\}$ converges at some point M of D_t, it converges uniformly everywhere within $D_{1t^o}^{t_1}$ to a parabolic function $u(P)$.

PROOF. The assertion is a corollary of the Pini–Hadamard lemma, which is an analog of Harnack's "large" inequality.

Lemma 18.1.1 (Pini–Hadamard inequality). *Let $u(P)$ be a nonnegative function continuous in $\bar{D}^t$ and parabolic inside D^t. Then there exists*

$\alpha(\delta) = \text{const} > 0$, *dependent only on the distance δ and the difference $t - t_1$ and consequently independent of u, such that*

$$0 \le u(Q) \le \alpha u(M,t) \quad \forall Q \in D_{1t_o}^{t_1} \tag{18.9}$$

where M is an arbitrary fixed point of D_t.[1]

Indeed, assume that the lemma has been proved and let $\{u_m\}$ be a nondecreasing sequence. By assumption, the sequence $\{u_k(M,t)\}$ is convergent, so that for every $\varepsilon = \text{const} > 0$ there exists $N \ge 0$ such that for all $n \ge N$ and $m \ge 0$,

$$0 \le u_{n+m}(M,t) - u_n(M,t) < \varepsilon/\alpha \tag{18.10}$$

Hence, by Lemma 18.1.1,

$$0 \le u_{n+m}(Q) - u_n(Q) \le \varepsilon \quad \forall Q \in D_{1t_o}^{t_1} \tag{18.11}$$

so that $\{u_k\}$ converges uniformly everywhere in $D_{t_o}^{t_1}$. By Theorem 18.1.1, this means that

$$u(Q) \overset{\text{def}}{=} \lim_{k \uparrow \infty} u_k(Q) \Rightarrow Lu(Q) = 0 \quad \forall Q \in D_{1t_o}^{t_1} \tag{18.12}$$

Q.E.D.

So let us prove the lemma, on the assumptions that there exists a Green's function $g(p,t,q,\tau)$ for the first boundary-value problem in D^t and that the Vyborny–Friedman Theorem is applicable at every point of the lateral boundary Σ^t. For every point $(p,T) \in D^t$ we have

$$u(p,T) = u_1(p,T) + u_2(p,T) \tag{18.13}$$

where

$$u_1(p,T) = -\int_0^T d\tau \int_{\Sigma_\tau} u(q,\tau) \frac{\partial}{\partial n_q} g(p,T,q,\tau) d\sigma_q \tag{18.14}$$

$$u_2(p,T) = \int_{D_0} u(q,0) g(p,T,q,0) d\sigma_q \tag{18.15}$$

Since $u(p,t)$ is of constant sign, the mean value theorem is applicable, so that

$$u_1(p,T) = -\frac{\partial}{\partial n_q} g(p,T,\tilde{q},\tilde{\tau}) \int_0^T d\tau \int_{\Sigma_\tau} u(q,\tau) d\sigma_q \tag{18.16}$$

$$u_2(p,T) = g(p,T,q^*,0) \int_{D_0} u(q,0) d\sigma_q \tag{18.17}$$

where $(\tilde{q}, \tilde{\tau})$ and q^* are intermediate points

$$(\tilde{q}, \tilde{\tau}) \in \Sigma^T, \quad q^* \in D_0. \tag{18.18}$$

---

[1] D_t is the upper base of D^t, that is, the section of D^t by a plane $\tau = t$.

Recall that, by the maximum principle,

$$g(p, T, q, \tau) > 0 \quad \forall \tau \in [0, T), \ \forall (p, T) \in D^T, \ \forall (q, \tau) \in D^T \tag{18.19}$$

Moreover, by definition,

$$g(p, \tau, q, \tau) = 0 \quad \forall (p, q) \in D_\tau, \ 0 \leq \tau \leq t, \ p \neq q \tag{18.20}$$

$$g(p, T, q, T) = 0 \quad \forall q \in \Sigma_\tau, \ \forall p \in D^t, \ \forall T \in [0, t] \tag{18.21}$$

Since $\mathbf{n}_q$ is the outward normal and the Vyborny–Friedman theorem is assumed to be applicable,[2] it follows from (18.17)–(18.19) that

$$-\frac{\partial}{\partial n_q} g(p, T, q^*, \tau^*) > 0 \quad \forall q^* \in \Sigma_{\tau^*}, \ 0 \leq \tau^* \leq T \tag{18.22}$$

Now let

$$N = \max_{(p,T) \in D_{1t_0}^{t_1}, (q,\tau) \in D^{t_1}} \left(-\frac{\partial}{\partial n_q} g(p, T, q^*, \tau^*) \right) \tag{18.23}$$

Then

$$-\frac{\partial}{\partial n_q} g(p, T, q^*, \tau^*) \leq N \tag{18.24}$$

Similarly, there exists

$$n = \max g(p, \tau, q, 0), \quad (p, \tau) \in D_{1t_0}^{t_1}, \ q \in D_0, \ 0 < t \leq \tau \leq T \leq t_1 < t \tag{18.25}$$

Equalities (18.16), (18.17), (18.24), and (18.25) imply

$$0 \leq u_1(p, T) \leq N \int_0^T d\tau \int_{\Sigma_\tau} u(q, \tau) d\sigma_q,$$
$$0 \leq u_2(p, T) \leq n \int_{D_0} u(q, 0) d\omega_q \tag{18.26}$$

Further, by (18.22),

$$u_1(M, t) = -\int_0^t d\tau \int_{\Sigma_\tau} u(q, \tau) \frac{\partial}{\partial n_q} g(M, t, q, \tau) d\sigma_q$$
$$> -\int_0^T d\tau \int_{\Sigma_\tau} u(q, \tau) \frac{\partial}{\partial n_q} g(M, t, q, \tau) d\sigma_q \tag{18.27}$$

Hence, by the mean value theorem,

$$u_1(M, t) \geq -\frac{\partial}{\partial n_q} g(M, t, q^*, \tau^*) \int_0^t d\tau \int_{\Sigma_\tau} u(q, \tau) d\sigma_q \tag{18.28}$$

---

[2]The Vyborny–Friedman theorem refers to a solution of the parabolic equation $Lu = 0$, whereas $g(p, t, q, \tau)$, considered as a function of the second pair (q, τ) of arguments, is a solution of the adjoint equation $Mv = 0$. Since the latter is transformed into the equation $Lv = 0$ by the substitution $\tau^* = t - \tau$, this means that the Vyborny–Friedman theorem is applicable to g for every $\tau < t$.

where

$$(q^*, \tau^*) \in \Sigma^T, \quad 0 < t_0 < T \le t_1 \tag{18.29}$$

which means that there exists $N_1 = \text{const} > 0$ such that

$$\min\left(-\frac{\partial}{\partial n_q} g(M, t, q^*, \tau^*)\right) = N_1 \tag{18.30}$$

Similarly,

$$u_2(M, t) = g(M, t, q^{**}, 0) \int_{D_0} u(q, 0) d\omega_q, \quad q^{**} \in D_0 \tag{18.31}$$

so that there exists $n_1 = \text{const} > 0$ such that

$$\min g(M, t, q^{**}, 0) = n_1 \tag{18.32}$$

Inequalities (18.26), (18.28), and (18.31) imply

$$0 \le u_1(p, T) \le \frac{N}{N_1} u_1(M, t), \quad 0 \le u_2(p, T) \le \frac{n}{n_1} u_2(M, t) \tag{18.33}$$

Here N_1 and n_1, as well N and n, depend on the distance δ between $\Sigma_{t_0}^{t_1}$ and Σ^t and on the differences $t - t_1$ and $t_0 - 0$, but not on $u(p, t)$. Hence, setting

$$\alpha = \max\left(\frac{N}{N_1}, \frac{n}{n_1}\right) \tag{18.34}$$

and taking (18.13) into account, we obtain

$$0 \le u(p, T) \le \alpha u(M, t) \tag{18.35}$$

Q.E.D.

Remark 18.1.1. Recall that (p, T) is an interior point of D^t not in the parabolic topology but in the usual topology of Euclidean space. Obviously, the inequalities $T < t_1 < t$, $t^o > 0$ are essential for the Pini–Hadamard theorem.

Remark 18.1.2. Recall that the proof of Harnack's second theorem given in Chapter 11 was based on the use of the Heine–Borel lemma. The proof of the Pini–Hadamard inequality based on Heine–Borel lemma may be also applied. If D^t is a noncylindrical region then this would require the rather complicated construction of a suitable cover. For cylindrical regions, however, the Pini–Hadamard theorem can be easily proved by reference to the Heine–Borel lemma without using assumptions that the region D^t is fundamental and that the Vyborny–Friedman theorem is applicable to every point of the lateral boundary Γ^t. Recall that these assumptions were used only in the proof of the Pini–Hadamard inequality. Therefore, we need only present a proof of this inequality relaying on the Heine–Borel lemma, as follows.

Assume that

$$D^t = D \times (0 < t < \infty) \tag{18.36}$$

where $D \in \mathbb{R}_3$. Let

$$D_0 \subset \bar{D}_0 \subset D, \quad D_{0,t_0}^{t_1} = D_0 \times (t_0 \leq \tau \leq t_1), \quad t_0 > 0, \ t_1 < t \tag{18.37}$$

Without loss of generality, we may assume that $M \in D_0$, since otherwise we need only enlarge D_0 and reduce δ. Note now that if

$$K_M^R \subset D \tag{18.38}$$

is an arbitrarily chosen sphere in D of radius R with center at M, and

$$C = K_M^R \times (0 < \tau \leq t) \tag{18.39}$$

is a parabolic cylinder, then C is a fundamental parabolic region since it satisfies the requirements of the existence theorem of Chapter 17, Section 1. Hence the Pini–Hadamard inequality is applicable to C. Thus, if $u(p, t) \geq 0$ is continuous in $\bar{C}$ and parabolic within C, then there exists $\alpha = \text{const} > 0$ dependent only on $\delta > 0$, $\eta > 0$, and $\varepsilon > 0$ and independent of $u(p, t)$, such that

$$0 \leq u(p, T) < \alpha u(M, t) \quad \forall (p, T) \in C_1 = K_M^{R-\delta} \times [\eta \leq T < t - \varepsilon] \tag{18.40}$$

Let $\Sigma = \bar{D} \setminus D$, $\Sigma_0 = \bar{D}_0 \setminus D_0$ be the boundaries of D and D_0, respectively, and let

$$4R = \min r_{pq}, \quad p \in \Sigma, \ q \in \Sigma_0 \tag{18.41}$$

Let q be an arbitrarily fixed point of D_0. Take a cover of $\bar{D}_0$ by a set of spheres K of radius R. By the Heine–Borel lemma, one may extract from this cover a finite number of spheres

$$K_i \overset{\text{def}}{=} K_{q_i}^R, \quad i = 0, 1, 2, \ldots, n \ q_0 = M, \ q_n = q \tag{18.42}$$

with centers at q_i. Then the set of parabolic regions

$$\tilde{C}_i = K_i \times (0 < \tau \leq t), \quad i = 0, 1, 2, \ldots, n \tag{18.43}$$

is a finite cover of $D_{0,t_0}^{t_1}$. Now take arbitrary $T \in [t_0, t_1]$ and set

$$\varepsilon = \frac{t-T}{n+1}, \quad \eta = \frac{T}{n+1}, \quad t_{1k} = t_1 + k\varepsilon, \quad t_{0k} = k\eta \tag{18.44}$$

so that

$$0 < t_{0n} < T, \ t_{1n} > t_1 \tag{18.45}$$

Hence the Pini–Hadamard inequality can be applied consecutively to all $k \in [0, n-1]$. Thus

$$u(q_{i+1}, t_{i+i}) < \alpha u(q_i, t_i), \quad i = 0, 1, 2, \ldots, n-1 \tag{18.46}$$

so that

$$0 \leq u(q, T) \prod_{i=1}^{n} u(q_i, t_i) \leq \alpha^{n+1} u(M, t) \prod_{i=1}^{n} u(q_i, t_i) \tag{18.47}$$

Hence

$$u(M, t) = 0 \Rightarrow u(q, T) = \beta u(M, t) = 0 \tag{18.48}$$

and

$$u(M, t) > 0 \Rightarrow u(q, T) \leq \beta u(M, t) \tag{18.49}$$

where

$$\beta = \alpha^{n+1} \tag{18.50}$$

which completes the proof of the Pini–Hadamard inequality for a cylindrical parabolic region.   **Q.E.D.**

2. Space of continuous super- and subparabolic functions

Recall that the spaces of twice differentiable super- and subharmonic functions are only subspaces of the spaces of continuous super- or subharmonic functions. The same is true of the space $\mathfrak{M}^{2,1}(D^t)$ of functions that are super- or subparabolic in a parabolic region D^t. Bearing in mind the formulation and proof of the parabolic analog of the basic Perron theorem, we now define the full spaces of super- and subparabolic functions in a normal parabolic region D^t. Recall that the spaces of super- or subharmonic functions are defined by introducing a function $\mathfrak{M}_u^k(p)$, continuous in the region D, equal to u identically outside K and harmonic within K. If $\mathfrak{M}_u^k(p) \leq u$ everywhere in D for any sphere $K \subset D$, then $u(p)$ is said to be *superharmonic* in D; if $\mathfrak{M}_u^k(p) \geq u$ for any $K \subset D$, then u is said to be *subharmonic* in D. The analogous definitions of super- and subparabolic functions in a normal parabolic region D^t must take into account that D^t may vary nonmonotonically, even degenerating at the initial time or at some time $t > 0$. This is so if one wants to be able to use the theory of super- and subparabolic functions to study, for example, phase transitions in a two-phase thermally conductive medium with prescribed nonmonotonic boundary temperature; in such situations one may observe the appearance of a new phase or the disappearance of an existing phase (see Figure 18.1).[3] Bearing this in mind, we introduce the following definitions.

Definition 18.2.1. Let $D^t \subset \mathcal{E}_{n+1} = \mathbb{R}_n \times (-\infty < t < \infty)$, $n = 1, 2, 3$, be a normal parabolic region and a region $D_t \subset \mathbb{R}_n$ its upper base. Then $\mathfrak{K}$ is a space of parabolic regions $K_{t_0}^t$, $0 < t_0 < t$, such that

(a) $K_{t_0}^t \subset \mathfrak{K} \Rightarrow K_{t_0}^t \subset D^t$

---

[3]Note that if D^T degenerates at $t = T > 0$, then the set D_T belongs to the parabolic boundary of D^T rather than to its interior. In this case the solutions of parabolic boundary-value problems may be constructed at $t = T$ only asymptotically as $t \uparrow T - 0$. Therefore we shall assume $t < T$ if such a time T of degeneration of the parabolic region exists; otherwise, we can assume $t \leq T$.

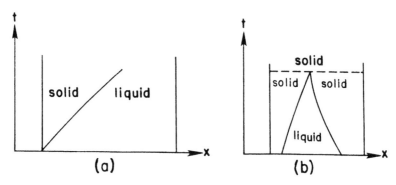

Figure 18.1.   Sketch of the generation of a new and degeneration of the existing parabolic regions.

(b) $K_{t_0}^t$ is fundamental with respect to a parabolic Dirichlet problems, that is, these problems are solvable in $K_{t_0}^t$ for any continuous boundary function $f(p,t)$.

In particular, if D^t is a parabolic cylinder then $\mathfrak{K}$ can be identified with the space of all parabolic cylinders $K_{t_0}^t = K \times (t_0 < \tau \leq t)$ belonging to D^t (K is a sphere in $\mathbb{R}_n$).

Definition 18.2.2. Given a function $u(p,t)$ continuous in D^t, a function $\mathfrak{M}_k(q, \tau \mid u, t, t_0)$ is defined by the conditions:

$$\mathfrak{M}_k(q, \tau \mid u, t, t_0) \in C(D^t) \cap C^{2,1}(K_{t_0}^t) \qquad (18.51)$$

$$\mathfrak{M}_k(q, \tau \mid u, t, t_0) = u(q, \tau) \quad \forall (q, \tau) \in D^t \setminus K_{t_0}^t \qquad (18.52)$$

$$L\mathfrak{M}_k(q, \tau \mid u, t, t_0) = 0 \quad \forall (q, t) \in K_{t_0}^t \qquad (18.53)$$

that is, $\mathfrak{M}_k(q, \tau \mid u, t_1, t_0)$ is parabolic in $\mathfrak{M}^{2,1}(K_{t_0}^t)$ in the sense defined in Chapter 14.[4]

Definition 18.2.2*. Given parabolic cylinder $C \overset{\text{def}}{=} K_M^R \times (t_0 < t \leq T)$, where $K_M^R \subset \mathbb{R}_n$ is a sphere with center at M, Γ is a parabolic boundary of C, and $u(p,t)$ is a function continuous on Γ, $\tilde{u}(p,t) \overset{\text{def}}{=} \mathfrak{M}(p, t \mid u, M, R, t_0, T)$ denotes a function such that

$$\tilde{u}(p,t) \in C(\tilde{C}) \cap C^{2,1}(C) \wedge L\tilde{u} = 0 \quad \forall (p,t) \in C \qquad (18.54)$$

Definition 18.2.3. $u(p,t)$ is said to be *superparabolic* in D^t if

$$\mathfrak{M}_k(q, \tau \mid u, t_1, t_0) \leq u(q, \tau) \quad \forall (q, \tau) \in K_{t_0}^t, \ \forall K_{t_0}^t \subset K \qquad (18.55)$$

---

[4]Recall that if D^t is a parabolic region then its lower base D belongs to the parabolic boundary Γ^t of D^t. Hence condition (18.52) means, in particular, that $\mathfrak{M}_c(q, t_0 \mid t, t_0) = u(q, t_0)$.

Definition 18.2.4. $u(p,t)$ is said to be *subparabolic* in D^t if

$$\mathfrak{M}_k(q,\tau \mid u,t,t_0) \geq u(q,\tau) \quad \forall (q,\tau) \in K_{t_0}^t, \ \forall K_{t_0}^t \subset K \tag{18.56}$$

Definition 18.2.5. $u(p,t)$ is *parabolic* in D^t if it is simultaneously superparabolic and subparabolic there.

As in the theory of sub- and superharmonic functions, the following lemmas are true.

Lemma 18.2.1. *If $v_i(p,t)$, $i=1,2$, are superparabolic (subparabolic) in D^t, then $v_1(p,t) + v_2(p,t)$ is superparabolic (subparabolic).*

Lemma 18.2.2. *If $u(p,t)$ is superparabolic (subparabolic) in D^t, then $-u(p,t)$ is subparabolic (superparabolic) in D^t.*

The validity of Lemmas 18.2.1–2 is obvious.

Lemma 18.2.3. *If $u(p,t) \in C^{2,1}(D^t)$ and*

$$Lu \equiv \Delta u - \frac{\partial}{\partial t}u = \psi(p,t) \quad \forall (p,t) \in D^t \tag{18.57}$$

then u is superparabolic (subparabolic) in D^t if

$$\psi(p,t) \leq 0 \ (\psi(p,t) \geq 0) \quad \forall (p,t) \in D^t \tag{18.58}$$

PROOF. Let $g(p,t,q,\tau)$ be the Green's function of the first boundary value problem for the heat-conduction equation in $K_{t_0}^t$.[5] Then, inside $K_{t_0}^t$,

$$u(p,t) = -\int_{t_0}^t d\tau \int_{S_\tau} u(q,\tau)\frac{\partial}{\partial n_q}g(p,t,q,\tau)d\sigma_q$$

$$+ \int_{K_{t_0}} u(q,t_0)g(p,t,q,t_0)d\omega_q$$

$$- \int_{t_0}^t d\tau \int_{K_\tau} \psi(q,\tau)g(p,t,\tau)d\omega_q \tag{18.59}$$

where $S_\tau = \bar{K}_\tau \setminus K_\tau$. On the other hand, inside $K_{t_0}^t$ we have

$$\mathfrak{M}_k(q,\tau \mid u,t,t_0) = -\int_{t_0}^t d\tau \int_{K_\tau} u(q,\tau)\frac{\partial}{\partial n_q}g(p,q,t-\tau)d\sigma_q$$

$$+ \int_{K_{t_0}} u(q,t_0)g(p,t,q,t_0)d\omega_q \tag{18.60}$$

---

[5]The Green's function g exists, since $K_{t_0}^t$ is fundamental by assumption.

so that, inside $K_{t_0}^t$,

$$\text{sign}\left(\mathfrak{M}_c(q,\tau \mid u,t_1,t_0) - u(p,t)\right) = \text{sign}\int_{t_0}^t d\tau \int_{K_\tau} \psi(q,\tau)g(p,t,q,\tau)d\omega_q$$

$$= \text{sign}\,\psi \qquad\qquad (18.61)$$

$$\textbf{Q.E.D.}$$

Lemma 18.2.4. *Let* $u(p,t) \in C^{2,1}(D^t)$ *be superparabolic (subparabolic) in* D^t. *Then*

$$Lu(p,t) \le 0 \quad (Lu(p,t) \ge 0) \quad \forall (p,t) \in D^t \qquad\qquad (18.62)$$

PROOF. Indeed, the fundamental identity (15.20) is applicable to u. Hence, as before,

$$u(p,t) = \mathfrak{M}_k(q,\tau \mid u,t,t_0) - \int_{t_0}^t d\tau \int_{K_\tau} Lu(q,\tau)g(p,t,q,\tau)d\omega_q \qquad\qquad (18.63)$$

By assumption, u is superparabolic in D^t, so that

$$\int_{t_0}^t d\tau \int_{K_\tau} Lu(q,\tau)g(p,t,q,\tau)d\omega_q \le 0 \qquad\qquad (18.64)$$

The mean value theorem yields

$$Lu(\tilde{q},\tilde{\tau})\int_{t_0}^t d\tau \int_{K_\tau} g(p,t,q,\tau)d\omega_q \le 0 \Rightarrow Lu(\tilde{q},\tilde{\tau}) \le 0 \qquad\qquad (18.65)$$

since $g \ge 0$. By assumption, Lu is continuous. Hence, contracting $K_{t_0}^t$ to the point (p,t), we obtain

$$Lu(p,t) \le 0 \qquad\qquad (18.66)$$

Since (p,t) is an arbitrarily chosen point of D^t, this proves the lemma.

$$\textbf{Q.E.D.}$$

Lemma 18.2.5. *Let* $u_i(p,t)$, $i=1,2$, *be superparabolic (subparabolic) in* D^t. *Then* $w=\min(u_1,u_2)$ $(w=\max(u_1,u_2))$ *is superparabolic (subparabolic) in* D^t.

PROOF. For any $K_{t_0}^t \subset D^t$ and any point $(p,t) \in K_{t_0}^t$,

$$\mathfrak{M}_k(q,\tau \mid u,t_1,t_0) = -\int_{t_0}^t d\tau \int_{S_\tau} \min[u_1(q,\tau),u_2(q,\tau)]\frac{\partial}{\partial n_q}g(p,t,q,\tau)d\sigma_q$$

$$+ \int_{K_{t_0}} \min[u_1(q,\tau),u_2(q,\tau)]g(p,t,q,t_0)d\omega_q \qquad\qquad (18.67)$$

Since $g \ge 0$ within $K_{t_0}^t$ for all $t \ge t_0$, $g=0$ in $S_{t_0}^t = \bar{K}_{t_0}^t \setminus K_{t_0}^t$ and n_q is an outward normal, it follows that $-\partial g/\partial n_q > 0$ for any $p \in K_t$ and $q \in S_\tau$.

This yields

$$\mathfrak{M}_k(q,\tau \mid u,t,t_0) = \min_{i=1,2}\left(-\int_{t_0}^t d\tau \int_{S_\tau} u_i(q,\tau)\frac{\partial}{\partial n_q}g(p,t,q,\tau)d\sigma_q\right.$$

$$\left. + \int_{K_{t_0}} u_i(q,\tau)g(p,t,q,t_0)d\omega_q\right)$$

$$= \min[\mathfrak{M}_k(q,\tau \mid u_1,t,t_0),\mathfrak{M}_k(q,\tau \mid u_2,t,t_0)]$$

$$\leq \min[u_1(p,t),u_2(p,t)] = u(p,t) \tag{18.68}$$

Thus

$$\mathfrak{M}_k(q,\tau \mid u,t,t_0) \leq u(p,t) \tag{18.69}$$

$$\textbf{Q.E.D.}$$

Lemma 18.2.6. *Let $u(p,t)$ be a given superparabolic (subparabolic) function in D^t, and let $K_{t_0}^t \subset \Re^t$. Then $v(q,\tau) \overset{\text{def}}{=} \mathfrak{M}_k(q,\tau \mid u,t,t_0)$ is superparabolic (subparabolic) in D^t.*

PROOF. Let $K_{t_o}^t \subset \Re^t$. To simplify the notation, denote

$$w(q,\tau) = \mathfrak{M}_k(q,\tau \mid v,t,t_o) \tag{18.70}$$

We have to prove that

$$w(q,\tau) \leq v(q,\tau) \quad \forall(q,\tau) \in D^t \tag{18.71}$$

The following possibilities must be considered (see Figure 18.2):

(a) $\qquad\qquad t_0 \leq t_o, K_{t_0}^t \cap K_{t_o}^t = \emptyset$ $\qquad\qquad$ (18.72)

(b) $\qquad\qquad t_0 \leq t_o, K_{t_0}^t \supset K_{t_o}^t$ $\qquad\qquad$ (18.73)

(c) $\qquad\qquad t_0 = t_o, K_{t_0}^t \subset K_{t_o}^t$ $\qquad\qquad$ (18.74)

(d) $\qquad\qquad t_0 \leq t_o, K_{t_0}^t \cap K_{t_o}^t \neq \emptyset \wedge K_{t_0}^t \setminus K_{t_o}^t \neq \emptyset$ $\qquad$ (18.75)

(a')–(c') $\qquad\qquad t_0 > t_o$ $\qquad\qquad\qquad\qquad$ (18.76)

(The assumptions in (a')–(c') concerning the location of $K_{t_0}^t$ and $K_{t_o}^t$ are the same as in (a)–(c).) We consider cases (a)–(c) in order.

Using the uniqueness theorem for solutions of the parabolic Dirichlet problem, we find

(a) (18.72) $\Rightarrow w \leq v \; \forall(q,\tau) \in K_{t_o}^t \wedge w = v \; \forall(q,\tau) \in D^t \setminus K_{t_o}^t$

$\qquad\qquad \Rightarrow w \leq v \; \forall(q,\tau) \in D^t$ $\qquad\qquad$ (18.77)

(b) (18.73) $\Rightarrow w \leq v \; \forall(q,\tau) \in K_{t_o}^t \wedge w = v \; \forall(q,\tau) \in D^t \setminus K_{t_o}^t$

$\qquad\qquad \Rightarrow w \leq v \; \forall(q,\tau) \in D^t$ $\qquad\qquad$ (18.78)

(c) (18.74) $\Rightarrow w = v \; \forall(q,\tau) \in D^t$ $\qquad\qquad$ (18.79)

(d) (18.75) $\Rightarrow w \leq v \; \forall(q,\tau) \in K_{t_o}^t \setminus K_{t_0}^t \wedge w = v \; \forall(q,\tau) \in D^t \setminus K_{t_o}^t$

$\qquad\qquad \Rightarrow w \leq v \; \forall(q,\tau) \in D^t \setminus (K_{t_0}^t \cup K_{t_o}^t)$ $\qquad$ (18.80)

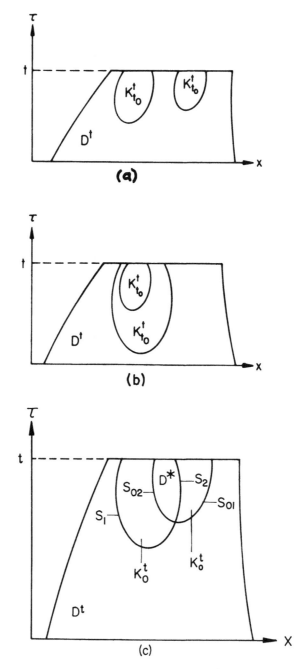

Figure 18.2.   The possible location of $K_{t_0}^t$ and $K_{t_o}^t$.

Hence we must only prove that

$$w \leq v \quad \forall (q, \tau) \in K_{t_0}^t \tag{18.81}$$

Let

$$D^* = K_{t_0}^t \cap K_{t_o}^t, \quad S^* = \bar{D}^* \setminus D^*, \quad S_1 = S_{t_0}^t \setminus K_{t_o}^t, \quad S_2 = S_{t_0}^t \cap K_{t_o}^t$$
$$S_{01} = S_{t_o}^t \setminus K_{t_0}^t, \quad S_{02} = S_{t_o}^t \cap K_{t_0}^t \Rightarrow S^* = S_2 \cup S_{02} \tag{18.82}$$

We have

$$w = v \quad \forall (q, \tau) \in D^t \setminus K_{t_o}^t \Rightarrow w = v \leq u \quad \forall (q, \tau) \in S_{02} \cup S_{01} \tag{18.83}$$

Hence, since v is parabolic and u is superparabolic in $K_{t_o}^t$

$$w \leq u = v \quad \forall (p, q) \in S_2 \Rightarrow w \leq v \quad \forall (q, \tau) \in S^* \tag{18.84}$$

By definition, v and w are parabolic functions within D^*. Hence, by the maximum principle, it follows from (18.84) that

$$w \leq v \quad \forall (q, \tau) \in \bar{D}^* \tag{18.85}$$

At the same time by (18.80), since u is superparabolic,

$$v = u \; \forall (q, \tau) \in D^t \setminus K_{t_0}^t \wedge v \leq u \; \forall (q, \tau) \in K_{t_0}^t$$
$$\Rightarrow v \leq u \; \forall (q, \tau) \in K_{t_0}^t$$
$$\Rightarrow w \leq v \; \forall (q, \tau) \in K_{t_0}^t \setminus K_{t_o}^t \tag{18.86}$$

Hence (18.85) and (18.86) imply

$$w \leq v \quad \forall (q, \tau) \in K_{t_o}^t \tag{18.87}$$

which proves that v is superparabolic, since $K_{t_o}^t$ is an arbitrarily chosen element of $\mathfrak{K}$. Cases (a')–(b') need no consideration, since they may be reduced to cases (a)–(b) by replacing (t_0, t) with (t_o, t) and vice versa. **Q.E.D.**

Lemma 18.2.7 (Maximum principle). *If $u(p, t)$ is superparabolic (subparabolic) in D^t and continuous in $\bar{D}^t$, then its minimum (maximum) is achieved on the parabolic boundary Γ^t of D^t.*

PROOF. (The proof of this lemma is an almost literal repetition of that of its superharmonic counterpart.) Assume that

$$\min_{(p, \tau) \in \Gamma^\tau} u(p, \tau) = M > \min_{(p, \tau) \in D^t} u(p, \tau) = m \tag{18.88}$$

Then there exists $(p_0, \tau) \in D^t$ such that

$$u(p_0, T) = m, \quad 0 < T \leq t \tag{18.89}$$

so that the set

$$\mathcal{A} = \left\{ q, \tau : u(q, \tau) = m, \; (q, \tau) \in D^t \right\} \neq \emptyset \tag{18.90}$$

is not empty. Let K be the subset of all $K_{t_0}^t \subset \mathfrak{K}$ such that

$$K_{t_0}^t \subset K \Rightarrow K_{t_0}^t \cap \mathcal{A} \neq \emptyset \tag{18.91}$$

Any function $\mathfrak{M}_c(q, \tau \mid)$ defined in $K_{t_0}^t \subset K$ is continuous in D^t, parabolic within $K_{t_0}^t$, and moreover belongs to $C^{2,1}(K_{t_0}^t)$, so that by the maximum principle for functions parabolic in $C^{2,1}(K_{t_0}^t)$, $u \equiv m$ everywhere in $K_{t_0}^t$. Hence K is a closed domain and u is equal to m at the boundary points of K. Obviously, by the assumption (18.88), the distance from K to the parabolic boundary of D^t is positive:

$$r_{K\Gamma^\iota} \overset{\text{def}}{=} R > 0 \tag{18.92}$$

so that

$$u(q, \tau) > m \quad \forall (q, \tau) \in D^t \setminus K \tag{18.93}$$

Hence there exists a region $\tilde{K} \subset D^t \setminus K$ that touches K at its boundary point $(\tilde{p}, \tilde{t})$ which is an inner point of some $K^* \overset{\text{def}}{=} K_{t_0}^t \subset \mathfrak{K}$. Hence, by the maximum principle for parabolic functions $v \in C^{2,1}(\tilde{K})$, $u(q, \tau) \equiv m$ within K^*, which contradicts (18.93). This contradiction shows that (18.88) cannot be true. **Q.E.D.**

3. Perron–Petrovsky's theorem. Parabolic barriers[6]

We now proceed to derive a parabolic analog of the basic Perron theorem concerning the existence of solutions to the Dirichlet problem for the Laplace equation. In what follows we restrict ourselves to proving only the most elementary relevant results; only the formulation of Petrovsky's fundamental theorem will be given.

Let D^t be a normal parabolic region and L the parabolic operator

$$Lu \equiv \Delta u - \frac{\partial}{\partial t} \tag{18.94}$$

Let us consider the Dirichlet problem: Find $u(p, t) \in C^{2,1}(D^t) \cap C(\bar{D}^t)$ such that

$$Lu = 0, \quad (p, t) \in D^t, \quad u = f(p, t) \in \Gamma^t \tag{18.95}$$

Definition 18.3.1. $u(p, t)$ is called an *upper (lower) function* with respect to the Dirichlet problem (or simply an upper (lower) function) if it is continuous in $\bar{D}^t$, superparabolic (subparabolic) in D^t, and satisfies the inequality

$$u \geq f \ \forall (p, t) \in \Gamma^t \quad (u \leq f \ \forall (p, t) \in \Gamma^t) \tag{18.96}$$

Definition 18.3.2. $\mathfrak{M}^+$ and $\mathfrak{M}^-$ are the classes of all upper and lower functions, respectively.

---

[6]A fundamental contribution to the theory of sub- and superparabolic functions has been made by I. G. Petrovsky [55], who studied the degeneration of parabolic regions (mainly in the space $\mathcal{E}_{1+1}$) at some time $t = T > 0$. In this connection he introduced (for comparison) regions that degenerated at $t = T$.

Note that, by the maximum principle (Lemma 18.2.7),

$$u \in \mathfrak{M}^- \wedge v \in \mathfrak{M}^+ \Rightarrow u \le v \quad \forall (p, t) \in D^t \tag{18.97}$$

Let

$$m \le f(p, t) \le M \tag{18.98}$$

Then

$$m \in \mathfrak{M}^-, \quad M \in \mathfrak{M}^+ \tag{18.99}$$

so that the classes of lower and upper functions are not empty:

$$\mathfrak{M}^- \ne \emptyset, \quad \mathfrak{M}^+ \ne \emptyset \tag{18.100}$$

Lemmas 18.2.1–7 may obviously be reformulated in terms of upper and lower functions as follows.

Lemma 18.3.1. *If* $v_i(p, t) \in \mathfrak{M}^+$, $(v_i(p, t) \in \mathfrak{M}^-)$, $i = 1, 2$, *then*

$$v_1(p, t) + v_2(p, t) \in \mathfrak{M}^+ \quad (v_1(p, t) + v_2(p, t) \in \mathfrak{M}^-) \tag{18.101}$$

Lemma 18.3.2. *If* $u(p, t) \in \mathfrak{M}^+$ $(u(p, t) \in \mathfrak{M}^-)$ *then*

$$-u(p, t) \in \mathfrak{M}^- \quad (-u(p, t) \in \mathfrak{M}^+) \tag{18.102}$$

Lemma 18.3.3. *Let* $u(p, t) \in C^{2,1}(D^t)$ *and*

$$Lu(p, t) \le 0 \quad (Lu(p, t) \ge 0) \quad \forall (p, t) \in D^t \tag{18.103}$$

Then

$$u(p, t) \in \mathfrak{M}^+ \quad (u(p, t) \in \mathfrak{M}^-) \tag{18.104}$$

Lemma 18.3.4. *Let* $u(p, t) \in C^{2,1}(D^t)$ *belong to* $\mathfrak{M}^+$ $(u(p, t) \in \mathfrak{M}^-)$. *Then*

$$Lu(p, t) \le 0 \ (Lu(p, t) \ge 0) \quad \forall (p, t) \in D^t \tag{18.105}$$

Lemma 18.3.5. *Let* $u_i(p, t) \in \mathfrak{M}^+$ $(u_i(p, t) \in \mathfrak{M}^-)$, $i = 1, 2$. *Then*

$$w = \min(u_1, u_2) \in \mathfrak{M}^+ \ (w = \max(u_1, u_2) \in \mathfrak{M}^-) \tag{18.106}$$

Lemma 18.3.6. *Let* $u(p, t) \in \mathfrak{M}^+$ $(u(p, t) \in \mathfrak{M}^-)$ *and* $K_{t_0}^{t_1} \subset D^t$. *Then*

$$v(q, \tau) \stackrel{\text{def}}{=} \mathfrak{M}_k(q, \tau \mid u, t_1, t_0)$$
$$\Rightarrow v(p, t) \in \mathfrak{M}^+ \ (u(p, t) \in \mathfrak{M}^-) \tag{18.107}$$

Lemma 18.3.7 (Maximum principle). *If* $u(p, t) \in \mathfrak{M}^+$ $(u(p, t) \in \mathfrak{M}^-)$, *then its minimum (maximum) is achieved on the parabolic boundary* Γ^t *of* D^t.

We can now formulate the following basic theorem.

Theorem 18.3.1 (Perron–Petrovsky). *Let*

$$u(p, t) = \inf \mathfrak{M}^+ \ (u(p, t) = \sup \mathfrak{M}^-) \tag{18.108}$$

Then

$$Lu = 0 \quad \forall (p,t) \in D^t \tag{18.109}$$

PROOF. Clearly, it is sufficient to prove that u is parabolic within any arbitrarily chosen parabolic cylinder

$$C_{M\eta}^{RT} = K_M^R \times (\eta < \tau \leq T < t] \in D^t \tag{18.110}$$

where $K_M^R \subset \mathbb{R}_3$ is the sphere of radius R with center at $M \in D_\eta$ and $t - T > 0$ is fixed arbitrarily small.

Take $R_1 > R$, and let t_o and T_1 be such that

$$C_{M\eta}^{Rt} \subset C_{M\eta}^{R_1 t} \subset D^t, \eta < t_o < T < T_1 < t \tag{18.111}$$

Let $\{\varepsilon_n\}$ be a numerical sequence such that

$$\varepsilon_1 > \varepsilon_2 > \varepsilon_3 > \cdots > \varepsilon_n > \cdots, \quad \lim_{n \uparrow \infty} \varepsilon_n = 0 \tag{18.112}$$

and let

$$v_0(p,\tau), \ v_1(p,\tau), \ v_3(p,\tau), \ldots, \ v_n(p,\tau), \ldots \tag{18.113}$$

be a sequence of upper functions such that

$$0 \leq v_n(M,T) - u(M,T) \leq \varepsilon_n, \quad n = 0,1,\ldots \tag{18.114}$$

Define a sequence $\{w_n\}|_{n=0}^\infty$ of functions as follows:

$$w_0 = v_0 \tag{18.115}$$

and, for all $n \geq 1$,

$$u_n = \mathfrak{M}(q, \tau \mid w_{n-1}, M, R_1, \eta, t), \quad w_n = \min(u_n, v_n) \tag{18.116}$$

By Lemmas 18.3.5–6,

$$u_n \in \mathfrak{M}^+ \quad \forall n \geq 1 \tag{18.117}$$

Furthermore, it follows from the maximum principle (Lemma 18.3.7) that

$$u_1 \geq u_2 \geq \cdots \geq u_n \geq \cdots > u \tag{18.118}$$

and

$$0 \leq u_n(M,T) - u(M,T) < \varepsilon_n \quad \forall n = 1,2,3,\ldots \tag{18.119}$$

Thus $\{u_n\}$ is a monotonically decreasing sequence of functions parabolic in the cylinder $C_{M\eta}^{R_1 t} \subset D_o^t$ and convergent at an interior (in the sense of the usual Euclidean topology) point $(M,T) \in C_{M\eta}^{Rt}$. Hence the Pini–Hadamard theorem is applicable, so that $\{u_n\}$ is convergent uniformly in $C_{Mt_o}^{RT_1} \subset C_{M\eta}^{R_1 t}$. Let

$$w(p,\tau) = \lim_{n \uparrow \infty} u_n(p,\tau) \quad \forall (p,\tau) \in C_{Mt_o}^{RT_1} \tag{18.120}$$

We must prove that

$$w(p,\tau) \equiv u(p,\tau) \quad \forall (p,\tau) \in C_{Mt_o}^{RT_1} \tag{18.121}$$

It is obvious that

$$w(p, \tau) \geq u(p, \tau) \quad \forall (p, \tau) \in C_{Mt_o}^{RT_1} \tag{18.122}$$

Indeed, if there exists (p, τ) such that

$$w(p, \tau) < u(p, \tau), \quad (p, \tau) \in C_{Mt_o}^{RT_1} \tag{18.123}$$

then, for some $n > 0$,

$$\mathfrak{M}^+ \ni w_n(p, \tau) < u(p, \tau) = \inf \mathfrak{M}^+ \tag{18.124}$$

which is impossible. Assume that, for some $(q, t^o) \in C_{Mt_o}^{RT_1}$ and $t_o < t^o$, we have[7]

$$w(q, t^o) > u(q, t^o) \tag{18.125}$$

Let

$$w(q, t^o) - u(q, t^o) = 4\delta > 0 \tag{18.126}$$

By the definition of u, there exists $v_0(p, t) \in \mathfrak{M}^+$ such that

$$v_0(q, t^o) - u(q, t^o) < \delta \tag{18.127}$$

Further, for all $n > 0$,

$$w(p, \tau) \leq u_n(p, \tau) \quad \forall (p, \tau) \in C_{Mn}^{Rt} \tag{18.128}$$

Hence (18.126)–(18.128) imply

$$\begin{aligned}
v_0(q, t^o) - u_n(q, t^o) &= [v_0(q, t^o) - u(q, t^o)] + [u(q, t^o) - w(q, t^o)] \\
&\quad + [w(q, t^o) - u_n(q, t^o)] \\
&\leq \delta - 4\delta + 0 = -3\delta
\end{aligned} \tag{18.129}$$

Note that v_0 is a continuous function and that the sequence $\{u_n\}$ is uniformly bounded and equicontinuous by the Arzela–Ascoli theorem, since $\{u_n\}$ is uniformly convergent. Hence there exists $r \in (0, R)$ such that if $r_{pq} < r$ then, for all $n \geq 1$,

$$|v_0(p, t^o) - v_0(q, t^o)| \leq \delta, \quad |u_n(p, t^o) - u_n(q, t^o)| \leq \delta \tag{18.130}$$

It is evident from (18.129) and (18.130) that for all $n \geq 1$ and all p such that $r_{pq} < \delta$,

$$\begin{aligned}
v_0(p, t) - u_n(p, t^o) &= [v_0(p, t^o) - v_0(q, t^o)] + [v_0(q, t^o) - u_n(q, t^o)] \\
&\quad + [u_n(q, t^o) - u_n(p, t^o)] \\
&< \delta - 3\delta + \delta = -\delta
\end{aligned} \tag{18.131}$$

Let

$$C_{Mt^o}^{rt} = K_M^\tau \times (t^o < \tau \leq t] \in C_{Mt_o}^{Rt} \tag{18.132}$$

and

$$w_n^o = \min(u_n, v_0), \quad \tilde{w}_n = \mathfrak{M}(p, \tau \mid w_n^o, r, t^o, t) \tag{18.133}$$

---

[7]Since T was chosen arbitrarily close to t, we may assume that $t^o < T$.

By Lemmas 18.3.5–6,

$$u_n \in \mathfrak{M}^+, \ v_0 \in \mathfrak{M}^+ \Rightarrow \tilde{w}_n \in \mathfrak{M}^+ \Rightarrow \tilde{w}_n(M,t) \geq u(M,t) = \inf \mathfrak{M}^+ \quad (18.134)$$

On the other hand, if $g(p,q,t-\tau)$ is the Green's function of the first boundary-value problem for the heat-conduction equation in the sphere $K \overset{\text{def}}{=} K_M^R \subset R_n$, then

$$\tilde{w}(M,t) - u(M,t) = u_n(M,t) - u(M,t) + [\tilde{w}_n(M,t) - u_n(M,t)]$$
$$= u_n(M,t) - u(M,t) + \Phi(M,t) \quad (18.135)$$

where

$$\Phi(M,t) = -\int_{t^o}^t d\tau \int_S [\tilde{w}_n(p,\tau) - u_n(p,\tau)] \frac{\partial}{\partial n_q} g(M,p,t-\tau) d\sigma_q$$

$$+ \int_K [\tilde{w}_n(p,t^o) - u_n(p,t^o)] g(M,p,t-t^o) d\sigma_q$$

$$= \Phi_0(M,t) + \Phi_1(M,t) \quad (18.136)$$

and $S = \bar{K} \setminus K$ is the boundary of K. Recall that

$$-\frac{\partial}{\partial n_q} g(M,p,t-\tau) > 0 \quad \forall p \in S, \quad g(M,p,t-\tau) > 0 \quad \forall t > t^o \quad (18.137)$$

Let

$$K^* = K_q^r \subset \mathbb{R}_3, \quad K_0 = K \cap K^*, \quad K_1 = K \setminus K^* \quad (18.138)$$

and

$$\int_{K_0} g(M,p,t-t^o) d\sigma_p = \alpha > 0 \quad (18.139)$$

We have

$$\tilde{w}_n - u_n \leq 0 \quad \forall(p,t_o) \in \bar{K}_1 \quad (18.140)$$
$$\tilde{w}_n - u_n \leq -\delta \quad \forall(p,t_o) \in \bar{K}_0 \quad (18.141)$$

$$\Phi_1 = \Phi_{10} + \Phi_{11}, \quad \Phi_{10} = \int_{K_0}, \quad \Phi_{11} = \int_{K_1} \quad (18.142)$$

Note that by (18.136)–(18.141),

$$\Phi_0 \leq 0, \ \Phi_{10} \leq -\alpha\delta, \ \Phi_{11} < 0$$
$$\Rightarrow \Phi(M,t) < -\alpha\delta \quad (18.143)$$

Hence

$$\tilde{w}_n(M,t) - u(M,t) = u_n(M,t) - u(M,t) + \Phi(M,t) \quad (18.144)$$

so that, by (18.136),

$$\tilde{w}_n(M,t) - u(M,t) < u_n(M,t) - u(M,t) - \alpha\delta \quad (18.145)$$

By construction, there exists $N > 0$ so large that for all $n \geq N$,

$$u_n(M,t) - u(M,t) < \varepsilon_n < \alpha\delta \qquad (18.146)$$

and

$$\tilde{w}_n(M,t) \leq u_n(M,t) \qquad (18.147)$$

Hence (18.145) yields

$$\tilde{w}_n(M,t) - u(M,t) < 0 \qquad (18.148)$$

which is impossible. Hence the assumption (18.125) implies a contradiction, which proves that

$$w(p,\tau) = u(p,\tau) \quad \forall (p,\tau) \in D^t \qquad (18.149)$$
$$\textbf{Q.E.D.}$$

As in Perron's treatment of the Dirichlet problem for the Laplace equation, we must consider the relationship between $u(p,t) = \inf \mathfrak{M}^+$ and the Dirichlet problem under consideration. We must therefore define regularity of a boundary point.

Definition 18.3.3. A point $(p,t) \in \Gamma^t$ is said to be *regular* if the Perron–Petrovsky function $u(q,\tau) = \inf \mathfrak{M}^+$ is continuous in a neighborhood $U \subset \bar{D}^t$ of (p,t) and if

$$\lim_{(q,\tau) \to (p,t)} u(q,\tau) = f(p,t) \quad (q,\tau) \in \Gamma \subset D^t \qquad (18.150)$$

for any continuous boundary function $f(p,t)$ and Γ–any smooth curve non-tangential to Γ^t.

Let $P = (p,t)$, and $Q = (q,\tau)$.

Theorem 18.3.2 (Perron–Petrovsky; global barriers). *A boundary point (p,t) is regular if and only if there exists a superparabolic function $\Omega_P(Q)$, called a* global barrier, *possessing the following properties:*

$$\Omega_P(Q) \in C(\bar{D}^t) \qquad (18.151)$$
$$\Omega_P(P) = 0 \qquad (18.152)$$
$$\Omega_P(Q) > 0 \quad \forall Q \in \Gamma^t \setminus P \qquad (18.153)$$

Remark. Equality (18.150) may hold for certain functions $f(Q)$ when the boundary point P is not regular. For example, if $f(Q) = 0$ for all $Q \in \Gamma^t$ then the Perron–Petrovsky function $\inf \mathfrak{M}^+$ vanishes identically, so that the equality (18.150) is valid regardless of the geometrical structure of D^t.

As emphasized in Section 4 of Chapter 11, the property of a boundary point's regularity does not depend on the properties of the whole boundary Γ of the region; rather, it is a local property. The same is true in the parabolic case, as the following theorem shows.

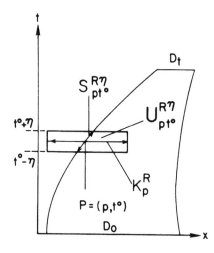

Figure 18.3.   Sketch of the Perron–Petrovsky theorem.

Theorem 18.3.3. *Let* $P = (p, t^o) \in \Gamma^t$, *and let* $U_M^{R\eta}$ *be a relative neighborhood of* P^o:

$$U_{pt^o}^{R\eta} = K_p^R \times (t^o - \eta < \tau < t^o + \eta) \quad (K_p^R \subset \mathbb{R}_3) \tag{18.154}$$

Let

$$S_{pt^o}^{R\eta} = U_{pt^o}^{R\eta} \cap \Sigma^t \tag{18.155}$$

be the part of the parabolic boundary Γ^t *cut out by* $U_{pt^o}^{R\eta}$ *(see Figure 18.3). Then a point* P *is regular if and only if there exists a local barrier* $\omega_P(Q)$, *that is, the function possessing the following properties.*

(a) $\omega_P(Q)$ *is continuous in the closure* $\bar{U}_{pt^o}^{R\eta}$ *of* $U_{pt^o}^{R\eta}$ *and is superparabolic in its interior.*

(b) *There exists* $k = \text{const} > 0$ *such that*

$$\omega_P(Q) \geq k \quad \forall Q \in U_P^{R\eta} \quad (Q = q, \tau) \tag{18.156}$$

(c)

$$\omega_P(P) = 0, \quad \omega_P(Q) > o \forall Q \in S_P^{R\eta} \setminus (P) \tag{18.157}$$

The proof of Theorems 18.3.2–3 coincides word-for-word with that of Perron's Theorems 11.4.1–2 (naturally, replacing the terms "superharmonic" and "harmonic"),[8] and we shall therefore not reproduce them.

The regularity of a boundary point is a local geometrical property. In noncylindrical regions $D^t \subset \mathcal{E}_{n+1}$ it depends on the structure of any of its sections D_τ by the planes $\tau = \text{const}$, as well on its variation in time. Very profound, (almost necessary and sufficient) conditions for a boundary point

---

[8]This also applies to problems considered in $\mathbb{R}_1$.

to be regular in the 1-dimensional case were established by Petrovsky in his fundamental work [55]. We shall formulate Petrovsky's theorem without proving it. Thus, let D^t be a 1-dimensional parabolic region;

$$D^t = \{x, t \colon \varphi_1(t) < x < \varphi_2(t), \ t > 0\} \tag{18.158}$$

and consider the Dirichlet problem

$$\frac{\partial^2}{\partial x^2} u(x,t) = \frac{\partial}{\partial t} u(x,t) \quad \forall (x,t) \in D^t \tag{18.159}$$

$$u[\varphi_i(t), t] = f_i(t), \quad i = 1, 2, \ t \geq 0 \tag{18.160}$$

$$u(x, 0) = \psi(x) \quad \forall x \in [\varphi_1(0), \varphi_2(0)] \tag{18.161}$$

Theorem 18.3.4 (Petrovsky's lateral boundary regularity).

(a) *Any point of the base $t = 0$ of the parabolic region D^t is regular.*[9]
(b) *A boundary point $[t, \varphi_1(t)]([t, \varphi_2(t)])$ is regular if there exists a continuous positive function $\rho_1(h)$ $(\rho_2(h))$ defined for $h < 0$ and tending to zero monotonically as $h \uparrow 0$, such that for any negative h of sufficiently small modulus,*

$$\varphi_1(t + h) - \varphi_1(t) \geq -2\sqrt{|h \ln[\rho_1(h)|},$$
$$\varphi_2(t + h) - \varphi_2(t) \leq 2\sqrt{|h \ln[\rho_2(h)]|} \tag{18.162}$$

and also

$$\lim_{\varepsilon \uparrow 0} \int_c^\varepsilon \frac{\rho_1(h)}{h} \sqrt{|h \ln[\rho_1(h)]|} dh = -\infty$$
$$\left(\lim_{\varepsilon \uparrow 0} \int_c^\varepsilon \frac{\rho_2(h)}{h} \sqrt{|h \ln[\rho_2(h)]|} dh = -\infty \right) \tag{18.163}$$

where c is some suitably chosen negative constant.

On the other hand, the boundary point is irregular if there is a function $\rho_1(h)$ $(\rho_2(h))$ such that inequality (18.162) holds and simultaneously the integral (18.163) is convergent:

$$\left| \lim_{\varepsilon \uparrow 0} \int_c^\varepsilon \frac{\rho_1(h)}{h} \sqrt{|h \ln[\rho_1(h)]|} dh \right| < \infty$$
$$\left(\left| \lim_{\varepsilon \uparrow 0} \int_c^\varepsilon \frac{\rho_2(h)}{h} \sqrt{|h \ln[\rho_2(h)]|} dh \right| < \infty \right) \tag{18.164}$$

Corollary. *Let $x = \varphi(t)$ be one (left or right) of the boundaries of the parabolic region D^t. Then*

(a) *A boundary point $[\varphi(t_0), t_0]$ is regular if*

$$|\varphi(t) - \varphi(t_0)|^2 \leq 4(t_0 - t) \ln \left(\frac{1}{t_0 - t} \right) \tag{18.165}$$

---

[9]See Lemma 18.4.1.

in a left semineighborhood of t_0.

(b) A boundary point $[\varphi(t_0), t_0]$ is irregular if

$$|\varphi(t) - \varphi(t_0)|^2 \geq (4 + \varepsilon)(t_0 - t) \ln\left(\frac{1}{t_0 - t}\right) \qquad (18.166)$$

in a left semineighborhood of t_0.[10]

4. Case of cylindrical region. Tichonov's theorem. Duhamel test

Recall that a parabolic region is said to be *fundamental* if the Dirichlet problem for the heat-conduction equation is solvable within the region, whatever the boundary function $f(p)$. Recall, moreover, that by Tichonov's theorem (Theorem 15.4.1*) any cylindrical region fundamental with respect to the Dirichlet problem for the heat-conduction equation is fundamental for the electrostatic Dirichlet problem. Our purpose is now to prove the converse. However, we first introduce a slightly weakened definition of fundamental parabolic region.

Definition 18.4.1. Let D^t be a parabolic region in $\mathcal{E}_{3+1}$ whose lower base D_0 is a simply connected region satisfying the conditions of Theorem 14.2.3. Consider the parabolic Dirichlet problem: Find $u(p,t)$ such that

(a) $u(p,t)$ is uniformly bounded in D^t;
(b) $u(p,t)$ is continuous in the closure $\bar{D}_\eta^t$ of D_η^t $\forall t \geq \eta > 0$;
(c) $u(p,t)$ is continuous within any region $G^t \subset \bar{G}^t \subset D^t$ $\forall t \geq 0$;
(d) $Lu \stackrel{\text{def}}{=} \Delta u(p,t) - \partial u(p,t)/\partial t = 0$ $\forall (p,t) \in D^t$, $\forall t > 0$;
(e) $\forall t_o \in (0, l]$ and $V(p_o, l_o) \in \Sigma^t, u(p_o, l_o) = f(p_o, t_o)$; and
(f) $u(p, 0) = \varphi(p)$ $\forall p \in D_0$.

Then D^t is fundamental if a solution $u(p,t)$ exists for any continuous $f(p,t)$ and $\varphi(p)$.

Remark 18.4.1. Without loss of generality, we may assume that the boundary function $f(p,t)$ is continuously differentiable with respect to t. Indeed, by the Weierstrass approximation theorem [13], any function $f(p,t)$ continuous in the segment $[0,t]$ may be uniformly approximated by a sequence $\{P_n(p,t)\}$ of polynomials. Hence, by the Pini–Hadamard theorem, if $u_n(p,t)$ is the solution of the problem (a)–(f) with f replaced by P_n, then the equality

$$\lim_{n \uparrow \infty} P_n(p, \tau) = f(p, \tau) \qquad \forall (p, \tau) \in D^t \qquad (18.167)$$

---

[10] A similar assertion is true in the multidimensional case. A detailed formulation may be found in Petrovsky's original paper [55].

implies that, for all $\eta > 0$ and $G^t \subset \bar{G}^t \subset D^t$,

$$\lim_{n \uparrow \infty} u_n(p, t) = u(p, \tau) \quad \forall (p, t) \in \bar{D}^t_\eta \cap \bar{G} \qquad (18.168)$$

Q.E.D.

Remark 18.4.2. It is obvious that Theorem 18.3.1. (Perron–Petrovsky) is applicable to the parabolic Dirichlet problem (a)–(f) provided the definition of the classes $\mathfrak{M}^+$ and $\mathfrak{M}^-$ are suitably altered (i.e., by adding the requirement that any function in one of the classes is uniformly bounded and continuous in the sense of (b)–(c)). In what follows we shall use this remark without special reference.

We can now formulate and prove Tichnov's theorem.

Theorem 18.4.1. *Any bounded region D that is fundamental with respect to the electrostatic Dirichlet problem is also fundamental with respect to the Dirichlet problem for the heat-conduction equation in a cylindrical region $D^t = D \times (0 < t < \infty)$.*

PROOF. The assertion is an obvious corollary of the following lemmas.

Lemma 18.4.1. *Any point p of the lower base D of D^t is regular.*

Lemma 18.4.2. *If $p \in \Sigma = \bar{D} \setminus D$ and $\Omega_p(q)$ is a barrier at p for the electrostatic Dirichlet problem, then $\Omega_P(Q)$, $Q = (q, \tau)$, is the barrier at $P = (p, t) \in \Sigma^t_\eta \; \forall \eta \in (0, t]$ for the parabolic Dirichlet problem $(a)-(f)$ in D^t.*

PROOF OF LEMMA 18.4.1. Let

$$p = (x_1, x_2, x_3), \quad P = (p, 0), \quad q = (\xi_1, \xi_2, \xi_3), \quad Q = (q, \tau)$$

$$\Omega_P(Q) = \sum_{i=1}^{3} (\xi_i - x_i)^2 + 6\tau \qquad (18.169)$$

Then

$$\Omega_P(Q) > 0 \quad \forall Q \in \bar{D}^t \setminus (P), \Omega_P(P) = 0$$

$$L\Omega_P(Q) \overset{\text{def}}{=} \left(\Delta - \frac{\partial}{\partial \tau} \right) \Omega_P(Q) = 0 \qquad (18.170)$$

so that $\Omega_P(Q)$ is a barrier at the point P.   **Q.E.D.**

PROOF OF LEMMA 18.4.2. Lemma 18.4.2 is in turn a corollary of the following two lemmas.

Lemma 18.4.3. *If D is the region fundamental for the electrostatic Dirichlet problem then the parabolic Dirichlet problem (a)–(f) is solvable for any boundary function f independent of t and for an initial function $\varphi(p) \equiv 0$.[11]*

Lemma 18.4.4 (Duhamel's principle). *Let*

$$P = (p, t), \quad f(P) \in C(\Sigma_0^t), \quad \frac{\partial}{\partial t} f(P) \in C(\Sigma_0^t), \quad f(p, 0) = 0 \qquad (18.171)$$

Let $w(p, t \mid \lambda)$ be a solution of the problem

$$Lw = 0 \quad \forall p \in D, \ \forall t \geq 0$$

$$w = 0 \quad \forall p \in D, \ t = 0, \qquad w = \frac{\partial}{\partial \lambda} f(p, \lambda) \quad \forall p \in \Sigma_0^t, \ t = 0 \qquad (18.172)$$

where λ is considered as a numerical parameter. Then

$$u(p, t) = \int_0^t w(p, t - \lambda) d\lambda \qquad (18.173)$$

is a solution of the problem (a)–(f) satisfying the requirements of Lemma 18.4.3.

PROOF. Assume that Lemma 18.4.3 has been proved. Then $w(p, t \mid \lambda)$ exists. We have

$$Lu(p, t) = \left(\Delta - \frac{\partial}{\partial t} \right) \int_0^t w(p, t - \lambda) d\lambda$$

$$\equiv -w(p, 0) + \int_0^t Lw(p, t - \lambda) d\lambda = 0 \qquad (18.174)$$

Note, further, that the definitions (18.172) and (18.173) and the assumption $f(p, 0) = 0$ imply

$$u(p, t)\big|_{p \in \Sigma} = \int_0^t w(p, t - \lambda)\big|_{p \in \Sigma} d\lambda = \int_0^t \frac{\partial}{\partial t} f(p, t - \lambda) d\lambda$$

$$= -\int_0^t \frac{\partial}{\partial \lambda} f(p, t - \lambda) d\lambda = -f(p, 0) + f(p, t) = f(p, t) \quad (18.175)$$

Thus the Duhamel principle is proved, provided that Lemma 18.4.3 is valid.

PROOF OF LEMMA 18.4.3. Consider a monotonic sequence of regions

$$D_1 \subset \bar{D}_1 \subset D_2 \subset \bar{D}_2 \subset \cdots \subset D_n \subset \bar{D}_n \subset \cdots \subset D \qquad (18.176)$$

such that

$$\lim_{n \uparrow \infty} D_n = D \qquad (18.177)$$

---

[11] The assumption $u(p, 0) = 0$ is not restrictive. Indeed, if $u(p, 0) = \varphi(p) \neq 0$, one can set $u(p, t) = u_1(p, t) + \Pi(p, t \mid \varphi)$ and $u_1(p, t) = f(p, t) - \Pi(p, t \mid \varphi)$ on the boundary Σ for all $t > 0$, where Π is the Poisson integral.

Let

$$\delta_n = D \setminus D_n, \quad \mu_n = \text{mes } \delta_n, \quad n = 1, 2, \ldots \tag{18.178}$$

so that by (18.177),

$$\lim_{n \uparrow \infty} \mu_n = 0 \tag{18.179}$$

Now define a sequence $\varphi_n(p) \in C(\bar{D})$ of functions continuous in $\bar{D}$ and such that

$$\varphi_n(p) = \begin{cases} f(p) & \forall p \in \Sigma \\ \varphi(p) & \forall p \in D_n \end{cases}, \quad \max[\max |\varphi_n|, \max |f|] = M \tag{18.180}$$

and a sequence $\{u_n(p,t)\}$, $n = 1, 2, \ldots$, of functions continuous in $\bar{D}^t = \bar{D} \times [0, t]$ for all $t > 0$ such that:

$$Lu_n = \Delta u_n - \frac{\partial}{\partial t} u_n = 0 \quad \forall (p, t) \in D^t \tag{18.181}$$

$$u_n = f(p) \quad \forall (p, t) \in \Sigma, u_n = \varphi_n \quad \forall p \in D, \ t = 0$$

Clearly $\{u_n\}$ is uniformly convergent in the region

$$\bar{D}_{q\rho} = \bar{D}_\eta^t \cup D^{\rho t} \tag{18.182}$$

where

$$\bar{D}_\eta^t = D \times (\eta < \tau \le t), \quad D^{\rho t} = D^\rho \times (0 < \tau \le t), \quad D^\rho \subset \bar{D}^\rho \subset D \tag{18.183}$$

Indeed, let

$$w_{mn} = u_n - u_{m+n} \quad \forall (m, n) > 0 \tag{18.184}$$

We have

$$\begin{aligned} Lw_{mn} &= 0 & \forall (p, t) \in D^t \\ w_{mn} &= 0 & \forall p \in \Sigma, \ \forall t > 0 \\ w_{mn} &= 0 & \forall p \in D_n, \ t = 0 \\ |w_{mn}| &\le 2M & \forall p \in \delta_n, \ t = 0 \end{aligned} \tag{18.185}$$

Now let

$$v_n = 2M \int_{\delta_n} E(p, q, t) d\omega_q = M \int_{\delta_n} \frac{\exp\left[-r_{pq}^2/4t\right]}{(4\pi t)^{3/2}} d\omega_q \tag{18.186}$$

The function v_n is defined everywhere in $\mathcal{E}_{3+1}$ and in particular in $\bar{D}^t$. Using the properties of the Poisson integral, we find that

$$\begin{aligned} Lv_n &= 0 & \forall (p, t) \in D^t \\ v_n &= 0 & \forall p \in D \setminus \delta_n \ t = +0 \\ v_n &= 2M & \forall p \in \delta_n, \ t = +0 \\ v_n &> 0 & \forall p \in \delta_n, \ t > 0 \end{aligned} \tag{18.187}$$

By the generalized maximum principle (Theorem 13.2.3), it follows from
(18.185) and (18.188) that

$$-v_n < w_{nm} < v_n \quad \forall (p, \tau) \in \bar{D}_\eta^t \cup \bar{D}^{\rho t} \tag{18.188}$$

Note that by (18.178), (18.179), and (18.186), for all $\varepsilon > 0$ there exists
$N > 0$ so large that

$$-\varepsilon < v_n < \varepsilon \quad \forall (p, \tau) \in \bar{D}_\eta^t \cup \bar{D}^{\rho t} \quad \forall n > N \tag{18.189}$$

Indeed, using the mean value theorem we obtain from (18.187) that

$$0 < v_n < 2ME(p, \tilde{q}, t)\mu_n \quad \forall n > N \tag{18.190}$$

where $\tilde{q} \in \delta_n$. Hence, since $E(p, \tilde{q}, t)$ is bounded in $\bar{D}_\eta^t \cup \bar{D}^{\rho t}$ if $\bar{D}^{\rho 0} \in D_n$, it
follows that (18.189) is valid. A comparison of (18.189) and (18.190) shows
that

$$|w_{nm}| < \varepsilon \quad \forall (p, \tau) \in \bar{D}_\eta^t \cup \bar{D}^{\rho t} \tag{18.191}$$

which proves that the sequence $\{u_n(p, \tau)\}$ is uniformly convergent in
$\bar{D}_\eta^t \cup \bar{D}^{\rho t}$.

Now let

$$u(p, t) = \lim_{n \uparrow \infty} u_n(p, t) \tag{18.192}$$

Since $|u_n(p, t)| < M$ for every $n > N$, (18.192) means that

$$|u(p, t)| \leq M \tag{18.193}$$

Hence, by Theorem 18.1.1, u is a bounded solution of the problem (a)–(f),
which proves Lemma 18.4.3 and thereby also Theorem 18.4.4.   **Q.E.D.**

5. Application of Schwarz alternating method to solution of Dirichlet problem for heat-conduction equation in noncylindrical region

As mentioned in Chapter 11, Section 5, one must distinguish between two
kinds of existence theorems: "pure" existence theorems and "constructive"
existence theorems.[12] Pure theorems establish the mere fact that a solution

---

[12](a) To say that an existence theorem is "constructive" means that the proof
outlines an explicit algorithm that produces the solution, but no more; it by
no means implies that the algorithm is also convenient as a practical tool for
constructing the solution. For example, finite-difference approximation is usually
far more suitable for practical work than numerical solution of a Volterra integral
equation, although very frequently (particularly in the numerical solution of
nonlinear problems) finite-difference methods are used without proving that the
approximations in fact converge to a solution. For a suitable numerical procedure,
the existence of a solution to an integral equation and the convergence to it are
guaranteed.

(b) All existence theorems based on application of the Shauder fixed-point theorem
or Tichonov's generalization of it [80] are pure theorems, since they do not refer
to any explicitly described, constructive algorithm (see the reference in Section 5
of Chapter 11 to Brouwer's degree of a map).

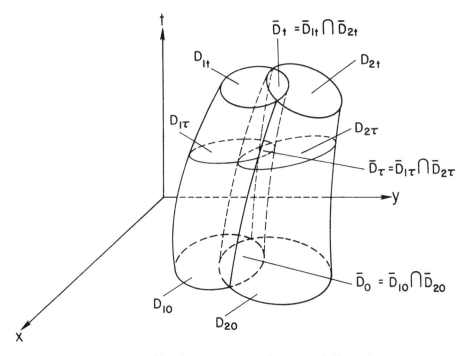

Figure 18.4.   Sketch of the union of two parabolic regions.

exists, without explicitly indicating any algorithm by which the solution
might be constructed in a finite or denumerable number of steps.[13]   In
this sense, the basic Perron and Perron–Petrovsky theorems are pure. On
the other hand, the Schwarz alternating method for the solution of the
electrostatic Dirichlet problem belongs to the class of constructive existence
theorems.   As we have seen, it yields an existence theorem of the same
generality as the pure Perron theorem.   Our intention is now to prove
that the Schwarz alternating iterations may be used to solve the parabolic
Dirichlet problem in a noncylindrical region. In order to avoid the difficulties
that arise in description of intersections of 3-dimensional regions varying
in time, we shall restrict our presentation to the 2-dimensional thermal
Dirichlet problem.

Given two bounded regions $D_i^t \in \mathcal{E}_{2+1}$, $i = 1, 2$, fundamental with respect
to the thermal Dirichlet problem, let D^t be their union

$$D^t = D_1^t \cup D_2^t \qquad (18.194)$$

(see Figure 18.4) Let $\Sigma_{i\tau}$, $i = 1, 2$, $0 \leq \tau \leq t$, be curves that intersect at

---

[13] Recall that we are counting the computation of any regular or singular integral as
an operation requiring the performance of one step only.

points a_τ and b_τ that divide $\Sigma_{i\tau}$ into two parts:

$$\bar{\Sigma}_{i\tau} = \bar{\Sigma}^1_{i\tau} \cup \bar{\Sigma}^2_{i\tau}, \quad i = 1, 2 \tag{18.195}$$

such that

$$\bar{\Sigma}^1_{1\tau} \cap D_{2\tau} = \emptyset, \Sigma^2_{1\tau} \subset D_{2\tau} \tag{18.196}$$

$$\Sigma^1_{2\tau} \subset D_{1\tau}, \Sigma^2_{2\tau} \cap D_{1\tau} = \emptyset \tag{18.197}$$

Let

$$D^{1t}_1 = D^t_1 \setminus \bar{D}^t_2, D^{2t}_2 = D^t_2 \setminus \bar{D}^t_1 \tag{18.198}$$

so that the boundaries of D^{1t}_1 and D^{2t}_2 are

$$\partial D^{1t}_1 = \Sigma^{1t}_1 \cup \Sigma^{2t}_1, \quad \partial D^{2t}_2 = \Sigma^{1t}_2 \cup \Sigma^{2t}_2 \tag{18.199}$$

Consider the Dirichlet problem: Find a function $u(p, t)$, continuous in $\bar{D}^t$ for all $t \geq 0$, such that

$$Lu = \Delta u - \frac{\partial}{\partial t} u = 0 \quad \forall (p, t) \in D^t \tag{18.200}$$

$$u(p, t) = F(p, t) \quad \forall (p, t) \in \Gamma^t, \ \forall t > 0 \tag{18.201}$$

where

$$F(p, t) \overset{\text{def}}{=} \begin{cases} f(p, t) & \forall (p, t) \in \Sigma^t \\ 0 & \forall p \in D, \ t = +0 \end{cases} \tag{18.202}$$

In what follows we assume that the curves $\Sigma_{1\tau}$ and $\Sigma_{2\tau}$ do not touch one another at the points a_τ and b_τ, $0 \leq \tau \leq t$, $t \geq 0$, but intersect there at angles $\omega_{a\tau}$ and $\omega_{b\tau}$; we assume also that the surfaces Σ^t_i, $i = 1, 2$, satisfy the conditions of Theorems 14.2.4 and 16.2.1. Under these conditions the Schwarz alternating process appears to be applicable. We recall its definition (see Chapter 11, Section 5). Let $u_{11}(p, t)$ be a solution of the problem

$$Lu_{11} = 0 \quad \forall (p, t) \in D^{1t}, \ \forall t > 0 \tag{18.203}$$

$$u_{11}(p, t) = \Phi_1(p, t) \quad \forall (p, t) \in \Sigma^{1t}_1 \tag{18.204}$$

where

$$\Phi_1(p, t) = \begin{cases} f(p, t) & \forall (p, t) \in \Sigma^{1t}_1 \\ \psi(p, t) & \forall (p, t) \in \Sigma^{2t}_1 \\ 0 & \forall p \in D^1_{10}, \ t = 0 \end{cases} \tag{18.205}$$

where $\psi(p, t)$ is an arbitrary continuous function equal to $f(p, \tau)$ at a_τ and b_τ.

Now define two families of functions $\{u_{2m}(p, t)\}\big|^\infty_{m=1}$ and $\{u_{2m+1}(p, t)\}\big|^\infty_{m=0}$ inductively, by the conditions

$$u_{2m} \in C(\bar{D}^t_2), \quad Lu_{2m} = 0 \quad \forall (p, t) \in D^t_2, u_{2m} = \Phi_{2m} \quad \forall (p, t) \in \Gamma^t_2 \tag{18.206}$$

where

$$\Phi_{2m}(p,t) = \begin{cases} f & \forall(p,t) \in \Sigma_2^{2t} \\ u_{2m-1} & \forall(p,t) \in \Sigma_2^{1t} \\ 0 & \forall p \in D_{10}^1, \ t = 0 \end{cases} \tag{18.207}$$

and

$$u_{2m+1} \in C(\bar{D}_1^t), \quad L u_{2m+1} = 0 \quad \forall(p,t) \in D_1^t, u_{2m+1} = \Phi_{2m+1} \quad \forall(p,t) \in \Gamma_1^t \tag{18.208}$$

and where

$$\Phi_{2m+1}(p,t) = \begin{cases} f & \forall(p,t) \in \Sigma_1^{1t} \\ u_{2m} & \forall(p,t) \in \Sigma_1^{2t} \\ 0 & \forall p \in D_{10}, \ t = 0 \end{cases} \tag{18.209}$$

Since $u_{11}(p,t)$ is defined by (18.203)–(18.205) and both regions D_1^{it}, D_2^{it}, $i = 1, 2$, are assumed to be fundamental with respect to the parabolic Dirichlet problem, all the functions u_{2m} and u_{2m-1} may be considered known. We can now establish a parabolic analog of Schwarz's theorem.

Theorem 18.5.1. *The sequences $\{u_{2m}(p,t)\}\big|_{m=1}^{\infty}$ and $\{u_{2m+1}(p,t)\}\big|_{m=0}^{\infty}$ converge uniformly in regions $\bar{D}_2^t$ and $\bar{D}_1^t$, respectively, to functions parabolic inside these regions. Their limits coincide at the intersection of D_1^t and D_2^t:*

$$\lim_{m\uparrow\infty} u_{2m} = \lim_{m\uparrow\infty} u_{2m+1} \quad \forall(P,t) \in D_1^t \cap D_2^t \tag{18.210}$$

so that

$$u(p,t) = \begin{cases} \lim_{m\uparrow\infty} u_{2m+1} & \forall(p,t) \in D_1^t \\ \lim_{m\uparrow\infty} u_{2m} & \forall(p,t) \in D_2^t \end{cases} \tag{18.211}$$

is a solution of problem (18.200)–(18.202).

PROOF. Recall that the proof that the Schwarz alternating iterations converge to a solution of the electrostatic Dirichlet problem (given in Chapter 11, Section 5), is based on the generalized maximum principle (Lemma 11.5.1) and on the validity of Schwarz's Lemma 11.5.2. These are the only elements in the proof that refer to specific features of the electrostatic Dirichlet problem. In other words, to prove Theorem 18.5.1 all we need are parabolic analogs of these lemmas. We already have a generalized maximum principle for the Fourier operator (see Theorems 14.2.3–4). Hence it remains only to formulate and prove a parabolic analog of Schwarz's lemma.

Lemma 18.5.1. *Let D^t be a bounded parabolic region satisfying the conditions of Theorem 16.2.1, and let*

$$S^t = \bigcup_{\tau=0}^{t} S_\tau \tag{18.212}$$

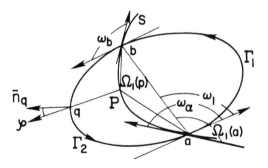

Figure 18.5.   Sketch of Lemma 18.5.2.

be a smooth surface dividing D^t into two parts D_1^t and D_2^t such that the
curves S_τ and $\Sigma_{i\tau}$ intersect at points a_τ and b_τ without touching, for all
$\tau \in [0,t]$ and all $t \geq 0$, at angles $\omega_{ia\tau}$ and $\omega_{ib\tau}$, $i = 1, 2$, corresponding to the
positive direction of the circuit. Furthermore, let

$$L_a^t = \bigcup_{\tau=0}^{t} (a_\tau), \quad L_b^t = \bigcup_{\tau=0}^{t} (a_\tau) \qquad (18.213)$$

be the curves of intersection of S^t and Σ^t, and assume that they satisfy the
assumptions of Theorem 14.2.4. Finally, let $u(p,t)$ be a function parabolic
within D^t, continuous in $\bar{D}_1^t$ and in $\bar{D}_2^t$, and such that

$$u(p,t) = \begin{cases} 0 & \forall (p,t) \in \Sigma_2^t \\ 1 & \forall (p,t) \in \Sigma_1^t \\ 0 & \forall p \in D_0, \ t = 0 \end{cases} \qquad (18.214)$$

Then there exists $q = \mathrm{const}$, $0 < q < 1$, such that for any $M = (m, \tau) \in L_a^t \cup L_b^t$

$$0 \leq \lim_{S^t \ni P \to M} u(P) \leq q \quad [P = (p, \tau)] \qquad (18.215)$$

PROOF. The lemma is in turn a corollary of the following generalization
of Theorem 16.2.2 on the discontinuities of the heat potential of a double
layer.

Lemma 18.5.2 (Discontinuities of the double-layer electrostatic and heat
potentials with piecewise continuous densities).
    (A) Let $D \subset \mathbb{R}_2$ be a convex region bounded by a smooth curve Γ of
finite curvature, which is divided by a smooth arc S into two parts D_1 and
D_2 intersecting Γ at points a and b without touching, at angles ω_{ia} and ω_{ib},
respectively. Let $f(p)$ be a function continuous on Γ_i, having at a one-sided
limits

$$\lim_{\Gamma_i \ni p \to a} f(p) = \alpha_i, \quad i = 1, 2 \qquad (18.216)$$

(see Figure 18.5). Finally, let $W(p \mid f)$ be the electrostatic potential of a *double layer*

$$W(p \mid f) = \int_\Gamma f(q) \frac{\cos \varphi}{2\pi r} ds_q \qquad (18.217)$$

Then

$$\lim_{S \ni p \to a} W(p \mid f) = \frac{\alpha_1 (\pi - \omega_{1a}) + \alpha_2 \omega_{2a}}{2\pi} + W(a \mid f) \qquad (18.218)$$

(B) *Let* D^t, D^t_1, D^t_2, *the boundaries* Σ^t_i, $i = 1, 2$, *the surface* S^t, *and the angles* $\omega_{a\tau}, \omega_{b\tau}$ *be defined in agreement with the conditions of Lemma 18.5.1. Let* $W[p, t \mid f(q, \tau)]$ *be the parabolic potential of a double layer*

$$W[p, t \mid f(q, \tau)] = \int_0^t d\tau \int_{\Sigma_\tau} f(q, \tau) \frac{\partial}{\partial n_q} E(p, q, t - \tau) d\sigma_q \qquad (18.219)$$

with density $f(q, \tau)$ *continuous in* Σ^t_i, $i = 1, 2$, *which has one-sided limits* $\alpha_i(\tau)$ *along the curve* L^t_a:

$$\lim_{\Sigma^t_i \ni P \to a(t)} f(p, t) = \alpha_i(t), \quad i = 1, 2 \quad [P = (p, t)] \qquad (18.220)$$

Then

$$\lim_{S^t \ni P \to a(t)} W[p, t \mid f(q, \tau)] = \frac{\alpha_1(t)[\pi - \omega_{1a}(t)] + \alpha_2(t) \cdot \omega_{2a}(t)}{2\pi}$$

$$+ W[a(t), t \mid f(q, \tau)] \qquad (18.221)$$

PROOF. (A) Let ω_i, $i = 1, 2$, be the angle subtended by the arc Γ_i from the point a. We have

$$W = W_1 + W_2 \qquad (18.222)$$

where

$$W_i \overset{\text{def}}{=} W_i(p \mid f) = \int_{\Gamma_i} f(q) \frac{\cos \varphi}{2\pi r_{pq}} ds_q \qquad (18.223)$$

If $p \in S \cap D$ (see Chapter 12, Section 7), then

$$W_i(p \mid 1) = \frac{\Omega_i(p)}{2\pi} \qquad (18.224)$$

where $\Omega_i(p)$ is the angle subtended by Γ_i from the interior point p of the arc $S \cap D$. It follows from (18.224) and from the continuity of f on the arc S that

$$\lim_{S \ni p \to a} W_i(p \mid f) = \frac{\Omega_i(a) - \omega_i}{2\pi} \cdot \alpha_i + W_i(a \mid f), \quad i = 1, 2 \qquad (18.225)$$

Using this and the equalities

$$\omega_a + \Omega_1 - \omega_1 = \pi, \quad \Omega_1 + \Omega_2 = 2\pi, \quad \omega_1 + \omega_2 = \pi \qquad (18.226)$$

(see Figure 18.5) we find that (18.222) and (18.225) imply (18.218).

(B) Again, reducing the theorem on discontinuities of the heat potential of a double layer to the theorem for an electrostatic potential of the double layer (see the proof of Theorem 16.2.1), we find that (A) implies (B), which completes the proof of Lemma 18.5.2. **Q.E.D.**

Remark. If simultaneously

$$S_t \ni (p,t) \to a(t), \quad t \downarrow 0 \tag{18.227}$$

then the right-hand side of (18.218) becomes indeterminate. It would be equal to

$$\lim_{S^t \ni P \to a(t), t \downarrow 0} W[p,t \mid f(q,\tau)] = \lim_{t \downarrow 0} \frac{\alpha^1 \alpha_1(t)\omega_{1a}(t) + \alpha^2 \alpha_2(t)\omega_{2a}(t)}{2\pi} \tag{18.228}$$

where α^1 and α^2 depend on $\lim_{t \downarrow 0}(r_{pa}/2\sqrt{t})$, but a posteriori[14]

$$0 \leq \alpha^1, \quad \alpha^2 \leq 1 \tag{18.229}$$

We can now prove the theorem. Indeed, let the contour Σ_τ be smooth for any $\tau \in [0,t]$, and any $t \geq 0$. Take

$$v(p,t) = W[p,t \mid f(q,\tau)] \tag{18.230}$$

with

$$f(q,\tau) = \begin{cases} 2 & \text{if } (q,\tau) \in \Sigma_{2\tau} \\ 0 & \text{if } (q,\tau) \in \Sigma_{1\tau} \end{cases} \quad \forall \tau \in [0,t] \tag{18.231}$$

Then, by Lemma 18.5.2,

$$\lim_{D^t \ni p \to M = (m,t) \in \Sigma_{1\tau}} v(p,t) = W[m,t \mid f(q,\tau)] \tag{18.232}$$

$$\lim_{D^t \ni p \to M = (m,t) \in \Sigma_{2\tau}} v(p,t) = 1 + W[M \mid f(q,\tau)] \tag{18.233}$$

if $l > 0$. Here $W[M \mid f(q,\tau)]$ is continuous on Σ^t. Hence $v(p,l)$ has the same jump at points of $L_a^t \cup L_b^t$ as $u(p,t)$. This means that

$$u(p,t) = v(p,t) + v_1(p,t) \tag{18.234}$$

where $v_1(p,t)$ is continuous in $\bar{D}^t \setminus (L_a^t \cup L_b^t)$, is uniformly bounded in D, has no jumps on $L_a^t \cap L_b^t$, and is parabolic within D^t. Hence, by the uniqueness theorem (see Theorem 14.2.4), $v_1(p,t)$ is continuous in $\bar{D}^t$, so that

$$\lim_{S^t \ni P \to a(t)} u(p,t) = \lim_{S^t \ni P \to a(t)} v(p,t) + v_1[a(t),t]$$

$$= \frac{\omega_2(t)}{\pi} + W[a(t),t \mid f(q,\tau)] + v_1[a(t),t] \tag{18.235}$$

But

$$1 = \lim_{D^t \ni p \to m \Sigma_{2t}} u(p,t) = \lim_{D^t \ni p \to m \in \Sigma_{2t}} v(p,t) + v_1(M_1,t)$$

$$= 1 + W[M \mid f(q,\tau)] + v_1(M,t) \tag{18.236}$$

---

[14]If $p \to a(t)$ and simultaneously $t \to 0$ then $\max(\omega_2/\pi)$ must be replaced by $\alpha \cdot \max(\omega_2)/\pi$, where $0 \leq \alpha \leq 1$, for the conclusion to remain in force.

Hence

$$W[M \mid f(q,\tau)] + v_1(M,t) = 0 \quad \forall M \in \Sigma_{2t} \tag{18.237}$$

By the continuity of $W[M \mid f(q,\tau)]$ and $v_1(M,t)$, this yields

$$W[a(t),t \mid f(q,\tau)] + v_1[a(t),t] = 0 \tag{18.238}$$

which implies

$$\lim_{S^t \ni P \to a(t)} u(p,t) = \frac{\omega_2(t)}{\pi} \tag{18.239}$$

Since by assumption $0 \leq \omega_2(t) < \pi$ for all $t \geq 0$, this means that

$$0 \leq q = \max \frac{\omega_2(t)}{\pi} < 1 \tag{18.240}$$

proving Lemma 18.7.2 and together with it Theorem 18.5.1.   **Q.E.D.**[15]

Problems

P18.1.1. Why it is essential for the validity of the Theorem 18.5.2 that it refer to the usual Euclidean topology instead of the parabolic topology?

P18.1.2. Prove the Pini–Hadamard inequality, referring to the Heine–Borel lemma for a noncylindrical parabolic region in $\mathcal{E}_{1+1}$.

P18.2–3.1. Is it possible to formulate and prove the analog of the Perron–Petrovsky basic theorem for the second boundary value problem for Laplace and Fourier equations? If possible, do so; if not, explain why.

P18.4.1. Compare the Duhamel test with the convolution theorem in the theory of the Laplace–Carson transform (see Appendix 4).

P18.4.2. Is the Duhamel test applicable to noncylindrical regions? If possible, prove that it is; if not, explain why not.

P18.5.1. Give a reasonable estimate for the convergence rate of the Schwarz alternating method and prove its validity.

---

[15]In 1941 V. Krylov [41] considered conditions under which the Schwarz alternating method may be applied to the solution of a boundary-value problem in a bounded region $D \subset \mathbb{R}_2$ for a nonlinear equation $F(x,y,u,u_x,u_y,u_{xx},u_{xy},u_{yy}) = 0$. These conditions essentially reproduce those under which the convergence of the Schwarz alternating method for the solution of electrostatic and parabolic Dirichlet problems was proved in Chapters 11 and 18.

19

Fourier method for bounded regions

1. Vibration of a bounded string. D'Alembert's solution and superposition of standing waves. Formal scheme of the method of separation of variables

The previous chapters were devoted to methods that are specific for the solution of problems of different types. We now proceed to consider a universal method, applicable equally well to the solution of linear problems of different kinds.

To begin with, let us return to the problem of free vibration of a finite string. The solution was constructed in Chapter 5, using D'Alembert's method, as a superposition of traveling waves. This solution can be transformed into a superposition of standing waves. Indeed, consider the boundary-value problem

$$\Box u \stackrel{\text{def}}{=} a^2 \frac{\partial^2 u}{\partial x^2} - \frac{\partial^2 u}{\partial t^2} = 0, \quad 0 < x < \ell, \ t > 0 \tag{19.1}$$

$$u(0, t) = 0, \quad u(\ell, t) = 0, \quad t > 0 \tag{19.2}$$

$$u(\ell, 0) = f(\ell), \quad \frac{\partial}{\partial t} u(\ell, 0) = \varphi(\ell), \quad 0 < x < \ell \tag{19.3}$$

D'Alembert's solution, obtained by the method of traveling waves, is

$$u = u_1 + u_2 \tag{19.4}$$

where

$$\Box u_1 = 0, \quad u_1(x, t) = \frac{f(x - at) + f(x + at)}{2}$$

$$u_1(x, 0) = f(x), \quad \frac{\partial}{\partial t} u_1(x, 0) = 0 \tag{19.5}$$

468

$$\Box u_2 = 0, \quad u_2(x,t) = \frac{1}{2a} \int_{x-at}^{x+at} \varphi(s)ds$$

$$u_2(x,0) = 0, \quad \frac{\partial}{\partial t} u_2(x,0) = \varphi(x)$$

(19.6)

and $f(x)$ and $\varphi(x)$ are extended outside of the segment $[0, \ell]$ as odd periodic functions of period 2ℓ. Now let

$$f(x) = \frac{2}{\ell} \sum_1^\infty a_m \sin \frac{m\pi x}{\ell}, \quad \varphi(x) = \frac{2}{\ell} \sum_1^\infty b_m \sin \frac{m\pi x}{\ell}$$

(19.7)

be the Fourier expansions of f and φ, so that

$$a_m = \int_0^\ell f(\xi) \sin \frac{m\pi\xi}{\ell} d\xi, \quad b_m = \int_0^\ell \varphi(\xi) \sin \frac{m\pi\xi}{\ell} d\xi$$

(19.8)

Substitution of the expansions (19.7) into (19.5) and (19.6), using the identities

$$\sin \frac{m\pi(x-at)}{\ell} + \sin \frac{m\pi(x+at)}{\ell} = 2 \sin \frac{m\pi x}{\ell} \cos \frac{m\pi at}{\ell}$$

$$\cos \frac{m\pi(x-at)}{\ell} - \sin \frac{m\pi(x+at)}{\ell} = 2 \sin \frac{m\pi x}{\ell} \sin \frac{m\pi at}{\ell}$$

(19.9)

yields

$$u_1(x,t) = \frac{2}{\ell} \sum_1^\infty a_m \sin \frac{m\pi x}{\ell} \cos \frac{m\pi at}{\ell}$$

(19.10)

$$u_2(x,t) = \frac{2}{\ell} \sum_1^\infty b_m \frac{\ell}{m\pi a} \sin \frac{m\pi x}{\ell} \sin \frac{m\pi at}{\ell}$$

(19.11)

Note now that each of the standing waves

$$u_{1m}(x,t) = \sin \frac{m\pi x}{\ell} \cos \frac{m\pi at}{\ell}, \quad u_{2m}(x,t) = \sin \frac{m\pi x}{\ell} \sin \frac{m\pi at}{\ell}$$

(19.12)

is a solution of the wave equation (19.1), satisfying the homogeneous boundary conditions (19.2). Thus D'Alembert's representation of the solution to problem (19.1)–(19.3) as a superposition of traveling waves is reduced to a superposition of denumerably many standing waves – particular solutions of the homogeneous wave equation with homogeneous boundary conditions – that satisfy the equation and the boundary and initial conditions formally (i.e., without the necessary verification that these infinite series are convergent in $[0, \ell]$ for $t > 0$ or inside the interval $(0, \ell)$ for $t \geq 0$; the same applies to the series obtained by repeated term-by-term differentiation with respect to t in $(0, \ell)$ for $t > 0$ or with respect to x for $t > 0$ and $x \in (0, \ell)$).

Clearly, such a representation of the solution of the problem as a superposition of particular solutions with "separated variables" may be obtained directly, without reference to a known solution in the form of traveling waves. The formal scheme of operations is as follows. We seek u_1 as a superposition of functions of the type

$$v(x, t) = X(x)T(t) \tag{19.13}$$

Substitution of v into equation (19.1) yields

$$a^2 T \frac{d^2}{dx^2} X = X \frac{d^2}{dt^2} T \tag{19.14}$$

or, separating variables,

$$\frac{\frac{d^2}{dx^2} X}{X} = \frac{\frac{d^2}{dt^2} T}{a^2 T} = -\lambda^2 \tag{19.15}$$

where[1]

$$\lambda = \lambda_1 + i\lambda_2 = \text{const} \tag{19.16}$$

Thus X must be determined by the conditions

$$\frac{d^2}{dx^2} X + \lambda^2 X = 0, \quad X(0) = X(\ell) = 0 \tag{19.17}$$

and T by the equation

$$\frac{d^2}{dx^2} T + a^2 \lambda^2 T = 0 \tag{19.18}$$

It follows from (19.17) that

$$X = a \sin(\lambda x) + b \cos(\lambda x) \tag{19.19}$$

The first boundary condition yields

$$b = 0 \tag{19.20}$$

Thus

$$X = a \sin(\lambda x) \tag{19.21}$$

We are of course looking for a nontrivial solution (i.e., not identically equal to zero). Therefore, we must assume that $a \neq 0$, which implies

$$X(\ell) = 0 \Rightarrow \sin(\lambda \ell) = 0 \tag{19.22}$$

Values of λ satisfying this condition are called *eigenvalues*, and the set of all eigenvalues is called the *spectrum* of the operator:

$$\mathcal{L}_1 u \overset{\text{def}}{=} \begin{cases} \dfrac{d^2}{dx^2} X + \lambda^2 X = 0, & 0 < x < \ell \\ X(0) = 0, & X(\ell) = 0 \end{cases} \tag{19.23}$$

---

[1] $\lambda = \text{const}$ since the left-hand side of (19.15) depends only on x whereas the right-hand side depends only on t.

Thus the spectrum of the operator $\mathcal{L}_1$ consists of the numbers

$$\lambda_n = \frac{n\pi}{\ell}, \quad n = 1, 2, \ldots \tag{19.24}$$

The nontrivial solutions corresponding to an eigenvalue λ_n, are called *eigenfunctions*.

The general solution of equation (19.18), corresponding to the nth eigenvalue, is

$$T_n(t) = A_n \sin \frac{n\pi a t}{\ell} + B_n \cos \frac{n\pi a t}{\ell} \tag{19.25}$$

Thus the general solution $v_n(x,t)$ of the equation $\square u_1 = 0$, corresponding to the nth eigenvalue, is

$$v_n(x,t) = \frac{2}{\ell} \left(A_n \cos \frac{n\pi a t}{\ell} + B_n \sin \frac{n\pi a t}{\ell} \right) \sin \frac{n\pi x}{\ell} \tag{19.26}$$

The solution $u_1(x,t)$ may now be determined as a series in the eigenfunctions v_n,

$$u_1(x,t) = \frac{2}{\ell} \sum_{n=1}^{\infty} \left(A_n \cos \frac{n\pi a t}{\ell} + B_n \sin \frac{n\pi a t}{\ell} \right) \sin \frac{n\pi x}{\ell} \tag{19.27}$$

where the coefficients A_n and B_n are chosen so that the initial conditions (19.5) are satisfied. This formally implies that

$$f(x) = \frac{2}{\ell} \sum_{n=1}^{\infty} A_n \sin \frac{n\pi x}{\ell}, \quad 0 = \frac{2}{\ell} \sum_{n=1}^{\infty} B_n \frac{n\pi a}{\ell} \sin \frac{n\pi x}{\ell} \tag{19.28}$$

so that

$$A_n = \int_0^\ell f(\xi) \sin \frac{n\pi \xi}{\ell} d\xi, \quad B_n = 0, \quad n = 1, 2, \ldots \tag{19.29}$$

Thus

$$u_1(x,t) = \frac{2}{\ell} \sum_{n=1}^{\infty} A_n \sin \frac{n\pi x}{\ell} \cos \frac{n\pi a t}{\ell} \tag{19.30}$$

Quite similarly, we find

$$u_2(x,t) = \frac{2}{\ell} \sum_{n=1}^{\infty} B_n \sin \frac{n\pi x}{\ell} \sin \frac{n\pi a t}{\ell} \tag{19.31}$$

where

$$B_n = \frac{l}{n\pi a} \int_0^\ell \varphi(\xi) \sin \frac{n\pi \xi}{\ell} d\xi \tag{19.32}$$

We have thus returned to the solution obtained from D'Alembert's solution (19.10) and (19.11) by converting traveling waves into standing waves.

Consider now the nonhomogeneous problem of the vibration of a loaded string, with the motion of its ends prescribed:

$$\Box u + F(x,t) = 0, \quad 0 < x < \ell, \quad t > 0$$

$$u(0,t) = f_1(t), \quad u(\ell,t) = f_2(t) \tag{19.33}$$

$$u(x,0) = \varphi_1(x), \quad \frac{\partial}{\partial t}u(x,0) = \varphi_2(x)$$

Assume that f_1 and f_2 are twice continuously differentiable for all $t > 0$. We can then seek $u(x,t)$ as the sum of a solution u_1 of the homogeneous equation with nonhomogeneous initial data and homogeneous boundary condition, a solution u_2 of the nonhomogeneous equation with homogeneous initial and boundary conditions, and a function u_3 that is linear in x and equal to f_1 and f_2 at $x = 0$ and $x = \ell$, respectively,

$$u = u_1 + u_2 + f_1(t) + \frac{x}{\ell}[f_2(t) - f_1(t)] \tag{19.34}$$

where

$$\Box u_2(x,t) + \Phi(x,t) = 0, \quad u_2(x,0) = 0, \quad \frac{\partial}{\partial t}u_2(x,0) = 0 \quad \forall x \in [0,\ell] \tag{19.35}$$

$$u_2(0,t) = u_2(\ell,t) = 0 \quad \forall t > 0$$

$$\Box u_1(x,t) = 0, \quad u_1(0,t) = u_2(\ell,t) = 0$$

$$u_1(x,0) = \varphi_1(x) - f_1(0) - \frac{x}{\ell}[f_2(0) - f_1(0)] \tag{19.36}$$

$$\frac{\partial}{\partial t}u_1(x,0) = \varphi_2(x) - \frac{x}{\ell}\frac{d}{dt}[f_2(0) - f_1(0)] \tag{19.37}$$

and

$$\Phi(x,t) = F(x,t) - \frac{x}{\ell}\frac{d^2}{dt^2}[f_2(t) - f_1(t)] \tag{19.38}$$

The function $u_1(x,t)$ may be considered as represented by a sum of series of type (19.30) and (19.31) or by the corresponding D'Alembert formula, so that it is only necessary to determine $u_2(x,t)$.

Let us expand $\Phi(x,t)$ as a series in the eigenfunctions of the homogeneous problem

$$\Phi(x,t) = \frac{2}{\ell}\sum_{n=1}^{\infty}C_n(t)\sin\frac{n\pi x}{\ell} \tag{19.39}$$

and seek $u_2(x,t)$ as the sum of a series

$$u_2(x,t) = \frac{2}{\ell}\sum_{n=1}^{\infty}\alpha_n(t)\sin\frac{n\pi x}{\ell} \tag{19.40}$$

On the assumption that repeated term-by-term differentiation of this series is permissible, we find that

$$\Box u_2 + \Phi = -\frac{2}{\ell}\sum_{n=1}^{\infty}\left[\frac{d^2}{dt^2}\alpha_n(t) + \left(\frac{an\pi}{\ell}\right)^2\alpha_n(t) - C_n(t)\right]\sin\frac{n\pi x}{\ell}$$

$$= 0 \tag{19.41}$$

Since the orthonormal system $\sin(n\pi x/\ell)$ is complete, we have

$$\frac{d^2}{dt^2}\alpha_n(t) + \left(\frac{n\pi a}{\ell}\right)^2 \alpha_n(t) - C_n(t) = 0 \tag{19.42}$$

Since

$$u_2(x,0) = \frac{\partial}{\partial t}u_2(x,0) = 0 \tag{19.43}$$

the solutions of equations (19.42) must satisfy the initial conditions

$$\alpha_n(0) = \frac{d}{dt}\alpha_n(0) = 0 \tag{19.44}$$

which implies

$$\alpha_n(t) = \frac{\ell}{n\pi a}\int_0^t C_n(\tau)\sin\left[\frac{n\pi a}{\ell}(t-\tau)\right]d\tau \tag{19.45}$$

Substitution of the functions $\alpha_n(t)$ thus determined into the series (19.40) accomplishes the formal construction of u_2 and together with this the formal solution of the nonhomogeneous problem under consideration.[2]

2. Heat transfer through a homogeneous slab

Let us consider the first boundary-value problem for the heat-conduction equation:

$$a^2\frac{\partial^2}{\partial x^2}u + f(x,t) = \frac{\partial}{\partial t}u \quad \forall x \in (0,\ell), \ \forall t > 0 \tag{19.46}$$

$$u(0,t) = \varphi_1(t), \quad u(\ell,t) = \varphi_2(t) \tag{19.47}$$

$$u(x,0) = \psi(x) \quad \forall x \in (0,\ell) \tag{19.48}$$

Assume first that the problem is homogeneous,[3] so that

$$f = \varphi_1(t) = \varphi_2(t) \equiv 0 \tag{19.49}$$

Then, applying the method of separation of variables, we seek a particular solution of the form

$$u(x,t) = X(x)T(t) \tag{19.50}$$

which implies

$$\frac{\frac{d^2}{dx^2}X}{X} = \frac{\frac{d}{dt}T}{a^2 T} = -\lambda^2 \tag{19.51}$$

---

[2]We emphasize that the procedure as described requires justification, that is, specification of conditions under which all the operations performed are admissible. This will be done in Appendix 3, where the general Sturm–Liouville theory is considered.

[3]Here and everywhere in this chapter a problem is called *homogeneous* if its solution satisfies the homogeneous equation and homogeneous boundary condition with nonhomogeneous initial conditions.

so that

$$T = \exp(-a^2\lambda^2 t), \quad X = a_\lambda \sin\frac{\lambda x}{\ell} + b_\lambda \cos\frac{\lambda x}{\ell} \tag{19.52}$$

Hence the spectrum of the operator

$$\mathfrak{L}X \overset{\text{def}}{=} \frac{d^2}{dx^2}X + \lambda^2 X = 0, \quad X = 0 \quad \text{at } x = 0, \; x = \ell \tag{19.53}$$

is the set of numbers

$$\lambda_n = \frac{n\pi}{\ell}, \quad n = 1, 2, \dots \tag{19.54}$$

and the corresponding eigenfunctions are

$$X_n(x) = \sin\frac{\lambda x}{\ell} \tag{19.55}$$

which means that the eigenfunctions of the operator

$$Lu = a^2\frac{\partial^2}{\partial x^2}u - \frac{\partial}{\partial t}u = 0, \quad u = 0 \quad \text{at } x = 0, \quad x = \ell \tag{19.56}$$

are

$$u_n(x, t) = \exp\left(-\frac{a^2 n^2 \pi^2 t}{\ell^2}\right)\sin\frac{\lambda x}{\ell} \tag{19.57}$$

Thus we are seeking a solution of the homogeneous problem as the sum of a series in eigenfunctions

$$u(x, t) = \frac{2}{\ell}\sum_{n=1}^{\infty} a_n \exp\left(-\frac{a^2 n^2 \pi^2 t}{\ell^2}\right)\sin\frac{n\pi x}{\ell} \tag{19.58}$$

where coefficients must be determined from the initial conditions. This formally implies that

$$\psi(x) = \frac{2}{\ell}\sum_{n=1}^{\infty} a_n \sin\frac{n\pi x}{\ell} \tag{19.59}$$

Hence

$$a_n = \int_0^\ell \psi(\xi)\sin\frac{n\pi\xi}{\ell}d\xi \tag{19.60}$$

Inserting (19.60) into (19.58) and changing the order of summation and integration, we obtain

$$u(x, t) = \int_0^\ell \psi(\xi)G(x, \xi, a^2 t)d\xi \tag{19.61}$$

where

$$G(x, \xi, a^2 t) = \frac{2}{\ell}\sum_{n=1}^{\infty}\exp\left(-\frac{a^2 n^2 \pi^2 t}{\ell^2}\right)\sin\frac{n\pi x}{\ell}\sin\frac{n\pi\xi}{\ell} \tag{19.62}$$

Theorem 19.2.1. *The function $G(x, \xi, a^2 t)$, defined by (19.62), is the Green's function $g(x, \xi, a^2 t)$ of the first boundary-value problem for the heat-conduction equation in the interval $(0, \ell)$.*

PROOF. Let $u(x, t)$ be a solution of problem (19.46)–(19.48), (19.4). Then (see (15.44))

$$u(x, t) = \int_0^\ell \psi(\xi) g(x, \xi, a^2 t) d\xi \qquad (19.63)$$

If formulas (19.61) and (19.62) indeed yield a solution of problem (19.46)–(19.48) and (19.4) then, whatever the continuous function $\psi(x)$, it follows from (19.61) and (19.63) that

$$\int_0^\ell \psi(\xi)[g(x, \xi, a^2 t) - G(x, \xi, a^2 t)] d\xi = 0 \qquad (19.64)$$

which in turn implies

$$g(x, \xi, a^2 t) \equiv G(x, \xi, a^2 t) \qquad (19.65)$$

Indeed, let $\varepsilon = \text{const} > 0$ be arbitrarily small. Define $\psi(x)$ by the conditions

$$\psi(x) \in C[0, \ell], \quad \psi(x) \equiv 0 \quad \forall x \notin [x_0 - \varepsilon, x_0 + \varepsilon] \subset (0, \ell)$$
$$\psi(x) > 0 \quad \forall x \in (x_0 - \varepsilon, x_0 + \varepsilon) \qquad (19.66)$$

Then, by the mean value theorem,

$$[g(x, \xi^*, a^2 t) - G(x, \xi^*, a^2 t)] \int_{x_0 - \varepsilon}^{x_0 + \varepsilon} \psi(\xi) d\xi = 0$$
$$\Rightarrow g(x, \xi^*, a^2 t) - G(x, \xi^*, a^2 t) = 0 \qquad (19.67)$$

where $\xi^* \in (x_0 - \varepsilon, x_0 + \varepsilon)$. Since $\varepsilon > 0$ is arbitrarily small, this means that

$$g(x, x_0, a^2 t) - G(x, x_0, a^2 t) = 0 \qquad (19.68)$$

Note that the requirement that $\psi(x)$ be an arbitrary continuous function may be replaced by the assumption that $\psi(x)$ be smooth enough – for example, that $\psi(x)$ be an arbitrary function that is twice continuously differentiable and equal to zero at $x = 0$, $x = \ell$. Indeed, in that case repeated integration by parts yields

$$a_n = \int_0^\ell \psi(\xi) \sin \frac{n\pi\xi}{\ell} d\xi$$

$$= -\frac{\ell}{n\pi} \psi(\xi) \cos \frac{n\pi\xi}{\ell} \Big|_{\xi=0}^\ell + \frac{\ell}{n\pi} \int_0^\ell \frac{d}{d\xi} \psi(\xi) \cos \frac{n\pi\xi}{\ell} d\xi$$

$$= \left(\frac{\ell}{n\pi}\right)^2 \frac{d}{d\xi} \psi(\xi) \sin \frac{n\pi\xi}{\ell} \Big|_{\xi=0}^\ell - \left(\frac{\ell}{n\pi}\right)^2 \int_0^\ell \frac{d^2}{d\xi^2} \psi(\xi) \sin \frac{n\pi\xi}{\ell} d\xi$$

$$= O(1) n^{-2} \qquad (19.69)$$

This rate of decrease of a_n guarantees the absolute and uniform convergence of the series (19.59). This in turn implies that the function $u(x,t)$ defined by (19.58) is in fact a solution of the problem (19.46)–(19.49).

Let us return to the nonhomogeneous problem (19.46)–(19.48). Its solution is

$$u(x,t) = \sum_{i=1}^{3} u_i(x,t) \tag{19.70}$$

where

$$u_1(x,t) = \int_0^\ell \psi(\xi)g(x,\xi,a^2 t)d\xi \tag{19.71}$$

$$u_2(x,t) = a^2 \int_0^t \varphi_1(\tau)\frac{\partial}{\partial\xi}g[x,0,a^2(t-\tau)]d\tau$$

$$- a^2 \int_0^t \varphi_2(\tau)\frac{\partial}{\partial\xi}g[x,\ell,a^2(t-\tau)]d\tau \tag{19.72}$$

$$u_3(x,t) = \int_0^t d\tau \int_0^\ell f(\xi,\tau)g[x,\xi,a^2(t-\tau)]d\xi \tag{19.73}$$

so that u_1 is a solution of the homogeneous problem with the prescribed initial data, u_2 is a solution of the homogeneous equation with homogeneous initial data and nonhomogeneous boundary data, and u_3 is the solution of the nonhomogeneous equation with homogeneous boundary and initial data. Inserting the representation (19.62) of g by its Fourier series expansion into these formulas and changing the order of summation and integration, we obtain the solution of the nonhomogeneous problem as an expansion in Fourier series, without the need for a preliminary formulation of problems whose solutions are u_1, u_2, and u_3 (i.e., without the procedure that was required for the solution of the nonhomogeneous wave equation). This essential distinction between the wave and the heat-conduction equations obviously reflects the fact that for hyperbolic equations there is no equivalent of the Green's functions, which constitutes one of the important achievements of the classical theory of partial differential equation of elliptic and parabolic types.[4]

3. Two-dimensional Dirichlet problem for Poisson equation in a rectangle

Consider the 2-dimensional Dirichlet problem in a rectangle,
$$D = (0 < x < a) \times (0 < y < b) \tag{19.74}$$

---

[4] An attempt in the theory of hyperbolic equations to define concepts that might play the role of the Green's functions led to the creation by Hadamard of the theory of integrals in the sense of "partie fini" [34], which, in parallel with Dirac's concept of the δ function, later became one of the essential elements motivating the creation of the theory of distributions [72].

$$\Delta u + f(x,y) = 0 \quad \forall (x,y) \in D \tag{19.75}$$
$$u(0,y) = f_1(y), \quad u(a,y) = f_2(y) \tag{19.76}$$
$$u(x,0) = \varphi_1(x), \quad u(x,b) = \varphi_2(y) \tag{19.77}$$

Let us first consider this problem on the assumption that the boundary conditions are homogeneous, that is, setting

$$f_i = \varphi_i \equiv 0, \quad i = 1,2 \tag{19.78}$$

Recall that in order to solve the analogous problem for the 1-dimensional wave equation we expanded the source term as a series in the eigenfunctions, that is, a Fourier sine series. Bearing this in mind, let us expand $f(x,y)$ as a double Fourier series:

$$f(x,y) = \frac{4}{ab} \sum_{m=1}^{\infty} \sum_{n=1}^{\infty} \alpha_{mn} \sin \frac{m\pi x}{a} \sin \frac{n\pi y}{b} \tag{19.79}$$

where

$$\alpha_{mn} = \int_0^a d\xi \int_0^b f(\xi,\eta) \sin \frac{m\pi\xi}{a} \sin \frac{n\pi\eta}{b} d\eta \tag{19.80}$$

Continuing the analogy, let us seek $u(x,y)$ as the sum of a double Fourier series:

$$u(x,y) = \frac{4}{ab} \sum_{m=1}^{\infty} \sum_{n=1}^{\infty} a_{mn} \sin \frac{m\pi x}{a} \sin \frac{n\pi y}{b} \tag{19.81}$$

Assume that repeated term-by-term differentiation of this series is permissible. Then by (19.75), (19.79), and (19.81),

$$\frac{4}{ab} \sum_{m=1}^{\infty} \sum_{n=1}^{\infty} \left\{ a_{mn} \left[\left(\frac{m\pi}{a}\right)^2 + \left(\frac{n\pi}{b}\right)^2 \right] - \alpha_{mn} \right\} \sin \frac{m\pi x}{a} \sin \frac{n\pi y}{b} = 0 \tag{19.82}$$

Since the orthonormal system $\{\sin(m\pi x/a) \cdot \sin(n\pi y/b)\}$ is complete in the rectangle D, we have

$$a_{mn} \left[\left(\frac{m\pi}{a}\right)^2 + \left(\frac{n\pi}{b}\right)^2 \right] = \alpha_{mn} \quad \forall (m,n) \geq 1 \tag{19.83}$$

Thus the formal solution of the problem (19.74)–(19.77) is

$$u(x,y) = \frac{4}{ab} \sum_{m=1}^{\infty} \sum_{n=1}^{\infty} \alpha_{mn} \left[\left(\frac{m\pi}{a}\right)^2 + \left(\frac{n\pi}{b}\right)^2 \right]^{-1}$$
$$\times \sin \frac{m\pi x}{a} \sin \frac{n\pi y}{b} = 0 \tag{19.84}$$

Inserting the definitions (19.80) into (19.81) and changing the order of integration and summation, we now obtain

$$u(x,y) = \int_0^a d\xi \int_0^b f(\xi,\eta) G(x,y,\xi,\eta) d\eta \tag{19.85}$$

where

$$G(x,y,\xi,\eta) = \frac{4}{ab} \sum_{m=1}^{\infty}\sum_{n=1}^{\infty} \left[\left(\frac{m\pi}{a}\right)^2 + \left(\frac{n\pi}{b}\right)^2 \right]^{-1}$$

$$\times \sin\frac{m\pi x}{a} \sin\frac{n\pi y}{b} \sin\frac{m\pi\xi}{a} \sin\frac{n\pi\eta}{b} \qquad (19.86)$$

Repeating the arguments of Section 2, one easily proves that $G(p,q)$ $[p = (x,y), q = (\xi,\eta)]$, as defined by (19.86), is the Fourier expansion of the Green's function of the electrostatic Dirichlet problem in the rectangle D.

It is easy to reduce the expansion (19.86) of the Green's function $G(p,q)$ to an ordinary Fourier expansion. To this end, note that

$$\frac{2}{\pi} \sum_{k=1}^{\infty} \frac{\cos kx}{k^2 + \alpha^2} = \frac{\cosh[\alpha(\pi - x)]}{\alpha\sinh(\alpha\pi)} - \frac{1}{\alpha^2\pi} \qquad \forall x \in [0, 2\pi] \qquad (19.87)$$

(see, e.g. [29]) and that the series (19.87) may be rewritten as

$$G(x,y,\xi,\eta) = \frac{2}{ab} \sum_{m=1}^{\infty} \sin\frac{m\pi x}{a} \sin\frac{m\pi\xi}{a}$$

$$\times \sum_{n=1}^{\infty} \left(\cos\frac{n\pi(y-\eta)}{b} - \cos\frac{n\pi(y+\eta)}{b} \right)$$

$$\times \left[\left(\frac{m\pi}{a}\right)^2 + \left(\frac{n\pi}{b}\right)^2 \right]^{-1} \qquad (19.88)$$

It follows from (19.87) and (19.88) that for all $\eta \in [0,y]$, $y \in [0,b]$,

$$G(p,q) = \frac{1}{\pi} \sum_{m=1}^{\infty} \left[\cosh\left(\frac{m\pi}{a}(b-y+\eta)\right) - \cosh\left(\frac{m\pi}{a}(b-y-\eta)\right) \right]$$

$$\times \left(m\sinh\frac{m\pi b}{a} \right)^{-1} \sin\frac{m\pi x}{a} \sin\frac{m\pi\xi}{a} \qquad (19.89)$$

or, what is the same,

$$G(p,q) = \frac{2}{\pi} \sum_{m=1}^{\infty} \sin\frac{m\pi x}{a} \sin\frac{m\pi\xi}{a} \sinh\frac{m\pi(b-y)}{a} \sinh\frac{m\pi\eta}{a}$$

$$\times \left(m\sinh\frac{m\pi b}{a} \right)^{-1}, \qquad \forall\eta \in [0,y], \ \forall y \in [0,b] \qquad (19.90)$$

Since $G(p,q)$ is an even function of $y - \eta$, it follows instantly from (19.89) that

$$G(p,q) = \frac{2}{\pi} \sum_{n=1}^{\infty} \sin\frac{m\pi x}{a} \sin\frac{m\pi\xi}{a} \sinh\frac{m\pi(b-\eta)}{a} \sinh\frac{m\pi y}{a}$$

$$\times \left(m\cdot\sinh\frac{m\pi b}{a} \right)^{-1} \qquad \forall y \in [0,\eta], \ \forall\eta \in [0,b] \qquad (19.91)$$

Applying the representation of the solution of the nonhomogeneous electrostatic Dirichlet problem in terms of the Green's function, we immediately obtain a representation of the solution as an expansion in Fourier series, provided that it is legitimate to change the order of summation and integration. This is yet another manifestation of similarity of properties of elliptic and parabolic operators.

Note that the standard method of separation of variables, applied to the problem

$$\Delta u = 0 \quad \forall (x, y) \in D \tag{19.92}$$

$$u(0, y) = f_1(y), \quad u(a, y) = f_2(y), \quad u(x, 0) = \varphi_1(x) \tag{19.93}$$

$$u(x, b) = \varphi_2(y) \tag{19.94}$$

requires the representation of u as the sum

$$u = u_1 + u_2 \tag{19.95}$$

of solutions of the problems

$$\Delta u_1 = 0 \quad \forall (x, y) \in D \tag{19.96}$$

$$u_1(0, y) = 0, \quad u_1(a, y) = 0, \quad u_1(x, 0) = \varphi_1(x) \tag{19.97}$$

$$u_1(x, b) = \varphi_2(x) \tag{19.98}$$

$$\Delta u_2 = 0 \quad \forall (x, y) \in D \tag{19.99}$$

$$u_2(x, 0) = 0, \quad u_1(x, b) = 0, \quad u_2(0, y) = f_1(y) \tag{19.100}$$

$$u_2(a, y) = f_2(y) \tag{19.101}$$

The solution $u_1(x, y)$ must be represented by its expansion in terms of the eigenfunctions of the 1-dimensional Sturm–Liouville problem, that is, by the series

$$u_1(x, y) = \frac{2}{a} \sum_{n=1}^{\infty} \left(\alpha_n \sinh \frac{n\pi y}{a} + \gamma_n \cosh \frac{n\pi y}{a} \right) \sin \frac{n\pi x}{a}$$

$$\equiv \frac{2}{a} \sum_{n=1}^{\infty} \left(a_n \sinh \frac{n\pi y}{a} + c_n \sinh \frac{n\pi (b - y)}{a} \right) \sin \frac{n\pi x}{a} \tag{19.102}$$

so that the coefficients a_n and b_n are determined by expanding the boundary data φ_1 and φ_2 in series,

$$\varphi_1(x) = \frac{2}{a} \sum_{n=1}^{\infty} c_n \sinh \frac{n\pi b}{a} \sin \frac{n\pi x}{a}$$

$$\varphi_2(x) = \frac{2}{a} \sum_{n=1}^{\infty} a_n \sinh \frac{n\pi b}{a} \sin \frac{n\pi x}{a} \tag{19.103}$$

The result is

$$a_n = \left(\sinh \frac{n\pi b}{a} \right)^{-1} \int_0^a \varphi_2(\xi) \sin \frac{n\pi \xi}{a} d\xi$$

$$c_n = \left(\sinh \frac{n\pi b}{a} \right)^{-1} \int_0^a \varphi_1(\xi) \sin \frac{n\pi \xi}{a} d\xi \tag{19.104}$$

Analogous Fourier expansions may be written out for $u_2(x,t)$.

It is easy to check that this procedure leads to the same result as the treatment using the expansion of the Green's function.

4. Vibration of circular membrane with rigidly fixed boundary under action of instant point impulse initially applied at an interior point of membrane

In suitably chosen dimensionless polar coordinates (ρ, ψ), the problem to be solved is

$$\Box u \overset{\text{def}}{=} \Delta u - \frac{\partial^2}{\partial t^2} u = 0 \quad \forall \rho \in (0,1), \quad \forall t > 0 \tag{19.105}$$

$$u(1,t) = 0, \quad u(\rho,0) = 0, \quad \frac{\partial}{\partial t} u(\rho,0) = \delta(p,q) \tag{19.106}$$

Here Δ is the 2-dimensional Laplace operator, that is,

$$\Delta \overset{\text{def}}{=} \left(\frac{\partial^2}{\partial \rho^2} + \frac{1}{\rho} \frac{\partial}{\partial \rho} + \frac{1}{\rho^2} \frac{\partial^2}{\partial \psi^2} \right) \tag{19.107}$$

and $\delta(p,q)$ is the 2-dimensional Dirac δ function in polar coordinates, that is,[5]

$$\delta(p,q) = \frac{1}{\rho} \delta(r - \rho)\delta(\varphi - \psi), \quad p = (r, \varphi), \quad q = (\rho, \psi) \tag{19.108}$$

Following the standard procedure for separation of variables, we seek a particular solution of equation (19.105) of the form

$$u(\rho, \psi, t) = w(\rho, \psi) T(t) \tag{19.109}$$

where

$$w(\rho, \varphi) = R(\rho)\Phi(\psi) \tag{19.110}$$

Substitution of (19.109) into (19.105) yields

$$\frac{\Delta w}{\rho^{-2} w} = \frac{\frac{d^2 T}{dt^2}}{T} = -\lambda^2 \tag{19.111}$$

---

[5] By the definition of the 2-dimensional Dirac δ function,

$$\int_0^\infty f(q)\delta(p,q)d\sigma_q = f(p) \quad \forall f(p) \in C(\mathbb{R}_2) \tag{α}$$

Since $d\sigma_q = \rho d\rho d\psi$, it follows from (α) that for any $f(p) = R(r)F(\varphi)$,

$$\int_0^\infty \rho R(\rho)\delta(r - \rho)dr \int_0^{2\pi} F(\varphi)\delta(\varphi - \psi)d\psi \tag{β}$$

so that in fact

$$\delta(p,q) = \frac{1}{\rho}\delta(r - \rho)\delta(\varphi - \psi) \tag{γ}$$

which implies

$$T = a_\lambda \cos(\lambda t) + b_\lambda \sin(\lambda t) \tag{19.112}$$

and

$$\Delta w + (\lambda/\rho)^2 w = 0 \tag{19.113}$$

Formulas (19.113) and (19.110) in turn yield

$$\left(\frac{\partial^2}{\partial \rho^2} + \frac{1}{\rho} \frac{\partial}{\partial \rho} + \frac{1}{\rho^2} \frac{\partial^2}{\partial \psi^2} + \lambda^2 \right) w = 0 \quad \forall \rho \in [0,1), \ \forall \psi \in [0, 2\pi] \tag{19.114}$$

The solution of this equation must satisfy a homogeneous boundary condition at $z = 1$ and be bounded in the neighborhood of the origin (see Appendix 3):

$$|w(0,\psi)| < \infty, \quad w(1,\psi) = 0 \quad \forall \psi \in [0, 2\pi], \quad w(\rho,0) = w(\rho, 2\pi) \tag{19.115}$$

where the condition on the right side of (19.115) is a corollary of the requirement that w be continuous. Insert (19.110) into equation (19.114). Separating variables, we obtain

$$W = R_{\lambda\mu} \Phi_\mu, \quad \frac{\left(\frac{d^2}{d\rho} + \frac{1}{\rho} \frac{d}{d\rho} + \lambda^2 \right) R_{\lambda\mu}}{\rho^{-2} R} = -\frac{\frac{d^2}{dx^2} \Phi}{\Phi} = \mu^2 \tag{19.116}$$

The requirement on the right side of (19.115) that w be periodic implies the periodicity condition

$$\Phi(0) = \Phi(2\pi) \tag{19.117}$$

which means that μ is an arbitrary integer

$$\mu = k, \quad k = 0, 1, 2, \dots \tag{19.118}$$

so that

$$\Phi(\psi) = a_k \cos(k\psi) + b_k \sin(k\psi) \tag{19.119}$$

On the other hand, the second of equations (19.116) yields

$$\left(\frac{d^2}{d\rho^2} + \frac{1}{\rho} \frac{d}{d\rho} + \lambda^2 - \frac{k^2}{\rho^2} \right) R_{\lambda k} = 0 \tag{19.120}$$

The substitution

$$z = \lambda\rho, \quad R_{\lambda k}(\rho) = J_k(z) \tag{19.121}$$

transforms equation (19.120) into

$$\left(\frac{d^2}{dz^2} + \frac{1}{z} \frac{d}{dz} + 1 - \frac{k^2}{z^2} \right) J_k(z) = 0 \tag{19.122}$$

This kth order Bessel equation must be solved subject to the homogeneous conditions

$$|J(0)| < \infty, \quad J_k(\lambda) = 0 \tag{19.123}$$

which yields

$$R_{\lambda k}(1) = J_k(\lambda) = 0 \tag{19.124}$$

This means that the spectrum of equation (19.120) consists of all the roots of the Bessel function $J_k(\lambda)$:

$$\lambda = \lambda_{km}, \quad m = 1, 2, 3, \ldots \tag{19.125}$$

(see Appendix 3, Section 5).

Thus the solution of the problem (19.105) and (19.106) must be sought as the sum of the double series

$$u = \sum_{k=0}^{\infty} \sum_{m=1}^{\infty} \{[a_{km} \cos(k\psi) + b_{km} \sin(k\psi)] J_k(\lambda_{km}\rho) \cos(\lambda_{km}t)$$

$$+ [c_{km} \cos(k\psi) + d_{km} \sin(k\psi)] J_k(\lambda_{km}\rho) \sin(\lambda_{km}t)\} \tag{19.126}$$

The initial conditions (19.106) imply

$$u(\rho, \psi, 0) = \sum_{k=0}^{\infty} \sum_{m=1}^{\infty} [a_{km} \cos(k\psi) + b_{km} \sin(k\psi)] J_k(\lambda_{km}\rho) = 0 \tag{19.127}$$

$$\frac{\partial}{\partial t} u(\rho, \psi, 0) = \sum_{k=0}^{\infty} \sum_{m=1}^{\infty} \lambda_{km} [c_{km} \cos(k\psi) + d_{km} \sin(k\psi)] J_k(\lambda_{km}\rho)$$

$$= \delta(p, q) \tag{19.128}$$

Since the orthogonal systems $\{\exp(ik\psi)\}$ and $\{J_k(\lambda_{km}\rho)\}$ are complete, it follows from (19.127) that

$$a_{km} = b_{km} = 0 \quad \forall (k, m) \geq 1 \tag{19.129}$$

At the same time, by (19.106), (19.108), and (19.128),

$$\frac{1}{\rho} \delta(r - \rho) \delta(\varphi - \psi)$$

$$= \sum_{k=0}^{\infty} \sum_{m=1}^{\infty} \lambda_{km} [c_{km} \cos(k\psi) + d_{km} \sin(k\psi)] J_k(\lambda_{km}\rho) \tag{19.130}$$

Hence

$$c_{km} = \frac{\cos(k\varphi) J_k(\lambda_{km}r)}{\lambda_{km} \||\cos(k\varphi)\| \cdot \|J_k(\lambda_{km}r)\|}$$

$$d_{km} = \frac{\sin(k\varphi) J_k(\lambda_{km}r)}{\lambda_{km} \|\sin(k\varphi)\| \cdot \|J_k(\lambda_{km}r)\|} \tag{19.131}$$

Thus (19.126), (19.129), and (19.131) determine a formal solution of the problem (19.105) and (19.106):

$$u(\rho, \psi) = \sum_{k=0}^{\infty} \sum_{m=1}^{\infty} \frac{\cos[k(\varphi - \psi)] J_k(\lambda_{km}r) J_k(\lambda_{km}\rho)}{\lambda_{km} \|\cos(k\varphi)\| \cdot \|J_k(\lambda_{km}r)\|} \sin(\lambda_{km}t) \tag{19.132}$$

5. Heat transfer through two-layer circular disk with Newtonian irradiation from medium of prescribed temperature

As a last example of the expansion of solutions of boundary-value problems as series of orthogonal functions, let us consider the following problem: Find $u(r, \varphi, t)$ such that

$$u \in C(\bar{D}_i), \quad \frac{\partial u}{\partial r} \in C(\bar{D}_i), \quad i = 1, 2 \tag{19.133}$$

where

$$
\begin{aligned}
D_1 &= \{r, \varphi \colon 0 < r < R_0, 0 \le \varphi \le 2\pi\} \\
D_2 &= \{r, \varphi \colon R_0 < r < R, 0 \le \varphi \le 2\pi\}
\end{aligned}
\tag{19.134}
$$

$$\operatorname{div}(K \operatorname{grad} u) + f(r, \varphi, t) = C \frac{\partial u}{\partial t} \quad \forall (r, \varphi) \in D_1 \cup D_2 \tag{19.135}$$

$$K \frac{\partial u}{\partial r} + hu = h\Phi(t, \varphi) \quad \forall \varphi \in [0, 2\pi], \ \forall t > 0, \ r = R \tag{19.136}$$

$$u(r, \varphi, 0) = \Psi(r, \varphi) \tag{19.137}$$

$$[u] = 0, \ \left[K \frac{\partial u}{\partial r} \right] = 0 \quad \forall \varphi \in [0, 2\pi], \ \forall t > 0, \ r = R_0 \tag{19.138}$$

Here, as before, $[f]$ denotes the jump of an arbitrary function f. We assume that

$$K = K_i = \text{const} > 0, \quad C = C_i = \text{const} > 0 \quad \forall (r, \varphi) \in D_i, \quad i = 1, 2 \tag{19.139}$$

Applying the construction of the Green's function as described in Section 2, we first consider the homogeneous problem by setting

$$f(r, \varphi, t) \equiv 0, \quad \Phi(\varphi, t) \equiv 0 \tag{19.140}$$

with the initial condition

$$\Psi(r, \varphi) = \frac{1}{\rho} \delta(\rho - r) \delta(\varphi - \psi) \tag{19.141}$$

Thus, we wish to find a particular solution of problem (19.133)–(19.141) satisfying all the conditions except the initial condition (19.141):

$$v(r, \varphi, t) = V(r, \varphi) T(t) \tag{19.142}$$

Substitution of v into (19.135) yields

$$\frac{\operatorname{div}[K \operatorname{grad} V]}{CV} = \frac{\frac{dT}{dt}}{T} = -\lambda^2 \tag{19.143}$$

Hence

$$T = \exp(-\lambda^2 t) \tag{19.144}$$

and V must be a solution of the problem

$$\operatorname{div}[K \operatorname{grad} V] + \lambda^2 CV = 0 \quad \forall (r, \varphi) \in D_1 \cup D_2 \tag{19.145}$$

$$K_2 \frac{\partial V}{\partial r} + hV = 0 \quad \text{at } r = R \ \forall \varphi \in [0, 2\pi] \tag{19.146}$$

$$[V] = 0, \ \left[K \frac{\partial V}{\partial r} \right] = 0 \quad \text{at } r = R_0 \ \forall \varphi \in [0, 2\pi] \tag{19.147}$$

Applying separation of variables, we let

$$V(r, \varphi) = F(r)\Phi(\varphi) \tag{19.148}$$

Then

$$\frac{K\left(\frac{d^2}{dr^2} + \frac{1}{r}\frac{d}{dr} + \lambda^2 C\right)F}{\frac{F}{r^2}} = \frac{\frac{d^2\Phi}{d\varphi^2}}{\Phi} = -\mu^2 \tag{19.149}$$

Hence

$$\Phi = \alpha\cos(\mu\varphi) + \beta\sin(\mu\varphi) \quad \forall \varphi \in (0, 2\pi) \tag{19.150}$$

and

$$K_1\left(\frac{d^2}{dr^2} + \frac{1}{r}\frac{d}{dr}\right)F_1 + \left(\lambda^2 C_1 - \frac{\mu^2}{r^2}\right)F_1 = 0 \quad \forall r \in (0, R_0) \tag{19.151}$$

$$|F| < \infty \tag{19.152}$$

$$K_2\left(\frac{d^2}{dr^2} + \frac{1}{r}\frac{d}{dr}\right)F_2 + \left(\lambda^2 C_2 - \frac{\mu^2}{r^2}\right)F_2 = 0 \quad \forall r \in (R_0, R) \tag{19.153}$$

$$F_1(R_0) = F_2(R_0), \quad K_1\frac{dF_1}{dr}\bigg|_{r=R_0} = K_2\frac{dF_2}{dr}\bigg|_{r=R_0} \tag{19.154}$$

$$K_2\frac{dF_2}{dr}\bigg|_{r=R} + hF_2\bigg|_{r=R} = 0 \tag{19.155}$$

Since Φ must be periodic with period 2π, it follows from (19.150) that

$$\mu = m, \quad m = 1, 2, 3, \ldots \tag{19.156}$$

The general solution of equation (19.151) and (19.153) is

$$F_i(r \mid \lambda) = \gamma_{i\mu_i}J_{\mu_i}\left(\frac{\lambda r}{a_i}\right) + \delta_{i\mu_i}Y_{\mu_i}\left(\frac{\lambda r}{a_i}\right), \quad i = 1, 2 \tag{19.157}$$

where J_{μ_i} and Y_{μ_i} are the Bessel and Neumann functions of order μ_i. Here

$$a_i = \sqrt{K_i/C_i}, \quad \mu_i = m/\sqrt{K_i}, \quad i = 1, 2 \tag{19.158}$$

Since

$$\lim_{|z|\downarrow 0} Y_\nu(z) = \infty \quad \forall \nu \tag{19.159}$$

(see Appendix 2) we have

$$\delta_{1\mu_1} = 0 \quad \forall m, \quad m = 1, 2, \ldots \tag{19.160}$$

Introducing equalities (19.157) and (19.160) into the matching conditions (19.154) and the boundary condition (19.155), we obtain a homogeneous system of algebraic equations[6]

$$\gamma_{1\mu_1} J_{\mu_1}\left(\frac{\lambda R_0}{a_1}\right) - \gamma_{2\mu_2} J_{\mu_2}\left(\frac{\lambda R_0}{a_2}\right) - \delta_{2\mu_2} Y_{\mu_2}\left(\frac{\lambda R_0}{a_2}\right) = 0 \qquad (19.161)$$

$$\gamma_{1\mu_1}\omega_1 \dot{J}_{\mu_1}\left(\frac{\lambda R_0}{a_1}\right) - \gamma_{2\mu_2}\omega_2 \dot{J}_{\mu_2}\left(\frac{\lambda R_0}{a_2}\right) - \omega_2\delta_{2\mu_2} \dot{Y}_{\mu_2}\left(\frac{\lambda R_0}{a_2}\right) = 0 \quad (19.162)$$

$$\gamma_{2\mu_2}\left(\lambda\omega_2 \dot{J}_{\mu_2}\left(\frac{\lambda R}{a_2}\right) + h J_{\mu_2}\left(\frac{\lambda R}{a_2}\right)\right)$$

$$+ \delta_{2\mu_2}\left(\lambda\omega_2 \dot{Y}_{\mu_2}\left(\frac{\lambda R}{a_2}\right) + h Y_{\mu_2}\left(\frac{\lambda R}{a_2}\right)\right) = 0 \qquad (19.163)$$

The eigenvalues λ are the roots of the determinant

$$\Delta = \begin{vmatrix} J_{\mu_1}\left(\frac{\lambda R_0}{a_1}\right) & -J_{\mu_2}\left(\frac{\lambda R_0}{a_2}\right) & -Y_{\mu_2}\left(\frac{\lambda R_0}{a_2}\right) \\ \omega_1 \dot{J}_{\mu_1}\left(\frac{\lambda R_0}{a_1}\right) & -\omega_2 \dot{J}_{\mu_2}\left(\frac{\lambda R_0}{a_2}\right) & -\omega_2 \dot{Y}_{\mu_2}\left(\frac{\lambda R_0}{a_2}\right) \\ 0 & \lambda\omega_2 \dot{J}_{\mu_2}\left(\frac{\lambda R}{a_2}\right) + h J_{\mu_2}\left(\frac{\lambda R}{a_2}\right) & \lambda\omega_2 \dot{Y}_{\mu_2}\left(\frac{\lambda R}{a_2}\right) + h Y_{\mu_2}\left(\frac{\lambda R}{a_2}\right) \end{vmatrix} = 0$$

$$(19.164)$$

The coefficients $\gamma_{2\mu_2}$ and $\delta_{2\mu_2}$ are expressed in terms of $\gamma_{1\mu_1}$ by

$$\gamma_{2\mu_2} = \gamma_{1\mu_1}\frac{\Delta_\gamma}{\Delta^*}, \qquad \delta_{2\mu_2} = \gamma_{1\mu_1}\frac{\Delta_\gamma}{\Delta^*} \qquad (19.165)$$

where

$$\Delta_\gamma = \begin{vmatrix} J_{\mu_1}\left(\frac{\lambda R_0}{a_1}\right) & Y_{\mu_2}\left(\frac{\lambda R_0}{a_2}\right) \\ \mu_1 \dot{J}_{\mu_1}\left(\frac{\lambda R_0}{a_1}\right) & \mu_2 \dot{Y}_{\mu_2}\left(\frac{\lambda R_0}{a_2}\right) \end{vmatrix}$$

$$(19.166)$$

$$\Delta_\delta = \begin{vmatrix} J_{\mu_2}\left(\frac{\lambda R_0}{a_2}\right) & J_{\mu_1}\left(\frac{\lambda R_0}{a_1}\right) \\ \mu_2 \dot{J}_{\mu_2}\left(\frac{\lambda R_0}{a_2}\right) & \mu_1 \dot{J}_{\mu_1}\left(\frac{\lambda R_0}{a_1}\right) \end{vmatrix}$$

and

$$\Delta^* = \begin{vmatrix} J_{\mu_2}\left(\frac{\lambda R_0}{a_2}\right) & Y_{\mu_2}\left(\frac{\lambda R_0}{a_2}\right) \\ \mu_2 \dot{J}_{\mu_2}\left(\frac{\lambda R_0}{a_2}\right) & \mu_2 \dot{Y}_{\mu_2}\left(\frac{\lambda R_0}{a_2}\right) \end{vmatrix} = \frac{2}{\mu_2\pi\lambda R_0} \qquad (19.167)$$

According to the general theory of the Sturm–Liouville equations (see Appendix 3), this determinant Δ has denumerable set of positive simple roots:

$$\lambda = \lambda_{mn}, \quad n = 1, 2, \ldots, \quad m = 0, 1, 2, \ldots \qquad (19.168)$$

---

[6]Throughout this section, $\dot{f}(x) \stackrel{\text{def}}{=} df/dx$.

to which correspond eigenfunctions

$$F_{mn}(r) = \begin{cases} F^1_{mn}(r) & 0 < r < R_0 \\ F^2_{mn}(r) & R_0 < r < r \end{cases} \tag{19.169}$$

which are orthogonal in the interval $(0, R)$ with weight $rC(r)$, so that

$$\int_0^R rC(r)F_{mn}(r)F_{mk}(r)dr = 0 \tag{19.170}$$

The functions $F^i_{mn}(r)$ are defined by

$$F^i_{mn}(r) = F_i(r \mid \lambda_{mn}), \quad i = 1, 2 \tag{19.171}$$

where F_i in turn are defined by (19.157) and (19.160). Since $F^i_{mn}(r)$ are by definition proportional to $\gamma_{1\mu_1}$, which is as yet an undetermined constant, we may choose it as a normalizing factor for F_{mn} so that

$$\|F_{mn}(r)\|^2 = \int_0^R \rho C(\rho)F^2_{mn}(\rho)d\rho = 1 \tag{19.172}$$

We now have to satisfy the initial condition, applying the expansion

$$\frac{1}{r}\delta(r - \rho)\delta(\varphi - \psi)$$

$$= \sum_{m=0}^\infty \sum_{n=0}^\infty F_n(\rho)[\alpha_{mn}\cos(m\psi) + \beta_{mn}\sin(\mu\psi)] \tag{19.173}$$

so that

$$\alpha_{mn} = \frac{1}{\pi}\int_0^R \rho C(\rho)d\rho \int_0^{2\pi} \frac{1}{\rho}\delta(r - \rho)\delta(\varphi - \psi)F_{mn}(\rho)\cos(m\psi)d\psi$$

$$= \frac{1}{\pi}F_{mn}(r)\cos(m\varphi) \tag{19.174}$$

$$\beta_{mn} = \frac{1}{\pi}F_{mn}(r)\sin(m\varphi) \quad \forall(m, n) \geq 1$$

and

$$\alpha_{0n} = 0, \quad \beta_{0n} = \frac{1}{\pi}F_{0n}(r) \quad \forall n \geq 1 \tag{19.175}$$

Thus

$$g(r, \rho, \varphi, \psi, t) = \frac{1}{2\pi}\sum_{n=1}^\infty \exp(-\lambda^2_{0n}t)F_{0n}(r)F_{0n}(\rho)$$

$$+ \frac{1}{\pi}\sum_{m=1}^\infty \sum_{n=1}^\infty \exp(-\lambda^2_{mn}t)F_{mn}(r)F_{mn}(\rho)(\cos[m(\varphi - \psi)]) \tag{19.176}$$

The solution of the nonhomogeneous problem (19.135)–(19.138) has an integral representation

$$u(r,\varphi,t) = h \int_0^t d\tau \int_0^{2\pi} \Phi(\tau,\psi)Rg(r,R,\varphi-\psi,t-\tau)d\psi$$

$$+ \int_0^R \rho d\rho \int_0^{2\pi} \Psi(\rho,\psi)g(r,\rho,t)d\psi$$

$$+ \int_0^t d\tau \int_0^R \rho d\rho \int_0^{2\pi} f(\rho,\psi,\tau)g(r,\rho,\varphi-\psi,t-\tau)d\psi \quad (19.177)$$

Inserting the expression (19.176) for g, changing the order of integration and summation, and evaluating the integrals, we obtain an expansion of the solution in double Fourier series. Note that this result is not only formal but true, since the inequalities $\lambda_{mn} > 0$ for all m,n guarantee the uniform convergence of all three integrals for all $t \geq t_0$ whatever $t_0 > 0$. That the initial condition is also uniformly satisfied follows from the filtering property of the Green's function.[7]

6. Application of Fourier method to solution of mixed problems. Reduction to denumerable system of algebraic equations. Perfect systems[8]

The Fourier method may sometimes be used to solve mixed boundary value problems, in which conditions of different kinds are imposed on different parts of the boundary. Recall that the Fourier method may be used to solve problems in a region bounded by coordinate lines in some curvilinear coordinate system (cartesian, cylindrical, spherical, etc.). There are two essentially different cases.

(a) If boundary conditions of different kinds are imposed on different coordinate lines, application of the Fourier method does not differ from the standard procedures described in the previous sections of this chapter.

(b) If boundary conditions of different kinds are imposed on different parts of the same coordinate line, the application of the Fourier method reduces the problem to the solution of a denumerable system of algebraic equations, that is, to a problem of incomparably greater complexity.

We consider this latter case, using as a characteristic example the following stationary 2-dimensional mixed problem.

Find a function $u(x,y)$, harmonic within the square D (see Figure 19.1),

$$D = \{x,y \colon 0 < x,y < 1\} \quad (19.178)$$

---

[7]See Section 2.

[8]See note 10.

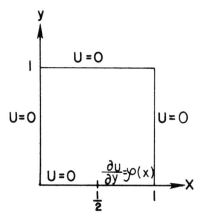

Figure 19.1. Boundary conditions in the problem (19.178)–(19.182).

continuous in the closure of D, and such that

$$u \equiv 0 \quad \text{at } x = 0 \text{ and } x = 1 \quad \forall y \in [0, 1] \tag{19.179}$$

$$u \equiv 0 \quad \text{at } y = 1 \quad \forall x \in [0, 1] \tag{19.180}$$

$$u \equiv 0 \quad \text{at } y = 0 \quad \forall x \in [0, 1/2], \quad \frac{\partial}{\partial x} u(1/2 + 0, 0) = 0 \tag{19.181}$$

$$\frac{\partial}{\partial y} u = \varphi(x) \quad \text{at } y = 0, \quad \forall x \in [1/2, 1) \tag{19.182}$$

Smoothness conditions will be imposed on φ in due course.

We first reduce the problem to a Fredholm integral equation of the first kind, defining

$$u(x, 0) \stackrel{\text{def}}{=} f(x) \quad \text{if } 1/2 \le x \le 1 \tag{19.183}$$

Recall that for $0 \le \eta \le y \le 1$ the electrostatic Green's function in D (see (19.90)) is:

$$g(x, y, \xi, \eta) = \frac{2}{\pi} \sum_{1}^{\infty} \frac{\sin(m\pi x) \sin(m\pi\xi) \sinh[m\pi(1-y)] \sinh(m\pi\eta)}{m \sinh(m\pi)} \tag{19.184}$$

Hence, if $u(x, y)$ is a solution of problem (19.179)–(19.182), then

$$u(x, y) = -\int_{\Gamma} u(q) \frac{\partial}{\partial n_q} g(p, q) ds_q = \int_{1/2}^{1} f(\xi) \frac{\partial}{\partial \eta} g(x, y, \xi, 0) d\xi$$

$$= 2 \sum_{1}^{\infty} \frac{\sin(m\pi x) \sinh[m\pi(1-y)]}{\sinh(m\pi)} \cdot \int_{1/2}^{1} f(\xi) \sin(m\pi\xi) d\xi \tag{19.185}$$

Repeated integration by parts, bearing in mind conditions (19.181) and the continuity of u on Γ, yields

$$u(x,y) = -2 \sum_1^\infty \frac{\sin(m\pi x) \sinh[m\pi(1-y)]}{(m\pi)^2 \sinh(m\pi)}$$

$$\times \int_{1/2}^1 \frac{d^2}{d\xi^2} f(\xi) \sin(m\pi\xi) d\xi \qquad (19.186)$$

Condition (19.182) now yields

$$\varphi(x) = \frac{2}{\pi} \sum_1^\infty \frac{\sin(m\pi x) \cosh(m\pi)}{m \sinh(m\pi)} \int_{1/2}^1 \frac{d^2}{d\xi^2} f(\xi) \sin(m\pi\xi) d\xi \qquad (19.187)$$

(If $d^2 f/dx^2$ is smooth enough – for example, if it is a function of bounded variation – then term-by-term differentiation of the series is permissible.) Let

$$\alpha_m = \int_{1/2}^1 \frac{d^2}{d\xi^2} f(\xi) \sin(m\pi\xi) d\xi, \quad b_m = \int_{1/2}^1 \varphi(\xi) \sin(m\pi\xi) d\xi \qquad (19.188)$$

By (19.187),

$$b_m = \frac{2}{\pi} \sum_1^\infty \frac{1}{n} \cdot \frac{\cosh(n\pi)}{\sinh(n\pi)} \cdot I_{mn} \alpha_n \qquad (19.189)$$

where

$$I_{mn} = \int_{1/2}^1 \sin(m\pi\xi) \sin(n\pi\xi) d\xi \qquad (19.190)$$

that is,

$$I_{2m,2n} = I_{2m+1,2n+1} = 0 \quad \forall (m,n) \text{ integer} \geq 0 \qquad (19.191)$$

$$I_{2m,2n+1} = \frac{(-1)^{m+n}}{2\pi} \left(\frac{1}{2m+2n+1} + \frac{1}{2m-2n-1} \right) \qquad (10.102)$$

$$I_{2m-1,2n} = \frac{(-1)^{m+n}}{2\pi} \left(\frac{1}{2m-2n-1} - \frac{1}{2m+2n-1} \right) \qquad (19.193)$$

In what follows we shall assume, for the sake of simplicity, that

$$b_{2m} = 0 \quad \forall m \geq 1 \qquad (19.194)$$

Then, by (19.189)–(19.193),

$$0 = \frac{2}{\pi} \sum_1^\infty \frac{1}{2n-1} \frac{\cosh[(2n-1)\pi]}{\sinh[(2n-1)\pi]} I_{2m,2n-1} \alpha_{2n-1} \qquad (19.195)$$

$$b_{2m-1} = \frac{2}{\pi} \sum_1^\infty \frac{1}{2n} \frac{\cosh(2n\pi)}{\sinh(2n\pi)} I_{2m-1,2n} \alpha_{2n} \qquad (19.196)$$

It follows from (19.195) that[9]

$$\alpha_{2m-1} = 0 \quad \forall m \geq 1 \tag{19.197}$$

Thus the problem is reduced to solution of the denumerable system of algebraic equations (19.196). Assume that the solution of this system exists, and that the series

$$\psi(x) = \frac{2}{\pi} \sum_{1}^{\infty} \alpha_{2m} \cdot \sin(2m\pi x) \tag{19.198}$$

is uniformly convergent in $[1/2, 1]$. Then

$$\frac{d^2}{dx^2} f(x) = \psi(x) \tag{19.199}$$

so that repeated integration, taking into account the conditions

$$f(1-0) = 0, \quad f(1/2+0) = 0, \quad \frac{d}{dx} f(1/2+0) = 0 \tag{19.200}$$

determines a solution of the integral equation (19.185) and together with it a solution of the original problem (19.179)–(19.182).

Thus the next steps in the consideration of the problem are as follows: Prove that system (19.196) is solvable; and study the order of decrease of the coefficients of the series.

To prove the solvability of a denumerable system of linear algebraic equations is, in general, a problem requiring complex analysis. There exists, however, a relatively simple case – perfect systems – that can be solved by Picard's method of successive approximation.

Definition 19.6.1. The system of linear algebraic equations

$$z_m = \varphi_m + \sum_{n=1}^{\infty} \delta_{mn} \cdot z_n, \quad m = 1, 2, 3, \ldots \tag{19.201}$$

is called "Perfect" if the series of free terms is absolutely convergent, that is, if there exists $M > 0$ such that

$$\sum_{m=1}^{\infty} |\varphi_m| < M \tag{19.202}$$

and there exists $\rho = \text{const} > 0$ such that[10]

$$\sum_{m=1}^{\infty} |\delta_{mn}| \leq \rho < 1, \quad n = 1, 2, 3, \ldots \tag{19.203}$$

---

[9]Statement (19.197) needs rigorous proof, since homogeneous denumerable systems of linear algebraic equations may, generally speaking, have a nontrivial solution, even when any subsystem of a finitely many equations has only trivial solutions. This may be the case if the solution of the denumerable system of equations is not the limit of the solutions of the truncated systems. The following analysis of system (19.196) clearly shows that this is impossible.

[10]Conditions (19.202) and (19.203) are stronger than necessary. L. Kantorovich, one of the authors of the theory of fully regular systems of linear algebraic equations

Definition 19.6.2. The norm in the sequence space

$$\Pi = \{x\}, \quad x = \{x_1, x_2, \dots, x_n, \dots\} \tag{19.204}$$

is defined as

$$\|x\| = \sum_{k=1}^{\infty} |x_k| \tag{19.205}$$

Theorem 19.6.1. *If conditions* (19.202) *and* (19.203) *are satisfied, then the transformation*

$$y = Tz \overset{\text{def}}{=} \left\{ y_m = \varphi_m + \sum_{n=1}^{\infty} \delta_{mn} z_n; \ m = 1, 2, \dots \right\} \tag{19.206}$$

is a contraction mapping of any bounded closed set $\mathfrak{M} \subset \Pi$ *into itself.*

PROOF. Let

$$y \in \Pi, \quad z \in \Pi, \quad \varphi \in \Pi \tag{19.207}$$

Then

$$\|y\| \le \|\varphi\| + \sum_{m=1}^{\infty} \sum_{n=1}^{\infty} |\delta_{mn}||z_n| = \|\varphi\| + \sum_{n=1}^{\infty} |z_n| \sum_{m=1}^{\infty} |\delta_{mn}|$$

$$\le M + \rho\|z\| \tag{19.208}$$

Hence

$$\|z\| \le N, \ N > \frac{M}{1-\rho} \implies \|y\| < N \tag{19.209}$$

so that T maps $\mathfrak{M}$ into itself.

Now let

$$z_i \in \mathfrak{M}, \quad y_i = Tz_i, \quad i = 1, 2 \tag{19.210}$$

---

[41], used the following definition: A system (19.201) is called *fully regular* if its free terms are uniformly bounded and there exists $\rho = \text{const} > 0$ such that

$$\sum_{n=1}^{\infty} |\delta_{mn}| \le 1 - \rho < 1, \quad m = 1, 2, 3, \dots \tag{α}$$

Kantorovich proved that if these conditions hold then system (19.201) is uniquely solvable by successive approximations, with arbitrarily chosen initial approximation. In the case of a regular system, when (α) is replaced by the weaker condition

$$\sum_{1}^{\infty} |\delta_{mn}| \le \rho < 1, \quad m = 1, 2, 3, \dots \tag{β}$$

the successive approximations converge to a solution of the system only for a special choice of the initial approximation. We use the term "perfect system" instead of "fully regular system" because of the difference in the definitions, which may sometimes be significant.

Then, by virtue of (19.203),

$$\|y_1 - y_2\| = \sum_{m=1}^{\infty}\left|\sum_{n=1}^{\infty}\delta_{mn}(z_{1n} - z_{2n})\right| \leq \sum_{n=1}^{\infty}|(z_1 - z_2)|\sum_{m=1}^{\infty}|\delta_{mn}|$$

$$\leq \rho\|z_1 - z_2\| < \|z_1 - z_2\| \tag{19.211}$$

Hence T is also a contraction mapping of $\mathfrak{M}$. By the Cacciopoli–Banach contracting mapping principle, this implies that T has one and only one fixed point in $\mathfrak{M}$, which can be found by successive approximations with an arbitrary initial approximation. **Q.E.D.**

We return now to the consideration of system (19.196), rewriting it as

$$b_{2m-1} = \frac{1}{8}\sum_{n=1}^{\infty}\beta_n \frac{I_{2m-1,2n}}{I_{2n-1,2n}} \tag{19.212}$$

by setting

$$\beta_m = \frac{16}{\pi}\alpha_{2m}\frac{\cosh(2m\pi)}{\sinh(2m\pi)}\cdot I_{2m-1,2m} \tag{19.213}$$

One could try to solve this system with the aid of Theorem 19.6.1. However, it is easy to see that the system is neither perfect nor regular, since

$$\sum_{m=1}^{\infty}\frac{1}{2n}\left|\frac{I_{2m-1,2n}}{I_{2n-1,2n}}\right| = \infty, \quad \forall n \geq 1 \tag{19.214}$$

On the other hand, the system (19.214) may be replaced by a perfect system. Indeed, by setting

$$C_{1n} = \frac{I_{1,2n}}{2nI_{2n-1,2n}} \tag{19.215}$$

$$C_{mn} = \frac{I_{2m-1,2n} + I_{2m+1,2n}}{2nI_{2n-1,2n}}$$

$$= \frac{(-1)^{m+n}m(4n+1)}{2n[(2m+1)^2 - 4n^2][(2m-1)^2 - 4n^2]}, \quad m = 2,3,\dots \tag{19.216}$$

then

$$\sum_{m=1}^{\infty}|C_{m1}| = \frac{1}{2} + \frac{1}{2}\sum_{m=2}^{\infty}\frac{5m}{(4m^2 - 3)^2 - 16m^2}$$

$$< \frac{1}{2} + \frac{5}{4}\int_4^{\infty}\frac{dx}{(4x-3)^2 - 16x} < \frac{2}{3} \tag{19.217}$$

$$\sum_{m=1}^{\infty}|C_{m2}| = \frac{7}{4\cdot 15} + \frac{1}{4\cdot 15} + \sum_{m=3}^{\infty}\frac{m}{[(2m+1)^2 - 16][(2m-1)^2 - 16]}$$

$$< \frac{2}{15} + \frac{9\cdot 5}{4\cdot 4}\int_3^{\infty}\frac{xdx}{(4x^2 - 15)^2 - 16x^2} < \frac{5}{24} \tag{19.218}$$

and, for every integer $n \geq 2$,

$$\sum_{m=1}^{\infty} |C_{mn}|$$

$$= |C_{1n}| + \sum_{m=2}^{n-1} |C_{mn}| + |C_{nn}| + \sum_{m=n+1}^{\infty} |C_{mn}| \Rightarrow$$

$$= \frac{1}{2n} + \frac{1}{2n^2} \sum_{m=2}^{n-1} \frac{m(4n+1)}{[4n^2 - (2m+1)^2][4n^2 - (2m-1)^2]^2}$$

$$+ \frac{1}{2n(4n-1)} + \frac{1}{2n^2} \sum_{m=n}^{\infty} \frac{m(4n+1)}{[4n^2 - (2m+1)^2][4n^2 - (2m-1)^2]}$$

$$< \frac{2}{4n-1} + \frac{4n+1}{2n^2} \left(\int_{2}^{n-1} + \int_{n}^{\infty} \frac{x\,dx}{|4n^2 - (2x+1)^2||4x^2 - (2m-1)^2|} \right)$$

$$< \frac{1}{2} \tag{19.219}$$

Thus, for every integer $n \geq 1$, there is a positive constant ρ such that

$$\sum_{m=1}^{\infty} |C_{mn}| \leq \rho < 1 \tag{19.220}$$

Q.E.D.

Note that, quite similarly, we can prove that

$$\sum_{n=1}^{\infty} |C_{mn}| \leq \rho < 1 \quad \forall m \geq 1 \tag{19.221}$$

It follows from (19.220) that if

$$f_m = b_{2m-1} + b_{2m+1}, \quad \sum_{m=1}^{\infty} |f_m| < \infty \tag{19.222}$$

then the system

$$f_m = \sum_{n=1}^{\infty} C_{mn} \beta_n \tag{19.223}$$

is perfect. But the convergence of the series on the right side of (19.222) may be guaranteed by the requirement that the prescribed function $\varphi(x)$ be sufficiently smooth. We assume that

$$\varphi(1/2) = \varphi(1) = 0, \qquad \left| \frac{d^2}{dx^2} \varphi \right| \leq M < \infty \tag{19.224}$$

Then the series

$$\sum_{m=1}^{\infty} |b_{2m-1}| \tag{19.225}$$

converges and together with it the series on the right side of (19.222).

It is obvious that the existence and uniqueness of the solution of the original system (19.212) imply the existence and uniqueness of the solution of system (19.215). However, the existence and uniqueness of the solution of system (19.212) is not an obvious corollary of the existence and uniqueness of the solution of system (19.223); a rigorous proof is necessary. This we now provide.[11]

Let β_{2n}^*, $n = 1, 2, 3, \ldots$, denote the solution of the system (19.215). The series obtained is absolutely convergent. Indeed, by the assumption (19.225),

$$|b_{2m-1}| < \frac{M}{m^3} \tag{19.226}$$

which implies

$$|\beta_{2m}^*| = 0(N^{-2}) \quad \forall m \geq N \tag{19.227}$$

since, by the definition (19.215) and (19.217),

$$|C_{mn}| < \frac{1}{2m(4m-1)} \tag{19.228}$$

Further, by (19.193),

$$\frac{1}{\pi m} \cdot \frac{\cosh(2m\pi)}{\sinh(2m\pi)} \cdot \left| \frac{I_{2m-1,2n}}{I_{2n+1,2n}} \right| < \frac{1}{4m^2} \tag{19.229}$$

Moreover, this shows that for any $\varepsilon = \text{const} > 0$ there exists $N > 0$ such that, for all $m > N$,

$$\left| b_{2m-1} - \frac{2}{\pi} \sum_{n=1}^{\infty} \frac{1}{2n} \cdot \frac{\cosh(2n\pi)}{\sinh(2n\pi)} \cdot \frac{I_{2m-1,2n}}{I_{2n-1,2n}} \cdot \beta_{2n}^* \right| < \varepsilon \tag{19.230}$$

This means that the equations of the system (19.212) for sufficiently large m are satisfied by β_{2m}^* with arbitrarily good accuracy.

We now use the method of descent. Let $m > 0$ be chosen so that equation (19.212) is satisfied by $\{\beta_{2n}^*\}$ to within ε. Now take the $(m-1)$th equation of system (19.223), for which $\{\beta_{2n}^*\}$ is an exact solution. Since this equation is the sum of equations of system (19.212) of the numbers m and $m-1$, this means that the last one is satisfied by $\{\beta_{2n}^*\}$ to within ε. Continuation of this descent procedure shows that all the equations of the system (19.212) are satisfied by $\{\beta_{2n}^*\}$ to within ε. Since $\varepsilon > 0$ is arbitrarily small, this means that $\{\beta_{2n}^*\}$ is the exact solution of the system (19.212). **Q.E.D.**

---

[11]The introduction of a perfect system on the right side of (19.223) instead of a nonregular system (19.212), and the use of the method of descent to prove that the solution of the latter coincides with the solution of the former, are borrowed from Kantorovich's monograph [41].

Problems

P19.1.1. For solving the problem (19.1)–(19.3) one could perform computations similar to those that led to the representation of the solution of problem (19.46)–(19.48) in the form (19.70)–(19.73). Do this and explain why these computations are not very useful. Do they make it possible to define the Green's functions for the wave equation?

P19.1.2. Use the method of standing waves to solve the problem of vibration of a loaded finite string with free right-hand end and an elastically fixed left-hand end.

P19.1.3. For a linear spring described by the equation (see Chapter 2, Section 6)

$$m\frac{d^2}{dt^2}u + 2\beta\frac{d}{dt}u + ku = 0 \tag{1}$$

where $u(t)$ is the deviation of the mass m from its equilibrium position, β is the friction coefficient and k is the elasticity constant of the spring, the transition from $\beta < \sqrt{mk}$ to $\beta > \sqrt{mk}$ corresponds to transition from the regime of decaying oscillations to that of monotonic decay. The distributed analog of equation (1) is the telegraph equation (see Problem 2.7.1),

$$m\frac{\partial^2}{\partial t^2}u + 2\beta\frac{\partial}{\partial t}u = k\frac{\partial^2}{\partial x^2}u \tag{2}$$

where $u(x,t)$ is the displacement of the spring at the point x and m is the linear mass density of the spring (β and k as previously). What physical and mathematical reality corresponds to the analog of the aforementioned transition for equation (2), and to the limit $m \downarrow 0$?

Hint: Consider, through separation of variables, the initial–boundary-value problem with homogeneous boundary conditions of Dirichlet type for equation (2) on a segment.

P19.2.1. Use the method of expansion in Fourier series to solve the heat-conduction equation in a homogeneous slab of a finite length, in close contact at its right-hand end with a medium at zero temperature and irradiated at the left-hand end by a similar medium. Assume that at the initial time an instantaneous unit source of heat is placed at some arbitrarily chosen point of the slab.

P19.3.1. Derive the Fourier expansion of the solution to problem (19.74)–(19.76) using expression (19.87) for the Green's function $G(x,y,\xi,\eta)$.

P19.4.1. Formulate sufficient conditions for the solution (19.132) of problem (19.105) and (19.106) to be rigorously valid (not only formal). Recall that to this end one must demonstrate that: (1) the series (19.132) and the series obtained by term-by-term differentiation are uniformly convergent in the closure of the region D under consideration for all $t \geq t_0$ and arbitrarily small $t_0 = \text{const} > 0$; (2) the series obtained by repeated term-by-term differentiation is uniformly convergent in every closed subregion of

D and (3) the sum of the series (19.132) satisfies the boundary and initial conditions.

P19.5.1. Estimate the order of decrease of eigenvalues λ_{mn} defined by equation (19.164).

P19.6.1. Provide the electrostatic and thermal interpretations for the problem (19.178)–(19.182).

P19.6.2. Solve a problem analogous to (19.178)–(19.182) in which the region is not a rectangle but rather a circle $r \leq 1$, where (r,φ) are polar coordinates in $\mathbb{R}_2$.

P19.6.3. As stated in Chapter 2, the problem of solving a Fredholm integral equation of the first kind is generally ill-posed in Hadamard's sense. Problem (19.178)–(19.182) was solved by reducing it to a Fredholm equation of the first kind. Determine whether the solution of this problem is stable with respect to small perturbations of the boundary data, assuming that the perturbed data are smooth enough and satisfy the conditions

$$u \equiv 0 \quad \text{at } y = 0, \quad x = {}^1\!/_2, \ x = 1, \qquad \frac{\partial}{\partial x} \cdot u({}^1\!/_2 \pm 0, 0) = 0$$

P19.6.4. The proof of solubility of the infinite system (19.196) of algebraic equations referred to the method of descent by L. B. Kantorovich [41]. Is it possible to transform this system to a perfect one by substitution of the type $\gamma_k = \alpha_k \ln^q(k+p)$ with suitable chosen $p > 0$ and $q > 0$, so that there will be no need for the use of the method of descent? If so, prove it; if not, explain why.

20

Integral transform method in unbounded regions

1. Integral transforms in solution of boundary-value problems in unbounded regions

We recall the first boundary-value problem for the heat-conduction equation in a finite slab:

$$Lu \overset{\text{def}}{=} \frac{\partial^2}{\partial x^2}u - \frac{\partial}{\partial t}u = 0 \quad \forall x \in (0, \ell), \ \forall t > 0 \tag{20.1}$$

$$u(0, t) = 0, \quad u(\ell, t) = 0 \quad \forall t > 0 \tag{20.2}$$

$$u(x, 0) = f(x) \quad \forall x \in (0, \ell) \tag{20.3}$$

The Fourier series expansion

$$u(x, t) = \frac{2}{\ell} \sum_{n=1}^{\infty} a_n \exp\left(-\frac{n^2\pi^2 t}{\ell^2}\right) \sin\left(\frac{n\pi x}{\ell}\right) \tag{20.4}$$

of the solution of this problem (obtained in Chapter 19, Section 2) is the expansion of the solution as a series in the eigenfunctions of the Sturm–Liouville problem:

$$\frac{d^2 X}{dx^2} + \lambda^2 X = 0, \quad 0 < x < \ell, \quad X(0) = 0, \quad X(\ell) = 0 \tag{20.5}$$

which has the discrete spectrum

$$\lambda = \frac{n\pi}{\ell}, \quad n = 1, 2, 3, \dots \tag{20.6}$$

As $\ell \uparrow \infty$, the difference

$$\Delta\lambda = \frac{n+1}{\ell}\pi - \frac{n}{\ell}\pi = \frac{\pi}{\ell} \tag{20.7}$$

between any two neighboring eigenvalues tends to zero, so that the discrete spectrum becomes a continuous spectrum $0 \leq \lambda < \infty$ and the series (20.4) becomes an integral

$$u(x, t) = \frac{2}{\pi} \int_0^{\infty} \alpha(\lambda) \exp(-\lambda^2 t) \sin(\lambda x) d\lambda \tag{20.8}$$

where[1]

$$\alpha(\lambda) = \int_0^\infty f(\xi) \sin(\lambda\xi) d\xi \tag{20.9}$$

Inserting (20.9) into (20.8), we obtain

$$u(x,t) = \frac{2}{\pi} \int_0^\infty \exp(-\lambda^2 t) \sin(\lambda x) d\lambda \int_0^\infty f(\xi) \sin(\lambda\xi) d\xi \tag{20.10}$$

Changing the order of integration, we get

$$u(x,t) = \int_0^\infty f(\xi) g(x,\xi,t) d\xi \tag{20.11}$$

where

$$g(x,\xi,t) = \frac{2}{\pi} \int_0^\infty \exp(-\lambda^2 t) \sin(\lambda x) \sin(\lambda\xi) d\lambda$$

$$= E(x - \xi, t) - E(x + \xi, t) \tag{20.12}$$

is the Green's function of the thermal Dirichlet problem in a half-space $x > 0$
(see (15.109)). Thus our heuristic arguments yield the rigorously verifiable
result.

Note that, when $\ell \uparrow \infty$, problem (20.1)–(20.3) becomes

$$\frac{\partial^2}{\partial x^2} u - \frac{\partial}{\partial t} u = 0 \quad \forall x \in (0,\infty), \ \forall t > 0 \tag{20.13}$$

$$u(0,t) = 0 \tag{20.14}$$

$$u(x,0) = f(x) \quad \forall x \in (0,\infty) \tag{20.15}$$

$$\lim_{x \uparrow \infty} u(x,t) = 0 \quad \forall t > 0 \tag{20.16}$$

Obviously, in order to obtain this result there is no need to first truncate
the half-space $x > 0$ to a finite interval $(0, \ell)$. Indeed, we can look for the
solution as a Fourier sine integral

$$u(x,t) = \frac{2}{\pi} \int_0^\infty v(\lambda, t) \sin(\lambda x) d\lambda \tag{20.17}$$

Inserting (20.17) into (20.13) and changing the order of differentiation and
integration, we obtain

$$Lu = \frac{2}{\pi} \int_0^\infty \left(\frac{\partial}{\partial t} + \lambda^2 \right) v(\lambda, t) \sin(\lambda x) d\lambda = 0 \tag{20.18}$$

---

[1]Recall that

$$a_m = \int_0^\ell f(\xi) \sin\left(\frac{m\pi\xi}{\ell} \right) d\xi$$

so that, in the limit of $\ell = \infty$, $\frac{2}{\ell} a_m$ becomes the integral

$$a(\lambda) = \frac{2}{\pi} \int_0^\infty f(\xi) \sin(\lambda\xi) d\xi$$

This equality becomes an identity if

$$\left(\frac{\partial}{\partial t} + \lambda^2\right) v(\lambda, t) \equiv 0 \quad \forall \lambda \in (0, \infty), \ \forall t > 0 \qquad (20.19)$$

Integration yields

$$v(\lambda, t) = \alpha(\lambda) \exp(-\lambda^2 t) \qquad (20.20)$$

so that

$$u(x, t) = \frac{2}{\pi} \int_0^\infty \alpha(\lambda) \exp(-\lambda^2 t) \sin(\lambda x) d\lambda \qquad (20.21)$$

The *spectral function* $\alpha(\lambda)$ is determined by the initial condition

$$f(x) = \frac{2}{\pi} \int_0^\infty \alpha(\lambda) \sin(\lambda x) d\lambda \qquad (20.22)$$

This condition is a singular integral equation of the first kind. The theory of Fourier integrals provides its solution:

$$\alpha(\lambda) = \int_0^\infty f(s) \sin(s\lambda) ds \qquad (20.23)$$

Substitution of (20.23) into (20.21) yields the previously obtained result (20.10).

The following terminology is customary in the theory of Fourier integrals: If

$$F_s(\lambda) = \sqrt{2/\pi} \int_0^\infty f(s) \sin(\lambda s) ds$$

$$\Rightarrow f(s) = \sqrt{2/\pi} \int_0^\infty F_s(\lambda) \sin(\lambda s) ds \qquad (20.24)$$

then $F_s(\lambda)$ is called the *sine transform* of $f(x)$; $f(x)$ is called the *source function*. In the problem (20.7)–(20.9) $f(x)$ is a prescribed function, and is therefore the source function.

Returning now to equality (20.17), let us consider u as the source function. Then its sine transform is

$$v(\lambda, t) = \sqrt{2/\pi} \int_0^\infty u(x, t) \sin(\lambda x) dx \qquad (20.25)$$

Assume that $u(x, t)$ is a solution of problem (20.1)–(20.2). Then, taking the sine transform of the identity

$$0 \equiv \left(\frac{\partial^2}{\partial x^2} - \frac{\partial}{\partial t}\right) u(x, t) \qquad (20.26)$$

we obtain

$$0 \equiv \sqrt{2/\pi} \int_0^\infty \left(\frac{\partial^2}{\partial \xi^2} - \frac{\partial}{\partial t}\right) u(\xi, t) \sin(\lambda \xi) d\xi \qquad (20.27)$$

Integration by parts, taking into account the boundary condition (20.14) and the change in the order of differentiation and integration, gives

$$\left(\frac{\partial}{\partial t} + \lambda^2 \right) v(\lambda, t) = 0 \tag{20.28}$$

so that again

$$v(\lambda, t) = \alpha(\lambda) \exp(-\lambda^2 t) \tag{20.29}$$

The spectral function $\alpha(\lambda)$ is determined by the initial condition (20.2), so that

$$\alpha(\lambda) = \sqrt{2/\pi} \int_0^\infty u(\xi, 0) \sin(\lambda\xi)d\xi = \sqrt{2/\pi} \int_0^\infty f(\xi)\sin(\lambda\xi)d\xi \tag{20.30}$$

Finally, the source function $u(x, t)$ is found by applying the inversion formula

$$u(x, t) = \sqrt{2/\pi} \int_0^\infty v(s, t)\sin(sx)ds \tag{20.31}$$

Thus there are two equivalent procedures:

(a) Treat $u(x, t)$ as the sine transform of the unknown function $v(\lambda, t)$. This leads to an ordinary differential equation for v. The general integral of this equation depends on the spectral function $\alpha(\lambda)$, which is determined from an additional condition (in the example under consideration, the initial condition of the problem) by using the inversion formula (20.24).

(b) Treat $v(\lambda, t)$ as the sine transform of the source function $u(x, t)$. Then the sine transform of the partial differential equation for u determines an ordinary differential equation for $v(\lambda, t)$, so that $v(\lambda, t)$ depends on the spectral function $\alpha(\lambda)$, which in turn is determined from an additional (initial) condition of the problem.

This is the idea and general scheme of the integral transform method, one of the strongest methods used in the theoretical investigation and numerical analysis of linear processes described by differential equations. It was originally developed in Fourier's famous treatise *La théorie analitique de la Chaleur* [19] where the integrals, now named for Fourier, appeared for the first time, and which led to the modern theory of integral transforms in general and of the Fourier integral in particular. This theory is one of the most profound and beautiful theories of mathematical analysis, providing tools for many different branches of mathematics and physics, including the descriptive theory of functions, theory of analytical functions, theory of partial differential equations, the general spectral theory of linear operators, theory of probability, and so on.

In what follows we restrict ourselves to formulation of the principal results, referring the reader to the special literature [74;86]. For the reader's convenience, some elementary information is presented in Appendix 4.

2. Fourier transform, sine and cosine Fourier transform. Double Fourier integral and Fourier–Lebesgue theorem. Fourier transform of derivatives

If

$$\varphi(x) = \int_{-\infty}^{\infty} f(s)K(xs)ds \Rightarrow f(x) = \int_{-\infty}^{\infty} \varphi(\lambda)H(\lambda x)d\lambda \qquad (20.32)$$

then $K(xs)$ and $H(\lambda x)$ are known as *reciprocal Fourier kernels*. The most important example of reciprocal kernels is

$$K(x) = \frac{1}{\sqrt{2\pi}} \exp(-ix), \quad H(x) = \frac{1}{\sqrt{2\pi}} \exp(ix) \qquad (20.33)$$

The integral transform based on these two kernels is called the *Fourier transform.*

Note that the pair (20.32) of equalities is equivalent to the identity

$$f(x) = \int_{-\infty}^{\infty} H(xs)ds \int_{\infty}^{\infty} f(\lambda)K(\lambda s)d\lambda \qquad (20.34)$$

Inserting the kernels (20.33), we obtain

$$f(x) = \frac{1}{2\pi} \int_{-\infty}^{\infty} \exp(ixs)ds \int_{-\infty}^{\infty} f(\lambda)\exp(-i\lambda s)d\lambda \qquad (20.35)$$

The integral in the right-hand side of (20.35) is called the *double Fourier integral.*

The truth of equality (20.35) is subject to certain conditions.

Theorem 20.2.1 (Dirichlet's conditions). *Let $f(x)$ satisfy the Dirichlet conditions:*

(a) *$f(x)$ is bounded and absolutely integrable for all $x \in (-\infty, \infty)$;*
(b) *$f(x)$ has at most a finite number of extremum points and discontinuities of the first kind.*

Then, for any $x \in (-\infty, \infty)$,

$$\frac{f(x+0) + f(x-0)}{2} = \frac{1}{2\pi} \int_{-\infty}^{\infty} \exp(ixs)ds \int_{-\infty}^{\infty} f(\lambda)\exp(-i\lambda s)d\lambda \qquad (20.36)$$

A proof of this theorem will be given in Appendix 4. More general conditions for the validity of this identity are given by the Lebesgue–Riemann theorem [86]: *If $f(x)$ is bounded and absolutely integrable in $(-\infty, \infty)$, then the equality (20.35) is valid at every point x where $f(x)$ has bounded variation.*

Now let $f(x)$ be an odd function satisfying the Dirichlet condition. Then the Fourier transform becomes a sine Fourier transform, so that

$$F_s(s) = \sqrt{2/\pi} \int_0^{\infty} f(\lambda)\sin(\lambda s)ds$$

$$\Rightarrow f(x) = \sqrt{2/\pi} \int_0^{\infty} F_s(s)\sin(xs)ds \qquad (20.37)$$

and together with this

$$\frac{f(x+0)+f(x-0)}{2} = \frac{1}{2\pi}\int_{-\infty}^{\infty} ds \int_{-\infty}^{\infty} f(\lambda)\sin[(s-\lambda)]d\lambda \qquad (20.38)$$

Finally, if $f(x)$ is an even function satisfying the Dirichlet condition, then the Fourier transform becomes a cosine Fourier transform, so that

$$F_c(s) = \sqrt{2/\pi}\int_0^{\infty} f(\lambda)\cos(\lambda s)d\lambda$$

$$\Rightarrow f(x) = \sqrt{2/\pi}\int_0^{\infty} F_c(s)\cos(xs)ds \qquad (20.39)$$

and together with this

$$\frac{f(x+0)+f(x-0)}{2} = \frac{1}{2\pi}\int_{-\infty}^{\infty} ds \int_{-\infty}^{\infty} f(\lambda)\cos[s(x-\lambda)]d\lambda \qquad (20.40)$$

In what follows, we denote the Fourier transform, cosine transform, and sine transform of f by $F(f)$, $F_c(f)$, and $F_s(f)$, respectively.

Assume that f and its derivative $df(x)/dx$ satisfy the Dirichlet conditions. Then integration by parts yields

$$F(f) = \frac{1}{is\sqrt{2\pi}}\int_{-\infty}^{\infty} \exp(-i\lambda s)\frac{d}{d\lambda}f(\lambda)d\lambda \quad \Rightarrow F\left(\frac{df}{ds}\right) = isF(f) \quad (20.41)$$

and similarly

$$F_c\left(\frac{df}{ds}\right) = sF_s(f) + \sqrt{2/\pi}f(0), \quad F_s\left(\frac{df}{ds}\right) = -sF_c(f) \qquad (20.42)$$

Hence, by induction, if $d^k f/dx^k$ satisfies the Dirichlet conditions for all $k \in [0, n]$, then

$$F\left(\frac{d^n}{dx^n}f\right) = s^n\exp\left(\frac{in\pi}{2}\right)F(f) \qquad (20.43)$$

$$F_c\left(\frac{d^n}{dx^n}f\right) = \begin{cases} (-1)^m s^{2m} F_c(f) - \sqrt{2/\pi}\sum_{k=0}^{m-1}(-1)^k s^{2k}\frac{d^{n-2k-1}}{ds^{n-2k-1}}f(0) \\ \quad \text{if } n = 2m \geq 2 \\ (-1)^m s^{2m+1} F_s(f) - \sqrt{2/\pi}\sum_{k=0}^{m}(-1)^k \frac{d^{n-2k}}{ds^{n-2k}}f(0)s^{2k} \\ \quad \text{if } n = 2m+1 \geq 1 \end{cases} \qquad (20.44)$$

$$F_s\left(\frac{d^n}{dx^n}\cdot f\right) = \begin{cases} (-1)^{m+1} s^n F_s(f) - \sqrt{2/\pi}\sum_{k=0}^{m}(-1)^k \frac{d^{n-2k}}{ds^{n-2k}}\cdot f(0)s^{2k-1} \\ \quad \text{if } n = 2m \geq 2 \\ (-1)^{m+1} s^n F_c(f) - \sqrt{2/\pi}\sum_{k=1}^{m}(-1)^k \frac{d^{n-2k}}{ds^{n-2}}f(0)s^{2k-1} \\ \quad \text{if } n = 2m+1 \geq 1 \end{cases} \qquad (20.45)$$

3. Use of Fourier transforms to solve Cauchy problem of heat conduction

A representation of the solution of the Cauchy problem

$$Lu \stackrel{\text{def}}{=} \frac{\partial^2 u}{\partial x^2} - \frac{\partial u}{\partial t} = 0 \quad \forall x \in (-\infty, \infty), \; \forall t > 0, u(x,0) = f(x) \qquad (20.46)$$

as a Poisson integral

$$u(x,t) = \int_{-\infty}^{\infty} f(\xi) E(x - \xi, t) d\xi \qquad (20.47)$$

has been obtained in Chapter 13 (Sections 1 and 2) by considering a similarity solution of the heat-conduction equation and differentiating it with respect to the spatial coordinate x. We now obtain the same solution by the Fourier transform method. This repeated solution of an "old" problem will demonstrate that application of a more general, universal method sometimes essentially involves a need for much more restrictive assumptions than a method based on specific properties of the problem under consideration.

Thus, let us assume that $f(x)$ satisfies the Dirichlet condition and introduce the Fourier transforms of u and f:

$$U(s,t) = \sqrt{\frac{1}{2\pi}} \int_{-\infty}^{\infty} u(\lambda, t) \exp(-i\lambda s) d\lambda \qquad (20.48)$$

Let us assume that it is legitimate to change the order of integration and differentiation, so that

$$\frac{\partial}{\partial t} U(s,t) = \sqrt{\frac{1}{2\pi}} \int_{-\infty}^{\infty} \frac{\partial}{\partial t} u(\lambda, t) \exp(-i\lambda s) d\lambda \qquad (20.49)$$

Since u is assumed to be a solution of problem (20.46), we have

$$\frac{\partial}{\partial t} U(s,t) + s^2 U(s,t) = 0 \qquad (20.50)$$

Integration yields

$$U(s,t) = \alpha(s) \exp(-s^2 t) \qquad (20.51)$$

This and the initial condition (20.46) give

$$\alpha(s) = \sqrt{\frac{1}{2\pi}} \int_{-\infty}^{\infty} f(\lambda) \exp(-i\lambda s) d\lambda \qquad (20.52)$$

Applying the inversion formula, we obtain

$$u(x,t) = \frac{1}{2\pi} \int_{-\infty}^{\infty} \exp(ixs - s^2 t) ds \int_{-\infty}^{\infty} f(\lambda) \exp(-i\lambda s) d\lambda \qquad (20.53)$$

Since $f(\lambda)$ is assumed to be absolutely integrable, we may change the order of integration; hence

$$u(x,t) = \int_{-\infty}^{\infty} f(\lambda) \mathcal{E}(x,t,\lambda) d\lambda \qquad (20.54)$$

where

$$\mathcal{E}(x,t,\lambda) = \int_{-\infty}^{\infty} \exp[i(x-\lambda)s - s^2 t]ds \qquad (20.55)$$

This integral is easily calculated. Indeed, let

$$z = s\sqrt{t} - i(x-\lambda)/2\sqrt{t} \qquad (20.56)$$

and consider the contour integral in the z plane

$$I(x,t,\lambda) = \frac{1}{2\pi\sqrt{t}} \exp\left(-\frac{(x-\lambda)^2}{4t}\right) \int_{\Gamma} \exp(-z^2)dz \qquad (20.57)$$

where Γ is the line

$$\Gamma = \left\{ z : \Re z \in (-\infty,\infty), \ \Im z = -(x-\lambda)/2\sqrt{t} \right\} \qquad (20.58)$$

Obviously,

$$\mathcal{E}(x,t,\lambda) = I(x,t,\lambda) \qquad (20.59)$$

Let

$$\Gamma^* \stackrel{\text{def}}{=} \Gamma_{abcda} \qquad (20.60)$$

be a rectangular contour with vertices at the points

$$a = \left[N, -\frac{x-\lambda}{2\sqrt{t}}\right], \quad b = [N,0], \quad c = [-N,0], \quad d = \left[-N, -\frac{x-\lambda}{2\sqrt{t}}\right] \quad (20.61)$$

By the Cauchy theorem,

$$\int_{\Gamma_{abcda}} e(-z^2)dz = 0 \qquad (20.62)$$

so that

$$\int_{\Gamma_{da}} = \int_{\Gamma_{cb}} - \int_{\Gamma_{ab}} + \int_{\Gamma_{dc}} \qquad (20.63)$$

We have

$$\lim_{N\uparrow\infty} \int_{\Gamma_{ab}} = \lim_{N\uparrow\infty} \int_{\Gamma_{dc}} = 0, \ \lim_{N\uparrow\infty} \int_{\Gamma_{da}} = \int_{\Gamma} \exp(-z^2)dz$$

$$\lim_{N\uparrow\infty} \int_{\Gamma_{cb}} = \int_{-\infty}^{\infty} \exp(-x^2)dx = \sqrt{\pi} \qquad (20.64)$$

Hence

$$I(t,\lambda) = \frac{1}{2\pi\sqrt{t}} \exp\left[-\frac{(x-\lambda)^2}{4t}\right] \int_{\Gamma} \exp(-z^2)dz \equiv E(x-\lambda,t) \qquad (20.65)$$

Thus, it follows from (20.54), (20.59), and (20.65) that

$$u(x,t) = \int_{-\infty}^{\infty} f(\xi)E(x-\xi,t)d\xi \qquad (20.66)$$

so that the Fourier transform indeed enables us to obtain the Poisson integral, not only heuristically but quite rigorously. However, this result

is the consequence of much more restrictive assumptions than are in fact
necessary. Indeed, the main restriction is imposed by the Dirichlet condition,
which requires that $f(x)$ be absolutely integrable; that is, it should tend
to zero sufficiently rapidly as $|x| \uparrow \infty$. In contrast, the representation of
$u(x,t)$ as a Poisson integral requires only that the rate of growth of $f(x)$ at
infinity not exceed $K \exp(ax^2)$ for any $K = \text{const} > 0$ and $a = \text{const} > 0$.[2]
This situation is typical. Very often the final result of a computation is
true under much less stringent conditions than required by the methods
employed. Moreover, it is usually not an easy task to check whether
the rigorous requirements for the method to be applicable are satisfied.
Therefore, the most desirable way to solve a problem is first to obtain a
formally valid result, and then to check whether this solution in fact meets
all the requirements; a check of this kind is usually much more feasible.

   Sometimes the mere possibility of using some other methods yields a
rigorous proof of the existence of a solution with the properties necessary
to justify application of the integral transform method. This is true, in
particular, with regard to properties of the solutions of Volterra integral
equations, the use of which in numerical computations is often very difficult,
but which provide simple and lucid tools for a theoretical analysis of the
problem.

4. Fourier–Bessel (Hankel) transform and solution of boundary-value problems with cylindrical symmetry. Fundamental solution of heat-conduction equation with forced convection, generated by continuously acting source of incompressible liquid

We shall now use the double Fourier–Bessel integral (See Appendix 2,
Section 5)[3]

$$f(x) = \int_0^\infty sJ_\nu(xs)ds \int_0^\infty f(\lambda)J_\nu(s\lambda)d\lambda \qquad (20.67)$$

or (what is the same) the pair of transforms

$$\varphi(s) = \int_0^\infty \lambda f(\lambda)J_\nu(s\lambda)d\lambda, \quad f(x) = \int_0^\infty s\varphi(s)J_\nu(sx)ds \qquad (20.68)$$

to solve boundary-value problems with cylindrical symmetry.

   As a typical problem, consider the following: Find a function $u(r,t)$
satisfying the conditions

$$\frac{\partial^2 u}{\partial r^2} + \frac{1-2\nu}{r} \cdot \frac{\partial u}{\partial r} - \frac{\partial u}{\partial t} + F(r,t) = 0 \quad \forall r \in (0,\infty), \ \forall t > 0 \qquad (20.69)$$

---

[2]The generalization, due to Titchmarsh [87], enables us to consider the Fourier
transforms of functions that go to infinity as $K \exp(a|x|)$ (see Appendix 4,
Section 3), which is much weaker than the natural restriction $|f| < K \exp(ax^2)$.

[3]The identity (20.67) is valid if $f(x)$ satisfies the same requirements as those listed
in the Dirichlet conditions (Theorem 20.2.1), provided that $\nu > -1/2$ (see, e.g.,
[74;86]).

$$u(r,0) = \varphi(r), \quad |u| < \infty \quad \forall r \in (0,\infty), \quad u(0,t) = f(t) \quad \forall t > 0 \qquad (20.70)$$

$$\varlimsup_{r\downarrow 0} |ur^{-\nu}| < \infty, \quad \varlimsup_{r\uparrow\infty} |ur^{-\nu}| < \infty \qquad (20.71)$$

where ν is assumed to be a positive constant

$$\nu = \mathrm{const} > 0 \qquad (20.72)$$

Note that if $\nu = 0$, problem (20.69), (20.70), (20.71) is the standard Cauchy problem of heat conduction in $\mathbb{R}_2$ with cylindrical symmetry and distributed sources of heat. In that case the boundary condition on the right side of (20.70) must be replaced by the usual condition that u be bounded. If $\nu < 0$, so that the axis $r = 0$ is a heat sink, then the temperature at $r = 0$ cannot be prescribed, either, and the requirement that u be bounded cannot guarantee the uniqueness of the solution. Sometimes the physical context of the problem allows one to introduce a boundary condition at $r = 0$, such as the heat-exchange condition. Namely, we consider the problem in the region $r > \rho$ and assume that heat exchange with a medium of a prescribed temperature takes place at the boundary $r = \rho$:

$$\frac{\partial}{\partial r} u(\rho,t) - hu(\rho,t) = -hf(t) \qquad (20.73)$$

Then the boundary condition on the right side of (20.70) may be replaced by the limiting relation

$$\lim_{\rho\downarrow 0} \left(\frac{\partial}{\partial r} u(\rho,t) - hu(\rho,t) \right) = -hf(t) \qquad (20.74)$$

We shall concentrate, however, on the case $\nu > 0$, treating the problem (20.69)–(20.72) only.[4]
We first consider the Cauchy problem

$$\frac{\partial^2 v}{\partial r^2} + \frac{1 - 2\nu}{r} \frac{\partial v}{\partial r} - \frac{\partial v}{\partial t} = 0 \quad \forall r \in (0,\infty), \ \forall t > 0 \qquad (20.75)$$

$$v(r,0) = \varphi(r), \quad |v| < \infty \quad \forall r \in (0,\infty) \qquad (20.76)$$

Following the standard procedure of the method of separation of variables, we seek a particular solution of equation (20.75) in the form

$$v(r,t) = R(r)\exp(-\lambda^2 t) \qquad (20.77)$$

where R is an integral of the equation

$$\frac{d^2 R}{dr^2} + \frac{1 - 2\nu}{r} \frac{dR}{dr} + \lambda^2 R = 0 \qquad (20.78)$$

---

[4]Problem (20.70)–(20.72) admits the following interpretation: Let hot water at temperature $f(t)$ be injected at a constant rate through a unit well of infinitely thin radius into a thin layer of a homogeneous porous medium at initial temperature $\varphi(r)$. The source term $F(r,t)$ describes the influx of heat from the surrounding rocks; $u(r,t)$ determines the redistribution of the temperature of the stratum due to the injection of heat.

The substitution

$$R = r^{\nu} R^* \tag{20.79}$$

gives

$$\frac{d^2 R^*}{dr^2} + \frac{1}{r}\frac{dR^*}{dr} + \left(\lambda^2 - \frac{\nu^2}{r^2}\right)R^* = 0 \tag{20.80}$$

The general solution of this equation,

$$R^* = \alpha(\lambda)J_{\nu}(\lambda r) + \beta(\lambda)Y_{\nu}(\lambda r) \tag{20.81}$$

satisfies the requirement (20.71) only if $\beta(\lambda) \equiv 0$. Thus we must take

$$R(r) = \alpha(\lambda)r^{\nu}J_{\nu}(\lambda r) \tag{20.82}$$

Since the spectrum is continuous, we must set

$$v(r,t) = \int_0^{\infty} \lambda r^{\nu}\alpha(\lambda)\exp(-\lambda^2 t)J_{\nu}(\lambda r)d\lambda \tag{20.83}$$

The spectral function $\alpha(\lambda)$ is determined from the initial condition

$$\varphi(r) = \int_0^{\infty} \lambda r^{\nu}\alpha(\lambda)J_{\nu}(\lambda r)d\lambda \tag{20.84}$$

which implies

$$\alpha(\lambda) = \int_0^{\infty} \rho^{1-\nu}\varphi(\rho)J_{\nu}(\lambda\rho)d\rho \tag{20.85}$$

Insertion of (20.85) in (20.83) and changing the order of integration yield

$$v(r,t) = \int_0^{\infty}\int_0^{\infty} \rho^1\varphi(\rho)d\rho$$
$$\times \int_0^{\infty} \lambda\exp(-\lambda^2 t)\left(\frac{r}{\rho}\right)^{\nu} J_{\nu}(\lambda\rho)J_{\nu}(\lambda r)d\lambda \tag{20.86}$$

The inner integral in the right-hand side of (20.86) is the second Weber integral (see (A2.202)):

$$\int_0^{\infty} \lambda\exp(-\lambda^2 t)\left(\frac{r}{\rho}\right)^{\nu} J_{\nu}(\lambda\rho)J_{\nu}(\lambda r)d\lambda = E_{\nu}(r,\rho,t) \tag{20.87}$$

where

$$E_{\nu}(r,\rho,t) \overset{\text{def}}{=} \left(\frac{r}{\rho}\right)^{\nu}\frac{\exp[-(r^2+\rho^2)/4t]}{2t}\cdot I_{\nu}\left(\frac{r\rho}{2t}\right) \tag{20.88}$$

Thus

$$v(r,t) = \int_0^{\infty} \rho\varphi(\rho)E_{\nu}(r,\rho,t)d\rho \tag{20.89}$$

This result has been obtained by a formal application of the rules of the Fourier–Bessel transform, and it therefore requires justification by direct means. It is, however, almost obvious that (20.89) is the solution of the Cauchy problem under consideration. Indeed, for any $t > 0$, $r > 0$, and $\rho > 0$,

$$LE_\nu \overset{\text{def}}{=} \left(\frac{\partial^2}{\partial r^2} + \frac{1 - 2\nu}{r} \frac{\partial}{\partial r} - \frac{\partial}{\partial t} \right) E_\nu = 0$$

$$ME_\nu \overset{\text{def}}{=} \left(\frac{\partial^2}{\partial \rho^2} + \frac{1 + 2\nu}{\rho} \frac{\partial}{\partial \rho} - \frac{\partial}{\partial t} \right) E_\nu = 0$$

(20.90)

In addition, it follows from (A2.97) and (A2.225) that

$$E_\nu(r, \rho, t) \asymp \left(\frac{r}{\rho} \right)^\nu \frac{\exp[-(r - \rho)^2/4t]}{2\sqrt{\pi t}} \cdot \frac{1}{\sqrt{r\rho}} \quad \text{as} \quad \frac{r\rho}{2t} \uparrow \infty \qquad (20.91)$$

$$E_\nu(r, \rho, t) \asymp \left(\frac{r}{2\sqrt{t}} \right)^\nu \frac{\exp[-(r^2 + \rho^2)/4t]}{2t} \cdot \frac{1}{\Gamma(\nu + 1)} \quad \text{as} \quad \frac{r\rho}{2t} \downarrow 0 \qquad (20.92)$$

and, moreover, for any $r, \rho, t > 0$,

$$0 < E_\nu(r, \rho, t) < \alpha_\nu \left(\frac{r}{2\sqrt{t}} \right)^\nu \cdot \frac{\exp[-(r - \rho)^2/4t]}{2t} \qquad (20.93)$$

where[5]

$$\alpha_\nu = \max \left(\frac{1}{\Gamma(\nu + 1)}, \frac{1}{\Gamma(\nu + 2)} \right) \qquad (20.94)$$

Hence the integral (20.89) is uniformly convergent and the asymptotic equality (20.91) is applicable if $\varphi(\rho)$ is bounded in the neighborhood of $\rho = 0$. But this means that $E_\nu(r, \rho, t)$ possesses the same filtering properties as the fundamental solution of the heat-conduction equation in $\mathbb{R}_2$. Thus the integral (20.89) is not only a formal but also a true solution of the Cauchy problem.

We can now construct an integral representation of the solution of problem (20.69)–(20.72). To this end, let $\delta > 0$ and $\varepsilon > 0$ be arbitrarily small. Let $u(r, t)$ be a solution of problem (20.69)–(20.72). Then the integral identity may be applied to u and E_ν:

$$0 \equiv \int_0^t d\tau \int_\delta^\infty \rho[E_\nu(r, \rho, t + \varepsilon - \tau)Lu(\rho, \tau) - u(\rho, \tau)ME_\nu(r, \rho, t + \varepsilon - \tau)]d\rho$$

$$= -\int_0^t \rho \left(u \frac{\partial}{\partial \rho} E_\nu - E_\nu \frac{\partial u}{\partial \rho} \right) \Big|_{\rho=\delta}^\infty d\tau - 2\nu \int_0^t u E_\nu \Big|_{\rho=\delta}^\infty d\tau$$

$$- \int_\delta^\infty \rho E_\nu u \Big|_{\tau=0}^t d\rho \qquad (20.95)$$

---

[5]If $\mu > -1/2$ then $\Gamma^-(1 + \mu)$ decreases monotonically. Therefore, if $z > 0$ then

$$0 < \left(\frac{z}{2} \right)^{-\nu} I_\nu(z) = \sum_{n=0}^\infty \frac{(z/2)^{2n}}{\Gamma(n + 1)\Gamma(n + \nu + 1)} < \alpha_\nu \left(\sum_{n=0}^\infty \frac{z/2}{n!} \right)^2 = \alpha_\nu \exp(2z)$$

By the filtering property of E_ν, this means that in the limit of $\delta = 0$ and $\varepsilon = 0$,

$$u(r,t) = 2\nu \int_0^t f(\tau)E_\nu(r,0,t-\tau)d\tau + \int_0^\infty \rho\varphi(\rho)E_\nu(r,\rho,t-\tau)d\rho$$

$$+ \int_0^t d\tau \int_0^\infty \rho F(\rho,\tau)E_\nu(r,\rho,t-\tau)d\rho = u_1 + u_2 + u_3 \qquad (20.96)$$

provided that

$$\lim_{\delta\downarrow 0} \int_0^t \rho\left(u\frac{\partial}{\partial\rho}E_\nu - E_\nu\frac{\partial u}{\partial\rho}\right)\bigg|_{\rho=\delta} d\tau = 0 \qquad (20.97)$$

The boundedness of u and the definition of E_ν imply

$$\lim_{\rho\downarrow 0} \rho u\frac{\partial}{\partial\rho}E_\nu = 0 \quad \forall r > 0, \ \forall t \geq 0 \qquad (20.98)$$

Hence, the integral representation (20.96) is valid if

$$\lim_{\rho\downarrow 0} \int_0^t \rho\frac{\partial u}{\partial\rho}E_\nu(r,\rho,t-\tau)d\tau = 0 \quad \forall t > 0, \ \forall r \geq 0 \qquad (20.99)$$

This integral representation has been obtained on the assumption that u is a solution of the problem and that condition (20.99) is valid. However, the following assertion is also true: If $f(t)$ and $F(r,t)$ are differentiable for all $r \geq 0$ and $t \in (0,T]$, and if there exists $c = \text{const} > 0$ such that

$$F(0,t) = 0, \quad |F(r,t)|\exp(-cr^2) < \infty \qquad (20.100)$$

$\varphi(r)$ is continuous for all $r \geq 0$, and if

$$|\varphi(r)|\exp(-cr^2) < \infty \qquad (20.101)$$

then the function $u(r,t)$ defined by (20.96) is a solution of problem (20.69)–(20.72).

Indeed, it follows from (20.92) that

$$E_\nu(r,0,t) \asymp \left(\frac{r}{2\sqrt{t}}\right)^\nu \frac{\exp[-(r^2/4t)]}{2t}\frac{1}{\Gamma(\nu+1)} \qquad (20.102)$$

which implies that, for all $t = \text{const} > 0$,

$$\lim_{r\downarrow 0} u_1(r,t) = \lim_{r\downarrow 0} 2\nu \int_0^t f(\tau)E_\nu(r,0,t-\tau)d\tau$$

$$= \frac{1}{\Gamma(\nu+1)}\lim_{r\downarrow 0}\int_{r/2\sqrt{t}}^\infty f\left(t-\frac{r^2}{2z^2}\right)\exp(-z^2)z^{2\nu-1}dz$$

$$= \frac{f(t)}{\Gamma(\nu)}\int_0^\infty \exp(-\zeta)\zeta^{\nu-1}d\zeta = f(t) \qquad (20.103)$$

Further, by (20.90) and the fact that

$$E_\nu(r,0,0) = 0 \quad \forall r > 0 \qquad (20.104)$$

we have, for all $r > 0$ and $t > 0$,

$$Lu_1 = 2\nu \int_0^t f(\tau)LE_\nu(r,0,t-\tau)d\tau - 2f(t)E_\nu(r,0,0) = 0 \qquad (20.105)$$

Finally,

$$\int_0^t \rho \frac{\partial}{\partial \rho} u_1(\rho,\tau)E_\nu(r,\rho,t-\tau)d\tau$$

$$= 2\nu \int_0^t \rho E_\nu(r,\rho,t-\tau)d\tau \int_0^\tau f(s)\frac{\partial}{\partial \rho}E_\nu(\rho,0,\tau-s)ds \qquad (20.106)$$

which clearly implies

$$\lim_{\rho \downarrow 0} \int_0^t \rho \frac{\partial}{\partial \rho} u_1(\rho,\tau)E_\nu(r,\rho,t-\tau)d\tau = 0 \qquad (20.107)$$

Similarly, the filtering property of E_ν, the definition (20.96) of u_2, and (20.92) imply that

$$Lu_2 = 0 \quad \forall r \geq 0, \ \forall t > 0 \qquad (20.108)$$

$$u_2(r,0) = \varphi(r) \quad \forall r > 0, \ u_2(+0,t) = 0 \quad \forall t > 0 \qquad (20.109)$$

$$\lim_{\rho \downarrow 0} \int_0^t \rho \frac{\partial}{\partial \rho} u_2(\rho,\tau)E_\nu(r,\rho,t-\tau)d\tau = 0 \qquad (20.110)$$

It remains to consider u_3. We omit the details of the arguments, restricting ourselves to the following remarks.

(a) Arguments identical to those used in connection with volume heat potentials (see Theorem 16.1.1) and use of the filtering property of E_ν enable one to demonstrate that

$$Lu_3 + F(r,t) = 0 \quad \forall r > 0, \quad \forall t > 0 \qquad (20.111)$$

(b) The equalities

$$u_3(r,+0) = 0 \quad \forall r > 0, \quad u_3(+0,t) = 0 \quad \forall t > 0 \qquad (20.112)$$

are obviously true.

(c) Statements (20.111) and (20.112) are valid without reference to the differentiability of $f(t)$ and the vanishing of F at the axis $r = 0$. These requirements are necessary because otherwise the equality

$$\lim_{\rho \downarrow 0} \int_0^t \left(\rho \frac{\partial}{\partial \rho} u_3\right) E_\nu(r,\rho,t-\tau)d\tau = 0 \qquad (20.113)$$

apparently cannot be proved. Even when these assumptions are made, the proof uses the properties of confluent hypergeometric functions [18]; since this class of special functions is beyond the scope of our book, the proof is omitted.[6]

---

[6] From the point of view of the preceding physical interpretation of problem (20.69)–(20.72), the assumption $F \equiv 0$ is the most natural, since $u(r,t)$ may be considered as the non-stationary deviation of the oil stratum from its initial stationary temperature, due to a local geothermal gradient, and this implies that $F \equiv 0$. In practical situations, $F \neq 0$ may occur only in extraordinary cases.

5. Laplace–Carson transform and its simplest properties[7]

We have used the Fourier and Fourier–Bessel transforms to solve boundary-value problems where all functions subject to transformation are considered as functions of spatial coordinates, and where the time appears in the "spectral functions" as a parameter. We now proceed to consider the Laplace integral transform, which is the most suitable one for solving linear non–steady-state boundary-value problems and where the parameter of the transformation is usually the time. Let us introduce some definitions.

Definition 20.5.1. $\mathfrak{L}(f)$ is the space of all functions $f(t) \in \mathbb{R}_1$ such that

(a) $f(t)$ is integrable in every interval of finite length;
(b) $f(t) \equiv 0$ for all $t < 0$;
(c) the rate of growth of $f(t)$ at infinity is at most exponential, which implies

$$f \in \mathfrak{L} \Rightarrow \exists c > 0 \text{ such that } |f(t)\exp(-ct)| < \infty \ \forall t \geq 0 \qquad (20.114)$$

Infimum c is called the *abscissa of convergence*.

Definition 20.5.2. Let $p = \sigma + i\tau$ and $f(t) \in \mathfrak{L}$. The function

$$F(p) = \int_0^\infty e^{-pt} f(t)\,dt \qquad (20.115)$$

is called the *Laplace transform* of f or the transform of f, and f may be called the *source function*. Note that by (20.114), the integral $F(p)$ is uniformly convergent if

$$\mathfrak{R}p = \sigma \geq \sigma_0 > c \qquad (20.116)$$

Since $\exp(-pt)$ is a holomorphic function of p and the integral (20.115) is uniformly convergent for all $\sigma \geq \sigma_0$, for any $\sigma_0 > c$ it follows that $F(p)$ is holomorphic in the half-plane $\mathfrak{R}p > c$. This explains why c is called the convergence abscissa.

Let $\eta(t)$ be the Heaviside unit function:

$$\eta(t) = \begin{cases} 0 & \text{if } t < 0 \\ 1/2 & \text{if } t = 0 \\ 1 & \text{if } t > 0 \end{cases} \qquad (20.117)$$

---

[7]Computations based on the use of the Laplace transform (or equivalently, the Laplace–Carson transform) are often referred to as *operational calculus*. This calculus, containing no reference to the integral transform and dealing only with a field of operators, was developed by Mikusinsky [46]. His approach has the advantage that its results need not be checked a posteriori for their validity. However, this is achieved at the cost of a certain loss in lucidity and a reduction in the range of application of the method.

and let $H(p)$ be its Laplace transform. We shall denote the correspondence between the source functions and their Laplace transforms by the symbol $\rightarrow$. We have

$$\eta(t) \rightarrow H(p) = 1/p \tag{20.118}$$

In order to make the transform of the Heaviside unit function equal to 1, Carson proposed the transform

$$F(p) = p \int_0^\infty e^{-pt} f(t)dt \tag{20.119}$$

We denote the Laplace–Carson transform by the symbol $\longrightarrow$ so that[8]

$$f(t) \longrightarrow F(p) \tag{20.120}$$

The distinction between the Laplace and Laplace–Carson transforms is not essential. Numerous tables are available giving the correspondence between various functions and their Laplace transforms or Laplace–Carson transforms. Which of these transforms is used is purely a matter of taste and habit; we shall consider only Laplace–Carson transforms.

Let us derive the simplest basic rules of the Laplace–Carson transforms. Straightforward computations give:

$$f(t) \longrightarrow F(p) \Rightarrow \int_0^t f(\tau)d\tau \longrightarrow \frac{1}{p}F \tag{20.121}$$

$$f(t) \longrightarrow F(p) \Rightarrow \frac{df}{dt} \longrightarrow pF - pf(0) \tag{20.122}$$

From (20.122), by induction we obtain

$$f \longrightarrow F \Rightarrow \frac{d^n}{dt^n}f \longrightarrow p^n F - \sum_{k=0}^{n-1} p^{n-k} \frac{d^k}{dt^k}f(0) \tag{20.123}$$

Further,

$$f(t) \longrightarrow F(p) \Rightarrow F(ap) \longleftarrow f(t/a) \tag{20.124}$$

$$f(t) \longrightarrow F(p) \Rightarrow e^{-\alpha p}F(p) \longleftarrow f(t-\alpha)\eta(t-\alpha) \tag{20.125}$$

Formula (20.125) is called the *lag theorem*;

$$f(t) \longrightarrow F(p) \Rightarrow \frac{p}{p+\beta}F(p+\beta) \longleftarrow e^{-\beta t}f(t) \tag{20.126}$$

formula (20.126) is known as the *displacement theorem*;

$$f(t) \longrightarrow F(p) \wedge \psi(t) \longrightarrow \Psi(p) \Rightarrow \int_0^t f(\tau)f(t-\tau)d\tau \longrightarrow \frac{1}{p}F(p)\Psi(p) \tag{20.127}$$

---

[8] The correspondence $\varphi \longrightarrow f$ will always mean that φ is the source function and f its transform; while $\varphi \longleftarrow f$ will mean that f is the source function and φ its transform. Arrows are always directed from the source function to the transform.

formula (20.127) is known as the *convolution* or *multiplication theorem*. Finally,

$$f(t) \longrightarrow F(p) \Rightarrow p \int_p^\infty \frac{F(q)}{q} \cdot dq \longleftarrow \frac{f(t)}{t} \tag{20.128}$$

$$f(t) \longrightarrow F(p) \Rightarrow p \frac{d}{dp}\left(\frac{F(p)}{p}\right) \longleftarrow -t f(t) \tag{20.129}$$

The following elegant theorem is due to A. Efros [93].

Theorem 20.5.1. *Let*

$$f(t) \longrightarrow F(p) \tag{20.130}$$

and

$$\Phi(\xi, t) \longrightarrow \exp[-\xi q(p)] u(p) q(p) \tag{20.131}$$

Then formally

$$\int_0^\infty \Phi(\xi, t) f(\xi) d\xi \longrightarrow u(p) F[q(p)] \tag{20.132}$$

PROOF. We have

$$p \int_0^\infty e^{-pt} dt \int_0^\infty f(\xi) \Phi(\xi, t) d\xi$$

$$= \int_0^\infty f(\xi) d\xi \, p \int_0^\infty e^{-pt} \Phi(\xi, t) dt$$

$$= u(p) q(p) \int_0^\infty \exp[-\xi q(p)] f(\xi) d\xi = u(p) F[q(p)] \tag{20.133}$$

Q.E.D.

Remark. If the change in the order of integration were permissible, the statement of Efros's theorem would be justified. But since (in nontrivial situations), practically the only – and the best – justification is a direct check of the final result, we shall not concern ourselves with a rigorous proof of Efros's theorem, in spite of its importance as a convenient tool for obtaining many other important rules of operational calculus. Such rules will be introduced as needed.

We first use the tables (20.121)–(20.133) of operational rules to obtain concrete correspondences between source functions and their transforms. First,

$$\eta(t) \longrightarrow 1 \wedge (20.126) \Rightarrow \frac{p}{p+\beta} \longleftarrow \exp(-\beta t) \tag{20.134}$$

Let

$$\beta = a + ib \tag{20.135}$$

If p is real then separation of the real and imaginary parts of (20.134) implies

$$p \cdot \frac{p+a}{(p+a)^2 + b^2} \longleftarrow \exp(-at)\cos(bt) \tag{20.136}$$

$$\frac{pb}{(p+a)^2 + b^2} \longleftarrow \exp(-at)\sin(bt) \tag{20.137}$$

Since both the left- and right-hand sides of these correspondences are analytic functions, formulas (20.136) and (20.137) remain true for complex p throughout the region

$$\Re(p + \beta) > 0 \tag{20.138}$$

Note also that

$$\frac{t^\nu}{\Gamma(\nu + 1)} \longrightarrow \frac{1}{p^\nu} \tag{20.139}$$

Using these correspondences we can prove the following theorem.

Theorem 20.5.2 (First Heaviside theorem). *Let the series*

$$p^{-\alpha} F(p) = \sum_{n=0}^{\infty} \beta_n p^{-n} \tag{20.140}$$

be uniformly convergent in the neighborhood of the point at infinity,

$$|p| > R > 0 \tag{20.141}$$

and let

$$\Re\alpha < 1 \tag{20.142}$$

Then

$$F(p) \longleftarrow f(t) \overset{\text{def}}{=} \sum_{n=0}^{\infty} \beta_n \frac{t^{n-\alpha}}{\Gamma(n - \alpha + 1)} \tag{20.143}$$

and, moreover, $t^\alpha f(t)$ is an entire function.

PROOF. By the D'Alembert test, the radius of convergence of the series (20.143) is infinite. Indeed, by (20.140) and (20.141),

$$\overline{\lim_{n\uparrow\infty}} \left| \frac{\beta_n}{\beta_{n+1}} \right| \le \frac{1}{R} \Rightarrow \lim_{n\uparrow\infty} \frac{\beta_n}{\beta_{n+1}} \cdot \frac{\Gamma(n+1-\alpha)}{\Gamma(n+2-\alpha)} = 0 \tag{20.144}$$

This means that it is permissible to change the order of integration and summation in the equality

$$p \sum_{n=0}^{\infty} \int_0^\infty \beta_n \frac{\exp(-pt)t^{n-\alpha}}{\Gamma(n+1-\alpha)} dt = \sum_{n=0}^{\infty} \beta_n \frac{1}{p^{n-\alpha}} \int_0^\infty \frac{\exp(-s)s^{n-\alpha}}{\Gamma(n+1-\alpha)} ds \tag{20.145}$$

so that indeed

$$\sum_{n=0}^{\infty} \beta_n \cdot \frac{t^{n-\alpha}}{\Gamma(n+1-\alpha)} \longrightarrow F(p) \tag{20.146}$$

Q.E.D.

Now let

$$F(p) = \exp\left(-\frac{\alpha}{p}\right) \equiv \sum_{n=0}^{\infty} (-1)^n \frac{\alpha^n}{p^n} \cdot \frac{1}{n!} \tag{20.147}$$

Then, according to the first Heaviside theorem,

$$\exp\left(-\frac{\alpha}{p}\right) \longleftarrow \sum_{n=0}^{\infty} (-1)^n \frac{(\alpha t)^n}{(n!)^2} = J_0(2\sqrt{\alpha t}) \tag{20.148}$$

and similarly

$$\exp\left(\frac{\alpha}{p}\right) \longleftarrow \sum_{n=0}^{\infty} \frac{(\alpha t)^n}{(n!)^2} = I_0(2\sqrt{\alpha t}) \tag{20.149}$$

Here J_0 and I_0 are (respectively) the Bessel function and the modified Bessel function of order zero.

Now consider the Lipschitz integral (A2.54):

$$\int_0^{\infty} e^{-rt} J_0(\rho t) dt = (r^2 + \rho^2)^{-1/2} \tag{20.150}$$

The change of variables

$$rt = p\tau, \quad \rho t = a\tau \tag{20.151}$$

yields

$$J_0(at) \longrightarrow p \cdot (p^2 + a^2)^{-1/2} \tag{20.152}$$

Both sides of (20.152) are analytic functions of a. Hence

$$I_0(at) \longrightarrow p \cdot (p^2 - a^2)^{-1/2} \tag{20.153}$$

Finally, let us determine the Laplace–Carson transform of the fundamental solution of the heat-conduction equation:

$$E(x,t) = \frac{\exp\left[-(x^2/4t)\right]}{2\pi^{1/2}t^{1/2}} \tag{20.154}$$

We have

$$E(x,t) \longrightarrow \mathcal{E}(p) = p \int_0^{\infty} \frac{\exp\left[-(x^2/4t)\right]}{2\pi^{1/2}t^{1/2}} \exp(-pt) dt \tag{20.155}$$

so that

$$\mathcal{E}(p) = \frac{\exp(-|x|\sqrt{p})}{2\sqrt{\pi}} p \int_0^{\infty} \exp\left[-\left(\frac{\sqrt{pt} - |x|}{2\sqrt{t}}\right)^2\right] \frac{dt}{\sqrt{t}} \tag{20.156}$$

Since

$$J \stackrel{\text{def}}{=} \frac{2}{\sqrt{\pi}} \int_0^{\infty} \exp\left[-\left(\lambda - \frac{\alpha}{\lambda}\right)^2\right] d\lambda = 1 \quad \forall \alpha \geq 0 \tag{20.157}$$

we see that

$$E(x,t) \longrightarrow \mathcal{E}(p) = (\sqrt{p}/2) \exp(-|x|\sqrt{p}) \tag{20.158}$$

Let us now apply Efros's Theorem 20.5.1. Take

$$\Phi(\xi,t) = 2E(\xi,t), \quad q(p) = \sqrt{p}, \quad u(p) = 1 \qquad (20.159)$$

This yields

$$f(t) \longrightarrow F(p) \Rightarrow F(\sqrt{p}) \longleftarrow 2\int_0^\infty E(\xi,t)f(\xi)d\xi \qquad (20.160)$$

Now take

$$\Phi(\xi,t) = J_0(2\sqrt{\xi t}), \quad q = 1/p, \quad u(p) = p \qquad (20.161)$$

This yields

$$f(t) \longrightarrow F(p) \Rightarrow p \cdot F\left(\frac{1}{p}\right) \longleftarrow 2\int_0^\infty f(\xi)J_0(2\sqrt{\xi t})d\xi \qquad (20.162)$$

We cite without proof three other operational rules that can be derived using Efros's theorem:

$$f(p) \longrightarrow F(p) \Rightarrow \frac{F(p+1/p)}{p+1/p} \longleftarrow \int_0^t f(\tau)J_0[2\sqrt{\tau(t-\tau)}]d\tau \qquad (20.163)$$

$$f(p) \longrightarrow F(p) \Rightarrow \frac{p}{p^2+1} \cdot F\left((p^2+1)^{1/2}\right)$$

$$\longleftarrow \int_0^t f(\tau)J_0\left((t^2-\tau^2)^{1/2}\right)d\tau \qquad (20.164)$$

$$f(p) \longrightarrow F(p) \Rightarrow \frac{p}{p^2-1} \cdot F\left((p^2-1)^{1/2}\right)$$

$$\longleftarrow \int_0^t f(\tau)I_0\left((t^2-\tau^2)^{1/2}\right)d\tau \qquad (20.165)$$

6. Relationship between Laplace and Fourier transforms. Bromwich integral and Jordan lemma

Until now we have restricted our attention to rules that make it possible to determine various functions from their transforms through prepared tables and the simplest rules of operational calculus. We now proceed to derive a general formula for inversion of transforms, one that follows from the theory of Fourier transforms. The following theorems are basic.

Theorem 20.6.1. *Let $f(t) \in \mathcal{L}$ and let c be its abscissa of convergence, so that the function*

$$\varphi(t) = f(t)\exp(-\sigma t) \quad \forall(\sigma \geq \sigma_0 > c) \qquad (20.166)$$

satisfies the Dirichlet conditions. Then the correspondence

$$f(t) \longrightarrow F(p) \qquad (20.167)$$

implies the following conclusions:

(a) *$F(p)$ is an analytic function of p without singularities in the half-plane $\sigma \geq \sigma_0 > c$.*

(b) *B: The identity*

$$f(t) = \frac{1}{2\pi i} \int_{\sigma-i\infty}^{\sigma+i\infty} e^{pt} F(p) \frac{dp}{p} \tag{20.168}$$

holds for all $\sigma \geq \sigma_0$.

Remark. The integral (20.168) is known as the *Bromwich integral* [6].

Theorem 20.6.2. *Let $F(p)$ be an analytic function of $p = \sigma + i\tau$ without singularities in a half-plane $\sigma \geq \sigma_0 > c$. Suppose that there exists $M > 0$ such that*

$$\int_{-\infty}^{\infty} \left| \frac{F(p)}{p} \right| d\tau < M \quad \forall \sigma \geq \sigma_0 \tag{20.169}$$

which means, in particular, that

$$\lim_{|\tau| \uparrow \infty} |F(p)| = 0 \tag{20.170}$$

and let

$$f(t) = \frac{1}{2\pi i} \int_{\sigma-i\infty}^{\sigma+i\infty} e^{pt} F(p) \frac{dp}{p} \tag{20.171}$$

Then $f(t) \in \mathcal{L}$, and $F(p)$ is its Laplace–Carson transform:

$$F(p) = p \int_0^{\infty} e^{-pt} f(t) dt \tag{20.172}$$

PROOF OF THEOREM 20.6.1. Let Φ be the Fourier transform of $\varphi(t)$, that is,

$$\Phi(p) = \int_{-\infty}^{\infty} e^{-i\tau t} f(t) e^{-\sigma t} dt = \int_0^{\infty} e^{-pt} f(t) dt \tag{20.173}$$

By the inversion formula,

$$f(t)e^{-\sigma t} = \frac{1}{2\pi} \int_{-\infty}^{\infty} e^{i\sigma\tau} \Phi(p) d\tau \Rightarrow f(t) = \frac{1}{2\pi i} \int_{\sigma-i\infty}^{\sigma+i\infty} e^{pt} F(\mu) \frac{dp}{p} \tag{20.174}$$

Q.E.D.

PROOF OF THEOREM 20.6.2. By assumption,

$$f(t) = \frac{1}{2\pi i} \int_{\sigma-i\infty}^{\sigma+i\infty} e^{pt} F(p) \frac{dp}{p} \tag{20.175}$$

Let us first show that $f(t)$ is independent of σ for all $\sigma \geq \sigma_0$. Consider a rectangular contour $ABCDA$ with vertices at the points

$$A = (\sigma - iT), \quad B = (\tilde{\sigma} - iT), \quad C = (\tilde{\sigma} + iT), \quad D = (\sigma + iT) \tag{20.176}$$

where $\tilde{\sigma} > \sigma$. The integrand in (20.176) has no singularities inside this contour. Hence, by the Cauchy theorem,

$$\int_{AB} + \int_{BC} + \int_{CD} + \int_{DA} = 0 \tag{20.177}$$

We have

$$\left| \int_{AB} e^{pt} F(p) \frac{dp}{p} \right| < \int_{\sigma}^{\tilde{\sigma}} e^{st} |F| \frac{ds}{T} \tag{20.178}$$

so that

$$\lim_{T \uparrow \infty} \int_{AB} = 0 \tag{20.179}$$

and similarly

$$\lim_{T \uparrow \infty} \int_{CD} = 0 \tag{20.180}$$

This and (20.176) and (20.177) yield

$$\lim_{T \uparrow \infty} \int_{AD} = \lim_{T \uparrow \infty} \int_{BC} \tag{20.181}$$

which proves that $f(t)$ is indeed independent of σ.

Let $t < 0$; then, by (20.169),

$$|f(t)| < M e^{\sigma t} \tag{20.182}$$

Since f is independent of σ and $t < 0$, this means that

$$|f(t)| \leq \lim_{\sigma \uparrow \infty} M e^{\sigma t} = 0 \tag{20.183}$$

so that

$$f(t) = 0 \quad \forall t < 0 \tag{20.184}$$

We can now prove that $F(p)$ is indeed the Laplace–Carson transform of $f(t)$.

Let $F^*(p) \longleftarrow f(t)$, so that (see (20.175))

$$F^*(p) = p \int_0^\infty e^{-pt} dt \frac{1}{2\pi i} \int_{\sigma - i\infty}^{\sigma + i\infty} e^{qt} F(q) \frac{dq}{q} \tag{20.185}$$

or, what is the same,

$$\frac{F^*(p)}{p} = \frac{1}{2\pi} \int_0^\infty e^{-i t \tau} dt \int_{-\infty}^\infty e^{ist} F(q) \frac{ds}{q} \tag{20.186}$$

Here $p = \sigma + i\tau$ and $q = \sigma + is$. The integral (20.186) is a double Fourier integral, and this means that

$$F^*(p) \equiv F(p) \tag{20.187}$$

Q.E.D.

The practical use of the Bromwich integral is based on the following theorem, which is known as Jordan's lemma.

Theorem 20.6.3 (Jordan's lemma). *Given a function $\Phi(p)$ and a sequence $\{C_n\}|_{n=1}^\infty$ of arcs of circles,*

$$C_n = \{|p| = R_n; \pi/2 - \varepsilon_n \leq \arg p \leq 3\pi/2 + \varepsilon_n\} \tag{20.188}$$

where

$$R_n < R_{n+1}, \quad \varepsilon_n = \arcsin(\sigma/R_n), \quad \lim_{n\uparrow\infty} R_n = \infty \tag{20.189}$$

Assume that there exist $A > 0$ and $k > 0$ such that

$$|\Phi(p)| < A|p^{-k}| \quad \forall p \in C_n, \ n = 1, 2, \dots \tag{20.190}$$

Then

$$\lim_{n\uparrow\infty} \int_{C_n} \Phi(p)e^{pt}dp = 0 \quad \forall t = \text{const} > 0 \tag{20.191}$$

PROOF. We have

$$\int_{C_n} \Phi(p)e^{pt}dp \leq AR_n^{1-k} \int_{\pi/2-\varepsilon_n}^{3\pi/2+\varepsilon_n} \exp(tR_n \cos\varphi)d\varphi \tag{20.192}$$

Define ψ by

$$\varphi = \begin{cases} \pi/2 + \psi & \text{if } \pi/2 < \varphi < \pi \\ 3\pi/2 - \psi & \text{if } \pi < \varphi < 3\pi/2 \\ \pi/2 - \psi & \text{if } \pi/2 - \varepsilon_n < \varphi < \pi/2 \\ 3\pi/2 + \psi & \text{if } 3\pi/2 < \varphi < 3\pi/2 + \varepsilon_n \end{cases} \tag{20.193}$$

Then

$$\int_{\pi/2-\varepsilon_n}^{3\pi/2+\varepsilon_n} \exp[tR_n \cos(\varphi)]d\varphi$$

$$= 2\int_0^{\pi/2} \exp[-tR_n \sin(\psi)]d\psi + 2\int_0^{\varepsilon_n} \exp[tR_n \sin(\psi)]d\psi$$

$$< 2\int_0^{\pi/2} \exp\left[-2tR_n\frac{\psi}{\pi}\right]d\psi + 2\int_0^{\varepsilon_n} \exp(tR_n\psi)d\psi \tag{20.194}$$

Hence

$$0 < \int_{\pi/2-\varepsilon_n}^{3\pi/2+\varepsilon_n} \exp[tR_n \cos(\varphi)]d\varphi$$

$$< \frac{\pi}{tR_n}\left(1 - \exp(-tR_n) + \frac{2}{\pi}\exp(tR_n\varepsilon_n)\right) \tag{20.195}$$

Since

$$0 < R_n\varepsilon_n = R_n \arcsin\left(\frac{\sigma}{R^n}\right) < \frac{\pi\sigma}{2} \tag{20.196}$$

it follows from (20.192) and (20.196) that

$$\left|\int_{C_n} \Phi(p)e^{pt}dp\right| \leq \frac{A\pi}{tR_n^k}\left(1 - \exp(-tR_n) + \frac{2}{\pi}\exp\left[t\sigma\frac{\pi}{2}\right]\right) \tag{20.197}$$

so that

$$\lim_{n\uparrow\infty} \int_{C_n} \Phi(p)e^{pt}dp = 0 \tag{20.198}$$

Q.E.D.

Now let $F(p)$ be an analytic function without essential singularities, having at most denumerable number of poles and branch points (which may simultaneously be poles) all located in the half-plane $\Re p < \sigma_0$, which have no points of accumulation in any final subregion of the complex plane $p = \sigma + i\tau$. Let $\{q_n\} \, |_{n=0}^{\infty}$ and $\{p_n\} \, |_{n=0}^{\infty}$ be the sequences of poles and branch points, respectively, enumerated so that

$$|p_n| \le |p_{n+1}|, \quad |q_n| \le |q_{n+1}|, \quad n = 1, 2, \ldots \tag{20.199}$$

Assume, to fix ideas, that $p = 0$ is a branch point of $F(p)$.

Let $\{C_m\} \, |_{m=0}^{\infty}$ be a sequence of arcs

$$C_m = \{p \colon |p| = R_m, \ \arg p_m \in [\pi/2 - \varepsilon_m, 3\pi/2 + \varepsilon_m]\} \tag{20.200}$$

where

$$R_m < R_{m+1}, \quad \varepsilon_m = \arcsin(\sigma/R_m), \quad \lim_{m \uparrow \infty} R_m = \infty \tag{20.201}$$

chosen so that no singular point of $F(p)$ lies on any of these arcs. Cut the half-plane $\Re p \le \sigma$ by branch cuts encircling all the branch points p_n of $F(p)$ in the counterclockwise sense, chosen so that no two such branch cuts intersect. Consider the region D_n bounded by the arc C_n, segments $\{\tilde{\Gamma}_m\}$ of all the cuts Γ_m emanating from the branch points p_m situated inside C_n, and the segment $[AB]$ of the straight line $p = \sigma + i\tau$ with its endpoints on C_n; let

$$L_n \overset{\text{def}}{=} \bar{D}_n \setminus D_n = C_n \cup \{\Gamma_m\} \cup [AB] \tag{20.202}$$

Inside D_n, $F(p)$ is a univalent function, so that by the residue theorem

$$\frac{1}{2\pi i} \int_{L_n} e^{pt} F(p) \frac{dp}{p} = \frac{1}{2\pi i} \left(\int_{AB} - \sum_m \int_{\tilde{\Gamma}_m} + \int_{C_n} \right) e^{pt} F(p) \frac{dp}{p}$$

$$= \sum_k \operatorname{res} \left(e^{pt} \frac{F(q_k)}{q_k} \right) \tag{20.203}$$

where the sum on the right extends over all poles inside D_n and the sum on the left over all segments of branch cuts located within D_n, on the assumption that $\tilde{\Gamma}_m$ is described in the counterclockwise sense (see Figure 20.1). Let $n \uparrow \infty$. By Jordan's lemma and Theorem 20.6.2, we find that if $f(t)$ is the function whose transform is $F(p)$ then

$$f(t) = \frac{1}{2\pi i} \int_{\sigma - i\infty}^{\sigma + i\infty} e^{pt} F(p) \frac{dp}{p}$$

$$= \sum_{k=0}^{\infty} \operatorname{res} \left(e^{pt} \frac{F(q_k)}{q_k} \right) + \sum_{n=0}^{\infty} \frac{1}{2\pi i} \int_{\Gamma_n} e^{pt} F(p) \frac{dp}{p} \tag{20.204}$$

Recall that

$$\int_{\Gamma_n} = \int_{\Gamma_n^+} - \int_{\Gamma_n^-} + \lim_{\delta \downarrow 0} \int_{C_n^\delta} \tag{20.205}$$

where C_n^δ is circle of radius δ with center at the branch point p_n, and where

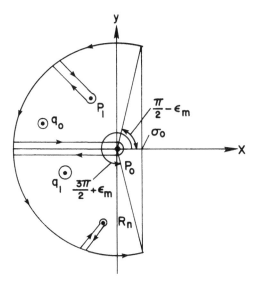

Figure 20.1. Sketch of the proof of Jordan's lemma.

Γ_n^+ and Γ_n^- are the upper and the lower borders of Γ_n, so that integration along them goes from p_n to infinity.

Remark. The only purpose of the branch cuts is to obtain a Riemann surface on which $F(p)/p$ is a univalent function. This means that they do not necessarily encircle the point at infinity. For example, if

$$F(p) = \frac{p}{\sqrt{p^2+1}} \tag{20.206}$$

then there are two branch points $p = \pm i$, located on the imaginary axis of the p plane. Therefore a cut encircling these two points makes $F(p)/p$ a univalent function in the cut plane.

As an example, let us calculate the Bromwich integral

$$f(t) = \frac{1}{2\pi i}\int_{\sigma-i\infty}^{\sigma+i\infty} \frac{e^{pt}}{\sqrt{\alpha+p}}\frac{dp}{p} \tag{20.207}$$

Jordan's lemma is obviously applicable. Here we have a simple pole at $p=0$ and a branch point $p=-\alpha$ that is simultaneously a pole. As the branch cut we take the ray $p+\alpha = \rho e^{i\pi}$, which gives

$$f(t) = \operatorname{res}\left(\frac{e^{pt}}{\sqrt{\alpha+p}}\cdot\frac{1}{p}\right)\bigg|_{p=0} + \frac{1}{2\pi i}\int_0^\infty \exp[-(\rho+\alpha)t]e^{-i\pi/2}\rho^{-1/2}\frac{d\rho}{\alpha+\rho}$$

$$- \frac{1}{2\pi i}\int_0^\infty \exp[-(\rho+\alpha)t]e^{i\pi/2}\rho^{-1/2}\frac{d\rho}{\alpha+\rho} \tag{20.208}$$

so that

$$f(t) = \frac{1}{\sqrt{\alpha}} - \frac{1}{\pi} \int_0^\infty \exp[-(\rho + \alpha)t] \rho^{-1/2} \frac{d\rho}{\alpha + \rho} \qquad (20.209)$$

Note that

$$f(0) = \frac{1}{\sqrt{\alpha}} \left(1 - \frac{2}{\pi} \int_0^\infty \frac{dx}{1 + x^2} \right) = 0 \qquad (20.210)$$

$$\frac{df}{dt} = \frac{1}{\pi} \int_0^\infty \exp[-(\rho + \alpha)t] \frac{d\rho}{\sqrt{\rho}} = \frac{2}{\sqrt{\pi}} \frac{\exp(-\alpha t)}{\sqrt{t}} \qquad (20.211)$$

and integration yields

$$f(t) = \frac{2}{\sqrt{\pi}} \int_0^t \exp(-\alpha t) \frac{dt}{\sqrt{t}} = \frac{1}{\sqrt{\alpha}} \operatorname{erf}(\sqrt{\alpha t}) \qquad (20.212)$$

7. Relationship between limits of functions and their transforms. Asymptotic expansion

Consider the Laplace–Carson transform of $f(t) \in \mathcal{L}$ for real values of the parameter p. There is an important relationship between the asymptotic behavior of a function and its transform in the neighborhood of $t = 0$ and $p = \sigma = \infty$, or $t = \infty$ and $\sigma = 0$.

Theorem 20.7.1. *Let $f(t) \in \mathcal{L}$ and assume that $\lim_{t \downarrow 0} f(t)$ exists. Then $\lim_{\sigma \uparrow \infty} F(\sigma)$ exists and*

$$\lim_{t \downarrow 0} f(t) = \lim_{\sigma \uparrow \infty} F(\sigma) \qquad (20.213)$$

PROOF. Let C be the abscissa of convergence of f. Then, for all $\sigma \geq \sigma_0 > C$, the integral

$$F(p) = p \int_0^\infty e^{-pt} f(t) dt, \quad p = \sigma \qquad (20.214)$$

is absolutely convergent. Assume that $\lim_{t \downarrow 0} f(t)$ exists, and take $\varepsilon > 0$ arbitrarily small and $\delta > 0$ such that

$$|f(t) - f(+0)| < \varepsilon \quad \text{if } 0 \leq t \leq \delta \qquad (20.215)$$

We have

$$F(\sigma) - f(+0)$$
$$= \sigma \int_0^\infty e^{-\sigma t} f(t) dt - f(+0) = \sigma \int_0^\infty e^{-\sigma t} [f(t) - f(+0)] dt$$
$$= \sigma \int_0^\delta e^{-\sigma t} [f(t) - f(+0)] dt + \sigma \int_\delta^\infty e^{-\sigma t} [f(t) - f(+0)] dt$$
$$= \sigma \int_0^\delta e^{-\sigma t} [f(t) - f(+0)] dt + \sigma e^{-\sigma \delta} \int_0^\infty e^{-\sigma s} [f(s + \delta) - f(+0)] ds$$
$$= I_1 + I_2 \qquad (20.216)$$

Let $\sigma > 2\sigma_0$. Then, since $f \in \mathfrak{L}$, there exists $M = \text{const} > 0$ such that

$$|I_2| < \sigma e^{-\sigma\delta} \int_0^\infty e^{-\sigma_0 t}[|f(t)| + |f(+0)|]dt < M\sigma e^{-\sigma\delta} \qquad (20.217)$$

At the same time,

$$|I_1| < \varepsilon[1 - e^{-\delta}] < \varepsilon \qquad (20.218)$$

Hence it follows from (20.215), (20.217), and (20.218) that there exists σ_1 so large that, for all $\sigma > \sigma_1$,

$$|F(\sigma) - f(+0)| < 2\varepsilon \qquad (20.219)$$

Since $\varepsilon > 0$ is arbitrarily small, this means that $\lim_{\sigma\uparrow\infty} F(\sigma)$ exists and that

$$\lim_{\sigma\uparrow\infty} F(\sigma) = \lim_{t\downarrow 0} f(t) \qquad (20.220)$$

Q.E.D.

Theorem 20.7.2. *If* $\lim_{t\uparrow\infty} f(t)$ *exists then* $\lim_{\sigma\downarrow 0} F(\sigma)$ *exists and*

$$\lim_{\sigma\downarrow 0} F(\sigma) = \lim_{t\uparrow\infty} f(t) \qquad (20.221)$$

PROOF. The proof is analogous to that of the Theorem 20.7.1. Indeed, let

$$f_\infty = \lim_{t\uparrow\infty} f(t) \qquad (20.222)$$

and let $N = \text{const} > 0$ be arbitrarily large. We have

$$I = \sigma \int_0^\infty e^{-\sigma t} f(t)dt - f_\infty = I_1 + I_2 \qquad (20.223)$$

where

$$I_1 = \sigma \int_0^N e^{-\sigma t}[f(t) - f_\infty]dt \qquad (20.224)$$

and

$$I_2 = \sigma \int_N^\infty e^{-\sigma t}[f(t) - f_\infty]dt \qquad (20.225)$$

By assumption, for any $\varepsilon = \text{const} > 0$ there exists $N > 0$ so large that

$$|f(t) - f_\infty| < \varepsilon \Rightarrow |I_2| < \varepsilon \qquad (20.226)$$

Fix such an $N > 0$ and take $\sigma_1 > 0$ so small that

$$|I_1| < \varepsilon \quad \forall \sigma \in [0, \sigma_1] \qquad (20.227)$$

This means that for all $\sigma \in [0, \sigma_1]$,

$$|I| < 2\varepsilon \Rightarrow \exists \lim_{\sigma\downarrow 0} F(\sigma) = \lim_{t\uparrow\infty} f(t) \qquad (20.228)$$

Q.E.D.

Remark. Theorems 20.7.1–2 assert only that the existence of the relevant limits of a function implies that of the corresponding limits of its transform for real values of the parameter. The converse is not true, as is seen from the following examples.

(1)

$$f(t) = \frac{\cos(1/2t)}{\sqrt{\pi t}} \longrightarrow F(\sigma) = \sqrt{\sigma}\exp(-\sqrt{\sigma})\cos(\sqrt{\sigma}) \qquad (20.229)$$

Hence

$$\lim_{\sigma\uparrow\infty} F(\sigma) = 0 \text{ exists} \qquad (20.230)$$

but

$$\lim_{t\downarrow 0} f(t) \text{ does not exist} \qquad (20.231)$$

(2)

$$f(t) = \sin(t) \longrightarrow F(\sigma) = \frac{\sigma}{\sigma^2+1} \qquad (20.232)$$

Hence

$$\lim_{\sigma\downarrow 0} F(\sigma) = 0 \text{ exists} \qquad (20.233)$$

but

$$\lim_{t\uparrow\infty} f(t) \text{ does not exist} \qquad (20.234)$$

We now state without proof two theorems that provide rather more important results. The first concerns the relationship between the average values of the source function in $0 \le t \le \infty$ and its transform; the second theorem concerns the asymptotic expansion of a function in the neighborhood of infinitely large t, expressed in terms of the asymptotic expansion of its transform in the neighborhood of all singular points with the maximal real part.

Theorem 20.7.3. *Let $f(t) \in \mathfrak{L}$. Assume that $f(t)$ is bounded from below, that is, assume there exists $M = \text{const} > 0$ such that*

$$f(t) + M \ge 0 \quad \forall t \in [0,\infty) \qquad (20.235)$$

Then

(a) *if $\lim_{T\uparrow\infty}(1/T)\int_0^T f(t)dt$ exists then $\lim_{\sigma\downarrow 0}\sigma F(\sigma)$ exists and also*

$$\lim_{T\uparrow\infty}\frac{1}{T}\int_0^T f(t)dt = \lim_{\sigma\downarrow 0}\sigma F(\sigma) \qquad (20.236)$$

and vice versa;

(b) *if $\lim_{T\downarrow 0}(1/T)\int_0^T f(t)dt$ exists then $\lim_{\sigma\uparrow\infty}\sigma F(\sigma)$ exists and also*

$$\lim_{T\downarrow 0}\frac{1}{T}\int_0^T f(t)dt = \lim_{\sigma\uparrow\infty}\sigma F(\sigma) \qquad (20.237)$$

and vice versa.

Theorem 20.7.4. *Let $f(t)$ be represented as a Bromwich integral,*

$$f(t) = \frac{1}{2\pi i} \int_{\sigma-\infty}^{\sigma+\infty} e^{pt} F(p) \frac{dp}{p} \quad \forall t > 0 \tag{20.238}$$

where

(a) $F(p)/p$ *satisfies the conditions of Jordan's lemma;*
(b) $F(p)/p$ *has at most a finite number n of singular points (poles or branch points);*
(c) $F(p)/p$ *can be expanded in series in a neighborhood of singular points $p = p_i$, $i = 1, 2, \dots, m \leq n$, with maximal real parts*

$$\frac{F(p)}{p} = \sum_{k=0}^{\infty} c_{ik}(p - p_i)^{\lambda_{ik}}, \quad -\infty < \lambda_{i0} < \lambda_{i1} < \cdots < \lambda_{is} < \cdots \tag{20.239}$$

where

$$\lim_{s \uparrow \infty} \lambda_{is} = \infty, \quad |p - p_i| < \ell_i, \quad i = 1, 2, \dots, m \tag{20.240}$$

Then $f(t)$ can be expanded in an asymptotic series

$$f(t) \asymp \sum_{i=1}^{m} e^{p_i t} \sum_{s=0}^{\infty} \frac{c_{is}}{\Gamma(-\lambda_{is})} \cdot t^{-\lambda_{is}-1} \quad \text{for } t \uparrow \infty \tag{20.241}$$

Problems

P20.2.1. Use equalities (20.43)–(20.45) to solve the nth order ordinary differential equations with constant coefficients

$$Lu = \sum_{k=0}^{n} \alpha_k \frac{d^k}{dx^k} u(x) = f(x), \quad x \in (-\infty, \infty) \tag{α}$$

$$\tilde{L}u = \sum_{k=0}^{m} \alpha_k \frac{d^{2k}}{dx^{2k}} u(x) = f(x), \quad 2m = n, \quad x \in (0, \infty) \tag{β}$$

assuming that f, u, and all its derivatives up to nth order vanish at infinity, and that in case (β) all derivatives of odd order (or even order) up to the $(m-1)$th order are specified at $x = 0$. Compare this method with the method of variation of constants. Is it possible to use this method to solve the Cauchy problem (case (β)) if the equation contains odd-order derivatives? If so, solve the problem; if not, explain why not. Is there any similarity with the problem (19.178)–(19.182) studied in Chapter 19, Section 6?

Hint: Apply the Fourier transform to solve problem (α), and cosine transform or sine transform, respectively, to solve problem (β).

P20.2.2. Solve the problem

$$\Box u \overset{\text{def}}{=} \frac{\partial^2 u}{\partial x^2} - \frac{\partial^2 u}{\partial t^2} = 0 \quad \forall x \in (0, 1), \quad \forall t > 0$$

$$\frac{\partial u}{\partial x} + hu = hf(t) \quad \text{at } x = 0, \quad u = 0 \quad \text{at } x = 1, \quad \forall t > 0$$

$$u(x, 0) = 0, \quad \frac{\partial}{\partial t} u(x, 0) = 0$$

P20.3.1. Is it possible to formulate the Fourier expansion method in terms of integral transforms in a bounded region? If so, explain; if not, why not?

P20.5.1. Prove the second Heaviside expansion theorem: Let

$$f(t) \Rightarrow F(p) = \frac{Q(p)}{P(p)}$$

where $\mathbf{Q}$ and $\mathbf{P}$ are (respectively) polynomials of degree m and n, $m \leq n$, without common roots. Let $p = p_k$, $k = 1, 2, \ldots, s$, be the roots of $P(p)$, of multiplicities n_k, so that

$$P(p) = \prod_{k=1}^{s} \left(p - p_k^{n_k}\right), \quad \sum_{k=1}^{s} n_k = n$$

Then

$$f(t) \Leftarrow \frac{Q(0)}{P(0)} + \sum_{k=1}^{s} \sum_{i=1}^{n_k} \left(\frac{d^{n_k-1}}{dp^{n_k-1}} C_k(p_k) \frac{1}{(n_k - i)!} \right) \frac{t^{i-1}}{(i-1)!} e^{p_k^t}$$

where

$$C_k(p) = \frac{Q(p)}{pP(p)}$$

Hint: The second Heaviside expansion theorem is an exact analog of the expansion theorem for the integrals of rational functions in integral calculus. It suffices to apply, in addition, the operational rules (20.139) and the lag theorem (20.126).

P20.5.2. Find the Laplace–Carson transform of the Dirac δ function

P20.5.3. What is the reason for the condition $m \leq n$ in the formulation of the second Heaviside expansion theorem? (See Problem 20.5.1.)

P20.5.4. Solve the problem of the dynamics of sorption (see Chapter 6, Section 3) by using Laplace–Carson transforms.

P20.5.5. Apply the operational rules (20.121)–(20.132) to the solution of the generalized Abel integral equation (15.149). Then derive the final expression (15.162) for the Green's function $g_3(x, \xi, t - \tau)$.

P20.5.6. Interpret and solve the b.v.p

$$\frac{\partial^2 u}{\partial r^2} + \frac{1}{r}\frac{\partial u}{\partial r} + \frac{\partial^2 u}{\partial z^2} = \frac{1}{a^2}\frac{\partial u}{\partial t} \quad 0 < z < \infty, \ 0 < r < \infty, \ t > 0$$

$$\frac{\partial^2 u}{\partial r^2} + \frac{1 - 2\nu}{r}\frac{\partial u}{\partial r} + \alpha\frac{\partial u}{\partial z} = \frac{\partial u}{\partial t} \quad z = 0, \ 0 < r < \infty, \ t > 0$$

$$u = 0, \ 0 \leq z < \infty, \ 0 < r < \infty, \ t = 0$$

$$u = T(t), \ z = 0, \ r = 0, \ t > 0$$

$$\lim_{z \to \infty} u = 0, \ \nu = \text{const} > 0, \ \alpha = \text{const} > 0$$

Hint: For interpretation see Chapter 2, Section 10 and Appendix 5, Section 1. For solution use the Fourier–Bessel integral transform and the Weber–Schafheitlin discontinuous integral.

21

Asymptotic expansions. Asymptotic solution of boundary-value problems

1. Solution of Cauchy problem for 1-dimensional Fourier equation. Short relaxation time asymptotics for solution of hyperbolic heat-conduction equation

In previous chapters we have often dealt with the asymptotic behavior of solutions of various nonstationary boundary-value problems, while considering the question of their approach to stationary limits. Quite another kind of problem arises when one is studying the behavior of solutions of boundary-value problems when the parameters appearing in the formulation of the problem are varied. The classical example of asymptotic analysis of this kind is represented by Oseen's treatment of the translational movement of a solid body in the $\tilde{x}$ direction at velocity U in an incompressible liquid of vanishingly small viscosity μ in the absence of external mass forces ([53]; see also [8]). Namely, let $(\tilde{x}, \tilde{y}, \tilde{z})$ be the laboratory coordinate system and (x, y, z) a system moving together with the body:

$$ x = \tilde{x} - Ut, \quad y = \tilde{y}, \quad z = \tilde{z} \tag{21.1} $$

Assuming the motion to be stationary in the coordinate system (x, y, z), we find that the system of equations to be considered is[1]

$$ -U\frac{\partial}{\partial x}\mathbf{v} - \mathbf{v} \times \operatorname{rot}\mathbf{v} = -\frac{1}{\rho}\operatorname{grad} q + \frac{\mu}{\rho}\Delta\mathbf{v} \tag{21.2} $$

$$ \operatorname{div}\mathbf{v} = 0 \tag{21.3} $$

---

[1]It can be easily shown (cf. [64]) that the material derivative can be rewritten as

$$ \frac{d}{dt}\mathbf{v} = \frac{\partial}{\partial t}\mathbf{v} - \mathbf{v} \times \operatorname{rot}\mathbf{v} + \operatorname{grad}\frac{v^2}{2} $$

where

$$q = p + \frac{\rho v^2}{2} \tag{21.4}$$

The non-slip boundary conditions state that the velocity of the liquid at the surface S of the solid body coincides with the velocity of the latter:

$$v_x = U, \quad v_y = 0, \quad v_z = 0, \quad (x, y, z) \in S \tag{21.5}$$

Oseen considered the simplified problem obtained from (21.2) and (21.3) by omitting the quadratic "vorticity" term $\mathbf{v} \times \operatorname{rot} \mathbf{v}$. Thus he considered the boundary-value problem

$$\mu \Delta \mathbf{v} + \rho U \frac{\partial}{\partial x} \mathbf{v} - \operatorname{grad} q = 0$$

$$\operatorname{div} \mathbf{v} = 0, \quad v_x = U, \quad v_y = 0, \quad v_z = 0 \quad \forall (x, y, z) \in S \tag{21.6}$$

Together with the solution $\mathbf{v}$ of this problem, let $\tilde{\mathbf{v}}$ be the solution of the corresponding problem for the irrotational flow of an ideal liquid:

$$\rho U \frac{\partial}{\partial x} \tilde{\mathbf{v}} - \operatorname{grad} q = 0, \quad \operatorname{div} \tilde{\mathbf{v}} = 0, \quad v_n = U_{n_x}(x, y, z) \in S \tag{21.7}$$

Oseen proved that

$$\lim_{\mu \downarrow 0} \mathbf{v} \neq \tilde{\mathbf{v}} \tag{21.8}$$

It turns out that the head resistance of the solid to the motion of the liquid does not vanish with vanishing viscosity, whereas the solid does not resist the motion of a perfect liquid, as stated by D'Alembert's paradox.

Oseen's discovery provided the impetus for the development of boundary-layer theory in ordinary and partial differential equations, in which the coefficient of the highest-order derivatives is a small parameter. (Recall that viscosity is the coefficient of the higher order derivatives in the Navier–Stokes equations.) This theory is, in turn, an instance of singular perturbations, which represent one of the central issues of asymptotic analysis.

Another, less studied, paradox is that of the infinite speed of redistribution of thermal or diffusion disturbances. In Chapter 13 we pointed out that this paradox is a consequence of the fact that Fourier's law neglects the time of fast (microscopic) relaxation; to allow for the finite relaxation time one must replace the parabolic Fourier heat-conduction equation by a hyperbolic equation, that is, an equation whose solutions possess the property that disturbances propagate at a finite speed. We now wish to demonstrate that the solutions of the Fourier heat-conduction equation are asymptotic limits of the solutions of the hyperbolic heat-conduction equation in the limit of vanishing relaxation time τ. An exact formulation of our assertion is given by the following theorem.

Theorem 21.1.1. *Let $u(x, t \mid F)$ be a solution of the parabolic Cauchy problem*

$$a^2 \frac{\partial^2}{\partial x^2} u(x, t \mid F) = \frac{\partial}{\partial t} u(x, t \mid F), \quad u(x, 0 \mid F) = u_0(x), \quad -\infty < x < \infty \tag{21.9}$$

and let $u(x, t \mid H)$ *be a solution of the corresponding hyperbolic Cauchy problem*

$$a^2 \frac{\partial^2}{\partial x^2} u(x, t \mid H) = \tau \frac{\partial^2}{\partial t^2} u(x, t \mid H) + \frac{\partial}{\partial t} u(x, t \mid H)$$

$$u(x, 0 \mid H) = u_0(x), \quad \frac{\partial}{\partial t} u(x, 0 \mid H) = u_1(x), \quad -\infty < x < \infty$$

(21.10)

Then, uniformly with respect to $x \in (-\infty, \infty)$ *for any* $t > 0$,

$$\lim_{\tau \downarrow 0} u(x, t \mid H) = u(x, t \mid F)$$

(21.11)

For the sake of simplicity we shall assume that u_0 and u_1 are bounded, smooth functions in $(-\infty, \infty)$ and that, by scaling, $a^2 = 1$.

PROOF. Note first that both $u(x, t \mid F)$ and $u(x, t \mid H)$ have explicit integral representations. Indeed, $u(x, t \mid F)$ is determined by the Poisson integral

$$u(x, t \mid F) = \int_{-\infty}^{\infty} u_0(\xi) E(x - \xi, t) d\xi$$

(21.12)

To find $u(x, t \mid H)$, note that the substitution

$$y = \frac{x}{2\sqrt{\tau}} - \frac{t}{2\tau}, \quad z = \frac{x}{2\sqrt{\tau}} + \frac{t}{2\tau}, \quad u(x, t) = v(y, z) \exp\left(-\frac{t}{2\tau}\right)$$

(21.13)

reduces the Cauchy problem (21.10) to the problem

$$4 \frac{\partial^2}{\partial y \partial z} v(y, z) + v(y, z) = 0, \quad -\infty < y < \infty, \ z - y > 0$$

(21.14)

$$v(y, y) = u_0(x)$$

(21.15)

$$\frac{\partial}{\partial y} v(y, y) = \sqrt{\tau} \frac{d}{dx} u_0(x) - u_0(x) - \tau u_1(x)$$

$$\frac{\partial}{\partial z} v(y, y) = \sqrt{\tau} \frac{d}{dx} u_0(x) + u_0(x) + \tau u_1(x)$$

(21.16)

The corresponding Riemann function (see Chapter 6, Sections 1 and 3) is

$$g(y, z, \eta, \zeta) = J_0(\sqrt{(y - \eta)(z - \zeta)})$$

(21.17)

where J_0 is Bessel function of order zero (see Appendix 2, Section 6), so that in the region $t > 0$ of the plane (x, t),

$$g = I_0(\lambda), \quad \lambda = (t/2\tau)[1 - (x - \xi)^2 \tau/t^2]^{1/2}$$

(21.18)

By (6.16) (see also Chapter 6, Section 3),

$$v \mid_M = \frac{1}{2} \left(vg \mid_P + vg \mid_Q - \int_{PQ} \left(g \left(\frac{\partial v}{\partial \zeta} - \frac{\partial v}{\partial \eta} \right) - v \left(\frac{\partial g}{\partial \zeta} - \frac{\partial g}{\partial \eta} \right) \right)_{\eta=\zeta} d\zeta \right)$$

$$P = (\eta, z), \quad Q = (y, \zeta)$$

(21.19)

where $M(y, z)$ is a given point of the region $z > y$ of the (y, z) plane and P, Q are points of intersection of the characteristic curves emanating from M with the carrier

$$\Gamma = \{y, z: y - z = 0, \; -\infty < z < \infty\} \tag{21.20}$$

In view of definitions (21.13) and equalities (21.15), (21.16), and (A2.101), it follows from (21.19) that

$$u(x, t \mid H) = \frac{1}{2} e^{-(t/2\tau)} [u_0(x^-) + u_0(x^+)]$$

$$+ \frac{1}{4\sqrt{\tau}} \int_{x^-}^{x^+} e^{(-t/2\tau)} \Bigg([u_0(\xi) + 2\tau u_1(\xi)] I_0(\lambda)$$

$$+ u_0(\xi) \left[1 - \frac{(x - \xi)^2 \tau}{t^2} \right]^{-1/2} I_1(\lambda) \Bigg) d\xi \tag{21.21}$$

where

$$x^+ = x + t/\sqrt{\tau}, \quad x^- = x - t/\sqrt{\tau} \tag{21.22}$$

We wish to let $\tau \to 0$ in (21.21), leaving x and $t > 0$ fixed. By assumption, u_0 and u_1 are bounded functions, and therefore the term outside the integral in the right-hand side of (21.21) tends to zero. Furthermore, the upper and lower limits of integration x^+ and x^- tend to $+\infty$ and $-\infty$, respectively. However, the integrand depends on the parameter τ. Hence, in order to be able to interchange the limit and the integral signs while at the same time letting x^+ and x^- go to infinity, we must verify that Tannery's theorem is applicable.[2] In order to demonstrate this, we first note that, as ξ increases from x^- to x, the argument of I_0 and I_1 increases from 0 to $t/2\tau$; and when ξ increases from x to x^+, the argument decreases from $t/2\tau$ to 0. Hence the main contribution to the

---

[2] We mean the following theorem [78]: Let

$$I(n) = \int_a^{\lambda_n} f(x, n) dx, \quad J = \int_a^{\infty} g(x) dx, \quad \lim_{n \uparrow \infty} \lambda_n = \infty$$

and assume that uniformly in any finite x interval,

$$\lim_{n \uparrow \infty} f(x, n) = g(x)$$

Then

$$\lim_{n \uparrow \infty} I(n) = J$$

if and only if there exists $M(x)$ such that

$$\int_a^{\infty} M(x) dx \text{ is convergent} \quad \text{and} \quad |f(x, n)| < M(x) \text{ for all } n$$

Concerning different tests for the uniform convergence of singular integrals, see [6].

integral comes from a neighborhood of the point $\xi = x$. This means that we may use the asymptotic values of I_0 and I_1, which are

$$I_k(\lambda) = (2\pi\lambda)^{-1/2}e^\lambda[1 + R_k(\lambda)], \quad k = 0, 1 \tag{21.23}$$

where $R_k(\lambda)$ are holomorphic functions of λ for all $|\lambda| \geq \rho > 0$:

$$R_k(\lambda) = \frac{4k^2 - 1}{8\lambda} + O(\lambda^{-2}) \tag{21.24}$$

(see Appendix 2, Sections 7 and 10).

We have

$$-\frac{t}{2\tau} + \lambda \equiv -\frac{t}{2\tau} + \left[\left(\frac{t}{2\tau}\right)^2 - \frac{(x-\xi)^2}{4\tau}\right]^{1/2}$$

$$= -(x-\xi)^2(2t)^{-1}\left(1 + \left[1 - \frac{(x-\xi)^2\tau}{t^2}\right]^{1/2}\right)^{-1} \tag{21.25}$$

In the neighborhood of $\xi = x$, with $t > 0$ fixed and $\tau \ll 1$, the binomial expansion of $[1 - (x-\xi)^2\tau/t]^{-1/2}$ is convergent (see Appendix 2, Section 1), so that

$$\left[1 - \frac{(x-\xi)^2\tau}{t^2}\right]^{-1/2} = \sum_0^\infty (-1)^n \left[\frac{(x-\xi)^2\tau}{t^2}\right]^n \Gamma\left(\frac{1}{2}\right)\Gamma^{-1}(n+1)\Gamma^{-1}\left(\frac{1}{2} - n\right)$$

$$= 1 + \frac{(x-\xi)^2}{2t^2}\tau + O(\tau^2) \tag{21.26}$$

Hence

$$-\frac{t}{2\tau} + \lambda = -\frac{(x-\xi)^2}{4t}\left(1 - \frac{(x-\xi)^2}{2t^2}\tau + O(\tau^2)\right) \tag{21.27}$$

Similarly, by the binomial series,

$$(2\pi)^{-1/2}\lambda^{-1/2} = \sqrt{\tau}(\pi t)^{-1/2}\left(1 + \frac{(x-\xi)^2}{2t^2}\tau + O(\tau^2)\right) \tag{21.28}$$

and

$$u_0(\xi)\left[1 - \frac{(x-\xi)^2\tau}{2t^2}\right]^{-1/2} = u_0(\xi)\left(1 - \frac{(x-\xi)^2\tau}{2t^2} + O(\tau^2)\right) \tag{21.29}$$

Inserting (21.23), (21.24), and (21.27)–(21.29) into (21.21), we find that

$$\lim_{\tau \downarrow 0} u(x, t \mid H) = \lim_{\tau \downarrow 0} \int_{x-}^{x+} [u_0(\xi) + 2\tau u_1(\xi)]E(x - \xi, t)(1 + O(\tau))d\xi \tag{21.30}$$

Hence the conditions of Tannery's theorem are satisfied, and therefore, in view of (21.12),

$$\lim_{\tau \downarrow 0} u(x,t \mid H) = u(x,t \mid F) \tag{21.31}$$

Q.E.D.

2. Asymptotic sequences. Expansions in asymptotic series. Definitions and preliminary statements

The example of Section 1 represents a very rare case, one in which an asymptotically valid approximation to a solution of a boundary-value problem with a small parameter is being worked at through a transition to a limit in the solution itself in the analytically closed form. Normally, the situation is the very opposite: One attempts to approximate the solution of the full problem through the solutions of various limit problems that are usually easier to solve. This strategy is known as the theory of *perturbations*. An important version of this theory concerns the problems where, as in Section 1, the small parameter is the coefficient of the highest-order derivative. Problems of this type, characterized by the occurrence in their solutions of narrow boundary or transition layers, require for their asymptotic treatment use of the method known as *matched asymptotic expansions*.

Matched asymptotic expansions is one of several methods designed to tackle *singular perturbation problems*, of which the boundary- or transition-layer problems are an important subclass. Singular perturbation problems are to be distinguished from *regular perturbation* ones, which may be addressed by regular perturbation methods.

Before outlining some of these notions, let us introduce a few definitions common to all asymptotic treatments. Let $f(p \mid \alpha)$ be a scalar function of p, defined in some region $D \subset \mathbb{R}_n$ and dependent on a scalar parameter α varying in an interval $I = (\alpha', \alpha'')$, and let $U(\alpha_0) \subset I$ be a neighborhood of a limit point $\alpha_0 \in I$.

Definition 21.2.1 (Large O). $\varphi(p \mid \alpha \to \alpha_0) = O[f(p \mid \alpha \to \alpha_0)]$ in $U(\alpha)$ for $p \in D$ if there exists $k(p)$ such that

$$|\varphi(p \mid \alpha)| \leq k(p)|f(p \mid \alpha)| \quad \forall \alpha \in U(\alpha_0) \tag{21.32}$$

If $f(p \mid \alpha) \neq 0$ for all $\alpha \in U(\alpha_0)$, then (21.32) means that $\varphi(p \mid \alpha)/f(p \mid \alpha)$ is bounded in $U(\alpha_0)$.

Definition 21.2.2. (Small o). $\varphi(p \mid \alpha \to \alpha_0) = o[f(p \mid \alpha \to \alpha_0)]$ as $\alpha \to \alpha_0$ if for any positive function $\delta(p)$ there exists $U(\alpha_o)$ such that

$$|\varphi(p \mid \alpha)| \leq \delta(p)|f(p \mid \alpha)| \quad \forall \alpha \in U(\alpha_0) \tag{21.33}$$

If $f(p \mid \alpha) \neq 0$ for all $\alpha \in U(\alpha_0)$, then (21.33) means that

$$\lim_{\alpha \uparrow \alpha_0} \frac{\varphi(p \mid \alpha)}{f(p \mid \alpha)} = 0 \tag{21.34}$$

Let us denote the *maximal* of $U(\alpha_0)$ as $dU(\alpha \mid \alpha_0, p)$.

Definition 21.2.3. If $dU(\alpha \mid \alpha_0, p)$, and $k(p)$ are independent of $p \in D$ then
$$\varphi(p \mid \alpha \to \alpha_0) = O[f(p \mid \alpha \to \alpha_0)]$$
or
$$\varphi(p \mid \alpha \to \alpha_0) = o[f(p \mid \alpha \to a_0)]$$
uniformly in D.

(21.35)

Definition 21.2.4. Let $\{\varphi_n(p \mid \alpha)\} \mid_{n=1}^{\infty}$, be a sequence defined in $D \times I$. Then $\{\varphi_n(p \mid \alpha)\}$ is an asymptotic sequence at $p \in D$ and $\alpha \to \alpha_0$ if, for all $n \geq 1$,
$$\varphi_{n+1}(p \mid \alpha) = o[\varphi_n(p \mid \alpha)] \quad \text{as } \alpha \to \alpha_0 \tag{21.36}$$
If (21.36) is valid uniformly in D, then $\{\varphi_n(p \mid \alpha)\}$ is said to be *uniformly asymptotic* in D.

Definition 21.2.5. Let $f(p, \alpha) \in D \times I$, $\varphi_n(p \mid \alpha) \in D \times I$, and $a_n(p) \in D$ for all $n \geq 1$. The finite sequence of partial sums
$$S_n(p \mid \beta) = \sum_{k=1}^{n} a_k(p)\varphi_k(p \mid \beta) \tag{21.37}$$
where $1 \leq n \leq N$, is called an *asymptotic expansion* of $f(p \mid \alpha)$ up to order N if
$$f(p \mid \alpha) - S_n(p \mid \alpha) = o[\varphi_n(p \mid \alpha)] \quad \forall n \in [1, N] \quad \text{as } \alpha \to \alpha_0 \tag{21.38}$$
Obviously, the relation (21.37) is equivalent to
$$f(p \mid \alpha) - S_n(p \mid \alpha) = O[\varphi_{n+1}(p \mid \alpha)] \quad \forall n \in [1, N] \quad \text{as } \alpha \to \alpha_0 \tag{21.39}$$
If (21.39) holds as $N \uparrow \infty$, one says that the series
$$\sum_{k=1}^{\infty} a_k(p)\varphi_k(p \mid \alpha) \tag{21.40}$$
is an asymptotic expansion of $f(p \mid \alpha)$, writing[3]
$$f(p \mid \alpha) \asymp \sum_{k=1}^{\infty} a_k(p)\varphi_k(p \mid \alpha) \tag{21.41}$$
If the relation (21.39) holds uniformly in D, then the series (21.41) is a uniform asymptotic expansion of $f(p \mid \alpha)$.

Theorem 21.2.1. *If* (21.40) *is an uniform asymptotic expansion of* $f(p \mid \alpha)$ *in* D, *and if* $\tilde{D}$ *is an arbitrary subset of* D, *then*
$$\int_{\tilde{D}} f(p \mid \alpha)d\omega_p \asymp \sum_{k=1}^{\infty} \int_{\tilde{D}} a_k(p)\varphi_k(p \mid \alpha)d\omega_p \tag{21.42}$$

---

[3]Often the symbol $\sim$ is used instead of $\asymp$.

so that an asymptotic series may be integrated term by term.

PROOF. Assume that D is a bounded region. Then, by uniformity for all $n \geq 1$,

$$\left| \int_{\tilde{D}} \left(f(p \mid \alpha) - \sum_{k=1}^{n} a_k(p)\varphi_k(p \mid \alpha) \right) d\omega_p \right|$$

$$= \left| \int_{\tilde{D}} (o[a_n(p)\varphi_n(p \mid \alpha)] d\omega_p \right|$$

$$= |o[a_n(q)\varphi_n(q \mid \alpha)]| \int_{\tilde{D}} d\omega_q \to 0 \quad \text{as } \alpha \to \alpha_0 \qquad (21.43)$$
$$q \in \tilde{D}$$

which proves the assertion if D is a bounded region. If D is unbounded, the assertion remains valid if all the integrals in (21.42) are uniformly convergent.

Remark 21.2.1. An asymptotic series may be divergent; in fact, this is so in most cases. Generally, the results of term-by-term differentiating of the asymptotic expansion of $f(p \mid \alpha)$ with respect to p are not the asymptotic expansion of the derivative of $f(p \mid \alpha)$. Nonetheless, we have the following theorems.

Theorem 21.2.2. *Results of term by term differentiation of the asymptotic power expansion of $f(x \mid \alpha)$ in α is the asymptotic expansion of $df(x \mid \alpha)/dx$.*

Theorem 21.2.3. *The sum and product of any finite number of asymptotic series is an asymptotic series.*

The validity of Theorems 21.2.2–3 is obvious.

Example. Let

$$f(x) \overset{\text{def}}{=} \text{erfc}(x) = \frac{2}{\sqrt{\pi}} \int_x^\infty e^{-\tau^2} d\tau, \quad x > 0 \qquad (21.44)$$

Integrating n times by parts, we obtain

$$f(x) = \frac{1}{\sqrt{\pi}x} e^{-x^2} \left(1 + \sum_{k=1}^{n} (-1)^k \frac{(2k-1)!}{(2x^2)^k} \cdot \frac{2^{1-k}}{(k-1)!} \right) + R_n(x) \qquad (21.45)$$

where the remainder term is

$$R_n(x) = \frac{2}{\sqrt{\pi}} (-1)^{n+1} \frac{(2n-1)!}{2^n} \cdot \frac{2^{1-n}}{(n-1)!} \int_x^\infty e^{-y^2} y^{-2n} dy \qquad (21.46)$$

Because the integral

$$\int_x^\infty e^{-y^2} dy \qquad (21.47)$$

is convergent,

$$\lim_{x\uparrow\infty} R_n(x) = 0 \quad \forall n = \text{const} \geq 0 \tag{21.48}$$

Hence the asymptotic expansion of $\text{erfc}(x)$ in the neighborhood of infinity is given by the series

$$\text{erfc}(x) = \frac{1}{\sqrt{\pi}x} e^{-x^2} \left(1 + \sum_{k=1}^{\infty} (-1)^k \frac{(2k-1)!}{(2x^2)^k} \cdot \frac{2^{1-k}}{(k-1)!} \right) \tag{21.49}$$

which is divergent for any fixed x.

3. Regular and singular perturbations. Differential equations depending on parameters. Scaling. Outer and inner expansions. Matching

In what follows we shall be concerned with deriving asymptotic solutions of boundary-value problems for partial differential equations dependent on one or more parameters. It is usually almost impossible to determine the exact solutions of such problems (certainly, nonlinear or even quasilinear problems), at least in a form that is easily handled – even when such properties of the problem as the existence of a solution in specific functional spaces, uniqueness, stability, and so forth are known in advance. As we have stated, a much more promising approach is to construct the solution to another problem, one that approximates the solution of the given problem with reasonable accuracy. One of the most powerful methods of this kind involves constructing an asymptotic expansion of the solution with respect to the parameters occurring in the formulation of the problem. This is the general methodology of the theory of perturbations.

Whenever the direct formal transition to the asymptotic limit in the original formulation yields no loss of any essential qualitative features of the solution of the full problem, the latter is said to be a problem of *regular* perturbations. Otherwise, it is said to be a problem of *singular* perturbations. To illustrate this, consider the quadratic equation

$$y^2 - 2\varepsilon y - 1 = 0, \quad \varepsilon \ll 1 \tag{21.50}$$

Both of its roots,

$$y_1 = \varepsilon + (1 + \varepsilon^2)^{1/2} \Rightarrow y_1 = 1 + \varepsilon + \tfrac{1}{2}\varepsilon^2 - \tfrac{1}{8}\varepsilon^4 + \cdots \tag{21.51}$$

$$y_2 = \varepsilon - (1 + \varepsilon^2)^{1/2} \Rightarrow y_2 = -1 + \varepsilon - \tfrac{1}{2}\varepsilon^2 + \tfrac{1}{8}\varepsilon^4 + \cdots \tag{21.52}$$

are fully recovered when seeking the solution as a power series

$$y \asymp \sum_{k=0}^{\infty} y_k \varepsilon^k \tag{21.53}$$

Substitution of (21.53) into (21.50) and equating the coefficients at each power of ε yields the following sequence of relations:

$$O(1) \qquad\qquad y_0^2 = 1 \Rightarrow y_0 = \pm 1 \tag{21.54}$$

$$O(\varepsilon) \qquad\qquad 2y_0 y_1 - 2y_0 = 0 \Rightarrow y_1 = 1 \tag{21.55}$$

Figure 21.1.   Sketch of a spring with the mass and damper at its end.

which in turn yields (21.51) and (21.52). This is a typical example of regular perturbations.

Now consider another quadratic equation, this time with a small parameter at the highest power of y:

$$\varepsilon y^2 - 2y - 1 = 0 \tag{21.56}$$

The roots of equation (21.56) are

$$y_1 = \varepsilon^{-1} + (\varepsilon^{-2} + \varepsilon^{-1})^{1/2} \Rightarrow y_1 = 2\varepsilon^{-1} + {}^1\!/_2 - {}^1\!/_8\varepsilon + \cdots \tag{21.57}$$

$$y_2 = \varepsilon^{-1} - (\varepsilon^{-2} + \varepsilon^{-1})^{1/2} \Rightarrow y_2 = -{}^1\!/_2 + {}^1\!/_8\varepsilon + \cdots \tag{21.58}$$

However, a direct substitution of the expansion (21.53) into (21.56) yields only the smallest (order 1), root (21.58). The large, order ε^{-1}, root (21.57) is lost in such a direct perturbation procedure, owing to degeneration of the quadratic equation (21.56) into a linear equation upon a formal transition to the limit $\varepsilon \to 0$ in the procedure. Such a qualitative degeneration upon a formal passage to a limit is typical for all singular perturbation problems. (In order to recover the root (21.57), the unknown should be rescaled so as to balance the first and second term in equation (21.56); the perturbation procedure may then be applied to the rescaled equation.)

Let us consider another example, this time for an ordinary differential equation. This example, borrowed from [44], concerns oscillations of a linear spring with a coefficient of stiffness K under the influence of an initial impulse I_0, on the assumption that one of the ends is fixed rigidly and the second is connected with some mass M and a damper B (see Figure 21.1). The equation and the initial conditions are

$$M\frac{d^2}{dT^2}Y + B\frac{d}{dT}Y + KY = 0, \quad t > 0, \tag{21.59}$$

$$Y(0) = 0, \quad \frac{d}{dT}Y(0) = \frac{I_0}{M} \tag{21.60}$$

We first introduce dimensionless variables

$$t = T/T_0, \quad y = Y/A \tag{21.61}$$

with scales T_0 and A undefined so far. Equation (21.59) and the initial conditions (21.60) become

$$\varepsilon^2 \frac{d^2}{dt^2} y(t) + 2\beta \frac{d}{dt} y(t) + y(t) = 0, \tag{21.62}$$

$$y(0) = 0, \quad \frac{d}{dt} y(0) = I \tag{21.63}$$

where

$$\varepsilon = MT_0^{-2}/K, \quad 2\beta = B/(T_0 K), \quad I = I_0 T_0/MA \tag{21.64}$$

The following two time scales are intrinsic to equation (21.59): free oscillation scale T_i,

$$T_i = (M/K)^{1/2} \tag{21.65}$$

and damping scale T_d,

$$T_d = M/B \tag{21.66}$$

of course, along with any suitable combination of those two (the one with time dimension). Thus

$$\varepsilon = T_i^2/T_0^2, \quad 2\beta = T_i^2/(T_d T_0) \tag{21.67}$$

Let us consider the following limiting cases.

A. Small damping

Choose

$$T_0 = T_i, \quad A = I_0(MK)^{-1/2} \tag{21.68}$$

so that equation (21.62) and the initial conditions (21.63) become

$$\frac{d^2}{dt^2} y + 2\beta \frac{d}{dt} y + y = 0, \quad y(0) = 0, \quad \frac{d}{dt} y(0) = 1 \tag{21.69}$$

Let the parameters of the system be such that

$$\beta = \frac{1}{2} \frac{B}{T_0 K} = \frac{T_i^2}{2 T_d T_0} = \frac{1}{2} \frac{T_i}{T_d} \ll 1 \tag{21.70}$$

so that equation (21.62) contains a small parameter at the first-order derivative. Thus the boundary-value problem (21.69) may be considered as if obtained by perturbation of the boundary-value problem

$$\frac{d^2}{dt^2} y + y = 0, \quad y(0) = 0, \quad \frac{d}{dt} y(0) = 1 \tag{21.71}$$

which describes the oscillation of a spring without damping owing to introduction of a damper of low efficiency. The introduction of the term $2\beta(du)/dt$ into the equation on the left side of (21.71) does not change the order of the equation, which means that the perturbation does not change the number and character of the boundary (in our case initial) conditions to be satisfied. The perturbation problem is regular in this sense (it is still singular in a sense to be specified later).

We approximate the exact solution by an asymptotic expansion in powers of the small parameter β:

$$y \asymp \sum_{i=0}^{\infty} y_i(t)\beta^i \qquad (21.72)$$

Substitution of the series (21.72) into (21.69) and comparison of coefficients of terms with like powers of β yield

$$\frac{d^2}{dt^2}y_0 + y_0 = 0, \quad y_0(0) = 0, \quad \frac{d}{dt}y_0(0) = 1 \qquad (21.73)$$

$$\frac{d^2}{dt^2}y_n + y_n = -2\frac{d}{dt}y_{n-1}, \quad y_n(0) = 0, \quad \frac{d}{dt}y_n(0) = 1, \quad n = 1,2,\dots \quad (21.74)$$

Integration yields

$$y_0 = \sin t, \quad y_1 = -t\sin t, \quad y_2 = O(\beta^2 t^2) \qquad (21.75)$$

Thus, up to quantities of second order in βt, we have

$$y(t \mid \beta) = \sin t - \beta t \sin t + O(\beta^2 t^2) \qquad (21.76)$$

Note that the exact solution of the problem (21.69) is

$$y(t \mid \beta) = e^{-\beta t}(1 - \beta^2)^{-1/2}\sin[(1 - \beta^2)^{1/2}t] \qquad (21.77)$$

Applying Lagrange's formula of finite increments, we find that for any fixed t in the neighborhood of $\beta = 0$,

$$y(t \mid \beta) = y(t \mid 0) + \beta\frac{\partial}{\partial\beta}y(t \mid 0) + \frac{1}{2}\beta^2\frac{\partial^2}{\partial\beta^2}y(t \mid \vartheta\beta)$$

$$= \sin t - \beta t \sin t + O(\beta^2 t^2) \qquad (21.78)$$

which shows that formula (21.76) yields an approximation of the exact solution y up to second order in β for $\beta \downarrow 0$ at any finite time interval.

On the other hand, the presence in (21.76) of secular terms that are proportional to powers of t but absent from the exact solution makes the expansion (21.72) invalid uniformly for the entire time half-axis. This is the sense (in contrast to that of the following subsection) in which (21.69) is a problem of singular perturbations.

B. Small mass–initial layer problem

If $M = 0$ ($\varepsilon = 0$), $\beta = O(1)$ then equation (21.62) becomes a first-order equation,

$$2\beta\frac{d}{dt}y(t) + y(t) = 0 \qquad (21.79)$$

Therefore the second-order equation (21.62) may be treated as the result of a perturbation that alters the order of the unperturbed equation. This is a typical boundary (initial) layer case, representing an important branch of singular perturbation problems that we will tackle via the following matched asymptotic expansions arguments. We consider three different kinds of asymptotic expansions, each corresponding to a different scaling of the variables.

1. Outer expansion (outer, bulk region)

Take

$$T_0 = T_i^2/T_d = B/K, \quad A = I_0/B, \quad \varepsilon = T_i^2/T_0^2 MK/B^2 \ll 1 \qquad (21.80)$$

so that ε is now a small parameter. The boundary value problem (21.62) and the left side of (21.63) becomes

$$\varepsilon \frac{d^2}{dt_o^2} y + \frac{d}{dt_o} y + y = 0, \quad y(0) = 0, \quad \frac{d}{dt_o} y(0) = \frac{1}{\varepsilon} \qquad (21.81)$$

where we have written t_o instead of t to distinguish this time scaling from the previous one.

We now approximate the solution of this problem by an asymptotic expansion:

$$y(t_o \mid \varepsilon) \asymp h(t_o, \varepsilon) \overset{\text{def}}{=} \sum_{i=1}^{\infty} h_i(t_o)\nu_i(\varepsilon) \qquad (21.82)$$

where $\{\nu_i(\varepsilon)\}$ is a temporarily undefined asymptotic sequence. The coefficients h_i of this expansion must be solutions of the problems

$$\frac{d}{dt_o} h_1 + h_1 = 0 \qquad (21.83)$$

$$\frac{d}{dt_o} h_2 + h_2 = \begin{cases} -\dfrac{d^2}{dt_o^2} h_1 & \text{if } \dfrac{\nu_1 \varepsilon}{\nu_2} = 1 \\[2ex] 0 & \text{if } \dfrac{\nu_1 \varepsilon}{\nu_2} \to 0 \end{cases} \qquad (21.84)$$

$$\vdots$$

Integration of equations (21.83) and (21.84) yields

$$h_1(t_o) = A_1 e^{-t_o}$$
$$h_2(t_o) = A_2 e^{-t_o} - A_1 t_o e^{-t_o} \quad \text{if } \nu_1 \varepsilon/\nu_2 = 1 \qquad (21.85)$$
$$h_2(t_o) = A_2 e^{-t_o} \quad \text{if } \nu_1 \varepsilon/\nu_2 \to 0$$

$$\vdots$$

The constants of integration remain temporarily undefined, since the initial conditions cannot be satisfied by any term of this outer expansion. These initial conditions and the orders of the various $\nu_i(\varepsilon)$ are unknown; they must be determined by matching with the inner expansion that is yet to be determined. These equalities apply at any fixed t_o outside a neighborhood of the origin $t_o = 0$; this justifies the use of the term outer expansion.

2. Inner asymptotic expansion (initial, boundary layer)

In order to handle the vicinity of $t_o = 0$ and to be able to satisfy the initial conditions, we have to introduce a new (fast) time so that the coefficient at the second derivative will not be small, despite the fact that M is small. This may be done by setting

$$t_\eta = t_o/\eta(\varepsilon) \tag{21.86}$$

where $\eta(\varepsilon)$ is such that

$$\lim_{\varepsilon \downarrow 0} \eta(\varepsilon) = 0 \tag{21.87}$$

With this scaling, equation (21.81) becomes

$$\frac{\varepsilon}{\eta^2(\varepsilon)} \frac{d^2}{dt_\eta^2} y + \frac{1}{\eta(\varepsilon)} \frac{d}{dt_\eta} y + y = 0 \tag{21.88}$$

There are three possibilities for the rate of decay of $\eta(\varepsilon)$.

(1) Inner–inner limit:

$$\lim_{\varepsilon \downarrow 0} \frac{\eta(\varepsilon)}{\varepsilon} = 0 \Rightarrow \frac{d^2}{dt_\eta^2} y = 0 \tag{21.89}$$

This limit, not of interest to us, corresponds to the very short time interval in which the process is dominated by the inertia and the damping or stiffness of the spring have not come yet into effect.

(2) Inner initial (boundary)-layer limit:

$$\eta(\varepsilon) = \varepsilon \tag{21.90}$$

(3) Intermediate limit:

$$\lim_{\varepsilon \downarrow 0} \frac{\varepsilon}{\eta(\varepsilon)} = 0, \quad \lim_{\varepsilon \downarrow 0} \eta(\varepsilon) = 0 \tag{21.91}$$

Thus there exists a class of limits and of their corresponding asymptotic expansions, intermediate between the inner and outer limits, which must be considered when the outer and inner asymptotic expansions are matched.

Thus, consider the inner (initial-layer) asymptotic expansion:

$$y(t \mid \varepsilon) \asymp g(t_i \mid \varepsilon) \stackrel{\text{def}}{=} \sum_{k=1}^{\infty} \mu_k(\varepsilon) g_k(t_i) \tag{21.92}$$

where $\{\mu_k(\varepsilon)\}$ is an asymptotic sequence to be defined and $t_i = t_o/\varepsilon$.

Inserting the expansion $g(t_i \mid \varepsilon)$ into (21.88) and then comparing terms of the same order of magnitude, we obtain

$$\frac{d^2}{dt_i^2} g_1 + \frac{d}{dt_i} g_1 = 0 \tag{21.93}$$

$$\frac{d^2}{dt_i^2} g_k + \frac{d}{dt_i} g_k = \begin{cases} -g_{k-1} & \text{if } \dfrac{\varepsilon \mu_{k-1}}{\mu_k} \to 1 \\ 0 & \text{if } \dfrac{\varepsilon \mu_{k-1}}{\mu_k} \to 0 \end{cases} \quad \forall k \geq 2 \tag{21.94}$$

The initial conditions determine $\mu_1(\varepsilon)$, since

$$\frac{d}{dt} y \asymp \sum_{k=1}^{\infty} \frac{\mu_k(\varepsilon)}{\varepsilon} \cdot \frac{d}{dt_i} g_k(t_i) = \frac{1}{\varepsilon} \text{ as } \varepsilon \downarrow 0 \Rightarrow \mu_1(\varepsilon) = 1 \tag{21.95}$$

and by induction,

$$\mu_k(\varepsilon) = \varepsilon^{k-1} \quad \forall k \geq 2 \tag{21.96}$$

Integration of equations (21.93) and (21.94), taking into account the initial conditions,

$$g_1(0) = 0, \quad \frac{d}{dt_i} g_1(0) = 1$$

$$\tag{21.97}$$

$$g_k(0) = 0, \quad \frac{d}{dt_i} g_k(0) = 0 \quad \forall k \geq 2$$

yields

$$g_1(t_i) = 1 - e^{-t_i} \tag{21.98}$$

$$g_2(t_i) = 2 - t_i - e^{-t_i}(2 + t_i) \tag{21.99}$$

$$\vdots$$

Thus the inner expansion to order $O(\varepsilon)$ is

$$g(t_i) \asymp 1 - e^{-t_i} + \varepsilon[2 - t_i - e^{-t_i}(2 + t_i)] + O(\varepsilon^2) \tag{21.100}$$

3. Intermediate expansion

The outer expansion (21.82) has a limit:

$$\lim_{t_o \downarrow 0} h(t_o \mid \varepsilon) = A_1 \nu_1(\varepsilon) + O\left(\nu_2(\varepsilon)\right) \tag{21.101}$$

On the other hand,

$$t_o = \text{const} > 0 \wedge \varepsilon \downarrow 0 \Rightarrow t_i = \frac{t_o}{\varepsilon} \uparrow \infty \tag{21.102}$$

The purpose of the intermediate asymptotic expansion is to match the inner and outer expansions to order $O(\varepsilon^k)$, where k is the number of terms of outer and inner expansions to be matched, in order to determine the unknown coefficients of the outer expansion. The matching procedure consists of finding a scaling and a corresponding expansion whose region of validity intersects the regions in which the outer and initial-layer (boundary-layer) limits are valid in a nonempty set.

We first match the first terms of these expansions up to order $O(1)$. This means that we must find $\eta(\varepsilon)$ such that, for fixed $t_\eta > 0$,

$$\lim_{\varepsilon \downarrow 0} \varepsilon / \eta(\varepsilon) = 0, \lim_{\varepsilon \downarrow 0} \left\{ \nu_1(\varepsilon) h_1(\eta t_\eta) - g_1\left(\frac{\eta t_\eta}{\varepsilon}\right) \right\} = 0 \tag{21.103}$$

(see (21.82) and (21.92)). Equalities (21.85) and (21.98) yield

$$\lim_{\varepsilon \downarrow 0} \left\{ \nu_1(\varepsilon) A_1 \exp(-\eta t_\eta) - [1 - \exp(-\eta t_\eta/\varepsilon)] \right\} = 0$$

$$\tag{21.104}$$

$$t_\eta = \text{const} > 0$$

Since $\exp(-\eta t_\eta/\varepsilon)$ tends to zero faster than any power of ε (we say that it is transcendentally small), we see that equality (21.104) holds to within $O(1)$ if and only if

$$\nu_1(\varepsilon) \equiv 1, \quad A_1 = 1 \tag{21.105}$$

and that the overlap region for matching to within $O(1)$ is determined by the inequalities

$$O(\varepsilon) < O\left(\eta(\varepsilon)\right) < O(1) \quad (\text{e.g.,}\ \varepsilon|\ln\varepsilon| \le \eta(\varepsilon) < O(1)) \tag{21.106}$$

Assume now that

$$\nu_2(\varepsilon) = \varepsilon\nu_1(\varepsilon) \equiv \varepsilon \tag{21.107}$$

(as we shall see below, this choice is implied by (21.85) and the matching requirements) and carry out matching to order $O(\varepsilon)$. We have to determine A_2 from the equality

$$\lim_{\varepsilon\downarrow 0}\frac{1}{\varepsilon}\left\{\nu_1(\varepsilon)h_1(\eta t_\eta) + \nu_2 h_2(\eta t_\eta) - g_1\left(\frac{\eta t_\eta}{\varepsilon}\right) - \varepsilon g_2\left(\frac{\eta t_\eta}{\varepsilon}\right)\right\} = 0$$

$$t_\eta = \text{const} \tag{21.108}$$

By (21.104)–(21.107), (21.85), (21.98), and (21.99),

$$\nu_1(\varepsilon)h_1(\eta t_\eta) - g_1\left(\frac{\eta t_\eta}{\varepsilon}\right) = A_1\exp(-t_\eta\eta) - 1 + \exp(\eta t_\eta/\varepsilon) = -t_\eta\eta + O(\eta^2)$$

$$\nu_2 h_2(\eta t_\eta) - \varepsilon g_2\left(\frac{\eta t_\eta}{\varepsilon}\right) = \varepsilon A_2 - 2\varepsilon + t_\eta\eta + (\varepsilon^2) \tag{21.109}$$

where transcendentally small terms are neglected. Thus, taking

$$A_2 = 2 \tag{21.110}$$

we find that matching has been carried out to order $O(\varepsilon)$ if

$$\lim_{\varepsilon\downarrow 0}\eta^2/\varepsilon = 0 \tag{21.111}$$

Hence the overlap region is determined by the inequalities

$$O(\varepsilon) < O(\eta(\varepsilon)) < O(\varepsilon^{1/2}) \quad \left(\text{e.g.,}\ \varepsilon|\ln\varepsilon| \le \eta(\varepsilon) < O(\varepsilon^{1/2})\right) \tag{21.112}$$

which is narrower than that previously found for matching up to order $O(1)$.

This procedure of matching may be continued, but an increase in the order of accuracy of the matching is accompanied by a reduction in the size of the region of overlap – the intersection of the regions in which the outer and inner expansions are simultaneously valid.

The matching implies that one can construct a composite approximation uniformly valid in any finite interval $0 \le t_o \le T$ on the time axis. Indeed, the outer expansion is

$$y(t_o \mid \varepsilon) \asymp e^{-t_o} + \varepsilon(2 - t_o)e^{-t_o} + \cdots \tag{21.113}$$

and the inner one

$$y(t_o \mid \varepsilon) \asymp 1 - e^{-t_o/\varepsilon} + \varepsilon(2 - t_o/\varepsilon)e^{-t_o/\varepsilon} + \cdots \tag{21.114}$$

Our matching procedure has determined the common part of these two expansions:

$$\text{cp} = 1 + \varepsilon(2 - t_o/\varepsilon) + \cdots \tag{21.115}$$

Adding the outer and inner expansions and subtracting the common part, we obtain the desired uniform composite asymptotic expansion:

$$y(t_o \mid \varepsilon) \asymp e^{-t_o} - e^{-t_o/\varepsilon} + \varepsilon[(2-t_o)e^{-t_o} - (2+t_o/\varepsilon)e^{-t_o/\varepsilon}] + \cdots \quad (21.116)$$

The technique described here is applicable in incomparably more complicated cases of linear, quasilinear, and even nonlinear boundary-value problems.

A few monographs devoted to a mathematically rigorous treatment of the method of singular perturbation are now available (see e.g. [17;42]). However, the application of this theory requires a deep (mainly intuitive), understanding of the physical essence of the problem to be solved. Hence references to these rigorous mathematical theories do not seem very promising for anyone interested in applications of the theory, which is the main task of mathematical physics. One of the best monographs on the perturbation method is that of Kevorkian and Cole [44], where the reader may find the information actually necessary for practical applications. A detailed consideration of the theory is beyond the scope of this book. We prefer, therefore, to refer the reader to Kevorkian and Cole, restricting our presentation to the preceding introductory remarks and to the following illustration of the application of the theory to a concrete problem of electrodiffusion.

4. Electrodiffusion and the nonequilibrium space charge in the 1-dimensional liquid junction [40;66]

By a liquid junction or *liquid junction potential* we mean an electric potential developing in an electrically insulated electrolyte solution with different ionic diffusivities and a discontinuity of the initial concentration. This is probably the simplest nonequilibrium electrodiffusion situation. Nevertheless, a detailed understanding of the dynamics of liquid junction is important for electrochemical and membrane transport applications.

Consider two compartments, represented by the regions $-1 < x < 0$ and $0 < x < 1$, filled with solutions of the same univalent electrolyte at concentrations 1 and Δ, respectively. Suppose that at some time $t = 0$ the wall separating the compartments at $x = 0$ is removed. The solution within the compartments is assumed to be immobilized, (say) with gelatin, so that the entire transport is due to electrodiffusion only. The initial concentrations of the electrolyte are maintained at the external walls, $x = \pm 1$, which are electrically insulated so that no electric current can pass through them.

The boundary-value problem is thus

$$\frac{\partial}{\partial t}p = \frac{\partial}{\partial x}\left(\frac{\partial}{\partial x}p + p\frac{\partial}{\partial x}\varphi\right), \quad -1 < x < 1, \ t > 0 \quad (21.117)$$

$$\frac{\partial}{\partial t}n = D\frac{\partial}{\partial x}\left(\frac{\partial}{\partial x}n - n\frac{\partial}{\partial x}\varphi\right) \quad (21.118)$$

$$\varepsilon\frac{\partial^2}{\partial x^2} = n - p \quad (21.119)$$

$$p(-1,t) = n(-1,t) = 1, \quad t > 0 \tag{21.120}$$

$$\varphi(-1,t) = 0, \quad t > 0 \tag{21.121}$$

$$p(1,t) = n(1,t) = \Delta > 0, \quad t > 0 \tag{21.122}$$

$$\left(\varepsilon \frac{\partial^2}{\partial x \partial t} \varphi + \frac{\partial}{\partial x} p - D \frac{\partial}{\partial x} n + (p + Dn) \frac{\partial}{\partial x} \varphi \right) \bigg|_{x=1} = 0 \tag{21.123}$$

$$p(x,0) = n(x,0) = \begin{cases} 1, & -1 < x < 0 \\ \Delta, & 0 < x < 1 \end{cases} \tag{21.124}$$

Here the variables are scaled in such a way that the diffusivity D_p of the cation p becomes 1, and D is the ratio of the anion diffusivity D_n to D_p. Condition (21.123) implies insulation, that is, vanishing of the total current (the difference of the two ionic fluxes plus the displacement current). ε is a small parameter equal to the square of the ratio of the Debye length[4] to the dimensional thickness L of the compartments.

Henderson's [37] classical treatment of liquid junction consists of setting $\varepsilon = 0$ in equations (21.119) and (21.123), which leads to a locally electroneutral boundary-value problem:

$$\frac{\partial}{\partial t} c = \frac{\partial}{\partial x} \left(\frac{\partial}{\partial x} c + c \frac{\partial}{\partial x} \varphi^h \right), \quad -1 < x < 1, \ t > 0 \tag{21.125}$$

$$\frac{\partial}{\partial t} c = D \frac{\partial}{\partial x} \left(\frac{\partial}{\partial x} c - c \frac{\partial}{\partial x} \varphi^h \right) \tag{21.126}$$

$$c(-1,t) = 1, \quad \varphi^h(-1,t) = 0 \quad \forall t > 0 \tag{21.127}$$

$$c(1,t) = \Delta, \quad \left((1-D) \frac{\partial}{\partial x} c + (1+D)c \frac{\partial}{\partial x} \varphi^h \right) \bigg|_{x=1} = 0 \tag{21.128}$$

$$c(x,0) = \begin{cases} 1, & -1 < x < 0 \\ \Delta, & 0 < x < 1 \end{cases} \tag{21.129}$$

Here

$$c(x,t) \stackrel{\text{def}}{=} p(x,t) \equiv n(x,t), \quad \varphi^h(x,t) \stackrel{\text{def}}{=} \varphi(x,t \mid \varepsilon) \tag{21.130}$$

Denote

$$a^2 \stackrel{\text{def}}{=} \frac{2D}{D+1}, \quad \delta \stackrel{\text{def}}{=} \frac{D-1}{D+1} \tag{21.131}$$

Equations (21.125) and (21.126) yield

$$a^2 \frac{\partial^2}{\partial x^2} c = \frac{\partial}{\partial t} c \tag{21.132}$$

$$\frac{\partial}{\partial x} \left(\delta \frac{\partial}{\partial x} c - c \frac{\partial}{\partial x} \varphi^h \right) = 0 \tag{21.133}$$

---

[4] A charged particle in an electrolyte solution is surrounded primarily by ions of opposite charge which leads to its "screening," so that the field of the source particle, beyond some short distance from it, becomes very small. This distance is known as the Debye length (Debye radius of screening).

Integration of (21.133), taking into account the boundary condition (21.127) and (21.128), yields

$$\varphi^h(x,t) = \delta \ln c(x,t) \qquad (21.134)$$

On the other hand, equation (21.132), the boundary conditions (21.127) and (21.128) imposed on $c(x,t)$, and the initial conditions (21.129) yield an integral representation of $c(x,t)$, in the form

$$c(x,t) = \int_{-1}^{0} g(x,\xi,t)d\xi + \Delta \int_{0}^{1} g(x,\xi,t)d\xi$$

$$+ a^2 \int_{0}^{t} \frac{\partial}{\partial \xi} g(x,-1,t-\tau)d\tau - a^2 \int_{0}^{t} \Delta \frac{\partial}{\partial \xi} g(x,1,t-\tau)d\tau \quad (21.135)$$

where

$$g(x,\xi,t) = \sum_{-\infty}^{\infty} \left[E\left(\frac{x-\xi}{2} + 2k, a^2(t-\tau) \right) \right.$$

$$\left. - E\left(\frac{x+\xi}{2} + 2k + 1, a^2(t-\tau) \right) \right] \qquad (21.136)$$

is the Green's function of the first boundary-value problem of heat conduction in $-1 < x < 1$.

Note that (21.129) and (21.134) imply

$$\varphi^h(1,t) = \delta \ln(\Delta) \qquad (21.137)$$

Finally, (21.134), (21.135), and (21.137) represent Henderson's solution to the problem of junction potential based on the assumption of local electroneutrality, that is, obtained by equating the small parameter ε to zero.

The purpose of this section is to prove the following statements.

(1) In terms of singular perturbation theory, Henderson's solution yields the leading term of the outer asymptotic expansion.
    Corrections of higher order in ε may be constructed by a straightforward perturbation procedure.
(2) Henderson's solution is discontinuous at $x = 0$ as $t \downarrow 0$ (corresponding to the initial discontinuity of the concentrations p, n and φ), whereas a solution of the full Poisson equation (21.119) is at least C^1 regular at $t = 0$; this means that the outer asymptotic expansion is valid not for $t \geq 0$ but for $t = O(1)$, and that there exists an initial layer.

A systematic singular perturbation treatment of the problem is carried out next, using matched asymptotic expansions.

Let $u(x,t,\varepsilon) = (p,n,\varphi)$ be the solution vector and $\tilde{u}(x,t,\varepsilon)$ the *outer* asymptotic expansion of u:

$$u(x,t,\varepsilon) \asymp \tilde{u}(x,t,\varepsilon) \overset{\text{def}}{=} \sum_{n=0}^{\infty} \tilde{u}_n(x,t)\varepsilon^n \qquad (21.138)$$

Substituting $\tilde{u}$ into (21.117)–(21.124) and equating terms of like order, for $\tilde{u}_k(x,t)$ $(k=0,1)$ we obtain[5]

$$\tilde{p}_0(x,t) = \tilde{n}_0(x,t) = c(x,t), \quad \tilde{\varphi}_0(x,t) = \varphi^h(x,t) \tag{21.139}$$

$$\frac{\partial}{\partial t}\tilde{p}_1 = \frac{\partial}{\partial x}\left(\frac{\partial}{\partial x}\tilde{p}_1 + \tilde{p}_0\frac{\partial}{\partial x}\tilde{\varphi}_1 + \tilde{p}_1\frac{\partial}{\partial x}\tilde{\varphi}_0\right) \tag{21.140}$$

$$\frac{\partial}{\partial t}\tilde{n}_1 = D\frac{\partial}{\partial x}\left(\frac{\partial}{\partial x}\tilde{n}_1 - \tilde{n}_0\frac{\partial}{\partial x}\tilde{\varphi}_1 - \tilde{n}_1\frac{\partial}{\partial x}\tilde{\varphi}_0\right) \tag{21.141}$$

$$\tilde{n}_1 - \tilde{p}_1 = \frac{\partial^2}{\partial x^2}\tilde{\varphi}_0 \tag{21.142}$$

$$\tilde{n}_1(-1,t) = \tilde{p}_1(-1,t) = \tilde{n}_1(1,t) = \tilde{p}_1(1,t) = 0 \tag{21.143}$$

$$\vdots$$

The initial conditions, corresponding to the boundary-value problem (21.140)–(21.143) and to the boundary-value problems for the other terms $\tilde{u}_k$, $k=2,3,\ldots$, must be worked out by matching with the initial-layer asymptotic approximation of u. It is easily inferred that these conditions are

$$\tilde{u}_k(x,0) = 0, \quad k=1,2,\ldots \tag{21.144}$$

The integration of system (21.140)–(21.144) (as well as all the resulting systems) is quite straightforward, analogous to integration of Henderson's system (21.132) and (21.133) together with the boundary and initial conditions. This confirms the first of our assertions.

We now turn to the construction of the *inner asymptotics*. We introduce fast time and space variables by setting

$$\tau = t/\varepsilon, \quad \xi = x/\sqrt{\varepsilon} \tag{21.145}$$

As $\varepsilon \downarrow 0$, the range of ξ becomes

$$-\infty < \xi < \infty \tag{21.146}$$

Substituting τ and ξ into equations (21.117)–(21.119), we obtain

$$\frac{\partial}{\partial \tau}p = \frac{\partial}{\partial \xi}\left(\frac{\partial}{\partial \xi}p + p\frac{\partial}{\partial \xi}\varphi\right) \tag{21.147}$$

---

[5]Note that since the small parameter ε does not occur in the ion diffusion equations (but only in the full Poisson equation and the boundary condition determining φ), and since the potential φ need not satisfy any initial condition, the zero-order outer asymptotic approximation of u satisfies all boundary and initial conditions of the full problem, so that the next terms of the outer asymptotic expansion need satisfy only homogeneous boundary conditions and initial conditions for the ion concentrations. This is precisely the aforementioned imaginary contradiction between the singular nature of the original problem and the possibility of satisfying all the boundary and initial conditions by the local electroneutrality approximation in Henderson's theory.

$$\frac{\partial}{\partial \tau} n = D \frac{\partial}{\partial \xi} \left(\frac{\partial}{\partial \xi} n - n \frac{\partial}{\partial \xi} \varphi \right) \tag{21.148}$$

$$\frac{\partial^2}{\partial \xi^2} \varphi = (n - p) \tag{21.149}$$

while the boundary conditions (21.120)–(21.123) and initial conditions (21.124) become

$$\lim_{\xi \downarrow -\infty} p(\xi, \tau) = \lim_{\xi \downarrow -\infty} n(\xi, \tau) = 1 \quad \forall \tau > 0 \tag{21.150}$$

$$\lim_{\xi \uparrow \infty} p(\xi, \tau) = \lim_{\xi \uparrow \infty} n(\xi, \tau) = \Delta \quad \forall \tau > 0 \tag{21.151}$$

$$\lim_{\xi \downarrow -\infty} \varphi(\xi, \tau) = 0 \tag{21.152}$$

$$\left(\frac{\partial^2}{\partial \xi \partial \tau} \varphi + \left(\frac{\partial}{\partial \xi} p - D \frac{\partial}{\partial \xi} n + (p + Dn) \frac{\partial}{\partial \xi} \varphi \right) \right) \Big|_{\xi = \infty} = 0 \tag{21.153}$$

$$p(\xi, 0) = n(\xi, 0) = \begin{cases} 1 & \forall \xi < 0 \\ \Delta & \forall \xi > 0 \end{cases} \tag{21.154}$$

Now let $v(\xi, \tau, \varepsilon) = (p, n, \varphi)$ be a solution vector. Let us seek v as an asymptotic series

$$v(\xi, \tau, \varepsilon) \asymp \sum_{k=0}^{\infty} v_k(\xi, \tau) \varepsilon^k \tag{21.155}$$

$v_0(\xi, \tau)$ is a solution of problem (21.147)–(21.152).[6] This means, in particular, that

$$\frac{d^2}{d\xi^2} \varphi_0(\xi, 0) = \begin{cases} 0 & \forall \xi < 0 \\ 0 & \forall \xi > 0 \end{cases} \tag{21.156}$$

Hence $d^2 \varphi_0(\xi, 0)/d\xi^2$ may be defined as zero at $x = 0$, which means that

$$\frac{d}{d\xi} \varphi_0(\xi, 0) \in C^1(-\infty < \xi < \infty) \tag{21.157}$$

On the other hand, it is easy to show that $v_0(\xi, \tau)$ matches the leading term $\tilde{u}_0(x, t)$ of the outer expansion of $u(x, t)$ in the neighborhood of $x = 0$. Indeed, in terms of the variables ζ, τ, where ζ is the similarity variable

$$\zeta = \xi / \sqrt{\tau} \tag{21.158}$$

problem (21.147)–(21.152) yields for $v_0(\xi, \tau)$

$$\frac{1}{\tau} \left[\frac{\partial^2}{\partial \zeta^2} p_0(\zeta, \tau) + \frac{\partial}{\partial \zeta} \left(p_0(\zeta, \tau) \frac{\partial}{\partial \zeta} \varphi_0(\zeta, \tau) \right) \right]$$

---

[6] For piecewise constant initial conditions (21.154), the inner asymptotic expansion (21.155) consists of only one term $v_0(\xi, \tau)$. This would not be the case for an arbitrary initial condition.

$$= \frac{\partial}{\partial \tau} p_0(\zeta, \tau) - \frac{\zeta}{2\tau} \frac{\partial}{\partial \zeta} p_0(\zeta, \tau) \tag{21.159}$$

$$\frac{D}{\tau} \left[\frac{\partial^2}{\partial \zeta^2} n_0(\zeta, \tau) - \frac{\partial}{\partial \zeta} \left(n_0(\zeta, \tau) \frac{\partial}{\partial \zeta} \varphi_0(\zeta, \tau) \right) \right]$$

$$= \frac{\partial}{\partial \tau} n_0(\zeta, \tau) - \frac{\zeta}{2\tau} \frac{\partial}{\partial \zeta} n_0(\zeta, \tau) \tag{21.160}$$

$$\frac{1}{\tau} \frac{\partial^2}{\partial \zeta^2} \varphi_0(\zeta, \tau) = \frac{1}{2} \left(n_0(\zeta, \tau) - p_0(\zeta, \tau) \right) \tag{21.161}$$

$$p_0(-\infty, \tau) = n_0(-\infty, \tau) = 1, \quad \varphi_0(-\infty, \tau) = 0 \tag{21.162}$$

$$p_0(\infty, \tau) = n_0(\infty, \tau) = \Delta, \quad \frac{\partial}{\partial \zeta} \varphi_0(\infty, \tau) = 0 \tag{21.163}$$

$$p_0(\zeta, 0) = n_0(\zeta, 0) = \begin{cases} 1 & \forall \zeta < 0 \\ \Delta & \forall \zeta > 0 \end{cases} \tag{21.164}$$

Equations (21.159)–(21.161) suggest that

$$\lim_{\tau \uparrow \infty} \frac{\partial}{\partial \tau} p_0(\zeta, \tau) = \lim_{\tau \uparrow \infty} \frac{\partial}{\partial \tau} n_0(\zeta, \tau) = 0$$

$$\lim_{\tau \uparrow \infty} [n_0(\zeta, \tau) - p_0(\zeta, \tau)] = 0 \tag{21.165}$$

Accordingly, we seek large-τ asymptotic formulas for $p_0(\zeta, \tau)$, $n_0(\zeta, \tau)$, and $\varphi_0(\zeta, \tau)$ as expansions in negative powers of τ:

$$p_0(\zeta, \tau) \asymp \sum_{k=0}^{\infty} p_0^k(\zeta) \tau^{-k}, \quad n_0(\zeta, \tau) \asymp \sum_{k=0}^{\infty} n_0^k(\zeta) \tau^{-k},$$

$$\varphi_0(\zeta, \tau) \asymp \sum_{k=0}^{\infty} \varphi_0^k(\zeta) \tau^{-k} \tag{21.166}$$

Substitution of (21.166) into (21.159)–(21.164) yields the following expressions for the leading terms of the asymptotic series (21.166):

$$\frac{d}{d\zeta} \left[\frac{d}{d\zeta} p_0^0(\zeta) + p_0^0(\zeta) \frac{d}{d\zeta} \varphi_0^0(\zeta) \right] = -\frac{\zeta}{2} \frac{d}{d\zeta} p_0 \tag{21.167}$$

$$\frac{d}{d\zeta} \left[\frac{d}{d\zeta} n_0^0(\zeta) - n_0^0(\zeta) \frac{d}{d\zeta} \varphi_0^0(\zeta) \right] = -\frac{\zeta}{2D} \frac{d}{d\zeta} n_0 \tag{21.168}$$

$$n_0^0(\zeta) - p_0^0(\zeta) = 0, \quad n_0^1(\zeta) - p_0^1(\zeta) = 2 \frac{d^2}{d\zeta^2} \varphi_0^0(\zeta) \tag{21.169}$$

$$p_0^0(-\infty) = n_0^0(-\infty) = 1, \quad \varphi_0^0(-\infty) = 0 \tag{21.170}$$

$$p_0^0(\infty) = n_0^0(\infty) = \Delta, \quad \varphi_0^0(\infty) = 0 \tag{21.171}$$

Obviously, it follows from (21.167)–(21.171) that

$$p_0^0(\zeta) = n_0^0(\zeta) = c_0^0 \overset{\text{def}}{=} 1 + (\Delta - 1) \operatorname{erf}(\zeta/a) \tag{21.172}$$

$$\varphi_0^0(\zeta) = \delta \ln c_0^0 \tag{21.173}$$

Here a^2 and δ are again defined by (21.131). Since, by (21.145) and (21.158),

$$\zeta = x/\sqrt{t} \tag{21.174}$$

the leading term of the initial-layer solution, with the fast-time τ asymptotic given by (21.172) and (21.173), obviously matches the leading term of the outer, slow-time solution (21.139) and (21.135).

The construction can be carried out in higher orders of ε as well. It can be seen that the composite solution (uniform asymptotic expansion) for the potential φ possesses, in terms of all orders in ε, the required C^1 regularity with respect to x, up to the initial time $t = 0$. Thus, obviously, the composite leading term for φ,

$$\varphi_0(x,t) = \tilde{\varphi}_0(x,t) + \varphi_0(x/\sqrt{\varepsilon}, t/\varepsilon) - \varphi_0^0(x/\sqrt{t}) \tag{21.175}$$

is in C^1 for $t \geq 0$.   **Q.E.D.**

Problems

P21.1.1. Construct the leading term of the inner asymptotic series for short relaxation times $\tau \downarrow 0$ of the solution of the Cauchy problem posed for the hyperbolic heat-conduction equation.

P21.4.1. Construct the outer asymptotic approximation for the solution of the liquid junction problem up to first order in ε.

Appendix 1: Elements of vector analysis

1. Definitions

D1. Flux of a vector through a surface

Let $\mathbf{v}(r)$ be a vector-valued function, defined and continuous at the points of a two-sided smooth surface Σ oriented by the normal n.[1] Define

$$\mathbf{d\sigma}_q = d\sigma_q \mathbf{n}_q^0 \tag{A1.1}$$

where $\mathbf{n}_q^0$ is the unit vector in the direction of $\mathbf{n}$ at the point q. Then the integral

$$Q = \int_\Sigma \mathbf{v}(q)\mathbf{d\sigma}_q \tag{A1.2}$$

is called the *total flux* (or simply flux) of v through the surface Σ.

D2. Divergence of a vector

Given a simply connected region D bounded by a surface Σ and a smooth vector-valued function $\mathbf{v}(p)$ defined in $\bar{D}$, let d be the diameter of D:

$$d = \max_{p,q \in D} r_{pq} \tag{A1.3}$$

and let $p \in D$ be an arbitrarily fixed point. Then the scalar $\operatorname{div} \mathbf{v}$ defined by

$$\operatorname{div} \mathbf{v}(p) = \lim_{d \downarrow 0} \frac{\int_\Sigma \mathbf{v}(q)\mathbf{d\sigma}_q}{\int_D d\omega} \tag{A1.4}$$

is called the *divergence* of $\mathbf{v}$ at p. Thus $\operatorname{div} \mathbf{v}(p)$ is the quotient of the flux of $\mathbf{v}$ through the boundary of a region of infinitesimal diameter containing p, to the volume of the region.

---

[1] If Σ is a closed surface then we assume $\mathbf{n}$ to be the outward normal. In what follows we shall say "outward normal" even if Σ is not closed, assuming in such cases that some direction of $\mathbf{n}$ has been chosen arbitrarily as the positive one.

D3. Gradient of a scalar

Let $\varphi(p)$ be a smooth scalar function defined in a region D. Consider the level surface

$$\Sigma = \{p: \varphi(p) = \text{const}, \ p \in \bar{D}\} \tag{A1.5}$$

and let $\mathbf{n}$ and $\boldsymbol{\tau}$ be the vectors of the normal and tangent to the surface Σ. The positive direction of $\mathbf{n}$ is assumed to coincide with that of increasing φ. Let $\mathbf{s}$ be a vector in an arbitrary direction, and let α_n, α_τ be the cosines of the angles between $\mathbf{s}$ and $\mathbf{n}, \boldsymbol{\tau}$, respectively. Then

$$\frac{\partial}{\partial s_p}\varphi(p) = \alpha_\tau \frac{\partial}{\partial \tau_p}\varphi(p) + \alpha_n \frac{\partial}{\partial n_p}\varphi(p) \equiv \alpha_n \frac{\partial}{\partial n_p}\varphi(p) \quad \forall p \in \Sigma \tag{A1.6}$$

Since $|\alpha_n| \leq 1$, this means that $\mathbf{n}_p$ is the direction in which φ increases most rapidly. The vector defined as[2]

$$\text{grad}\,\varphi \overset{\text{def}}{=} \frac{\partial}{\partial n_p}\varphi(p)\mathbf{n}_p^0 \tag{A1.7}$$

where n_p^0 is the unit vector in the direction $\mathbf{n}_p$, is called the *gradient* of $\varphi(p)$.

D4. Laplacian

Let $\varphi(p)$ be a twice continuously differentiable function. Then

$$\Delta\varphi(p) \overset{\text{def}}{=} \text{div}[\text{grad}\,\varphi(p)] \tag{A1.8}$$

is called the *Laplacian* of φ, and the operator

$$\Delta \overset{\text{def}}{=} \text{div}(\text{grad}) \tag{A1.9}$$

is called the *Laplace operator* or simply *Laplacian*.

D5. Rotor

Let $D \subset \mathbb{R}_3$ be a simply connected region of the diameter d, bounded by a piecewise smooth surface Σ; let $\mathbf{v}(p) \in C^1(\bar{D})$. Let $p \in D$ be an arbitrary

---

[2]Let (x, y, z) be the orthogonal cartesian coordinates and $\mathbf{i}, \mathbf{j}, \mathbf{k}$ the basis vectors. Let ∇ be the Hamilton operator

$$\nabla = \frac{\partial}{\partial x}\mathbf{i} + \frac{\partial}{\partial y}\mathbf{j} + \frac{\partial}{\partial z}\mathbf{k}$$

Let $\varphi(x, y, z)$ be a smooth scalar function, and let $\mathbf{A} = X\mathbf{i} + Y\mathbf{j} + Z\mathbf{k}$ be a smooth vector-valued function. Then

$$\text{div}\,\mathbf{A} = \nabla \cdot \mathbf{A}, \quad \text{grad}\,\varphi = \nabla\varphi$$

For the definition of div and grad operators in any orthogonal curvilinear coordinate system, see Section 3. Clearly it follows from these definitions that the notion of the Hamilton operator is useful only in a cartesian coordinate system. Therefore we usually avoid using it.

interior point of D. Then the vector $\operatorname{rot} \mathbf{v}(p)$ is defined by

$$\operatorname{rot} \mathbf{v}(p) = \lim_{d \downarrow 0} \frac{\int_{\Sigma} [\mathbf{v}(q) \cdot d\boldsymbol{\sigma}_q]}{\int_D d\omega_q} \tag{A1.10}$$

where $[\mathbf{a} \cdot \mathbf{b}]$ is the vector product of $\mathbf{a}$ and $\mathbf{b}$.

Remark. One often writes $\operatorname{curl} \mathbf{v}(p)$ instead of $\operatorname{rot} \mathbf{v}(p)$.

D6. Circulation

Let γ be a closed contour without self-intersections, taken with some definite sense of description. Let $\mathbf{v}(p) \in C(\gamma)$ be a vector defined and continuous on γ. Then the scalar

$$\Gamma = \int_{\gamma} \mathbf{v}(q) \cdot d\mathbf{s}_q \tag{A1.11}$$

where $d\mathbf{s}_q$ is the element of the tangent vector to γ in the positive sense of description, is called the *circulation* of v along γ.

2. Gauss divergence theorem and Stokes's theorem

The following theorems, which we state without proof, are of fundamental importance.

Theorem A1.2.1 (Gauss's divergence theorem). *Let (x, y, z) be an orthogonal coordinate system, and let $D \subset \mathbb{R}_3$ be a region bounded by a piecewise smooth surface Σ such that every plane parallel to the coordinate planes divides Σ into at most finitely many disconnected pieces. Let $\mathbf{f}(p)$ be a vector-valued function such that*

$$\mathbf{f}(p) \in C^1(\bar{D}) \cap C^2(D) \tag{A1.12}$$

Then

$$\int_D \operatorname{div} \mathbf{f}(q) d\omega_q = \int_{\Sigma} \mathbf{f}(q) \cdot d\boldsymbol{\sigma}_q \tag{A1.13}$$

Theorem A1.2.2 (Stoke's theorem). *Let $\gamma \subset \mathbb{R}_3$ be a closed curve, taken with some positive sense of description, and let Σ be a surface bounded by Γ with the positive direction of the normal $\mathbf{n}_p$ to Σ at any point p of Σ chosen so as to form a right-hand screw together with the positive sense of γ. Then*

$$\int_{\gamma} \mathbf{v}(q) \cdot d\mathbf{s}_q = \int_{\Sigma} \operatorname{rot} \mathbf{v}(q) \cdot d\boldsymbol{\sigma}_q \tag{A1.14}$$

Corollary. $\operatorname{rot} \mathbf{v}$ *is a solenoidal vector, that is,* $\operatorname{div} \operatorname{rot} \mathbf{v}(p) = 0$.

3. Orthogonal curvilinear coordinate systems. Lamé coefficients. Basic operators of vector analysis

Let D be a region in $\mathbb{R}_3$ and (x_1, x_2, x_3) a rectangular cartesian coordinate system; let

$$q_i = f_i(x_1, x_2, x_3) \in C^1(D), \quad i = 1, 2, 3 \tag{A1.15}$$

such that the Jacobian is different from zero everywhere in D:

$$\frac{D(f_1, f_2, f_3)}{D(x_1, x_2, x_3)} \neq 0 \quad \forall p = (x_1, x_2, x_3) \in D \tag{A1.16}$$

By the implicit function theorem, the x_i are smooth functions of f_j, $i, j = 1, 2, 3$, in D. The system (q_1, q_2, q_3) is a general curvilinear coordinate system in D, with coordinate surfaces $q_i = \text{const}$, $i = 1, 2, 3$, and coordinate lines Γ_i, $i = 1, 2, 3$; the latter are the lines of intersection of coordinate surfaces $(q_j = \text{const}, q_k = \text{const}, j \neq k, j, k \neq i)$. A coordinate system (q_1, q_2, q_3) is *orthogonal* if every two coordinate lines Γ_i and Γ_j intersect at right angles. Obviously, a curvilinear coordinate system (q_1, q_2, q_3) is orthogonal if

$$\sum_{i=1}^{3} \frac{\partial x_i}{\partial q_k} \frac{\partial x_i}{\partial q_m} = 0, \quad k \neq m, \quad k, m = 1, 2, 3 \tag{A1.17}$$

Indeed, let $\mathbf{i}_k$ and $\mathbf{e}_k$, $k = 1, 2, 3$, be the basic vectors of the cartesian and curvilinear coordinate systems, respectively, and let $\mathbf{ds}_n$ be the line element at the coordinate line Γ_n, $n = 1, 2, 3$.[3] Then

$$\mathbf{ds}_k = \sum_{i=1}^{3} \frac{\partial x_i}{\partial q_k} dq_k \mathbf{i}_i, \quad \mathbf{ds}_m = \sum_{j=1}^{3} \frac{\partial x_j}{\partial q_m} dq_m \mathbf{i}_j \tag{A1.18}$$

Hence

$$\mathbf{ds}_k \cdot \mathbf{ds}_m = \sum_{i=1}^{3} \sum_{j=1}^{3} \frac{\partial x_i}{\partial q_k} \frac{\partial x_j}{\partial q_m} dq_k dq_m \mathbf{i}_i \cdot \mathbf{i}_j = \sum_{i=1}^{3} \frac{\partial x_i}{\partial q_k} \frac{\partial x_i}{\partial q_m} dq_k dq_m \tag{A1.19}$$

so that

$$\mathbf{ds}_k \cdot \mathbf{ds}_m = 0 \Rightarrow \sum_{i=1}^{3} \frac{\partial x_i}{\partial q_k} \frac{\partial x_i}{\partial q_m} = 0, \quad k \neq m \tag{A1.20}$$

$$\textbf{Q.E.D.}$$

Let Γ be an arbitrary smooth curve and $\mathbf{ds}$ the line element on Γ. Then

$$ds^2 = \sum_{i=1}^{3} dx_i^2 = \left(\sum_{i=1}^{3} \left(\sum_{k=1}^{3} \left(\frac{\partial x_i}{\partial q_k} \right) dq_k \right) \cdot \left(\sum_{m=1}^{3} \left(\frac{\partial x_i}{\partial q_m} \right) dq_m \right) \right) \tag{A1.21}$$

---

[3] That is, $\mathbf{de}_n$ is an infinitesimal vector tangent to Γ_n at a point (q_1, q_2, q_3) on Γ_n.

By changing the order of summation and using the fact that the curvilinear coordinate system is orthogonal, one obtains

$$ds^2 = \sum_{k=1}^{3} H_k^2 dq_k^2 \qquad (A1.22)$$

where

$$H_k = \left(\sum_{i=1}^{3} \left(\frac{\partial x_i}{\partial q_k} \right)^2 \right)^{1/2} \qquad (A1.22)$$

In particular,

$$ds_k = H_k dq_k \qquad (A1.24)$$

so that

$$\mathbf{ds}_k = H_k dq_k \mathbf{e}_k \qquad (A1.25)$$

The H_k are called the *Lamé coefficients*.

We now develop expressions for the basic operators of vector analysis in orthogonal curvilinear coordinate systems.

A. Gradient of φ

Let S be a level surface of a function φ and T the tangent plane to S at a point p. Then, by definition (A1.24), any vector $\mathbf{ds}$ in T can be expressed as

$$\mathbf{ds} = \sum_{i=1}^{3} H_i dq_i \mathbf{e}_i \qquad (A1.26)$$

and simultaneously

$$\sum_{i=1}^{3} \frac{\partial \varphi}{\partial q_i} dq_i = 0 \qquad (A1.27)$$

Hence the vector

$$\mathbf{N} = \sum_{i=1}^{3} \frac{1}{H_i} \frac{\partial \varphi}{\partial q_i} \mathbf{e}_i \qquad (A1.28)$$

is orthogonal to $\mathbf{ds}$; that is, the unit vector $\mathbf{n}^0$ of a normal to S is

$$\mathbf{n}^0 = \frac{\sum_{i=1}^{3} \frac{1}{H_i} \frac{\partial \varphi}{\partial q_i} \mathbf{e}_i}{\left(\sum_{i=1}^{3} \left(\frac{1}{H_i} \frac{\partial \varphi}{\partial q_i} \right)^2 \right)^{1/2}} \qquad (A1.29)$$

Thus the direction cosines β_k of $\mathbf{n}^0$ in the curvilinear coordinate system under consideration are

$$\beta_k = \mathbf{n}^0 \cdot \mathbf{e}_k = \frac{\frac{1}{H_k} \frac{\partial \varphi}{\partial q_k}}{\left(\sum_{i=1}^{3} \left(\frac{1}{H_i} \frac{\partial \varphi}{\partial q_i} \right)^2 \right)^{1/2}} \qquad (A1.30)$$

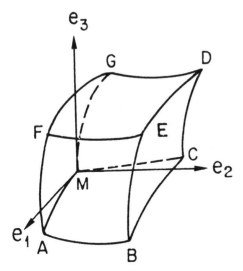

Figure A1.1.   Sketch of the definition of div.

Hence

$$\operatorname{grad}\varphi = \sum_{k=1}^{3}\frac{\partial\varphi}{\partial s_k}\beta_k\mathbf{n}^0 = \sum_{k=1}^{3}\frac{1}{H_k}\frac{\partial\varphi}{\partial q_k}\beta_k\mathbf{n}^0$$

$$= \sum_{k=1}^{3}\frac{1}{H_k}\frac{\partial\varphi}{\partial q_k}\frac{\frac{1}{H_k}\frac{\partial\varphi}{\partial q_k}}{\left(\sum_{i=1}^{3}\left(\frac{1}{H_i}\frac{\partial\varphi}{\partial q_i}\right)^2\right)^{1/2}}\frac{\sum_{i=1}^{3}\frac{1}{H_i}\frac{\partial\varphi}{\partial q_i}\mathbf{e}_i}{\left(\sum_{i=1}^{3}\left(\frac{1}{H_i}\frac{\partial\varphi}{\partial q_i}\right)^2\right)^{1/2}}$$

$$= \sum_{i=1}^{3}\frac{1}{H_i}\frac{\partial\varphi}{\partial q_i}\frac{\sum_{k=1}^{3}\left(\frac{1}{H_k}\frac{\partial\varphi}{\partial q_k}\right)^2}{\sum_{i=1}^{3}\left(\frac{1}{H_i}\frac{\partial\varphi}{\partial q_i}\right)^2}\mathbf{e}_i = \sum_{i=1}^{3}\frac{1}{H_i}\frac{\partial\varphi}{\partial q_i}\mathbf{e}_i \qquad (A1.31)$$

Thus

$$\operatorname{grad}\varphi = \sum_{i=1}^{3}\frac{1}{H_i}\frac{\partial\varphi}{\partial q_i}\mathbf{e}_i \qquad (A1.32)$$

B. Divergence

Consider an infinitesimal parallelepiped P spanned by vectors $\mathbf{ds}_k$ emanating from a point M (see Figure A1.1), and a vector

$$\mathbf{v} = \sum_{k=1}^{3}v_k\mathbf{e}_k \qquad (A1.33)$$

Let S_i^{-} and S_i^{+} denote the faces of P normal to the vectors $\mathbf{ds}_i$, $i = 1, 2, 3$, where the superscript $^{+}$ $(^{-})$ indicates that the direction of the outward normal to the face coincides with the positive (negative) direction of the corresponding basis vector. The flux F of $\mathbf{v}$ through the surface S bounding P is

$$
\int_S \mathbf{v}(q)d\boldsymbol{\sigma}_q = \sum_{i=1}^3 \left(\int_{S_i^+} - \int_{S_i^-} \mathbf{v}(q)d\sigma_q \mathbf{e}_i \right)
$$

$$
= \sum_{i=1}^3 \left(\int_{S_i^+} - \int_{S_i^-} v_i(q)d\sigma_q \right) \tag{A1.34}
$$

We have

$$
d\sigma_i = H_k H_m dq_k dq_m, \quad k,m \neq i, \; k \neq m, \; i,k,m = 1,2,3 \tag{A1.35}
$$

and, up to second-order infinitesimals,

$$
H_k H_m v_i \mid_{S_i^+} - H_k H_m v_i \mid_{S_i^-} = \frac{\partial}{\partial q_i}(H_k H_m v_i) \bigg|_M dq_i \tag{A1.36}
$$

Hence, up to the same order,

$$
\int_S \mathbf{v}(q)d\boldsymbol{\sigma}_q = \sum_{i=1}^3 \frac{\partial}{\partial q_i}(H_k H_m v_i) \bigg|_M dq_i dq_k dq_m \tag{A1.37}
$$

But again up to second-order infinitesimals,

$$
\int_P d\omega_q = \prod_{k=1}^3 H_k dq_k \tag{A1.38}
$$

The definition (A1.4) and equalities (A1.30) and (A1.31) now yield

$$
\text{div } \mathbf{v} = \frac{\frac{\partial}{\partial q_1}(H_2 H_3 v_1) + \frac{\partial}{\partial q_2}(H_3 H_1 v_2) + \frac{\partial}{\partial q_3}(H_1 H_2 v_3)}{H_1 H_2 H_3} \tag{A1.39}
$$

C. Laplacian

Definition (A1.8) and equalities (A1.32) and (A1.39) yield

$$
\Delta\varphi = \frac{1}{H_1 H_2 H_3} \left(\frac{\partial}{\partial q_1} \left(\frac{H_2 H_3}{H_1} \frac{\partial \varphi}{\partial q_1} \right) \right.
$$
$$
\left. + \frac{\partial}{\partial q_2} \left(\frac{H_3 H_1}{H_2} \frac{\partial \varphi}{\partial q_2} \right) + \frac{\partial}{\partial q_1} \left(\frac{H_1 H_2}{H_3} \frac{\partial \varphi}{\partial q_3} \right) \right) \tag{A1.40}
$$

D. Rotor v

Let Γ be a contour bounding an infinitesimal rectangle spanned by vectors $\mathbf{ds}_1$ and $\mathbf{ds}_2$. Applying Stoke's theorem to this contour, we obtain

$$
\text{rot } \mathbf{v} \cdot \mathbf{e}_3 H_1 H_2 dq_1 dq_2 = \left(v_2 H_2 \mid_{q_1 + dq_1} - v_2 H_2 \mid_{q_1} \right) dq_2
$$
$$
- \left(v_1 H_1 \mid_{q_2 + dq_2} - v_1 H_1 \mid_{q_2} \right) dq_1 \tag{A1.41}
$$

so that

$$
\text{rot } \mathbf{v} \mid_3 = \frac{1}{H_1 H_2} \left(\frac{\partial}{\partial q_1}[H_2 v_2] - \frac{\partial}{\partial q_2}[H_1 v_1] \right) \tag{A1.42}
$$

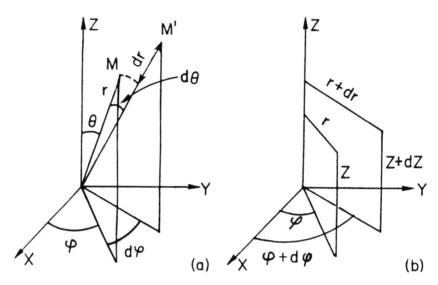

Figure A1.2. Lamé coefficients in spherical and cylindrical coordinate systems.

Cyclic permutation yields

$$\text{rot}\,\mathbf{v}\,|_2 = \frac{1}{H_3 H_1}\left(\frac{\partial}{\partial q_3}[H_1 v_1] - \frac{\partial}{\partial q_1}[H_3 v_3]\right) \tag{A1.43}$$

$$\text{rot}\,\mathbf{v}\,|_1 = \frac{1}{H_2 H_3}\left(\frac{\partial}{\partial q_2}[H_3 v_3] - \frac{\partial}{\partial q_3}[H_2 v_2]\right) \tag{A1.44}$$

E. Lamé coefficients in cartesian, cylindrical, and spherical coordinate systems

1. cartesian coordinates

$$H_1 = H_2 = H_3 = 1 \tag{A1.45}$$

2. Cylindrical coordinates (r, φ, z) (see Figure A1.2(b))

$$x = r\cos\varphi, \quad y = r\sin\varphi, \quad z = z$$
$$q_1 = r, \quad q_2 = \varphi, \quad q_3 = z \Rightarrow ds_1 = dr, \quad ds_2 = rd\varphi, \quad ds_3 = dz$$
$$\Rightarrow H_1 = 1, \quad H_2 = r, \quad H_3 = 1 \tag{A1.46}$$

3. Spherical coordinates (r, φ, ϑ) (see Figure A1.2(a))

$$x = r\cos\varphi\sin\vartheta, \quad y = r\sin\varphi\sin\vartheta, \quad z = \cos\vartheta$$
$$\Rightarrow q_1 = r, \quad q_2 = \varphi, \quad q_3 = \vartheta$$
$$\Rightarrow ds_1 = dr, \quad ds_2 = r\sin\vartheta d\varphi, \quad ds_3 = rd\vartheta$$
$$\Rightarrow H_1 = 1, \quad H_2 = r\sin\vartheta, \quad H_3 = r \tag{A1.47}$$

F. Gradient, divergence, and Laplacian in cartesian, cylindrical, and spherical coordinates

Equalities (A1.32), (A1.39), (A1.40), and (A1.45)–(A1.47) yield the following.

1. cartesian coordinates

$$\operatorname{grad} F = \frac{\partial}{\partial x} F \mathbf{i} + \frac{\partial}{\partial y} F \mathbf{j} + \frac{\partial}{\partial z} F \mathbf{k} \tag{A1.48}$$

where $\mathbf{i}, \mathbf{j}, \mathbf{k}$ are basis vectors.

$$\mathbf{F} = X\mathbf{i} + Y\mathbf{j} + Z\mathbf{k} \Rightarrow \operatorname{div} \mathbf{F} = \frac{\partial}{\partial x} X + \frac{\partial}{\partial y} Y + \frac{\partial}{\partial z} Z \tag{A1.49}$$

$$\Delta F = \frac{\partial^2}{\partial x^2} F + \frac{\partial^2}{\partial y^2} F + \frac{\partial^2}{\partial x^2} F \tag{A1.50}$$

2. Cylindrical coordinates

$$\operatorname{grad} F = \frac{\partial}{\partial r} F \mathbf{e}_1 + \frac{1}{r} \frac{\partial}{\partial \varphi} F \mathbf{e}_2 + \frac{\partial}{\partial z} F \mathbf{e}_3 \tag{A1.51}$$

$$\mathbf{F} = R\mathbf{e}_1 + \Phi\mathbf{e}_2 + Z\mathbf{e}_3 \Rightarrow \operatorname{div} \mathbf{F} = \frac{1}{r} \frac{\partial}{\partial r} (rR) + \frac{1}{r} \frac{\partial}{\partial \varphi} \Phi + \frac{\partial}{\partial z} Z \tag{A1.52}$$

$$\Delta F = \frac{1}{r} \frac{\partial}{\partial r} \left(r \frac{\partial}{\partial r} F \right) + \frac{1}{r^2} \frac{\partial^2}{\partial \varphi^2} F + \frac{\partial^2}{\partial z^2} F \tag{A1.53}$$

3. Spherical coordinates

$$\operatorname{grad} F = \frac{\partial}{\partial r} F \mathbf{e}_1 + \frac{1}{r \sin \vartheta} \frac{\partial}{\partial \varphi} F \mathbf{e}_2 + \frac{1}{r} \frac{\partial}{\partial \vartheta} F \mathbf{e}_3 \tag{A1.54}$$

$$\mathbf{F} = R\mathbf{e}_1 + \Phi\mathbf{e}_2 + \Theta\mathbf{e}_3$$

$$\Rightarrow \operatorname{div} \mathbf{F} = \frac{1}{r^2} \frac{\partial}{\partial r} (r^2 R) + \frac{1}{r \sin \vartheta} \frac{\partial}{\partial \varphi} \Phi + \frac{1}{r \sin \vartheta} \frac{\partial}{\partial \vartheta} (\sin \vartheta \Theta) \tag{A1.55}$$

$$\Delta F = \frac{1}{r^2} \frac{\partial}{\partial r} \left(r^2 \frac{\partial}{\partial r} F \right) + \frac{1}{r^2 \sin^2 \vartheta} \frac{\partial^2}{\partial \varphi^2} F + \frac{1}{r^2 \sin \vartheta} \frac{\partial}{\partial \vartheta} \left(\sin \vartheta \frac{\partial}{\partial \vartheta} F \right) \tag{A1.56}$$

Appendix 2: Elements of theory of Bessel functions[1]

1. Introduction. Euler's gamma function

Before presenting the elements of the theory of Bessel functions, we cite without proof the necessary information about Euler's gamma function.

Definitions.

$$\Gamma(z) = \int_0^\infty e^{-t} t^{z-1} dz \Rightarrow \Gamma(z+1) = z\Gamma(z) \quad \forall z, \ \Re z > 0 \qquad (A2.1)$$

so that for every integer $n > 0$,

$$\Gamma(n+1) = n! \qquad (A2.2)$$

For arbitrary z, $\Gamma(z)$ is defined recursively by

$$\Gamma(z) = \frac{\Gamma(z+1)}{z} \qquad (A2.3)$$

Equation (A2.3) implies that $\Gamma(z)$ is a holomorphic function, with simple poles at the points $z = -n$, $n = 0, 1, 2, \ldots$.

Duplication formula.

$$\Gamma(2z) = 2^{2z-1} \pi^{-1/2} \Gamma(z)\Gamma(z+1/2) \qquad (A2.4)$$

Functional equations (Reflection formula).

$$\Gamma(z)\Gamma(1-z) = \pi \csc(\pi z) \Rightarrow \Gamma(1/2+z)\Gamma(1/2-z) = \pi \sec(\pi z) \qquad (A2.5)$$

Asymptotic behavior for large $|z|$ (Stirling's formula).

$$\Gamma(z) \asymp e^{-z} \cdot e^{(z-1/2)\ln(z)} (2\pi)^{1/2} \left(1 + \frac{z^{-1}}{12} + O(z^{-2})\right), \quad |\arg(z)| < \pi \qquad (A2.6)$$

---

[1] Only those elements of the theory of Bessel functions that are used in the main part of the book are presented here. Roots of Bessel functions are considered in Appendix 3, Section 5.

Figure A2.1. Contour for $\Gamma(z)$ definition in a complex plane.

Representation of $\Gamma(z)$ by a contour integral. Let C be a contour in the ζ plane as shown in Figure A2.1; that is, a contour with initial and final points at $|\zeta| = \infty$, $\arg\zeta = 0$, and $\arg\zeta = 2\pi$, encircling $\zeta = 0$ in the positive sense. Then

$$\Gamma(z) = (\exp(2\pi z i) - 1)^{-1} \int_C e^{-\zeta} \zeta^{z-1} d\zeta \qquad (A2.7)$$

Logarithmic derivative of $\Gamma(z)$.

$$\psi(z) \stackrel{\text{def}}{=} \frac{d}{dz} \ln[\Gamma(z)] = \lim_{n\uparrow\infty} \left(\ln(n) - \sum_{m=0}^{n} \frac{1}{m+z} \right)$$

$$\equiv -\gamma - \frac{1}{z} + \sum_{m=1}^{\infty} \frac{z}{m(m+z)} \qquad (A2.8)$$

where

$$\gamma = \lim_{n\uparrow\infty} \left(\sum_{m=1}^{n} \frac{1}{m} - \ln(n) \right) = \lim_{n\uparrow\infty} [h_n - \ln(n)] \qquad (A2.9)$$

Beta function $B(p,q)$. The beta function is defined as follows:

$$B(p,q) = \frac{\Gamma(p)\Gamma(q)}{\Gamma(p+q)} \qquad (A2.10)$$

Integral representation of $B(p,q)$ for $\Re p > 0$, $\Re q > 0$.

$$B(p,q) = \int_0^1 z^{p-1}(1-z)^{q-1} dz \qquad (A2.11)$$

2. Generating functions and Bessel functions of first kind. Neumann functions

Let $\varphi(t,z)$ be a function of two complex variables, represented by its Laurent expansion in powers of t in the neighborhood of $t = 0$:

$$\varphi(t,z) = \sum_{-\infty}^{\infty} f_n(z) t^n \qquad (A2.12)$$

Then $\varphi(t, z)$ is called a *generating function* for the family $f_n(z)$, $n = 0, \pm 1, \pm 2, \ldots$.

Let

$$\varphi(t, z) = \exp\left[\frac{z}{2}(t - t^{-1})\right] \tag{A2.13}$$

Then $f_n(z)$ is called the *Bessel function of the first kind* of order n, denoted by $J_n(z)$. Thus,

$$\varphi(t, z) \overset{\text{def}}{=} \exp\left[\frac{z}{2}(t - t^{-1})\right] = \sum_{-\infty}^{\infty} J_n(z)t^n \tag{A2.14}$$

Multiplying the series

$$e^{zt/2} = \sum_0^{\infty} \left(\frac{z}{2}\right)^m \frac{t^m}{m!}, \qquad e^{-z/2t} = \sum_0^{\infty} \left(\frac{z}{2}\right)^k (-1)^k \frac{t^{-k}}{k!} \tag{A2.15}$$

and setting $k - m = n$, we obtain for any $n \geq 0$

$$J_n(z) \equiv \sum_0^{\infty} \frac{(-1)^m (z/2)^{2m+n}}{m!(n+m)!} \tag{A2.16}$$

Let ν be an arbitrary complex number. Define

$$J_\nu(z) \equiv \sum_0^{\infty} \frac{(-1)^m (z/2)^{2m+\nu}}{\Gamma(m+1)\Gamma(\nu+m+1)} \tag{A2.17}$$

Since

$$\Gamma(p+1) = p! \tag{A2.18}$$

for every integer p, $J_\nu(z)$ is the Bessel function of order ν of the first kind if ν is an integer. This justifies defining the Bessel functions not as the coefficients of the Laurent series (A2.14), but as the sum of the series (A2.17) for any constant (real or complex) ν. In what follows, however, we restrict our attention to real indices ν.

Obviously $J_\nu(z)$ is an entire function of z, since the radius of the convergence of the series (A2.17) is equal to infinity – this follows instantly from D'Alembert's test. At the same time, $J_\nu(z)$ is holomorphic with respect to ν, since $\Gamma^{-1}(p)$ is a holomorphic function with simple roots at the points

$$p = 0, -1, -2, \ldots \tag{A2.19}$$

as follows from (A2.3).

We now return to the generating function $\varphi(t, z)$. Differentiation of $\varphi(t, z)$ with respect to z yields

$$\frac{\partial}{\partial z} \varphi = \frac{1}{2}[t - t^{-1}]\varphi = \sum_{-\infty}^{\infty} \frac{d}{dz} J_n(z) t^n \tag{A2.20}$$

On the other hand, definition (A2.14) yields

$$\frac{\partial}{\partial z} \varphi = \frac{1}{2} \left(\sum_{-\infty}^{\infty} J_n(z) t^{n+1} - \sum_{-\infty}^{\infty} J_n(z) t^{n-1} \right) \tag{A2.21}$$

Hence

$$\frac{1}{2} \sum_{-\infty}^{\infty} J_n(z)(t^{n+1} - t^{n-1}) = \sum_{-\infty}^{\infty} \frac{d}{dz} J_n(z) t^n \tag{A2.22}$$

which implies

$$\sum_{-\infty}^{\infty} \left(2\frac{d}{dz} J_n(z) - J_{n-1}(z) + J_{n+1}(z) \right) t^n = 0 \tag{A2.23}$$

The uniqueness of the Laurent expansions implies that all the terms of this series must vanish identically, whence we derive the following recurrence formula:

$$J_{n-1}(z) - J_{n+1}(z) = 2\frac{d}{dz} J_n(z) \tag{A2.24}$$

for any n.

Differentiation of φ with respect to t gives

$$\frac{\partial}{\partial t} \varphi = \frac{z}{2}[1 + t^{-2}]\varphi \equiv \sum_{-\infty}^{\infty} n J_n(z) t^{n-1} \tag{A2.25}$$

which implies

$$\sum_{-\infty}^{\infty} J_n(z) \left[nt^{n-1} - \frac{z}{2}(t^n + t^{n-2}) \right] \equiv 0 \tag{A2.26}$$

This identity may be written as

$$\sum_{-\infty}^{\infty} \left\{ n J_n(z) - \frac{z}{2}[J_{n-1}(z) + J_{n+1}(z)] \right\} t^{n-1} \equiv 0 \tag{A2.27}$$

which yields, for every n,

$$J_{n-1}(z) + J_{n+1}(z) = \frac{2n}{z} J_n(z) \tag{A2.28}$$

Furthermore, by comparing (A2.28) and (A2.24), we find that

$$\frac{d}{dz}J_n(z) + \frac{n}{z}J_n(z) = J_{n-1}(z), \qquad \frac{d}{dz}J_n(z) - \frac{n}{z}J_n(z) = -J_{n+1}(z) \quad \text{(A2.29)}$$

These equations may be rewritten as

$$\frac{d}{dz}[z^n J_n(z)] = z^n J_{n-1}(z), \qquad \frac{d}{dz}[z^{-n} J_n(z)] = -z^{-n} J_{n+1}(z) \qquad \text{(A2.30)}$$

so that, for all $m \geq 0$,

$$\left(\frac{1}{z}\frac{d}{dz}\right)^m [z^n J_n(z)] = z^{n-m} J_{n-m}(z)$$

$$\left(\frac{1}{z}\frac{d}{dz}\right)^m [z^{-n} J_n(z)] = (-1)^m z^{-n-m} J_{n+m}(z) \qquad \text{(A2.31)}$$

By eliminating J_{n-1} from identities (A2.29), we easily obtain

$$\frac{d^2}{dz^2}J_n(z) + \frac{1}{z}\frac{d}{dz}J_n(z) + \left(1 - \frac{n^2}{z^2}\right)J_n(z) = 0 \qquad \text{(A2.32)}$$

Equation (A2.32) is the *Bessel differential equation of the first kind* of order n. We have derived it here for an integer n. However, by direct term-by-term differentiation of the series (A2.17), using the recurrent properties (A2.3) of the gamma function, one sees that the Bessel functions $J_\nu(z)$ of arbitrary order ν are solutions of the Bessel equation

$$LJ_\nu(z) \stackrel{\text{def}}{=} \left(\frac{d^2}{dz^2} + \frac{1}{z}\frac{d}{dz} + \left(1 - \frac{\nu^2}{z^2}\right)\right)J_\nu(z) = 0 \qquad \text{(A2.33)}$$

Every linear differential equation of second order has two linearly independent particular solutions. One of the solutions of the Bessel equation (A2.33) is $J_\nu(z)$. Since replacing ν by $-\nu$ leaves equation (A2.33) invariant, the second solution is $J_{-\nu}(z)$, called the *Bessel function of the second kind* of order ν.

If $\nu = n$ an integer, then by (A2.17) we have

$$J_{-n}(z) = \sum_0^\infty \frac{(-1)^m (z/2)^{-n+2m}}{\Gamma(m+1)\Gamma(m-n+1)}$$

$$\equiv \sum_n^\infty \frac{(-1)^m (z/2)^{-n+2m}}{\Gamma(m+1)\Gamma(m-n+1)} \Rightarrow \qquad \text{(A2.34)}$$

$$(-1)^n \sum_0^\infty \frac{(-1)^m (z/2)^{n+2m}}{\Gamma(m+1)\Gamma(m+n+1)} = (-1)^n J_n(z)$$

since $1/\Gamma(z)$ is an entire function with simple zeros at points $z = 0, -1, -2, \ldots$. Hence J_ν and $J_{-\nu}$ are linearly dependent if ν is an integer; if ν is not an integer then J_ν and $J_{-\nu}$ are linearly independent. Indeed, it follows from (A2.17) that

$$\lim_{z \to 0} z^{-\nu} J_\nu(z) = 1/(\nu+1), \qquad \lim_{z \to 0} z^\nu J_{-\nu}(z) = 1/(1-\nu) \qquad \text{(A2.35)}$$

so that

$$\lim_{z \to 0} J_\nu(z) = 0, \qquad \lim_{z \to 0} J_{-\nu}(z) = \infty \qquad (A2.36)$$

Thus, if ν is not an integer then J_ν and $J_{-\nu}$ are indeed linearly independent.

In order to determine a second solution of the Bessel equation which is linearly independent of J_ν for all values of ν, we note that any linear combination of n linearly independent solutions of a linear differential equation of order n is also a solution of the equation. Taking this into account, we define the *Neumann function* by

$$Y_\nu(z) = [J_\nu(z) \cos(\nu\pi) - J_{-\nu}(z)] / \sin(\nu\pi) \qquad (A2.37)$$

If ν is an integer, the right hand side of (A2.37) becomes an indeterminate expression of the type $0/0$, which can be determined with L'Hospital's rule:

$$Y_n(z) = \lim_{\nu \to n} \frac{\cos(\nu\pi) \frac{d}{d\nu} J_\nu(z) - \pi \sin(\nu\pi) J_\nu(z) - \frac{d}{d\nu} J_{-\nu}(z)}{\pi \cos(\nu\pi)}$$

$$= \pi^{-1} \frac{d}{dn} J_n(z) - (-1)^n \pi^{-1} \frac{d}{d\nu} J_{-\nu}(z) \Big|_{\nu=n} \qquad (A2.38)$$

Using the formula for the logarithmic derivative of the Gamma function (see (A2.8)), we obtain a series representation of $Y_n(z)$[2]:

$$\pi Y_n(z) = 2 \left[\gamma + \ln\left(\frac{z}{2}\right) \right] J_n(z) - \sum_{m=0}^{n-1} \frac{(z/2)^{2m-n}(n-m-1)!}{m!}$$

$$- \sum_{m=0}^{\infty} (-1)^m \frac{(z/2)^{n+2m}(h_{m+n} + h_m)}{m!(n+m)!}, \qquad n = 1, 2, \ldots \qquad (A2.39)$$

$$\pi Y_0(z) = 2 \left[\gamma + \ln\left(\frac{z}{2}\right) \right] J_0(z) - 2 \sum_{m=0}^{\infty} (-1)^m \left(\frac{z}{2}\right)^{2m} (m!)^{-2} h_m \qquad (A2.40)$$

Thus the general solution of the Bessel equation is a linear combination of J_ν and Y_ν:

$$u(z) = c_1 J_\nu(z) + c_2 Y_\nu(z) \qquad (A2.41)$$

Note that $Y_0(z)$ has a logarithmic singularity at the origin $z = 0$ and $Y_n(z)$ has a pole of order n at $z = 0$ for every $n > 0$ (see Appendix 3, Section 4). Hence any solution of the Bessel equation $L_n u = 0$ which is bounded at $z = 0$ is proportional to $J_n(z)$.[3]

---

[2] Recall that

$$h_n = \sum_{k=1}^{n} \frac{1}{k} \quad \text{and} \quad \gamma = \lim_{n \uparrow \infty} (h_n - \ln(n))$$

(see (A2.8)).

[3] If ν is not an integer then $z = 0$ is a branch point of the algebraic or logarithmic type.

The Wronskians of the solutions of Bessel equation are:

$$W(J_\nu, J_{-\nu}) = -2(\pi z)^{-1}\sin(\pi\nu), \qquad W(J_\nu, Y_\nu) = 2(\pi z)^{-1} \qquad \text{(A2.42)}$$

This follows easily from (A2.35) and (A2.37).

3. Bessel and Lipschitz integrals

Let us return to definition (A2.13). The residue theorem yields

$$J_n(z) = (2\pi i)^{-1} \oint_C t^{-1-n} \exp\left[\frac{z}{2}(t - t^{-1})\right] dt \qquad \text{(A2.43)}$$

where C is an arbitrary closed contour surrounding the origin $t = 0$. Letting

$$C = \left\{|t| = 1; \; t = e^{i\varphi}\right\} \qquad \text{(A2.44)}$$

we obtain the *Bessel integral*:

$$\begin{aligned} J_n(z) &= (2\pi)^{-1}\int_{-\pi}^{\pi} \exp[i(z\sin\varphi - n\varphi)]d\varphi \\ &\equiv \frac{1}{\pi}\int_0^{\pi} \cos(z\sin\varphi - n\varphi)d\varphi \end{aligned} \qquad \text{(A2.45)}$$

We now derive the *Lipschitz integral*. Expansion of the binomial series yields

$$(1 + z^2)^{-1/2} = \sum_0^\infty C_{-1/2}^m z^{2m} \qquad \text{(A2.46)}$$

where the binomial coefficient $C_{-1/2}^m$ is

$$C_{-1/2}^m = \frac{\Gamma(1/2)}{\Gamma(m+1)\Gamma(1/2 - m)} \qquad \text{(A2.47)}$$

By (A2.5),

$$\Gamma(m + 1/2)\Gamma(1/2 - m) = (-1)^m \pi, \qquad \Gamma(1/2) = \sqrt{\pi} \qquad \text{(A2.48)}$$

so that

$$(1 + z^2)^{-1/2} = \sum_0^\infty (-1)^m z^{2m} \frac{\Gamma(m+1/2)}{\Gamma(m+1)\sqrt{\pi}} \qquad \text{(A2.49)}$$

On the other hand, (A2.17) yields

$$\begin{aligned} \int_0^\infty e^{-t} J_0(zt)dt &\equiv \sum_0^\infty \frac{(-1)^m (z/2)^{2m}}{\Gamma(m+1)\Gamma(m+1)} \int_0^\infty e^{-t} t^{2m} dt \\ &= \sum_0^\infty \frac{(-1)^m (z/2)^{2m}}{\Gamma(m+1)\Gamma(m+1)} \Gamma(2m+1) \end{aligned} \qquad \text{(A2.50)}$$

Using the duplication formula (A2.4), we obtain

$$\int_0^\infty e^{-t} J_0(zt)dt = \sum_0^\infty \frac{(-1)^m z^{2m}\Gamma(m+1/2)}{\Gamma(m+1)}\pi^{-1/2} \qquad \text{(A2.51)}$$

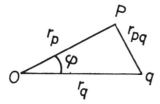

Figure A2.2. Sketch of Neumann's addition theorem.

Comparison with (A2.49) shows that

$$\int_0^\infty e^{-t} J_0(zt)dt = (1+z^2)^{-1/2} \qquad (A2.52)$$

This is the *Lipschitz integral*. It is convenient to change variables:

$$z = \frac{\rho}{r}, \quad t = r\tau \qquad (A2.53)$$

This gives

$$(r^2 + \rho^2)^{-1/2} = \int_0^\infty e^{-rt} J_0(\rho t)dt \qquad (A2.54)$$

4. Neumann's addition theorem

Let p and q be two given points at distances r and ρ from the origin, and let $\varphi = \pi - \psi$ be the angle between the radius vectors $\mathbf{r}$ and $\boldsymbol{\rho}$ (see Figure A2.2). Then

$$r_{pq}^2 = r^2 + \rho^2 + 2r\rho\cos(\psi) \qquad (A2.55)$$

Define angles α and β by

$$r_{pq} = re^{i\alpha} + \rho e^{i\beta} = r\cos\alpha + \rho\cos\beta, \quad 0 = r\sin\alpha + \rho\sin\beta \qquad (A2.56)$$

We have

$$\exp\left[\frac{re^{i\alpha} + \rho e^{i\beta}}{2}\left(t - \frac{1}{t}\right)\right]$$

$$= \exp\left[\frac{re^{i\alpha}}{2}\left(t - \frac{1}{t}\right)\right]\exp\left[\frac{\rho e^{i\beta}}{2}\left(t - \frac{1}{t}\right)\right] \qquad (A2.57)$$

Hence, by (A2.14), we have

$$\sum_{-\infty}^{\infty} t^n J_n[r\exp(i\alpha) + \rho\exp(i\beta)]$$

$$= \sum_{-\infty}^{\infty} t^m J_m[r\exp(i\alpha)]\sum_{-\infty}^{\infty} t^k J_k[r\exp(i\alpha)] \qquad (A2.58)$$

On the other hand,

$$\frac{r}{2}\exp(i\alpha)\left(t - \frac{1}{t}\right) = -\frac{ir\sin\alpha}{t} + \frac{r}{2}[t\exp(i\alpha) - t^{-1}\exp(-i\alpha)] \qquad (A2.59)$$

so that

$$\exp\left[\frac{r}{2}\exp(i\alpha)[t - t^{-1}]\right] = \exp[-irt^{-1}\sin\alpha]$$

$$\times \exp\left(\frac{r}{2}[\exp(i\alpha)t - \exp(-i\alpha)t^{-1}]\right)$$

$$= \exp[-irt^{-1}\sin\alpha]\sum_{-\infty}^{\infty}\exp(im\alpha)t^m J_m(r) \quad (A2.60)$$

and

$$\exp\left[\frac{\rho}{2}\exp(i\beta)[t - t^{-1}]\right] = \exp[-i\rho t^{-1}\sin\beta]\sum_{-\infty}^{\infty}\exp(im\beta)t^m J_m(\rho) \quad (A2.61)$$

Hence

$$\exp\left(\frac{re^{i\alpha} + \rho e^{i\beta}}{2}\left(t - \frac{1}{t}\right)\right) = \exp[-it^{-1}(r\sin\alpha + \rho\sin\beta)]$$

$$\times \sum_{-\infty}^{\infty}\exp(im\alpha)t^m J_m(r)\sum_{-\infty}^{\infty}\exp(ik\beta)t^k J_k(\rho) \quad (A2.62)$$

Let $n = k + m$. Then, multiplying the series on the right of (A2.62) and using (A2.58), we obtain

$$\sum_{\infty}^{\infty}t^n J_n[r\exp(i\alpha) + \rho\exp(i\beta)]$$

$$= \sum_{-\infty}^{\infty}t^n\sum_{-\infty}^{\infty}\exp\{i[m\alpha + (n - m)\beta]\}\,J_m(r)\,J_{n-m}(\rho) \quad (A2.63)$$

so that

$$J_n(r_{pq}) = J_n[r\exp(i\alpha) + \rho\exp(i\beta)]$$

$$= \sum_{-\infty}^{\infty}\exp\{i[m\alpha + (n - m)\beta]\}\,J_m(r)\,J_{n-m}(\rho) \quad (A2.64)$$

In particular,

$$J_0(r_{pq}) = \sum_{-\infty}^{\infty}\exp[im(\alpha - \beta)]J_m(r)\,J_{-m}(\rho) \quad (A2.65)$$

By (A2.55) and (A2.56),

$$r_{pq}^2 = r^2 + \rho^2 + 2r\rho\cos(\psi)$$

$$= r^2\cos^2(\alpha) + \rho^2\cos^2(\beta) + 2r\rho\cos(\alpha)\cos(\beta) \quad (A2.66)$$

$$0 = r^2\sin^2(\alpha) + \rho^2\sin^2(\beta) + 2r\rho\sin(\alpha)\sin(\beta)$$

so that

$$r^2 + \rho^2 + 2r\rho\cos(\psi) = r^2 + \rho^2 + 2r\rho\cos(\alpha - \beta) \tag{A2.67}$$

Hence

$$\alpha - \beta = \psi = \pi - \varphi \tag{A2.68}$$

so that

$$\cos[m(\alpha - \beta)] = (-1)^m \cos(m\varphi), \qquad \sin[m(\alpha - \beta)] = \sin(m\varphi) \tag{A2.69}$$

Taking this and (A2.34) into account, we obtain

$$J_0(r_{pq}) = \sum_{-\infty}^{\infty} J_m(r) J_{-m}(\rho)(-1)^m \cos(m\varphi)$$

$$+ i \sum_{-\infty}^{\infty} J_m(r) J_{-m}(\rho)(-1)^m \sin(m\varphi) \tag{A2.70}$$

or, by (A2.34),

$$J_0(r_{pq}) = J_0(r) J_0(\rho) + 2 \sum_{1}^{\infty} J_m(r) J_m(\rho) \cos(m\varphi) \tag{A2.71}$$

since

$$\sum_{-\infty}^{\infty} J_m(r) J_{-m}(\rho) \sin(m\varphi)$$

$$= \sum_{1}^{\infty} J_m(r) J_m(\rho)(-1)^m[\sin(m\varphi) + \sin(-m\varphi)] = 0 \tag{A2.72}$$

Equality (A2.71) is known as the *Neumann addition theorem*.

5. Potential of double layer of dipoles distributed with unit density along surface of infinitely long circular cylinder. Discontinuous Weber–Schafheitlin integral. Fourier–Bessel double integral

Let (r, φ, z) be the cylindrical coordinate system and C an infinitely long circular cylinder of radius R,

$$C = \{\rho = R; \ 0 \le \varphi \le 2\pi; \ -\infty < z < \infty\}, \tag{A2.73}$$

$$p = (r, 0, 0), \qquad q = (\rho, \varphi, \zeta) \tag{A2.74}$$

$$R_{pq}^2 = r_{pq}^2 + \zeta^2 \equiv r^2 + \rho^2 - 2r\rho\cos(\varphi) + \zeta^2 \tag{A2.75}$$

so that R_{pq} is the distance between the points p and q and r_{pq} the distance between p and the projection of q on the plane $z = 0$.

Let $W(p, q)$ be the potential at p of a double layer of dipoles distributed over C with unit density:

$$W(p, q) = -\frac{1}{4\pi} \int_C \frac{\partial}{\partial \rho} \frac{1}{R_{pq}} d\sigma_q \tag{A2.76}$$

By the theorem on discontinuities of a double layer electrostatic potential,

$$W(p,q) = \begin{cases} 1 & r < R \\ 1/2 & r = R \\ 0 & r > R \end{cases} \tag{A2.77}$$

On the other hand, use of the Lipschitz integral and the Neumann addition theorem yields

$$\frac{1}{R_{pq}} = \int_0^\infty \exp(-\lambda|\zeta|)\, J_0(\lambda r_{pq})d\lambda$$

$$\equiv \int_0^\infty \exp(-\lambda|\zeta|)$$

$$\times \left(J_0(\lambda r)\, J_0(\lambda \rho) + 2\sum_1^\infty J_m(\lambda r)\, J_m(\lambda \rho)\cos(m\varphi) \right) d\lambda \tag{A2.78}$$

and, consequently,[4]

$$4\pi W(p,q) = -\int_{-\infty}^\infty d\zeta \int_0^{2\pi} R\, d\varphi \int_0^\infty \exp(-\lambda|\zeta|)$$

$$\times \frac{\partial}{\partial \rho}\left(J_0(\lambda r)\, J_0(\lambda \rho) + 2\sum_1^\infty J_m(\lambda r)\, J_m(\lambda \rho) \right)\Bigg|_{\rho=R} d\lambda$$

$$= -R\int_0^\infty d\lambda \frac{2}{\lambda}\int_{-\infty}^\infty \left(2\pi J_0(\lambda r)\frac{\partial}{\partial \rho}J_0(\lambda \rho) \right.$$

$$\left. + 2\sum_1^\infty J_m(\lambda r)\frac{\partial}{\partial \rho}J_m(\lambda \rho)\Bigg|_{\rho=R} \int_0^{2\pi}\cos(m\varphi)d\varphi \right)$$

$$\times \exp(-\lambda|\zeta|)d\zeta \tag{A2.79}$$

Thus, by (A2.31), we obtain

$$W(p,q) = R\int_0^\infty J_0(\lambda r)\, J_1(\lambda R)d\lambda \tag{A2.80}$$

Comparing (A2.80) and (A2.77), we obtain

$$R\int_0^\infty J_0(\lambda r)\, J_1(\lambda R)d\lambda = \begin{cases} 1 & r < R \\ 1/2 & r = R \\ 0 & r > R \end{cases} \tag{A2.81}$$

This integral is a particular case of the general *Weber–Schafheitlin discontinuous integral*:

$$\int_0^\infty J_\mu(\lambda a)\, J_\nu(\lambda b)\lambda^{\mu-\nu+1}d\lambda = f(a,b) \tag{A2.82}$$

---

[4]Term-by-term differentiation of the series under the sign of the double integral in (A2.78) is easily justified by using asymptotic estimates for Bessel functions of large arguments (see Section 7 of this Appendix).

where

$$f(a,b) = \begin{cases} 2^{\mu-\nu+1}a^\mu b^{-\nu}(b^2-a^2)^{\nu-\mu-1}/\Gamma(\nu-\mu) & \text{if } b > a \\ 0 & \text{if } b < a \end{cases} \quad \text{(A2.83)}$$

and

$$\Re\nu > \Re\mu > -1 \quad \text{(A2.84)}$$

The proof of (A2.82)–(A2.84), unlike that of (A2.81), is based on properties of hypergeometric series and on a very delicate analysis of the convergence of the relevant series and integrals [90].

The Weber–Schafheitlin discontinuous integrals yield an easy derivation of the *Fourier–Bessel double integrals*. Let $f(r)$ be a continuous function of bounded variation, absolutely integrable on the half-line. Then, by (A2.81),

$$\int_\rho^\infty f(R)dR = \int_0^\infty Rf(R)dR \int_0^\infty J_0(\lambda\rho) J_1(\lambda R)d\lambda \quad \text{(A2.85)}$$

so that

$$f(\rho) = -\frac{d}{d\rho}\int_0^\infty Rf(R)dR \int_0^\infty J_0(\lambda\rho) J_1(\lambda R)d\lambda \quad \text{(A2.86)}$$

Formally inverting the order of differentiation and integration and using (A2.29), we get

$$f(\rho) = \int_0^\infty Rf(R)dR \int_0^\infty \lambda J_1(\lambda\rho) J_1(\lambda R)d\lambda \quad \text{(A2.87)}$$

This is the Fourier–Bessel integral of order 1. Using (A2.82), we can derive the Fourier–Bessel integral of order ν:

$$f(\rho) = \int_0^\infty \lambda J_\nu(\lambda\rho)d\lambda \int_0^\infty Rf(R) J_\nu(\lambda R)dR \quad \text{(A2.88)}$$

To do this, set

$$f(\rho) = \varphi(\rho)\rho^{-\nu-1} \quad \text{(A2.89)}$$

It follows from (A2.82) and (A2.83) that

$$\int_0^\infty J_\nu(\rho t) J_{\nu+1}(Rt)dt = \begin{cases} \rho^\nu R^{-1-\nu} & \text{if } \rho < R \\ 0 & \text{if } \rho > R \end{cases} \quad \text{(A2.90)}$$

Hence

$$\int_0^R \varphi(\rho)d\rho = \int_0^\infty \varphi(\rho)\rho^{-\nu}R^{\nu+1}d\rho \int_0^\infty J_\nu(\rho t) J_{\nu+1}(Rt)dt \quad \text{(A2.91)}$$

which formally implies

$$\varphi(R) = \int_0^\infty \rho^{-\nu}\varphi(\rho)d\rho \int_0^\infty J_\nu(\rho t)\frac{\partial}{\partial R}\left(R^{\nu+1}J_{\nu+1}(Rt)\right)dt \quad \text{(A2.92)}$$

Since

$$\frac{\partial}{\partial R}\left(R^{\nu+1}J_{\nu+1}(Rt)\right) = tR^{\nu+1}J_\nu(Rt) \quad \text{(A2.93)}$$

(A2.89) and (A2.91) imply

$$f(R) = \int_0^\infty \rho f(\rho) d\rho \int_0^\infty J_\nu(\rho t) J_\nu(Rt) t dt \qquad (A2.94)$$

or, changing the order of integration,

$$f(R) = \int_0^\infty t J_\nu(Rt) dt \int_0^\infty \rho f(\rho) J_\nu(\rho t) d\rho \qquad (A2.95)$$

which coincides with (A2.88) up to notation.

This result has been obtained without justifying the change in the order of integration and differentiation or examining the behavior of the integrand at zero and infinity. However, use of the asymptotic properties of the Bessel functions (see the subsequent sections) yields a rigorous proof.

6. Bessel functions of imaginary argument. Spherical Bessel functions

The function

$$\psi(t, z) = \exp\left(\frac{z}{2}(t + t^{-1})\right) = \sum_{-\infty}^\infty I_n(z) t_n \qquad (A2.96)$$

generates the Bessel functions $I_n(z)$ of the first kind of an imaginary argument or the modified Bessel functions of the first kind of order n:

$$I_n(z) = \sum_{n=0}^\infty \frac{(z/2)^{2m+n}}{m!(m+n)!} \qquad (A2.97)$$

Comparison with the expansion (A2.17) of $J_n(z)$ yields

$$I_n(z) = \exp(-in\pi/2) J_n[\exp(i\pi/2)z] \qquad (A2.98)$$

which explains the first of the aforementioned names. The identity (A2.97) instantly implies

$$I_{n-1}(z) - I_{n+1}(z) = \frac{2n}{z} I_n(z) \qquad (A2.99)$$

and

$$2\frac{d}{dz} I_n(z) = I_{n-1}(z) - I_{n+1}(z) \qquad (A2.100)$$

By (A2.98) and (A2.31),

$$\frac{d}{dz}[z^n I_n(z)] = z^n I_{n-1}(z), \frac{d}{dz}[z^{-n} I_n(z)] = z^{-n} I_{n+1}(z) \qquad (A2.101)$$

so that

$$\left(\frac{1}{z}\frac{d}{dz}\right)^m [z^n I_n(z)] = z^{n-m} I_{n-m}(z)$$

$$\left(\frac{1}{z}\frac{d}{dz}\right)^m [z^{-n} I_n(z)] = z^{-n-m} I_{n+m}(z)$$

$$\qquad (A2.102)$$

Finally, the Bessel equation (A2.33) becomes the *modified Bessel equation:*

$$\frac{d^2}{dx^2} I_n(z) + \frac{1}{z}\frac{d}{dz} I_n(z) - \left(1 + \frac{n^2}{z^2}\right) I_n(z) = 0 \qquad \text{(A2.103)}$$

Here again, n may be replaced by any real or complex ν. Thus

$$I_\nu(z) = \sum_0^\infty \frac{(z/2)^{2m+\nu}}{\Gamma(m+1)\Gamma(m+\nu+1)} \qquad \text{(A2.104)}$$

and

$$\frac{d^2}{dx^2} I_\nu(z) + \frac{1}{z}\frac{d}{dz} I_\nu(z) - \left(1 + \frac{\nu^2}{z^2}\right) I_\nu(z) = 0 \qquad \text{(A2.105)}$$

The modified Bessel functions of the second kind $I_{-n}(z)$ are related to $I_n(z)$ by the equality

$$I_{-n}(z) = I_n(z), \quad n = 0,1,2,\ldots \qquad \text{(A2.106)}$$

A second solution of the modified Bessel equation, linearly independent of $I_\nu(z)$, called the *Macdonald function,* is defined by

$$K_\nu(z) = \frac{\pi}{2}\frac{I_{-\nu}(z) - I_\nu(z)}{\sin(\pi\nu)} \qquad \text{(A2.107)}$$

Using L'Hospital's rule, for integers ν we obtain

$$K_\nu(z) = (-1)^{n+1}I_n(z)\ln\left(\frac{z}{2}\right) + \frac{1}{2}\sum_0^{n-1}(-1)^n\left(\frac{z}{2}\right)^{2m-n}\frac{(n-m-1)!}{m!}$$

$$+ \sum_0^\infty \left(\frac{z}{2}\right)^{2m+n}\frac{\psi(n+m+1) + \psi(m+1)}{m!} \qquad \text{(A2.108)}$$

where $\psi(k)$ is the logarithmic derivative of $\Gamma(k)$.

Recursive relations (A2.102) and definition (A2.107) imply

$$\left(\frac{1}{z}\frac{d}{dz}\right)^m [z^\nu K_\nu(z)] = (-1)^m z^{\nu-m} K_{\nu-m}(z)$$

$$\left(\frac{1}{z}\frac{d}{dz}\right)^m [z^{-\nu} K_\nu(z)] = (-1)^m z^{-\nu-m} K_{\nu+m}(z) \qquad \text{(A2.109)}$$

We now proceed to consider the spherical Bessel functions or Bessel functions of order a half-integer. We have

$$J_{1/2}(z) \equiv \sum_0^\infty \frac{(-1)^m (z/2)^{2m+1/2}}{\Gamma(m+1)\Gamma(m+3/2)} \qquad \text{(A2.110)}$$

By the duplication formula (A2.4),

$$\Gamma(m+1)\Gamma(m+3/2) = \sqrt{\pi}\,\Gamma[2(m+1)]2^{-1-2m} \qquad \text{(A2.111)}$$

Hence

$$J_{1/2}(z) = \left(\frac{2}{\pi z}\right)^{1/2}\sum_0^\infty \frac{(-1)^m z^{2m+2}}{\Gamma[2(m+1)]} \equiv \left(\frac{2}{\pi z}\right)^{1/2}\sin(z) \qquad \text{(A2.112)}$$

Similarly,

$$J_{-1/2}(z) = \left(\frac{2}{\pi z}\right)^{1/2} \cos(z) \qquad (A2.113)$$

Using the recurrence formulas

$$\left(\frac{1}{z}\frac{d}{dz}\right)^m [z^n J_n(z)] = z^{n-m} J_{n-m}(z)$$

$$\left(\frac{1}{z}\frac{d}{dz}\right)^m [z^{-n} J_n(z)] = (-1)^m z^{-n-m} J_{n+m}(z) \qquad (A2.114)$$

we obtain

$$J_{n-1/2}(z) = (-1)^n \sqrt{2/\pi} z^{n-1/2} \left(\frac{1}{z}\frac{d}{dz}\right)^n \cos(z) \qquad (A2.115)$$

and

$$J_{1/2-n}(z) = \sqrt{2/\pi} z^{n-1/2} \left(\frac{1}{z}\frac{d}{dz}\right)^n \sin(z) \qquad (A2.116)$$

Quite similarly, using definitions (A2.37), (A2.112), (A2.113), (A2.98), and (A2.107), we find

$$Y_{1/2}(z) = -J_{-1/2}(z) = -\left(\frac{2}{\pi z}\right)^{1/2} \cos(z) \qquad (A2.117)$$

$$Y_{-1/2}(z) = J_{1/2}(z) = \left(\frac{2}{\pi z}\right)^{1/2} \sin(z) \qquad (A2.118)$$

$$I_{1/2}(z) = \exp\left(-\frac{i\pi}{4}\right) J_{1/2}\left[\exp\left(\frac{i\pi}{2}\right)z\right] = \left(\frac{2}{\pi z}\right)^{1/2} \sinh(z) \qquad (A2.119)$$

$$I_{-1/2}(z) = \exp\left(-\frac{i\pi}{4}\right) J_{-1/2}\left[\exp\left(\frac{i\pi}{2}\right)z\right] = \left(\frac{2}{\pi z}\right)^{1/2} \cosh(z) \qquad (A2.120)$$

$$K_{1/2}(z) = K_{-1/2}(z) = (2\pi z)^{-1/2} e^{-z} \qquad (A2.121)$$

Finally, similarly to (A2.115) and (A2.116), we have

$$I_{n-1/2}(z) = \sqrt{2/\pi} z^{n-1/2} \left(\frac{1}{z}\frac{d}{dz}\right)^n \cosh(z) \qquad (A2.122)$$

$$I_{1/2-n}(z) = \sqrt{2/\pi} z^{n-1/2} \left(\frac{1}{z}\frac{d}{dz}\right)^n \sinh(z) \qquad (A2.123)$$

and

$$K_{1/2-n}(z) = (-1)^n \sqrt{\frac{\pi}{2}} z^{n-1/2} \left(\frac{1}{z}\frac{d}{dz}\right)^n e^{-z} \qquad (A2.124)$$

$$K_{-1/2-n}(z) = (-1)^n \sqrt{\frac{\pi}{2}} z^{-n+1/2} \left(\frac{1}{z}\frac{d}{dz}\right)^n e^{-z} \qquad (A2.125)$$

7. Asymptotic behavior of Bessel functions

In order to derive asymptotic expressions for the Bessel functions as $z \uparrow \infty$, we write the Bessel equation as

$$\frac{d^2}{dz^2}u + \frac{1}{z}\frac{d}{dz}u + \left(1 - \frac{1/4}{z^2}\right)u = \left(\frac{\nu^2 - 1/4}{z^2}\right)u \tag{A2.126}$$

Applying the method of variation of constants, we find that the general solution of this equation can be represented as

$$Z_\nu(z) = \alpha(z)\,J_{1/2}(z) + \beta(z)Y_{1/2}(z) \tag{A2.127}$$

where

$$\frac{d\alpha}{dz}J_{1/2}(z) + \frac{d\beta}{dz}Y_{1/2}(z) = 0 \tag{A2.128}$$

$$\frac{d\alpha}{dz}\frac{d}{dz}J_{1/2}(z) + \frac{d\beta}{dz}\frac{d}{dz}Y_{1/2}(z) = \left(\frac{\nu^2 - 1/4}{z^2}\right)Z_\nu(z) \tag{A2.129}$$

In view of (A2.42) we find that, for all $z_0 > 0$,

$$Z_\nu(z) = \alpha_0 J_{1/2}(z) + \beta_0 Y_{1/2}(z)$$
$$+ \left(\nu^2 - \frac{1}{4}\right)\int_{z_0}^z \frac{Z_\nu(\zeta)}{z^{1/2}\zeta^{3/2}}\sin(z - \zeta)d\zeta \tag{A2.130}$$

The constants α_0 and β_0 depend on values $Z_\nu(z_0)$ and $dZ_\nu(z_0)/dz$.

It is easy to see that the solution of the integral equation (A2.130) is bounded for all $z \geq z_0$. Indeed, let z_1 be fixed arbitrarily large,

$$\max_{z_0 < x < z_1}|Z_\nu(z)| = |Z_\nu(\tilde{z})| = M > 0 \tag{A2.131}$$

and

$$N = |\alpha_0| + |\beta_0| \tag{A2.132}$$

Majorizing (A2.130), we get

$$M < N\tilde{z}^{-1/2} + 2M(\nu^2 - 1/4)(z_0\tilde{z})^{-1/2} \tag{A2.133}$$

so that

$$M < N[1 - 2(\nu^2 - 1/4)(z_0\tilde{z})^{-1/2}]\tilde{z}^{-1/2} \tag{A2.134}$$

Hence, taking

$$z_0 > 4\nu^2 - 1 \tag{A2.135}$$

we obtain

$$M < 2N\tilde{z}^{-1/2} < 2Nz_0^{-1/2} \tag{A2.136}$$

Thus

$$Z_\nu(z) = O(z^{-1/2}) \tag{A2.137}$$

Moreover, differentiation of (A2.130) shows that also

$$\frac{d}{dz}Z_\nu(z) = O(z^{-1/2}) \tag{A2.138}$$

Inserting these estimates into (A2.130) we find that, asymptotically as $z \uparrow \infty$,

$$Z_\nu(z) \asymp [\alpha \sin(z) + \beta \cos(z)]z^{-1/2} + (\nu^2 - 1/4) \int_z^\infty Z_\nu(\zeta)(z\zeta^3)^{-1/2}d\zeta$$

$$(A2.139)$$

Thus we finally conclude that, asymptotically as $z \uparrow \infty$,[5]

$$Z_\nu(z) \asymp [\alpha \sin(z) + \beta \cos(z)]z^{-1/2} + O(z^{-3/2}) \qquad (A2.140)$$

We now consider Bessel functions $I_\nu(z)$ of an imaginary argument. Let

$$u(z) \overset{\text{def}}{=} I_\nu(z) = v(z)w(z), \qquad w(z) \overset{\text{def}}{=} e^z z^{-1/2} \qquad (A2.141)$$

Recall that, for any μ,

$$L_\mu(u) \overset{\text{def}}{=} \frac{d^2}{dz^2}u + \frac{1}{z}\frac{d}{dz}u - (1 + \mu^2 z^{-2})u \qquad (A2.142)$$

We have

$$L_\nu(u) = L_{1/2}(u) - (\nu^2 - 1/4)z^{-2}u$$

$$\equiv \left(2\frac{dw}{dz} + \frac{1}{z}w\right)\frac{dv}{dz} - \left(\frac{\nu^2 - 1/4}{z^2}\right)wv + w\frac{d^2v}{dz^2} = 0 \qquad (A2.143)$$

since

$$L_{1/2}w = 0 \qquad (A2.144)$$

Thus v must be a solution of the equation

$$\frac{d^2v}{dz^2} + \left(\frac{1}{z} + \frac{2}{w}\frac{dw}{dz}\right)\frac{dv}{dz} - \left(\frac{\nu^2 - 1/4}{z^2}\right)v = 0 \qquad (A2.145)$$

Fix some $z_0 > 0$. Integration of (A2.145) shows that v must be a solution of the integral equation

$$v(z) = \alpha + \beta e^{-2z} + \frac{4\nu^2 - 1}{8} \int_{z_0}^z v(\zeta)\zeta^{-2}(1 - \exp[-2(z - \zeta)])d\zeta \qquad (A2.146)$$

with integration constants α and β dependent on v at $z = z_0$.

This shows that, asymptotically as $z \uparrow \infty$,

$$v(z) \asymp \alpha = \text{const} \qquad (A2.147)$$

so that, by (A2.141),

$$I_\nu(z) \asymp \alpha e^z z^{-1/2} \quad \text{as } z \uparrow \infty \qquad (A2.148)$$

---

[5]The constants α and β here remain undetermined. Their determination requires a more precise representation of the Bessel functions, in the form of contour and singular integrals, referred to in Section 9 and used in Section 10 in connection with the asymptotic behavior of the Macdonald function $K_\nu(z)$.

Quite similarly, it can be shown that

$$K_\nu(z) \asymp \beta e^{-z} z^{-1/2} \quad \text{as } z \uparrow \infty \tag{A2.149}$$

As in the case of $J_\nu(z)$ and $Y_\nu(z)$, the constants α and β can be determined by using more accurate computations:[6]

$$\alpha = \sqrt{1/2\pi}, \qquad \beta = \sqrt{\pi/2} \tag{A2.150}$$

Remarks. (1) The asymptotic representation (A2.148) shows that the Lipschitz integral (A2.52) is applicable for complex z if $\Re z < 1$, since this inequality guarantees the convergence of the integral at infinity. Thus

$$\int_0^\infty e^{-t} I_0(zt)\,dt = (1 - z^2)^{-1/2} \quad \forall z^2 < 1 \tag{A2.151}$$

(2) The Bessel integral (A2.45) is applicable for complex z:

$$I_0(z) = \frac{2}{\pi} \int_0^1 \cosh(\lambda z)(1 - \lambda^2)^{-1/2} d\lambda$$

$$\equiv (\pi z)^{-1} \int_0^z (e^s + e^{-s})(1 - s^2 z^{-2})^{-1/2} ds \tag{A2.152}$$

Using the binomial expansion

$$(1 - s^2 z^{-2})^{-1/2} = \sum_0^\infty (-1)^n \left(s^2 z^{-2}\right)^n \frac{\Gamma(1/2)}{\Gamma(1/2 - n)\Gamma(n+1)} \tag{A2.153}$$

we see from (A2.152) that, for $I_0(z)$, equalities (A2.150) are indeed true.

8. Method of averaging. Weber's integrals [30]

In Chapter 18, Section 4, the Weber integral

$$(2t)^{-1} \exp\left[-\frac{(\lambda^2 + \mu^2)}{4t}\right] I_\nu\left(\frac{\lambda\mu}{2t}\right) = \int_0^\infty e^{-t\rho^2} \rho J_\nu(\lambda\rho) J_\mu(\mu\rho)\,d\rho \tag{A2.154}$$

Is used to construct a fundamental solution of the axially symmetric heat-conduction equation with forced convection. We now prove this identity.

Following Weber, consider the solution of the wave equation

$$\Delta u = \frac{\partial^2}{\partial t^2} u \tag{A2.155}$$

of the type of a stationary oscillation:

$$u(\tilde{p}, t) = v(\tilde{p}) e^{ikt}, \quad \tilde{p} \in \mathbb{R}_3 \tag{A2.156}$$

The amplitude $v(\tilde{p})$ of the oscillations is the integral of the Helmholtz equation

$$\Delta v(\tilde{p}) + k^2 v(\tilde{p}) = 0 \tag{A2.157}$$

---

[6]See note 5.

Let

$$p = (x, y, z), \qquad q = (\xi, \eta, \zeta) \tag{A2.158}$$

be two given points in a cartesian coordinate system. We shall also use a cylindrical coordinate system (ρ, φ, z) so that

$$p = (r \cos\varphi, \, r \sin\varphi, \, z), \qquad q = (\rho \cos\psi, \, \rho \sin\psi, \, \zeta) \tag{A2.159}$$

and a polar coordinate system (r, φ, ϑ), with pole at p, such that

$$\xi = x + r \cos\varphi \cdot \sin\vartheta, \quad \eta = y + r \sin\varphi \cdot \sin\vartheta, \quad \zeta = z + r \cos\vartheta \tag{A2.160}$$

Let v be an arbitrary solution of equation (A2.157) which is bounded at p, and let S_p^r be the surface of the sphere of radius r with center at p. Let

$$w(r \mid p) = \frac{1}{4\pi r^2} \int_{S_p^r} v(q) d\sigma_q \tag{A2.161}$$

so that $w(r \mid p)$ is the average value of v over the surface of the sphere S_p^r. Then (see Chapter 8, Section 1)

$$v(p) = \lim_{r \downarrow 0} w(r \mid p) \tag{A2.162}$$

and

$$\frac{d^2}{dr^2}(rw) + k^2 rw = 0 \tag{A2.163}$$

An integral of this equation bounded at p is

$$w = A \frac{\sin(kr)}{r} \tag{A2.164}$$

Using (A2.162), we find

$$A = \frac{v(p)}{k} \tag{A2.165}$$

so that

$$w(r \mid p) = v(p) \frac{\sin(kr)}{kr} \tag{A2.166}$$

Now let

$$\Omega(x, y, z) \stackrel{\text{def}}{=} \Omega(p) = \frac{1}{4\pi} \int\!\!\!\int\!\!\!\int_{-\infty}^{\infty} \exp(-tr_{pq}^2) v(q) d\xi d\eta d\zeta$$

$$= \frac{1}{4\pi} \int_0^{\infty} dr \int_{S_p^r} \exp(-tr_{pq}^2) v(q) d\sigma_q \tag{A2.167}$$

or, what is the same,

$$\Omega(p) = \int_0^{\infty} \exp(-tr^2) r^2 w(r \mid p) dr \tag{A2.168}$$

Using (A2.166), we obtain

$$\Omega(p) = v(p) \, J(k) \tag{A2.169}$$

where

$$J(k) = \frac{1}{k} \int_0^\infty \exp(-tr^2) r \sin(kr) dr = \frac{1}{2ki} \int_{-\infty}^\infty \exp(-tr^2 + ikr) r dr \quad (A2.170)$$

Let

$$s = k/2\sqrt{t}, \quad z = r\sqrt{t} - is \quad (A2.171)$$

Then we find

$$J(k) = \frac{e^{-k^2/4t}}{2t} \left(\frac{1}{ki} \int_{-\infty-is}^{\infty-is} e^{-z^2} z dz + \frac{1}{2\sqrt{t}} \int_{-\infty-is}^{\infty-is} e^{-z^2} dz \right) \quad (A2.172)$$

Consider the contour $ABB'A'A$ in the complex plane z, where

$$A = N, \quad B = -N$$
$$B' = -N - ik/2\sqrt{t}, A' = N - ik/2\sqrt{t} \quad (A2.173)$$

and the two integrals

$$I_N^1 = \int_{ABB'A'A} z e^{-z^2} dz, \quad I_N^2 = \int_{ABB'A'A} e^{-z^2} dz \quad (A2.174)$$

By Cauchy's theorem,

$$I_N^1 = 0, \quad I_N^2 = 0 \quad (A2.175)$$

On the other hand,

$$I_N^i = \sum_{j=1}^4 I_{N_j}^i \quad (A2.176)$$

where

$$I_{N_1}^i = \int_{AB}, \quad I_{N_2}^i = \int_{BB'}, \quad I_{N_3}^i = \int_{B'A'}, \quad I_{N_4}^i = \int_{A'A} \quad (A2.177)$$

Since $z = x + iy$,

$$|I_{N_2}^1| \leq \frac{|k|}{2\sqrt{t}} \int_0^{|s|} |z| |e^{-z^2}| dy, \quad \left(x = -N, \ s = -\frac{k}{2\sqrt{t}} \right) \quad (A2.178)$$

so that

$$\lim_{N\uparrow\infty} I_{N_2}^1 = 0 \quad (A2.179)$$

and similarly

$$\lim_{N\uparrow\infty} I_{N_4}^1 = 0, \quad \lim_{N\uparrow\infty} I_{N_2}^2 = 0, \quad \lim_{N\uparrow\infty} I_{N_4}^2 = 0 \quad (A2.180)$$

Hence

$$J(k) = -\lim_{N\uparrow\infty} \left(\frac{e^{-k^2/4t}}{2ik\sqrt{t}} I_{N_1}^1 + \frac{e^{-k^2/4t}}{4t^{3/2}} I_{N_1}^2 \right) = \frac{\sqrt{\pi}}{4t^{3/2}} e^{-k^2/4t} \quad (A2.181)$$

Thus, by (A2.169),

$$\Omega(p) = \frac{\sqrt{\pi}}{4t^{3/2}} e^{-k^2/4t} v(p) \quad (A2.182)$$

Recall that $v(p)$ is an arbitrary bounded solution of the Helmholtz equation (A2.157). Taking this into account, assume that $v(p)$ is independent of z. Then v must be a solution of the two-dimensional Helmholtz equation:

$$\left(\frac{\partial^2}{\partial r^2} + r^{-1}\frac{\partial}{\partial r} + r^{-2}\frac{\partial^2}{\partial \varphi^2} + k^2\right) v(r, \varphi) \qquad (A2.183)$$

This equation admits a solution with separable variables,

$$v(r, \varphi) = R(r)\Phi(\varphi) \qquad (A2.184)$$

where $\Phi(\varphi)$ is 2π-periodic, since only under these conditions can $v(r, \varphi)$ be continuous. Substituting (A2.184) into (A2.183) and separating variables,[7] we obtain

$$\frac{\frac{d^2}{dr^2}R + \frac{1}{r}\frac{d}{dr}R + k^2 R}{Rr^{-2}} = -\frac{\frac{d^2}{d\varphi^2}\Phi}{\Phi} = \lambda^2 \qquad (A2.185)$$

Integration yields

$$\Phi = a\cos(\lambda\varphi) + b\sin(\lambda\varphi) \qquad (A2.186)$$

Using the periodicity condition, we find

$$\lambda = n, \quad n = \text{integer} \qquad (A2.187)$$

Hence, by (A2.185),

$$\frac{d^2}{dr^2}R + \frac{1}{r}\frac{d}{dr}R + [k^2 - n^2 r^{-2}]R = 0 \qquad (A2.188)$$

so that the bounded solution R is proportional to $J_n(kr)$. Thus

$$v(r, \varphi) = J_n(kr)[a\cos(n\varphi) + b\sin(n\varphi)] \qquad (A2.189)$$

Taking

$$a = 1, \quad b = i \qquad (A2.190)$$

we get

$$v(r, \varphi) = J_n(kr)\exp(in\varphi) \qquad (A2.191)$$

Inserting (A2.191) into (A2.167) and using (A2.182), we obtain

$$\frac{\sqrt{\pi}}{4t^{3/2}}\exp(-k^2/4t + in\varphi)J_n(kr) = \frac{1}{4\pi}\int_{-\infty}^{\infty}d\xi\int_{-\infty}^{\infty}\exp(in\psi)J_n(k\rho)d\eta$$

$$\times \int_{-\infty}^{\infty}\exp\left\{-t[(x - \xi)^2 + (y - \eta)^2 + (z - \zeta)^2]\right\}d\zeta \qquad (A2.192)$$

Recall that here

$$(x - \xi)^2 + (y - \eta)^2 + (z - \zeta)^2 = r_{pq}^2 = r^2 + \rho^2 - 2r\rho\cos(\varphi - \psi)$$
$$x = r\cos(\varphi), \quad y = r\sin(\varphi), \quad \xi = \rho\cos(\psi), \quad \eta = \rho\sin(\psi) \qquad (A2.193)$$

---

[7] Concerning the method of separation of variables, see Chapter 19 and Appendix 3.

Besides,

$$\int_{-\infty}^{\infty} \exp[-t(\zeta - z)^2]d\zeta = (\pi/t)^{1/2} \qquad (A2.194)$$

Hence

$$\frac{1}{4}\sqrt{\pi}t^{-3/2} \exp\left(-k^2/4t + in\varphi\right) J_n(kr)$$

$$= \frac{1}{4\pi}\left(\frac{\pi}{t}\right)^{1/2} \int_0^{\infty} \rho d\rho \int_{-\pi}^{\pi} \exp(in\psi) J_n(k\rho)$$

$$\times \exp\left\{-t[(r^2 + \rho^2) + 2tr\rho\cos(\varphi - \psi)]\right\} d\psi \qquad (A2.195)$$

Taking

$$\varphi = \pi/2 \qquad (A2.196)$$

we obtain

$$\frac{1}{t}\exp(-k^2/4t + in\pi/2) J_n(kr) = \frac{1}{\pi}\int_0^{\infty} \rho J_n(k\rho)\exp[-t(r^2 + \rho^2)]d\rho$$

$$\times \int_{-\pi}^{\pi} \exp[in\psi + 2tr\rho\sin(\psi)]d\psi \quad (A2.197)$$

Using the Bessel integral (A2.45), we find that

$$\frac{1}{\pi}\int_{-\pi}^{\pi} \exp[in\psi + 2tr\rho\sin(\psi)]d\psi = 2J_n(2itr\rho)$$

$$= 2\exp\left(\frac{in\pi}{2}\right) I_n(2tr\rho) \qquad (A2.198)$$

which yields

$$\frac{1}{t}e^{-k^2/4t} J_n(kr) = 2\int_0^{\infty} \rho J_n(k\rho)I_n(2tr\rho)\exp[-t(r^2 + \rho^2)]d\rho \qquad (A2.199)$$

Set

$$2tr = \mu, \qquad k = \lambda \qquad (A2.200)$$

This implies

$$\frac{1}{2t}e^{-(\lambda^2 - \mu^2)/4t} J_n\left(\frac{\lambda\mu}{2t}\right) = \int_0^{\infty} \rho J_n(\lambda\rho)I_n(\mu\rho)\exp(-tr^2)d\rho \qquad (A2.201)$$

We have proved this equality for positive real μ. But since both sides are analytic functions of μ throughout the complex plane, it follows by the principle of analytic continuation that we may replace μ by $i\mu$, which yields

$$\frac{1}{2t}e^{-(\lambda^2 + \mu^2)/4t} I_n\left(\frac{\lambda\mu}{2t}\right) = \int_0^{\infty} \rho J_n(\lambda\rho) J_n(\mu\rho)\exp(-t\rho^2)d\rho \qquad (A2.202)$$

Q.E.D.

Using another method [90], one may prove that the identity (A2.202) is valid not only for integer n but for any complex n, provided only that $\Re n > -1/2$.

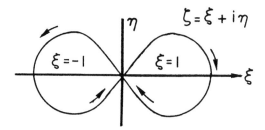

Figure A2.3. Contour for $J_0(z)$ definition by a contour integral.

Remark. Dividing (A2.202) by $I_n(\lambda\mu/2t)$ and letting $\mu \downarrow 0$, we obtain *Weber's first integral*:

$$\int_0^\infty \rho^{n+1} J_n(\lambda\rho) \exp(-t\rho^2) d\rho = \frac{\lambda^n}{(2t)^{n+1}} e^{-\lambda^2/4t} \qquad (A2.203)$$

9. Representation of Bessel functions by contour and singular integrals [30, 90]

A. *Contour integral representing $J_n(z)$*

Let C be the contour in the complex plane ζ shown in Figure A2.3, i.e., with initial point at $\Re\zeta = 0$, $\Im\zeta = 0$, encircling the point $\zeta = -1$ in the positive direction, crossing the real axis at $|\zeta| = 0$, $\arg\zeta = 2\pi$, encircling the point $|\zeta| = 1$, $\arg\zeta = \pi$ in the negative direction, and returning to $|\zeta| = 0$, $\arg\zeta = 0$. Then

$$J_n(z) = \frac{e^{i\pi/2}}{2\pi\sqrt{\pi}} \Gamma\left(\frac{1}{2} - n\right) \left(\frac{z}{2}\right)^n \int^{(-1+,+1-)} e^{-iz\zeta}(\zeta^2 - 1)^{n-1/2} d\zeta \quad (A2.204)$$

B. *Contour integral representing $I_n(z)$*

We have

$$I_n(z) = \frac{e^{i\pi/2}}{2\pi\sqrt{\pi}} \Gamma\left(\frac{1}{2} - n\right) \left(\frac{z}{2}\right)^n \int^{(-1+,+1-)} e^{z\zeta}(\zeta^2 - 1)^{n-1/2} d\zeta \quad (A2.205)$$

C. *Singular integrals representing $J_\nu(z)$ and $I_\nu(z)$*

If

$$\Re\nu > -\frac{1}{2} \qquad (A2.206)$$

then

$$J_\nu(x) = \frac{2(x/2)^\nu}{\sqrt{\pi} \cdot \Gamma(n+1/2)} \int_0^1 \cos(x\xi)(1 - \xi^2)^{\nu-1/2} d\xi \qquad (A2.207)$$

and

$$I_\nu(x) = \frac{2(x/2)^\nu}{\sqrt{\pi} \cdot \Gamma(n+1/2)} \int_0^1 \cosh(x\xi)(1-\xi^2)^{\nu-1/2} d\xi \qquad \text{(A2.208)}$$

D. Integral representation of $K_\nu(z)$

If

$$-\pi \ll \arg z < \pi, \quad z \neq 0, \quad \Re\nu > -1/2 \qquad \text{(A2.209)}$$

then

$$K_\nu(z) = \frac{(\pi/2z)^{1/2}}{\Gamma(\nu+1/2)} e^{-z} \int_0^\infty e^{-\xi}\xi^{\nu-1/2}\left(1+\frac{\xi}{2z}\right)^{\nu-1/2} d\xi \qquad \text{(A2.210)}$$

For the proof of (A2.204)–(A2.210) see [30; 90].

10. Asymptotic representation of Bessel functions in complex plane[8]

The integral (A2.210) can be used to obtain asymptotic representations of K_ν and I_ν.[9] To that end, we shall find an expression for the remainder term in the binomial expansion of $(1+\zeta/2z)^{\nu-1/2}$ in (A2.210). We have

$$\int_0^1 (1+zt)^{m-1} dt = \frac{(1+z)^m - 1}{mz} \qquad \text{(A2.211)}$$

Thus

$$(1+z)^m = 1 + mz \int_0^1 (1+zt)^{m-1} dt$$

$$= 1 + mz \int_0^1 [1+z(1-\tau)]^{m-1} d\tau \qquad \text{(A2.212)}$$

Integration by parts yields

$$(1+z)^m = 1 + mz + m(m-1)z^2 \int_0^1 [1+z(1-\tau)]^{m-2}\tau^2 d\tau \qquad \text{(A2.213)}$$

---

[8] As pointed out in Section 7 (see note 5), more accurate asymptotic expansions of the Bessel functions can be obtained by using their representations by contour and singular integrals. This is very important because such representations yield exact expressions for remainder terms of asymptotic expansions, which is sometimes decisive for the solution of nontrivial problems of mathematical physics; a characteristic example is the Stefan problem with cylindrical symmetry. The main tool for deriving these integrals is the theory of analytic functions and, in particular, the technique of transformation of contour integrals. However, since the theory of analytical functions is beyond the scope of this text, we restrict ourselves to references to the final results without their derivation, where this is possible within the context of our asymptotic analysis.

[9] Concerning the asymptotic behavior of J_ν and Y_ν, see the problems at the end of this appendix.

By induction we find that, for every $n \geq 1$,

$$(1+z)^m = \sum_0^{n-1} \frac{\Gamma(m+1)\,z^k}{\Gamma(k+1)\Gamma(m-k+1)} + R_n^*(z \mid m) \qquad (A2.214)$$

where

$$R_n^*(z \mid m) = \frac{\Gamma(m+1)\,z^n}{\Gamma(n+1)\Gamma(m-n+1)} \int_0^1 n\tau^{n-1}[1+z(1-\tau)]^{m-n}d\tau$$

$$= \frac{\Gamma(m+1)\,z^n}{\Gamma(n+1)\Gamma(m-n+1)} \int_0^1 n(1-\tau)^{n-1}(1+z\tau)^{m-n}d\tau \quad (A2.215)$$

The only condition for the validity of this equality is

$$1+z\tau \neq 0 \quad \forall \tau \in [0,1], \ \forall(z,m) \text{ real or complex} \qquad (A2.216)$$

taking the branch of $(1+zt)^{m-1}$ that is equal to 1 at $z=0$.

Inserting (A2.213)–(A2.216) into (A2.210), we obtain

$$K_\nu(z) = \frac{(\pi/2z)^{1/2}}{\Gamma(\nu+1/2)}e^{-z}$$

$$\times \int_0^\infty e^{-\xi}\xi^{\nu-1/2}\left(\sum_0^{n-1} \frac{\Gamma(\nu+1/2)(\xi/2z)^k}{\Gamma(k+1)\Gamma(\nu-k+1/2)}\right.$$

$$\left. + R_n^*\left(\frac{\xi}{2z}\,\bigg|\,\nu-\frac{1}{2}\right)\right)d\xi \qquad (A2.217)$$

or, by (A2.215),

$$K_\nu(z) = \frac{(\pi/2z)^{1/2}}{\Gamma(\nu+1/2)}e^{-z}$$

$$\times \left(\sum_0^{n-1} \frac{\Gamma(\nu+1/2)\Gamma(\nu-1/2+k)}{\Gamma(k+1)\Gamma(\nu-k+1/2)}(2z)^{-k} + R_n(z \mid \nu)\right) \quad (A2.218)$$

where

$$R_n(z \mid \nu) = \frac{\Gamma(\nu+1/2)(2z)^{-n}}{\Gamma(n+1)\Gamma(\nu-n+1/2)} \int_0^\infty e^{-\zeta}\zeta^{\nu-1/2+n}d\zeta$$

$$\times \int_0^1 n(1-t)^{n-1}(1+\zeta t/2z))^{\nu-n-1/2}dt \qquad (A2.219)$$

Note that, in view of conditions (A2.209) and (A2.221), the integrand is a holomorphic function of $\zeta = re^{i\varphi}$ without singularities, which tends uniformly to zero as $r \uparrow \infty$ and if $-\pi/2+\varepsilon \leq \phi \leq \pi/2-\varepsilon$ for every $\varepsilon \in (0,\pi/2)$. Therefore, by Cauchy's theorem, instead of integrating along the positive real axis one can integrate along any ray within this region. Thus

$$R_n(z \mid \nu) = \frac{\Gamma(\nu+1/2)}{\Gamma(n+1)\Gamma(\nu-n+1/2)} \int_0^{\infty e^{i\gamma}} e^{-\zeta}\zeta^{\nu-1/2+n}d\zeta$$

$$\times \int_0^1 n(1-t)^{n-1}(1+\zeta t/2z))^{\nu-n-1/2}dt \qquad (A2.220)$$

$$-\pi/2+\varepsilon \leq \gamma = \text{const} \leq \pi/2-\varepsilon, \quad 0 < \varepsilon < \pi/2$$

In what follows we will deal only with the principal term of this representation, that is, with the equality

$$K_\nu(z) = \frac{e^{-z}}{\Gamma(\nu+1/2)} \left(\frac{\pi}{2z}\right)^{1/2} \left(1 + \frac{\Gamma(\nu+1/2)}{\Gamma(\nu-1/2)} \cdot \int_0^{\infty e^{i\gamma}} e^{-\zeta}\zeta^{\nu+1/2}d\zeta \right.$$

$$\left. \times \int_0^1 \left(1 + \frac{\zeta t}{2z}\right)^{\nu-3/2} dt \right) \tag{A2.221}$$

Consider now the asymptotic behavior of $I_\nu(z)$. Note, first of all, that by (A2.104),

$$I_\nu(e^{i\pi}z) = e^{i\pi\nu}I_\nu(z), \qquad I_{-\nu}(e^{i\pi}z) = e^{-i\pi\nu}I_{-\nu}(z) \tag{A2.222}$$

so that, according to (A2.107),

$$K_\nu(e^{i\pi}z) = \frac{e^{-i\pi\nu}I_{-\nu}(z) - e^{i\pi\nu}I_\nu(z)}{\sin(\pi\nu)}\frac{\pi}{2}$$

$$e^{-i\pi\nu}K_\nu(z) = e^{-i\pi\nu} \cdot \frac{I_{-\nu}(z) - I_\nu(z)}{\sin(\pi\nu)}\frac{\pi}{2} \tag{A2.223}$$

which yields

$$\pi i I_\nu(z) = e^{-i\pi\nu}K_\nu(z) - K_\nu(e^{i\pi}z) \tag{A2.224}$$

This, together with (A2.221), implies

$$I_\nu(z) = \frac{1}{\sqrt{2\pi z}}\left(\exp(z)\left(1 + \frac{\Gamma(\nu+1/2)}{\Gamma(\nu-1/2)(2\pi z)^{1/2}}R_1(ze^{i\pi}|\nu)\right)\right.$$

$$+ \exp\left[-z - i\pi\left(\nu + \frac{1}{2}\right)\right]$$

$$\left. \times \left(1 + \frac{\Gamma(\nu+1/2)}{\Gamma(\nu-1/2)(2\pi z)^{1/2}}R_1(z|\nu)\right)\right) \tag{A2.225}$$

Note that $R_1(ze^{i\pi}|\nu)$ and $R_1(z\mid\nu)$ are integrals of holomorphic functions which converge uniformly if

$$-\frac{3\pi}{2} + \varepsilon \le \arg z \le \frac{\pi}{2} - \varepsilon \quad \forall \varepsilon \in (0, \pi/2)$$

$$\varepsilon \le |z| < \infty \quad \forall \varepsilon = \text{const} > 0 \tag{A2.226}$$

and γ satisfies condition (A2.220). Hence (A2.225) is an exact integral representation of $I_\nu(z)$ for every $\nu > -1/2$. At the same time, the first term on the right of (A2.225) is the principal term of the asymptotic expansion of $I_\nu(z)$. There is a very essential distinction between the two terms in this formula. Recall that asymptotic expansions of any function $f(z)$ may be integrated term by term, so that the resulting expansion is the asymptotic expansion of $f(z)$. But term-by-term differentiation of the asymptotic expansion of arbitrary $f(z)$ does not, in general, yield the asymptotic expansion of df/dz (see Chapter 21, Section 2). Therefore, whenever it is necessary to differentiate expressions in order to derive asymptotic representations, exact rather than asymptotic formulas must be used.

11. Hint for solution of cylindrical Stefan problem

In view of the remark at the end of Section 10, we shall now develop an integral representation of the fundamental solution of the problem of heat conduction with forced convection, studied in Chapter 20, Section 4. This representation will enable us to determine the asymptotic behavior of the solution for r and ρ bounded away from zero. The fundamental solution is

$$E_\nu(r,\rho,t-\tau) = \frac{(r/\rho)^\nu}{2(t-\tau)} \exp\left(\frac{r^2+\rho^2}{4(t-\tau)}\right) I_\nu\left(\frac{r\rho}{2(t-\tau)}\right), \quad \nu > 0 \quad \text{(A2.227)}$$

Using (A2.225), we obtain

$$E_\nu(r,\rho,t-\tau) = \frac{1}{\sqrt{\pi r \rho}} \cdot \frac{(r/\rho)^\nu}{2(t-\tau)^{1/2}}$$

$$\times \left(\exp\left(-\frac{(r-\rho)^2}{4(T-\tau)}\right)\left[1 + R_1\left(\frac{r\rho e^{i\pi}}{2(t-\tau)}\,\middle|\,\nu\right)\right]\right.$$

$$+ \exp\left(-\frac{(r+\rho)^2}{4(t-\tau)}\right)$$

$$\left.\times \exp\left[-i\pi\left(+\frac{1}{2}\right)\right]\left(1 + R_1\left(\frac{r\rho}{2(t-\tau)}\,\middle|\,\nu\right)\right)\right) \quad \text{(A2.228)}$$

Again, although this equality is an exact representation of $E_\nu(r,\rho,t-\tau)$, the first term on its right is the leading term of asymptotic representation of $E_\nu(r,\rho,t-\tau)$ for large r and finite t.

In the particular case of heat conduction without forced convection

$$E_0(r,\rho,t-\tau) = \frac{1}{\sqrt{\pi r \rho}} \cdot \frac{1}{2(t-\tau)^{1/2}}$$

$$\times \left(\exp\left(-\frac{(r-\rho)^2}{4(t-\tau)}\right) \left[1 + R_1\left(\frac{r\rho e^{i\pi}}{2(t-\tau)} \,\Big|\, 0\right)\right] \right.$$

$$+ \exp\left(-\frac{(r+\rho)^2}{4(t-\tau)}\right)$$

$$\left. \times e^{-i\pi/2}\left(1 + R_1\left(\frac{r\rho}{2(t-\tau)} \,\Big|\, 0\right)\right) \right) \tag{A2.229}$$

Using identities (A2.229) and (A2.102), we find that[10]

$$\frac{\partial}{\partial r}E_0 = -\frac{\partial}{\partial \rho}E_1 - \frac{E_1}{\rho} = -\frac{r}{2(t-\tau)}E_0 + \frac{\rho}{2(t-\tau)}E_1$$

$$\frac{\partial^2}{\partial r \partial \rho}E_0 = -\frac{\partial}{\partial t}E_1 = \frac{r\rho}{2(t-\tau)}E_0 - \frac{r^2+\rho^2}{4(t-\tau)}E_1 \tag{A2.230}$$

$$\frac{\partial^2}{\partial r^2}E_0 + \frac{1}{r}\frac{\partial}{\partial r}E_0 = \frac{\partial}{\partial t}E_0, \qquad E_i(r,\rho,t-\tau) = E_i(\rho,r,t-\tau)$$

This, together with (A2.229), implies

$$\frac{\partial}{\partial r}E_0(r,\rho,t) = -\frac{r-\rho}{4\sqrt{\pi r \rho}t^{3/2}}\exp\left(-\frac{(r-\rho)^2}{4t}\right)[1 + R_1\left(\zeta \,|\, 0\right)]$$

$$- \frac{r+\rho}{4\sqrt{\pi r \rho}t^{3/2}}\exp\left(-\frac{(r+\rho)^2 - i\pi/2}{4t}\right)[1 + R_1\left(z \,|\, 0\right)]$$

$$+ \frac{\sqrt{\rho}}{4t^{3/2}\sqrt{\pi r}}\exp\left(-\frac{(r-\rho)^2 + i\pi}{4t}\right)\frac{d}{d\zeta}R_1\left(\zeta \,|\, 0\right)$$

$$+ \frac{\sqrt{\rho}}{4t^{3/2}\sqrt{\pi r}}\exp\left(-\frac{(r+\rho)^2 - i\pi/2}{4t}\right)\frac{d}{dz}R_1\left(z \,|\, 0\right)$$

$$- \frac{1}{4r^{3/2}\sqrt{\pi \rho t}}\exp\left(-\frac{(r-\rho)^2}{4t}\right)\left[1 + R_1\left(ze^{i\pi} \,|\, 0\right)\right]$$

$$- \frac{1}{4r^{3/2}\sqrt{\pi \rho t}}\exp\left(-\frac{(r+\rho)^2 - i\pi/2}{4t}\right)[1 + R_1\left(\zeta \,|\, 0\right)] \tag{A2.231}$$

where

$$z = \frac{r\rho}{2t}, \qquad \zeta = \frac{r\rho}{2t}e^{i\pi} \tag{A2.232}$$

---

[10]It is assumed that the variables are scaled in such a way that the coefficient of thermal diffusivity is 1.

Recall that $R(z \mid n)$ and $R(\zeta \mid n)$ are uniformly and absolutely convergent integrals of holomorphic functions for every $n \geq -1/2$. Hence, for all $n \geq 0$ and $m \geq 0$, there exists $C_{nm} = \text{const} > 0$ such that

$$\left| \frac{d^m}{dz\,m} R(z \mid n) \right|, \quad \left| \frac{d^m}{dz\,m} R(z \mid n) \right| < C_{nm} \qquad \text{(A2.233)}$$

The results of this section are easily used to prove local existence, uniqueness, and stability theorems for the solution of the Stefan problem with cylindrical symmetry in the region

$$D: \{ r > R = \text{const} > 0 \} \quad \forall R = \text{const} > 0 \qquad \text{(A2.234)}$$

employing the method developed in Chapter 17 to solve the Stefan problem.

Problems

P.A2.1. Derive asymptotic representations of $J_\nu(z)$ and $Y_\nu(z)$ in the complex plane.

P.A2.2. Can Neumann's addition theorem be used to obtain an expansion of $I_0(r_{pq})$? If so, derive the expansion; if not, why not?

P.A2.3. Consider the following boundary-value problem: Find a function $u(r)$ satisfying the conditions

$$\frac{d^2}{dr^2}u + \frac{1}{r}\frac{d}{dr}u - u = 0, \quad 0 < r < \infty \qquad (\alpha)$$

$$\lim_{r \downarrow 0} \frac{u(r)}{\ln(1/r)} = 1, \qquad \lim_{r \uparrow \infty} u(r) = 0 \qquad (\beta)$$

Give a physical interpretation of the boundary conditions (β).   *Hint:* The equation is the stationary equation of diffusion of a radioactive gas.

Appendix 3: Fourier's method and Sturm–Liouville equations

1. Separation of variables and eigenvalue problem

The method of separation of variables was used in Chapter 18 in the solution of a number of boundary-value problems relating to the wave, heat-conduction, and Laplace equations, with reference to the general theory of the Sturm–Liouville equations. We now present a short description of the basic facts of this theory, a systematic study of which is provided in the next sections.

We shall restrict attention to 2-dimensional problems. Let the equation

$$Lu \stackrel{\text{def}}{=} \frac{\partial}{\partial x}\left(k(x)\frac{\partial u}{\partial x}\right) - q(x)u = -r(x)\mathfrak{L}u \qquad (A3.1)$$

be given in the region

$$G = \{x, y : 0 < x < a; \ 0 < y < b \leq \infty\} \qquad (A3.2)$$

where $k(x)$ is a piecewise continuous, piecewise smooth function, $q(x)$ and $r(x)$ are piecewise continuous, and also

$$k(x), \quad r(x) > 0, \quad q(x) > 0 \quad \forall x \in (0, a], \quad k(0), \quad r(0) > 0 \qquad (A3.3)$$

To fix ideas and simplify the exposition, let us assume that k, dk/dx, q, and r have only one point of discontinuity, $x = x_0$. If

$$k(0) = 0, \quad r(0) = 0 \qquad (A3.4)$$

then we assume

$$k(x) = x\tilde{k}(x), \quad r(x) = x\tilde{r}(x), \quad q(x) = \frac{\tilde{q}(x)}{x} \qquad (A3.5)$$

$$\tilde{k}(0) > 0, \quad \tilde{r}(0) > 0, \quad \tilde{q}(0) \geq 0$$

The case $K(0) > 0$, $r(0) > 0$ is called *regular*, whereas the case $K(0) = 0$, $r(0) = 0$ is called *singular*.

Concerning the operator $\mathfrak{L}$ we assume that

$$\mathfrak{L}u = -\frac{\partial^2 u}{\partial y^2} \vee \left(-\frac{\partial u}{\partial y}\right) \vee \frac{\partial^2 u}{\partial y^2} \qquad (A3.6)$$

so that equation (A3.1) is assumed to be of the hyperbolic, parabolic, or elliptic type. In the case of a hyperbolic or parabolic equation we assume that $b = \infty$, so that in these cases the region G is a half-strip.

For a regular equation (A3.1) we consider the problems

$$\alpha \frac{\partial u}{\partial x} - \beta u = 0 \quad \text{at } x = 0, \quad \gamma \frac{\partial u}{\partial x} + \delta u = 0 \quad \text{at } x = a$$

$$\alpha^2 + \beta^2 = \text{const} > 0, \quad \gamma^2 + \delta^2 > 0, \quad \alpha, \beta, \gamma, \delta \geq 0 \tag{A3.7}$$

or, in the singular case,

$$u \text{ is bounded in the neighborhood of } x = 0$$

$$\gamma \frac{\partial u}{\partial x} + \delta u = 0 \quad \text{at } x = a \tag{A3.8}$$

Furthermore:

(1) in the *elliptic* case,

$$c_1 \frac{\partial u}{\partial y} - d_1 u = f_1(x) \quad \text{if } y = 0, \quad c_2 \frac{\partial u}{\partial y} + d_2 u = f_2(x) \quad \text{if } y = b$$

$$c_i^2 + d_i^2 = \text{const} > 0, \quad c_i, \quad d_i \geq 0, \quad i = 1, 2 \tag{A3.9}$$

(2) in the *hyperbolic* case, u remains bounded as $y \uparrow \infty$ and

$$u = f_1(x), \quad \frac{\partial u}{\partial y} = f_2(x) \quad \text{if } y = 0 \tag{A3.10}$$

(3) in the *parabolic* case, u remains bounded as $y \uparrow \infty$ and

$$u = f(x) \quad \text{if } y = 0 \tag{A3.11}$$

At the line of discontinuity $x = x_0$ of $k(x)$, $q(x)$, and $r(x)$, the functions u and $k(\partial u/\partial x)$ are required to be continuous:

$$[u] = \left[k \frac{\partial u}{\partial x} \right] = 0 \quad \text{if } x = x_0 \quad \forall y \in [0, b] \tag{A3.12}$$

Thus all the problems considered in Chapter 18 belong to this "admissible" type.

Recall that in the method of separation of variables one looks for particular solutions of the equations, of the form

$$u(x, y) = X(x) \cdot Y(y) \tag{A3.13}$$

that satisfy the homogeneous conditions (A3.7) or, in the singular case, conditions (A3.8).

Inserting (A3.13) into (A3.1), one finds

$$\frac{1}{r(x)X(x)} \left(\frac{d}{dx} \left[k(x) \frac{dX}{dx} \right] - q(x)X(x) \right) = \frac{\mathcal{L}Y(y)}{Y} = -\lambda \quad \forall x \neq x_0 \tag{A3.14}$$

Since the left-hand side of this identity depends only on x, and the right-hand side only on y, it follows that $\lambda = \text{const}$, which yields

$$L^* X \stackrel{\text{def}}{=} \frac{d}{dx} \left[k(x) \frac{dX}{dx} \right] - q(x)X(x) + \lambda r(x)X(x) = 0 \quad \text{if } x \neq x_0 \tag{A3.15}$$

$$[X] = \left[k(x) \frac{dX}{dx} \right] = 0 \quad \text{at } x = x_0 \tag{A3.16}$$

and

$$\alpha \frac{dX}{dx} - \beta X = 0 \quad \text{if } x = 0, \quad \gamma \frac{dX}{dx} + \delta X = 0 \quad \text{if } x = a$$

$$\alpha^2 + \beta^2 = \text{const} > 0, \quad \gamma^2 + \delta^2 > 0, \quad \alpha, \beta, \gamma, \delta \geq 0 \tag{A3.17}$$

in the regular case, or

$$X \text{ is bounded in the neighborhood of } x = 0$$

$$\gamma \frac{dX}{dx} + \delta X = 0 \quad \text{at } x = a \tag{A3.18}$$

in the singular case.

To fix ideas, we assume that

$$\alpha = \beta = 1, \quad \gamma = 0, \quad \delta = 1 \tag{A3.19}$$

$Y(y)$ in turn must be determined from the equation

$$\mathcal{L}Y + \lambda Y = 0 \tag{A3.20}$$

From now on, Y will denote the general solution of this equation.

Definition A3.1.1. Values of λ for which the boundary-value problem (A3.15)–(A3.18) has nontrivial solutions (i.e., not identically equal to zero) are called *eigenvalues*.

Definition A3.1.2. The set of all eigenvalues is called the *spectrum* of the Sturm–Liouville operator L^*.

Definition A3.1.3. Any nontrivial solution of the problem (A3.15)–(A3.18), corresponding (belonging) to the eigenvalue λ is called an *eigenfunction*.

The following theorems are proved in the theory of Sturm–Liouville equations.

Theorem A3.1.1. *The spectrum* $\Lambda = \{\lambda_n\}$ *of the Sturm–Liouville operator satisfying conditions* (A3.16)–(A3.18) *is discrete and denumerable. All the eigenvalues are positive:*

$$0 < \lambda_1 < \lambda_2 < \cdots < \lambda_n < \cdots, \quad \lim_{n \uparrow \infty} \lambda_n = \infty \tag{A3.21}$$

Theorem A3.1.2. *Eigenfunctions* X_n *and* X_m *belonging to different eigenvalues* λ_n *and* λ_m *are orthogonal with weight* r:

$$\int_0^a r(x) X_n(x) X_m(x) dx = 0 \tag{A3.22}$$

Theorem A3.1.3. *To every eigenvalue* λ_n *there corresponds only one eigenfunction* X_n; *that is, any two eigenfunctions belonging to the same eigenvalue are linearly dependent.*

Theorem A3.1.4. *Let $v(x)$ be a function satisfying the following conditions in the interval $(0, a)$:*

(1) $v(x)$ *is continuous in* $[0, a]$;
(2) $v(x)$ *has at most finitely many extrema in* $[0, a]$;
(3) $v(x)$ *satisfies the boundary conditions* (A3.17) *in the regular case or* (A3.18) *in the singular case.*

Then $v(x)$ can be expanded in terms of the eigenfunctions into a series

$$\sum_{n=1}^{\infty} a_n X_n \tag{A3.23}$$

that converges uniformly over every segment in the interval $(0, a)$ and whose sum satisfies the boundary conditions (A3.17) *or* (A3.18)*, respectively.*

Corollary. *The coefficients of the general solution $Y_n(y)$ of equation* (A3.19) *(with $\lambda = \lambda_n$) may be chosen so that the sum of the series*

$$u(x, y) = \sum_{n=1}^{\infty} X_n(x) Y_n(y) \tag{A3.24}$$

satisfies conditions (A3.9)*,* (A3.10)*, or* (A3.11)*, depending on the type of problem under consideration. This means that if the series* (A3.24) *can be differentiated term by term the necessary number of times, then its sum represents the solution of problem* (A3.1)–(A3.12) *not only formally but truly.*

2. Elementary theory of regular Sturm–Liouville equations

We now present the proofs of Theorems A3.1.1–4 in the regular case. Let $\mathfrak{M}$ and $\mathfrak{M}^*$ be sets of functions satisfying the following conditions.

Definition A3.2.1. $u(x) \in \mathfrak{M}$ if

(1) $u(x)$ is defined and continuous in $[0, a]$;
(2) du/dx and d^2u/dx^2 are piecewise continuous in $[0, a]$;
(3) $k(x)du/dx$ is continuous in $[0, a]$;
(4) $du/dx - u = 0$ if $x = 0$ and $u = 0$ if $x = a$.

Definition A3.2.2. $u(x) \in \mathfrak{M}^*$ if

(1) $u(x)$ is defined and continuous in $[0, a]$;
(2) $du(x)/dx$ is piecewise continuous in $[0, a]$;
(3) $k(x)du/dx$ is continuous in $[0, a]$.

Obviously,

$$\mathfrak{M} \subset \mathfrak{M}^* \tag{A3.25}$$

Definition A3.2.3. Lu is the *linear operator* defined in $\mathfrak{M}$ by

$$Lu = \frac{d}{dx}\left(k(x)\frac{du}{dx}\right) - q(x)u(x) \quad \forall u(x) \in \mathfrak{M} \tag{A3.26}$$

Definition A3.2.4. $F(u, v)$ is the *bilinear operator* defined in $\mathfrak{M}^*$ by

$$F(u, v) = k(0)u(0)v(0) + \int_0^a \left(k(x)\frac{du}{dx}\frac{dv}{dx} + q(x)u(x)v(x) \right) dx \qquad (A3.27)$$

$F(u)$ is the *quadratic operator*

$$F(u) = F(u, u) \qquad (A3.28)$$

Definition A3.2.5. The *scalar product* in $\mathfrak{M}^*$ with weight r is

$$(u \cdot v) = \int_0^a r(x)u(x)v(x)dx \quad \forall u, v \in \mathfrak{M}^* \qquad (A3.29)$$

The *norm* in $\mathfrak{M}^*$ is defined as

$$\|u\| = (u \cdot u)^{1/2} \qquad (A3.30)$$

We can now prove the theorems formulated in Section 1.

Theorem A3.2.1. *If λ and μ are different eigenvalues of the operator L, and if u and v are eigenfunctions belonging to λ and μ, respectively, then $(u \cdot v) = 0$.*

PROOF. We have

$$vL(u) - uL(v) = \frac{d}{dx}\left(k(x)\left[v(x)\frac{du}{dx} - u(x)\frac{dv}{dx} \right] \right)$$

$$= (\mu - \lambda)r(x)u(x)v(x) \qquad (A3.31)$$

Integrating over $[0, a]$ and taking into account the continuity of w and $k(dw/dx)$ for all $w \in \mathfrak{M}$, we obtain

$$(\mu - \lambda)(u \cdot v) = 0 \Rightarrow (u \cdot v) = 0 \qquad (A3.32)$$
$$\textbf{Q.E.D.}$$

Theorem A3.2.2. *All the eigenvalues of $L(u)$ are real.*

PROOF. Suppose that there exists an eigenvalue

$$\lambda = \alpha + i\beta \qquad (A3.33)$$

with $\beta \neq 0$. Since the functions k, q, and r in the definition of the operator L are real, any eigenfunction u belonging to λ must have nonzero imaginary part:

$$u = v + iw, \quad w \not\equiv 0 \qquad (A3.34)$$

Hence the complex conjugates $\bar{\lambda}$ and $\bar{u}$,

$$\bar{\lambda} = \alpha - i\beta, \qquad \bar{u} = v - iw \qquad (A3.35)$$

are also an eigenvalue and an eigenfunction, respectively. Hence, by Theorem A3.2.1,

$$|u|^2 = (u \cdot \bar{u}) \overset{\text{def}}{=} \int_0^a r(x)|u(x)|^2 dx = 0 \qquad (A3.36)$$

Since by assumption $r(x) > 0$ for all $x \in (0, b)$ and u is continuous, it follows from (A3.36) that

$$u(x) \equiv 0 \tag{A3.37}$$

so that (A3.34) is false. **Q.E.D.**

Theorem A3.2.3. *Let λ be an eigenvalue and u an eigenfunction belonging to λ. Then*

$$\lambda(u \cdot u) = F(u) \tag{A3.38}$$

PROOF. Along with the scalar product with weight r, consider the scalar product $(u \cdot v)_1$ with the weight 1:

$$(u \cdot v)_1 = \int_0^a u(x)v(x)dx \tag{A3.39}$$

By the definitions of L, M and M^* we have

$$(u \cdot Lu)_1 = \int_0^a u(x) \left[\frac{d}{dx}\left(k(x)\frac{du}{dx} \right) - q(x)u(x) \right] dx$$

$$= ku\frac{du}{dx}\bigg|_{x=0}^a - \int_0^a \left[k(x)\left(\frac{du}{dx} \right)^2 + q(x)u(x)^2 \right] dx \tag{A3.40}$$

By (A3.17), (A3.19) and the definitions (A3.27) and (A3.28), this yields

$$(u \cdot Lu)_1 = -F(u) \tag{A3.41}$$

Since by assumption λ is an eigenvalue, we have

$$Lu = -\lambda r(x)u(x) \tag{A3.42}$$

which implies

$$F(u) = \lambda(u \cdot u) \tag{A3.43}$$

Q.E.D.

Corollary. *All the eigenvalues of L are positive.*

Indeed, by definition $(u \cdot u) > 0$, $F(u) > 0$, so that (A3.43) implies

$$\lambda > 0 \tag{A3.44}$$

Q.E.D.

In what follows we assume that all the eigenfunctions are normalized (i.e., have norm 1), so that

$$\|u\|^2 = \int_0^a r(x)u^2(x)dx = 1 \tag{A3.45}$$

The following theorem is fundamental.

Theorem A3.2.4. *The minimum of the functional $F(u)$ on the set $\mathfrak{M}^*$ is attained at an element $u_0(x) \in \mathfrak{M}$; it is equal to the smallest eigenvalue of the operator L.*

PROOF. Since we are assuming that all functions in $\mathfrak{M}^*$ are normalized, we must find a conditional extremum of the functional F. Using the Lagrange method of undetermined multipliers, we see that u_0 is an unconditional minimum of the functional

$$\Phi(u) \equiv F(u) - \lambda(u \cdot u)$$

$$= k(0)u^2(0) + \int_0^a \left[k(x)\left(\frac{du}{dx}\right)^2 + [q(x) - \lambda r(x)]u(x)^2 \right] dx \tag{A3.46}$$

Hence $u_0(x)$ is an integral of the Euler equation

$$\left(\frac{\partial}{\partial u} - \frac{d}{dx}\frac{\partial}{\partial u_1} \right) \left[k(x)\left(\frac{du}{dx}\right)^2 + [q(x) - \lambda r(x)]u(x)^2 \right]$$

$$\equiv -2(Lu + \lambda r u) = 0, \quad \left(u_1 \stackrel{\text{def}}{=} \frac{du(x)}{dx} \right) \tag{A3.47}$$

satisfying the "natural" boundary condition[1]

$$\frac{d}{dx}u(0) - u(0) = 0 \tag{A3.48}$$

---

[1] We reproduce the derivation of (A3.47) and (A3.48). Note, first of all, that for all $\alpha, \beta, \Phi(\alpha u + \beta v) = \alpha^2 \Phi(u) + \alpha\beta\Phi(u,v) + \beta^2\Phi(\beta)$, where the functionals $\Phi(u)$ and $\Phi(u,v)$ are defined by analogy with $F(u)$ and $F(u,v)$, respectively. Now let u_0 be an extremal and η an admissible function. The minimum of the function $\Phi(u_0 + t\eta), \eta$ being arbitrary, is attained at a point where $d\Phi/dt = 0$. Thus

$$\frac{d\Phi(0)}{dt} = 2\Phi(0) = k(0)u_0(0)\eta(0) + \int_0^\pi \left[\frac{kdu}{dx}\frac{d\eta}{dx} + (q - \lambda r u) \cdot \eta \right] dx = 0$$

By the Du Bois–Reymond lemma, the extremal $u_0(x)$ has piecewise continuous second derivatives. Integration by parts yields

$$(u_0 \cdot \eta) = k(0)\left(u_0(0) - \frac{d}{dx}u_0(0) \right)\eta(0) - \int_0^a [Lu_0 + \lambda r u_0]\eta dx$$

and λ is the smallest eigenvalue of the operator L.   **Q.E.D.**

Theorem A3.2.5. *Let*

$$\lambda_1 < \lambda_2 < \cdots < \lambda_n \tag{A3.49}$$

be the first n eigenvalues and

$$u_1, u_2, \ldots, u_n \tag{A3.50}$$

the corresponding eigenfunctions. Then the conditional extremum of the functional $F(u)$ on the class $\mathfrak{M}^*$ of normalized functions u orthogonal to $u_1, u_2, \ldots, u_n$ is equal to the $(n+1)$th eigenvalue λ_{n+1}, and is achieved at the eigenfunction u_{n+1} belonging to λ_{n+1}.

PROOF. By the Lagrange method of undetermined multipliers we must find an unconditional extremum of the functional

$$\Psi(u) = F(u) - \lambda(u, u) - 2\sum_{k=0}^{n} \mu_k(u \cdot u_k) \tag{A3.51}$$

The Euler equation is

$$Lu + \lambda r u + \sum_{k=0}^{n} \mu_k r u_k = 0 \tag{A3.52}$$

and its integral must satisfy the natural boundary conditions

$$\frac{d}{dx}u(0) - u(0) = 0 \tag{A3.53}$$

The Lagrange multipliers are determined by the condition that u is normalized and is orthogonal to all the u_k, $k = 1, 2, \ldots, n$. Thus,

$$(u \cdot u) = 1 \tag{A3.54}$$

and

$$(u \cdot u_k) = 0 \tag{A3.55}$$

Since all the u_k are normalized and orthogonal to one another, we see from (A3.52) that

$$(u_k \cdot Lu)_1 + \mu_k = 0 \quad \forall k = 1, 2, \ldots, n \tag{A3.56}$$

On the other hand,

$$Lu_m + \lambda_m r u_m = 0 \quad \forall m = 1, 2, \ldots, n \tag{A3.57}$$

which implies

$$(u \cdot Lu_m)_1 = 0, \quad m = 1, 2, \ldots, n \tag{A3.58}$$

(We have used the fact that $\eta(a) = u(a) = 0$.) Since $\eta(x)$ is arbitrary, we conclude that $Lu_0 + \lambda r u_0 = 0$ and $du_0(0)/dx = u_0(0)$.   **Q.E.D.**

In addition, (A3.56) and (A3.58) yield

$$\mu_k = (u \cdot Lu_k)_1 - (u_k \cdot Lu)_1$$

$$= \int_0^a \frac{d}{dx}\left[k\left(u\frac{d}{dx}u_k - u_k\frac{d}{dx}u \right)\right] dx = k\left(u\frac{d}{dx}u_k - u_k\frac{d}{dx}u \right)\Big|_{x=0}^{a} \quad \text{(A3.59)}$$

Since $u_k \in \mathfrak{M}^*$ and satisfy the natural boundary conditions, it follows from (A3.59) that

$$\mu_k = 0, \quad k = 1, 2, \ldots, n \quad \text{(A3.60)}$$

so that (A3.52) yields

$$Lu + \lambda ru = 0 \quad \text{(A3.61)}$$
Q.E.D.

Remark. Obviously, $\lambda_{\pi+1} > \lambda_n$ is the eigenvalue nearest to λ_n.

Theorem A3.2.6. *Let* $\{\lambda_n\}\,|_{n=1}^{\infty}$ *be the set of eigenvalues. Then*

$$\lim_{n\uparrow\infty} \lambda_n = \infty \quad \text{(A3.62)}$$

PROOF. Let λ_n be the nth eigenvalue of L and u_n an eigenfunction belonging to it, so that

$$(u_n \cdot u_n) = \int_0^a ru_n^2 dx = 1 \quad \text{(A3.63)}$$

and

$$F(u_n) = k(0)u_n^2(0) + \int_0^a \left[k(x)\left(\frac{du_n}{dx}\right)^2 + q(x)u_n^2(x) \right] dx$$

$$= \lambda_n(u_n \cdot u_n) \quad \text{(A3.64)}$$

Suppose that $\{\lambda_n\}$ is bounded. Then

$$\lim_{m\uparrow\infty} \lambda_m \overset{\text{def}}{=} \mu > \lambda_n \quad \forall n \geq 1 \quad \text{(A3.65)}$$

By Theorem A3.2.3 and (A3.65),

$$F(u_n) < \mu \quad \text{(A3.66)}$$

Since k and q are positive, it follows from (A3.66) that there exists $M = \text{const} > 0$ such that

$$\int_0^a \left(\frac{d}{dx}u_n\right)^2 dx < M \quad \text{(A3.67)}$$

which means that $\{u_k\}$ is uniformly bounded and equicontinuous. Indeed, by virtue of the Schwarz inequality, for all $x, y \in [0, a]$ we have

$$
\begin{aligned}
|u_n(x) - u_n(y)| &= \left| \int_y^x \left(\frac{d}{dx} u_n \right) dx \right| \\
&\leq \left| \int_y^x \left(\frac{d}{dx} u_n \right)^2 dx \int_y^x dx \right|^{1/2} \\
&< M|x - y|^{1/2} \longrightarrow
\end{aligned}
\tag{A3.68}
$$

$$
|u_n(x)| \equiv |u_n(x) - u_n(a)| < Ma^{1/2}
$$

This means that, by the Arzela–Ascoli theorem, $\{u_n\}$ contains a uniformly convergent subsequence. To simplify the notation, let us assume that this subsequence is $\{u_n\}$. Thus for any $\varepsilon > 0$ there exists $N = \text{const} > 0$ such that for all $n, m > N$,

$$
|u_n(x) - u_m(x)| < \varepsilon
\tag{A3.69}
$$

Hence

$$
\|u_n(x) - u_m(x)\|^2 = \int_0^a r(x)|u_n(x) - u_m(x)|^2 dx < \varepsilon^2 M^*
\tag{A3.70}
$$

where $M^* \geq ar(x)$ for all $x \in [0, a]$. Hence $\varepsilon > 0$ may be chosen so small that

$$
\|u_n(x) - u_m(x)\| < 1
\tag{A3.71}
$$

On the other hand, by the orthogonality of u_n and u_m,

$$
\begin{aligned}
\|u_n(x) - u_m(x)\|^2 &= \int_0^a r(x)|u_n(x) - u_m(x)|^2 dx \\
&= (u_n \cdot u_n) - 2(u_n \cdot u_m) + (u_m \cdot u_m) = 2
\end{aligned}
\tag{A3.72}
$$

which contradicts (A3.71). Thus the assumption that $\{\lambda_n\}$ is bounded leads to a contradiction. **Q.E.D.**

The following theorem is useful in studying how variation of the coefficients of the operator L affects the distribution of the eigenvalues.

Theorem A3.2.7 (Courant). *Let* $\varphi_1(x), \varphi_2(x), \ldots, \varphi_{n-1}(x)$ *be arbitrary functions belonging to* $\mathfrak{M}^*$ *and*

$$
\lambda(\varphi_1, \varphi_2, \ldots, \varphi_{n-1}) = \min F(u)
\tag{A3.73}
$$

where the minimum is over all functions u *such that*

$$
u \in \mathfrak{M}^*, \quad (u \cdot \varphi_k) = 0, \quad k = 1, 2, \ldots, n-1, \quad \|u\| = 1
\tag{A3.74}
$$

Let

$$
\bar{\lambda} = \sup_{\varphi_k \in \mathfrak{M}^*, k=1,2,\ldots,n-1} \lambda(\varphi_1, \varphi_2, \ldots, \varphi_{n-1})
\tag{A3.75}
$$

Then the nth eigenvalue of the operator L is equal to $\bar{\lambda}$:

$$\lambda_n = \bar{\lambda} \qquad (A3.76)$$

PROOF. Let $u, v \in \mathfrak{M}$. Then, by Definitions A3.2.1–4,

$$F(u,v) = -\int_0^a vLu\,dx \qquad (A3.77)$$

Therefore, if u_i and u_j are eigenfunctions belonging to the eigenvalues λ_i and λ_j, $i \neq j$, then

$$F(u_i, u_j) = 0, \quad i \neq j, \ i,j = 1,2,\dots,n \qquad (A3.78)$$

Now let

$$u = \sum_{i=1}^{n} c_i u_i \qquad (A3.79)$$

be a linear combination of the first n eigenfunctions with undetermined coefficients. The identity

$$F(\alpha u + \beta v) = \alpha^2 F(u) + 2\alpha\beta F(u,v) + \beta^2 F(v) \qquad (A3.80)$$

and (A3.79) imply

$$F(u) = \sum_{i=1}^{n} c_i^2 \cdot \lambda_i \qquad (A3.81)$$

since $F(u_i) = \lambda_i(u_i \cdot u_i) = \lambda_i$.

Now let $\varphi_1(x), \dots, \varphi_{n-1}$ be an arbitrary sequence of functions belonging to $\mathfrak{M}^*$. Let us look for a function u satisfying the conditions

$$(u \cdot u) = 1, \quad (u \cdot \varphi_i) = 0, \quad i = 1,2,\dots,n-1 \qquad (A3.82)$$

By the orthogonality of $\{u_k\}$,

$$\sum_{i=1}^{n} c_i(u_i \cdot \varphi_k) = 0 \quad \forall k = 1,2,\dots,n-1 \qquad (A3.83)$$

$$\sum_{i=1}^{n} c_i^2 = 1 \qquad (A3.84)$$

System (A3.83) is a system of $n-1$ linear homogeneous equations with n unknowns. Hence it is always solvable. Equation (A3.84) is used only to normalize u, ensuring that it is an admissible function. Since $\lambda_k < \lambda_{k+1}$ for all k, it follows from (A3.81) and (A3.84) that

$$F(u) < \lambda_n \Rightarrow \lambda(\varphi_1, \varphi_2, \dots, \varphi_{n-1}) \leq \lambda_n \Rightarrow \bar{\lambda} \leq \lambda_n \qquad (A3.85)$$

On the other hand, by Theorem A3.2.5,

$$\lambda_n = \lambda(u_1, u_2, \dots, u_{n-1}) \qquad (A3.86)$$

whence

$$\bar{\lambda} \geq \lambda_n \qquad (A3.87)$$

Hence (A3.85) and (A3.87) yield

$$\lambda_n = \bar{\lambda} \tag{A3.88}$$

Q.E.D.

Remark. The fact that the functional $F(u)$ has a minimum, adopted in the proofs of Theorems A3.2.4, A3.2.5, and A3.2.7, is proved by the direct methods of variational calculus [11; 12].

Theorem A3.2.8. *Two eigenfunctions u and v belonging to the same eigenvalue λ cannot be linearly independent.*[2]

PROOF. Suppose that there are two linearly independent eigenfunctions u and v belonging to the same eigenvalue λ. Then every solution w of the equation

$$Lw = 0 \tag{A3.89}$$

is a linear combination of u and v:

$$w = \alpha u + \beta v \tag{A3.90}$$

Chose α and β so that

$$w(a) = 1, \quad \frac{d}{dx} w(a) = 0 \tag{A3.91}$$

This is possible since the Cauchy problem (A3.89), (A3.91) certainly has a solution. Thus

$$w(a) = \alpha u(a) + \beta v(a) = 1 \tag{A3.92}$$

which contradicts the equalities

$$u(a) = v(a) = 0 \tag{A3.93}$$

in the definition of eigenfunctions. Thus two eigenfunctions belonging to the same eigenvalue cannot be linearly independent. **Q.E.D.**

3. Expansion of functions in $\mathfrak{M}^*$ in series of eigenfunctions of regular Sturm–Liouville operator

A. Approximation of functions by orthogonal polynomials. Bessel inequality and Parseval equality

Given a family $\{u_m\}$ of orthogonal and normalized functions in $\mathfrak{M}^*$, let

$$f(x) \in L_2(0,a) \tag{A3.94}$$

be a square integrable function defined in $[0,a]$. Let us associate with f a polynomial in u_m with undetermined coefficients:

$$f(x) \sim \sum_{m=1}^{n} c_m u_m(x) \tag{A3.95}$$

---

[2] Theorem A3.2.8 is identical to Theorem A3.1.3.

Lemma A3.3.1. *Let* a_m, $m = 1, 2, \ldots, a_n$, *be the Fourier coefficients of* $f(x)$; *that is,*

$$a_m = (f \cdot u_m) \stackrel{\text{def}}{=} \int_0^a r(x) f(x) u_m(x) dx, \quad m = 1, 2, \ldots n \quad \text{(A3.96)}$$

Let $\mathcal{M}[J(c_1, c_2, \ldots, c_n)]$ *be the set of quadratic functionals*

$$J(c_1, \ldots, c_n) = \int_0^a r(x) \left(f(x) - \sum_{m=1}^n c_m u_m(x) \right)^2 dx \quad \text{(A3.97)}$$

Then

$$\min \mathcal{M}[J(c_1, \ldots, c_n)] = J(a_1, \ldots, a_n) \quad \text{(A3.98)}$$

PROOF. By the orthogonality of the eigenfunctions, we have

$$J(c_1, \ldots, c_n) = (f \cdot f) - 2 \sum_{m=1}^n c_m a_m + \sum_{m=1}^n c_m^2$$

$$= (f \cdot f) + \sum_{m=1}^n c_m^2 - 2 \sum_{m=1}^n c_m a_m + \sum_{m=1}^n a_m^2 - \sum_{m=1}^n a_m^2$$

$$= (f \cdot f) + \sum_{m=1}^n (c_m - a_m)^2 - \sum_{m=1}^n a_m^2 \quad \text{(A3.99)}$$

Hence

$$J(c_1, \ldots, c_n) = \min \mathcal{M} \quad \text{(A3.100)}$$

if

$$c_k = a_k, \quad k = 1, 2, \ldots, n \quad \text{(A3.101)}$$
Q.E.D.

Corollary (Bessel Inequality). *The series of squares of the Fourier coefficients of any function* $f \in \mathfrak{M}^*$ *is convergent:*

$$\sum_{m=1}^\infty a_m^2 \leq (f \cdot f) \quad \text{(A3.102)}$$

Definition A3.3.1. A normalized system $\{u_n(x)\}_{n=1}^\infty$ of functions orthogonal with weight $r(x)$ is called a *complete orthonormal system* if the sum of squares of the Fourier coefficients converges to the square of the norm of f, whatever the function $f \in L_2(0, a)$:

$$\sum_{m=1}^\infty a_m^2 = (f, f) \quad \forall f \in L_2(0, a) \quad \text{(A3.103)}$$

Remark A3.3.1. Equality (A3.103) is known as the *Parseval equality*; it means that the Fourier series of any $f \in L_2(0, a)$ converges in mean to f:

$$\lim_{n \uparrow \infty} \int_0^a \left(f(x) - \sum_{m=0}^n a_m u_m(x) \right)^2 dx = 0 \quad \text{(A3.104)}$$

The following theorem is of fundamental importance.

Theorem A3.3.1 (Completeness Theorem). *An orthonormal system $\{u_m(x)\}$ of eigenfunctions of the Sturm–Liouville operator L is complete in $L_2(0, a)$.*

PROOF. Let us first prove that the Parseval equality holds for every $f \in \mathfrak{M}$. Let

$$\varphi_n(x) = f(x) - \sum_{m=1}^{n} c_m u_m(x), \quad c_m = \int_0^a r f u_m \, dx \quad \forall m = 1, 2, \ldots \quad \text{(A3.105)}$$

We have to prove that

$$\lim_{n \uparrow \infty} \|\varphi_n(x)\| = 0 \tag{A3.106}$$

Since the system of eigenfunctions is orthonormal, we have

$$(\varphi_n \cdot u_m) = \begin{cases} 0 & m = 1, 2, \ldots, n \\ c_m & m = n+1, n+2, \ldots \end{cases} \tag{A3.107}$$

Since $\varphi_n \in \mathfrak{M} \subset \mathfrak{M}^*$, the function

$$\psi_n = \frac{\varphi_n}{\|\varphi_n\|} \tag{A3.108}$$

belongs to the class of functions over which the functional $F(u)$ is minimized in order to determine the $(n+1)$th eigenvalue λ_{n+1}. Hence

$$F(\psi_n) \geq \lambda_{n+1} \tag{A3.109}$$

which implies

$$\|\varphi_n\|^2 \leq F(\varphi_n)/\lambda_{n+1} \tag{A3.110}$$

Hence, if $F(\varphi_n)$ is uniformly bounded,

$$F(\varphi_n) < K = \text{const} > 0 \quad \forall n = 1, 2, \ldots \tag{A3.111}$$

then

$$\lim_{n \uparrow \infty} \|\varphi_n\|^2 = 0 \tag{A3.112}$$

since

$$\lim_{n \uparrow \infty} \lambda_n = \infty \tag{A3.113}$$

Thus our claim (A3.106) will be true if there exists K satisfying (A3.111). In order to prove this we note that by (A3.41),

$$F(\varphi_n) = -(\varphi_n \cdot L\varphi_n)_1 \tag{A3.114}$$

so that by (A3.105) and the equalities

$$Lu_m = -\lambda_m r \cdot u_m, \quad (u_k \cdot u_m) = \begin{cases} 0 & k \neq m \\ 1 & k = m \end{cases} \tag{A3.115}$$

$$c_m = (f \cdot u_m), \quad (f \cdot Lu_m)_1 = -\lambda_m(f \cdot u_m) = -\lambda_m c_m$$

we have

$$F(\varphi_n) = -\left(\left(f(x) - \sum_{m=1}^n c_m u_m(x)\right) \cdot \left(Lf(x) - \sum_{m=1}^n c_m L u_m(x)\right)\right)_1$$

$$= -(f \cdot Lf)_1 - \sum_{m=1}^n c_m(u_m \cdot Lf)_1 - \sum_{m=1}^n c_m(f \cdot L u_m)_1 - \sum_{m=1}^n \lambda_m c_m^2$$

$$\Rightarrow F(\varphi_n) = -(f \cdot Lf)_1 - \sum_{m=1}^n c_m(u_m \cdot Lf)_1 \qquad (A3.116)$$

Let

$$\Psi = \frac{Lf}{r}, \quad d_m = (\Psi \cdot u_m) \qquad (A3.117)$$

We have

$$\sum_{m=1}^n c_m(Lf \cdot u_m)_1 = \sum_{m-1}^n c_m d_m \qquad (A3.118)$$

Using the Cauchy inequality, we find

$$\left|\sum_{m-1}^n c_m d_m\right|^2 \le \sum_{m-1}^n c_m^2 \cdot \sum_{m-1}^n d_m^2 \qquad (A3.119)$$

By assumption $r \ge r_0 > 0$ and $Lf \in L_2(0, a)$, so that

$$\Psi \in L_2(0, a) \qquad (A3.120)$$

Hence the Bessel inequality is applicable to Ψ. This yields

$$0 \le F(\varphi_n) \le -(f \cdot Lf)_1 + \left(\sum_{m=1}^n c_m^2 \sum_{m=1}^n d_m^2\right)^{1/2}$$

$$\le (f \cdot Lf)_1 + [(f \cdot f)(\Psi \cdot \Psi)]^{1/2} \stackrel{\text{def}}{=} K \qquad (A3.121)$$

Since K is independent of n, our assertion is proved.

We now drop the assumption that $f \in \mathfrak{M}$ and consider an arbitrary function $f \in L_2(0, a)$. Since $\mathfrak{M}$ is everywhere dense in $L_2(0, a)$, the Parseval equality is valid for every function $f \in L_2(0, a)$. Indeed, let $\varepsilon > 0$ be arbitrarily small. Let $f \in L_2(0, a)$. Since $\mathfrak{M}$ is dense in $L_2(0, a)$, there exists $\varphi \in \mathfrak{M}$ such that

$$\|f - \varphi\|^2 = \int_0^a r(f - \varphi)^2 dx < \varepsilon \qquad (A3.122)$$

Let c_m and a_m be the Fourier coefficients of f and φ, respectively. Then, by Lemma A3.3.1,

$$\int_0^a r\left(f - \sum_{m=1}^n c_m u_m\right)^2 dx \le \int_0^a r\left(f - \sum_{m=1}^n a_m u_m\right)^2 dx$$

$$= \int_0^a r \left(f - \varphi + \varphi - \sum_{m=1}^n a_m u_m \right)^2 dx$$

$$= \left\| (f - \varphi) + \left(\varphi - \sum_{m=1}^n a_m u_m \right) \right\|^2$$

$$\leq 2 \left(\|f - \varphi\|^2 + \left\| \varphi - \sum_{m=1}^n a_m u_m \right\|^2 \right) \qquad (A3.123)$$

By the Parseval equality, which has been proved for $\varphi \in \mathfrak{M}$, there exists $N > 0$ so large that for all $n \geq N$,

$$\left\| \varphi - \sum_{m=1}^n a_m u_m \right\|^2 < \varepsilon \qquad (A3.124)$$

Together with (A3.122), this gives

$$\left\| f - \sum_{m=1}^n c_m u_m \right\|^2 < 4\varepsilon \qquad (A3.125)$$

Since $\varepsilon > 0$ is arbitrarily small, this proves that the Parseval equality is valid for all $f \in L_2(0, a)$. **Q.E.D.**

Theorem A3.3.2. *A system* $\{u_m\}$ *of eigenfunctions of the operator L is complete if and only if one cannot enlarge it by addition of a normalized function $v \in \mathfrak{M}$ orthogonal to all u_m.*

PROOF. Indeed, suppose there exists a function

$$v \in \mathfrak{M}, \quad \|v\| = 1 \qquad (A3.126)$$

such that

$$c_m = (v \cdot u_m) = 0 \quad \forall m = 1, 2, \ldots \qquad (A3.127)$$

Then

$$\int_0^a r \left(v - \sum_{m=1}^\infty c_m u_m \right)^2 dx = \int_0^a r v^2 dx = 1 \qquad (A3.128)$$

contrary to the Parseval equality.

Conversely, assume that the Parseval equality is valid for all $f \in \mathfrak{M}$, so that

$$\|v\|^2 - \sum_{m=1}^\infty c_m^2 = 1 - \sum_{m=1}^\infty c_m^2 = 0 \qquad (A3.129)$$

Hence there exists m such that

$$c_m^2 > 0 \qquad (A3.130)$$

which means that v cannot be orthogonal to all eigenfunctions. **Q.E.D.**

B. Rate of increase of the eigenvalues

The basis for the forthcoming estimates is provided by the following theorem.

Theorem A3.3.3 (Sturm Comparison Theorem). *Together with the operator L, functionals $F(u,v)$, $F(u)$, and the scalar product $(u \cdot v)$, consider an operator L^*, functionals $F^*(u,v)$, $F^*(u)$, and the scalar product $(u \cdot v)^*$ defined in the same way as $L, \ldots, (u \cdot v)$ but with k, q, r, λ replaced by k^*, q^*, r^*, λ^* (respectively), where k^*, q^*, and r^* belong to the same classes of functions as k, q, and r.[3] Under these assumptions, if*

$$0 < k_0 \le k^* < k, \quad 0 \le q_0 \le q^* \le q, \quad 0 < r_0 \le r \le r^* \tag{A3.131}$$

then

$$\lambda_n^* \le \lambda_n \quad \forall n = 1, 2, \ldots \tag{A3.132}$$

PROOF. Sturm's theorem is a corollary of Courant's Theorem A3.2.7. Indeed, by that theorem,

$$\lambda_n = \sup \lambda(\varphi_1, \varphi_2, \ldots, \varphi_{n-1}), \quad \varphi_k \in \mathfrak{M}^*, \quad k = 1, 2, \ldots, n-1 \tag{A3.133}$$

where

$$\lambda(\varphi_1, \varphi_2, \ldots, \varphi_{n-1}) = \min_{u \in \mathfrak{M}} F(u), \quad (u, u) = 1$$
$$(u, \varphi_k) = 0 \quad \forall k = 1, 2, \ldots, n-1 \tag{A3.134}$$

Similar definitions give $\lambda^*(\varphi_1, \varphi_2, \ldots, \varphi_{n-1})$ and λ_n^*.

---

[3]In particular, this means that if k, q, and r have a point of discontinuity x_0, then k^*, q^*, and r^* cannot have any point of discontinuity other than x_0.

Obviously, the minimum of the functional $F(u)$ over the set of normalized functions $u \in \mathfrak{M}$ coincide with the minimum of the functional

$$\tilde{F}(u) = \frac{F(u)}{(u \cdot u)} \tag{A3.135}$$

over the class of not necessarily normalized functions $u \in \mathfrak{M}$.

Take an arbitrary set

$$\varphi_1, \varphi_2, \dots, \varphi_{n-1}, \quad \varphi_k \in \mathfrak{M}, \quad k = 1, 2, \dots, n-1 \tag{A3.136}$$

and define

$$\psi_k = \frac{r}{r^*} \varphi_k, \quad k = 1, 2, \dots, n-1 \tag{A3.137}$$

Since $0 < r_0 \leq r \leq r^*$, the set of all admissible functions ψ coincides with the set of all admissible functions φ. We have

$$\lambda(\varphi_1, \varphi_2, \dots, \varphi_{n-1}) = \min \tilde{F}(u) \tag{A3.138}$$

where the minimum is over all u such that

$$(u \cdot \varphi_k) = 0 \quad \forall k = 1, 2, \dots, n-1 \tag{A3.139}$$

By (A3.137) and (A3.139),

$$(u \cdot \psi_k)^* = \int_0^a r^* u \cdot \left(\frac{\varphi_k r}{r^*} \right) dx = \int_0^a r u \varphi_k dx = 0 \quad \forall k = 1, 2, \dots, n-1 \tag{A3.140}$$

At the same time, it follows from (A3.131) that

$$F(u) \geq F^*(u), \quad (u \cdot u) \leq (u, u)^* \tag{A3.141}$$

which means that

$$\begin{aligned}
\lambda(\varphi_1, \varphi_2, \dots, \varphi_{n-1}) &= \min\{\tilde{F}(u), \ (u \cdot \varphi_k) = 0, \ k = 1, 2, \dots, n-1\} \\
&\geq \min\{\tilde{F}^*(u), \ (u \cdot \varphi_k) = 0, \ k = 1, 2, \dots, n-1)\} \\
&= \lambda^*(\varphi_1, \varphi_2, \dots, \varphi_{n-1})
\end{aligned} \tag{A3.142}$$

Since the set of all functions φ coincides with the set of functions ψ, this implies

$$\lambda_n = \sup \lambda(\varphi_1, \varphi_2, \dots, \varphi_{n-1}) \geq \sup \lambda^*(\varphi_1, \varphi_2, \dots, \varphi_{n-1}) = \lambda_n^* \tag{A3.143}$$

Q.E.D.

Theorem A3.3.4.

$$\lambda_n = O(n^2) \quad \text{as } n \uparrow \infty \tag{A3.144}$$

PROOF. Take

$$k^* = k_0, \quad q^* = 0$$
$$r^* = R = \text{const} > r(x) \quad \forall x \in (0, a) \tag{A3.145}$$

By the Sturm theorem,

$$\lambda_n^* \leq \lambda_n \quad \forall n \geq 1 \tag{A3.146}$$

It easy to see that

$$\lambda_n^* = O(n^2) \quad \text{for } n \uparrow \infty \qquad (A3.147)$$

Indeed, by (A3.145),

$$L^* u = k_0 \frac{d^2}{dx^2} u + \lambda^* R u = 0 \quad \forall x \in (0, a)$$

$$\frac{d}{dx} u(0) - u(0) = 0, \quad u(a) = 0 \qquad (A3.148)$$

Let

$$\mu^{*2} = \frac{\lambda^* R}{k_0} \qquad (A3.149)$$

The general solution of the problem is

$$u = \alpha \sin[\mu^*(a - x)] + \beta \cos[\mu^*(a - x)] \qquad (A3.150)$$

so that the boundary conditions (A3.148) imply $\beta = 0$ and

$$\mu^* = \tan(\mu^* \cdot a) \qquad (A3.151)$$

This equation has a denumerable set of roots,

$$\mu_n^* \asymp \frac{n\pi}{a} + \frac{\pi}{2a} \quad \text{as } n \uparrow \infty \qquad (A3.152)$$

Thus

$$\lambda_n^* \asymp O(n^2) \quad \text{as } n \uparrow \infty \qquad (A3.153)$$

Now take

$$k_1 = \text{const} \geq k(x), \quad q_1 = \text{const} \geq q(x),$$

$$\tilde{R} = \text{const} \leq r(x) \quad \forall x \in (0, a) \qquad (A3.154)$$

and define

$$\tilde{L} u = k_1 \frac{d^2}{dx^2} u + q_1 u + \tilde{\lambda} \tilde{R} u = 0$$

$$\frac{d}{dx} u(0) - u(0) = 0, \quad u(a) = 0 \qquad (A3.155)$$

Then, by the Sturm theorem, we find that

$$\tilde{\lambda}_n \geq \lambda_n \quad \forall n \geq 1 \qquad (A3.156)$$

At the same time, setting

$$\nu^2 = \frac{\tilde{\lambda} \tilde{R} + q_1}{k_1} \qquad (A3.157)$$

we again find that

$$\nu_n \asymp O(n) \quad \text{as } n \uparrow \infty \qquad (A3.158)$$

Hence (A3.146), (A3.153), (A3.156), and (A3.158) yield

$$\lambda_n = O(n^2) \quad \text{as } n \uparrow \infty \qquad (A3.159)$$

Q.E.D.

C. Rate of decrease of Fourier coefficients of functions $f \in \mathfrak{M}$

Theorem A3.3.5. *Let $f \in \mathfrak{M}$ and let c_m be its Fourier coefficients,*

$$c_m = (f \cdot u_m) \quad \forall m = 1, 2, \ldots \tag{A3.160}$$

with respect to the eigenfunctions of the Sturm–Liouville operator L. Then[4]

$$|c_m| = o(m^{-5/2}) \quad \text{for } m \uparrow \infty \tag{A3.161}$$

PROOF. Recall that for all $f \in \mathfrak{M}$ and $u_m \in \mathfrak{M}$,

$$\int_0^a (fLu_m - u_m Lf)dx = 0 \tag{A3.162}$$

Let

$$\varphi = Lf, \quad \psi = \varphi/r \tag{A3.163}$$

Since by assumption $r \geq r_0 > 0$ and $\varphi \in L_2(0, a)$,

$$\psi \in L_2(0, a) \tag{A3.164}$$

so that the Bessel inequality is applicable. Hence

$$a_m = (\psi \cdot u_m) \Rightarrow \sum_{m=1}^{\infty} a_m^2 \leq (\psi \cdot \psi) \tag{A3.165}$$

Comparison with the divergent harmonic series

$$\sum_{m=1}^{\infty} \frac{1}{m} \tag{A3.166}$$

shows that

$$|a_m| = o(1/\sqrt{m}) \tag{A3.167}$$

It follows from (A3.162), the fact that $Lu_m = -\lambda_m r u_m$, and (A3.167) that

$$\int_0^a fLu_m dx = -\lambda_m c_m = a_m \tag{A3.168}$$

Hence (A3.167) and (A3.159) yield

$$c_m = o(m^{-5/2}) \quad \text{as } m \uparrow \infty \tag{A3.169}$$

Q.E.D.

Remark A3.3.2. Assume that

$$\psi = \frac{1}{r} Lf \in \mathfrak{M} \tag{A3.170}$$

Then, by (A3.169),

$$a_m = (\psi \cdot u_m) = o(m^{-5/2}) \quad \text{for } m \uparrow \infty \tag{A3.171}$$

---

[4]Recall that the symbols O and o are defined in Chapter 21, Section 2.

Hence, by (A3.168) and (A3.161),

$$c_m = (f \cdot u_m) = o(m^{-9/2}) \tag{A3.172}$$

D. Asymptotic order of eigenfunctions and their derivatives

Assume for the sake of simplicity that k and r are twice continuously differentiable functions, and that q is continuous in $[0, a]$.

Theorem A3.3.6. *Let* λ_n *and* u_n, $n = 1, 2, \ldots$, *be the eigenvalues and corresponding eigenfunctions of the Sturm–Liouville problem*

$$Lu \stackrel{\text{def}}{=} \frac{d}{dx}\left(k\frac{du}{dx}\right) + qu = -\lambda ru, \quad \frac{d}{dx}u(0) - u(0) = 0, \quad u(a) = 0 \tag{A3.173}$$

Then

$$u_n(x) = \sqrt{2/a}\sin[\sqrt{\lambda_n}(x-a)] + o(\lambda_n^{-1/2}) \tag{A3.174}$$

$$\frac{d}{dx}u_n(x) = \sqrt{2\lambda_n/a}\cos[\sqrt{\lambda_n}(x-a)] + o(1) \tag{A3.175}$$

$$\frac{d^2}{dx^2}u_n(x) = \sqrt{2/a}\lambda_n\sin[\sqrt{\lambda_n}(x-a)] + o(\lambda_n^{1/2}) \tag{A3.176}$$

PROOF. First let

$$k = 1, \quad r = 1, \quad \|u\| = 1, \quad u(a) = 0 \tag{A3.177}$$

Then

$$\frac{d}{dx}u(a) \neq 0 \tag{A3.178}$$

This means that u is proportional to the solution of the Cauchy problem

$$\frac{d^2}{dx^2}v + \lambda_n v = -q(x)v(x), \quad v(a) = 0, \quad \frac{d}{dx}v(a) = \sqrt{\lambda_n} \tag{A3.179}$$

Clearly $v(x)$ is a solution of the integral equation

$$v(x) = -\sin[\sqrt{\lambda_n}(a-x)]$$
$$- \left(\frac{1}{\lambda_n}\right)^{1/2}\int_x^a q(\xi)v(\xi)\sin\left[\sqrt{\lambda_n}(x-\xi)\right]d\xi \tag{A3.180}$$

Let

$$0 \leq q \leq M \tag{A3.181}$$

and take $N > 0$ so large that for all $n \geq N$,

$$\lambda_n > 4M^2 a^2 \tag{A3.182}$$

Then, by (A3.180),

$$\max|v| < 1 + \frac{1}{2Ma}\int_0^a M\max|v(x)|dx \Rightarrow |v| < 2 \tag{A3.183}$$

Together with (A3.180), this gives

$$v(x) = -\sin[\sqrt{\lambda_n}(a-x)] + O(1)/\sqrt{\lambda_n} \qquad \text{(A3.184)}$$

and

$$(v \cdot v) = \frac{a}{2}\left(1 + \frac{O(1)}{\sqrt{\lambda_n}}\right) \qquad \text{(A3.185)}$$

Hence

$$u_n = \frac{v}{(v \cdot v)^{1/2}} = \sqrt{2/a}\sin\left[\sqrt{\lambda_n}(a-x)\right] + \frac{O(1)}{\sqrt{\lambda_n}} \qquad \text{(A3.186)}$$

which proves (A3.174). The validity of (A3.175) and (A3.176) follows from (A3.180) by differentiation.

We now return to the general case. The substitution

$$u = v \cdot \varphi, \quad y = \int_0^x \frac{ds}{\psi(s)} \qquad \text{(A3.187)}$$

reduces the problem (A3.173) to a problem of type (A3.179) if

$$\varphi = (kr)^{-1/2}, \quad \psi = \sqrt{k/r} \qquad \text{(A3.188)}$$

The truth of estimates (A3.174)–(A3.176) now follows. **Q.E.D.**

Remark A3.3.3. As stated previously, the assumption that k and r are twice continuously differentiable functions and that q is continuous in $[0, a]$ was adopted only to simplify the exposition. Theorems A3.3.4–6 and Remark A3.3.2 are true for all admissible k, q, and r.

Corollary of Theorems A3.3.4–6 and Remark A3.3.2. *Theorems A3.3.4–A3.3.6 show that the Fourier series expansion of any function $f \in \mathfrak{M}$ and its first derivative converge absolutely and uniformly. However, it does not follow from these theorems that the repeated term-by-term differentiation of this series is always admissible. The estimate (A3.167) was obtained by a very coarse method – comparison of a convergent series with the divergent harmonic series. Using a more delicate approach, one can prove that the coefficients of repeatedly differentiated Fourier series of $f \in \mathfrak{M}$ are of order $O(1/m)$, so that the series is conditionally convergent within the interval $(0, a)$ [57]. However, if the conditions of the Remark A3.3.2 are valid, then the series obtained by repeated differentiation of the Fourier series expansion is absolutely and uniformly convergent in $[0, a]$.*

4. Remarks on case of singular operator

The case of a singular operator appears in the analytical theory of ordinary differential equations and in particular in the theory of special functions, among the most important of which are the Bessel functions (see Appendix 2). In what follows we present some basic information about this case, without proofs.[5]

---

[5]For proofs see, e.g., [39] and [85].

Let L be a singular Sturm–Liouville operator, that is, an operator

$$LX \overset{\text{def}}{=} \frac{d}{dx}\left[k(x)\frac{dX}{dx}\right] - q(x)X(x) + \lambda r(x)X(x) = 0 \quad \forall x \in (0, a) \quad \text{(A3.189)}$$

where

$$X \text{ is bounded in the neighborhood of } x = 0$$

$$\gamma \frac{dX}{dx} + \delta X = 0 \quad \text{at } x = a \tag{A3.190}$$

Again (for the sake of simplicity only), k, dk/dx, q, and r are assumed to be continuous in $[0, a]$, and moreover

$$k(x) = x\tilde{k}(x), \quad r(x) = x\tilde{r}(x), \quad q(x) = \frac{\tilde{q}(x)}{x}$$

$$\tilde{k}(0) > 0, \quad \tilde{r}(0) > 0, \quad \tilde{q}(0) \geq 0 \tag{A3.191}$$

Lemma A3.4.1. *Let u and v be two linearly independent solutions of the equation*

$$Lw + \lambda rw = 0 \quad \forall x \in (0, a) \tag{A3.192}$$

and assume that u is bounded in the neighborhood of $x = 0$. Then v has a logarithmic singularity at $x = 0$ if $u(0) \neq 0$ or a pole of order ν if (x) has at $x = 0$ a root of order ν.

Lemma A3.4.2. *If u is the solution of the equation (A3.192) which is bounded in the neighborhood of $x = 0$, then*

$$\lim_{x \downarrow 0} \left(k(x)\frac{d}{dx}u(x)\right) = 0 \tag{A3.193}$$

Lemma A3.4.3. *Let $u(x)$ be a bounded solution of $L(u) = 0$. Let $q(x)$ be continuous in the neighborhood of $x = 0$ and*

$$u(0) \neq 0, \frac{d}{dx}u(0) = \frac{q(0)u(0)}{\tilde{k}(0)} \tag{A3.194}$$

If

$$q(x) = \frac{\tilde{q}(x)}{x}, \quad \tilde{q}(0) \neq 0 \tag{A3.195}$$

then

$$u(x) = x^\nu z(x), \quad z(0) \neq 0, \quad \nu = \sqrt{\tilde{q}(0)/\tilde{k}(0)} \tag{A3.196}$$

Note now that the assumption that the operator L is regular was not used in the proofs of Theorems A3.2.1–2. Thus these theorems remain valid in the singular case. Moreover, by Lemma A3.4.2,

$$ku\frac{d}{dx}u \Big|_{x=0}^{a} = 0 \tag{A3.197}$$

and by Lemma A3.4.3 the integral

$$\int_0^a \left[k(x) \left(\frac{du}{dx} \right)^2 + q(x)u(x)^2 \right] dx \qquad (\text{A3.198})$$

is convergent. Thus Theorem A3.2.3 is also valid in the singular case.

Let us now augment the definition of the classes $\mathfrak{M}$ and $\mathfrak{M}^*$, adding the condition that any function in either class has the rate of growth specified in Lemma A3.4.3. Then Theorems A3.2.4–8 remain in force, provided that the natural boundary condition is replaced by the condition that u is bounded in the neighborhood of $x = 0$. Moreover, all the results of the Section 3 remain valid. Indeed, in the proof of the Sturm theorem one need only use as a comparison operator not the harmonic oscillator $d^2u/dx^2 + \lambda u$ but the Bessel operator

$$Lu = \frac{d^2}{dx^2}u + \frac{1}{x}\frac{d}{dx}u + \left(1 - \frac{\nu^2}{x^2}\right)u \qquad (\text{A3.199})$$

where ν is defined by (A3.196), and take into account that the asymptotic of eigenvalues of the Bessel operator is similar to that of the eigenvalues of the regular operator (see Appendix 2, Section 7); that is, they are of order n^2. In particular, this justifies the expansion in the Fourier–Bessel series used in Chapter 19, Section 5, to solve the problem of vibration of a circular membrane with rigidly fixed boundary.

5. Expansions into Fourier–Bessel and Dini series

We now state without proof a few facts related to the expansion of functions in Fourier–Bessel and Dini series (for proofs, see [90]).

A. Roots of Bessel functions

Let

$$\mathcal{E}_\nu(z) = J_\nu(z)\cos\alpha - Y_\nu(z)\sin\alpha \qquad (\text{A3.200})$$

where J_ν and Y_ν are Bessel and Neumann functions of order ν. Assume that

$$\alpha, \beta, \nu \text{ are real,} \quad \nu > -1/2 \qquad (\text{A3.201})$$

Theorem A3.5.1. *Any real[6] $\mathcal{E}_\nu(z)$ has a denumerable number of roots $\lambda_{\nu n}$. All the roots $\lambda_{\nu n}$ are simple and real.*

Theorem A3.5.2. *Let $\{\lambda_{\nu n}\}$ be the set of all positive roots of a real $\mathcal{E}_\nu$ enumerated in increasing order. Then the positive roots of $\mathcal{E}_{\nu+1}$ alternate with the positive roots of $\mathcal{E}_\nu$:*

$$0 < \lambda_{\nu,1} < \lambda_{\nu+1,1} < \lambda_{\nu,2} < \lambda_{\nu+1,2} < \cdots \qquad (\text{A3.202})$$

---

[6] $\mathcal{E}_\nu(z)$ is called a *real* Bessel function if α, β, and ν are real and $z > 0$.

Theorem A3.5.3. *All the positive roots $\lambda_{\nu,m}$ of $J_\nu(z)$ lie in the interval $(m\pi + \pi/4,\ m\pi + 3\pi/4)$.*

Theorem A3.5.4. *Let $\nu > 1/2$ and let $\lambda_{\nu,m}$ be a root of $\mathcal{E}_\nu$. If*

$$\lambda_{\nu,m} > (2\nu + 1)(2\nu + 3)/\pi \tag{A3.203}$$

then

$$m\pi - \alpha + \frac{\nu\pi}{2} + \frac{\pi}{2} < \lambda_{\nu,m} < m\pi - \alpha + \frac{\nu\pi}{2} + \frac{3\pi}{4} \tag{A3.204}$$

B. Expansion in Fourier–Bessel series

Definition A3.5.1. The series

$$f(z) \sim \sum_{m=1}^{\infty} \alpha_m J_\nu(\lambda_m z), \quad \lambda_m \overset{\text{def}}{=} \lambda_{\nu,m} \tag{A3.205}$$

is called the *Fourier–Bessel* series of $f(z)$ if, for all $m = 1, 2, \ldots$,

$$\alpha_m = \frac{\int_0^t t f(t)\, J_\nu(\lambda_m t)\, dt}{\|J_\nu(\lambda_m t)\|^2} \tag{A3.206}$$

Let $\nu > -1/2$.

Theorem A3.5.5. *For all $m \neq n$, $J_\nu(\lambda_m z)$ and $J_\nu(\lambda_n z)$ are orthogonal with weight z:*

$$\int_0^1 z J_\nu(\lambda_m z) \cdot J_\nu(\lambda_n z)\, dz = 0 \tag{A3.207}$$

Theorem A3.5.6.

$$\|J_\nu(z)\|^2 = {}^1\!/_2\, J_{\nu+1}(z) \tag{A3.208}$$

Theorem A3.5.7 (Analog of the Riemann Lebesgue lemma). *Let $f(x)$ be defined in $(0,1]$ and assume that the integral*

$$\int_0^1 t^{1/2} f(t)\, dt \tag{A3.209}$$

is absolutely convergent. Let

$$0 < x \le 1, \quad (a,b) \subset (0,1), \quad x \notin (a,b) \tag{A3.210}$$

and

$$T_n(x,t) = 2 \sum_{m=1}^{n} \frac{J_\nu(\lambda_m x)\, J_\nu(\lambda_m t)}{J_{\nu+1}^2(\lambda_m)} \tag{A3.211}$$

Then

$$\lim_{n\uparrow\infty} \int_a^b t f(t) T_n(x,t)\, dt = 0 \tag{A3.212}$$

Theorem A3.5.8. *Under the assumptions of Theorem A3.5.7, if $f(x)$ is of bounded variation in an interval (a, b) then its Fourier–Bessel series converges and*

$$\sum_{m=1}^{\infty} \alpha_m J_\nu(\lambda_m x) = \frac{1}{2}[f(x-0) + f(x+0)] \quad \forall x \in (a, b) \tag{A3.213}$$

Theorem A3.5.9. *Let $f(x)$ be continuous and of bounded total variation in an interval $(a, b) \subset [0, 1]$. If the integral (A3.209) is absolutely convergent, then the Fourier–Bessel series of $f(x)$ is uniformly convergent in the segment*

$$a + \Delta \leq x \leq b - \Delta \quad \forall \Delta > 0 \tag{A3.214}$$

If, moreover,

$$f(1 - 0) = 0 \tag{A3.215}$$

then the Fourier–Bessel series of $f(x)$ is uniformly convergent in the segment

$$a + \Delta \leq x \leq 1 \quad \forall \Delta > 0 \tag{A3.216}$$

Finally, if $x^{1/2} f(x)$ is of bounded total variation in a segment $[0, b]$, $0 < b < 1$, then the series

$$x^{1/2} \sum_{m=1}^{\infty} \alpha_m J_\nu(\lambda_m x) \tag{A3.217}$$

converges uniformly to $x^{1/2} f(x)$ for all $x \in [0, b]$.

The rate of increase of the terms of the Fourier–Bessel series of $f(x)$ is defined by the following theorem.

Theorem A3.5.10. *If $x^{1/2} f(x)$ is of bounded total variation in $[0, 1]$, then*

$$2 J_\nu(\lambda_m x) \, J_{\nu+1}^{-2}(\lambda_m) \int_0^1 t f(t) \, J_\nu(\lambda_m t) dt = O\left(\frac{1}{\lambda_m}\right) \tag{A3.218}$$

Term-by-term differentiation of Fourier–Bessel series is justified by the following theorem.

Theorem A3.5.11. *Let $f(x)$ be continuous in $0 < x < 1$,*

$$\lim_{x \downarrow 0} x^{\nu+2} f(x) = 0, \quad \lim_{x \uparrow 1} f(x) = 0 \tag{A3.219}$$

and let

$$\varphi(x) \overset{\text{def}}{=} \frac{d}{dx} f(x) - \frac{\nu}{x} f(x) \tag{A3.220}$$

have bounded total variation and absolutely integrable in a segment $[a, b] \subset [0, 1]$. Then the identity

$$f(x) = \sum_{n=1}^{\infty} \alpha_n J_\nu(\lambda_n x) \tag{A3.221}$$

implies

$$\frac{d}{dx}f(x) = \sum_{n=1}^{\infty} \alpha_n \frac{d}{dx} J_\nu(\lambda_n x) \tag{A3.222}$$

C. Expansion in Dini series

Definition A3.5.2. Let $f(x)$ be defined in $(0,1)$ and satisfy a homogeneous condition of the third kind at $x = 1$:

$$\frac{d}{dx}f(1-0) + f(1-0) = 0 \tag{A3.223}$$

Further, let h be a given constant and

$$\nu > -1 \tag{A3.224}$$

The series

$$f(x) \sim \sum_{n=1}^{\infty} \beta_n J_\nu(\gamma_n x) \tag{A3.225}$$

where $\{\gamma_m\}$ is the monotonically increasing sequence of positive roots of the equation

$$\left(z\frac{d}{dz}J_\nu(z) + hJ_\nu(z) \right) = 0 \tag{A3.226}$$

is called a *Dini series* if the coefficients β_n are defined by

$$\beta_n = \frac{2\gamma_n^2 \int_0^1 tf(t) J_\nu(\gamma_n t)dt}{[\gamma_n^2 - \nu^2]J_\nu^2(\gamma_n) + \gamma_n^2 \left(\frac{d}{dz}J_\nu(z)\right)^2}, \quad n = 1,2,\dots z = \gamma_n \tag{A3.227}$$

Theorem A3.5.12. *The functions $J_\nu(\gamma_n x)$ and $J_\nu(\gamma_m x)$ are orthogonal with weight x:*

$$\int_0^1 xJ_\nu(\gamma_n x) J_\nu(\gamma_m x)dx = 0 \quad \forall m \neq n \tag{A3.228}$$

and

$$\|J_\nu(\gamma_n x)\|^2 = \frac{[\gamma_n^2 - \lambda_n^2]J_\nu^2(\gamma_n) + \gamma_n^2 \left(\frac{d}{dx}J_\nu(\gamma_n)\right)}{2\gamma_n^2} \tag{A3.229}$$

Theorem A3.5.13. *Let*

$$x^{1/2}f(x) \sim \sum_{n=1}^{\infty} \alpha_n J_\nu(\lambda_n x) \tag{A3.230}$$

and

$$x^{1/2}f(x) \sim \sum_{n=1}^{\infty} \beta_n J_\nu(\gamma_n x) \tag{A3.231}$$

be the Fourier–Bessel and Dini series corresponding to $f(x)$. If the Fourier–Bessel series (A3.230) is convergent in some point x then the Dini series (A3.231) is also convergent at that point, and the two series converge to the same sum there. If the Fourier–Bessel series (A3.230) is uniformly convergent in a segment $0 < a \le x \le b < 1$, then the same is true for the Dini series (A3.231). In particular, if $f(x)$ is of bounded variation in a segment $0 \le a \le x \le b < 1$ then, for all $\Delta > 0$ and $x \in (a + \Delta, \, b - \Delta)$,

$$\frac{f(x+0) + f(x-0)}{2} = \sum_{n=1}^{\infty} \beta_n J_\nu(\gamma_n x) \qquad (A3.232)$$

and the convergence is uniform if $f(x)$ is continuous. If $f(x)$ is of bounded total variation in $(a, 1]$, then

$$\lim_{x \uparrow 1} \sum_{n=1}^{\infty} \beta_n J_\nu(\gamma_n x) = f(1 - 0) \qquad (A3.232)$$

If $f(x)$ is also continuous in $(0, 1]$ and the integral

$$\int_0^1 x^{1/2} f(x) dx \qquad (A3.234)$$

is absolutely convergent, then the Dini series is uniformly convergent in $(a + \Delta, 1]$, for all $\Delta > 0$.

Problems

PA3.3.1. Prove the following complement to the Sturm comparison Theorem A3.3.3: *The eigenvalues of the Strum–Liouville operator are not increasing functions of the length of the interval of its definition.*

PA3.4.1. Prove all the theorems of Section 4.

Appendix 4: Fourier integral

1. Riemann–Lebesgue lemma

Definition A4.1.1. $\mathfrak{M}(f)$ will denote the set of functions defined in $-\infty < x < \infty$ that satisfy the Dirichlet conditions:

(a) $f(x)$ is absolutely integrable in $-\infty < x < \infty$;
(b) $f(x)$ has at most finite number n of extrema or points of discontinuity of the first kind.

Lemma A4.1.1 (Riemann–Lebesgue).

$$f(x) \in \mathfrak{M} \Rightarrow \lim_{\lambda \uparrow \infty} \int_{-\infty}^{\infty} f(x) e^{i\lambda x} dx = 0 \qquad \text{(A4.1)}$$

PROOF. Since f is absolutely integrable in $-\infty < x < \infty$, it follows that for every $\varepsilon > 0$ there exist $N > 0$ and $M = \text{const} > 0$ so large that

$$\int_{-\infty}^{-N} |f(x)| dx < \frac{\varepsilon}{3}, \quad \int_{N}^{\infty} |f(x)| dx < \frac{\varepsilon}{3}, \quad |f(x)| < M, \quad \forall |x| \qquad \text{(A4.2)}$$

Let

$$x_1 < x_2 < \cdots < x_n \qquad \text{(A4.3)}$$

be the set of all the discontinuity or extremum points of $f(x)$; $f(x)$ is continuous and monotonic in every interval (x_k, x_{k+1}). Let

$$x_0 = -N, \quad x_{n+1} = N, \quad I_N = \sum_{k=0}^{n} I_k \qquad \text{(A4.4)}$$

where

$$I_k = \int_{x_k}^{x_{k+1}} f(x) e^{i\lambda x} dx \qquad \text{(A4.5)}$$

To fix ideas, assume that $f(x)$ decreases in (x_k, x_{k+1}). Then, by the second mean value theorem of integral calculus, there exists $\xi_k \in (x_k, x_{k+1})$ such

617

that

$$I_k = f(x_k + 0) \int_{x_k}^{\xi_k} e^{i\lambda x} dx \;\Rightarrow\; |I_k| < \frac{M}{|\lambda|}(x_{k+1} - x_k) \tag{A4.6}$$

Hence

$$|I_N| < \frac{2MN}{|\lambda|} \;\Rightarrow\; \exists \Lambda > 0 \text{ so large that } |I_N| < \frac{\varepsilon}{3} \;\forall |\lambda| > \Lambda \tag{A4.7}$$

Since $\varepsilon > 0$ is arbitrarily small, it follows from (A4.2) and (A4.7) that

$$\left| \int_{-\infty}^{\infty} f(x) e^{i\lambda x} dx \right| \leq \int_{-\infty}^{-N} |f(x)| dx + \int_{N}^{\infty} |f(x)| dx + |I_N| < \varepsilon$$

$$\Rightarrow \lim_{\lambda \uparrow \infty} \int_{-\infty}^{\infty} f(x) e^{i\lambda x} dx = 0 \tag{A4.8}$$

$$\text{Q.E.D.}$$

Remark. The requirement that f satisfy the Dirichlet condition is unnecessarily strong. The Riemann–Lebesgue lemma is valid if $f(x)$ is absolutely Lebesgue-integrable [86].

2. Fundamental Fourier theorem

Theorem A4.2.1 (Double Fourier integral). *Let*

$$f(x) \in \mathfrak{M} \tag{A4.9}$$

Then, for all $x \in (-\infty, \infty)$,

$$\lim_{N \uparrow \infty} \frac{1}{2\pi} \int_{-N}^{N} d\lambda \int_{-\infty}^{\infty} f(\xi) \exp[i\lambda(x - \xi)] d\xi$$

$$= \frac{f(x - 0) + f(x + 0)}{2} \tag{A4.10}$$

Proof. Let

$$I = \frac{1}{2\pi} \int_{-N}^{N} d\lambda \int_{-\infty}^{\infty} f(\xi) \exp[i\lambda(x - \xi)] d\xi \tag{A4.11}$$

Let x be fixed. Since $f(x)$ is absolutely integrable, it is permissible to change the order of integration in (A4.11), so that

$$I = \frac{1}{2\pi} \int_{-\infty}^{\infty} f(\xi) d\xi \int_{-N}^{N} \exp[i\lambda(x - \xi)] d\lambda = \sum_{k=1}^{4} I_k(x, \lambda) \Big|_{\lambda = -N}^{N} \tag{A4.12}$$

where

$$I_1 = \frac{1}{2\pi} \int_{-\infty}^{x-\delta}, \quad I_2 = \frac{1}{2\pi} \int_{x-\delta}^{x}, \quad I_3 = \frac{1}{2\pi} \int_{x}^{x+\delta}, \quad I_4 = \frac{1}{2\pi} \int_{x+\delta}^{\infty} \tag{A4.13}$$

and $\delta > 0$ is arbitrarily small.

Consider first I_4. Integration yields

$$I_4 = \frac{1}{2\pi} \int_{x+\delta}^{\infty} f(\xi) \frac{\exp[i\lambda(x-\xi)]}{i(x-\xi)} \bigg|_{-N}^{N} d\xi \qquad (A4.14)$$

Since

$$f(\xi) \in \mathfrak{M} \quad \forall \xi \in (x+\delta, \infty) \qquad (A4.15)$$

the Riemann–Lebesgue lemma is applicable, so that

$$\lim_{N \uparrow \infty} I_4 = 0 \qquad (A4.16)$$

Clearly the same is true for I_1:

$$\lim_{N \uparrow \infty} I_1 = 0 \qquad (A4.17)$$

Since $\delta > 0$ is arbitrarily small and the number of extremum or discontinuity points of f is finite, we may assume that x is the only such point in $[x - \delta, x]$ and $[x, x + \delta]$, so that $f(x)$ is continuous and monotonic in these segments. Consider I_2, assuming to fix ideas that $f(x)$ increases in $[x - \delta, x]$. We have

$$I_2 = f(x-0)I_{21} + I_{22} \qquad (A4.18)$$

where

$$I_{21} = \frac{1}{2\pi} \int_{x-\delta}^{x} \frac{\exp[i\lambda(x-\xi)]}{i(x-\xi)} \bigg|_{\lambda=-N}^{N} d\xi = \frac{1}{\pi} \int_{0}^{\delta N} \frac{\sin(s)}{s} ds \qquad (A4.19)$$

and

$$I_{22} = \frac{1}{\pi} \int_{x-\delta}^{x} [f(\xi) - f(x-0)] \frac{\sin[N(x-\xi)]}{x-\xi} d\xi \qquad (A4.20)$$

Since $\delta = \text{const} > 0$, the limit

$$\lim_{N \uparrow \infty} I_{21} = \frac{1}{2} \qquad (A4.21)$$

exists [86]. Furthermore, $f(x)$ is monotonic in $[x-\delta, x]$; hence, by the second mean value theorem, there exists $\eta \in (0, \delta)$ such that

$$I_{22} = [f(x-0) - f(x-\delta)] \frac{1}{\pi} \int_{x-\eta}^{x} \frac{\sin[N(x-\xi)]}{x-\xi} d\xi \qquad (A4.22)$$

The convergence of the integral

$$\int_{0}^{\infty} \frac{\sin s}{s} ds \qquad (A4.23)$$

means that for some sufficiently large $L > 0$ and all $N > 0$,

$$\left| \frac{1}{\pi} \int_{x-\eta}^{x} \frac{\sin[N(x-\xi)]}{x-\xi} d\xi \right| < L \qquad (A4.24)$$

Hence, by the continuity of $f(\xi)$ in $[x - \delta, x]$, for all $\varepsilon_1 > 0$ and $N > 0$ there exists $\delta > 0$ so small that

$$|I_{22}| < \varepsilon_1 \qquad (A4.25)$$

Combining (A4.18), (A4.21), and (A4.25), we conclude that for all $\varepsilon > 0$ and fixed x there exists $\delta > 0$ so small and $N_1 > 0$ so large that

$$\left| I_2(x, N) - \frac{f(x-0)}{2} \right| < 2\varepsilon_1 \tag{A4.26}$$

Similarly,

$$\left| I_3(x, N) - \frac{f(x+0)}{2} \right| < 2\varepsilon_1 \tag{A4.27}$$

Since ε_1 is arbitrary it follows from (A4.23), (A4.24), and (A4.13) that for all $x = \text{const}$ and $\varepsilon > 0$ there exists $N_1 > 0$ so large that, for all $N > N_1$,

$$\left| I(x, N) - \frac{f(x-0) + f(x+0)}{2} \right| < \varepsilon \tag{A4.28}$$

Thus, for all $x = \text{const}$,

$$\lim_{N \uparrow \infty} I(x, N) = \frac{f(x-0) + f(x+0)}{2} \tag{A4.29}$$

Q.E.D.

3. Fourier transform of function of exponential growth at infinity. Relationship between double Fourier integral and Fourier series

We now present Titchmarsh's generalization of the theory of Fourier transforms to functions of exponential growth at infinity [86]. Let $f(x)$ be defined in $(-\infty, \infty)$ and possess the following properties: There exist $\alpha_0 > 0$ and $\beta_0 > 0$ such that if

$$\varphi(x) = \begin{cases} f(x)\exp(-\alpha x) & \forall x > 0 \\ f(x)\exp(\beta x) & \forall x < 0 \end{cases} \tag{A4.30}$$

and

$$\varphi_+(x) = \varphi(x)\eta(x) \quad \forall x > 0, \quad \varphi_-(x) = \varphi(x)\eta(-x) \quad \forall x < 0 \tag{A4.31}$$

where $\eta(x)$ is the Heaviside unit function

$$\eta(x) = \begin{cases} 1 & \forall x > 0 \\ \tfrac{1}{2} & x = 0 \\ 0 & \forall x < 0 \end{cases} \tag{A4.32}$$

then $\varphi(x)$ satisfies the Dirichlet conditions.[1]

---

[1]See the remark in Section 1 of this appendix.

Denote

$$F_+(w) = \frac{1}{\sqrt{2\pi}} \int_{-\infty}^{\infty} \varphi_+(\xi) \exp(is\xi)d\xi$$

$$\equiv \frac{1}{\sqrt{2\pi}} \int_{0}^{\infty} f(\xi) \exp(iw\xi)d\xi, \quad w = s + i\alpha \tag{A4.33}$$

$$F_-(w) = \frac{1}{\sqrt{2\pi}} \int_{-\infty}^{\infty} \varphi_-(\xi) \exp(is\xi)d\xi$$

$$\equiv \frac{1}{\sqrt{2\pi}} \int_{0}^{\infty} f(\xi) \exp(iw\xi)d\xi, \quad w = s + i\beta \tag{A4.34}$$

Using the inversion formula, we obtain

$$\varphi_\pm(x) = \frac{1}{\sqrt{2\pi}} \int_{-\infty}^{\infty} F_\pm(\xi) \exp(-isx)ds \tag{A4.35}$$

or, what is the same,

$$f(x)\eta(x) = \frac{1}{\sqrt{2\pi}} \int_{-\omega+\alpha}^{\omega+i\alpha} F_+(\xi) \exp(-iwx)ds \tag{A4.36}$$

$$f(x)\eta(-x) = \frac{1}{\sqrt{2\pi}} \int_{-\omega+\beta}^{\omega+i\beta} F_-(\xi) \exp(-iwx)ds \tag{A4.37}$$

Hence

$$f(x) = \frac{1}{\sqrt{2\pi}} \int_{-\omega+i\alpha}^{\omega+i\alpha} F_+(w) \exp(-iwx)ds$$

$$+ \frac{1}{\sqrt{2\pi}} \int_{-\omega+i\beta}^{\infty+i\beta} F_-(w) \exp(-iwx)ds \tag{A4.38}$$

This is Titchmarsh's inversion formula for the Fourier transform of a function of exponential growth at infinity.

We can now present a rigorous reduction of a Fourier double integral to a Fourier series of periodic functions. Indeed, let $f(x)$ be a periodic function with period 2π, satisfying the Dirichlet conditions in $[0, 2\pi]$. Then $f(x)$, considered as defined in $-\infty < x < \infty$, satisfies Titchmarsh's assumptions. This means that whenever $\Im w > 0$,

$$F_+(w) = \frac{1}{\sqrt{2\pi}} \int_{0}^{\infty} f(\xi) \exp(iw\xi)d\xi \tag{A4.39}$$

or, equivalently (because $f(x)$ is periodic),

$$F_+(w) = \sum_{k=0}^{\infty} \frac{1}{\sqrt{2\pi}} \int_{2k\pi}^{2(k+1)\pi} f(\xi) \exp(iw\xi)d\xi$$

$$= \frac{1}{\sqrt{2\pi}} \int_{0}^{2\pi} f(\xi) \exp(iw\xi) \sum_{k=0}^{\infty} \exp(2k\pi wi)d\xi$$

$$= \frac{1}{\sqrt{2\pi}} \int_{0}^{2\pi} f(\xi) \frac{\exp(iw\xi)}{1 - \exp(2\pi wi)} d\xi \quad \forall \Im w > 0 \tag{A4.40}$$

Similarly,

$$F_-(w) = \frac{1}{\sqrt{2\pi}} \int_0^{2\pi} f(\xi) \frac{\exp(iw\xi)}{1 - \exp(2\pi wi)} d\xi \quad \forall \Im w < 0 \qquad (A4.41)$$

The inversion formula (A4.38) yields

$$f(x) = \frac{1}{\sqrt{2\pi}} \int_{-\omega+i\alpha}^{\omega+i\alpha} \varphi(w) \frac{\exp(-iwx)}{1 - \exp(2\pi wi)} dw$$

$$+ \frac{1}{\sqrt{2\pi}} \int_{-\omega+i\beta}^{\omega+i\beta} \varphi(w) \frac{\exp(-iwx)}{1 - \exp(2\pi wi)} dw \qquad (A4.42)$$

where

$$\varphi(w) = \int_0^{2\pi} f(\xi) \exp(iw\xi) d\xi \qquad (A4.43)$$

Note that $\varphi(w)$ is an entire function. Hence the residue theorem is applicable to the integrals in the right-hand side of (A4.43). This yields

$$f(x) = \frac{1}{2\pi} \sum_{-\infty}^{\infty} \varphi(n) \exp(inx) \qquad (A4.44)$$

In view of (A4.43), this is the Fourier series expansion of $f(x)$.

4. Convolution theorem and evaluation of definite integrals

As shown in Chapter 19, Section 5, the convolution theorem plays an important role in the theory and application of Laplace–Carson transforms. An analogous theorem is valid in the theory of Fourier transforms.

Definition A4.4.1. Given two functions $f(x) \in \mathfrak{M}$ and $g(x) \in \mathfrak{M}$, the integral

$$\Phi(x) = \frac{1}{\sqrt{2\pi}} \int_{-\infty}^{\infty} g(\xi) f(x - \xi) d\xi \equiv \frac{1}{\sqrt{2\pi}} \int_{-\infty}^{\infty} f(\xi) g(x - \xi) d\xi \qquad (A4.45)$$

is called the *convolution* of the functions f and g.[2]

Let us find the Fourier transform Φ_f of the convolution $\Phi(x)$. Let $F(x)$ and $G(x)$ be the Fourier transforms of f and g, respectively. We have

$$\Phi_f(x) = \frac{1}{\sqrt{2\pi}} \int_{-\infty}^{\infty} e^{ixs} ds \int_{-\infty}^{\infty} g(\xi) f(s - \xi) d\xi \qquad (A4.46)$$

---

[2] The convolution of two functions f and g is often denoted by $f * g$. This notation is applicable in the theory of Fourier transforms as well as Laplace–Carson or any other integral transforms. Naturally, the specific definition of the convolution depends on the type of integral transform under consideration.

By changing the order of integration, one obtains

$$\Phi_f(x) = \frac{1}{2\pi} \int_{-\infty}^{\infty} \exp(i\xi x) g(\xi) d\xi \int_{-\infty}^{\infty} \exp[i(s-\xi)x] f(s-\xi) ds$$

$$= \frac{1}{2\pi} \int_{-\infty}^{\infty} \exp(i\xi x) g(\xi) d\xi \int_{-\infty}^{\infty} \exp(isx) f(s) ds$$

$$= F(x) G(x) \tag{A4.47}$$

Thus, formally, the Fourier transform of a convolution is the product of the Fourier transform of its factors. Clearly, this result is not only formally but rigorously true if all the integrals involved are convergent and if it is legitimate to change the order of integration.[3]

If $x = 0$, formula (A4.47) yields the *Parseval equality*:

$$\int_{-\infty}^{\infty} f(s) g(-s) ds = \int_{-\infty}^{\infty} F(s) G(s) ds \tag{A4.48}$$

If f and g are both even or both odd functions then the Parseval equality becomes the analogous identity for cosine or sine transforms, respectively, so that

$$\int_{-\infty}^{\infty} f(s) g(-s) ds = \int_{0}^{\infty} F_s(s) G_s(s) ds \tag{A4.49}$$

or

$$\int_{-\infty}^{\infty} f(s) g(-s) ds = \int_{0}^{\infty} F_c(s) G_c(s) ds \tag{A4.50}$$

The convolution theorem and the Parseval equality may be used to evaluate definite integrals, as the following examples show.

1. Let

$$f(x) = \exp(-\alpha x), \qquad g(x) = \exp(-\beta x) \tag{A4.51}$$

Then

$$\int_{0}^{\infty} f(x) g(x) dx = \frac{1}{\alpha + \beta} \tag{A4.52}$$

We have

$$F_c(s) = \left(\frac{2}{\pi}\right)^{1/2} \int_{0}^{\infty} \exp(-\alpha x) \cos(sx) dx = \frac{\sqrt{2}}{\sqrt{\pi}} \frac{\alpha}{\alpha^2 + s^2}$$

$$G_c(s) = \left(\frac{2}{\pi}\right)^{1/2} \int_{0}^{\infty} \exp(-\beta x) \cos(sx) dx = \frac{\sqrt{2}}{\sqrt{\pi}} \frac{\beta}{\beta^2 + s^2} \tag{A4.53}$$

Hence, by the Parseval equality,

$$\frac{2}{\pi} \cdot \int_{0}^{\infty} \frac{\alpha \beta ds}{(\alpha^2 + s^2)(\beta^2 + s^2)} = \frac{1}{\alpha + \beta} \tag{A4.54}$$

---

[3] A number of sufficient conditions for validity of the equality (A4.47) are provided by Titchmarsh [86].

2. Let

$$f(x) = \eta(a - x), \qquad g(x) = \eta(b - x) \tag{A4.55}$$

where $\eta(x)$ is the Heaviside unit. Then

$$
\begin{aligned}
F_c(x) &= \frac{\sqrt{2}}{\sqrt{\pi}} \int_0^a \cos(xs)ds = \frac{\sqrt{2}}{\sqrt{\pi}} \cdot \frac{\sin(ax)}{x} \\
G_c(x) &= \frac{\sqrt{2}}{\sqrt{\pi}} \int_0^b \cos(xs)ds = \frac{\sqrt{2}}{\sqrt{\pi}} \cdot \frac{\sin(bx)}{x}
\end{aligned} \tag{A4.56}
$$

Hence

$$\frac{2}{\pi} \int_0^\infty \frac{\sin(ax)\sin(bx)}{x^2}dx = \int_0^{\min(a,b)} dx = \min(a,b) \tag{A4.57}$$

5. Abel-summable integrals and solution of problems with concentrated capacity

One way of giving a meaning to the concept of the sum of a divergent series or integral is provided by the theory of summable series and integrals. A detailed account of the theory may be found in many papers and monographs; see, for example, [36; 86]. In what follows we shall restrict ourselves to a very short and even superficial introduction to the definition of Abel-summable integrals, and demonstrate how the concept can be applied to solution of boundary-value problems by means of a characteristic example concerning the solution of a concrete problem with concentrated capacity.[4]

Definition A4.5.1. A function $f(x)$ is said to be *Abel-summable* in $(0,\infty)$ if the limit

$$\lim_{\alpha \downarrow 0} \int_0^\infty e^{-\alpha x} f(x)dx \overset{\text{notation}}{=\!=} \int_0^\infty {}^a f(x)dx \tag{A4.58}$$

exists.

The following trivial theorem justifies the introduction of the definition.

Theorem A4.5.1. *If $f(x)$ is absolutely integrable in $(0,\infty)$ then*

$$\int_0^\infty {}^a f(x)dx = \int_0^\infty f(x)dx \tag{A4.59}$$

We exhibit two important examples of functions that are not integrable but are Abel-summable. We claim that

$$I(s) \overset{\text{def}}{=} \int_0^\infty e^{isx}dx \tag{A4.60}$$

---

[4]For the definition of concentrated capacity, see Chapter 2, Section 10.

is divergent, whereas

$$I^a(s) \stackrel{\text{def}}{=} \int_0^\infty {}^a e^{isx} dx = \begin{cases} 0 + i/s & \text{if } s > 0 \\ \infty + i \cdot 0 & \text{if } s = 0 \end{cases} \tag{A4.61}$$

Indeed, if $s > 0$ then

$$\int_0^N e^{isx} dx = \frac{\sin(Ns)}{s} + \frac{i[1 - \cos(Ns)]}{s} \tag{A4.62}$$

Since $\lim_{N \uparrow \infty} \sin(Ns)$ and $\lim_{N \uparrow \infty} \cos(Ns)$ do not exist, the integral (A4.60) is indeed divergent. At the same time,

$$\int_0^N e^{-\alpha x + isx} dx = \frac{\alpha + is}{s^2 + \alpha^2} [1 - \exp(is - \alpha)N]$$

$$\Rightarrow \lim_{N \uparrow \infty} \int_0^N e^{-\alpha x + ix} dx = \frac{\alpha + is}{s^2 + \alpha^2} \quad \forall \alpha > 0, \ \forall s > 0 \tag{A4.63}$$

Hence $\exp(ix)$ is Abel-summable and

$$\int_0^\infty {}^a e^{ix} dx = \lim_{\alpha \downarrow 0} \frac{\alpha + is}{s^2 + \alpha^2} = \frac{i}{s} \tag{A4.64}$$

Thus we have

$$\int_0^\infty {}^a \cos(sx) dx = 0, \quad \int_0^\infty {}^a \sin(sx) dx = \frac{1}{s} \quad \forall s > 0 \tag{A4.65}$$

We now apply this result to the solution of a problem with concentrated capacity. Let (r, φ, z) be cylindrical coordinate in $\mathbb{R}_3$, and let D be the region

$$D^* = D \times [0 < \varphi \le 2\pi], \quad D = \{r, z: 1 < r < \infty, \ 0 < z < \infty\} \tag{A4.66}$$

The problem is: Find $u(r, z)$ satisfying the conditions

$$\left[\frac{\partial^2}{\partial r^2} + \frac{1}{r} \frac{\partial}{\partial r} + \frac{\partial^2}{\partial z^2} \right] u(r, z) = 0 \quad \forall (r, z) \in D \tag{A4.67}$$

$$\left[\frac{\partial^2}{\partial z^2} + 2\beta \frac{\partial}{\partial r} \right] u(1, z) = 0 \quad \forall z > 0 \tag{A4.68}$$

$$u(1, 0) = 1 \tag{A4.69}$$

$$\frac{\partial}{\partial z} u(r, 0) = 0 \quad \forall r > 1 \tag{A4.70}$$

$$\lim_{r \uparrow \infty} u(r, z) = 0 \quad \forall z \ge 0 \tag{A4.71}$$

$$\lim_{z \uparrow \infty} u(r, z) = 0 \quad \forall r \ge 1 \tag{A4.72}$$

This problem may be interpreted as determining the stationary distribution of temperature in a thermally anisotropic medium occupying the region

$$\bar{D} = \{r, z: 0 < r < \infty, \ 0 \le \varphi \le 2\pi, \ 0 < z < \infty\} \tag{A4.73}$$

when the medium possesses infinitely large thermal conductivity in the cylinder $D_0 = \{r, z: 0 \le r < 1, 0 \le \varphi \le 2\pi, 0 < z < \infty\}$ and finite thermal conductivity in the cylinder $D_1 = \{r, z\colon 1 < r < \infty, 0 \le \varphi \le 2\pi, 0 < z < \infty\}$, being thermally isotropic there. Condition (A4.68) is the concentrated capacity condition. Condition (A4.69) is added in order to guarantee the unique solvability of the problem.[5]

Let us seek the transform $F_c(r, \lambda)$ of $u(r, z)$:

$$v(r, \lambda) \stackrel{\text{def}}{=} F_c(r, \lambda) = \left(\frac{2}{\pi}\right)^{1/2} \int_0^\infty u(r, z) \cos(\lambda z) dz \qquad (\text{A4.74})$$

Then, by (A4.67),

$$\left[\frac{d^2}{dr^2} + \frac{1}{r}\frac{d}{dr} - \lambda^2\right] v(r, \lambda) = 0 \quad \forall r > 1 \qquad (\text{A4.75})$$

A solution of this equation, bounded at infinity, is (see Appendix 2)

$$v(r, \lambda) = c(\lambda) K(\lambda r) \qquad (\text{A4.76})$$

where $K(z)$ is the Macdonald function. The inversion formula for cosine Fourier transforms yields

$$u(r, z) = \left(\frac{2}{\pi}\right)^{1/2} \int_0^\infty c(\lambda) K(\lambda r) \cos(\lambda z) d\lambda \qquad (\text{A4.77})$$

In regard to satisfaction of the boundary condition (A4.68), we formally obtain

$$\int_0^\infty c(\lambda)[\lambda^2 K_0(\lambda) + 2\beta\lambda K_1(\lambda)] \cos(\lambda z) d\lambda = 0 \qquad (\text{A4.78})$$

If this integral were convergent, it would follow that

$$c(\lambda) \equiv 0 \qquad (\text{A4.79})$$

which in turn implies

$$u(r, z) \equiv 0 \qquad (\text{A4.80})$$

Hence convergence of the integral (A4.78) would contradict the assumption that the solution of the problem may indeed be represented as a cosine Fourier integral. However, if we interpret the integral (A4.78) as Abel-summable, then we can use the representation of zero in the form of the trigonometrical integral on the left side of (A4.65):

$$\int_0^\infty {}^a c(\lambda)[\lambda^2 K_0(\lambda) + 2\beta\lambda K_1(\lambda)] \cos(\lambda z) d\lambda = A \int_0^\infty {}^\alpha c(\lambda z) d\lambda = 0 \quad (\text{A4.81})$$

This gives

$$c(\lambda) = A/[\lambda^2 K_0(\lambda) + 2\beta\lambda K_1(\lambda)] \qquad (\text{A4.82})$$

---

[5]This problem, as well as some other problems with concentrated capacities, has been solved by Antimirov [2] using this method.

where A is a constant, to be determined from the (as yet unused) condition (A4.69):

$$A = \left(\frac{2}{\pi}\right)^{1/2} \int_0^\infty \alpha \frac{K_0(\lambda)}{\lambda^2 K_0(\lambda) + 2\beta\lambda K_1(\lambda)} d\lambda \qquad \text{(A4.83)}$$

Thus, finally,

$$u(r,z) = \frac{\int_0^\infty \alpha \frac{K_0(\lambda r)\cos(\lambda z)}{\lambda^2 K_0(\lambda) + 2\lambda\beta K_1(\lambda)} d\lambda}{\int_0^\infty \alpha \frac{K_0(\lambda)}{\lambda^2 K_0(\lambda) + 2\lambda\beta K_1(\lambda)} d\lambda} \qquad \text{(A4.84)}$$

It is easy to see that the integrals on the right of (A4.84) converge in the usual sense, so that[6]

$$u(r,z) = \frac{\int_0^\infty \frac{K_0(\lambda r)\cos(\lambda z)}{\lambda^2 K_0(\lambda) + 2\lambda\beta K_1(\lambda)} d\lambda}{\int_0^\infty \frac{K_0(\lambda)}{\lambda^2 K_0(\lambda) + 2\lambda\beta K_1(\lambda)} d\lambda} \qquad \text{(A4.85)}$$

Problems

P.A4.4.1. (a) Prove that

$$\frac{2}{\pi} \int_0^\infty \alpha^{-1} \cos(\alpha x) \sin(\alpha x) d\alpha = \begin{cases} 1 & \text{if } 0 < x < \alpha \\ 0 & \text{if } x \geq \alpha \end{cases}$$

(b) Prove that

$$\frac{2}{\pi} \int_0^\infty \sin(\alpha x) \left(\frac{\alpha\cos(\alpha a) - b\cos(\alpha b)}{\alpha} + \frac{\sin(\alpha b) - \sin(\alpha a)}{\alpha^2}\right) d\alpha$$

$$= \begin{cases} 0 & \text{if } 0 < x < a \\ x & \text{if } a < x < b \\ 0 & \text{if } x > b \end{cases}$$

Hint: Use cosine and sine Fourier transforms.

P.A4.5.1. (a) Prove the correctness of assertion (A4.85). (b) Prove that the solution (A4.84) indeed satisfies all the conditions (A4.67)–(A4.72).
Hint: See the solution of problem P10.2.3.

---

[6]See Problem A4.5.1.

Appendix 5: Examples of solution of nontrivial engineering and physical problems

In this appendix we shall show how the methods studied in the main part of the book can be used to resolve far-from-trivial engineering or physical problems. All purely technical details will be omitted, although references to the corresponding sections of the main text will be provided.

1. Heat loss in injection of heat into oil stratum [67]

A.

The injection of heat into oil strata is one of the tertiary methods of oil recovery, and has been extensively discussed in the technological literature. Although the most effective method of thermally influencing oil production seems to be the injection of superheated steam into production wells, the injection of a hot incompressible liquid is discussed in the literature primarily because it is much more amenable to analysis than steam injection, which involves consideration of the very complicated phenomenon of phase transition in porous media. Analysis of the injection of a hot incompressible liquid is incomparably easier, and provides useful information from the engineering point of view.

One of the basic problems in analyzing the process of heat injection is determining the ratio of the amount of heat used efficiently to improve oil recovery to the heat lost due to the unavoidable heat exchange between the productive stratum and surrounding unproductive rocks. In order to calculate this ratio, one need not know the spatial distribution of temperature within the productive stratum but only the overall effect of the temperature distribution. In what follows we shall present the respective computations.

Let $(\tilde{r}, \varphi, \tilde{z})$ be a cylindrical coordinate system, and consider a homogeneous productive porous layer occupying the region

$$D_0 = \{\tilde{r}, \varphi, \tilde{z} \colon 0 < \tilde{r} < \infty,\ 0 \le \varphi \le 2\pi,\ -h < \tilde{z} < 0\} \qquad \text{(A5.1)}$$

Suppose that a hot incompressible liquid is injected into D_0 through an injection well of infinitesimal radius whose axis coincides with the $\tilde{z}$ axis. Subscripts $i = 0$, 1, and -1 will refer to quantities defined in D_0 and in the layers D_1 and D_{-1} covering and underlying D_0. We assume that the productive stratum D_0 is located so deep under the earth's surface that, throughout the heat injection process, the influence of thermal conditions at this surface on the temperature distribution in D_0 is negligible. We may therefore assume that

$$D_1 = \{\tilde{r}, \varphi, \tilde{z} \colon 0 < \tilde{r} < \infty,\ 0 \le \varphi \le 2\pi,\ 0 < \tilde{z} < \infty\}$$
$$D_{-1} = \{\tilde{r}, \varphi, \tilde{z} \colon 0 < \tilde{r} < \infty,\ 0 \le \varphi \le 2\pi,\ -\infty < \tilde{z} < -h\}$$
(A5.2)

Thus, the subscripts $i = 0, 1, -1$ will refer to the regions D_i, $i = 0, 1, -1$. All the physical parameters characterizing the properties of the materials filling these regions are assumed to be constant; as usual, c_i, k_i, and a_i^2 will denote (respectively) specific heat capacity per unit volume, thermal conductivity, and thermal diffusivity. For simplicity, we shall assume that the rocks covering and underlying the productive layer D_0 have the same physical properties. The local geothermal gradient is assumed to be constant and the initial temperature distribution $\vartheta_i \big|_{\tilde{t}=0}$ in D_i to be the stationary one due to the local geothermal gradient. Since the porous layer is homogeneous we may assume that the system is axially symmetric, so that the temperatures ϑ_i, $i = 0, 1, -1$, are independent of φ.

Denote the temperature of the injected liquid by $T^*(\tilde{z}, \tilde{t})$ at the productive layer level[1] and assume, for simplicity, that

$$T^*(\tilde{z}, \tilde{t}) = T(\tilde{t}) + \vartheta_0(\tilde{z}, 0)$$
(A5.3)

Let $\tilde{u}_i$ be the perturbation of the temperature in D_i caused by the injection of hot liquid to the productive layer D_0, so that $\tilde{u}_i$ is equal to the difference between ϑ_i at time $\tilde{t} > 0$ and the initial temperature $\vartheta_i(\tilde{z}, 0)$. Let

$$\tilde{w}(\tilde{t}) = 2\pi \int_0^\infty k_0 \left(\frac{\partial}{\partial z} \tilde{u}_0 \bigg|_{\tilde{z}=-h} - \frac{\partial}{\partial z} \tilde{u}_0 \bigg|_{z=0} \right) \tilde{r}\, d\tilde{r}$$
(A5.4)

and

$$\tilde{w}_0(\tilde{t}) = \frac{\int_0^{\tilde{t}} \tilde{w}(\tau)\, d\tau}{Q c_0 \int_0^{\tilde{t}} \tilde{T}(\tau)\, d\tau}$$
(A5.5)

where Q is the output of the injection well in unit time.

Since $\tilde{w}$ is the total amount of heat delivered per unit time by the whole productive layer to the surrounding rocks through the floor and foot of the layer, $\tilde{w}_0(\tilde{t})$ is equal to the ratio of the total amount of heat lost in time $\tilde{t}$, measured from the beginning of the injection process, to the total

---

[1] See Chapter 20, Section 4, concerning the possibility of identifying the temperature at the productive layer level with the temperature within the injection well at the same level, i.e., with the temperature that can be prescribed.

amount of heat injected into the productive layer during this time. In other words, it is this quantity that we wish to calculate.

We introduce dimensionless variables by the following scaling:

$$2\nu = \frac{Qc_0}{2\pi h k_0}, \quad \lambda = \frac{k_0}{k_1}, \quad a^2 = \frac{a_1^2}{a_0^2}, \quad r = \frac{2\tilde{r}}{h}$$

$$z = \frac{2\tilde{z}}{h}, \quad t = \frac{4a_0^2 \tilde{t}}{h^2}, \quad u_i = \frac{\tilde{u}_i}{T_m}, \quad T = \frac{\tilde{T}}{T_m} \qquad (A5.6)$$

$$v(r,t) = \frac{\partial}{\partial z} u_0 \bigg|_{z=0}, \quad w(t) = -4\pi \int_0^\infty v(r,t) r \, dr$$

Here

$$T_m = \max_{0 \le t \le \infty} T(t) \qquad (A5.7)$$

Since by assumption the rocks covering and underlying the productive layer have the same physical properties, and moreover the initial temperature distribution is stationary, it follows that the temperature $u_i(r, z, t)$ is symmetric with respect to the plane $z = -1$, so that

$$\frac{\partial}{\partial z} u_0 \bigg|_{z=-1} = 0, \quad \frac{\partial}{\partial z} u_0 \bigg|_{z=0} = -\frac{\partial}{\partial z} u_0 \bigg|_{z=-2} \qquad (A5.8)$$

which implies

$$\tilde{w} = \frac{k_0 h}{2} T_m w, \quad w_0 = \frac{\int_0^t w(\tau) d\tau}{8\pi \nu \int_0^t T(\tau) d\tau} \qquad (A5.9)$$

B.

The temperature distribution in D_0 and D_1 must satisfy the following conditions

$$\frac{\partial^2}{\partial r^2} u_0 + \frac{1-2\nu}{r} \frac{\partial}{\partial r} u_0 + \frac{\partial^2}{\partial z^2} u_0 = \frac{\partial}{\partial t} u_0 \qquad (A5.10)$$

$$(r, z) \in \tilde{D}_0 = \{r, z: -1 < z < 0, \ 0 < r < \infty\}$$

$$a^2 \left[\frac{\partial^2}{\partial r^2} u_1 + \frac{1}{r} \frac{\partial}{\partial r} u_1 + \frac{\partial^2}{\partial z^2} u_1 \right] = \frac{\partial}{\partial t} u_1 \qquad (A5.11)$$

$$(r, z) \in \tilde{D}_1 = \{r, z: 0 < z < \infty, \ 0 < r < \infty\}$$

$$u_1 \big|_{t=0} = 0, \quad u_0 \big|_{r=0} = T(t), \quad \frac{\partial}{\partial z} u_0 \bigg|_{z=-1} = 0 \qquad (A5.12)$$

$$u_1 \big|_{z=0} = u_0 \big|_{z=0}, \quad \lambda \frac{\partial}{\partial z} u_0 \bigg|_{z=0} = \frac{\partial}{\partial z} u_1 \big|_{z=0} \equiv \lambda v(r,t) \qquad (A5.13)$$

Let $g_0(r, z, \rho, \zeta, t)$ and $g_1(r, z, \rho, \zeta, t)$ be the Green's functions of the second boundary-value problems of heat conduction in regions $\tilde{D}_0$ and $\tilde{D}_1$, respectively. Recall, first, that by Theorem 13.5.1 the fundamental solution of the heat-conduction equation in a multidimensional region which is a topological product of the 1-dimensional regions is the product of the corresponding 1-dimensional fundamental solutions. Obviously this theorem remains true if the product of fundamental solutions is replaced by the product of the corresponding Green's functions. Moreover, this theorem remains true if in these regions we deal not with Fourier equations but with the equation of heat conduction with forced convection. Bearing this in mind and using Theorem 13.5.1 and equalities (15.116) and (20.88), we find that

$$g_0(r, z, \rho, \zeta, t - \tau) = E_\nu(r, \rho, t - \tau) G(z, \zeta, t - \tau) \qquad (A5.14)$$

where

$$E_\nu = \left(\frac{r}{\rho}\right)^\nu \frac{\exp\left(-(r^2 + \rho^2)/4(t - \tau)\right)}{2(t - \tau)} I_\nu\left(\frac{r\rho}{2(t - \tau)}\right) \qquad (A5.15)$$

is a fundamental solution of the equation

$$\frac{\partial^2}{\partial r^2}u + \frac{1 - 2\nu}{r}\frac{\partial}{\partial r}u = \frac{\partial}{\partial t}u \qquad (A5.16)$$

(see (20.88)), and G is the Green's function of the parabolic Neumann problem in the strip $-1 < z < 0$; i.e.,

$$G = \sum_{-\infty}^{\infty}[E(z - \zeta - 2n, \tau - \tau) + E(z - \zeta + 2n, \tau - \tau)] \qquad (A5.17)$$

(see (15.116)), whereas

$$g_1 = E_0(r, \rho, a^2(t - \tau)) \cdot (E(z - \zeta, a^2(t - \tau)) + E(z + \zeta, a^2(t - \tau))) \qquad (A5.18)$$

Recall that here $E(x - \xi, t - \tau)$ is a fundamental solution of the heat-conduction equation:

$$E(x - \xi, t - \tau) = \frac{\exp\left[-(x - \xi)^2/4(t - \tau)\right]}{2\pi^{1/2}(t - \tau)^{1/2}} \qquad (A5.19)$$

Note that the following limiting equalities hold

$$\lim_{\rho\downarrow 0}\rho g_0 = \lim_{\rho\uparrow\infty}\rho g_0 = \lim_{\rho\downarrow 0}\rho\frac{\partial}{\partial\rho}g_0 = \lim_{\rho\uparrow\infty}\rho\frac{\partial}{\partial\rho}g_0 = 0 \qquad (A5.20)$$

$$\lim_{\rho\downarrow 0}g_0 = \frac{1}{\Gamma(\nu + 1)}\left[\frac{r^2}{4(t - \tau)}\right]^\nu \cdot \frac{\exp[-r^2/4(t - \tau)]}{2(t - \tau)}G(z, \zeta, t - \tau) \qquad (A5.21)$$

where $\Gamma(\nu + 1)$ is the gamma function (see Appendix 2, Sections 1 and 6).

By conditions (A5.10)–(A5.13) (see Chapter 15, Sections 1 and 3, and equality (20.98),

$$u_0(r, z, t) = 2\nu\int_0^t T(\tau)d\tau\int_{-1}^0 g_0(r, 0, z, \zeta, t - \tau)d\zeta$$

$$+ \int_0^t d\tau\int_0^\infty v(\rho, \tau)g_0(r, z, 0, \zeta, t - \tau)\rho d\rho \qquad (A5.22)$$

$$u_1(r, z, t) = -\lambda a^2\int_0^t d\tau\int_0^\infty v(\rho, \tau)g_1(r, z, \rho, 0, t - \tau)\rho d\rho \qquad (A5.23)$$

Assuming now the first of conditions (A5.13) is satisfied, we obtain the following integral equality that will in turn be used for computing $w(t)$ and $w_0(t)$ defined by (A5.6) and (A5.9), respectively:

$$\int_0^t d\tau \int_0^\infty v(\rho, \tau) K(r, \rho, t - \tau) \rho d\rho$$

$$= -2\nu \int_0^t T(\tau) d\tau \int_{-1}^0 g_0(r, 0, 0, \zeta, t - \tau) d\zeta \qquad (A5.24)$$

where

$$K = K_1 + K_2 + K_3 \qquad (A5.25)$$

$$K_1 = \lambda \frac{\exp\left[-(r^2 + \rho^2)/4a^2(t - \tau)\right]}{2a^3 \pi^{1/2}(t - \tau)^{3/2}} \cdot I_0 \left(\frac{r\rho}{2a^2(t - \tau)} \right) \qquad (A5.26)$$

$$K_2 = \left(\frac{r}{\rho} \right)^\nu \frac{\exp\left[-(r^2 + \rho^2)/4(t - \tau)\right]}{2\pi^{1/2}(t - \tau)^{3/2}} \cdot I_\nu \left(\frac{r\rho}{2(t - \tau)} \right) \qquad (A5.27)$$

$$K_3 = 2K_2 \sum_1^\infty \exp\left(-\frac{n^2}{t - \tau} \right) \qquad (A5.28)$$

Multiplying (A5.24) by $-4\pi r$, changing the order of integration, and integrating with respect to r, we obtain

$$-4\pi \int_0^t d\tau \int_0^\infty v(\rho, \tau) \rho d\rho \int_0^\infty K(r, \rho, t - \tau) r dr$$

$$= 8\pi\nu \int_0^t T(\tau) d\tau \int_{-1}^0 d\zeta \int_0^\infty g_0(r, 0, 0, \zeta, t - \tau) r dr \qquad (A5.29)$$

Define

$$J_i = \int_0^\infty K_i(r, \rho, t - \tau) r dr, \quad i = 1, 2, 3$$

$$\qquad (A5.30)$$

$$g(\zeta, t - \tau) = \int_0^\infty g_0(r, 0, 0, \zeta, t - \tau) r dr$$

It is easy to show that

$$\tilde{g} \overset{\text{def}}{=} \int_{-1}^0 g(\zeta, t) d\zeta \equiv 1 \qquad (A5.31)$$

Indeed, by (A5.21) and (A5.30) we have

$$\tilde{g} = \tilde{g}_1 \tilde{g}_2 \qquad (A5.32)$$

where

$$\tilde{g}_1 = \int_0^\infty \frac{1}{\Gamma(\nu + 1)} \left[\frac{r^2}{4(t - \tau)} \right]^\nu \frac{\exp[-r^2/4(t - \tau)]}{2(t - \tau)} r dr$$

$$\qquad (A5.33)$$

$$\tilde{g}_2 = \int_{-1}^0 G(0, \zeta, t - \tau) d\zeta$$

It follows from the definition of G that

$$F \overset{\text{def}}{=} \int_{-1}^{0} G(z, \zeta, t - \tau) d\zeta \tag{A5.34}$$

is the solution of the following boundary-value problem:

$$\frac{\partial^2}{\partial z^2} F = \frac{\partial}{\partial t} F, \quad -1 < z < 0, \quad t > \tau$$

$$\frac{\partial}{\partial z} F = 0 \quad \text{at } z = -1, \ z = 0, \quad F = 1 \quad \text{at } t = \tau \tag{A5.35}$$

By the uniqueness theorem,

$$F \equiv 1 \tag{A5.36}$$

Hence

$$\tilde{g} \equiv \tilde{g}_1 \tag{A5.37}$$

On the other hand

$$\tilde{g}_1 \equiv \int_0^\infty \frac{1}{\Gamma(\nu+1)} \left[\frac{r^2}{4(t-\tau)} \right]^\nu \frac{\exp\left[-r^2/4(t-\tau)\right]}{2\pi(t-\tau)} r \, dr$$

$$= \frac{1}{\Gamma(\nu+1)} \int_0^\infty s^\nu e^{-s} ds = 1 \tag{A5.38}$$

which completes the proof of (A5.31).

Now consider J_i, $i = 1, 2, 3$. Using the first Weber integral (A2.203), we obtain

$$J_1 = \lambda a [\pi(t - \tau)]^{-1/2}, \quad J_2 = [\pi(t - \tau)]^{-1/2}$$

$$J_3 = 2[\pi(t-\tau)]^{-1/2} \sum_1^\infty \exp\left(-\frac{n^2}{t-\tau}\right) \tag{A5.39}$$

Now we can compute $w_0(t)$, defined by (A5.6) and (A5.9). Inserting (A5.30), (A5.31), and (A5.39) into (A5.29), we obtain

$$\int_0^t w(\tau) R(t - \tau) d\tau = 8\pi\nu \int_0^t T(\tau) d\tau \tag{A5.40}$$

where

$$R(t) = (\pi t)^{-1/2} \left(1 + 2\lambda + 2 \sum_{n=1}^\infty \exp\left(-\frac{n^2}{t}\right)\right) \tag{A5.41}$$

The integral equation (A5.40) can be solved by using the Laplace–Carson transform. For every admissible function $F(t)$ we denote

$$F^*(p) \stackrel{\text{def}}{=} p \int_0^\infty e^{-pt} F(t)dt \tag{A5.42}$$

and apply the convolution theorem (20.127). This gives

$$w^*(p) = \frac{8\pi\nu T^*(p)}{R^*(p)} \tag{A5.43}$$

where

$$R^*(p) = p^{1/2}\left(1 + \lambda a + 2\sum_{n=1}^\infty e^{-2np^{1/2}}\right)$$

$$\equiv (1+\lambda a)p^{1/2}\frac{1 - \alpha e^{-2p^{1/2}}}{1 - e^{-2p^{1/2}}} \tag{A5.44}$$

and

$$\alpha = \frac{\lambda a - 1}{\lambda a + 1} \tag{A5.45}$$

Thus

$$w^*(p) = \frac{8\pi\nu}{1+\lambda a}\frac{1}{p}T^*(p)\psi^*(p) \tag{A5.46}$$

where

$$\psi^*(p) = p^{-1/2}\frac{1 - e^{-2p^{1/2}}}{1 - \alpha e^{-2p^{1/2}}} \tag{A5.47}$$

Again using the convolution theorem, we obtain

$$w(t) = \frac{8\pi\nu}{1+\lambda a}\int_0^t T(t-\tau)\psi(\tau)d\tau \tag{A5.48}$$

Let us assume that

$$T(t) = \begin{cases} 1 & \forall t < t_0 \\ 0 & \forall t > t_0 \end{cases} \tag{A5.49}$$

Then

$$w(t) = \begin{cases} w_1(t) \stackrel{\text{def}}{=} \dfrac{8\pi\nu}{1+\lambda a}\cdot\displaystyle\int_0^t \psi(\tau)d\tau & \forall t \le t_0 \\ w_1(t) - w_1(t-t_0) & \forall t > t_0 \end{cases} \tag{A5.50}$$

The original $\psi(t)$ of $\psi^*(p)$ can be obtained by means of the first Heaviside theorem 20.5.2, which is applicable if $\Re(p^{1/2}) > \sigma > 0$. By (A5.46),

$$\psi^*(p) = p^{1/2}\left(1 + (\alpha-1)\sum_{n=1}^\infty \alpha^{n-1}\exp(-2np^{1/2})\right) \tag{A5.51}$$

so that (see (20.154) and (20.158))

$$\psi(t) = (\pi t)^{-1/2} \left(1 + (\alpha - 1) \sum_{n=1}^{\infty} \alpha^{n-1} \exp\left(-\frac{n^2}{t}\right) \right) \qquad (A5.52)$$

Inserting (A5.52) into (A5.50) and changing the order of integration and summation, we finally obtain

$$w_0(t) = \frac{1}{8\pi\nu t} \int_0^t w_1(\tau) d\tau$$

$$= \begin{cases} \tilde{w}_0(t) & \forall t \le t_0 \\ \tilde{w}_0(t) - \dfrac{t-t_0}{t} \cdot \tilde{w}_0(t-t_0) & \forall t > t_0 \end{cases} \qquad (A5.53)$$

where

$$\tilde{w}_0(t) = \frac{1-\alpha}{2} \left\{ \frac{4}{3} \left(\frac{t}{\pi}\right)^{1/2} \left(1 - (1-\alpha) \sum_{n=1}^{\infty} \alpha^{n-1} \left(1 + \frac{n^2}{t}\right) \exp\left(-\frac{n^2}{t}\right) \right) \right.$$

$$\left. + (1-\alpha) \sum_{n=1}^{\infty} 2n\alpha^{n-1} \left(1 + \frac{2n^2}{3t}\right) \operatorname{erfc} \frac{n}{t^{1/2}} \right\} \qquad (A5.54)$$

This result had been compared (G.E. Malofeev) with experimental data obtained on a planar model of an oil layer. Surprisingly, this comparison showed the agreement of the above result (corresponding to the axially symmetric set-up) with experimental data for a planar heat injection (in the experiment, hot water was injected not through an isolated injection well but through an entire planar injection gallery). This coincidence is due to the fact (proved by M. Antimirov) that the total loss of heat depends only on the total amount of heat injected into the productive layer and is independent of the number and location of the injection wells. (For references see [68].)

2. Nonlinear effects in electrodiffusion equilibrium. Saturation of force of repulsion between two symmetrically charged spheres in electrolyte solution [65]

A. Nonlinear Poisson–Boltzmann equation. Formulation of problem

Subject to appropriate scaling, the flux of cations and anions in a completely dissociated univalent electrolyte is given by the Nernst–Planck equations (see Chapter 2, Section 2):

$$\mathbf{J}_+ = -\operatorname{grad} c_+ - c_+ \operatorname{grad} \varphi, \quad \mathbf{J}_- = -\operatorname{grad} c_- + c_- \operatorname{grad} \varphi \qquad (A5.55)$$

This and conservation of ions yield, for a steady state

$$\operatorname{div}[\operatorname{grad} c_+ + c_+ \operatorname{grad} \varphi] = 0, \quad \operatorname{div}[\operatorname{grad} c_- - c_- \operatorname{grad} \varphi] = 0 \qquad (A5.56)$$

with the electric potential φ satisfying the Poisson equation

$$\Delta\varphi = c_- - c_+ \tag{A5.57}$$

The fluxes $\mathbf{J}_+$ and $\mathbf{J}_-$ vanish at equilibrium (which is defined as a steady state without macroscopic fluxes). Hence, by (A5.55),

$$c_+ = e^{-\varphi}, \; c_- = e^{\varphi} \Rightarrow \Delta\varphi = e^{\varphi} - e^{-\varphi} \tag{A5.58}$$

Now consider a metal particle occupying a convex region $D_i \subset \mathbb{R}_3$, bounded by a smooth surface Σ_e charged to an electric potential $\zeta > 0$ and at equilibrium with an infinite solution. Then the equilibrium potential φ is the solution of the following boundary-value problem:

$$\Delta\varphi = e^{\varphi} - e^{-\varphi} \quad \forall p \in D_e = \mathbb{R}_3 \setminus \bar{D}_i \tag{A5.59}$$

$$\varphi(p) = \zeta \quad \forall p \in \Sigma_e \tag{A5.60}$$

$$\lim_{p \to \infty} \varphi = 0 \tag{A5.61}$$

The existence, uniqueness, and analyticity of a positive solution to this problem are easily proved by the method of successive approximations:

$$\Delta\varphi_n = e^{\varphi_{n-1}} - e^{-\varphi_{n-1}}, \quad \lim_{p \to \infty} \varphi_n = 0, \quad \varphi_n(p) = \zeta$$
$$\forall p \in \Sigma_e, \; n = 1, 2, \dots \tag{A5.62}$$

with due attention to the properties of electrostatic volume potential and the existence and uniqueness theorems for a solution of the outer Dirichlet problem (see Chapter 12, Sections 4–8).

Along with the potential φ in the electrolyte, which is the solution of the nonlinear problem (A5.59)–(A5.61), let us consider the potential $\tilde{\varphi}$ in vacuum or in a dielectric, defined as the solution of the *linear* problem

$$\Delta\tilde{\varphi} = 0 \quad \forall p \in D_e = \mathbb{R}_3 \setminus \bar{D}_i \tag{A5.63}$$

$$\tilde{\varphi}(p) = \zeta \quad \forall p \in \Sigma_e \tag{A5.64}$$

$$\lim_{p \to \infty} \tilde{\varphi} = 0 \tag{A5.65}$$

The force exerted by the electric field upon the particle is, accordingly,

$$\mathbf{F} = \frac{1}{2} \int_{\Sigma_e} \left(\frac{\partial}{\partial n_q} \varphi(q) \right)^2 d\sigma_q, \quad \tilde{\mathbf{F}} = \frac{1}{2} \int_{\Sigma_e} \left(\frac{\partial}{\partial n_q} \tilde{\varphi}(q) \right)^2 d\sigma_q \tag{A5.66}$$

Let $p \in D_e$ be fixed. Then

$$\lim_{\zeta \uparrow \infty} \tilde{\varphi}(p) = \infty, \quad \lim_{\zeta \uparrow \infty} |\tilde{F}| = \infty \tag{A5.67}$$

which is an obvious corollary of Coulomb's law. This unbounded increase of the electric field and force on a particle, due to the unbounded increase of its charge, is a consequence of the linearity of the Laplace equation. As opposed to this, in the *nonlinear* setting (A5.59)–(A5.61) a saturation of the electric potential and force occurs due to "screening" of the particle charge by the ions of the opposite sign.[2] In other words, there exist $K_1(p) > 0$ and $K_2 > 0$ such that

$$\left|\lim_{\varsigma\uparrow\infty} \varphi(p)\right| < K_1(p) < \infty \quad \forall p \notin \Sigma_e, \quad \lim_{\varsigma\uparrow\infty} |F| < K_2 \qquad (A5.68)$$

In the 1-dimensional case (i.e., for parallel plates) this follows from direct computations [15]. In what follows, we shall prove that the field and force saturation also occurs for parallel cylinders or spheres.

B. Auxiliary lemmas. Saturation of electric field

To prove the central Theorem A5.2.1 of this section, we shall need some simple lemmas.

Lemma A5.2.1. *Let $u_1(p)$ be a solution of the boundary-value problem*

$$\Delta u_i = f(u_i) \quad \forall p \in D_e \qquad (A5.69)$$
$$u_i(p) = g_i(p) \quad \forall p \in \Sigma_e \qquad (A5.70)$$
$$\lim_{p\to\infty} u_i(p) = 0, \quad i = 1, 2 \qquad (A5.71)$$

where $f(u)$ is a monotonically increasing function and

$$g_2(p) \geq g_1(p) \quad \forall p \in \Sigma_e \qquad (A5.72)$$

Then

$$u_2(p) \geq u_1(p) \quad \forall p \in D_e \qquad (A5.73)$$

PROOF. Let $G(p,q)$ be the Green's function of the Dirichlet problem in D_e. Applying the fundamental identity (9.112), taking into account that D_e is the outer region and the normal to Σ_e is directed outwards with respect to D_i, we find that

$$u_i(p) = \int_{\Sigma_e} g_i(q) \frac{\partial}{\partial n_q} G(p,q) d\sigma_q$$
$$- \int_{D_e} f[u_i(q)] G(p,q) d\omega_q, \quad i = 1, 2 \qquad (A5.74)$$

Suppose that

$$u_2(p) < u_1(p) \quad \forall p \in D_e \qquad (A5.75)$$

Since

$$g_2(p) \geq g_1(p) \quad \forall p \in \Sigma_e \quad \text{and} \quad \frac{\partial}{\partial x} f(x) > 0 \ \forall x \qquad (A5.76)$$

---

[2]The linearized version of the Poisson–Boltzmann equation is the Debye–Hückel equation. The electric field in this case is weaker than in a vacuum or a dielectric, owing to the linear screening effect of mobile ions. However, this screening effect is not sufficiently strong to cause field saturation.

it follows from the inequalities

$$G(p,q) > 0 \quad \forall (p,q) \in D_e$$
$$G(p,q) = 0, \quad \frac{\partial}{\partial n_q} G(p,q) > 0 \quad \forall p \in D_e, \ \forall q \in \Sigma_e \tag{A5.77}$$

that

$$u_2(p) - u_1(p) = \int_{\Sigma_e} [g_2(q) - g_1(q)] \frac{\partial}{\partial n_q} G(p,q) d\sigma_q$$
$$- \int_{D_e} [f(u_2(q)) - f(u_1(q))] \cdot G(p,q) \cdot d\omega_q > 0 \tag{A5.78}$$

contrary to (A5.75). Thus (A5.75) is impossible.

Suppose that inequality (A5.75) holds not everywhere in D_e but in some set $\mathfrak{M}$ of subregions $D \subset D_e$:

$$\mathfrak{M} = \cup D, \ p \in D \Rightarrow u_2(p) < u_1(p) \ \forall p \in D \tag{A5.79}$$
$$u_2(p) \geq u_1(p) \quad \forall p \in D_e \setminus \mathfrak{M} \tag{A5.80}$$

Let S be the boundary of $D \subset \mathfrak{M}$. By the continuity of $u_i(p)$, $i = 1, 2$,

$$u_2(p) = u_1(p) \quad \forall p \in S \tag{A5.81}$$

Applying the fundamental identity (A5.74) to D (where G is now the Green's function of the Dirichlet problem in D), we obtain

$$u_2(p) - u_1(p) = - \int_D [f(u_2) - f(u_1)] G(p,q) d\omega_q > 0 \tag{A5.82}$$

which again contradicts the assumption (A5.79). Hence the lemma is true. **Q.E.D.**[3]

Lemma A5.2.2. *Let $u_1(p)$ be defined by (A5.69)–(A5.71) and $u_2(p)$ by the conditions*

$$\Delta u_2 = f(u_2) + \delta(u,p) \quad \forall p \in D_e \tag{A5.83}$$
$$u_2(p) = g_1(p) \quad \forall p \in \Sigma_e \tag{A5.84}$$
$$\lim_{p \to \infty} u_2(p) = 0, \quad i = 1, 2 \tag{A5.85}$$

Then

$$\delta(u,p) \geq 0 \ \forall p \in D_e \Rightarrow u_2(p) \leq u_1(p)$$
$$\delta(u,p) \leq 0 \ \forall p \in D_e \Rightarrow u_2(p) \geq u_1(p) \ \forall p \in D_e \tag{A5.86}$$

PROOF. The proof is quite analogous to that of Lemma A5.2.1.

---

[3]This proof is based on the implicit assumption that if $D \in \mathfrak{M}$ then the fundamental identity is applicable, which in turn is true only if S is sufficiently smooth. For a more rigorous proof one must analyze the properties of the level surface $u_2 - u_1 = 0$.

Lemma A5.2.3. *Let $u(x \mid \zeta)$ be a solution of the problem*

$$\frac{d^2}{dx^2}u = e^u - e^{-u}, \quad 0 < x < \infty$$

$$\lim_{x\uparrow\infty} u = 0, \quad u(+0) = \zeta > 0$$

(A5.87)

Then

$$u(x \mid \zeta) = 2\ln\left(\frac{e^{\zeta/2} + 1 + (e^{\zeta/2} - 1) \cdot e^{-\sqrt{2}x}}{e^{\zeta/2} + 1 - (e^{\zeta/2} - 1) \cdot e^{-\sqrt{2}x}}\right)$$

(A5.88)

and also

$$\lim_{\zeta\uparrow\infty} u(x \mid s) \stackrel{\text{def}}{=} \tilde{u}(x) = 2\ln\left(\frac{1 + e^{-\sqrt{2}x}}{1 - e^{-\sqrt{2}x}}\right)$$

(A5.89)

$$J_1^* \stackrel{\text{def}}{=} \int_0^\infty (\exp[u(x \mid \zeta)] + \exp[-u(x \mid \zeta)] - 2)dx = \frac{\zeta}{\sqrt{2}}$$

(A5.90)

PROOF. It follows from (A5.87) (for the choice of sign of the square root, see the proof of Lemma A5.2.4) that

$$\frac{d}{dx}u = -\sqrt{2}(c + e^u + e^{-u})^{1/2}$$

(A5.91)

Since u tends to zero at infinity,

$$\lim_{x\uparrow\infty} \frac{d}{dx}u = 0$$

(A5.92)

so that (A5.91) implies $c = -2$. Thus

$$\frac{d}{dx}u = -\sqrt{2}(e^{u/2} - e^{-u/2})$$

(A5.93)

Integration, with due attention to the boundary condition at $x = 0$, yields (A5.88), whence (A5.89) and (A5.90) follow immediately. **Q.E.D.**

Lemma A5.2.4. *Let $\underline{u}(x \mid \zeta, \Delta)$ be a solution of the problem*

$$\frac{d^2}{dx^2}\underline{u} = e^{\underline{u}} - e^{-\underline{u}} \quad \forall x \in (0, \Delta)$$

(A5.94)

$$\underline{u}\big|_{x=0} = \zeta > 0, \quad \frac{\partial}{\partial x}\underline{u}\big|_{x=\Delta} = 0$$

Then, for any fixed $x = $ const in $(0, \Delta]$, there exists $M(x) > 0$ such that

$$0 < \underline{u}(x \mid \zeta) < M(x) < \infty \quad \forall \zeta > 0$$

(A5.95)

PROOF. First of all, we note that $\underline{u}(x)$ is a positive monotonically decreasing function. Indeed, by (A5.94),

$$x \in (0, \Delta] \wedge \underline{u}(x) > 0 \Rightarrow \frac{d^2}{dx^2}\underline{u}(x) > 0$$

(A5.96)

so that $\underline{u}(x)$ cannot have an inner maximum; by Hopf's lemma it cannot have a positive maximum at $x = \Delta$, since there $d\underline{u}(x)/dx = 0$. Neither can it have a negative minimum in $(0, \Delta)$ since $\underline{u}(x) < 0$ implies $d^2\underline{u}(x)/dx^2 < 0$. Finally, $\underline{u}(x)$ cannot have a positive inner minimum, since in that case $\underline{u}(x)$ would be convex in the neighborhood of $x = \Delta$, contrary to (A5.96). Thus $d\underline{u}/dx < 0$, so that, as in the case of (A5.91),

$$\frac{d}{dx}\underline{u} = -2 \cdot [\cosh \underline{u}(x \mid \zeta) - \alpha(\zeta)]^{1/2} \tag{A5.97}$$

where

$$\alpha(\zeta) = \cosh \underline{u}_0, \quad \underline{u}_0 \stackrel{\text{def}}{=} \underline{u}(\Delta \mid \zeta) \tag{A5.98}$$

is a constant of integration.

We claim that $\alpha(\zeta)$ remains bounded as $\zeta \uparrow \infty$. Indeed, integration of (A5.97) yields

$$\int_{\underline{u}_0}^{\zeta} \frac{dy}{[\cosh y - \alpha(\zeta)]^{1/2}} = 2\Delta \tag{A5.99}$$

The substitution

$$\cosh y = \frac{1}{t} \tag{A5.100}$$

yields

$$2\Delta = \alpha^{-1/2}(\zeta) \int_{1/\cosh(\zeta)}^{1/\alpha(\zeta)} (1 - t^2)^{-1/2} [\alpha^{-1}(\zeta) - t]^{-1/2} t^{-1/2} dt \tag{A5.101}$$

Suppose that

$$\lim_{\zeta \uparrow \infty} \alpha(\zeta) = \infty \tag{A5.102}$$

Then, using the mean value theorem, we find

$$0 < \Delta = \lim_{\zeta \uparrow \infty} \alpha^{-1/2}(\zeta) \left[\frac{\pi}{2} - \arcsin \left(\frac{\alpha^{1/2}}{\cosh^{1/2} \zeta} \right) \right]^{1/2} = 0 \tag{A5.103}$$

This contradiction shows that

$$\exists \lim_{\zeta \uparrow \infty} \alpha(\zeta) \stackrel{\text{def}}{=} \alpha_\infty < \infty \tag{A5.104}$$

Now fix $x \in (0, \Delta)$ and let $\zeta \uparrow \infty$ in the equality

$$2x = \int_{u(x|\zeta)}^{\zeta} \frac{dy}{[\cosh y - \alpha(\zeta)]^{1/2}}$$

$$= \alpha^{-1/2}(\zeta) \int_{1/\cosh(\zeta)}^{1/\cosh[u(x|\zeta)]} (1 - t^2)^{-1/2} [\alpha^{-1}(\zeta) - t]^{-1/2} t^{-1/2} dt \tag{A5.105}$$

Supposing that

$$\lim_{\zeta \uparrow \infty} u(x \mid \zeta) = \infty \tag{A5.106}$$

we again obtain

$$2x = 0 \tag{A5.107}$$

contradicting the condition $x > 0$. Thus $u(x \mid \zeta)$ is indeed bounded for every fixed $x > 0$. **Q.E.D.**

Theorem A5.2.1. *Let* $\varphi(p)$ *be a solution of problem* (A5.59)–(A5.61). *Let* $p \in D_e$ *and*

$$s = r_{p\Sigma} \overset{\text{def}}{=} \min_{q \in \Sigma} r_{pq} \tag{A5.108}$$

Then for $s > 0$, *the function* $\varphi(p \mid \zeta)$ *is nonnegative and remains bounded as* $\zeta \uparrow \infty$. *Moreover,*

$$0 \leq \varphi(p \mid \zeta) < 4 \exp(-\sqrt{2}s) + O[\exp(-2\sqrt{2}s)] \tag{A5.109}$$

PROOF. The theorem is a corollary of the following lemma.

Lemma A5.2.5. *Let* $\varphi(p)$ *be some solution of the problem*

$$\Delta\varphi = f(\varphi) \quad \forall p \in D \subset \mathbb{R}_n, \quad \varphi \mid_\Sigma = \zeta > 0 \tag{A5.110}$$

where

$$f(0) = 0, \quad \frac{d}{d\varphi} f(\varphi) > 0 \tag{A5.111}$$

and f *is an analytic function of* φ. *Then*

$$\varphi(p) > 0 \quad \forall p \in D \tag{A5.112}$$

PROOF. The proof of this lemma is almost identical to the proof in Lemma A5.2.4 that $\underline{u}$ is positive. Indeed, let D_m denote a subregion of D with boundary Σ_m such that $\varphi(p) < 0$ in D_m. By the continuity of φ,

$$\varphi(p) = 0 \quad \forall p \in \Sigma_m \tag{A5.113}$$

so that 0 is the maximum of φ in D_m. Consequently, $f(\varphi) \leq 0$ in D_m. Thus $\varphi(p)$ is superharmonic in D_m, so that its minimum is achieved on Σ_m. But this is impossible unless $\varphi(p) \equiv 0$ in D_m. Hence $\varphi(p) > 0$ for all $p \in D$. **Q.E.D.**

Now let $p \in \Sigma_e$, so that

$$\varphi(p) = \zeta \tag{A5.114}$$

Note that by (A5.59)–(A5.61) and Lemma A5.2.5,

$$\varphi(q) < \zeta \quad \forall q \in D_e \tag{A5.115}$$

Indeed, denote

$$f(\varphi) \overset{\text{def}}{=} e^\varphi - e^{-\varphi} \tag{A5.116}$$

so that

$$f(0) = 0, \quad \frac{d}{d\varphi} f(\varphi) > 0 \tag{A5.117}$$

and consequently Lemma A5.2.5 is applicable. Thus $\varphi(p)$ is subharmonic in D_e and hence φ achieves its maximum on Σ_e since $\lim_{p\to\infty} \varphi = 0$, which implies (A5.115).

Now let T_p be the tangent plane to Σ_e at p; let D_t be the region bounded by T_p in the exterior of D_i. Let (x, y, z) be a rectangular cartesian coordinate system with origin at p and x axis along the inward normal $\mathbf{n}_p$ to Σ_e at p. Consider the boundary-value problem

$$\Delta\psi(q) = e^{\psi(q)} - e^{-\psi(q)} \quad \forall q \in D_t \lim_{q\to\infty} \psi(q) = 0$$
$$\psi(q) = \zeta \quad \forall q \in T_p \tag{A5.118}$$

If $\psi(q)$ is a solution of this problem, then by (A5.115) and Lemma A5.2.1 we have

$$\varphi(q) \leq \psi(q) \quad \forall q \in \bar{D}_t \tag{A5.119}$$

Obviously $\psi(q)$ depends only on x, so that

$$\psi(q) \equiv u(x) \tag{A5.120}$$

where $u(x)$ is defined by (A5.88). Hence, by Lemma A5.2.4, $\psi(q)$ is bounded for all $x > 0$. Since p is an arbitrarily chosen point of Σ_e it follows from (A5.119) that

$$r_{q\Sigma} = s > 0 \Rightarrow 0 < \varphi(q) < u(s) < \infty \tag{A5.121}$$

The last assertion of the theorem follows immediately from (A5.119) and (A5.120), since, by (A5.88) and (A5.89),

$$u(s \mid \zeta) = 4e^{-2^{1/2}s} + O(e^{-2^{3/2}s}) \quad \text{for } \zeta \uparrow \infty, \quad s \gg 1 \tag{A5.122}$$

Remark A5.2.1. The estimate (A5.122) of the rate of decay at infinity is valid for an arbitrary system of N bodies D_{ik} with boundaries Σ_k, $k = 1, 2, \dots, N$, lying together with their boundaries in a convex region D_e bounded by a smooth surface Σ_e.

Indeed, let

$$\zeta = \sup_{p\in\cup_k\Sigma_k} \varphi(p) > 0, \quad \inf_{p\in\cup_k\Sigma_k} \varphi(p) > 0 \tag{A5.123}$$

Then, repeating the argument of Lemma A5.2.5, we conclude that $\varphi(p)$ is nonnegative in D_e. By Lemma A5.2.1 and the maximum principle,

$$\sup_{p\in\Sigma_e} \varphi(p) \leq \zeta \tag{A5.124}$$

Thus Lemma A5.2.1 and the arguments of Lemma A5.2.5 yield the desired estimate.

C. Saturation of repulsion force between two symmetrically charged spheres in electrolyte solution

The question to be discussed is whether saturation of the electric field, as asserted in Theorem A5.2.1, implies saturation of the interparticle force of

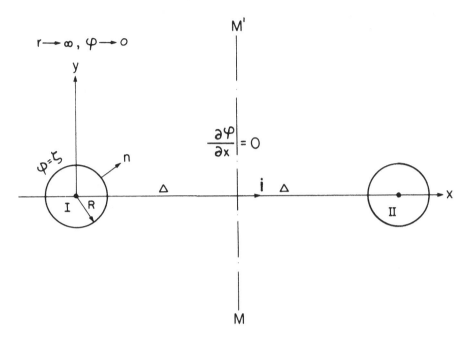

Figure A5.1. Location of two charged particles.

interaction. To fix ideas, let us consider repulsion between two symmetrically charged particles in a symmetric electrolyte solution. In the 1-dimensional case (repulsion between two parallel plates), the answer is known to be positive: the force of repulsion per unit of area of the plates saturates. This follows by direct integration of the Poisson–Boltzmann equation, as has been done in numerous publications, primarily in the context of colloid stability (see, e.g., [15]). Recall that in a vacuum, a dielectric, or an ionic system with linear screening, the repulsive force grows without bound with the charging of the particles.[4]

In the general, multidimensional case, it remains unsolved whether the repulsive force between noncongruent bodies symmetrically charged and immersed in an electrolyte solution saturates. In what follows we shall prove that saturation occurs for two spheres c_1 and c_2 in $\mathbb{R}_2$ of radius R, located at a fixed distance 2Δ from one another and charged to a constant potential ζ (see Figure A5.1). Let (x, y) be a rectangular cartesian coordinate system in $\mathbb{R}_2$, and let

$$\gamma_1 \overset{\text{def}}{=} \partial c_1 = \left\{ x, y \colon r^2 \overset{\text{def}}{=} x^2 + y^2 = R^2 \right\} \qquad \text{(A5.125)}$$

We assume that the x axis passes through the centers of c_1 and c_2 and the

_____
[4]See note 2, p. 637.

y axis through the center of c_1. The distribution of electric potential in the electrolyte surrounding the bodies is symmetric about the line $x = \Delta$. Hence this potential is a solution of the following boundary value problem:

$$\Delta\varphi = e^\varphi - e^{-\varphi}, \qquad r = (x^2 + y^2)^{1/2} > R, \quad -\infty < x < \Delta \qquad \text{(A5.126)}$$

$$\varphi\big|_{r=R} = \varsigma, \quad \frac{\partial}{\partial x}\varphi\bigg|_{x=\Delta} = 0, \quad \lim_{r\uparrow\infty}\varphi = 0 \qquad \text{(A5.127)}$$

In order to evaluate the total force acting on the particle, we note that it comprises two components: the total pressure exerted by the solution and the electrical force. It is easy to see that at equilibrium the total pressure force on the particles in the electrolyte solution vanishes. Indeed, at equilibrium the gradients of pressure p and electric potential φ satisfy the equation

$$-\operatorname{grad} p - \rho \operatorname{grad}\varphi = 0 \qquad \text{(A5.128)}$$

where ρ is the space charge density:

$$\rho = e^{-\varphi} - e^\varphi \qquad \text{(A5.129)}$$

Let $p_0 = \text{const}$ be the pressure at infinity,

$$\lim_{r\uparrow\infty} p = p_0 \qquad \text{(A5.130)}$$

Substitution of (A5.129) into (A5.128) yields

$$\operatorname{grad}\left(-p + e^\varphi + e^{-\varphi}\right) = 0 \Rightarrow -p + e^\varphi + e^{-\varphi} = \text{const} \qquad \text{(A5.131)}$$

so that by (A5.127) and (A5.130),

$$-p + e^\varphi + e^{-\varphi} - -p_0 + 2 \qquad \text{(A5.132)}$$

Hence

$$p\big|_{r=R} = p_0 + e^\varsigma + e^{-\varsigma} - 2 \qquad \text{(A5.133)}$$

so that the total pressure force on the particle c_1 vanishes:

$$\oint_{\gamma_1} p(q)\mathbf{n}_q^0 ds_q = \left(p_0 + e^\varsigma + e^{-\varsigma} - 2\right)\oint_{\gamma_1}\mathbf{n}_q^0 ds_q = 0, \quad \gamma = \bar{c}_1 \setminus c_1 \qquad \text{(A5.134)}$$

Thus we need only evaluate the electrical force of repulsion.

Theorem A5.2.2. *There exists $K = K(\Delta)$ such that*

$$|\mathbf{F}^{\text{el}}| = \frac{1}{2}\left|\oint_{\gamma_1}\left(\frac{\partial}{\partial n_q}\varphi(q)\right)^2\mathbf{n}_q^0 ds_q\right| < K(\Delta) \quad \text{as } \varsigma\uparrow\infty \qquad \text{(A5.135)}$$

PROOF. Let D^L be the doubly connected region bounded by the rectangle

$$Q = ABCDA, \quad A = (\Delta, -Y)$$
$$B = (\Delta, Y), \quad C = (-X, Y), \quad D = (-X, -Y) \qquad \text{(A5.136)}$$

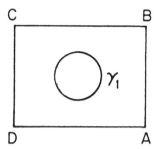

Figure A5.2. Contour $Q = ABCDA$.

and the circle γ_1 (see Figure A5.2). Let φ be the solution of the problem (A5.126) and (A5.127). Multiply the identity (A5.126) by $\operatorname{grad}\varphi$, integrate over D^L, and use the integral identity

$$\int_{D^L} \operatorname{grad}\varphi(q) \cdot \Delta\varphi(q) d\omega_q$$

$$= \int_{\Sigma^L} \left[\frac{\partial}{\partial n_q}\varphi(q) \operatorname{grad}\varphi(q) - \frac{1}{2}\operatorname{grad}^2[\varphi(q)]\mathbf{n}_q^0 \right] d\sigma_q \qquad \text{(A5.137)}$$

which is a corollary of the identity $\operatorname{div}(a\mathbf{B}) = a \cdot \operatorname{div}(\mathbf{B}) + \mathbf{B} \cdot \operatorname{grad}(a)$ and of the Gauss divergence theorem.[5] Denote

$$\mathbf{J}_1 \stackrel{\text{def}}{=} \int_{D^L} \operatorname{grad}\varphi \cdot \Delta\varphi \cdot d\omega_q, \quad \mathbf{J}_2 \stackrel{\text{def}}{=} -\int_{D^L} \operatorname{grad}\left(e^\varphi + e^{-\varphi}\right) d\omega_q \qquad \text{(A5.138)}$$

so that

$$\mathbf{J}_1 + \mathbf{J}_2 = 0 \qquad \text{(A5.139)}$$

Let $\mathbf{i}, \mathbf{j}$ be the basis unit vectors of the coordinate system (x, y) and α_q, β_q the direction cosines of the normal $\mathbf{n}_q$ to the circle c_1 at its point q. By (A5.137),

$$\mathbf{J}_1 = \sum_{i=1}^{5} \mathbf{J}_{1i} \qquad \text{(A5.140)}$$

---

[5] The proof of (A5.137) is as follows. Let $\mathbf{i}, \mathbf{j}, \mathbf{k}$ be the basis vectors of the coordinate system (x, y, z). Let ∇ be the Hamilton operator so that $\nabla\varphi \equiv \operatorname{grad}\varphi$. We have

$$\varphi_x \operatorname{div}(\nabla\varphi) + (\nabla\varphi_x)\nabla\varphi = \operatorname{div}(\varphi_x\nabla\varphi)$$

$$\Rightarrow \varphi_x\Delta\varphi = \operatorname{div}(\varphi_x\nabla\varphi) - \frac{1}{2}((\nabla\varphi)^2)_x$$

$$\Rightarrow \int_D \varphi_x\Delta\varphi d\omega = \int_\Sigma \varphi_x\frac{\partial}{\partial n}\varphi d\sigma - \frac{1}{2}\int_D ((\nabla\varphi)^2)_x d\omega$$

$$\Rightarrow \int_\Sigma \varphi_x\frac{\partial}{\partial n}\varphi d\sigma - \frac{1}{2}\cdot\int_D \operatorname{div}((\nabla\varphi)^2\mathbf{i})d\omega$$

$$\Rightarrow \int_D \varphi_x\Delta\varphi d\omega = \int_\Sigma \varphi_x\frac{\partial}{\partial n}\varphi d\sigma - \frac{1}{2}\int_\Sigma (\nabla\varphi)^2 n_x d\sigma$$

Similar identities hold for φ_y and φ_z. By summing we obtain (A5.137).

where

$$\mathbf{J}_{11} = \int_{AB} \frac{\partial}{\partial x} \varphi(\Delta, y) \operatorname{grad} \varphi(\Delta, y) dy - \frac{\mathbf{i}}{2} \int_{AB} \operatorname{grad}^2 \varphi(\Delta, y) dy$$

$$\mathbf{J}_{12} = -\frac{1}{2} \int_{CD} \operatorname{grad}^2 \varphi(q) \mathbf{n}_q^0 d\sigma_q + \int_{CD} \frac{\partial}{\partial n_q} \varphi(q) \operatorname{grad} \varphi(q) d\sigma_q$$

$$\mathbf{J}_{13} = \frac{1}{2} \int_{\gamma_1} \operatorname{grad}^2 \varphi(q) \mathbf{n}_q^0 d\sigma_q - \int_{\gamma_1} \frac{\partial}{\partial n_q} \varphi(q) \operatorname{grad} \varphi(q) d\sigma_q$$

$$\mathbf{J}_{14} = \int_{BC}, \quad \mathbf{J}_{15} = \int_{DA}$$

(A5.141)

(Recall that $\mathbf{n}_q^0$ in $\mathbf{J}_{13}$ is the outward normal to γ_1 with respect to c_1.) By the boundary conditions (A5.127),

$$\mathbf{J}_{11} = -\frac{\mathbf{i}}{2} \int_{AB} \left[\frac{\partial}{\partial y} \varphi(\Delta, y) \right]^2 dy \qquad (A5.142)$$

Next, since γ_1 is a level curve for φ, we have

$$\mathbf{J}_{13} = -\frac{1}{2} \oint_{\gamma_1} \operatorname{grad}^2 \varphi(q) \mathbf{n}_q^0 d\sigma_q = -\mathbf{F}^{\mathrm{el}} \qquad (A5.143)$$

Finally, by the symmetry of φ about the x axis,

$$\mathbf{J}_{14} + \mathbf{J}_{15} = 0 \qquad (A5.144)$$

Now, using the Gauss divergence theorem, we find that

$$\mathbf{J}_2 = -\int_{D^L} \left(\mathbf{i} \frac{\partial}{\partial x} + \mathbf{j} \frac{\partial}{\partial y} \right) \psi(q) d\omega_q = -\int_{D^L} [\mathbf{i} \operatorname{div}(\psi(q)\mathbf{i}) + \mathbf{j} \operatorname{div}(\psi(q)\mathbf{j})] d\omega_q$$

$$= \sum_{i=1}^{5} \mathbf{J}_{2i} \qquad (A5.145)$$

where

$$\psi = e^\varphi + e^{-\varphi}, \quad \mathbf{J}_{21} = -\mathbf{i} \int_{AB} \psi(\Delta, y) dy, \quad \mathbf{J}_{22} = \mathbf{i} \int_{CD} \psi(q) d\sigma_q$$

$$\mathbf{J}_{23} = \int_{\gamma_1} \psi(q) \mathbf{n}_q^0 d\sigma_q, \quad \mathbf{J}_{24} + \mathbf{J}_{25} = 0$$

(A5.146)

By the boundary condition (A5.127)

$$\mathbf{J}_{23} = (e^\varsigma + e^{-\varsigma}) \int_{\gamma_1} \mathbf{n}_q^0 d\sigma_q = 0 \qquad (A5.147)$$

Now let $X, Y \uparrow \infty$. Since φ is symmetric about the x axis,

$$\mathbf{I}_{11} \stackrel{\text{def}}{=} \lim_{Y \uparrow \infty} \mathbf{J}_{11} = -\frac{\mathbf{i}}{2} \int_{-\infty}^{\infty} \left(\frac{\partial}{\partial y} \varphi(\Delta, y) \right)^2 dy$$

$$= -\mathbf{i} \int_0^\infty \left(\frac{\partial}{\partial y} \varphi(\Delta, y) \right)^2 dy \qquad (A5.148)$$

Since φ is uniformly bounded at infinity,

$$\lim_{X\uparrow\infty} \operatorname{grad} \varphi(q) = 0 \qquad (A5.149)$$

so that

$$\mathbf{I}_{12} \stackrel{\text{def}}{=} \lim_{X\uparrow\infty} \mathbf{J}_{12} = 0 \qquad (A5.150)$$

J_{13} is independent of L, so that

$$\mathbf{I}_{13} = \lim_{X,Y\uparrow\infty} \mathbf{J}_{13} = -\mathbf{F}^{\text{el}} \qquad (A5.151)$$

And, since φ vanishes at infinity,

$$\lim_{X\uparrow\infty} \mathbf{J}_{22} = 2\mathbf{i} \int_0^Y 2dy \qquad (A5.152)$$

so that, by (A5.145)–(A5.147),

$$2\mathbf{I}_2 \stackrel{\text{def}}{=} \lim_{X,Y\uparrow\infty} \mathbf{J}_2 = \lim_{Y\uparrow\infty} \left(\mathbf{J}_{21} + \lim_{X\uparrow\infty} \mathbf{J}_{22} \right)$$

$$= -2\mathbf{i} \int_0^\infty (e^\varphi + e^{-\varphi} - 2) \Big|_{x=\Delta} dy \qquad (A5.153)$$

Thus it follows from the equality

$$\lim_{X,Y\uparrow\infty} (\mathbf{J}_1 + \mathbf{J}_2) = 0 \qquad (A5.154)$$

that the only nonvanishing x component of the electric repulsion force acting on the particle c_1 is

$$-F_x^{\text{el}} = I_1 + 2I_2 \qquad (A5.155)$$

where

$$I_1 = \int_0^\infty \left(\frac{\partial}{\partial y} \varphi(\Delta, y) \right)^2 dy, \quad I_2 = \int_0^\infty (e^\varphi + e^{-\varphi} - 2) \Big|_{x=\Delta} dy \qquad (A5.156)$$

It is easy to see that integrals I_1 and I_2 remain bounded as $\zeta \uparrow \infty$. First we note that, by arguments identical to those used in theorem A5.2.1, φ and hence also the right-hand side of equation (A5.126) are not negative.

To evaluate I_1 we note that $\varphi(x,y)$ is a monotonically decreasing function of y in the upper half plane $y > 0$. Indeed, let

$$v(x,y) = \frac{\partial}{\partial y} \varphi(x,y)$$
$$(x,y) \in D_+ = \{x,y: r > R, \ -\infty < x < \Delta, \ y > 0\} \qquad (A5.157)$$

By symmetry,

$$v(x,0) = 0 \qquad (A5.158)$$

Also φ is positive and thus by (A5.126) subharmonic in D_+. But φ and hence also v tend to zero at infinity, which means that φ has a nonnegative maximum at the boundary $r = R$, $y > 0$. Hence, by the Hopf lemma

$$v\big|_{r=R,y>0} < 0 \qquad (A5.159)$$

But by (A5.126) and (A5.157),

$$\Delta v(x,y) = \left(e^{\varphi} + e^{-\varphi}\right)v(x,y), \quad \forall(x,y) \in D_{+} \tag{A5.160}$$

Hence

$$v(x,y) \leq 0 \quad \forall(x,y) \in \bar{D}_{+} \tag{A5.161}$$

Thus $\varphi(x,y)$ is a monotonically decreasing function of y in D_{+}.   **Q.E.D.**

Using this result and the mean value theorem, we find that

$$0 \leq I_1 \leq \sup_{y>0}\left|\frac{\partial}{\partial y}\cdot\varphi(\Delta,y)\right|\cdot\left|\int_0^{\infty}\frac{\partial}{\partial y}\cdot\varphi(\Delta,y)dy\right|$$

$$= \sup_{y>0}\left|\frac{\partial}{\partial y}\cdot\varphi(\Delta,y)\right|\varphi(\Delta,0\,|\,\zeta) \tag{A5.162}$$

Denote

$$\Phi(\zeta) \stackrel{\text{def}}{=} \sup_{y>0}\left|\frac{\partial}{\partial y}\varphi(\Delta,y)\right| \tag{A5.163}$$

Arguments identical to those of Lemma A5.2.4 show that $\varphi(x,y\,|\,\zeta)$ is bounded in the region $r > R$, $R \leq x \leq \Delta$ as $\zeta \uparrow \infty$. Hence there exists $M(\Delta) > 0$ such that

$$0 \leq \varphi(\Delta,0\,|\,\zeta) < M(\Delta) < \infty \quad \text{for } \zeta \uparrow \infty \quad \forall\Delta > 0 \tag{A5.164}$$

The boundedness of φ implies in turn that there exists $\tilde{M}(\Delta) > 0$ such that

$$\Phi(\zeta) < \tilde{M}(\Delta) < \infty \quad \text{for } \zeta \uparrow \infty \tag{A5.165}$$

Thus it follows from (A5.162)–(A5.165) that I_1 is bounded:

$$I_1 < K(\Delta) = M(\Delta)\cdot\tilde{M}(\Delta) < \infty \tag{A5.166}$$

The proof that I_2 is bounded is quite analogous. Indeed, consider the straight line

$$L = \{x,y\colon y = y_0 > R,\ -\infty < x \leq \infty\} \tag{A5.167}$$

We have

$$I_2 = I_{21} + I_{22} \tag{A5.168}$$

where

$$I_{21} = \int_0^{y_0} (e^\varphi + e^{-\varphi} - 2) \bigg|_{x=\Delta} dy, \quad I_{22} = \int_{y_0}^\infty (e^\varphi + e^{-\varphi} - 2) \bigg|_{x=\Delta} dy \quad \text{(A5.169)}$$

The same arguments that led to the estimate (A5.164) show that there exists $M(\Delta, y_0) > 0$ such that

$$0 < I_{21} < M(\Delta, y_0) < \infty \quad \text{for } \zeta \uparrow \infty \quad \text{(A5.170)}$$

In order to evaluate I_{22}, we note that for $y = y_0$, $\varphi(x, y)$ is a smooth positive function of x vanishing at infinity. Therefore $\varphi(x, y_0)$ achieves its maximum at some x_0,

$$\varphi_m(\zeta) \overset{\text{def}}{=} \max \varphi(x, y_0) = \varphi(x_0, y_0) \quad \text{(A5.171)}$$

As in the proof of Theorem A5.2.1, we find that there exists $\varphi_m^\infty = \text{const} > 0$ such that

$$\varphi_m(\zeta) < \varphi_m^\infty \quad \text{for } \zeta \uparrow \infty \quad \text{(A5.172)}$$

On the other hand, let $\psi(y)$ be a solution of the 1-dimensional boundary-value problem

$$\frac{d^2}{dy^2} \psi(y) = e^\psi - e^{-\psi} \quad \forall y > y_0 \quad \text{(A5.173)}$$

$$\psi(y_0) = \varphi_m(\zeta), \quad \lim_{y\uparrow\infty} \psi = 0 \quad \text{(A5.174)}$$

so that, by Lemma A5.2.3,

$$\psi(y) = 2\ln\left(\frac{e^{\tilde\zeta/2} + 1 + (e^{\tilde\zeta/2} - 1) \cdot e^{-\sqrt{2}y}}{e^{\tilde\zeta/2} + 1 - (e^{\tilde\zeta/2} - 1) \cdot e^{-\sqrt{2}y}}\right), \tilde\zeta = \varphi_m(\zeta) \quad \text{(A5.175)}$$

Applying Lemma A5.2.1, we find that

$$0 < \varphi(x, y) \leq \psi(y) \quad \forall x \in (-\infty, \Delta), \quad \forall y > y_0 \quad \text{(A5.176)}$$

Hence

$$0 < I_{22} < \int_{y_0}^\infty (e^\psi + e^{-\psi} - 2)dy = \frac{\varphi_m(\zeta)}{\sqrt{2}} \quad \text{(A5.177)}$$

Hence, by virtue of (A5.172), we have

$$0 < I_{22} < \frac{\varphi_m^\infty}{\sqrt{2}} \quad \text{(A5.178)}$$

Thus formulas (A5.166), (A5.168), (A5.170), (A5.178), and (A5.155) imply that the repulsive force $-F_x^{\text{el}}$ remains bounded as $\zeta \uparrow \infty$, proving the assertion (A5.135) of Theorem A5.2.2. **Q.E.D.**

Remark A5.2.2. We have formulated and proved Theorem A5.2.2 for a system of two congruent spheres in $\mathbb{R}_2$. It is obvious, however, that the same arguments will establish the theorem for any two congruent, convex, symmetrically charged bodies in $\mathbb{R}_3$, provided that one of them is a mirror reflection of the other in some plane. This is obvious since the crucial point in the proof is a comparison of the solution of the problem in $\mathbb{R}_3$ with the solution of the 1-dimensional problem, which is known in explicit form, and this comparison is also valid in the more general cases. Intuitively, we expect the repulsive force to saturate in the general case of finitely many (not necessarily convex) bodies in arbitrary position and with arbitrary charges. However, this conjecture remains unproved.

3. Linear stability of Neumann's solution of two-phase Cauchy–Stefan problem

A. Introduction

The problem of the stability (linear asymptotic and global) of solutions of different versions of the Stefan problem has been extensively discussed in the mathematical and physical literature (see, e.g. [88] and [92].[6] In most papers, stability of the 1-dimensional planar Cauchy–Stefan problem is considered on the assumption that the reference front is moving at a constant rate, which implies the assumption that the initial temperature of at least one of the phases grows exponentially at infinity. It seems more appropriate from the physical point of view to assume that the initial temperature is bounded. In what follows we consider the problem of linear stability of the classical Neumann solution of the two-phase Cauchy–Stefan problem.

---

[6]Four different definitions of stability may be introduced. Let U be the solution of a perturbed problem. Then:

(a) u is called *locally stable* if there exists $T > 0$ such that for any $\varepsilon > 0$ there exists $\delta > 0$ such that $\|U - u\| < \delta$ at $t = 0 \Rightarrow \|U - u\| < \varepsilon$ for all $t \in [0, T]$.

(b) u is called *globally stable* (uniformly in x) if, for any $T > 0$ and $\varepsilon > 0$, there exists $\delta > 0$ such that $\|U - u\| < \delta$ at $t = 0 \Rightarrow \|U - u\| < \varepsilon$ for all $t \in [0, T]$.

(c) u is called *asymptotically stable* (uniformly in x) if, for any $\varepsilon > 0$, there exists $T = T(\varepsilon) > 0$ such that $t > T \Rightarrow \|U - u\| < \varepsilon$.

(d) u is called *weakly asymptotically stable* (uniformly in x) if, for any $\varepsilon > 0$ there exist $\delta > 0$ such that if $\|U - u\| < \delta$ at $t = 0$ then there exists $T > 0$ such that $\|U - u\| < \varepsilon$ for all x and $t > T$.

Here the norm is an appropriate norm defined in the 1-dimensional space $(-\infty < t < \infty)$.

In this spirit, by linear stability analysis one means the investigation of global and asymptotic behavior in the solution of a linear problem, obtained by linearization of the original problem with respect to perturbation around the solution whose stability is being studied.

Let (x, y, z) be a rectangular cartesian coordinate system. Let

$$D^1 = \{x: -\infty < x < s(t)\}, \quad D^2 = \{x: s(t) < x < \infty\}$$

$$D_t^i = D^i \times (0 < t < \infty), \quad i = 1, 2$$

$$D_3^1 = \{x, y, z: -\infty < x < S(y, z, t), \ -\infty < y, z < \infty, \ t = \text{const}\}$$

$$D_{3t}^1 = D_3^1 \times (0 \le t < \infty)$$

$$D_3^2 = \{x, y, z: S(y, z, t) < x < \infty, \ -\infty < y, z < \infty\}$$

$$D_{3t}^2 = D_3^2 \times (0 < t < \infty)$$

$$\text{(A5.179)}$$

Consider the problem

$$u_i(x, t) \in C(\bar{D}^i) \ \forall t > 0 \ \wedge \ u_i(x, t) \in C^{2,1}(D_t^i), \quad i = 1, 2 \tag{A5.180}$$

$$Lu_1(x, t) = 0 \quad \forall (x, t) \in D_t^1,$$

$$u_1(x, 0) = -\alpha_1 = \text{const} < 0 \quad \forall x \in D^1 \tag{A5.181}$$

$$L_a u_2(x, t) = 0 \quad \forall (x, t) \in D_t^2, \quad u_2(x, 0) = \alpha_2 = \text{const} > 0 \tag{A5.182}$$

$$u_1(s(t), t) = u_2(s(t), t) = 0 \quad \forall t > 0 \tag{A5.183}$$

$$\frac{ds}{dt} = \frac{\partial}{\partial x} u_1(s(t), t) - \frac{\partial}{\partial x} u_2(s(t), t) \quad \forall t > 0, \quad s(0) = 0 \tag{A5.184}$$

where we have used the notation

$$L = \frac{\partial^2}{\partial x^2} - \frac{\partial}{\partial t}, \quad L_a = a^2 \frac{\partial^2}{\partial x^2} - \frac{\partial}{\partial t} \tag{A5.185}$$

in D_t^1 and D_t^2, respectively, or

$$L_3 = \Delta - \frac{\partial}{\partial t}, \quad L_{3a} = a^2 \Delta - \frac{\partial}{\partial t} \tag{A5.186}$$

in D_{3t}^1 and D_{3t}^2, respectively. This is the problem of determining the temperature distribution in a two-phase thermally conductive medium, where the region D^1 is occupied by a solid phase and the region D^2 by a liquid one, on the assumption of a negligible contribution of convection to the temperature distribution and the rate of the motion of the phase interface due to the jump of density at the latter.

Problem (A5.180)–(A5.184) has the well-known Neumann similarity solution

$$u_1(x, t) = -\alpha_1 \frac{\text{erf}\,\gamma - \text{erf}(x/2t^{1/2})}{1 + \text{erf}(\gamma)} \tag{A5.187}$$

$$u_2(x, t) = \alpha_2 \frac{\text{erf}(x/2at^{1/2}) - \text{erf}(\gamma/a)}{1 - \text{erf}(\gamma/a)} \tag{A5.188}$$

$$s(t) = 2\gamma t^{1/2} \tag{A5.189}$$

where γ is the root of the equation

$$\gamma = \frac{\alpha_1 \exp(-\gamma^2)}{\pi^{1/2}(1 + \text{erf}\,\gamma)} - \frac{\alpha_2 \exp(-\gamma^2/a^2)}{a\pi^{1/2}[1 - \text{erf}(\gamma/a)]} \tag{A5.190}$$

In what follows we consider the problem of stability of the Neumann solution with respect to small 3-dimensional perturbations of the type

$$F(x,t)\exp[i(\alpha y + \beta z)] \qquad (A5.191)$$

B. Differential equations for perturbations

Along with problem (A5.180)–(A5.184), let us consider the 3-dimensional Stefan problem:

$$L_3 U_1 = 0 \quad \forall (x,y,z,t) \in D_{3t}^1$$
$$L_{a3} U_2 = 0 \quad \forall (x,y,z,t) \in D_{3t}^2 \qquad (A5.192)$$

$$U_i(x,y,z,0) = U_{i0}(x,y,z), \quad i = 1,2 \qquad (A5.193)$$

$$U_i(S(y,z,t),t) = 0, \quad i = 1,2 \qquad (A5.194)$$

$$\frac{\partial S}{\partial t} = \left[\frac{\partial}{\partial x} U_1 - \frac{\partial}{\partial x} U_2 - \frac{\partial S}{\partial y} \left(\frac{\partial}{\partial y} U_1 - \frac{\partial}{\partial y} U_2 \right) \right.$$
$$\left. - \frac{\partial S}{\partial z} \left(\frac{\partial}{\partial z} U_1 - \frac{\partial}{\partial z} U_2 \right) \right] \Bigg|_{x=S(y,z,t)} \qquad \forall (y,z), \ \forall t > 0 \quad (A5.195)$$

$$S(y,z,0) = S_0(y,z) \quad \forall (y,z) \qquad (A5.196)$$

Denote

$$\xi = x - \sigma(\eta,\zeta,\tau), \quad \eta = y, \quad \zeta = z, \quad \tau = t$$
$$\sigma(\eta,\zeta,\tau) = S(y,z,t) - s(t) \qquad (A5.197)$$

and write $U_i(\xi,\eta,\zeta,\tau)$ instead of $U_i(\xi + \sigma, \eta, \zeta, \tau)$, $i = 1,2$.
     Let

$$v_i(\xi,\eta,\zeta,\tau) = U_i(\xi,\eta,\zeta,\tau) - u_i(\xi,\tau), \quad i = 1,2 \qquad (A5.198)$$

Since

$$\frac{\partial S}{\partial y} = \frac{\partial \sigma}{\partial \eta}, \quad \frac{\partial S}{\partial z} = \frac{\partial \sigma}{\partial \zeta} \qquad (A5.199)$$

it follows from (A5.192) and (A5.197) that

$$\frac{\partial^2}{\partial \xi^2} U_1 \left[1 + \left(\frac{\partial \sigma}{\partial \eta} \right)^2 + \left(\frac{\partial \sigma}{\partial \zeta} \right)^2 \right] + \frac{\partial^2}{\partial \eta^2} U_1 + \frac{\partial^2}{\partial \zeta^2} U_1$$
$$- 2 \left[\frac{\partial^2}{\partial \xi \partial \eta} U_1 \frac{\partial \sigma}{\partial \eta} + \frac{\partial^2}{\partial \xi \partial \zeta} U_1 \frac{\partial \sigma}{\partial \xi} \right]$$
$$- \frac{\partial}{\partial \xi} U_1 \left[\frac{\partial^2}{\partial \eta^2} \sigma + \frac{\partial^2}{\partial \zeta^2} \sigma - \frac{\partial}{\partial \tau} \sigma \right] = \frac{\partial}{\partial \tau} U_1 \qquad (A5.200)$$

Write ξ, τ instead of x, t in (A5.181). Subtracting (A5.181) from (A5.199) and taking (A5.198) into account, we obtain

$$\Delta v_1 + \left(\frac{\partial^2}{\partial \xi^2} v_1 + \frac{\partial^2}{\partial \xi^2} u_1 \right) \left[\left(\frac{\partial \sigma}{\partial \eta} \right)^2 + \left(\frac{\partial \sigma}{\partial \zeta} \right)^2 \right]$$

$$- 2 \left[\frac{\partial^2}{\partial \xi \partial \eta} v_1 \frac{\partial \sigma}{\partial \eta} + \frac{\partial^2}{\partial \xi \partial \zeta} v_1 \frac{\partial \sigma}{\partial \xi} \right]$$

$$- \left(\frac{\partial}{\partial \xi} u_1 + \frac{\partial}{\partial \xi} \sigma \right) \left[\frac{\partial^2}{\partial \eta^2} \sigma + \frac{\partial^2}{\partial \zeta^2} \sigma - \frac{\partial}{\partial \tau} \sigma \right] = \frac{\partial}{\partial \tau} v_1 \qquad \text{(A5.201)}$$

In linear stability theory, all quadratic terms in the perturbations v_1, σ, and their derivatives must be dropped. Thus we obtain

$$\Delta v_1 - \frac{\partial}{\partial \xi} u_1 \left[\frac{\partial^2}{\partial \eta^2} \sigma + \frac{\partial^2}{\partial \zeta^2} \sigma - \frac{\partial}{\partial \tau} \sigma \right] = \frac{\partial}{\partial \tau} v_1 \qquad \text{(A5.202)}$$

$$-\infty < \xi < s(\tau), \quad -\infty < \eta, \zeta < \infty$$

Similarly,

$$a^2 \Delta v_2 - \frac{\partial}{\partial \xi} u_2 \left[a^2 \left(\frac{\partial^2}{\partial \eta^2} \sigma + \frac{\partial^2}{\partial \zeta^2} \sigma \right) - \frac{\partial}{\partial \tau} \sigma \right] = \frac{\partial}{\partial \tau} v_2 \qquad \text{(A5.203)}$$

$$-\infty < s(\tau) < \xi, \quad -\infty < \eta, \zeta < \infty$$

Now the Stefan condition (A5.195), together with (A5.197) and (A5.198), implies

$$\frac{\partial \sigma}{\partial \tau} = \frac{\partial v_1}{\partial \xi} - \frac{\partial v_2}{\partial \xi} + \left(\frac{\partial v_1}{\partial \xi} - \frac{\partial v_2}{\partial \xi} + \frac{\partial u_1}{\partial \xi} - \frac{\partial u_2}{\partial \xi} \right) \left(\left(\frac{\partial \sigma}{\partial \eta} \right)^2 + \left(\frac{\partial \sigma}{\partial \eta} \right)^2 \right)$$

$$- \left(\frac{\partial v_1}{\partial \eta} - \frac{\partial v_2}{\partial \eta} \right) \frac{\partial \sigma}{\partial \eta} - \left(\frac{\partial v_1}{\partial \zeta} - \frac{\partial v_2}{\partial \zeta} \right) \frac{\partial \sigma}{\partial \zeta}$$

$$\xi = s(t), \quad -\infty < \eta, \zeta < \infty \qquad \text{(A5.204)}$$

or, omitting nonlinear terms,

$$\frac{\partial \sigma}{\partial \tau} = \frac{\partial v_1}{\partial \xi} - \frac{\partial v_2}{\partial \xi}, \quad \xi = s(t), \quad -\infty < \eta, \zeta \leq \infty \qquad \text{(A5.205)}$$

Besides (A5.202), (A5.203), and (A5.205) we add the conditions

$$v_i(s(\tau), \tau) = 0, \quad i = 1, 2 \qquad \text{(A5.206)}$$

$$v_i(\xi, \eta, \zeta, 0) = v_{i0}(\xi, \eta, \zeta), \quad i = 1, 2 \qquad \text{(A5.207)}$$

In what follows we write x, y, z, t instead of ξ, η, ζ, τ. Let

$$v_i(x, y, x, t) = w_i(x, t) \exp[i(\alpha y + \beta z)], \quad i = 1, 2$$

$$\sigma(y, z, t) = f(t) \exp[i(\alpha y + \beta z)] \qquad \text{(A5.208)}$$

Then

$$\frac{\partial^2}{\partial x^2}w_1 - k^2 w_1 + \frac{\partial}{\partial x}u_1\left(\frac{d}{dt}f + k^2 f\right)$$

$$= \frac{\partial}{\partial t}w_1, \quad -\infty < x < s(t) \tag{A5.209}$$

$$\frac{\partial^2}{\partial x^2}w_2 - k^2 w_2 + \frac{\partial}{\partial x}u_2\left(a^{-2}\frac{d}{dt}f + k^2 f\right)$$

$$= a^{-2}\frac{\partial}{\partial t}w_1, \quad s(t) < x < \infty \tag{A5.210}$$

$$\frac{d}{dt}f = \frac{\partial}{\partial x}w_1 - \frac{\partial}{\partial x}w_2, \quad x = s(t) \tag{A5.211}$$

$$w_i[s(t),t] = 0, \quad w_i(x,0) = w_{i0}(x), \quad i = 1,2, \quad f(0) = f_0 \tag{A5.212}$$

where

$$k^2 = \alpha^2 + \beta^2 \tag{A5.213}$$

Setting

$$w_1 = \psi_1(x,t)e^{-k^2 t}, \quad w_2 = \psi_2(x,t)e^{-a^2 k^2 t}, \quad f(t) = \varphi(t)e^{-k^2 t} \tag{A5.214}$$

we find that

$$\frac{\partial^2}{\partial x^2}\psi_1 + \frac{\partial}{\partial x}u_1\frac{\partial}{\partial t}\varphi = \frac{\partial}{\partial t}\psi_1, \quad -\infty < x < s(t)$$

$$a^2\frac{\partial^2}{\partial x^2}\psi_2 + \frac{\partial}{\partial x}u_2\left[\frac{\partial}{\partial t}\varphi + k^2(a^2-1)\varphi\right]\exp[-k^2(a^2-1)] \tag{A5.215}$$

$$= \frac{\partial}{\partial t}\psi_2, \quad s(t) < x < \infty$$

$$\psi_i = 0 \text{ at } x = s(t), \quad \psi_i(x,0) = w_{i0}(x), \quad i = 1,2 \tag{A5.216}$$

and

$$\frac{d}{dt}f = \frac{\partial}{\partial x}\psi_1 \cdot e^{-k^2 t} - \frac{\partial}{\partial x}\psi_2 \cdot e^{-a^2 k^2 t} \text{ at } x = s(t), \quad f(0) = f_0 \tag{A5.217}$$

C. Reducing problem (A5.215) **and** (A5.216) **to a system of integral equations**

System (A5.215) and (A5.216) has a global solution (ψ_1, ψ_2, f) which is an analytic function of t for all $t > 0$ and of x for $x \neq s(T)$, if w_{i0} and f_0 are continuous functions.[7] Taking this into account, let T be an arbitrarily

---

[7]See [25]. We refer to Gevrey's treatise because it provides the most profound information concerning properties of parabolic operators. In particular, it is proved there that solutions of nonlinear parabolic equations are analytic in the regions indicated if the input data are analytic. However, the assertion of the existence of the global solution of the system (A5.215) and (A5.216) follows from the obvious applicability of results of Chapter 17, Section 1, not only to a single equation but to a system of an arbitrary finite number of equations. (See also [82].)

large fixed number and

$$\Psi_i(x) \overset{\text{def}}{=} \psi_i(x, T), \quad i = 1, 2, \quad f^0 \overset{\text{def}}{=} f(T) \tag{A5.218}$$

Define

$$q_i(t) = \frac{\partial}{\partial x} \psi_i[s(t), t], \quad t \geq T, \; i = 1, 2 \tag{A5.219}$$

The functions $\psi_i(x, t)$ have the integral representation (see (15.22))

$$\psi_1(x, t) = \int_{-\infty}^{s(T)} \Psi_1(\xi) E(x - \xi, t - T) d\xi + \int_T^t q_1(\tau) E(x - s(\tau), t - \tau) d\tau$$

$$+ \int_T^t d\tau \int_{-\infty}^{s(\tau)} \frac{\partial}{\partial \xi} u_1(\xi, \tau) \frac{d}{d\tau} \varphi(\tau) E(x - \xi, t - \tau) d\xi \tag{A5.220}$$

$$\psi_2(x, t) = \int_{s(T)}^{\infty} \Psi_2(\xi) E[x - \xi, a^2(t - T)] d\xi$$

$$- a^2 \int_T^t q_2(\tau) E[x - s(\tau), a^2(t - \tau)] d\tau$$

$$+ \int_T^t d\tau \int_{s(\tau)}^{\infty} \frac{\partial}{\partial \xi} u_2(\xi, \tau) \left(\frac{d}{d\tau} \varphi(\tau) + k^2(a^2 - 1) \varphi(\tau) \right)$$

$$\times \exp[k^2(a^2 - 1)] E[x - \xi, a^2(t - \tau)] d\xi \tag{A5.221}$$

Taking into account that

$$\psi_i[s(t), t] = 0, \quad i = 1, 2 \tag{A5.222}$$

and applying Theorem 16.2.1 (discontinuity of double-layer heat potentials), we find that

$$q_1(t) = \tilde{\Phi}_1(t) + 2 \int_T^t q_1(\tau) \frac{\partial}{\partial x} E[s(t) - s(\tau), t - \tau] d\tau$$

$$q_2(t) = \tilde{\Phi}_2(t) - 2a^2 \int_T^t q_2(\tau) \frac{\partial}{\partial x} E[s(t) - s(\tau), a^2(t - \tau)] d\tau \tag{A5.223}$$

where

$$\tilde{\Phi}_1(t) = \Phi_1(t) + 2 \int_T^t \frac{d}{d\tau} \varphi(\tau) d\tau \int_{-\infty}^{s(\tau)} \frac{\partial}{\partial \xi} u_1(\xi, \tau) \frac{\partial}{\partial x} E[s(t) - \xi, t - \tau] d\xi$$

$$\tilde{\Phi}_2(t) = \Phi_2(t) + 2 \int_T^t \left(\frac{d}{d\tau} \varphi(\tau) + k^2(a^2 - 1) \varphi(\tau) \right) \exp[k^2(a^2 - 1)\tau] d\tau$$

$$\times \int_{s(\tau)}^{\infty} \frac{\partial}{\partial \xi} u_2(\xi, \tau) \frac{\partial}{\partial x} E[x - \xi, a^2(t - \tau)] d\xi \tag{A5.224}$$

and

$$\Phi_1(t) = 2 \int_{-\infty}^{s(T)} \frac{d}{d\xi} \Psi_1(\xi) E(s(\tau) - \xi, t - T) d\xi$$

$$\Phi_2(t) = 2 \int_{s(T)}^{\infty} \frac{d}{d\xi} \Psi_2(\xi) E[s(\tau) - \xi, a^2(t - T)] d\xi \tag{A5.225}$$

Let Γ_i, $i = 1, 2$, be the resolvents of equations (A5.223).[8] Then

$$q_1(t) = \tilde{\Phi}_1(t) + \int_T^t \tilde{\Phi}_1(\tau)\Gamma_1(t, \tau)d\tau$$

$$q_2(t) = \tilde{\Phi}_2(t) + \int_T^t \tilde{\Phi}_2(\tau)\Gamma_2(t, \tau)d\tau \qquad \text{(A5.226)}$$

which may be rewritten as

$$q_1(t) = Q_1(t) + \int_T^t \frac{d}{d\tau}\varphi_1(\tau)R_1(t, \tau)d\tau \qquad \text{(A5.227)}$$

$$q_2(t) = Q_2(t) + \int_T^t \left(\frac{d}{d\tau}\varphi(\tau) + k^2(a^2 - 1)\varphi(\tau)\right)$$
$$\times \exp[k^2(a^2 - 1)\tau]R_2(\tau)d\tau \qquad \text{(A5.228)}$$

where

$$Q_i(t) = \Phi_i(t) + \int_T^t \Phi_i(\tau)\Gamma_i(t, \tau)dt, \quad i = 1, 2 \qquad \text{(A5.229)}$$

and

$$R_1(t, \tau) = 2 \int_{-\infty}^{s(\tau)} \frac{\partial}{\partial \xi} u_1(\xi, \tau)\frac{\partial}{\partial x}E[s(t) - \xi, t - \tau]d\xi + \int_\tau^t \Gamma_1(t, \mu)d\mu$$
$$\cdot 2 \int_{-\infty}^{s(\tau)} \frac{\partial}{\partial \xi} u_1(\xi, \tau)\frac{\partial}{\partial x}E[s(\mu) - \xi, \mu - \tau]d\xi \qquad \text{(A5.230)}$$

$$R_2(t, \tau) = 2 \int_{s(\tau)}^{\infty} \frac{\partial}{\partial \xi} u_2(\xi, \tau)\frac{\partial}{\partial x}E[s(t) - \xi, a^2(t - \tau)]d\xi + \int_\tau^t \Gamma_2(t, \mu)d\mu$$
$$\cdot 2 \int_{s(\tau)}^{\infty} \frac{\partial}{\partial \xi} u_2(\xi, \tau)\frac{\partial}{\partial x}E[s(\mu) - \xi, a^2(\mu - \tau)]d\xi \qquad \text{(A5.231)}$$

Equalities (A5.214), (A5.217), (A5.219), (A5.227), and (A5.228) finally yield

$$\frac{d}{dt}f(t) = \tilde{Q}(t) + \int_T^t \left(\frac{d}{dt}f(\tau) + k^2 f(\tau)\right)\exp[-k^2(t - \tau)]R_1(t, \tau)d\tau$$
$$- \int_T^t \left(\frac{d}{dt}f(\tau) + a^2 k^2 f(\tau)\right)$$
$$\times \exp[-a^2 k^2(t - \tau)]R_2(t, \tau)d\tau \qquad \text{(A5.232)}$$

where

$$\tilde{Q}(t) = Q_1(t)e^{-k^2 t} - Q_2 e^{-a^2 k^2 t} \qquad \text{(A5.233)}$$

D. Upper and lower bounds for $R_i(t, \tau)$ and $Q_i(t)$, $i = 1, 2$

Consider first the kernels R_1 and R_2 defined by (A5.230) and (A5.231). It follows from the definition of the resolvents Γ_1 and Γ_2 that R_1 and R_2 are

---

[8] We refer to the elementary theory of integral equations (see e.g. [28]).

solutions of the integral equations

$$R_1(t,\tau) = 2 \int_{-\infty}^{s(\tau)} \frac{\partial}{\partial \xi} u_1(\xi,\tau) \frac{\partial}{\partial x} E[s(t) - \xi, t - \tau] d\xi$$

$$+ 2 \int_{\tau}^{t} \frac{\partial}{\partial x} E[s(t) - s(\mu), t - \mu] R_1(\mu,\tau) d\mu \qquad \text{(A5.234)}$$

$$R_2(t,\tau) = 2a^2 \int_{s(\tau)}^{\infty} \frac{\partial}{\partial \xi} u_2(\xi,\tau) \frac{\partial}{\partial x} E[s(t) - \xi, a^2(t - \tau)] d\xi$$

$$+ 2a^2 \int_{\tau}^{t} \frac{\partial}{\partial x} E[s(t) - s(\mu), a^2(t - \mu)] R_2(\mu,\tau) d\mu \qquad \text{(A5.235)}$$

In turn, this means the following. Let U_1 and U_2 be uniformly bounded solutions of the boundary-value problems

$$\frac{\partial^2}{\partial x^2} U_1 = \frac{\partial}{\partial t} U_1, \quad -\infty < x < s(t), \quad t > \tau$$

$$U_1(s(t), t) = 0, \qquad U_1(x,\tau) = \frac{\partial}{\partial x} u_1(x,\tau) \qquad \text{(A5.236)}$$

$$a^2 \frac{\partial^2}{\partial x^2} U_2 = \frac{\partial}{\partial t} U_1, \quad s(t) < x < \infty, \quad t > \tau$$

$$U_2(s(t), t) = 0, \qquad U_2(x,\tau) = \frac{\partial}{\partial x} u_2(x,\tau) \qquad \text{(A5.237)}$$

Then

$$R_i(t,\tau) = \frac{\partial}{\partial x} U_i(s(t), t), \quad i = 1, 2 \qquad \text{(A5.238)}$$

Now consider the following auxiliary problems. Take an arbitrary $t > \tau$ and define

$$\bar{s}(T) = s(\tau) + [s(t) - s(\tau)] \frac{T - \tau}{t - \tau}$$

$$\underline{s}(T) = s(t) - \frac{d}{dt} s(t)(t - T) \qquad \text{(A5.239)}$$

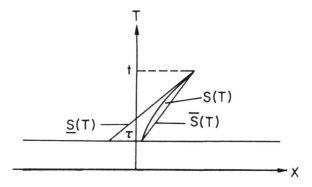

Figure A5.3.   Auxiliary contours $\underline{s}(T)$ and $\bar{s}(T)$.

(see Figure A5.3). Since $s(T)$ is convex,

$$\underline{s}(T) < s(T) < \bar{s}(T) \quad \forall T \in (\tau, t), \quad s(t) = \underline{s}(t) = \bar{s}(t) \tag{A5.240}$$

Now let $\underline{U}_1, \bar{U}_1, \underline{U}_2, \bar{U}_2$ be uniformly bounded solutions of the problems

$$\frac{\partial^2}{\partial x^2} \underline{U}_1 = \frac{\partial}{\partial T} \underline{U}_1, \quad -\infty < x < \underline{s}(T), \quad \tau < T \le t \tag{A5.241}$$

$$\underline{U}_1(\underline{s}(T), T) = 0, \quad \underline{U}_1(x, \tau) = U_1(x, \tau), \quad -\infty < x < \underline{s}(\tau)$$

$$\frac{\partial^2}{\partial x^2} \bar{U}_1 = \frac{\partial}{\partial T} \bar{U}_1, \quad -\infty < x < \bar{s}(T), \quad \tau < T \le t \tag{A5.242}$$

$$\bar{U}_1(\bar{s}(T), T) = 0, \quad \bar{U}_1(x, \tau) = U_1(x, \tau), \quad -\infty < x < \bar{s}(\tau)$$

$$a^2 \frac{\partial^2}{\partial x^2} \bar{U}_2 = \frac{\partial}{\partial T} \bar{U}_2, \quad \bar{s}(T) < x < \infty, \quad \tau < T \le t \tag{A5.243}$$

$$\bar{U}_2(\bar{s}(T), T) = 0, \quad \bar{U}_2(x, \tau) = U_2(x, \tau), \quad \bar{s}(\tau) < x < \infty$$

$$a^2 \frac{\partial^2}{\partial x^2} \underline{U}_2 = \frac{\partial}{\partial T} \underline{U}_2, \quad \underline{s}(T) < x < \infty, \quad \tau < T \le t \tag{A5.244}$$

$$\underline{U}_2(\underline{s}(T), T) = 0, \quad \bar{U}_2(x, \tau) = 0, \quad \underline{s}(\tau) < x < \bar{s}(\tau)$$

$$\underline{U}_2(x, \tau) = U_2(x, \tau), \quad \bar{s}(\tau) < x < \infty$$

Note that, by (A5.187) and (A5.188),

$$U_1(x, \tau) = \alpha_1 (\pi \tau)^{-1/2} (1 + \operatorname{erf} \gamma)^{-1} \exp(-x^2/4\tau) \tag{A5.245}$$

$$U_2(x, \tau) = \alpha_2 (a(\pi \tau)^{1/2})^{-1} (1 - \operatorname{erf}(\gamma/a))^{-1} \exp(-x^2/4a^2\tau) \tag{A5.246}$$

Since U_i, $\underline{U}_i$ and $\bar{U}_i$, $i = 1, 2$, tend uniformly to zero at infinity and are uniformly bounded, we see, by the maximum principle, that

$$0 < \underline{U}_1 < U_1, \quad -\infty < x < \underline{s}(T), \quad \tau < T \le t$$

$$0 < U_1 < \bar{U}_1, \quad -\infty < x < \bar{s}(T), \quad \tau < T \le t \tag{A5.247}$$

$$0 = \underline{U}_1 = U_1 = \bar{U}_2 \quad \text{at } T = t, \ x = s(t)$$

Hence

$$\underline{R}_1 < R_1 < \bar{R}_1 < 0 \tag{A5.248}$$

where

$$\bar{R}_1 = \frac{\partial}{\partial x}U_1, \ \underline{R}_1 = \frac{\partial}{\partial x}\bar{U}_1 \quad \text{at } x = s(t), \ T = t \tag{A5.249}$$

Quite similarly,

$$0 < \underline{R}_2 < R_2 < \bar{R}_2 \tag{A5.250}$$

where

$$\bar{R}^2 = \frac{\partial}{\partial x}U_2, \ \underline{R}_2 = \frac{\partial}{\partial x}\bar{U}_2 \quad \text{at } x = s(t), \ T = t \tag{A5.251}$$

Using the integral representations (15.54) and (15.55) for solutions of the Dirichlet problem, after substitution of the Green's function (15.139) we find (after straightforward, though rather lengthy, computations) that

$$\begin{aligned} \underline{R}_1 &= A(R_1^* + \underline{R}_1^*), & \bar{R}_1 &= A(R_1^* + \bar{R}_1^*) \\ \underline{R}_2 &= B(R_2^* + \underline{R}_2^*), & \bar{R}_2 &= B(R_2^* + \bar{R}_2^*) \end{aligned} \tag{A5.252}$$

where

$$\begin{aligned} A &= -\alpha_1 \exp(-\gamma^2)[\pi^{1/2}(1 + \operatorname{erf}\gamma)]^{-1} \\ B &= \alpha_2 \exp(-\gamma^2/a^2)[a\pi^{1/2}(1 - \operatorname{erf}\gamma/a)]^{-1} \end{aligned} \tag{A5.253}$$

$$R_1^* = \tau^{1/2}t^{-1}[\pi(t-\tau)]^{-1/2}\exp\left[-\frac{\gamma^2(t^{1/2}-\tau^{1/2})}{t^{1/2}+\tau^{1/2}}\right] \tag{A5.254}$$

$$R_2^* = \tau^{1/2}(at)^{-1}[\pi(t-\tau)]^{-1/2}\exp\left[-\frac{(\gamma/a)^2(t^{1/2}-\tau^{1/2})}{t^{1/2}+\tau^{1/2}}\right] \tag{A5.255}$$

$$\underline{R}_1^* = \gamma(2t)^{-1}\left(1 + \operatorname{erf}\left[\frac{\gamma(t^{1/2}-\tau^{1/2})^{1/2}}{(t^{1/2}+\tau^{1/2})^{1/2}}\right]\right) \tag{A5.256}$$

$$\bar{R}_1^* = \gamma\tau^{1/2}t^{-1}(t^{1/2}+\tau^{1/2})^{-1}\left(1 + \operatorname{erf}\left[\frac{\gamma(t^{1/2}-\tau^{1/2})^{1/2}}{(t^{1/2}+\tau^{1/2})^{1/2}}\right]\right) \tag{A5.257}$$

$$\underline{R}_2^* = -\gamma(2a^2t)^{-1}\operatorname{erfc}\left[\frac{\gamma a^{-1}(t^{1/2}-\tau^{1/2})^{1/2}}{(t^{1/2}+\tau^{1/2})^{1/2}}\right] \tag{A5.258}$$

$$\bar{R}_2^* = -\gamma\tau^{1/2}(a^2t)^{-1}(t^{1/2}+\tau^{1/2})^{-1}\operatorname{erfc}\left[\frac{\gamma a^{-1}(t^{1/2}-\tau^{1/2})^{1/2}}{(t^{1/2}+\tau^{1/2})^{1/2}}\right] \tag{A5.259}$$

We now estimate the free terms Q_1 and Q_2 of equations (A5.227) and (A5.228). In view of their definition (A5.229) and that of the resolvents Γ_i, $i = 1, 2$, we see that Q_1 and Q_2 are solutions of the integral equations

$$Q_1(t) = \Phi_1(t) + 2\int_T^t Q_1(\tau)\frac{\partial}{\partial x}E[s(t) - s(\tau), t - \tau]d\tau \tag{A5.260}$$

$$Q_2(t) = \Phi_2(t) - 2a^2\int_T^t Q_2(\tau)\frac{\partial}{\partial x}E[s(t) - s(\tau), a^2(t-\tau)]d\tau \tag{A5.261}$$

By (A5.225), (A5.226), and (A5.222), this implies the following assertion.

Let V_1 and V_2 be solutions of the problems

$$\frac{\partial^2}{\partial x^2}V_1 = \frac{\partial}{\partial t}V_1, \quad -\infty < x < s(t), \ t > T$$

$$V_1[s(t), t] = 0, \quad V_1(x, T) = \Psi_1(x), \quad |V_1| < \infty$$

(A5.262)

$$a^2\frac{\partial^2}{\partial x^2}V_2 = \frac{\partial}{\partial t}V_2, \quad s(t) < x < \infty, \ t > T$$

$$V_2[s(t), t] = 0, \quad V_2(x, T) = \Psi_2(x), \quad |V_2| < \infty$$

(A5.263)

Then

$$Q_i(t) = \frac{\partial}{\partial x}V_i[s(t), t], \quad i = 1, 2$$

(A5.264)

Recalling that $\Psi_i(x)$ are values of $\psi_i(x, T)$, $T > 0$, and that the solution of the parabolic Dirichlet problem is uniquely determined by the input data, we may assume that $V_i(x, t)$ is defined for $0 \le t < \infty$ and equal to $\psi_i(x, 0)$ at $t = 0$. Assume that

$$\psi_i(x, 0) = 0 \quad \forall x \notin (\alpha_i, \beta_i), \quad \alpha_1 < \beta_1 < \alpha_2 < \beta_2$$

$$|\psi_i(x, 0)| < N_i \quad \forall x \in (\alpha_i, \beta_i), \quad i = 1, 2$$

(A5.265)

Constructing similar bounds for Q_i, using (15.54) and (15.55) with the Green's function (15.139), differentiating with respect to x, and majorizing the results according to (A5.265), we obtain (again after straightforward but lengthy computations)

$$|Q_1| < N_1(2\pi^{1/2})^{-1}\int_{\alpha_1}^{\beta_1} \exp\left[\frac{-(2\gamma t^{1/2} - \tau)^2}{4t}\right]\tau t^{-1/2}d\tau$$

(A5.266)

$$|Q_2| < N_2(2a^3\pi^{1/2})^{-1}\int_{\alpha_2}^{\beta_2} \exp\left[-\frac{(2\gamma t^{1/2} - \tau)^2}{4a^2 t}\right]\tau t^{-1/2}d\tau$$

(A5.267)

Thus there is a constant $c > 0$ such that

$$|Q_i| < cNt^{-1/2}, \quad i = 1, 2, \quad N = \max(N_1, N_2)$$

(A5.268)

This result was obtained under the assumption that the temperature values lie in the strips (α_i, β_i), $i = 1, 2$. Now suppose instead that they are distributed along the x axis, so that

$$\psi_i(x, 0) = O(1)|x|^{-\alpha} \quad \text{as } |x| \uparrow \infty$$

$$0 < \alpha < 2, \quad \operatorname{sign}\psi_i(x, 0) = \text{const}, \quad i = 1, 2$$

(A5.269)

Then we obtain, instead of (A5.268),

$$|Q_i| < cNt^{-(1+\alpha)/2}$$

(A5.270)

Finally, if α in (A5.269) is greater than 2, then (A5.268) holds.

E. Asymptotic solution of equation (A5.232) for $k \neq 0$

Using (A5.247), (A5.249), and (A5.252)–(A5.259), we see that (A5.232) may be rewritten as

$$\frac{d}{dt} f(t) = \tilde{Q}(t) + 2A \int_0^{t-T/t} \exp(-k^2 t s^2) \left[h_0(1 - s^2) + s h_1[t(1 - s^2)] \right]$$

$$\times \left(\frac{d}{dt} f[t(1 - s^2)] + k^2 f[t(1 - s^2)] \right) ds$$

$$- 2B \int_0^{t-T/t} \exp(-k^2 a^2 t s^2) \left[h_2(1 - s^2) + s h_3[t(1 - s^2)] \right]$$

$$\times \left(\frac{d}{dt} f[t(1 - s^2)] + k^2 a^2 f[t(1 - s^2)] \right) ds \tag{A5.271}$$

where

$$h_0(x) = \pi^{-1/2} x^{1/2} \exp[-\gamma^2 (1 - x^{1/2})/(1 + x^{1/2})]$$
$$h_2(x) = a^{-1} \pi^{-1/2} x^{1/2} \exp[-\gamma^2 a^{-2}(1 - x^{1/2})/(1 + x^{1/2})] \tag{A5.272}$$

and h_1, h_3 are such that

$$x^{1/2}(1 + x^{1/2})^{-1} \operatorname{erfc}\left[-\frac{\gamma(1 - x^{1/2})^{1/2}}{(1 + x^{1/2})^{1/2}} \right]$$

$$< h_1(tx) < \frac{1}{2} \gamma \operatorname{erfc}\left[-\frac{\gamma(1 - x^{1/2})^{1/2}}{(1 + x^{1/2})^{1/2}} \right] \tag{A5.273}$$

$$\gamma a^{-1} x^{1/2}(1 + x^{1/2})^{-1} \operatorname{erfc}\left[\frac{\gamma a^{-1}(1 - x^{1/2})^{1/2}}{(1 + x^{1/2})^{1/2}} \right]$$

$$< h_3(tx) < \frac{1}{2} \gamma a^{-1} \operatorname{erfc}\left[\frac{\gamma a^{-1}(1 - x^{1/2})^{1/2}}{(1 + x^{1/2})^{1/2}} \right] \tag{A5.274}$$

Note that the solution $df(t)/dt$ of equation (A5.232) is globally bounded (see Chapter 17, Section 1). This means that the asymptotic behavior of the solution of the integrodifferential equation (A5.271) is determined by the behavior of the integrand near the point $t = 0$. Hence we can replace (A5.271) by the differential equation

$$\frac{d}{dt} f(t) = \tilde{Q}(t) + A k^{-1} t^{-1/2} \left(\frac{d}{dt} f(t) + k^2 f(t) \right)$$

$$- B a^{-2} k^{-1} t^{-1/2} \left(\frac{d}{dt} f(t) + k^2 a^2 f(t) \right) \tag{A5.275}$$

Write

$$c = (Ba^{-2} - A)/k, \quad d = k(B - A) \tag{A5.276}$$

By (A5.253),

$$c > 0, \quad d > 0 \tag{A5.277}$$

Equation (A5.275) becomes

$$\frac{d}{dt}f(t) + d(t^{1/2} + c)^{-1}f(t) = \tilde{Q}(t)t^{1/2}(t^{1/2} + c)^{-1} \tag{A5.278}$$

so that

$$f(t) = f(T)\exp[-2d(t^{1/2} - T^{1/2})][(c + t^{1/2})(c + T^{1/2})^{-1}]^{2cd}$$
$$+ \int_T^t \tilde{Q}(\tau)\tau^{1/2}(c + \tau^{1/2})^{-1-2cd}(c + t^{1/2})^{2cd}$$
$$\times \exp[-2d(t^{1/2} - \tau^{1/2})]d\tau \tag{A5.279}$$

By (A5.268)–(A5.270),

$$\tilde{Q}(t) = \exp(-\beta^2 t)t^{-\delta}q(t) \tag{A5.280}$$

where

$$\beta^2 = k^2 \min(a^2, 1), \quad \delta = \begin{cases} 3/2 & \text{if (A5.265) holds} \\ (1 + \alpha)/2 & \text{if (A5.269) holds} \end{cases} \tag{A5.281}$$
$$q(t) = O(1) \quad \text{as } t \uparrow \infty$$

Hence

$$f(t) = f(T)\exp[-2d(t^{1/2} - T^{1/2})]$$
$$\times [(c + t^{1/2})(c + T^{1/2})^{-1}]^{2cd} + I(t) \tag{A5.282}$$

where

$$I(t) = \int_T^t q(\tau)\tau^{1/2-\delta}(c + t^{1/2})^{2cd}(c + \tau^{1/2})^{-1-2cd}$$
$$\times \exp(-\beta^2 \tau + 2d\tau^{1/2} - 2dt^{1/2})d\tau \tag{A5.283}$$

By (A5.283),

$$|I(t)| < \max_{T \le \tau \le t} |q(\tau)|T^{-1-\delta}(c + T^{1/2})^{-2cd-1}$$
$$\times \int_T^t \exp(-\beta^2 \tau + 2d\tau^{1/2} - 2dt^{1/2})2^{cd}t^{cd}\tau^{-1/2}d\tau$$
$$= \exp(-2dt^{1/2})t^{cd}O(1) \quad \text{as } t \uparrow \infty \tag{A5.284}$$
$$|I(t)| > \min_{T \le \tau \le t} |q(\tau)|(2T)^{-1-\delta}(c + (2T)^{1/2})^{-2cd-1}$$
$$\times \int_T^{2T} \exp(-\beta^2 \tau + 2d\tau^{1/2} - 2dt^{1/2})t^{cd}\tau^{-1/2}d\tau$$
$$= \exp(-2dt^{1/2})t^{cd}O(1) \quad \text{as } t \uparrow \infty \tag{A5.285}$$

Hence

$$f(t) = \exp(-2dt^{1/2})t^{cd}O(1) \quad \text{as } t \uparrow \infty \tag{A5.286}$$

F. Additional remarks

1. Global stability and weakly asymptotic stability of the 1-dimensional Cauchy–Stefan problem under 1-dimensional perturbations

Let $u(p,t)$, $s(t)$ be the solution of the Cauchy–Stefan problem

$$Lu_1(x,t) = 0, \quad -\infty < x < s(t)$$
$$L_a u_2(x,t) = 0, \quad s(t) < x < \infty$$
$$u_1(x,0) = f_1(x) < 0, \quad -\infty < x < 0$$
$$u_2(x,0) = f_2(x) > 0, \quad s(t) < x < \infty \qquad \text{(A5.287)}$$
$$\lim_{|x|\uparrow\infty} u_i(x,t) = \alpha_i, \quad u_i[s(t),t] = 0, \quad i = 1,2$$

$$\frac{d}{dt}s(t) = \frac{\partial}{\partial x}u_1[s(t),t] - \frac{\partial}{\partial x}u_2[s(t),t], \quad t > 0, \quad s(0) = 0$$

The existence and uniqueness of the global solution to this problem were proved in [69], where it was also shown that the solution tends uniformly to a Neumann self-similar solution as $t \uparrow \infty$ in the exact (i.e. nonlinearized) setting. The global stability of the solution to this problem in the context of linear stability theory follows from the fact that the stability problem can be reduced to linear integral equations. Indeed, equations (A5.223) are valid for any reference front $s(t)$, and not only as a solution of the self-similar Neumann problem.

Assume that

$$\left|\frac{d}{dx}\Psi_i(x)\right| < \delta, \quad |x| < \infty, \quad i = 1,2 \qquad \text{(A5.288)}$$

Then (A5.225) implies

$$|\Phi_i| < \delta, \quad i = 1,2 \qquad \text{(A5.289)}$$

The resolvents Γ_i of equations (A5.223) have the majorants

$$|\Gamma_i| < \sum_{n=0}^{\infty} \Gamma\left(\frac{1}{2}\right)^{n+1} \frac{N^{n+1}t^{n/2-1/2}}{\Gamma[(n+1)/2]} \qquad \text{(A5.290)}$$

where N is a suitably chosen constant. This means that for every $T^* > 0$ there exists $L > 0$ such that

$$|\Gamma_i| < L \quad \text{for } 0 \le \tau \le t \le T^* \qquad \text{(A5.291)}$$

Hence, for any $\eta > 0$ there exists $\delta > 0$ (see (A5.260), (A5.261), and (A5.289)) such that

$$|Q_i| < \eta \quad \text{for } 0 \le \tau \le t \le T^* \qquad \text{(A5.292)}$$

Assume now that $M > 0$ is chosen so large that

$$\left|\frac{\partial}{\partial x}u_i\right| < M \quad \forall |x| < \infty, \quad 0 \le t \le T^* \qquad \text{(A5.293)}$$

Then it follows from (A5.234), (A5.235), and (A5.290) that there exists $A = A(T, M)$ such that

$$|R_i| < A(t - \tau)^{-1/2}, \quad 0 \leq \tau < t \leq T^* \tag{A5.294}$$

Taking this into account, we see that $df(t)/dt$ is the solution of a linear integral equation whose kernel is majorized by the expression

$$B(t - \tau)^{-1/2}, \quad B = B(T^*, M) \tag{A5.295}$$

which in turn means that for any $\varepsilon > 0$ there exists $\eta = \eta(T^*, M)$ such that

$$|Q_i| < \eta \Rightarrow \left| \frac{d}{dt} f(t) \right| < \varepsilon \tag{A5.296}$$

Combining (A5.296) and (A5.292), we see that for all $T^* > 0$ and any $\varepsilon > 0$ there exists $\delta > 0$ such that

$$\left| \frac{d}{dt} f(t) \right| < \varepsilon \tag{A5.297}$$

if (A5.288) holds. Hence the solution of problem (A5.287) is globally stable (in the sense of definition (b); see note 7). This result obviously implies that the free boundary is weakly asymptotically stable, since this has been proved for the Neumann solution.

2. An unsolved problem

The most natural formulation of the stability problem is as follows. Let σ and v_i be defined by (A5.197) and (A5.198), respectively. Prove that for any $\varepsilon > 0$ there exists $\delta > 0$ such that

$$\sup |v_i(\xi, \eta, \zeta, 0)| < \delta, \; \sup |\sigma(\eta, \zeta, 0)| < \delta$$
$$\Rightarrow |v_i(\xi, \eta, \zeta, \tau)| < \varepsilon, \; |\sigma(\eta, \zeta, \tau)| < \varepsilon \; \forall \tau > 0 \tag{A5.298}$$

Let w_i and f be defined by (A5.208). Then they are the Fourier transforms of v_i and σ. Clearly, since w_i and f are absolutely integrable and

$$|w_i| < \delta(\varepsilon), \quad |f| < \delta(\varepsilon) \quad \forall t \geq 0, \; \forall x, \; \forall k \geq 0 \tag{A5.299}$$

it follows that

$$|v_i| < \varepsilon, \quad |\sigma| < \varepsilon \tag{A5.300}$$

everywhere in their region of definition. Hence the problem is reduced to proving that the Fourier transforms of the perturbations are uniformly small with respect to the wave number k. This fact would be a corollary of the already proved results if one could prove the inequality (A5.299) for $k = 0$. In other words, one must prove that the free boundary is globally stable under 1-dimensional (in the x direction) perturbations of the initial data, in the sense of linear stability theory.

As mentioned previously, weak asymptotic stability of the Neumann self-similar solution was proved in [69] in a nonlinearized setting. However, the stability of Neumann similarity solution under 1-dimensional perturbation (in the sense of linear stability theory) remains unproved.

3. Distinction between rate of decay obtained here and in local theories, assuming a constant rate of motion of the reference front

As mentioned in the introduction to this section, most of the literature on stability of solutions to the Cauchy–Stefan problem assumes that the rate of the reference front is constant. Under this assumption, the coefficients of the equations determining the perturbations are constant. As a result, the time dependence of the perturbations is exponential, implying a rate of decay of order $O[\exp(-\mu_2 t)]$, $\mu_2 > 0$. In the theory developed here, the reference front moves at a rate $\gamma t^{-1/2}$; as a result, the coefficients of the respective equations are not constant, which implies the rate of decay $O[\exp(-\mu_1 t^{1/2})]$.

References

[1] D'Adhemar, R. 1904, Sur une classe d'equations aux dérivées partielles, du second ordre, du type hyperbolique, à 3 and 4 variables. *Journal de Liouville*, **5**, serie 10, 131–207; 1905, Sur une équation aux derivées partielles du type hyperbolic. *Rend. Circ. Mat. Palermo* **20**, 142–59.

[2] Antimirov, M. 1959, The solution of certain two-dimensional thermal problems of the mixed type by the method of divergent integrals and series. *Latvian Mathematical Annual*, v. 6. Riga: Zinatre (Russian)

[3] Benson, Sidney W. 1960, *Chemical Kinetics*. New York: McGraw-Hill.

[4] Bernstein, C. N. 1904, Sur la nature analytique des solutions des équations aus dérivées partielles des second ordre. *Math. Ann.* **59**, 20–76.

[5] Bluman, G. W., and Cole, J. D. 1974, *Similarity Methods for Differential Equations*. New York: Springer.

[6] Bromwich, T. I. 1908, *An Introduction to the Theory of Infinite Series*. London: Macmillan.

[7] Buckley, S. F., and Leverett, M. C. 1942, Mechanism of fluid displacement in sands. *Trans. A.I.M.E.* **146**, 107.

[8] Burgers, J. M. 1928, On Oseen's theory for the approximate determination of the flow of a fluid with very small friction along a body. *Proc. Kon. Akad. von Wetenshappen* **31**(4, 5), Amsterdam.

[9] Cannon, J. H. R. 1984, The one-dimensional heat equation. *Encyclopedia of Mathematics and Applications*, v. 23. Menlo Park, CA: Addison Wesley.

[10] Chandrasekhar, S. 1943, Stochastic problems in physics and astronomy, *Reviews of Modern Physics* **15**(1), 1–89.

[11] Courant, R. 1925, Über direkte Metoden bei Variations und Randvertproblemen, *Jber. Deutsch. Math. Verein.* **34**, 90–117.

[12] Courant, R. 1927, Über direkte Metoden in der Variationsrechnungh und über verwandte Fragen, *Math. Ann.*, **97**, 711–36.

[13] Courant, H., and Hilbert, D. 1962, *Methods of Mathematical Physics*, v. I, II. New York: Interscience.

[14] Crank, J. 1984, *Free and Moving Boundary Problems*. Oxford: Oxford University Press.

[15] Deriagin, B., and Landau, L. D. 1967, A theory of the stability of strongly charged lyophobic sols and the coalescence of strongly charged particles in electrolytic solutions. In L. D. Landau and D. Ter-Haar (eds.), *Collected Papers*. New York: Gordon and Breach.

[16] Dirac, P. D. 1935, *The Principles of Quantum Mechanics*. Oxford: Clarendon Press.

[17] Eckhaus, W. 1979. *Asymptotic Analysis and Singular Perturbations*. Amsterdam: North Holland.

[18] Erdély, A., Magnus, W., Oberhettinger, F., Tricomi, F. G. 1953, Higher transcendental functions. Bateman manuscript project; 1954, *Tables of Integral Transforms*. New York: McGraw-Hill.

[19] Fourier, J. R. J. 1822, La théorie analitique de la Chaleur. Paris. English translation: F. W. Gehring (ed.). 1878. *Analytic Theory of Heat*. Cambridge: Cambridge University Press.

[20] Frank, Ph., and Mises, R. (eds.). 1935, Differential- und Integralgleichungen der Mechanik und Physik, Bd. II. Braunschweig: Druck und Verlag von Fiedr. Vieweg & Sohn Akt. Ges.

[21] Frank-Kamenetski, D. A. 1947, *Diffusion and Heat Transfer in Chemical Kinetics*. Moscow: Akad. Nauk SSSR.

[22] Friedman, A. A. 1934, Study on Hydrodynamics of Compressible Fluid. Moscow: GTTI (Russian).

[23] Friedman, A. 1964, *Partial Differential Equations of Parabolic Type*. Englewood Cliffs, NJ: Prentice-Hall.

[24] Gel'fand, I. M. 1959, Some problems of the theory of quasilinear differential equations. *Uspekhi Mat. Nauk.* **14**, 2, 87–158. (See also Collins, R. E., 1961, *Flow of Fluids through Porous Materials*. New York: Reinhold, pp. 142–9.

[25] Gevrey, M. 1913, Equation aux dérivées partielles du type parabolique. *J. Mat. Purco ct Appl.* (6), **0**, 305–471.

[26] Glansdorf, P., and Prigogine, I. 1971, *Thermodynamic theory of structure, stability and fluctuations*. New York: Interscience.

[27] Goursat, E. 1923, *Course d'analyse mathematique*, v. 1, 5th ed. Paris: Gauthier-Villars.

[28] Goursat, E. 1923, *Course d'analyse mathematique*, v. 3, 4th ed. Paris: Gauthier-Villars.

[29] Gradstein, I. S., and Rizhik, I. M. 1980, *Tables of Integrals Series and Products*. New York: Academic Press.

[30] Gray, A., and Mathews, G. B. 1922, *A Treatise on Bessel Functions and their Application to Physics*, 2nd ed. London: Macmillan.

[31] Grew, K. E., and Ibbs, T. I. 1952, *Thermal Diffusion in Gases*. Cambridge: Cambridge University Press.

[32] Groot, C., and Masur, P. 1962, *Non-Equilibrium Thermodynamics*. Amsterdam: North Holland.

[33] Gunter, N. M. 1934, *La theorie du potentials et ses applications aux problèmes fondamentaux de la physique mathematique*. Paris: Gauthier-Villars.

[34] Hadamard, J. 1923, *Lectures on Cauchy's Problem in Linear Partial Differential Equations*. New Haven, CT: Yale University Press.

[35] Hadamard, J. 1954, Extension à l'équation de la chaleur d'un théorème de A. Harnack. *Rend. Circ. Mat. Palermo* (2), 337–46.

[36] Hardy, G. H. 1949, *Divergent Series*. Oxford: Oxford University Press.

[37] Henderson, P. 1907, Zur Thermodynamik der Flüssig Keitsketten. *Z. Phys. Chem.* **59**, 118; 1908, Zur Thermodynamik der Flüssig Keitsketten. *Z. Phys. Chem.* **63**, 325.

[38] Hopf, E. 1927, Elementare Bemerkungen über die Lösungen partielle Differentialgleichungen zweiter order vom elliptischen Typus. *Sitzungsber. Preuss. Akad.* **19**, 147–52.

[39] Ince, E. L. 1956, *Ordinary differential equations*. New York: Dover.

[40] Jackson, J. L. 1974, Charge neutrality in electrolyte solutions and the liquid junction potential. *J. Phys. Chem.* **78**, 2060–4.

[41] Kantorovich, I. V., and Krylov, I. 1958, *Approximate Methods of Higher Analysis*. New York: Interscience.

[42] Kaplun, S. 1967, Fluid mechanics and singular perturbations. Edited by P. A. Lagerstrom, L. N. Howard, and C. S. Liu. New York: Academic Press.

[43] Keller, J. B. 1958, Propagation of a magnetic field into a superconductor, *Phys. Review* **111**(6), 1497–9.

[44] Kevorkian, J., and Cole, J. D. 1981, Perturbation methods in applied mathematics. *Applied Mathematical Sciences*, v. 34. New York: Springer.

[45] Ladizhenskaja, O. A., Solonnikov, V. A., and Ural'ceva, N. N. 1968, Linear and quasilinear equations of parabolic type. *Tran. Math. Monographs*, v. 23. Providence, RI: Amer. Math. Soc.

[46] Mikusinski, J. 1959, *Operational Calculus*. New York: Pergamon Press.

[47] Miln-Tomson, L. M. 1960, *Theoretical Hydrodynamics*. London: Macmillan.

[48] Morrey, C, 1958, On the analyticity of the solutions of analytic nonlinear systems of partial differential equations. *Amer. J. Math.* **80**, 198–237.

[49] Nirenberg, L. 1953, A strong maximum principle for parabolic equations. *Comm. Pure Appl. Math.* **6**, 167–77.

[50] Oleinik, O. A. 1959, On uniqueness and stability of the generalized solution of the Cauchy problem for a quasilinear equation. *Uspekhi Mat. Nauk* **14**, 2, 165–70.

[51] Oleinik, O. A. 1959, On construction of a generalized solution of the Cauchy problem for a quasilinear equation of the first order by means of introducing the vanishing viscosity. *Uspekhi Mat. Nauk* **14**, 2, 159–64.

[52] Oleinik, O. A. 1964, On discontinuous solutions of non-linear differential equations. In *The First Mathematical School*, v. 2. Kiev: Naukova Dumka. (Russian)

[53] Oseen, C. W. 1927, *Neuere Methoden und Ergebnisse in der Hydrodynamik.* Leipzig: Akadem Verlagsgesellschaft.

[54] Perron, D. 1923, Eine neue Behandlung der Randwertaufgabe für $\Delta u = 0$, *Math. Z.* **1**, 42–54.

[55] Petrovsky, I. G. 1935, Zur ersten Randwertaufgabe der Wärmeleitungsgleihung. *Compositio Math.* **1**, 383–419.

[56] Petrovsky, I. G. 1939, Sur l'analyticité des solutions des systèmes d'équations differentielles, *Mat. Sb.* (N.S.) **5**(47), 3–68.

[57] Petrovsky, I. G. 1954, *Lectures on Partial Differential Equations.* New York: Interscience.

[58] Petrovsky, I. G. 1966, *Ordinary Differential Equations*, rev. ed. Englewood Cliffs, NJ: Prentice-Hall.

[59] Pini, B. 1954, Maggioranti e minoranti delle soluzioni equazioni paraboliche. *Ann. Mat. Pura. Appl.* (4) **3**, 249–64; 1954, Sulla soluzione generalizzata de Weiner per il primo problema di valore al conttorno nel caso parabolico, *Rend. Sem. Mat. Univ. Padova* **23**, 422–34.

[60] Prigogine, I. and Defay, R. 1954, *Chemical Thermodynamics.* London: Longmans, Green.

[61] Privalov, I. I. 1935, *Integral Equations.* Moscow: GTTI. (Russian)

[62] Protter, M. H., and Weinberger, H. F. 1967, *Maximum Principle in Differential Equations.* New York: Springer.

[63] Rapoport, L. A., and Leas, W. J. 1953, Properties of linear waterfloods, *Trans. A.I.M.E.* **196**, 139.

[64] Rose, N. B., Kibel, I. A., and Kotchin, N. E. 1937, *Theoretical Hydromechanics*, part 2. Moscow: ONTI (Russian) part 1, 1955, Moscow GTTI. (Russian)

[65] Rubinstein, I. 1988, Field and force saturation in ionic equilibrium, *SIAM J. Appl. Math.* **4**, 0.

[66] Rubinstein, I. 1990, Electro-diffusion of ions. Philadelphia: SIAM.

[67] Rubinstein, L. 1959, On the integral value of the losses of heat at the heat injection of a hot liquid into a stratum. *Izv. Vyssh. Uchebn. Zaved.* Oil and Gas, **9**, 41–8.

[68] Rubinstein, L. I. 1972, *Temperature Fields in Oil Layers.* Moscow: Nedra. (Russian)

[69] Rubinstein, L. I. 1971, The Stefan problem. *Trans. Math. Monographs.* v. 27, Providence, RI: American Mathematics Society.

[70] Rubinstein, L. 1980, A model axially symmetric problem of heat conduction with a free boundary moving within the concentrated capacity, In E. Magenes (ed.), *Free Boundary Problems* (proceedings of a seminar held in Pavia, Italy, September–October 1979), pp. 417–36. Roma: Instituto Nazionale di Alta Matematica Francesco Severi.

[71] Rubinstein, L. 1983, Free boundary problem for a nonlinear system of parabolic equations, including one with reversed time. *Ann. Mat. Pura Appl.* (4) **135**, 29–72.

[72] Schwarz, L. 1950, *Théorie des distributions*. Paris: Hermann.

[73] Semenov, N. N. 1934, *Chain Reactions*. Moscow: Gostoptekhisdat. (Russian)

[74] Sneddon, I., 1951, *Fourier Transform*. New York: McGraw-Hill.

[75] Sobolev, S. 1963, Applications of functional analysis in mathematical physics. *Transl. Math. Monographs*, v. 7. Providence, RI: American Mathematics Society.

[76] Sommerfeld, A. 1949, Electrodynamik, *Vorlesungen über Theoreriche Physik*, v. 3. Leipzig: Akademische Verlagsgesellschaft Geest und Portig K.-G.

[77] Tamm, I. E. 1979, *Fundamentals of the Theory of Electricity*. Moscow: Mir.

[78] Tannery, J. 1904, *Introduction à la théorie des fonction d'une variables*. Paris: Gauthier-Villars.

[79] Tichonoff (Tichonov), A. N. 1935, Théorèmes d'uniticitè pour l'équatien de la chaleur. *Mat. Sb.* (N.S.) **42**, 199–216.

[80] Tichonoff, A. N. 1935, Ein Fixpunktsatz. *Math. Ann.* **3**, 767–76.

[81] Tikhonoff, A. N. 1937, On the cooling of bodies by radiation, *Izv. Akad. Nauk SSSR Ser. Geofiz.* **3**, 461–79. (Russian)

[82] Tichonoff, A. 1938. Sur l'équation fonctionelle du type de Volterra. *Bulletin de l'Univeristé d'état de Moscow*, serie international, Section A, v. 1, Fasc. 8. Moscow: GONTI.

[83] Tichonoff, A. 1938. Sur l'équation de la chaleur à plusieurs variables. *Bulletin de l'Univeristé d'état de Moscow*, serie international, Section A, v. 1, Fasc. 9. Moscow: GONTI. (Russian)

[84] Tichonoff, A. N. 1950, Boundary conditions containing derivatives of order higher than the order of the equation. *Mat. Sb.* (N.S.) **26**(68), 35–56; English translation in *Amer. Math. Soc. Transl.*, v. 1 (1966). Providence, RI: Amer. Math. Soc., pp. 440–66.

[85] Tichonoff, A. N., and Samarski, A. A. 1959, *Differentialgleichumgem der Mathematischen Physik*. Berlin: Deutschland Verlag der Wissenschaften.

[86] Titchmarsh, E. C. 1937, *Introduction to the Theory of Fourier Integrals*. Oxford: Oxford University Press.

[87] Titchmarsh, E. C. 1946, *Eigen-function Expansions Associated with Second Order Differential Equations*. Oxford: Clarendon Press.

[88] Turland, B. D. and Peckover, R. S. 1980, The stability of planar melting fronts in two phase thermal Stefan problem, *J. Inst. Math. Appl.* **25**, 1–15.

[89] Vyborny, R. 1957, On properties of solutions of some boundary value problems for equations of parabolic type. *Dokl. Akad. Nauk SSSR* (N.S.) **117**, 563–5.

[90] Watson, G. N. 1948, *A Treatise on the Theory of Bessel Functions*, 2nd ed. Cambridge: Cambridge University Press.

[91] Widder, D. V. 1975, *The Heat Equation*. New York: Academic Press.

[92] Wolkind, D. J., and Notestine, R. D. 1981, A non-linear stability analysis of the solidification of a pure substance. *IMA J. Appl. Math.* **27**, 85–104.

[93] Efros, A. M., and Danilevski, A. M. 1937, *Operational Calculus and Contour Integrals.* Kharkov: DNTVU. (Russian)

Index

Printed in the United States
By Bookmasters